W9-CPO-109

(*continued on back*)

Empirical Model-Building
and Response Surfaces

Empirical Model-Building and Response Surfaces

GEORGE E. P. BOX

NORMAN R. DRAPER

John Wiley & Sons

New York . Chichester . Brisbane . Toronto . Singapore

Library of Congress Cataloging-in-Publication Data:

Box, George E. P.
  Empirical model-building and response surfaces.

  (Wiley series in probability and mathematical
statistics. Applied probability and statistics)
  Bibliography: p.
  Includes indexes.
  1. Experimental design. 2. Response surfaces
(Statistics) I. Draper, Norman Richard. II. Title.
III. Series.

QA279.B67 1986   519.5   86-4064
ISBN 0-471-81033-9

Printed in the United States of America

10 9 8 7 6 5 4

Preface

The experimenter frequently faces the task of exploring the relationship between some response y and a number of predictor variables $\mathbf{x} = (x_1, x_2, \ldots, x_k)'$. Thus, for example, the response y might be the reaction time of a subject and three variables x_1, x_2, x_3 might be hours of sleep deprivation, amount of exercise, and dose of a certain drug. Or, in a chemical experiment, the response y might be the yield of product and x_1, x_2, x_3 might be the temperature, time, and pressure at which the reaction was conducted.

Various degrees of knowledge or ignorance may exist about the nature of such relationships. At one extreme it may be that the relationship is "exactly known" from physical considerations and consequently a mathematical function can be written down which, apart from experimental error ε, is believed to exactly represent the relationship between the response y and the variables $\mathbf{x}$. We can then write down a *mechanistic model* of the form $y = f(\mathbf{x}) + \varepsilon$. When such knowledge of the physical mechanism underlying a relationship is known or even suspected, it should be used. A number of interesting problems concerning checking of model fit, model discrimination, and choice of experimental design occur when mechanistic models are being considered but these are not a major topic of discussion here.

Many examples exist, however, where knowledge of the mechanistic model is lacking and all that is known is that, apart from experimental error, the relationship between y and the various x's is likely to be smooth. Such relationships can be explored experimentally and useful conclusions drawn. In particular, if we have only one variable x, observations of y at various levels of x may be plotted and a smooth curve drawn through the points by eye. Alternatively, some smooth function such as a straight line or a quadratic curve may be fitted, for example, by least squares. If we denote such an approximating (or graduating) function by $g(x)$ we may refer to the relationship $y = g(x) + \varepsilon$ as an *empirical model*.

It is well known that the usefulness of a simple function such as a straight line or a quadratic curve is greatly enhanced if we allow for the possibility that, before making the plot, an appropriate transformation in y or in x or in both is made. For example, a relationship might be simplified by plotting log y against x, y against log x, or log y against log x; in other examples, a square root or reciprocal transformation, rather than a log, might prove valuable.

It is possible to use statistical estimation procedures both to choose the adjustable constants in $g(x)$ (determining, for example, the slope and constant of the fitted line) as well as the appropriate transformation for y or for x. Such fitting processes can be thought of as a means of matching a (possibly transformed) low degree polynomial $g(x)$ to the data in much the same way that a French curve is adjusted to the data to obtain a smooth fit. In this way a "mathematical French curve" $g(x)$ is made to substitute locally for the true but unknown function $f(x)$.

When there is more than one x variable, simple graphical plotting is not available. However, we can extend the idea of the fitting of a "mathematical French curve" to more than one dimension. A class of such graduating functions explored in this book consists of the polynomials of degree one and two in $x_1, x_2, \ldots, x_k$, again allowing for the possibility that both response y and each of the variables $x_1, x_2, \ldots, x_k$ might first be subjected to transformation. The response surfaces so generated may be put to a number of somewhat different uses:

1. To show how a particular response y is affected by a set of variables **x** over some specific region of interest.
2. To discover what settings of the x's will give a product simultaneously satisfying specifications for a number of responses $y_1, y_2, \ldots, y_m$ (e.g., yield, impurity, color, texture, and so on).
3. To explore the space of the x variables to define the maximum response and to determine the nature of this maximum.

In problems 2 and 3 it is very likely that movement away from the initial experimental region may be necessary before the objective is obtained. The method of steepest ascent is used to move from a region of low response to one of higher response using a local first-order approximation to the response function (a first degree polynomial in the x's, possibly transformed). Possibly after one or more applications of steepest ascent, a region of improved response may be found where second-order effects predominate and further exploration will then be possible in terms of a second degree polynomial in the x's, again possibly transformed. When this region contains a maximum it is important not only to estimate its location but also to determine its local geography. In particular, maxima are frequently associated with diagonally orientated ridges. The exploration of such ridges can provide alternative near-optimal processes and thus allow the near optimization of more than one response. The examination of such ridge systems is therefore of great economic importance. Such elucidation of ridge systems is also of importance because it may provide clues to the mechanistic (physical) nature of the phenomenon under study.

The book contains 15 chapters. Chapters 1 and 2 are concerned with the general philosophy of response surface methodology. Chapter 3 provides a summary of important results in least squares with the geometry necessary to provide intuitive understanding. First-order experimental designs derived from two-level factorials and fractionals are discussed in Chapters 4 and 5 and their application to steepest ascent is developed in Chapter 6. Chapter 7 introduces the fitting and checking of second-order models and Chapter 8 is concerned with the uses of transformation and with associated estimation problems. Chapter 9 introduces the exploration of maxima with second-order designs and Chapters 10 and 11 deal with the elucidation of ridge systems. The manner in which such systems can be used to form a link between empirical and theoretical models is covered in Chapter 12. Theoretical aspects of the choice of experimental designs are explored in Chapters 13 and 14. In Chapter 13, the influence of variance, bias, lack of fit, and other factors in the choice of experimental designs is explored. In Chapter 14 variance optimal designs are discussed and the alphabetic optimality approach is critically considered. Chapter 15 presents designs that provide useful compromises for various practical situations.

Numerous exercises for the procedures employed are given, together with answers. A bibliography of selected references is also included.

The original plan for this book was that it would be a three-author work by George Box, Norman Draper, and Stu Hunter. Initially this seemed entirely feasible, but with the two of us in Madison, and Stu in Princeton, it gradually became clear that the practicalities would not work out. George and Norman met every week that both were free for many years while writing this volume. Stu contributed by mail whenever possible. Eventually, Stu decided that he should not officially be a coauthor and formally withdrew from the project. Nevertheless, he continued to take a strong interest in the book, read much of the manuscript, and contributed commentary and exercises. We are extremely grateful for his help.

We are also grateful to the many colleagues and graduate students who have contributed commentary and criticism and, in some instances, helped with diagrams and calculations. Special thanks are due to Ray Niznick, Jake Sredni, Conrad Fung, and Stephen Jones.

We also wish to thank Mary Ann Clark, Mary Esser, Sally Ross, and Wanda Gray for their careful typing of the manuscript in all of its various manifestations, and Helga Fack for accurate and painstaking drawing of the diagrams.

We are *especially* grateful to Stephen Jones, who read the entire set of proofs most carefully. Thanks are also due to José Ramírez and to Søren Bisgaard for their help with checking proofs; José also provided emergency computing help at the galley stage.

Finally, we thank Bea Shube of John Wiley & Sons for her persistent good advice and her extraordinary patience with us.

GEORGE E. P. BOX
NORMAN R. DRAPER

Madison, Wisconsin
September 1986

Contents

Empirical Model-Building
and Response Surfaces

CHAPTER 1

Introduction to Response Surface Methodology

In this chapter we discuss, in a preliminary way, some general philosophy necessary to the understanding of the theory and practice of response surface methodology (RSM). Many of the points made here receive more detailed treatment in later chapters.

1.1. RESPONSE SURFACE METHODOLOGY (RSM)

The mechanisms of some scientific phenomena are understood sufficiently well that useful mathematical models that flow directly from the physical mechanism can be written down. Although a number of important statistical problems arise in the building and study of such models they are not considered in this book. Instead, the methods we discuss will be appropriate to the study of phenomena that are presently not sufficiently well understood to permit the mechanistic approach.

Response surface methodology comprises a group of statistical techniques for empirical model building and model exploitation. By careful design and analysis of experiments, it seeks to relate a *response*, or *output* variable to the levels of a number of *predictors*, or *input* variables, that affect it.

The variables studied will depend on the specific field of application. For example, response in a chemical investigation might be *yield* of sulfuric acid and the input variables affecting this yield might be the *pressure* and *temperature* of the reaction. In a psychological experiment, an investigator might want to find out how a test *score* (output) achieved by certain subjects depended upon the *duration* (input 1) of the period during which they studied the relevant material, and the *delay* (input 2) between study and test. In mathematical language, in this latter case, we can say that the

1

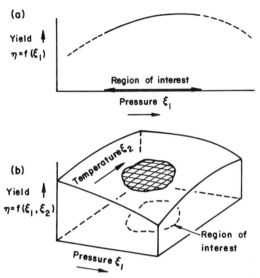

FIGURE 1.1.   (*a*) A response curve. (*b*) A response surface.

investigator is interested in a presumed *functional relationship*

$$\eta = f(\xi_1, \xi_2), \qquad (1.1.1)$$

that expresses the response or output score, η (Greek letter eta), as a function of two input variables *duration*, ξ_1, and *delay*, ξ_2 (Greek letters xi). Both our examples involve two inputs but, in general, we shall have not two, but k input variables $\xi_1, \xi_2, \ldots, \xi_k$, and the functional relationship between the mean response and the levels of the k inputs can then be written

$$\eta = f(\xi_1, \xi_2, \ldots, \xi_k). \qquad (1.1.2)$$

More compactly, if $\boldsymbol{\xi}$ denotes a column vector* with elements $\xi_1, \xi_2, \ldots, \xi_k$, the mean response function may be written as

$$\eta = f(\boldsymbol{\xi}). \qquad (1.1.3)$$

If there is only one input variable ξ_1, we can relate an output η to a single input ξ_1 by a *response curve* such as that in Figure 1.1a. If we have two inputs ξ_1 and ξ_2 and we draw, in three-dimensional space, a graph of η

---

*The word vector is used here only as a shorthand word referring to all k elements.

against ξ_1 and ξ_2, we obtain a *response surface* such as that shown in Figure 1.1*b*. When k ξ's are involved and k is greater than 2, we shall still talk of a response surface in the $k + 1$ dimensional space of the variables even though only sectional representation of the surface is possible in the three-dimensional space actually available to us.

The operation of the system with the k inputs adjusted to some definite set of levels ξ is referred to as an *experimental run*. If repeated runs are made at the same conditions ξ, the measured response will vary because of measurement errors, observational errors, and basic variability in the experimental material. We regard η therefore as the mean response at particular conditions $(\xi_1, \ldots, \xi_k)$. An actual observed response result, which we call y, would fall in some statistical distribution about its mean value η. We say that the expected value of y equals η, that is, $E(y) = \eta$. In any particular run, we shall refer to the discrepancy $y - \eta$ between the observed value y and the hypothetical mean value η as the *error* and denote it by ε (Greek epsilon). In general then, our object is to investigate certain aspects of a functional relationship affected by error and expressed as

$$y = f(\xi) + \varepsilon. \tag{1.1.4}$$

Extending the Scope of Graphical Procedures

The underlying relationship between η and a single-input variable ξ can often be represented by a smooth curve like that in Figure 1.1*a*. Even though the exact nature of the function f was unknown, a smooth curve such as a polynomial acting as a kind of mathematical French curve might be used to represent the function *locally*. Similarly, the relationship between η and two inputs ξ_1 and ξ_2 can often be represented by a smooth surface like that in Figure 1.1*b* and a suitable polynomial, acting as a two-dimensional French curve could be used to approximate it locally. Mathematical French curves, like other French curves, are, of course, not expected to provide a global representation of the function but only a local approximation over some limited region of current interest.

The empirical use of graphical techniques is an important tool in the solution of many scientific problems. The basic objective of the use of response surface methods is to extend this type of empirical procedure to cases where there are more than just one or two inputs. In these higher dimensional cases, simple graphical methods are inadequate, but we can still fit graduating polynomials and, by studying the features of these fitted empirical functions, greatly extend the range of application of empirical procedures.

So that the reader can better appreciate where response surface methodology fits in the overall picture of scientific experimentation, we now briefly survey some aspects of scientific method. A more extensive review of the role of statistics in the scientific method is given by Box (1976).

1.2. INDETERMINANCY OF EXPERIMENTATION

At first sight, the conduct of an experimental investigation seems to be a highly arbitrary and uncertain process. For example, suppose that we collected 10 teams* of experimenters competent in a particular field of science or technology, locked each team in a separate room, presented them all with the same general scientific problem, and asked each team to submit a plan that could lead to the problem's solution. It is virtually certain that no two teams would present the same plan. Consider, in particular, some of the questions which would arise in planning an initial set of experiments, but on which the teams would be unlikely to agree.

1. Which input variables $\xi_1, \xi_2, \ldots$, should be studied?

If it were a chemical reaction that was being studied, for example, it might be that most investigators would regard, say, temperature and pressure as being important but that there might be a diversity of opinion about which should be included among other input variables such as the initial rate of addition of the reactant, the ratio of certain catalysts, the agitation rate, and so on. Similar and perhaps even stronger disagreement might occur in a psychological experiment.

2. Should the input variables ξ be examined in their original form, or should transformed input variables be employed?

An input variable ξ, such as energy, may produce a linear increase in a response η, such as perceived loudness of noise, when ξ is varied on a *geometrical* scale—or, equivalently, when $\ln \xi$ is varied on a linear scale. It is simpler to express such a relationship therefore by first transforming the input ξ to its logarithm. Another input might be related to the response η by an inverse square law, suggesting an inverse square transformation ξ^{-2}. Other examples can be found leading to transformations such as the square root $\xi^{1/2}$, the inverse square root $\xi^{-1/2}$, and the reciprocal ξ^{-1}. A choice of a transformation for a single variable is often called a *choice of metric* for that variable. More generally, a transformation on the input variables can involve two or more of the original inputs. Suppose, for example, that the

---

*Or 10 individual experimenters.

amounts ζ_1 and ζ_2 (Greek zeta) of two nitrogenous fertilizers were being investigated. Rather than employing ζ_1 and ζ_2 themselves as the input variables, their sum $\xi_1 = \zeta_1 + \zeta_2$, the total amount of nitrogenous fertilizer, and their ratio $\xi_2 = \zeta_1/\zeta_2$ might be used if it were likely that the response relationship in terms of ξ_1 and ξ_2 could be more simply expressed. In some instances, the theory of dimensionless groups can be used to indicate appropriate transformations of this kind, but usually the best choice of metrics and transformations is not clearcut and, initially at least, will be the subject of conflicting opinion.

3. How should the response be measured?

It is often far from clear how the response should be defined. For example, in a study to improve a water treatment plant, the biochemical oxygen demand (BOD) of the effluent would often be regarded by experts as a natural measure of purity. However, it is possible to have an effluent with zero BOD which is lethal to both bacteria and humans. Thus the appropriate response depends on the end use of the response and will often be the subject of debate.

In (2), we discussed transformation of the input variables. Transformation of the response variable is also possible, and the appropriate metric for the response is not always clear. In subsequent chapters we show how transformation in the response can lead both to simpler representation and simpler assumptions.

4. At which levels of a given input variable ξ should experiments be run?

Suppose temperature is an important input. One experimenter might feel that, for the particular system under study, experiments covering temperatures over the range 100–140°C should be made. Another experimenter, believing the system to be very sensitive to temperature, might choose a range from 115 to 125°C. A third experimenter, believing that considerably higher temperatures would be needed might pick a range of from 140 to 180°C. In practice, of course, not one but several inputs must be considered simultaneously. If the temperature is to be changed over a 20° range, for example, what is a suitable* "commensurate" change for concentration?

In brief, the investigator must choose not only a location for his experiments but also an appropriate scaling for each of the variables.

---

*A "rule of thumb" sometimes used by chemists is that a 10°C change in temperature roughly doubles the rate of a chemical reaction. Thus, a commensurate change for concentration in this example would consist of quadrupling the concentration. However, here again, the effect of varying concentration differs for different chemical systems and, in particular, for the reaction order.

5. How complex a model is necessary in a particular situation?

This question is, of course, related to questions (2), (3), and (4). By definition, the more appropriately transformations and metrics are chosen, the simpler is the model that may be employed. Also, the more extensive the region of interest is in the ξ space, the more complex will be the model needed. Thus again we see that this question is a hazy one and is intimately bound up with questions considered under (2), (3), and (4).

6. How shall we choose qualitative variables?

The foregoing discussion is posed entirely in terms of *quantitative* inputs such as temperature and concentration. Similar indeterminacies occur in experimentation with *qualitative* inputs such as type of raw material, type of catalyst, identity of operator, and variety of seed. For example, if we are to compare three kinds of seed, which varieties should be employed? Should the tested varieties include the cheapest, that believed to give the greatest yield, that which is most popular? The answer depends on the objectives of the investigation, and that, partially at least, depends on opinion.

7. What experimental arrangement (experimental design) should be used?

This question is intimately bound up with all of the questions already studied. For instance, consider the hypothetical experiment of question (2). If the amounts ζ_1 and ζ_2 of the two nitrogen fertilizers are each tested at three levels in all nine possible combinations, the experimental arrangement is that of Figure 1.2a. However, if the same experimental plan is set out in terms of total nitrogen $\xi_1 = \zeta_1 + \zeta_2$, and nitrogen ratio $\xi_2 = \zeta_1/\zeta_2$, we

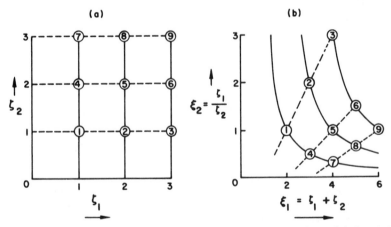

FIGURE 1.2. The effect of a transformation on the shape of an experimental design: (*a*) A balanced design in the (ζ_1, ζ_2) space. (*b*) The same design in the $\{(\zeta_1 + \zeta_2), \zeta_1/\zeta_2\}$ space.

obtain the very unbalanced arrangement of Figure 1.2*b*. In general, the question of "design optimality" is intimately bound up with arbitrary choices of metric, transformation, and the size and shape of the region to be studied.

In brief then, the investigator deals with a number of entities whose natures are necessarily matters of opinion. Among these are (a) the identity of the space of the inputs and outputs in which the experiments should be conducted; (b) the scales, metrics, and transformations in which the variables should be measured, and (c) the location of the region of interest, the specification of the model over it, and the experimental arrangement that should be used to explore the region of interest.

1.3. ITERATIVE NATURE OF THE EXPERIMENTAL LEARNING PROCESS

Faced with so many indeterminacies and uncertainties, the reader might easily despair of a successful outcome of any kind. His spirits may be sustained, however, by the thought that these difficulties have nothing to do with statistical methods per se, they are hurdles that confront and have always confronted all experimenters. In spite of them, practical experimentation is frequently successful. How does this come about? The situation appears more hopeful when we remember that any group of experimental runs is usually only *part of an iterative sequence*, and that an investigational strategy should be aimed at the overall furthering of knowledge rather than just the success of any individual group of trials. Our problem is to so organize matters that we are likely in due course to be led to the right conclusions even though our initial choices of the region of interest, the metrics, the transformations, and levels of the input variables may not all be good. Our strategy must be such as will allow any poor initial choices to be rectified as we proceed. Obviously, the path to success is not unique (although it may seem so to the first investigator in a field). It is not *uniqueness* of the path that we should try to achieve, therefore, but rather the probable and rapid convergence of an iterative sequence to the right conclusions.

This iterative process of learning by experience can be roughly formalized. It consists essentially of the successive and repeated use of the sequence

CONJECTURE—DESIGN—EXPERIMENT—ANALYSIS

as illustrated in Figure 1.3. Here, the words "design" and "analysis" do not

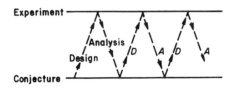

FIGURE 1.3. The iterative nature of experimentation.

refer only to *statistical* design and analysis. By "design," we mean the synthesis of a suitable experiment to test, estimate, and develop a current conjectured model. By "analysis" we mean the treatment of the experimental results, leading either to verification of a postulated model and the working out of its consequences, or to the forming of a new or modified conjecture. These complementary processes are used many times during an investigation and, by their intelligent alternation, the experimenter's knowledge of the system becomes steadily greater.

The indeterminancy of the learning process, but the ultimate possibility of its convergence, is familiarly exemplified by the playing of the game of 20 questions. Ten players engaged independently in playing 20 questions all might succeed eventually in discovering that the object in question was the left ear of the Statue of Liberty but they would almost certainly reach that conclusion by different routes. Let us consider how they do it.

A player may be supplied initially with the information that the object is "mineral." This limits consideration to a certain class of objects which he has to reduce to one as rapidly as possible. When the game is played so that each question concerns a choice between two specific alternatives, the best design of experiment (each "experiment" here is a question asked) is one which will classify the possible objects into two approximately equi-probable groups. A good (but not uniquely good) experiment would thus consist of asking "Is it metal or stone?;" a poor experiment would consist of asking "Is it the stone in my ring?." The experiment having been performed and the data (in the form of an answer) having become available, the player would now analyze the reply in the light of any relevant prior knowledge he possessed and would then form some new or modified conjecture as to the nature of the object in question. To the question, "Is it metal or stone?" the answer "stone" might conjure up in his mind, on the one hand, buildings, monuments, mountains, and so on, and, on the other hand, small stones, both precious (such as diamonds, rubies) and nonprecious (such as pebbles). The player might feel that a good way to discriminate between these two types of objects in the light of information then available would be the question "Is it larger or smaller than this room?." The answer to this question would now result in new analysis which would raise new conjectures and give rise to a new question and so the game would continue.

As another example of iterative investigation, consider a detective employed in solving a mystery. Some data in the form of a body, and the facts that the door was locked on the inside and that Mr. X would benefit from the will, are already known, or have been discovered. These data lead to conjectures on the part of the investigator which in turn lead to certain data-producing actions, or "experiments." As readers of detective novels will realize, the sequence of events which finally leads to the detection of the culprit is by no means unique. Alternative good (but by no means unique) experiments might be a visit to Baker Street subway station to question the ticket collector about a one-armed man with a scar under his left eye, or a visit to the Manor House to find out if the flower bed beneath Lady Cynthia's window shows a tell-tale footprint. The skill of the detective is partially measured by his ability to conceive, at each stage, appropriate conjectures and to select those experiments which, in the light of these conjectures, will best illuminate current aspects of the case. He pursues a strategy which (he hopes) will cause him to follow *one* of the paths leading to the unmasking of the assassin (preferably before the latter has had time to flee the country).

It is through the iterative process of "learning as we go" that the problem of indeterminancy which we mention may be resolved. We shall find that, in the application of the investigative processes described in this book, a multidimensional iteration occurs in which modification in the *location* of the experimental runs, in *scaling*, in *transformation*, and in the *complexity* of the contemplated model, all can occur. While we cannot, and should not, attempt to ensure uniqueness for the investigation, we can ensure that the crucial components of good design and analysis on which convergence depends are organized so as to illuminate and stimulate the ideas of the experimenter as effectively as possible, and so lead him quickly along one (nonunique) path to the truth. Figure 1.4 illustrates this general idea.

The truly iterative nature of some investigations may sometimes be obscured by the length of time taken by each iterative cycle. In this case it may be possible to see the wider iteration only by "stepping back" and examining what occurs over months or years. In this wider context, iteration may skip from one investigator to another, even from one country to another, and its phases may be very long. Even in this situation, however, it is important to bear in mind that the important consideration is the *overall* acquisition of knowledge.

What is being said is not, of course, peculiar to response surface methodology. "Traditional" designs such as randomized blocks, latin squares, and factorial designs have, since their inception, been used by statisticians as building blocks in iterative learning sequences. The possibility of rapid convergence of such a sequence depends, to an important

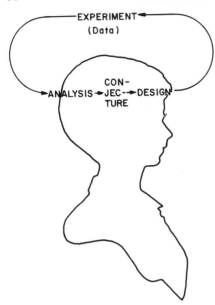

FIGURE 1.4. The iterative experimental process in relation to the experimenter.

extent, on the efficiency of these designs and their associated analyses, and their ability to illuminate and stimulate the ideas of the investigator. This notion was inherent in R. A. Fisher's* general attitude toward the use of these designs in scientific investigation.

1.4. SOME CLASSES OF PROBLEMS (WHICH, HOW, WHY)

Within the foregoing iterative context, we now consider some specific classes of scientific problems and the characteristics of models used in solving these problems.

Mechanistic and Empirical Models

It is helpful in this discussion to distinguish between empirical models and mechanistic models. We consider first what we might mean by a purely mechanistic model. Suppose that, in the study of some physical phenomenon, we know enough of its physical mechanism to *deduce* the form of the

---

*Sir Ronald Fisher (1890–1962), the famous British statistician, was responsible for many of the basic ideas of experimental design. His books *Statistical Methods for Research Workers* and *The Design of Experiments* are classic.

functional relationship linking the mean value η of the output to the levels ξ of the inputs via an expression

$$E(y) = \eta = f(\xi, \theta), \qquad (1.4.1)$$

where $\xi = (\xi_1, \xi_2, \ldots, \xi_k)'$ is a set of input variables, measuring, for example, initial concentrations of reactants, temperatures, and pressures, and where $\theta = (\theta_1, \theta_2, \ldots, \theta_p)'$ represents a set of physical parameters measuring such things as activation energies, diffusion coefficients, and thermal conductivities. Then we should say that Eq. (1.4.1) represented a mechanistic model.

In practice some or all of the θ's would need to be estimated from the data. Also, mechanistic knowledge might most naturally be expressed by a set of differential equations or integral equations for which (1.4.1) was the solution.

Often, however, the necessary physical knowledge of the system is absent or incomplete and consequently no mechanistic model is available. In these circumstances, it could often be realistically assumed that the relationship between η and ξ would be smooth and, consequently, that $f(\xi, \theta)$ could be locally *approximated* (over limited ranges of the experimental variables ξ) by an interpolation function $g(\xi, \beta)$, such as a polynomial. In this latter expression, the β's, which are the elements of β, would be coefficients in the interpolation function. They would be related to, but be distinct from, the parameters (the θ's) of the physical system. The interpolation function $g(\xi, \beta)$ could provide a local empirical model for the system and, as we have said, would act simply as a mathematical French curve.

Now the theoretical mechanistic model $\eta = f(\xi, \theta)$, and the purely empirical model $\eta \simeq g(\xi, \beta)$, as defined above, represent extremes. The former would be appropriate in the extreme case where a great deal was accurately known about the system, and the latter would be appropriate in the other extreme case, where nothing could be assumed except that the response surface was locally smooth. The situation existing in most real investigations is somewhere in between and, as experimentation proceeds, and we gain information, the situation can change. Because real problems occur at almost all points between the extremes mentioned above, a variety of statistical tools is needed to cope with them.

Both the state of ignorance in which we begin our experimental work, and the state of comparative knowledge to which we wish to be brought, will determine our approach. It must be realized, of course, that no real problem ever *quite* fits any prearranged category. With this proviso in mind, it is nevertheless helpful to distinguish the basic *types* of problems shown in Table 1.1. These are categorized in terms of what is unknown about the

TABLE 1.1. Some scientific problems

Supposed unknown	Objective	Descriptive name	Stage
f ξ θ	Determine the subset ξ of important variables from a given larger set Ξ of potentially important variables	Screening variables	Which
f θ	Determine empirically the effects of the known input variables ξ	Empirical model building	How (Response surface methodology)
f θ	Determine a local interpolation approximation $g(\xi, \beta)$ to $f(\xi, \theta)$	(Response surface methodology)	
f θ	Determine f	Mechanistic model building	Why
θ	Determine θ	Mechanistic model fitting	

"true mechanistic model." For reference purposes we have given each type of problem a descriptive name, as well as a briefer but more pointed "stage" name which makes it clear what stage of investigation is involved.

At the WHICH stage, our object is to determine *which* ones of all the suggested input variables have a significant and important effect upon the response variable. At the HOW stage we would like to learn more about the pattern of the response behavior as changes are made in these important variables. At the WHY stage we attempt to find the mechanistic reasons for the behavior we have observed in the previous stage. (In practice, the stages would usually overlap.)

We now briefly sketch some of the procedures that may be employed to deal with these problems. See also Box, Hunter, and Hunter (1978, Chapters 13–16). Our book will then discuss more fully the use of response surface methods for empirical model building.

Screening (WHICH Stage)

It often happens at the beginning of an investigation that there is rather a long list of variables $\xi_1, \xi_2, \ldots$ which could be of importance in affecting η. One way to reduce the list to manageable size is to sit down with the investigator (the biologist, chemist, psychologist, etc.) and to ask him to pick out the variables he believes to be most important. To press this too far, however, is dangerous because, not infrequently, a variable initially believed unimportant turns out to have a major effect. A good compromise is to employ a preliminary screening design such as a two-level fractional factorial (see Chapter 5) to pick out variables worthy of further study.

In one investigation, for instance, the original list of variables $\xi_1, \xi_2, \ldots$ that might have affected the response η contained 11 candidates. Three of these were, after careful thought, eliminated as being sufficiently unimportant to be safely ignored. A 16-run two-level fractional factorial design was run on the remaining eight variables and four of the eight were designated as probably influential over the ranges studied. Three of these four had already been selected by the investigator as likely to be critical, confirming his judgement. The fourth was unexpected and turned out to be of great importance. Screening designs are often carried out sequentially in small blocks and are very effective when performed in this way. For additional details see the papers by Box and Hunter (1961a, b, especially pp. 334–337) or see Box, Hunter, and Hunter (1978, Chapter 13).

Empirical Model Building (HOW Stage)

When input variables are quantitative, and the experimental error is not too large compared with the range covered by the observed responses (see Section 8.2), it may be profitable to attempt to *estimate* the response function within some area of immediate interest. In many problems, the form of the true response function $f(\xi, \theta)$ is unknown and cannot economically be obtained but may be capable of being locally approximated by a polynomial or some other type of graduating function, $g(\xi, \beta)$, say. Suitable experimental designs for this purpose have been developed. The essentially iterative nature of *response surface methodology* (RSM) would ensure that, as the investigation proceeded, it would be possible to learn about (a) the amount of replication needed to achieve sufficient precision, (b) the location of the experimental region of most interest, (c) appropriate scalings and transformations for the input and output variables, and (d) the degree of complexity of an approximating function, and hence of the designs, needed at various stages. These matters are discussed in their Chapters 14 and 15 by Box, Hunter, and Hunter (1978), and, more fully, in this book.

Mechanistic Model Building (WHY Stage)

If it were possible, we might like to use the *true* functional form $f(\xi, \theta)$ to represent the response, rather than to approximate it by a graduating function. In some problems, we can hope to achieve useful working mechanistic models which, at least, take account of the *principal* features of the mechanism. These models often are most naturally expressed in terms of differential equations or other nonexplicit forms, but modern developments in computing facilities and in the theory of nonlinear design and estimation have made it possible to cope with the resulting problems. A mechanistic model has the following advantages:

1. It contributes to our scientific understanding of the phenomenon under study.
2. It usually provides a better basis for extrapolation (at least to conditions worthy of further experimental investigation if not through the entire ranges of all input variables).
3. It tends to be parsimonious (i.e., frugal) in the use of parameters and to provide better estimates of the response.

Results from fitting mechanistic models have sometimes been disappointing because not enough attention has been given to discovering what *is* an appropriate model form. It is easy to collect data that never "place the postulated model in jeopardy" and so it is common (e.g., in chemical engineering) to find different research groups each advocating a different model for the same phenomenon and each proffering data that "prove" their claim. In such cases, methods that discriminate between the various candidate models must be applied. See Box, Hunter, and Hunter (1978, Chapter 16) for additional discussion.

Comment

As we have said, the lines between the WHICH, the HOW, and the WHY stages of investigation are often ill-defined in practice. Thus, toward the end of a (WHICH) screening investigation, for example, we may employ a fractional factorial design that is then used as the first of a series of building blocks in a response surface study (HOW). In other circumstances (see, e.g., Box and Youle, 1955) careful study of an empirical fitted response surface at the HOW stage can generate ideas about the possible underlying mechanism (WHY). The present text is concerned mostly with empirical model building, that is, the HOW stage of experimentation, and specifically with what has become to be called *response surface methodology* or RSM. The various aspects of RSM will be dealt with in the chapters that follow.

1.5. NEED FOR EXPERIMENTAL DESIGN

It sometimes happens (e.g., in investigations of industrial plant processes) that a large amount of past operating data is available. It may then be urged that no experimentation is actually needed because it ought to be possible to extract information relating changes in a response of interest to changes that have occurred *naturally* in the input variables. Such investigations are often valuable as preliminary studies, but the existence of such data rarely eliminates the need for further planned experimentation. There are several reasons for this. For example:

1. Important input variables affecting the response are often the very ones that are not varied.

2. Relations between the response variable and various input variables may be induced by unrecorded "lurking" variables that affect both the response and the input variables. These can give rise to "nonsense correlations."

3. Historical operating data often contain gaps and omit important ancillary information.

Matters such as these are more fully discussed in Box, Hunter, and Hunter (1978, pp. 487–498).

1.6. GEOMETRIC REPRESENTATION OF RESPONSE RELATIONSHIPS

Our ability to think about systems with multiple inputs and outputs is greatly enhanced by geometrical representation and, in particular, by the use of contour diagrams. Some examples are given in Figure 1.5. The curves of Figure 1.5a, for example, represent a relationship between yield of product η, and reaction ξ_1 at three different levels of catalyst concentration ξ_2.

In Figure 1.5b, a three-dimensional representation is given, and the three curves in Figure 1.5a are now sections of the response surface in Figure 1.5b.

If we were to slice off portions of the response surface at various horizontal levels of the mean yield, η, and then project the outlines of the slices onto the (ξ_1, ξ_2) base of Figure 1.5b, we would obtain the *contours* (or *contour lines*) in Figure 1.5c. We call Figure 1.5c a *contour diagram*. A

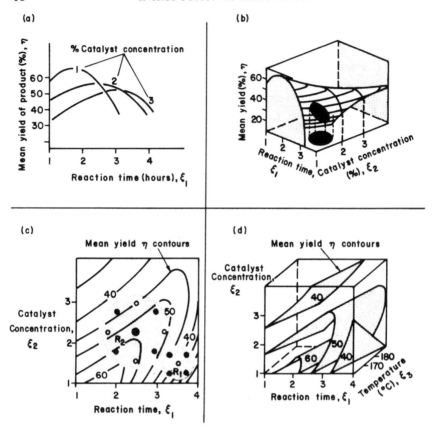

FIGURE 1.5. Geometrical representations of a relationship connecting the mean yield of product (%), η, with reaction time, ξ_1, catalyst concentration (%), ξ_2, and temperature (°C), ξ_3.

contour diagram has the advantage that a relationship between a response η and two predictors ξ_1 and ξ_2 (i.e., a relationship that involves three variables) can be represented in only two-dimensional space. Figure 1.5d is also a contour diagram but one in which a third dimension, representing a third predictor variable, temperature, has been added, and three-dimensional *contour surfaces* drawn. If we took a section, at a selected, fixed temperature, of the three-dimensional representation in Figure 1.5d, we would obtain a two-dimensional contour diagram of the form of Figure 1.5c. We see that, with only *three* dimensions at our disposal, we can actually think about the relationships between *four* variables $\eta, \xi_1, \xi_2, \xi_3$. Because the contour diagrams are drawn in the space of the input variables, the ξ's, each potential experimental run will appear simply as a point in this input space. A special selected pattern of points chosen to investigate a response function relationship is called an *experimental design*. Two such experimental design patterns are shown in the regions R_1 and R_2 in Figure 1.5c.

1.7. THREE KINDS OF APPLICATIONS

Response surface methods have proved valuable in the solution of a large variety of problems. We now outline three common classes of such problems using an industrial example for illustration.

A particular type of extruded plastic film should possess a number of properties, one of which is high transparency. This property, regarded as a response η_1, will be affected by such input variables as the screw speed of extrusion ξ_1, and the temperature of the extruder barrel, ξ_2.

1. Approximate Mapping of a Surface Within a Limited Region

Suppose we were exploring the capability of an extruding machine used in actual manufacture. Such a machine would usually be designed to run at a particular set of conditions, but moderate adjustments in the speed ξ_1, and temperature ξ_2, might be possible. If the unknown response function, $f(\xi_1, \xi_2)$ could be graduated by, say, a polynomial $g(\xi_1, \xi_2)$ in the limited region $R(\xi_1, \xi_2)$ over which the machine could be run, we could approximately predict the transparency of the product for any specified adjustment of the machine.

2. Choice of Operating Conditions to Achieve Desired Specifications

It frequently happens that we are interested in more than one property of the product. For example, in the case of the extruded film, we might be interested in tear strength η_2 in addition to transparency η_1.

It is convenient to visualize the situation by thinking of the contours of η_1 and η_2 superimposed in the ξ_1 and ξ_2 space. (See Figure 1.6) For clarity, we can think of the contours of η_1 as being drawn in red ink and those of η_2 drawn in green ink. Actual diagrams and models in which two or more sets of contours are superimposed are in fact often used in elucidating possibilities for industrial systems.

Suppose we desired a high transparency $\eta_1 > \eta_{10}$ and a high tear strength $\eta_2 > \eta_{20}$ where η_{10}, η_{20} are specified values. Then, if the responses

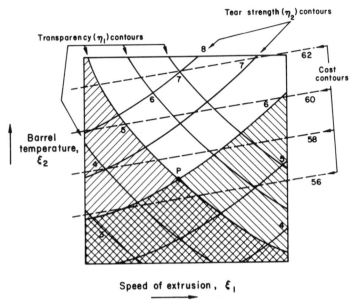

FIGURE 1.6.   The unshaded region contains values (ξ_1, ξ_2) in which the desired product with transparency $\eta_1 > 5$ and tear strength $\eta_2 > 6$ is obtained.

transparency and tear strength were known to be graduated, respectively, by

$$\eta_1 = g_1(\xi_1, \xi_2), \qquad \eta_2 = g_2(\xi_1, \xi_2),$$

we could draw the approximate critical contours $\eta_1 = \eta_{10}, \eta_2 = \eta_{20}$ on the same figure and hence determine the region in the (ξ_1, ξ_2) space in which *both* inequalities were satisfied and so *both* responses attained satisfactory values. Figure 1.6 illustrates how this might be done. The shading indicates the regions in the space of the input variables in which an undesirable product would be obtained. Transparencies greater than 5 and tear strengths greater than 6 can be obtained within the unshaded region.

In practice, of course, there may be more than two predictors and/or more than two response variables. One response that would often need to be examined is the overall average cost associated with any particular set of operating conditions. Thus we might seek to obtain a product which satisfied the specification $\xi_1 > \xi_{10}, \xi_2 > \xi_{20}$ *and* had the smallest possible cost η_3. We see from Figure 1.6 that manufacture at conditions corresponding to the point P would just satisfy the specifications and at the same time, lead to a minimum cost.

The formal description for the problem above has many aspects closely related to linear and nonlinear programming problems. The situation we consider here however, is more complicated, because the relationships between the η's and ξ's are not given but must be estimated.

3. Search for Optimal Conditions

In the examples above, we imply that reasonably satisfactory manufacturing conditions have already been determined. However, at an earlier stage of process development, work would have been carried out in the laboratory and pilot plant to determine suitable conditions. Frequently, there would be a large region in the space of the variables ξ_1 and ξ_2 where the process was operable. An important problem is that of finding, within this operability region, the *best* operating conditions. Let us suppose that the merit of any particular set (ξ_1, ξ_2) of input conditions can be assessed in terms of a response η representing, say, an overall measure of profitability. The problem would be that of locating, and determining the characteristics of, the optimum conditions.

In the chapters that follow, we shall discuss further the details of response surface methodology and how the various applications of it can be made.

CHAPTER 2

The Use Of Graduating Functions

2.1. APPROXIMATING RESPONSE FUNCTIONS

Basic Underlying Relationship

We have said that investigations of physical, chemical, and biological systems are often concerned with the elucidation of some functional relationship

$$E(y) = \eta = f(\xi_1, \xi_2, \ldots, \xi_k)$$

$$= f(\boldsymbol{\xi}) \tag{2.1.1}$$

connecting the expected value (i.e., the mean value) of a response y such as the yield of a product with k quantitative variables $\xi_1, \xi_2, \ldots, \xi_k$ such as temperature, time, pressure, concentration, and so on. In some problems, a number of different responses may be of interest, as for example, percentage yield y_1, purity y_2, viscosity y_3, and so on. For each of these, there will be a separate functional relationship. In this book we assume that the natures of these various functional relationships are not known, and that local polynomial approximations are used instead.

In what follows it is convenient not to have to deal with the actual numerical measures of the variables ξ_i, but instead to work with coded or "standardized" variables x_i. For example, if at some stage of an investigation we defined the current region of interest for ξ_i to be $\xi_{io} \pm S_i$, where ξ_{io} is the center of the region, then it would be convenient to define an equivalent working variable x_i where

$$x_i = \frac{\xi_i - \xi_{io}}{S_i}.$$

Thus, for example, if ξ_i is temperature and the current region of interest is $115 \pm 10°$, then for any setting of temperature we have the equivalent coded value $x_i = (\xi_i - 115)/10$. We note that the coded quantities x_i are simply convenient linear transformations of the original ξ_i, and so expressions containing the x_i can always be readily rewritten in terms of the ξ_i.

Polynomial Approximations

In general, a polynomial in the coded inputs $x_1, x_2, \ldots, x_k$ is a function which is a linear aggregate (or combination) of powers and products of the x's. A term in the polynomial is said to be of order j (or degree j) if it contains the product of j of the x's (some of which may be repeated). Thus terms involving x_1^3, $x_1 x_2^2$, and $x_1 x_2 x_3$ would all be said to be of order 3 (or degree 3). A polynomial is said to be of order d, or degree d, if the term(s) of highest order in it is (are) of order or degree d. Thus, if $k = 2$ and if x_1 and x_2 denote two coded inputs, the general polynomial can be written

$$g(\mathbf{x}, \boldsymbol{\beta}) = \beta_0 + (\beta_1 x_1 + \beta_2 x_2) + (\beta_{11} x_1^2 + \beta_{22} x_2^2 + \beta_{12} x_1 x_2)$$

$$+ (\beta_{111} x_1^3 + \beta_{222} x_2^3 + \beta_{112} x_1^2 x_2 + \beta_{122} x_1 x_2^2) + (\beta_{1111} x_1^4 \ldots)$$

$$+ \text{etc.,} \tag{2.1.2}$$

where terms of the same order are bracketed for convenience. Note that the subscript notation is chosen so that each β coefficient can be easily identified with its corresponding x term. For example, β_{122} is the coefficient of $x_1 x_2 x_2$, that is, $x_1 x_2^2$. In the expression (2.1.2), the β's are coefficients or (empirical) *parameters* which, in practice, have to be estimated from the data.

As is seen from Table 2.1, the number of such parameters increases rapidly as the number, k, of the input variables and the degree, d, of the polynomial are both increased.

A polynomial expression of degree d can be thought of as a Taylor's series expansion of the true underlying theoretical function $f(\boldsymbol{\xi})$ truncated after terms of dth order. The following will usually be true:

1. The higher the degree of the approximating function, the more closely the Taylor series can approximate the true function.
2. The smaller the region R over which the approximation needs to be made, the better is the approximation possible with a polynomial function of given degree.

TABLE 2.1. Number of coefficients in polynomials of degree d involving k inputs

| Number of inputs, k | Degree of polynomial, d | | | |
| | 1 | 2 | 3 | 4 |
	Planar	Quadratic	Cubic	Quartic
2	3	6	10	15
3	4	10	20	35
4	5	15	35	70
5	6	21	56	126

In practice, we can often proceed by supposing that, over limited regions of the factor space, a polynomial of only first or second degree might adequately represent the true function.

First Degree (or First-Order) Approximation

In Eq. (2.1.2) the first set of parentheses contains first-order terms. If we truncated the expression at this point, we should have the first degree polynomial approximation for $k = 2$ predictor variables, x_1 and x_2,

$$g(\mathbf{x}, \boldsymbol{\beta}) = \beta_0 + \beta_1 x_1 + \beta_2 x_2 \qquad (2.1.3)$$

capable of representing a tilted plane. The height and tilt of the plane are determined by the coefficients β_0, β_1, and β_2. Specifically (as in Figure 2.1a) β_0 is the intercept of the plane with the g axis at the origin of x_1 and x_2 and β_1 and β_2 are the gradients (slopes) in the directions x_1 and x_2. The height contours of such a plane would be equally spaced parallel straight lines. Some of these contours are shown on the plane in Figure 2.1a, and projected onto the (x_1, x_2) plane in Figure 2.1b. (Some possible height readings are attached to the contours to help the visualization.) We shall discuss such approximations in considerably more detail in Chapter 6.

Second Degree (or Second Order) Approximation

If we truncated the expression (2.1.2) at the second set of parentheses, we would have the polynomial approximation of second degree for $k = 2$

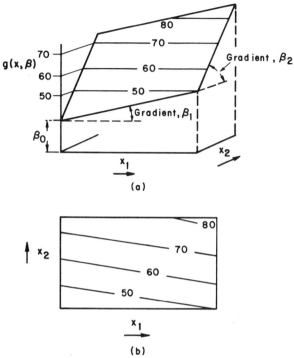

FIGURE 2.1. (*a*) Geometrical representation of a first degree polynomial (planar approxima-
tion) $g(\mathbf{x}, \boldsymbol{\beta}) = \beta_0 + \beta_1 x_1 + \beta_2 x_2$. (*b*) The equal height contours of $g(\mathbf{x}, \boldsymbol{\beta})$ projected onto
the (x_1, x_2) plane.

predictor variables, x_1, and x_2,

$$g(\mathbf{x}, \boldsymbol{\beta}) = \beta_0 + \beta_1 x_1 + \beta_2 x_2 + \beta_{11} x_1^2 + \beta_{22} x_2^2 + \beta_{12} x_1 x_2. \quad (2.1.4)$$

This defines what is called a general second order (or quadratic) surface,
here in two variables x_1 and x_2 only.

Figure 2.2 illustrates how, by suitable choices of the coefficients, the
second order surface in x_1 and x_2 can take on a variety of useful shapes.
Both contour plots and associated surfaces are shown.

A simple maximum is shown in Figure 2.2*a*, and a stationary or flat ridge
in Figure 2.2*b*. Figure 2.2*c* shows a rising ridge, and Figure 2.2*d* shows
what is variously called a col, saddle, or minimax. Although even locally the
true underlying model cannot be expected to correspond *exactly* with such
forms, nevertheless the main features of a true surface could often be
approximated by one of these forms. We shall discuss approximations in
considerably more detail in Chapters 9–11.

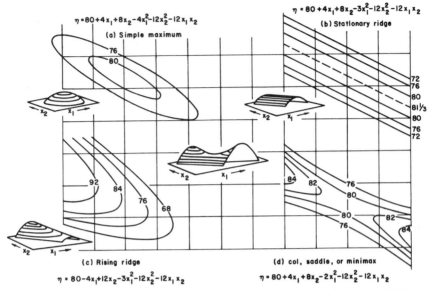

FIGURE 2.2. Some examples of types of surfaces defined by second-order polynomials in two predictor variables x_1 and x_2.

Relationship Between the Approximating Polynomial and the "True" Underlying Response Surface

To appreciate the potential usefulness of such empirical approximations, consider Figure 2.3. This shows a theoretical response surface in which the yield of a product η is represented as a function of time of reaction in hours, ξ_1, and the absolute temperature of reaction, ξ_2, in degrees Kelvin. A contour diagram corresponding to this surface is shown in Figure 2.4. [The theoretical function shown in Figures 2.3 and 2.4 arises from a particular physical theory about the manner in which the reaction of interest occurred. This theory leads to a set of differential equations whose solution is the function shown. The function itself, $\eta = f(\xi_1, \xi_2; \theta_1, \theta_2, \theta_3, \theta_4)$, which depends on two predictor variables ξ_1 and ξ_2 and on four physical constants, $\theta_1, \theta_2, \theta_3, \theta_4$, is given in Appendix 2A. For our present purposes, it is sufficient to appreciate that, underlying any system, there is a mechanism which, if fully understood, could be represented by a theoretical relationship $f(\xi, \theta)$.]

If we *knew* the *form* of the theoretical relationship (that is, the nature of the function f) but did not know the values of the parameters θ, then usually it would be best to fit this theoretical function directly to the data.

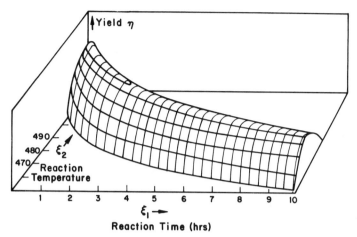

FIGURE 2.3. A theoretical response surface showing a relationship between response η (yield of product) as a function of predictors ξ_1 (reaction time in hours), and ξ_2 (reaction temperature in degrees Kelvin).

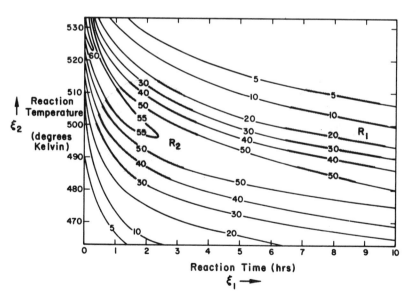

FIGURE 2.4. Yield contours of a theoretical response surface, showing the relationship between response η (yield of product) as a function of predictors ξ_1 (reaction time in hours), and ξ_2 (reaction temperature in degrees Kelvin).

25

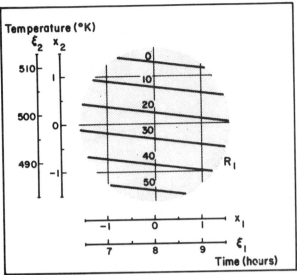

FIGURE 2.5. Local representation in region R_1 of the response surface in Figure 2.4. The contours plotted are those of the first degree polynomial $24.5 - 2x_1 - 18.5x_2$ where $x_1 = \xi_1 - 8$, and $x_2 = (\xi_2 - 498)/10$.

If, as we suppose here, we *do not* know the form of the function, then the relationship can often be usefully approximated, over a limited region, by a simple polynomial function. The purpose of this approximation is not to represent the true underlying relationship everywhere, but merely to graduate it locally. This is illustrated in Figure 2.5 which should be compared with Figure 2.4. We see that it is possible to represent the main features of the true relationship moderately well over the region labeled R_1 by the first degree polynomial

$$\eta \simeq 24.5 - 2x_1 - 18.5x_2,$$

in which the input variables are represented through the convenient coding

$$x_1 = (\xi_1 - 8)/1, \qquad x_2 = (\xi_2 - 498)/10.$$

The approximation is far from perfect. In particular, it is of too simple a kind to represent the uneven spacing of the theoretical contours. However, it does convey the correct impression that the main feature of the surface in the region is a large negative gradient in temperature. Furthermore, the inevitable experimental error met in practice might make dubious the usefulness of greater elaboration at this stage.

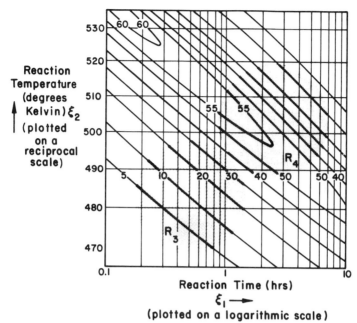

FIGURE 2.6. The contours of Figure 2.4 after transformation; ξ_1 is plotted on a logarithmic scale, and ξ_2 is plotted on a reciprocal scale.

Over the region R_2 in Figure 2.4, a second degree polynomial representing a surface like that of Figure 2.2c could, again, very approximately represent the function locally.

Improvement in Approximation Obtainable by Transformation

As we mentioned in Chapter 1, a considerable improvement in representational capability can often be obtained by allowing the possibility of transformations in the inputs and outputs. For illustration, we reconsider the kinetic example introduced earlier. Figure 2.6 is a contour diagram of the same function as plotted in Figure 2.4. However, in Figure 2.6, reaction time ξ_1 is plotted on a log scale, and temperature in degrees Kelvin ξ_2 is plotted on a reciprocal scale.* It is easy to see that considerably closer approximations by first and second degree polynomials would be possible

---

*The reciprocal axis is reversed to allow more ready comparison between Figures 2.4 and 2.6. *Over the range studied*, reciprocal temperature is almost a linear function of temperature so that, for this example, the improved approximation is almost totally due to plotting reaction time on a log scale.

TABLE 2.2.   Data on worsted yarn: Cycles to failure under various loading conditions

Length of test specimen (mm), ξ_1	Amplitude of load cycle (mm), ξ_2	Load (g), ξ_3	Cycles to failure, Y	Antilog of $\hat{y}$
250	8	40	674	692
350	8	40	3636	3890
250	10	40	170	178
350	10	40	1140	1000
250	8	50	292	309
350	8	50	2000	1738
250	10	50	90	94
350	10	50	360	447

after these transformations had been made. In particular, a fitted plane in the region R_3 and a fitted quadratic surface in R_4 would now represent the transformed function quite closely.

2.2.   AN EXAMPLE

We now give an example which has been chosen (a) to illustrate, in an elementary way, some of the considerations already mentioned, and (b) to motivate the work of the chapters which follow.

An Experiment in Worsted Yarn

Table 2.2 shows a partial* listing of data from an unpublished report to the Technical Committee, International Wool Textile Organization, by Dr. A. Barella and Dr. A. Sust. These numbers were obtained in a textile investigation of the behavior of worsted yarn under cycles of repeated loading. The final column should be ignored for the moment. Because of the wide range of variation of Y (from 90 to 3636 in this data set), it is more natural to consider an analysis in terms of $y = \log Y$ (the base 10 is used) rather than in terms of Y itself. Also the input or predictor variables are

---

*The complete set of data, and a more comprehensive analysis, are given in Chapter 7. These data were quoted and analyzed by Box and Cox (1964).

TABLE 2.3. Coded data on worsted yarn

Specimen length (coded), x_1	Amplitude (coded), x_2	Load (coded), x_3	Log cycles to failure, y	Fitted value log cycles, $\hat{y}$
-1	-1	-1	2.83	2.84
1	-1	-1	3.56	3.59
-1	1	-1	2.23	2.25
1	1	-1	3.06	3.00
-1	-1	1	2.47	2.49
1	-1	1	3.30	3.24
-1	1	1	1.95	1.90
1	1	1	2.56	2.65

more conveniently used in the coded forms

$$x_1 = (\xi_1 - 300)/50, \qquad x_2 = (\xi_2 - 9)/1, \qquad x_3 = (\xi_3 - 45)/5.$$

$$(2.2.1)$$

The transformed data are shown in Table 2.3. (Again, ignore the $\hat{y}$ column for the moment.) We shall show later that the transformed data can be fitted very well by a simple first degree polynomial in x_1, x_2, and x_3 of form

$$g(\mathbf{x}, \boldsymbol{\beta}) = \beta_0 + \beta_1 x_1 + \beta_2 x_2 + \beta_3 x_3, \qquad (2.2.2)$$

where we estimate the coefficients $\beta_0, \beta_1, \beta_2, \beta_3$ from the data. If we denote the respective estimates by b_0, b_1, b_2, b_3 and the *fitted value*, that is, the response obtained from the fitted equation at a general point (x_1, x_2, x_3) by $\hat{y}$, then

$$\hat{y} = b_0 + b_1 x_1 + b_2 x_2 + b_3 x_3. \qquad (2.2.3)$$

For this particular set of eight data values, we show later that estimates of the β coefficients, evaluated via the method of least squares, are

$$b_0 = 2.745 \pm 0.025,$$

$$b_1 = 0.375 \pm 0.025,$$

$$b_2 = -0.295 \pm 0.025,$$

$$b_3 = -0.175 \pm 0.025. \qquad (2.2.4)$$

(The numbers following the $\pm$ signs are the standard errors of the estimates.) Thus the fitted equation derived from the data is

$$\hat{y} = 2.745 + 0.375x_1 - 0.295x_2 - 0.175x_3. \qquad (2.2.5)$$

For comparison, the $\hat{y}$'s evaluated from this fitted equation by substituting the appropriate x values are shown beside the actual observed y's in Table 2.3. The antilogs of these $\hat{y}$'s are shown in Table 2.2, opposite the corresponding observed values of the Y's. It will be seen that there is a close agreement between each Y, the actual number of cycles to failure observed, and the corresponding antilog $\hat{y}$ predicted by the fitted equation.

Figure 2.7a shows contours of the fitted equation for $\hat{y}$ in the x_1, x_2, x_3 space computed from the linear expression Eq. (2.2.5). In Figure 2.7b, these same scales and contours have been relabeled in terms of the original variables $\xi_1, \xi_2,$ and ξ_3, and the original observations are shown at the corners of the cube.

A first inspection of Figure 2.7 certainly suggests that a first degree approximation in terms of the log response provides an excellent representation of the relationship between the response variable, cycles to failure, and the input variables, length, amplitude, and load.

If this is so, then the relationship could be of great practical value.

1. It could indicate the direction and magnitude of change in response when changes are made in each of the three input variables.
2. It could allow calculations to be made of the response at intermediate conditions not actually tested.
3. It could show the direction in which to move if we wish to change the input levels so as to increase the response "cycles to failure" as much as possible.

However, this example raises a number of questions. Some of these we now set out, and we also indicate where in this book these questions will be answered.

SOME QUESTIONS THAT ARISE

1. *Least squares.* The estimated expression $\hat{y} = 2.745 + 0.375x_1 - 0.295x_2 - 0.175x_3$ was obtained by fitting a first degree polynomial of general form (2.2.2) to the data of Table 2.3 by the method of least squares. What *is* the method of least squares? What are its assumptions, and how may it be generally applied to the fitting of such relationships? (Chapter 3.)

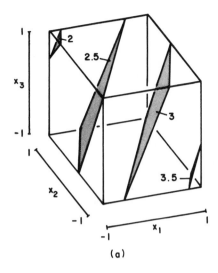

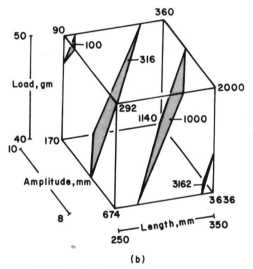

FIGURE 2.7. (*a*) Contours of fitted surface $\hat{y} = 2.745 + 0.375x_1 - 0.295x_2 - 0.175x_3$, with the transformation $y = \log Y$ applied to the original Y observations. (*b*) Contours of fitted surface as in (*a*) but labeled in original units of Y; the actual observations are shown at the cube corners.

2. *Standard errors of coefficients.* If relationships of this kind are to be intelligently employed, we need measures of precision of the estimated coefficients. In (2.2.4), standard errors, which are estimates of the standard deviations of the coefficients, were obtained from the data. How can such estimates be calculated? (Chapter 3.)

3. *Using first-order response models.* How may we use such expressions to ascertain a direction of increasing response, perhaps pointing the way toward a maximum response? (Chapter 6.)

4. *Higher order expressions.* How may we fit and use higher-order polynomial models, in particular, general second-order expressions? How may fitted expressions such as in (1) be used to indicate sets of conditions for which *several* responses achieve desired specifications. (Chapters 7, 9, 10, and 11.)

5. *Adequacy of fit.* In approximating an unknown theoretical function empirically, we need to be able to check whether or not a given degree of approximation is adequate. How can analysis of variance (ANOVA) and the examination of residuals (the discrepancies between observed and fitted values) help to check adequacy of fit? (Chapters 3 and 7.)

6. *Designs.* A simple arrangement of experiments was employed in the above example in which all eight combinations of two levels of the input variables were run. This is called a 2^3 *factorial design*. What designs are suitable for fitting polynomials of first and second degrees? (Chapters 4, 5, 15, and 13.)

7. *Transformations.* By making a log transformation of the output variable Y above, a very simple first degree relationship could be employed. How can such transformations be found in general? (Chapters 7 and 8.)

The motivation for topics discussed in chapters not specifically mentioned above will become clearer as we proceed.

APPENDIX 2A. A THEORETICAL RESPONSE FUNCTION

The function plotted in Figures 2.3 and 2.4 is given by

$$\eta = k_1\{\exp(-k_1\xi_1) - \exp(-k_2\xi_2)\}/(k_1 - k_2), \qquad (2A.1)$$

where

$$k_1 = \theta_1\exp\{-\theta_2/\xi_2\}, \qquad k_2 = \theta_3\exp\{-\theta_4/\xi_2\},$$

$$\xi_1 = \text{reaction time (h)},$$

and $\xi_2 = \text{reaction temperature}(^\circ\text{K}) = T + 273, \qquad (2A.2)$

where T is temperature in degrees Celsius. Expression (2A.1) arises if η is the fractional yield of the intermediate product B in a consecutive chemical reaction $A \rightarrow B \rightarrow C$ subject to first-order kinetics, with the dependence of the rates k_1, k_2 on temperature following the Arrhenius law. The actual parameter values which were used to give Figures 2.3 and 2.4 are

$$\theta_1 = 5.0259 \times 10^{16}, \qquad \theta_2 = 1.9279 \times 10^4,$$

$$\theta_3 = 1.3862 \times 10^{14}, \qquad \theta_4 = 1.6819 \times 10^4. \qquad (2A.3)$$

CHAPTER 3

Least Squares for Response Surface Work

Readers with an understanding of least squares (regression analysis) may wish to skip this chapter entirely, or simply dip into it as necessary. The chapter does not, of course, represent a complete treatment of least squares. For that, the reader should consult one of the standard texts.

3.1. THE METHOD OF LEAST SQUARES

We saw in Chapter 2 that some method was needed for fitting empirical functions to data. The fitting procedure we employ is the *method of least squares*. Given certain assumptions which are later discussed and which often approximately describe real situations, the method has a number of desirable properties.

We shall present the basic results needed, omitting proofs and illustrating with examples as we proceed. Because matrices are necessary to develop these methods efficiently, we provide a brief review of the required matrix algebra in Appendix 3C.

Least-Squares Estimates

We have said in Chapter 2 [see Eq. (2.1.1)] that the investigator often wishes to elucidate some model

$$y = f(\xi, \theta) + \varepsilon, \tag{3.1.1}$$

where

$$E(y) = \eta = f(\xi, \theta) \tag{3.1.2}$$

is the mean level of the response y which is affected by k variables $(\xi_1, \xi_2, \ldots, \xi_k) = \boldsymbol{\xi}'$. The model involves, in addition, p parameters $(\theta_1, \theta_2, \ldots, \theta_p) = \boldsymbol{\theta}'$, and ε is an experimental error. To examine this model, the experimenter may make a series of experimental runs at n different sets of conditions $\boldsymbol{\xi}_1, \boldsymbol{\xi}_2, \ldots, \boldsymbol{\xi}_n$, observing the corresponding values of the response, $y_1, y_2, \ldots, y_n$. Two important questions that arise are:

1. Does the postulated model adequately represent the data?
2. Assuming the model does adequately represent the data, what are the best estimates of the parameters in $\boldsymbol{\theta}$?

Somewhat paradoxically we need to study the second of these two questions first, leaving the first question until later.

Suppose some specific values were chosen for the parameters on some basis or another. This would enable us to compute $f(\boldsymbol{\xi}, \boldsymbol{\theta})$ for each of the experimental runs $\boldsymbol{\xi}_1, \boldsymbol{\xi}_2, \ldots, \boldsymbol{\xi}_n$ and hence to obtain the n discrepancies $\{y_1 - f(\boldsymbol{\xi}_1, \boldsymbol{\theta})\}, \{y_2 - f(\boldsymbol{\xi}_2, \boldsymbol{\theta})\}, \ldots, \{y_n - f(\boldsymbol{\xi}_n, \boldsymbol{\theta})\}$. The method of least squares selects, as the best estimate of $\boldsymbol{\theta}$, the value that makes the sum of squares of these discrepancies, namely

$$S(\boldsymbol{\theta}) = \sum_{u=1}^{n} \{y_u - f(\boldsymbol{\xi}_u, \boldsymbol{\theta})\}^2, \qquad (3.1.3)$$

as small as possible. $S(\boldsymbol{\theta})$ is called the *sum of squares function*. For any given choice of the p parameters in $\boldsymbol{\theta}$, there will be a specific value of $S(\boldsymbol{\theta})$. The minimizing choice of $\boldsymbol{\theta}$, its *least-squares estimate*, is denoted by $\hat{\boldsymbol{\theta}}$; the corresponding (minimized) value of $S(\boldsymbol{\theta})$ is thus $S(\hat{\boldsymbol{\theta}})$.

Are the least-squares estimates of the θ's "good" estimates? In general, their goodness depends on the nature of the distribution of the errors. Specifically, we shall see later that least-squares estimates would be appropriate if it could be assumed that the experimental errors $\varepsilon_u = y_u - \eta_u$, $u = 1, 2, \ldots, n$ were *statistically independent*, with *constant variance*, and were *normally distributed*. These "standard assumptions" and their relevance are discussed later; for the moment, we concentrate on the problem of how to calculate least-squares estimates.

3.2. LINEAR MODELS

A great simplification in the computation of the least-squares estimates occurs when the response function is linear in the parameters, that is, of the form

$$\eta = f(\boldsymbol{\xi}, \boldsymbol{\theta}) = \theta_1 z_1 + \theta_2 z_2 + \cdots + \theta_p z_p. \qquad (3.2.1)$$

In this expression, the z's are known constants and, in practice, are known functions of the experimental conditions $\xi_1, \xi_2, \ldots, \xi_k$ or, equivalently, of their coded forms $x_i = (\xi_i - \xi_{i0})/S_i$, $i = 1, 2, \ldots, k$, where ξ_{i0} and S_i are suitable location and scale factors, respectively. Adding the experimental error $\varepsilon = y - \eta$ we thus have the model

$$y = \theta_1 z_1 + \theta_2 z_2 + \cdots + \theta_p z_p + \varepsilon. \qquad (3.2.2)$$

Any model of this form is said to be a *linear model*, that is, *linear in the parameters*. Linear models are more widely applicable than might at first appear, as will be clear from the following examples.

EXAMPLE 3.1. An experimenter believes that the electrical conductivity y of cotton fibre depends on the humidity ξ and that, over the range of humidity of interest, an approximately linear relationship, obscured by experimental error, will exist. It is convenient to transform linearly a variable such as humidity ξ to coded form $x = (\xi - \xi_0)/S$, where ξ_0 is some convenient origin and S is a scale factor. The model may then be written

$$\eta = \beta_0 + \beta_1 x, \qquad (3.2.3)$$

where β_0 represents the conductivity at $x = 0$ (i.e., at $\xi = \xi_0$) and β_1 the increase in conductivity per unit increase in x (i.e., per S units of humidity). In this example, there is $k = 1$ predictor variable, namely the humidity x, and there are $p = 2$ parameters, β_0 and β_1. The response function can be written in the general form (3.2.1) as

$$\eta = \theta_1 z_1 + \theta_2 z_2 \qquad (3.2.4)$$

by setting $z_1 = 1$, $z_2 = x = (\xi - \xi_0)/S$, $\theta_1 = \beta_0$, and $\theta_2 = \beta_1$.

EXAMPLE 3.2. Suppose, in the foregoing example, that the relationship between the mean electrical conductivity and the humidity was expected to be curved. This might be provided for by adding a term in x^2 so that the postulated response function was

$$\eta = \beta_0 + \beta_1 x + \beta_{11} x^2. \qquad (3.2.5)$$

Identification with the general form

$$\eta = \theta_1 z_1 + \theta_2 z_2 + \theta_3 z_3 \qquad (3.2.6)$$

is provided by setting $z_1 = 1$, $z_2 = x$, $z_3 = x^2$, $\theta_1 = \beta_0$, $\theta_2 = \beta_1$, and $\theta_3 = \beta_{11}$. Note that this is an example of a function which is linear in the parameters but nonlinear (in fact quadratic) in x. It is clear that we can cover any of the polynomial models discussed in Chapter 2 using the general linear model formulation.

EXAMPLE 3.3. It is postulated that the presence of a certain additive in gasoline increases gas mileage y. A number of trials are run on a standard Volkswagen engine *with* and *without* the additive. The experimenter might employ the response function

$$\eta = \beta_0 + \beta_1 x \qquad (3.2.7)$$

in which x is used simply to denote *presence* or *absence* of the additive. Thus

$$x = 0 \qquad \text{if additive absent, and}$$

$$x = 1 \qquad \text{if additive present} \qquad (3.2.8)$$

and then β_0 represents the mean gas mileage with no additive while β_1 is the incremental effect of the additive. Here $k = 1$; the single predictor variable x does not have a continuous scale but can take only the values 0 or 1. Such a variable is sometimes called an *indicator*, or *dummy*, *variable*. There are $p = 2$ parameters, β_0 and β_1, and the model function in the general form (3.2.1) is as in Eq. (3.2.4).

Dummy variables are very useful to represent the effects of qualitative variables such as operator, day on which experiment is conducted, type of raw material, and so on.

An Algorithm for All Linear Least-Squares Problems

Suppose that we are in a situation where the response function

$$\eta = \theta_1 z_1 + \theta_2 z_2 + \cdots + \theta_p z_p \qquad (3.2.9)$$

has been postulated and experimental conditions $(z_{1u}, z_{2u}, \ldots, z_{pu})$, $u = 1, 2, \ldots, n$ have been run, yielding observations $y_1, y_2, \ldots, y_n$. Then the model relates the observations to the known z_{iu}'s and the unknown θ_i's by n

equations

$$y_1 = \theta_1 z_{11} + \theta_2 z_{21} + \cdots + \theta_p z_{p1} + \varepsilon_1,$$

$$y_2 = \theta_1 z_{12} + \theta_2 z_{22} + \cdots + \theta_p z_{p2} + \varepsilon_2,$$

$$\cdots$$

$$\cdots$$

$$y_n = \theta_1 z_{1n} + \theta_2 z_{2n} + \cdots + \theta_p z_{pn} + \varepsilon_n. \qquad (3.2.10)$$

These can be written in matrix form as

$$\mathbf{y} = \mathbf{Z}\boldsymbol{\theta} + \boldsymbol{\varepsilon}, \qquad (3.2.11)$$

where

$$\mathbf{y} = \begin{bmatrix} y_1 \\ y_2 \\ \vdots \\ y_n \end{bmatrix}, \quad \mathbf{Z} = \begin{bmatrix} z_{11} & z_{21} & \cdots & z_{p1} \\ z_{12} & z_{22} & \cdots & z_{p2} \\ \vdots & \vdots & & \vdots \\ z_{1n} & z_{2n} & \cdots & z_{pn} \end{bmatrix}, \quad \boldsymbol{\theta} = \begin{bmatrix} \theta_1 \\ \theta_2 \\ \vdots \\ \theta_p \end{bmatrix}, \quad \boldsymbol{\varepsilon} = \begin{bmatrix} \varepsilon_1 \\ \varepsilon_2 \\ \vdots \\ \varepsilon_n \end{bmatrix},$$
$$\quad\ n \times 1 \qquad\qquad\quad n \times p \qquad\qquad\qquad\ p \times 1 \qquad\ n \times 1$$

$$(3.2.12)$$

and where the dimensions of the vectors and matrices are given beneath them. The sum of squares function Eq. (3.1.3) is

$$S(\boldsymbol{\theta}) = \sum_{u=1}^{n} \left(y_u - \theta_1 z_{1u} - \theta_2 z_{2u} - \cdots - \theta_p z_{pu} \right)^2, \qquad (3.2.13)$$

or in matrix format,

$$S(\boldsymbol{\theta}) = (\mathbf{y} - \mathbf{Z}\boldsymbol{\theta})'(\mathbf{y} - \mathbf{Z}\boldsymbol{\theta}). \qquad (3.2.14)$$

We show later that the least-squares estimates $(\hat{\theta}_1, \hat{\theta}_2, \ldots, \hat{\theta}_p) = \hat{\boldsymbol{\theta}}'$ which minimize $S(\boldsymbol{\theta})$ are given by the solutions of the p *normal equations*

$$\hat{\theta}_1 \Sigma z_1^2 + \hat{\theta}_2 \Sigma z_1 z_2 + \cdots + \hat{\theta}_p \Sigma z_1 z_p = \Sigma z_1 y,$$

$$\hat{\theta}_1 \Sigma z_2 z_1 + \hat{\theta}_2 \Sigma z_2^2 + \cdots + \hat{\theta}_p \Sigma z_2 z_p = \Sigma z_2 y,$$

$$\cdots$$

$$\cdots$$

$$\hat{\theta}_1 \Sigma z_p z_1 + \hat{\theta}_2 \Sigma z_p z_2 + \cdots + \hat{\theta}_p \Sigma z_p^2 = \Sigma z_p y, \qquad (3.2.15)$$

where the abbreviated notation indicates that, for example,

$$\Sigma z_1 z_2 = \sum_{u=1}^{n} z_{1u} z_{2u}, \quad \text{and} \quad \Sigma z_1 y = \sum_{u=1}^{n} z_{1u} y_u. \qquad (3.2.16)$$

EXAMPLE 3.4. Consider the following data which relate to Example 3.1.

Humidity percent, ξ	20	30	40	50
Coded humidity, $x = (\xi - 35)/5$	-3	-1	1	3
Observed electrical conductivity, y	8	23	28	34

We postulate the relationship $y = \beta_0 + \beta_1 x + \varepsilon$ or $y = \theta_1 z_1 + \theta_2 z_2 + \varepsilon$, where $z_1 = 1$, $z_2 = x$, $\theta_1 = \beta_0$, and $\theta_2 = \beta_1$. Then $\Sigma z_1^2 = 4$, $\Sigma z_1 z_2 = 0$, $\Sigma z_2^2 = 20$, $\Sigma z_1 y = 93$, $\Sigma z_2 y = 83$, and the normal equations are

$$4\hat{\theta}_1 \qquad\qquad = 93,$$

$$20\hat{\theta}_2 = 83, \qquad (3.2.17)$$

whence $\hat{\theta}_1 = 23.25$, and $\hat{\theta}_2 = 4.15$. The fitted least-squares straight line is thus given by

$$\hat{y} = 23.25 + 4.15x, \qquad (3.2.18)$$

where $\hat{y}$ denotes a *fitted* or *estimated value*. The data points and the fitted line are plotted in Figure 3.1. Note that the line looks sensible in relation to the points, providing a modest, but important, visual check that the calculations are correct.

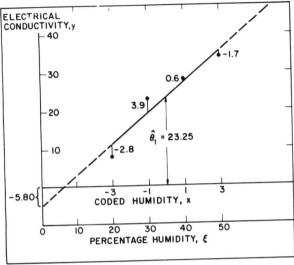

FIGURE 3.1.   Conductivity—humidity example. Plot of the data and the fitted line.

3.3.   MATRIX FORMULAS FOR LEAST SQUARES

In the general case, the p normal equations (3.2.15) are conveniently written in matrix form as

$$\mathbf{Z}'\mathbf{Z}\hat{\boldsymbol{\theta}} = \mathbf{Z}'\mathbf{y} \tag{3.3.1}$$

where $\mathbf{Z}'$ (of dimension $p \times n$) is the transpose of $\mathbf{Z}$. Thus, for the conductivity experiment,

$$\mathbf{y} = \begin{bmatrix} 8 \\ 23 \\ 28 \\ 34 \end{bmatrix}, \qquad \mathbf{Z} = \begin{bmatrix} 1 & -3 \\ 1 & -1 \\ 1 & 1 \\ 1 & 3 \end{bmatrix},$$

$$\mathbf{Z}'\mathbf{y} = \begin{bmatrix} 1 & 1 & 1 & 1 \\ -3 & -1 & 1 & 3 \end{bmatrix} \begin{bmatrix} 8 \\ 23 \\ 28 \\ 34 \end{bmatrix} = \begin{bmatrix} 93 \\ 83 \end{bmatrix},$$

$$\mathbf{Z}'\mathbf{Z} = \begin{bmatrix} 1 & 1 & 1 & 1 \\ -3 & -1 & 1 & 3 \end{bmatrix} \begin{bmatrix} 1 & -3 \\ 1 & -1 \\ 1 & 1 \\ 1 & 3 \end{bmatrix} = \begin{bmatrix} 4 & 0 \\ 0 & 20 \end{bmatrix}. \tag{3.3.2}$$

Thus the normal equations in matrix form (3.3.1) become

$$\begin{bmatrix} 4 & 0 \\ 0 & 20 \end{bmatrix} \begin{bmatrix} \hat{\theta}_1 \\ \hat{\theta}_2 \end{bmatrix} = \begin{bmatrix} 93 \\ 83 \end{bmatrix}, \tag{3.3.3}$$

which, of course, are the same equations as (3.2.17).

Conditions on the Experimental Design Necessary to Ensure that the θ's Can Be Separately Estimated

A schedule of experimental conditions is referred to as an *experimental design*. Thus, in the conductivity example, the set of conditions for humidity $\xi = 20, 30, 40, 50$, is the experimental design. The design determines the Z matrix. We here consider some elementary points about the choice of an experimental design. Consider again the linear model

$$y = \theta_1 z_1 + \theta_2 z_2 + \varepsilon, \tag{3.3.4}$$

where y might, for example, refer to the rate of reaction and z_1 and z_2 might be percentages of two catalysts A and B affecting this rate. Suppose an experimental design were chosen in which the levels of z_1 and z_2 happened to be proportional to one another, so that, for every one of the runs made, $z_2 = \delta z_1$. Thus, for $\delta = 2$, for example, the percentage of catalyst B would be twice that of catalyst A, in every run. We might then have

$$Z = \begin{bmatrix} 1 & 2 \\ 2 & 4 \\ 3 & 6 \\ 4 & 8 \end{bmatrix}. \tag{3.3.5}$$

In this case the model could equally well be written as

$$y = \theta_1 z_1 + \theta_2 \delta z_1 + \varepsilon$$

$$= (\theta_1 + \delta \theta_2) z_1 + \varepsilon$$

$$= \delta^{-1}(\theta_1 + \delta \theta_2) z_2 + \varepsilon. \tag{3.3.6}$$

The normal equations for θ_1 and θ_2 can be written down but they would not provide a unique solution for estimates of θ_1 and θ_2, which could not be estimated separately. We could estimate only the linear combination $\theta_1 + \delta \theta_2$. The reason is that, when $z_2 = \delta z_1$, changes associated with z_1 (catalyst

A) are completely indistinguishable from changes associated with z_2 (catalyst B). Now $z_2 = \delta z_1$ implies that $z_2 - \delta z_1 = 0$. In general, the pathological case occurs whenever there is an exact linear relationship of the form $\alpha_1 z_1 + \alpha_2 z_2 = 0$ (for our example, $\alpha_1 = -\delta$, $\alpha_2 = 1$) linking the columns of the matrix $\mathbf{Z}$. In general, an $n \times p$ matrix $\mathbf{Z}$ (where $n \geq p$) is said to have *full* (column) *rank* p if there are no linear relationships of the form

$$\alpha_1 z_1 + \alpha_2 z_2 + \cdots + \alpha_p z_p = 0 \qquad (3.3.7)$$

linking the elements of its columns, and the matrix $\mathbf{Z'Z}$ is then said to be *nonsingular*. If, instead, there are $q > 0$ independent linear relationships, among the columns of $\mathbf{Z}$, then $\mathbf{Z}$ is said to have rank $p - q$ and $\mathbf{Z'Z}$ is said to be *singular*. Whenever $\mathbf{Z}$ is of rank $p - q < p$, it will not be possible to estimate the p θ's separately but only $p - q$ linear functions of them.

To enable the p parameters of the model to be estimated separately, we must therefore employ an experimental design that provides an $n \times p$ matrix $\mathbf{Z}$ of full column rank. The matrix $\mathbf{Z'Z}$ will then be nonsingular and will possess an inverse. In such a case, the solution to the normal equations (3.3.1) may be written

$$\hat{\boldsymbol{\theta}} = (\mathbf{Z'Z})^{-1}\mathbf{Z'y}. \qquad (3.3.8)$$

Using our example once again to illustrate this, we find

$$(\mathbf{Z'Z})^{-1} = \begin{bmatrix} \frac{1}{4} & 0 \\ 0 & \frac{1}{20} \end{bmatrix} \qquad (3.3.9)$$

$$\begin{bmatrix} \hat{\theta}_1 \\ \hat{\theta}_2 \end{bmatrix} = \begin{bmatrix} \frac{1}{4} & 0 \\ 0 & \frac{1}{20} \end{bmatrix}\begin{bmatrix} 93 \\ 83 \end{bmatrix} = \begin{bmatrix} 23.25 \\ 4.15 \end{bmatrix}. \qquad (3.3.10)$$

Meaning of the Fitted Constants

Consider now the meaning of the estimated parameters in the fitted line

$$\hat{y} = 23.25 + 4.15x. \qquad (3.3.11)$$

Because $x = (\xi - 35)/5$, we can rewrite this equally well in terms of ξ, the percentage humidity, as

$$\hat{y} = -5.80 + 0.83\xi. \qquad (3.3.12)$$

TABLE 3.1. Calculation of residuals for our examples

z_1	$z_2 = x$	y	$\hat{y}$	$e = y - \hat{y}$
1	-3	8	10.8	-2.8
1	-1	23	19.1	3.9
1	1	28	27.4	0.6
1	3	34	35.7	-1.7
Sums		93	93.0	0

Notice that, in Figure 3.1, the estimated constant 23.25 in Eq. (3.3.11) is the value of $\hat{y}$ (the intercept) when $x = 0$. Similarly, in the form of Eq. (3.3.12), the intercept when $\xi = 0$ is -5.80. Outside the range of humidities actually tested (20 to 50%) the fitted line is shown as a broken line in Figure 3.1, because we do not have data to check its approximate validity outside that range. In particular, the intercept -5.80 should be regarded merely as a construction point in drawing the line over the experimental range. (Obviously the idea that electrical conductivity has a negative value at zero humidity is nonsense.)

The second estimated parameter in the fitted equation measures the gradient of the line in the units of measurement employed. Thus $\hat{y}$ increases by 4.15 units for each unit change in x or, equivalently, by $0.83 = 4.15/5$ units for each unit change in ξ.

Calculation of Residuals

Using Eq. (3.3.11) we can calculate, for each x value, corresponding values of $\hat{y}$ and of $e = y - \hat{y}$ as shown in Table 3.1. The reader should note that $\Sigma ez_1 = 0$ and $\Sigma ez_2 = 0$, namely, $-2.8 + 3.9 + 0.6 - 1.7 = 0$ and $-3(-2.8) - 1(3.9) + 1(0.6) + 3(-1.7) = 0$. These results provide a check on the calculations and are fundamental properties which are, in theory, exactly true. (In practice, they are valid only to the level of the rounding error in the calculations.) As we shall see in the next section, all least-squares estimates are such that the residuals have zero sum of products with each one of the z's. It is this property, formally stated, which produces *normal* equations.

The residuals are marked in Figure 3.1; they are the vertical distances between the observed data points y and the corresponding values $\hat{y}$ on the fitted straight line. As we discuss later, they provide all the information available from the data on the adequacy of fit of the model.

EXAMPLE 3.5. For a further illustration of least-squares calculations, consider Table 3.2 which shows again the eight observations already discussed in Section 2.2. To these were fitted the model

$$y = \beta_0 + \beta_1 x_1 + \beta_2 x_2 + \beta_3 x_3 + \varepsilon. \tag{3.3.13}$$

In the notation of Eq. (3.1.4), $p = 4$, $z_1 = 1$, $z_2 = x_1$, $z_3 = x_2$, $z_4 = x_3$, $\theta_1 = \beta_0$, $\theta_2 = \beta_1$, $\theta_3 = \beta_2$, and $\theta_4 = \beta_3$. The model can then be written as

$$y = Z\theta + \varepsilon \tag{3.3.14}$$

with

$$y = \begin{bmatrix} 2.83 \\ 3.56 \\ 2.23 \\ 3.06 \\ 2.47 \\ 3.30 \\ 1.95 \\ 2.56 \end{bmatrix}, \quad Z = \begin{bmatrix} 1 & -1 & -1 & -1 \\ 1 & 1 & -1 & -1 \\ 1 & -1 & 1 & -1 \\ 1 & 1 & 1 & -1 \\ 1 & -1 & -1 & 1 \\ 1 & 1 & -1 & 1 \\ 1 & -1 & 1 & 1 \\ 1 & 1 & 1 & 1 \end{bmatrix},$$
$$\qquad 8 \times 1 \qquad\qquad\qquad 8 \times 4$$

$$\theta = \begin{bmatrix} \beta_0 \\ \beta_1 \\ \beta_2 \\ \beta_3 \end{bmatrix}, \quad \varepsilon = \begin{bmatrix} \varepsilon_1 \\ \varepsilon_2 \\ \varepsilon_3 \\ \varepsilon_4 \\ \varepsilon_5 \\ \varepsilon_6 \\ \varepsilon_7 \\ \varepsilon_8 \end{bmatrix}, \quad Z'Z = \begin{bmatrix} 8 & 0 & 0 & 0 \\ 0 & 8 & 0 & 0 \\ 0 & 0 & 8 & 0 \\ 0 & 0 & 0 & 8 \end{bmatrix},$$
$$4 \times 1 \qquad\qquad 8 \times 1$$

$$(Z'Z)^{-1} = \begin{bmatrix} \frac{1}{8} & 0 & 0 & 0 \\ 0 & \frac{1}{8} & 0 & 0 \\ 0 & 0 & \frac{1}{8} & 0 \\ 0 & 0 & 0 & \frac{1}{8} \end{bmatrix}, \quad Z'y = \begin{bmatrix} 21.96 \\ 3.00 \\ -2.36 \\ -1.40 \end{bmatrix}.$$

Thus

$$\hat{\theta} = \begin{bmatrix} \frac{1}{8} & 0 & 0 & 0 \\ 0 & \frac{1}{8} & 0 & 0 \\ 0 & 0 & \frac{1}{8} & 0 \\ 0 & 0 & 0 & \frac{1}{8} \end{bmatrix} \begin{bmatrix} 21.96 \\ 3.00 \\ -2.36 \\ -1.40 \end{bmatrix} = \begin{bmatrix} 2.745 \\ 0.375 \\ -0.295 \\ -0.175 \end{bmatrix}. \tag{3.3.15}$$

TABLE 3.2. **A portion of the Barella and Sust data on failure of worsted yarn; the response is log (cycles to failure)**

x_1	x_2	x_3	y	$\hat{y}$	$e = y - \hat{y}$
-1	-1	-1	2.83	2.84	-0.01
1	-1	-1	3.56	3.59	-0.03
-1	1	-1	2.23	2.25	-0.02
1	1	-1	3.06	3.00	0.06
-1	-1	1	2.47	2.49	-0.02
1	-1	1	3.30	3.24	0.06
-1	1	1	1.95	1.90	0.05
1	1	1	2.56	2.65	-0.09

The fitted equation is therefore

$$\hat{y} = 2.745 + 0.375x_1 - 0.295x_2 - 0.175x_3. \qquad (3.3.16)$$

The contours of the fitted plane in terms of the coded variables x_1, x_2, x_3 were shown in Figure 2.7a; in terms of the original variables ξ_1 (length), ξ_2 (amplitude), ξ_3 (load), the contours appear in Figure 2.7b. The fitted values and residuals are shown in the last two columns of Table 3.2. We shall discuss this example further as we proceed.

Exercise. Confirm that the set of residuals is orthogonal to the set of elements in every column of the **Z** matrix, namely, that $\Sigma ez_1 = 0$, $\Sigma ez_2 = 0$, $\Sigma ez_3 = 0$, and $\Sigma ez_4 = 0$.

EXAMPLE 3.6.   Table 3.3 contains data obtained in a laboratory experiment to examine the change in percentage yield y of a certain dyestuff as the reaction temperature was changed. The experimenter believed that the yield would pass through a maximum in this temperature range, and that the true relationship could perhaps be graduated by a quadratic (or second-order) expression in the variable temperature,

$$y = \beta_0 + \beta_1 x + \beta_{11}x^2 + \varepsilon,$$

where $x = $ (temperature -60). Thus, for this example, $p = 3$, $z_1 = 1$, $z_2 = x$, $z_3 = x^2$, $\theta_1 = \beta_0$, $\theta_2 = \beta_1$, and $\theta_3 = \beta_{11}$. The model is of the form

TABLE 3.3. **Dyestuff yields at various reaction temperatures, fitted values, and residuals**

Reaction temperature	x	Yield (%), y	$\hat{y}$	e
56	-4	45.9	46.0	-0.1
60	0	79.8	78.6	1.2
61	1	78.9	80.5	-1.6
63	3	77.1	76.6	0.5
65	5	62.5	62.6	-0.1

(3.2.11) with

$$
y = \begin{bmatrix} 45.9 \\ 79.8 \\ 78.9 \\ 77.1 \\ 62.5 \end{bmatrix}, \qquad
Z = \begin{bmatrix} 1 & x & x^2 \\ 1 & -4 & 16 \\ 1 & 0 & 0 \\ 1 & 1 & 1 \\ 1 & 3 & 9 \\ 1 & 5 & 25 \end{bmatrix}, \qquad
\theta = \begin{bmatrix} \beta_0 \\ \beta_1 \\ \beta_{11} \end{bmatrix}, \qquad
\varepsilon = \begin{bmatrix} \varepsilon_1 \\ \varepsilon_2 \\ \varepsilon_3 \\ \varepsilon_4 \\ \varepsilon_5 \end{bmatrix},
$$

(3.3.17)

$$
Z'Z = \begin{bmatrix} 5 & 5 & 51 \\ 5 & 51 & 89 \\ 51 & 89 & 963 \end{bmatrix},
$$

$$
(Z'Z)^{-1} = \begin{bmatrix} 0.435321 & -0.002917 & -0.022785 \\ -0.002917 & 0.023398 & -0.002008 \\ -0.022785 & -0.002008 & 0.002431 \end{bmatrix},
$$

$$
Z'y = \begin{bmatrix} 344.2 \\ 439.1 \\ 3069.7 \end{bmatrix}, \qquad
\hat{\theta} = (Z'Z)^{-1}Z'y = \begin{bmatrix} 78.614709 \\ 3.106313 \\ -1.263149 \end{bmatrix}. \quad (3.3.18)
$$

The fitted equation is thus

$$
\hat{y} = 78.615 + 3.106x - 1.263x^2. \tag{3.3.19}
$$

A plot of the original data and the fitted second-order curve is shown in Figure 3.2. The fitted values and residuals appear in the last two columns of Table 3.3.

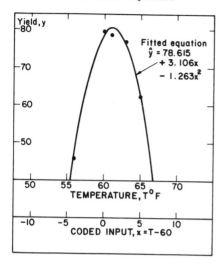

FIGURE 3.2. Plot of dyestuff data and fitted second-order equation.

3.4. GEOMETRY OF LEAST SQUARES

A deep understanding of least squares and its associated analysis is possible using elementary ideas of coordinate geometry.

Least Squares with One Regressor

Consider the simple model

$$y = \beta + \varepsilon, \tag{3.4.1}$$

which expresses the fact that y is varying about an unknown mean β. Suppose we have just three observations of y, $\mathbf{y} = (4, 1, 1)'$. Then, as we have seen, the model can be written

$$\mathbf{y} = \mathbf{z}_1 \theta + \varepsilon \tag{3.4.2}$$

with $\mathbf{z}_1 = (1, 1, 1)'$ and $\theta = \theta = \beta$. That is,

$$\begin{bmatrix} 4 \\ 1 \\ 1 \end{bmatrix} = \begin{bmatrix} 1 \\ 1 \\ 1 \end{bmatrix} \theta + \begin{bmatrix} \varepsilon_1 \\ \varepsilon_2 \\ \varepsilon_3 \end{bmatrix}. \tag{3.4.3}$$

Now the vectors $\mathbf{y}$ and $\mathbf{z}_1$ may be represented by points in three-dimensional space. When there are n observations, in general, we shall need an

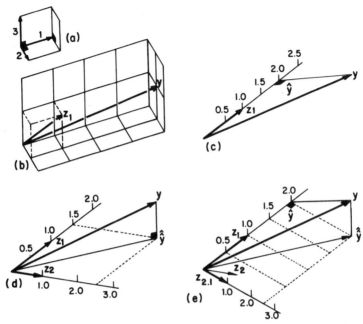

FIGURE 3.3. The geometry of least-squares estimation for a simple example involving three observations.

n-dimensional space but it is possible to argue directly by analogy. Suppose we agree that the first, second, and third coordinates of every vector will be measured along the three axes marked $1, 2$, and 3, respectively, in Figure 3.3a. Then we can construct a three-dimensional grid representing "three-dimensional graph paper" like that shown in Figure 3.3b with unit cubes being the three-dimensional analogy of unit squares on ordinary paper. On such a grid, we can represent a three-element vector $\mathbf{y} = (y_1, y_2, y_3)'$, say, by a line* joining the origin $(0, 0, 0)$ to the point with coordinates (y_1, y_2, y_3). In Figure 3.3b we show the vectors $\mathbf{z}_1 = (1, 1, 1)'$ and $\mathbf{y} = (4, 1, 1)'$ which appear in Eq. (3.4.3). Recall three basic facts of coordinate geometry:

1. The squared distance between any two points $U = (u_1, u_2, u_3)$ and $V = (v_1, v_2, v_3)$ is $(v_1 - u_1)^2 + (v_2 - u_2)^2 + (v_3 - u_3)^2$, the sum of squares of the differences of their coordinates. Thus, in particular, the squared length of a vector is equal to the sum of squares of its elements. Thus, the

---

*A vector has magnitude and direction but holds no particular *position* in space. Any other line having the same length and pointing in the same direction represents the same vector.

length of the vector $y = (4, 1, 1)'$ is $|y| = (4^2 + 1^2 + 1^2)^{1/2} = (18)^{1/2} = 3(2)^{1/2}$ and the length of $z = (1, 1, 1)'$ is $|z| = (1^2 + 1^2 + 1^2)^{1/2} = 3^{1/2}$.

2. If ϕ is the angle between two vectors $u = (u_1, u_2\, u_3)'$ and $v = (v_1, v_2, v_3)'$,

$$\cos \phi = \frac{u_1 v_1 + u_2 v_2 + u_3 v_3}{\left(u_1^2 + u_2^2 + u_3^2\right)^{1/2}\left(v_1^2 + v_2^2 + v_3^2\right)^{1/2}}$$

$$= \frac{\Sigma uv}{\left\{(\Sigma u^2)(\Sigma v^2)\right\}^{1/2}}$$

$$= \frac{u'v}{\left\{(u'u)(v'v)\right\}^{1/2}}. \qquad (3.4.4)$$

It follows that, if $u'v = 0$, the vectors u and v are at right angles to each other, that is, are orthogonal.

3. Multiplication of a vector by a constant multiplies all the elements of that vector by the constant. Thus, for example, $2.5z_1$ is the vector $(2.5, 2.5, 2.5)'$ whose endpoint is marked 2.5 in Figure 3.3c and which is in the same direction as z_1 but is 2.5 times as long.

Least-Squares Estimates

Let us now look at Figure 3.3c. The vector $\hat{y} = z_1\hat{\theta}$ is the point in the direction of z_1 which makes

$$\Sigma e^2 = e_1^2 + e_2^2 + e_3^2 = (y_1 - \hat{y}_1)^2 + (y_2 - \hat{y}_2)^2 + (y_3 - \hat{y}_3)^2 \quad (3.4.5)$$

as small as possible, that is, the point that makes the distance $|e| = |y - \hat{y}|$ as small as possible. Thus $\hat{\theta}$ must be chosen so that the vector $y - \hat{y}$ is at right angles to the vector z_1, as indicated in Figure 3.3c. It follows that

$$z_1'(y - \hat{y}) = 0,$$

or

$$z_1'(y - z_1\hat{\theta}) = 0, \qquad (3.4.6)$$

that is,

$$\Sigma z_1 y - \hat{\theta}\Sigma z_1^2 = 0.$$

This is the single normal equation for this example and corresponds to Eqs. (3.2.15) and (3.3.1) for $p = 1$. The solution is clearly

$$\hat{\theta} = \left(\Sigma z_1^2 \right)^{-1} \Sigma z_1 y. \qquad (3.4.7)$$

For our numerical example, $\Sigma z_1^2 = 3$, $\Sigma z_1 y = 6$, and hence $\hat{\theta} = \bar{y} = 2$. Thus the fitted model obtained by the method of least squares is

$$\hat{\mathbf{y}} = 2\mathbf{z}_1 \qquad (3.4.8)$$

as is evident by inspection of Figure 3.3c. The relationships between the original observations, the fitted values given by Eq. (3.4.8), and the residuals $y - \hat{y}$ are expressed by

$$\mathbf{y} = \hat{\mathbf{y}} + (\mathbf{y} - \hat{\mathbf{y}}),$$

$$\begin{bmatrix} 4 \\ 1 \\ 1 \end{bmatrix} = \begin{bmatrix} 2 \\ 2 \\ 2 \end{bmatrix} + \begin{bmatrix} 2 \\ -1 \\ -1 \end{bmatrix}. \qquad (3.4.9)$$

3.5. ANALYSIS OF VARIANCE FOR ONE REGRESSOR

Suppose that special interest were associated with some particular value of θ, say $\theta_0 = 0.5$, and we wanted to check whether the null hypothesis $H_0 : \theta = \theta_0 = 0.5$ was plausible. If this null hypothesis were true, the mean observation vector would be given by

$$\boldsymbol{\eta}_0 = \mathbf{z}_1 \theta_0, \qquad (3.5.1)$$

or

$$\begin{bmatrix} 0.5 \\ 0.5 \\ 0.5 \end{bmatrix} = \begin{bmatrix} 1 \\ 1 \\ 1 \end{bmatrix} 0.5. \qquad (3.5.2)$$

The appropriate observation breakdown

$$\mathbf{y} - \boldsymbol{\eta}_0 = (\hat{\mathbf{y}} - \boldsymbol{\eta}_0) + (\mathbf{y} - \hat{\mathbf{y}}) \qquad (3.5.3)$$

would then be

$$\begin{bmatrix} 3.5 \\ 0.5 \\ 0.5 \end{bmatrix} = \begin{bmatrix} 1.5 \\ 1.5 \\ 1.5 \end{bmatrix} + \begin{bmatrix} 2 \\ -1 \\ -1 \end{bmatrix}. \qquad (3.5.4)$$

Associated with this breakdown is the analysis of variance in Table 3.4.

TABLE 3.4. **Analysis of variance associated with the null hypothesis** $H_0: \theta = 0.5 \, (= \theta_0)$

Source	Degrees of freedom (df)	Sum of squares $(= \text{length}^2)$ (SS)	Mean square (MS)	F	Expected value of mean square, $E(\text{MS})$
Model	1	$\|\hat{\mathbf{y}} - \boldsymbol{\eta}_0\|^2 = (\hat{\theta} - \theta_0)^2 \Sigma z_1^2$ $= 6.75$	6.75		$\sigma^2 +$ $(\theta - \theta_0)^2 \Sigma z_1^2$
				$F_0 = 2.25$	
Residual	2	$\|\mathbf{y} - \hat{\mathbf{y}}\|^2 = \Sigma(y - \hat{\theta}z_1)^2$ $= 6.00$	3.00		σ^2
Total	3	$\|\mathbf{y} - \boldsymbol{\eta}_0\|^2 = \Sigma(y - \eta_0)^2$ $= 12.75$			

We shall discuss the use of this table later. For the moment, notice that it is essentially a bookkeeping analysis. The sums of squares represent the squared lengths of the vectors and the corresponding degrees of freedom indicate the number of dimensions in which the various vectors are free to move. To see this, consider the situation *before the data became available.* The data $\mathbf{y}$ could lie anywhere in the three-dimensional space so that $\mathbf{y} - \boldsymbol{\eta}_0$ has three degrees of freedom (3 df). However, the model (3.4.2) says that, whatever the data, the systematic part $\hat{\mathbf{y}} - \boldsymbol{\eta}_0 = (\hat{\theta} - \theta_0)\mathbf{z}_1$ of $\mathbf{y} - \boldsymbol{\eta}_0$ *must* lie in the direction of $\mathbf{z}_1$ so that $\hat{\mathbf{y}} - \boldsymbol{\eta}_0$ has only one degree of freedom (1 df). Finally, whatever the data, the residual vector must be perpendicular to $\mathbf{z}_1$ and is therefore free to move in two dimensions and thus has two degrees of freedom (2 df). The sum of the degrees of freedom for model ($\hat{\mathbf{y}} - \boldsymbol{\eta}_0$) and for residual ($\mathbf{y} - \hat{\mathbf{y}}$) is equal to n (i.e., to that for total $\mathbf{y} - \boldsymbol{\eta}_0$).

The sums of squares in Table 3.4 are the squared lengths of the vectors $\hat{\mathbf{y}} - \boldsymbol{\eta}_0$, $\mathbf{y} - \hat{\mathbf{y}}$, and $\mathbf{y} - \boldsymbol{\eta}_0$. It follows from Pythagoras's theorem that the sums of squares for the model and residual add up to give the total sum of squares.

Now what can be said about the null hypothesis that $\theta = \theta_0 = 0.5$? Look at the analysis of variance table, Table 3.4 and at Figure 3.4. The component $\|\hat{\mathbf{y}} - \boldsymbol{\eta}_0\|^2 = (\hat{\theta} - \theta_0)^2 \Sigma z^2$ is a measure of the discrepancy between the postulated model $\boldsymbol{\eta}_0 = \mathbf{z}_1\theta_0$ and the estimated model $\hat{\mathbf{y}} = \mathbf{z}_1\hat{\theta}$. Under the standard assumptions mentioned earlier, it can be shown that the expected value of the sum of squares for the model component assuming the model true is $(\theta - \theta_0)^2 \Sigma z_1^2 + \sigma^2$, while that for the residual component is $2\sigma^2$ (or, in general, $\nu_2\sigma^2$ where ν_2 is the number of degrees of freedom of the residuals). A natural measure of discrepancy from the null hypothesis

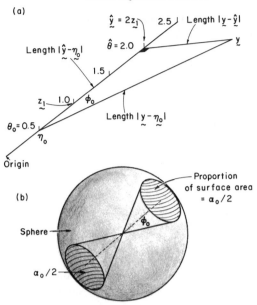

FIGURE 3.4. Geometry connected with check of the hypothesis $\theta = 0.5$ for the model $y = \theta z + \varepsilon$. ($a$) The sum of squares breakup. (b) The significance level as a surface area.

$\theta = \theta_0$ is therefore

$$F_0 = \frac{|\hat{\mathbf{y}} - \boldsymbol{\eta}_0|^2/1}{|\mathbf{y} - \hat{\mathbf{y}}|^2/2} = \frac{\text{mean square for model component}}{\text{mean square for residual}}. \quad (3.5.5)$$

If the null hypothesis $\theta = \theta_0$ were true, then both the numerator and the denominator of F_0 would estimate the same σ^2. A value of F_0 greater than 1 suggests the existence of $\theta - \theta_0$ and hence of a possible departure from the null hypothesis that $\theta = \theta_0$. Now consider the angle ϕ_0, shown on Figure 3.4a, between the vector $\mathbf{y} - \boldsymbol{\eta}_0$ and the $\mathbf{z}_1$ vector. We see that

$$F_0 = 2\cot^2\phi_0. \quad (3.5.6)$$

Thus the smaller is ϕ_0 (and the larger is F_0) the more we shall be led to doubt the null hypothesis. Now on the standard assumptions, the errors

$\varepsilon = y - \eta$ are identically, independently, and normally distributed, so that

$$p(y - \eta) = \text{constant } \sigma^{-3} \exp\left\{\frac{-\frac{1}{2}\sum_u (y_u - \eta_u)^2}{\sigma^2}\right\}$$

$$= f\left\{\sum_u (y_u - \eta_u)^2\right\}. \tag{3.5.7}$$

Thus, the three-dimensional distribution for the y's has spherical contours and, if the null hypothesis is true and $\eta = \eta_0$, the contours of probability density will be spheres centered at η_0. To determine how often an angle as small as ϕ_0 will occur by chance is evidently equivalent to asking "How large are the two caps of a sphere subtended by a cone of rotation of angle ϕ_0 compared to the area of the sphere's surface?" (see Figure 3.4b).

Calculations equivalent to the above have been made to obtain the tables of the percentage points of the F distribution given in this book. For a series of upper tail reference probabilities, denoted by α, and called *significance levels*, the fixed numbers $F_\alpha(\nu_1, \nu_2)$ are given such that $\text{prob}\{F(\nu_1, \nu_2) \geq F_\alpha(\nu_1, \nu_2)\} = \alpha$ where $F(\nu_1, \nu_2)$ denotes the random variable of an F distribution with ν_1 and ν_2 df. The F_0 value obtained in the analysis of variance table is compared to this set of reference $F_\alpha(\nu_1, \nu_2)$'s. For our example $F_0 = 2.25$. The tables show that, for $\nu_1 = 1$, $\nu_2 = 2$ df, this value would be exceeded by chance with a probability exceeding 25% (i.e., with an $\alpha > 0.25$) so that the value $\theta = \theta_0 = 0.5$ is not discredited by the data at even the $\alpha = 0.25$ level.

An important special case arises when the null hypothesis is $H_0: \theta = \theta_0 = 0$, implying that there is no relationship between y and the regressor z_1 or, equivalently in this case, that the mean of the data is zero.

Exercise. Compute the analysis of variance table for the data when $\theta_0 = 0$ (i.e., when $\eta_0 = 0$). (Answer: $12 + 6 = 18$, same df.)

3.6. LEAST SQUARES FOR TWO REGRESSORS

Our previous model Eq. (3.4.1) said that y could be represented by a mean value $\beta = \theta$ plus an error. We now suppose instead that, for the same observations $y' = (4, 1, 1)$, it was believed that there might be systematic deviations from the mean associated with the humidity in the laboratory. Suppose the coded levels of humidity were 0.8, 0.2, and -0.4 in the three experimental runs. Thus, the model would now be the equation of a straight line

$$y = \beta_0 + \beta_1 x + \varepsilon, \tag{3.6.1}$$

or

$$y = z_1\theta_1 + z_2\theta_2 + \varepsilon \tag{3.6.2}$$

with $y = (4, 1, 1)'$, $\theta_1 = \beta_0$, $\theta_2 = \beta_1$, $z_1 = (1, 1, 1)'$, and $z_2 = (0.8, 0.2, -0.4)'$.

Least-Squares Estimates

Now examine Figure 3.3d. Whereas the previous model function $\eta = z_1\theta_1$, with $z_1 = (1, 1, 1)'$ said that the vector η lay on the equiangular line, the revised model

$$\eta = z_1\theta_1 + z_2\theta_2, \tag{3.6.3}$$

says that $\eta = E(y)$ is in the plane defined by linear combinations of the vectors z_1, z_2. Note that (because $z_1'z_2 = \Sigma z_1 z_2 = 0.6 \neq 0$) the vectors z_1 and z_2 are *not* at right angles. However, a point on the plane corresponding to any choice of θ_1 and θ_2 could be found by imagining the parameter plane to be covered by an oblique "graph paper" grid on which the basic element was not a square as on ordinary graph paper but a parallelogram having z_1 and z_2 for two of its sides.

The least-squares values $\hat{\theta}_1$, $\hat{\theta}_2$ which produce a vector

$$\hat{y} = z_1\hat{\theta}_1 + z_2\hat{\theta}_2 \tag{3.6.4}$$

in the parameter plane are those which make the squared length $\Sigma(y - \hat{y})^2 = |y - \hat{y}|^2$ of the residual vector as small as possible. Thus $\hat{y}$ is the foot of the perpendicular from the end of the vector y onto the plane defined by z_1 and z_2. The normal equations now express the fact that $y - \hat{y}$ must be perpendicular (i.e., normal) to both z_1 and z_2. Therefore,

$$\left.\begin{array}{l} z_1'(y - \hat{y}) = 0 \\ z_2'(y - \hat{y}) = 0 \end{array}\right\} \tag{3.6.5}$$

or, equivalently,

$$\left.\begin{array}{l} \Sigma z_1(y - \hat{\theta}_1 z_1 - \hat{\theta}_2 z_2) = 0 \\ \Sigma z_2(y - \hat{\theta}_1 z_1 - \hat{\theta}_2 z_2) = 0 \end{array}\right\} \tag{3.6.6}$$

yielding the normal equations (3.2.15) with $p = 2$. In matrix format,

$$y = \begin{bmatrix} 4 \\ 1 \\ 1 \end{bmatrix}, \quad Z' = \begin{bmatrix} 1 & 1 & 1 \\ 0.8 & 0.2 & -0.4 \end{bmatrix}, \quad \hat{\theta} = \begin{bmatrix} \hat{\theta}_1 \\ \hat{\theta}_2 \end{bmatrix}. \tag{3.6.7}$$

The normal equations are then

$$\mathbf{Z}'(\mathbf{y} - \mathbf{Z}\hat{\boldsymbol{\theta}}) = \mathbf{0} \tag{3.6.8}$$

which yields the equations (3.3.1) with solution (3.3.8). For our example

$$\mathbf{Z}'\mathbf{Z} = \begin{bmatrix} 3.00 & 0.60 \\ 0.60 & 0.84 \end{bmatrix}, \quad (\mathbf{Z}'\mathbf{Z})^{-1} = \tfrac{1}{18}\begin{bmatrix} 7 & -5 \\ -5 & 25 \end{bmatrix},$$

$$\mathbf{Z}'\mathbf{y} = \begin{bmatrix} 6 \\ 3 \end{bmatrix}, \quad \hat{\boldsymbol{\theta}} = \begin{bmatrix} \hat{\theta}_1 \\ \hat{\theta}_2 \end{bmatrix} = \begin{bmatrix} 1.5 \\ 2.5 \end{bmatrix}. \tag{3.6.9}$$

The least-squares fit resulting from a model of form Eq. (3.6.1) is thus

$$\hat{y} = 1.5 + 2.5x. \tag{3.6.10}$$

Inserting successively the three values of x in the data provides fitted values $3.5, 2.0, 0.5$ and leads to the observation breakdown, exemplified in Figure $3.3d$,

$$\mathbf{y} = \hat{\hat{\mathbf{y}}} + (\mathbf{y} - \hat{\mathbf{y}}) \tag{3.6.11}$$

$$\begin{bmatrix} 4 \\ 1 \\ 1 \end{bmatrix} = \begin{bmatrix} 3.5 \\ 2.0 \\ 0.5 \end{bmatrix} + \begin{bmatrix} 0.5 \\ -1.0 \\ 0.5 \end{bmatrix}. \tag{3.6.12}$$

Let us now consider the two fitted models together.

$$p = 1, \quad \hat{y} = 2.0z_1, \quad \text{i.e., } \hat{y} = 2,$$

$$p = 2, \quad \hat{y} = 1.5z_1 + 2.5z_2, \quad \text{i.e., } \hat{y} = 1.5 + 2.5x. \tag{3.6.13}$$

The introduction of the second explanatory variable $x = z_2$ has resulted in the coefficient of z_1 changing from 2.0 to 1.5. To see why, look at Figures 3.3c and 3.3d. The projection of $\mathbf{y}$ on the line of z_1 is $\hat{y} = 2z_1$, but the projection of $\mathbf{y}$ on the plane of z_1 and z_2 is $\hat{y} = 1.5z_1 + 2.5z_2$. However, had z_1 and z_2 been orthogonal, that is, $z_1'z_2 = 0$, the coefficient of z_1 would *not* have changed. This is easily seen from the geometry. Alternatively, we can see that it is true algebraically, for in this case the off-diagonal element of $\mathbf{Z}'\mathbf{Z}$ would have been zero and the normal equations for $\hat{\theta}_1$ and $\hat{\theta}_2$ could have been solved independently. We should not be surprised by this change because the coefficient of z_1 has a different meaning in the two cases.

3.7. GEOMETRY OF THE ANALYSIS OF VARIANCE FOR TWO REGRESSORS

Now suppose special interest were associated with a null hypothesis involving particular values of the parameters, say $\theta_1 = \theta_{10} = 0.5$, $\theta_2 = \theta_{20} = 1.0$. If this null hypothesis were true, the mean observation vector η_0 would be representable as

$$\eta_0 = \theta_{10}z_1 + \theta_{20}z_2, \tag{3.7.1}$$

or

$$\begin{bmatrix} 1.3 \\ 0.7 \\ 0.1 \end{bmatrix} = 0.5 \begin{bmatrix} 1 \\ 1 \\ 1 \end{bmatrix} + 1.0 \begin{bmatrix} 0.8 \\ 0.2 \\ -0.4 \end{bmatrix}. \tag{3.7.2}$$

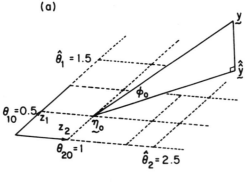

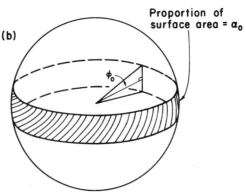

FIGURE 3.5. Geometry connected with check of the hypothesis $\theta_1 = 0.5$, $\theta_2 = 1.0$ for the model $y = z_1\theta_1 + z_2\theta_2 + \varepsilon$. (a) The sum of squares breakup. (b) The hypothesis probability as surface area.

TABLE 3.5.   Analysis of variance split-up appropriate for testing the null hypothesis
$\theta_1 = \theta_{10} = 0.5$, $\theta_2 = \theta_{20} = 1.0$

Source	df	SS	MS	F
Model z_1 and z_2	2	$\lvert\hat{\mathbf{y}} - \mathbf{\eta}_0\rvert^2 = \Sigma\{(\theta_1 - \theta_{10})z_1 + (\theta_2 - \theta_{20})z_2\}^2 = 6.69$	3.345	2.23
Residual	1	$\lvert\mathbf{y} - \hat{\mathbf{y}}\rvert^2 = \Sigma(y - \hat{\theta}_1 z_1 - \hat{\theta}_2 z_2)^2 \qquad\qquad = 1.50$	1.50	
Total	3	$\lvert\mathbf{y} - \mathbf{\eta}_0\rvert^2 = \Sigma(y - \eta_0)^2 \qquad\qquad\qquad = 8.19$		

Thus we could write

$$\mathbf{y} - \mathbf{\eta}_0 = (\hat{\mathbf{y}} - \mathbf{\eta}_0) + (\mathbf{y} - \hat{\mathbf{y}}),$$

$$\begin{bmatrix} 2.7 \\ 0.3 \\ 0.9 \end{bmatrix} = \begin{bmatrix} 2.2 \\ 1.3 \\ 0.4 \end{bmatrix} + \begin{bmatrix} 0.5 \\ -1.0 \\ 0.5 \end{bmatrix}. \qquad (3.7.3)$$

The geometry of this breakdown will be clear from Figure 3.5a. The associated analysis of variance is given in Table 3.5. Again, the plausibility of the null hypothesis $\theta_1 = 0.5$, $\theta_2 = 1.0$ can be assessed by comparing mean squares via the F statistic

$$F_0 = \frac{\lvert\hat{\mathbf{y}} - \mathbf{\eta}_0\rvert^2/2}{\lvert\mathbf{y} - \hat{\mathbf{y}}\rvert^2/1} = 2.23. \qquad (3.7.4)$$

As before, $F_0 = (\nu_2/\nu_1)\cot^2\phi_0$, where $\nu_1 = 2$, $\nu_2 = 1$ are the degrees of freedom for numerator and denominator, respectively. If the null hypothesis is true, then the data are generated by a spherical normal distribution centered at $\mathbf{\eta}_0$ and the probability that $F > F_0$ is the ratio of the area of the shaded band in Figure 3.5b compared with the area of the whole sphere.

3.8.   ORTHOGONALIZING THE SECOND REGRESSOR, EXTRA SUM OF SQUARES PRINCIPLE

In this example, the vector z_2 is not orthogonal to z_1. However, we can find the component $z_{2\cdot1}$ of z_2 that *is* orthogonal to z_1 and rewrite the response function in terms of z_1 and $z_{2\cdot1}$. To obtain $z_{2\cdot1}$, we use the least-squares property that a residual vector is orthogonal to the space in which the predictor variables lie. Temporarily regarding z_2 as the "response" vector and z_1 as the predictor variable, we obtain $\hat{z}_2 = 0.2z_1$ by least squares and hence the residual vector is

$$z_{2\cdot1} = z_2 - \hat{z}_2 = z_2 - 0.2z_1. \qquad (3.8.1)$$

Thus,

$$z_{2\cdot1} = (0.6, 0, -0.6)'. \qquad (3.8.2)$$

Now rewrite the model function in the form

$$\eta = (\theta_1 + 0.2\theta_2)z_1 + \theta_2(z_2 - 0.2z_1),$$

that is,

$$\eta = \theta z_1 + \theta_2 z_{2\cdot1}. \qquad (3.8.3)$$

For this form of the model

$$(Z'Z)^{-1} = \begin{bmatrix} 3.00 & 0 \\ 0 & 0.72 \end{bmatrix}^{-1} = \tfrac{1}{18} \begin{bmatrix} 6 & 0 \\ 0 & 25 \end{bmatrix},$$

$$Z'y = \begin{bmatrix} 6 \\ 1.8 \end{bmatrix}, \quad \text{and} \quad \hat{\theta} = \begin{bmatrix} \hat{\theta} \\ \hat{\theta}_2 \end{bmatrix} = \begin{bmatrix} 2 \\ 2.5 \end{bmatrix}. \qquad (3.8.4)$$

We can now compare our various fitted equations:

1. $\hat{y} = 2z_1$, for the model with one parameter
2a. $\hat{y} = 1.5z_1 + 2.5z_2$ $\quad\left.\begin{array}{l}\\\\\end{array}\right\}$ for the two forms of the model with
2b. $\hat{y} = 2.0z_1 + 2.5z_{2\cdot1}$ $\quad$ two parameters

Note the following points:

1. Because z_1 and $z_{2\cdot1}$ are orthogonal, the coefficient of z_1 in (2b) is the same as the coefficient 2 of z_1 in model (1), which uses only a single regressor.
2. The coefficient of $z_{2\cdot1}$ in (2b) is the same as that for z_2 in (2a).

These facts repay study because they generalize directly in a manner to be discussed shortly.

The geometry of what we have just done is illustrated in Figure 3.3e. The vector z_1 and the component $z_{2\cdot1}$ of z_2 which is orthogonal to z_1 can be used just as well as z_1 and z_2 to define the plane that constitutes the response function locus. In the nonorthogonal coordinates defined by the basis vectors z_1 and z_2, the point $\hat{y}$ has coordinates $\hat{\theta}_1 = 1.5$, $\hat{\theta}_2 = 2.5$; in the orthogonal coordinates defined by basis vectors z_1 and $z_{2.1}$, the *same point* $\hat{y}$ has coordinates $\hat{\theta} = 2$, $\hat{\theta}_2 = 2.5$.

Analysis of Variance Associated with Augmentation of the Model

Problems often arise where the efficacy of including an additional variable z_2 (or, as we shall see later, of including several additional variables) is under consideration. The appropriate null hypothesis is then that $\theta_2 = 0$. It is immaterial what value we choose for θ. Let it be such that $\eta_0 = z_1\theta$. Then the data may be broken down as follows:

$$\mathbf{y} - \mathbf{\eta}_0 = \hat{\mathbf{y}} - \mathbf{\eta}_0 + (\hat{\hat{\mathbf{y}}} - \hat{\mathbf{y}}) + (\mathbf{y} - \hat{\hat{\mathbf{y}}}). \tag{3.8.5}$$

More specifically we may as well set $\theta = 0$ so that $\mathbf{\eta}_0 = \mathbf{0}$ and, for our example,

$$\begin{bmatrix} 4 \\ 1 \\ 1 \end{bmatrix} = \begin{bmatrix} 2 \\ 2 \\ 2 \end{bmatrix} + \begin{bmatrix} 1.5 \\ 0 \\ -1.5 \end{bmatrix} + \begin{bmatrix} 0.5 \\ -1.0 \\ 0.5 \end{bmatrix}. \tag{3.8.6}$$

The three vectors on the right are all orthogonal to one another leading to the analysis of variance of Table 3.6. The relationship in the sum of squares column may be arrived at by noting that, because

$$\hat{\mathbf{y}} = z_1\hat{\theta} \quad \text{and} \quad \hat{\hat{\mathbf{y}}} - \hat{\mathbf{y}} = z_1\hat{\theta} + z_{2.1}\hat{\theta}_2 - z_1\hat{\theta} = z_{2.1}\hat{\theta}_2, \tag{3.8.7}$$

TABLE 3.6. Analysis of variance table for our example with $\theta = 0$ ($\mathbf{\eta}_0 = 0$), showing the orthogonal contribution produced by augmentation of the model

Source	df	SS
Response function with z_1 only	1	$\|\hat{\mathbf{y}} - \mathbf{\eta}_0\|^2 = (\hat{\theta} - \theta_0)^2\Sigma z_1^2 = 12.0$
Extra due to z_2 (given z_1)	1	$\|\hat{\hat{\mathbf{y}}} - \hat{\mathbf{y}}\|^2 = \hat{\theta}_2^2\Sigma z_{2.1}^2 = 4.5$
Residual	1	$\|\mathbf{y} - \hat{\hat{\mathbf{y}}}\|^2 = \Sigma(y - \hat{\hat{y}})^2 = 1.5$
Total	3	$\|\mathbf{y} - \mathbf{\eta}_0\|^2 = \Sigma(y - \eta_0)^2 = 18.0$

the equations $\mathbf{y} = \hat{\mathbf{y}} + (\hat{\hat{\mathbf{y}}} - \hat{\mathbf{y}}) + (\mathbf{y} - \hat{\hat{\mathbf{y}}})$ can be written

$$\mathbf{y} = \mathbf{z}_1\hat{\theta} + \mathbf{z}_{2\cdot 1}\hat{\theta}_2 + (\mathbf{y} - \hat{\hat{\mathbf{y}}}). \tag{3.8.8}$$

The three vectors on the right are orthogonal, so that

$$|\mathbf{y}|^2 = \hat{\theta}^2|\mathbf{z}_1|^2 + \hat{\theta}_2^2|\mathbf{z}_{2\cdot 1}|^2 + |\mathbf{y} - \hat{\hat{\mathbf{y}}}|^2. \tag{3.8.9}$$

and the sum of squares split-up of Table 3.6 is immediately obtained.

The Extra Sum of Squares Principle

Consider the extra sum of squares due to z_2 in the analysis of variance, Table 3.6. We have

$$|\hat{\hat{\mathbf{y}}} - \hat{\mathbf{y}}|^2 = |\hat{\hat{\mathbf{y}}}|^2 - |\hat{\mathbf{y}}|^2$$
$$= (\text{SS for } z_1 \text{ and } z_2) - (\text{SS for } z_1 \text{ alone}). \tag{3.8.10}$$

Alternatively and equivalently,

$$|\hat{\hat{\mathbf{y}}} - \hat{\mathbf{y}}|^2 = |\mathbf{y} - \hat{\mathbf{y}}|^2 - |\mathbf{y} - \hat{\hat{\mathbf{y}}}|^2$$
$$= (\text{residual SS with } z_1 \text{ only}) - (\text{residual SS with } z_1 \text{ and } z_2). \tag{3.8.11}$$

We see that, to obtain the "extra SS in the analysis of variance associated with the orthogonal sum of squares," we do not need to calculate $\mathbf{z}_{2\cdot 1}$ explicitly. All that is needed is to fit each model in turn and obtain the difference in the residual sums of squares. More generally, if we have a linear model including any set of regressors defined by the columns of matrix $\mathbf{Z}_1$, say, and we are considering a more elaborate model with additional regressors represented by the columns of $\mathbf{Z}_2$, then a sum of squares associated with the additional regressors, given the others, can always be found by first fitting the model with both sets of regressors $(\mathbf{Z}_1, \mathbf{Z}_2)$ and then fitting the simpler model with $\mathbf{Z}_1$ only. The required extra sum of squares is then obtained either as

$$(\text{regression SS for model with } \mathbf{Z}_1, \mathbf{Z}_2)$$
$$- (\text{regression SS for model with } \mathbf{Z}_1 \text{ only}) \tag{3.8.12}$$

or as

$$(\text{residual SS for model with } \mathbf{Z}_1 \text{ only})$$
$$- (\text{residual SS for model with } \mathbf{Z}_1 \text{ and } \mathbf{Z}_2). \tag{3.8.13}$$

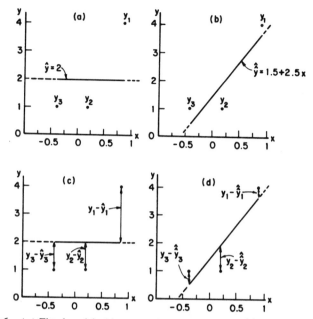

FIGURE 3.6. (*a*) Fitted model with mean value only. (*b*) Fitting sloping straight line model. (*c*) Splitup of the observations into fitted values and residuals for the mean only model. (*d*) Splitup for the sloping straight line model.

Another Way to Consider the Analysis

To see what has been done from a slightly different point of view, we can examine Figure 3.6 which shows a plot of the data in relation to the two different models. The first fitted model is the horizontal straight line $\hat{y} = 2$ in Figure 3.6*a*. The second fitted model is the sloping line $\hat{y} = 1.5 + 2.5x$ in Figure 3.6*b*; $\hat{\theta}_1 = 1.5$ is the value of $\hat{y}$ when $x = 0$ and $\hat{\theta}_2 = 2.5$ is the gradient (or slope) of the line. Figures 3.6*c* and 3.6*d* show the partitioning of the observations into fitted values and residuals.

Consider next the meaning associated with the breakdown

$$|y|^2 = |\hat{y}|^2 + |\hat{\hat{y}} - \hat{y}|^2 + |y - \hat{\hat{y}}|^2, \qquad (3.8.14)$$

that is,

$$\Sigma y^2 = \Sigma \hat{y}^2 + \Sigma(\hat{\hat{y}} - \hat{y})^2 + \Sigma(y - \hat{\hat{y}})^2. \qquad (3.8.15)$$

The analysis of variance of Table 3.4 but with $\theta_0 = 0$ says that the total sum of squares $y_1^2 + y_2^2 + y_3^2 = \Sigma y^2$ can be split into a model component

$\hat{y}_1^2 + \hat{y}_2^2 + \hat{y}_3^2$ and a residual component $(y_1 - \hat{y}_1)^2 + (y_2 - \hat{y}_2)^2 + (y_3 - \hat{y}_3)^2$. Because $\hat{y} = z_1\hat{\theta} = \hat{\theta} = \bar{y}$, the model component measures the deviation of the sample mean from zero. For the analysis of the sloping line model, the model component $\hat{y}_1^2 + \hat{y}_2^2 + \hat{y}_3^2$ measures the deviation of the fitted straight line values from zero and leaves a residual component $(y_1 - \hat{y}_1)^2 + (y_2 - \hat{y}_2)^2 + (y_3 - \hat{y}_3)^2$. In the analysis of variance of Table 3.6, the component $\hat{y}$ is split up into the parts $\hat{y} + (\hat{y} - \hat{y}) = \bar{y} + \hat{\theta}_2 z_{2\cdot 1} = \bar{y} + \hat{\theta}_2(x - \bar{x})$. The first represents the effect of the average and the second the deviations of the fitted values given by the straight line from the fitted average.

3.9. GENERALIZATION TO p REGRESSORS

All these ideas can be readily generalized to any number of observations n and any number of parameters p. The development is most easily carried out using matrix algebra but the reader should keep in mind, as a guide, the geometrical interpretations set out for $n = 3$ above.

Least Squares and the Normal Equations

The n relations implicit in the response function may be written

$$\boldsymbol{\eta} = \mathbf{Z}\boldsymbol{\theta} \tag{3.9.1}$$

where it is assumed that $\mathbf{Z}$ is of full column rank p and hence that $\mathbf{Z}'\mathbf{Z}$ is positive definite and invertible. Let $\hat{\boldsymbol{\theta}}$ be the vector of estimates given by the normal equations

$$(\mathbf{y} - \hat{\mathbf{y}})'\mathbf{Z} = \mathbf{0}, \tag{3.9.2}$$

that is,

$$(\mathbf{y} - \mathbf{Z}\hat{\boldsymbol{\theta}})'\mathbf{Z} = \mathbf{0}. \tag{3.9.3}$$

It is clear that $\hat{\boldsymbol{\theta}}$ provides least-squares estimates from the following argument. The sum of squares function is

$$S(\boldsymbol{\theta}) = (\mathbf{y} - \boldsymbol{\eta})'(\mathbf{y} - \boldsymbol{\eta})$$
$$= (\mathbf{y} - \hat{\mathbf{y}})'(\mathbf{y} - \hat{\mathbf{y}}) + (\hat{\mathbf{y}} - \boldsymbol{\eta})'(\hat{\mathbf{y}} - \boldsymbol{\eta}) \tag{3.9.4}$$

because the cross-product is zero from the normal equations. Thus

$$S(\boldsymbol{\theta}) = S(\hat{\boldsymbol{\theta}}) + (\hat{\boldsymbol{\theta}} - \boldsymbol{\theta})'\mathbf{Z}'\mathbf{Z}(\hat{\boldsymbol{\theta}} - \boldsymbol{\theta}). \tag{3.9.5}$$

TABLE 3.7. Analysis of variance table for the case of general regression

Source	df	SS
Response function	p	$\|\hat{\mathbf{y}} - \mathbf{\eta}\|^2 = (\hat{\mathbf{\theta}} - \mathbf{\theta})'\mathbf{Z}'\mathbf{Z}(\hat{\mathbf{\theta}} - \mathbf{\theta})$
Residual	$n - p$	$\|\mathbf{y} - \hat{\mathbf{y}}\|^2 = \Sigma(y - \hat{y})^2$
Total	n	$\|\mathbf{y} - \mathbf{\eta}\|^2 = \Sigma(y - \eta)^2$

However, because $\mathbf{Z}'\mathbf{Z}$ is positive definite, $S(\mathbf{\theta})$ is minimized when $\mathbf{\theta} = \hat{\mathbf{\theta}}$. Thus the normal equations solution

$$\hat{\mathbf{\theta}} = (\mathbf{Z}'\mathbf{Z})^{-1}\mathbf{Z}'\mathbf{y} \qquad (3.9.6)$$

will always produce least-squares estimates.

Analysis of Variance

The breakdown (3.9.4) can be conveniently set out in the form of Table 3.7.

Orthogonal Breakdown

Suppose that the p regressors represented by the p columns of $\mathbf{Z}$ split naturally into two sets $\mathbf{Z}_1$, $\mathbf{Z}_2$ of p_1, p_2 columns, respectively, so that the response function $\mathbf{\eta} = \mathbf{Z}\mathbf{\theta}$ can also be written

$$\mathbf{\eta} = \mathbf{Z}_1\mathbf{\theta}_1 + \mathbf{Z}_2\mathbf{\theta}_2. \qquad (3.9.7)$$

The simpler model function $\mathbf{\eta} = \mathbf{Z}_1\mathbf{\theta}_1$ might be one which it was hoped might be adequate and $\mathbf{Z}_2\mathbf{\theta}_2$ might represent further terms which perhaps would have to be added if the terms of $\mathbf{Z}_1\mathbf{\theta}_1$ were inadequate to represent the response. Suppose further (at first) that $\mathbf{Z}_1$ and $\mathbf{Z}_2$ were orthogonal so that $\mathbf{Z}_1'\mathbf{Z}_2 = \mathbf{0}$. Then the equation $\hat{\mathbf{\theta}} = (\mathbf{Z}'\mathbf{Z})^{-1}\mathbf{Z}'\mathbf{y}$ which provides the least-squares estimator $\hat{\mathbf{\theta}}$ of $\mathbf{\theta}$ splits into the two parts

$$\hat{\mathbf{\theta}}_1 = (\mathbf{Z}_1'\mathbf{Z}_1)^{-1}\mathbf{Z}_1'\mathbf{y} \quad \text{and} \quad \hat{\mathbf{\theta}}_2 = (\mathbf{Z}_2'\mathbf{Z}_2)^{-1}\mathbf{Z}_2'\mathbf{y}. \qquad (3.9.8)$$

Moreover, the sum of squares for the response function of the more

TABLE 3.8. Analysis of variance table when model splits up into orthogonal parts, that is, $Z_1'Z_2 = 0$

Source	df	SS
Response function (Z₁ only)	p_1	$(\hat{\theta}_1 - \theta_1)'Z_1'Z_1(\hat{\theta}_1 - \theta_1)$
Extra for Z_2	p_2	$(\hat{\theta}_2 - \theta_2)'Z_2'Z_2(\hat{\theta}_2 - \theta_2)$
Residual	$n - p_1 - p_2$	$(y - \hat{y})'(y - \hat{y}) = \Sigma(y - \hat{y})^2$
Total	n	$(y - \eta)'(y - \eta) = \Sigma(y - \eta)^2$

elaborate model may be written as

$$(\theta - \hat{\theta})'Z'Z(\theta - \hat{\theta}) = (\theta_1 - \hat{\theta}_1)Z_1'Z_1(\theta_1 - \hat{\theta}_1) + (\theta_2 - \hat{\theta}_2)Z_2'Z_2(\theta_2 - \hat{\theta}_2) \tag{3.9.9}$$

and the associated analysis of variance is thus as in Table 3.8.

Orthogonalization When $Z_1'Z_2 \neq 0$

Consider the model function $\eta = Z_1\theta_1$. An important fact about least-squares estimates, which follows from the normal equations, is that the vector of residuals

$$y - \hat{y} = y - Z_1\hat{\theta} \tag{3.9.10}$$

is orthogonal to every one of the columns of Z_1. This thus provides a general method for determining that component of a vector y which is orthogonal to a space of p_1 dimensions whose basis vectors are the columns of a matrix Z_1. The required orthogonal component, $y_{.1}$ say, is given by

$$y_{.1} = y - \hat{y} = y - Z_1\hat{\theta}_1$$

$$= y - Z_1(Z_1'Z_1)^{-1}Z_1'y$$

$$= (I - R_1)y \tag{3.9.11}$$

say, where

$$R_1 = Z_1(Z_1'Z_1)^{-1}Z_1'. \tag{3.9.12}$$

Now, instead of the single-column vector y, consider the p_2 columns of Z_2

in the same role. A matrix

$$\mathbf{Z}_{2\cdot1} = \mathbf{Z}_2 - \hat{\mathbf{Z}}_2 \qquad (3.9.13)$$

with every one of its p_2 columns orthogonal to every one of the p_1 columns of $\mathbf{Z}_1$ can be obtained via

$$\mathbf{Z}_{2\cdot1} = (\mathbf{I} - \mathbf{R}_1)\mathbf{Z}_2$$

$$= \mathbf{Z}_2 - \mathbf{Z}_1\mathbf{A} \qquad (3.9.14)$$

where

$$\mathbf{A} = (\mathbf{Z}_1'\mathbf{Z}_1)^{-1}\mathbf{Z}_1'\mathbf{Z}_2 \qquad (3.9.15)$$

is a $p_1 \times p_2$ matrix of coefficients obtained by regressing each of the p_2 columns of $\mathbf{Z}_2$ onto all the p_1 columns of $\mathbf{Z}_1$. This matrix $\mathbf{A}$ is sometimes referred to as the *alias matrix*, or the *bias matrix*.

The Orthogonalized Model and the Extra Sum of Squares Procedure in This Context

In the situation where $\mathbf{Z}_1$ and $\mathbf{Z}_2$ are not orthogonal, we can write the response function in the form

$$\mathbf{y} = \mathbf{Z}_1(\boldsymbol{\theta}_1 + \mathbf{A}\boldsymbol{\theta}_2) + (\mathbf{Z}_2 - \mathbf{Z}_1\mathbf{A})\boldsymbol{\theta}_2 + \boldsymbol{\varepsilon}$$

$$= \mathbf{Z}_1\boldsymbol{\theta} + \mathbf{Z}_{2\cdot1}\boldsymbol{\theta}_2 + \boldsymbol{\varepsilon} \qquad (3.9.16)$$

where $\boldsymbol{\theta} = \boldsymbol{\theta}_1 + \mathbf{A}\boldsymbol{\theta}_2$. The corresponding analysis of variance table is shown in Table 3.9.

TABLE 3.9. **Analysis of variance for orthogonalized general model**

Source	df	SS
Response function $\mathbf{Z}_1$ only	p_1	$(\boldsymbol{\theta} - \hat{\boldsymbol{\theta}})'\mathbf{Z}_1'\mathbf{Z}_1(\boldsymbol{\theta} - \hat{\boldsymbol{\theta}})$
Extra for $\mathbf{Z}_2$	p_2	$(\boldsymbol{\theta}_2 - \hat{\boldsymbol{\theta}}_2)\mathbf{Z}_{2\cdot1}'\mathbf{Z}_{2\cdot1}(\boldsymbol{\theta}_2 - \hat{\boldsymbol{\theta}}_2)$
Residual	$n - p_1 - p_2$	$(\mathbf{y} - \hat{\mathbf{y}})'(\mathbf{y} - \hat{\mathbf{y}}) = \Sigma(y - \hat{y})^2$
Total	n	$(\mathbf{y} - \boldsymbol{\eta})'(\mathbf{y} - \boldsymbol{\eta}) = \Sigma(y - \eta)^2$

The following points should be noticed:

1. The vector of estimates $\hat{\theta}$ obtained by fitting $y = Z_1\theta + Z_{2.1}\theta_2 + \varepsilon$ will be identical to the one that would be obtained by fitting the model $y = Z_1\theta + \varepsilon$ which contained only Z_1.

2. If the additional regressors in Z_2 are desired in the model, the vector of estimates $\hat{\theta}_2$ will be the same if they are obtained by least squares from the full model (3.9.7), or from the model $y = Z_{2.1}\theta_2 + \varepsilon$.
    (Points 1 and 2 are ensured by the orthogonality of the matrices Z_1 and $Z_{2.1}$, i.e., $Z_1'Z_{2.1} = 0$.)

3. If the additional regressors in Z_2 are required, that is, if $\theta_2 \neq 0$, then $\hat{\theta}$ will not provide an unbiased estimate of θ_1 but, rather, of the combination $\theta_1 + A\theta_2$.

4. The need for the regressors in Z_2 to be in the response function will be indicated by the size of the sum of squares "extra for Z_2" with p_2 df, when θ_2 is set equal to 0 in that sum of squares.

5. The "extra for Z_2" sum of squares may be obtained by first fitting the simpler model $y = Z_1\theta_1 + \varepsilon$ and then the more elaborate model $y = Z_1\theta_1 + Z_2\theta_2 + \varepsilon$ and calculating the difference in the model sums of squares or, alternatively, in the residual sums of squares.

3.10. BIAS IN LEAST-SQUARES ESTIMATORS ARISING FROM AN INADEQUATE MODEL

Suppose that it had been decided to fit the model* $y = Z_1\theta_1 + \varepsilon$ but the true model which should have been fitted was $y = Z_1\theta_1 + Z_2\theta_2 + \varepsilon$. We would have estimated θ_1 by

$$\hat{\theta}_1 = (Z_1'Z_1)^{-1}Z_1'y \qquad (3.10.1)$$

but, under the true model,

$$E(\hat{\theta}_1) = (Z_1'Z_1)^{-1}Z_1'E(y)$$

$$= (Z_1'Z_1)^{-1}Z_1'(Z_1\theta_1 + Z_2\theta_2)$$

$$= \theta_1 + A\theta_2 \qquad (3.10.2)$$

*As before we use the notation θ for the parameters of a general model and β's for specific cases.

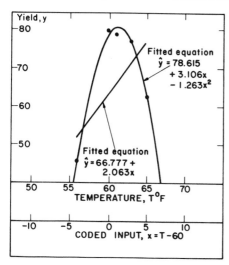

FIGURE 3.7. Plot of dyestuff data and fitted first- and second-order equation.

where

$$\mathbf{A} = (\mathbf{Z}_1'\mathbf{Z}_1)^{-1}\mathbf{Z}_1'\mathbf{Z}_2 \qquad (3.10.3)$$

is the so-called *bias* or *alias* matrix. Equation (3.10.2) tells us that, unless $\mathbf{A} = \mathbf{0}$, $\hat{\boldsymbol{\theta}}_1$ will estimate not $\boldsymbol{\theta}_1$, but a combination of $\boldsymbol{\theta}_1$ and $\boldsymbol{\theta}_2$. Now $\mathbf{A} = \mathbf{0}$ only if $\mathbf{Z}_1'\mathbf{Z}_2 = \mathbf{0}$, when all the regressors in $\mathbf{Z}_1$ are orthogonal to all the regressors in $\mathbf{Z}_2$.

EXAMPLE 3.7.   For illustration, consider again the data of Table 3.3, but suppose that, instead of fitting the "true" quadratic model $y = \beta_0 + \beta_1 x + \beta_{11} x^2 + \varepsilon$, we had fitted the straight line model $y = \beta_0 + \beta_1 x + \varepsilon$. Figure 3.7 shows the least-squares fitted straight line $\hat{y} = 66.777 + 2.063x$, which fits the data very poorly. It is obvious that the intercept $b_0 = 66.777$ which is the predicted value of y at $x = 0$, and the slope $b_1 = 2.063$ will have expectations contaminated by the true value of the omitted curvature parameter, β_{11}. The exact nature of this contamination can be found by calculating the matrix $\mathbf{A}$. We find

$$\mathbf{Z}_1 = \begin{bmatrix} 1 & -4 \\ 1 & 0 \\ 1 & 1 \\ 1 & 3 \\ 1 & 5 \end{bmatrix}, \qquad \boldsymbol{\theta}_1 = \begin{bmatrix} \beta_0 \\ \beta_1 \end{bmatrix}, \qquad \mathbf{Z}_2 = \begin{bmatrix} 16 \\ 0 \\ 1 \\ 9 \\ 25 \end{bmatrix} \qquad \boldsymbol{\theta}_2 = \beta_{11}. \quad (3.10.4)$$

Thus the bias matrix is

$$\mathbf{A} = (\mathbf{Z}_1'\mathbf{Z}_1)^{-1}\mathbf{Z}_1'\mathbf{Z}_2 = \begin{bmatrix} 5 & 5 \\ 5 & 51 \end{bmatrix}^{-1} \begin{bmatrix} 51 \\ 89 \end{bmatrix}$$

$$= \frac{1}{230} \begin{bmatrix} 51 & -5 \\ -5 & 5 \end{bmatrix} \begin{bmatrix} 51 \\ 89 \end{bmatrix}$$

$$= \frac{1}{230} \begin{bmatrix} 2156 \\ 190 \end{bmatrix}$$

$$= \begin{bmatrix} 9.374 \\ 0.826 \end{bmatrix}. \tag{3.10.5}$$

Hence

$$E(b_0) = \beta_0 + 9.374\beta_{11},$$

$$E(b_1) = \beta_1 + 0.826\beta_{11}. \tag{3.10.6}$$

We see that $b_0 = 66.777$ is not an unbiased estimate of β_0 in the true quadratic model but is instead an estimate of $\beta_0 + 9.374\beta_{11}$. Similarly, $b_1 = 2.063$ is an estimate of $\beta_1 + 0.826\beta_{11}$, rather than of β_1.

When, as in this example, it is possible to estimate the coefficients of the true model separately, the linear relationships indicated by the alias structure will also hold for the least-squares estimates.* For example, recall that the estimated quadratic equation was

$$\hat{y} = 78.615 + 3.106x - 1.263x^2 \tag{3.10.7}$$

while the estimated straight line equation was

$$\hat{y} = 66.777 + 2.063x.$$

It is readily confirmed that (to within rounding error)

$$66.777 = 78.615 + 9.374(-1.263),$$

$$2.063 = 3.106 + 0.826(-1.263). \tag{3.10.8}$$

---

*This fact is a direct consequence of the Gauss–Markov theorem, see Appendix 3B.

EXAMPLE 3.8. Aliases for a 2^{5-2} design. Consider the 2^{5-2} fractional factorial design* with generating relation $I = 124 = 135$ whose levels are given by the second through sixth columns of the $\mathbf{Z}_1$ matrix below. Suppose the fitted first degree polynomial was

$$\hat{y} = b_0 + b_1 x_1 + b_2 x_2 + b_3 x_3 + b_4 x_4 + b_5 x_5 \qquad (3.10.9)$$

when, in fact, to obtain an adequate representation over the region covered by the x's, we would need a second degree polynomial model

$$y = \beta_0 + \beta_1 x_1 + \beta_2 x_2 + \beta_3 x_3 + \beta_4 x_4 + \beta_5 x_5 \qquad (3.10.10)$$

$$+ \beta_{11} x_1^2 + \beta_{22} x_2^2 + \beta_{33} x_3^2 + \beta_{44} x_4^2 + \beta_{55} x_5^2$$

$$+ \beta_{12} x_1 x_2 + \beta_{13} x_1 x_3 + \beta_{14} x_1 x_4 + \beta_{15} x_1 x_5 + \beta_{23} x_2 x_3$$

$$+ \beta_{24} x_2 x_4 + \beta_{25} x_2 x_5 + \beta_{34} x_3 x_4 + \beta_{35} x_3 x_5 + \beta_{45} x_4 x_5 + \varepsilon.$$

What will be the expected values of $b_0, b_1, \ldots, b_5$? We have

$$
\mathbf{Z}_1 =
\begin{array}{c}
0\ \ 1\ \ 2\ \ 3\ \ 4\ \ 5 \\
\left[\begin{array}{rrrrrr}
1 & -1 & -1 & -1 & 1 & 1 \\
1 & 1 & -1 & -1 & -1 & -1 \\
1 & -1 & 1 & -1 & -1 & 1 \\
1 & 1 & 1 & -1 & 1 & -1 \\
1 & -1 & -1 & 1 & 1 & -1 \\
1 & 1 & -1 & 1 & -1 & 1 \\
1 & -1 & 1 & 1 & -1 & -1 \\
1 & 1 & 1 & 1 & 1 & 1
\end{array}\right]
\end{array},
$$

$$
\mathbf{Z}_2 =
\begin{array}{c}
\scriptstyle 11\ \ 22\ \ 33\ \ 44\ \ 55\ \ \ 12\ \ \ 13\ \ \ 14\ \ \ 15\ \ \ 23\ \ \ 24\ \ \ 25\ \ \ 34\ \ \ 35\ \ \ 45 \\
\left[\begin{array}{rrrrrrrrrrrrrrr}
1 & 1 & 1 & 1 & 1 & 1 & 1 & -1 & -1 & 1 & -1 & -1 & -1 & -1 & 1 \\
1 & 1 & 1 & 1 & 1 & -1 & -1 & -1 & -1 & 1 & 1 & 1 & 1 & 1 & 1 \\
1 & 1 & 1 & 1 & 1 & -1 & 1 & 1 & -1 & -1 & -1 & 1 & 1 & -1 & -1 \\
1 & 1 & 1 & 1 & 1 & 1 & -1 & 1 & -1 & -1 & 1 & -1 & -1 & 1 & -1 \\
1 & 1 & 1 & 1 & 1 & 1 & -1 & -1 & 1 & -1 & -1 & 1 & 1 & -1 & -1 \\
1 & 1 & 1 & 1 & 1 & -1 & 1 & -1 & 1 & -1 & 1 & -1 & -1 & 1 & -1 \\
1 & 1 & 1 & 1 & 1 & -1 & -1 & 1 & 1 & 1 & -1 & -1 & -1 & -1 & 1 \\
1 & 1 & 1 & 1 & 1 & 1 & 1 & 1 & 1 & 1 & 1 & 1 & 1 & 1 & 1
\end{array}\right]
\end{array},
$$

$$\boldsymbol{\theta}_1 = (\beta_0, \beta_1, \beta_2, \beta_3, \beta_4, \beta_5)',$$

$$\boldsymbol{\theta}_2 = (\beta_{11}, \beta_{22}, \beta_{33}, \beta_{44}, \beta_{55}, \beta_{12}, \beta_{13}, \beta_{14}, \beta_{15}, \beta_{23}, \beta_{24}, \beta_{25}, \beta_{34}, \beta_{35}, \beta_{45})',$$

*Designs of this type are discussed in Chapter 5.

$$\mathbf{A} = (\mathbf{Z}_1'\mathbf{Z}_1)^{-1}\mathbf{Z}_1'\mathbf{Z}_2$$

$$= \begin{bmatrix} 1 & 1 & 1 & 1 & 1 & & & & & & & & \\ & & & & & & & & & & 1 & & 1 \\ & & & & & & 1 & & & & & & \\ & & & & & & & 1 & & & & & \\ & & & & & 1 & & & & & & & \\ & & & & & & 1 & & & & & & \end{bmatrix},$$

$$(3.10.11)$$

where empty spaces represent zeros. Then

$$E(b_0) = \beta_0 + \beta_{11} + \beta_{22} + \beta_{33} + \beta_{44} + \beta_{55},$$

$$E(b_1) = \beta_1 + \beta_{24} + \beta_{35},$$

$$E(b_2) = \beta_2 + \beta_{14},$$

$$E(b_3) = \beta_3 + \beta_{15},$$

$$E(b_4) = \beta_4 + \beta_{12},$$

$$E(b_5) = \beta_5 + \beta_{13}. \qquad (3.10.12)$$

Thus all estimators are aliased unless the β_{ij} shown are all zero.

Notice that the aliasing which occurs in this design cannot be further elucidated unless additional experimental runs are performed. For example, the columns in $\mathbf{Z}_1$ and $\mathbf{Z}_2$ associated with x_1, $x_2 x_4$, and $x_3 x_5$ are *identical*. Here we can only say, for instance, that unless the unknown coefficients β_{24} and β_{35} are zero, then b_1 will be a biased estimate of β_1. Because of the inadequacy of the design, no possibility exists here to estimate the second degree model and so to unbias (or de-alias) the estimates. (In the previous example this could be done.)

3.11. PURE ERROR AND LACK OF FIT

Genuine Replicates

It is frequently useful to employ experimental arrangements in which two or more runs are made at an identical set of values of the input variables

$x_1, x_2, \ldots, x_k$. If these runs can be made in such a way that they are subject to all the sources of error that beset runs made at different conditions, we call them *genuine replicates*.*

Genuine replicates are very valuable because differences in the response y between them can provide an estimate of the error variance no matter what the true model may be. For r genuine replicates in which x is fixed, the model can be written

$$y_u = f(\mathbf{x}, \boldsymbol{\theta}) + \varepsilon_u, \qquad u = 1, 2, \ldots, r, \qquad (3.11.1)$$

implying

$$\bar{y} = f(\mathbf{x}, \boldsymbol{\theta}) + \bar{\varepsilon}, \qquad (3.11.2)$$

and

$$\sum_{u=1}^{r} (y_u - \bar{y})^2 = \sum_{u=1}^{r} (\varepsilon_u - \bar{\varepsilon})^2, \qquad (3.11.3)$$

whatever the model function $f(\mathbf{x}, \boldsymbol{\theta})$ may be. Thus, if the ε_u are independent random variables with variance σ^2,

$$E\left\{ \sum_{u=1}^{r} (y_u - \bar{y})^2 \right\} = E\left\{ \sum_{u=1}^{r} (\varepsilon_u - \bar{\varepsilon})^2 \right\} = (r-1)\sigma^2 \quad (3.11.4)$$

and, furthermore, if the $\varepsilon_u \sim N(0, \sigma^2)$ and are independent, both sides of (3.11.4) are distributed as $\sigma^2 \chi^2_{r-1}$ variables.

If we have m such sets of replicated runs with r_i runs in the ith set made at $\mathbf{x}_i$, the individual internal sums of squares may be pooled together to form a *pure error sum of squares* having, as its degrees of freedom, the sum of the separate degrees of freedom. Thus the (total) *pure error sum of squares* is

$$\sum_{i=1}^{m} \sum_{u=1}^{r_i} (y_{iu} - \bar{y}_i)^2 \qquad (3.11.5)$$

with degrees of freedom

$$\sum_{i=1}^{m} (r_i - 1) = \sum_{i=1}^{m} r_i - m. \qquad (3.11.6)$$

---

*Some care is needed to achieve genuine replicated runs. In particular, a group of such runs should normally not be run consecutively but should be randomly ordered. Replicate runs must be subject to all the usual setup errors, sampling errors, and analytical errors which affect runs made at different conditions. Failure to achieve this will typically cause underestimation of the error and will invalidate the analysis.

When $r_i = 2$, the simpler formula

$$\sum_{u=1}^{2} (y_{iu} - \bar{y}_i)^2 = \frac{1}{2}(y_{i1} - y_{i2})^2 \qquad (3.11.7)$$

can be used; such a component has 1 df.

When any model is fitted to data, the pure error sum of squares is always *part* of the residual sum of squares. The residual from the uth observation at $\mathbf{x}_i$ is

$$y_{iu} - \hat{y}_i = (y_{iu} - \bar{y}_i) - (\hat{y}_i - \bar{y}_i); \qquad (3.11.8)$$

on squaring both sides and summing,

$$\sum_{i=1}^{m} \sum_{u=1}^{r_i} (y_{iu} - \hat{y}_i)^2 = \sum_{i=1}^{m} \sum_{u=1}^{r_i} (y_{iu} - \bar{y}_i)^2 + \sum_{i=1}^{m} r_i(\hat{y}_i - \bar{y}_i)^2, \qquad (3.11.9)$$

that is,

$$\text{residual SS} = \text{pure error SS} + \text{lack of fit SS}, \qquad (3.11.10)$$

the cross-product vanishing in the summation over u for each i. The last term on the right is called the lack of fit sum of squares because it is a measure of the discrepancy between the model prediction $\hat{y}_i$ and the average $\bar{y}_i$ of the replicated runs made at the ith set of experimental conditions.

The corresponding equation for degrees of freedom is

$$\sum_{i=1}^{m} r_i - p = \sum_{i=1}^{m} (r_i - 1) + m - p \qquad (3.11.11)$$

where p is the number of parameters in the model. Obviously, we must have $m \geq p$, that is, there must be at least as many or more *distinct* $\mathbf{x}$ values at which y's are observed than there are parameters to be estimated. If $m = p$, there will be no sum of squares nor degrees of freedom for lack of fit, since the model will always "fit perfectly."

An analysis of variance table may now be constructed as in Table 3.10. In this and similar tables, the mean squares, denoted by MS, are obtained by dividing each sum of squares by the corresponding number of degrees of freedom (df).

TABLE 3.10. Analysis of variance table showing the entry of pure error

Source	SS	df	MS
b_0	$n\bar{y}^2$	1	
Additional model terms	$\hat{\theta}'Z'y - n\bar{y}^2$	$p - 1$	MS_M
Lack of fit	$\displaystyle\sum_{i=1}^{m} r_i(\hat{y}_i - \bar{y}_i)^2$	$m - p$	MS_L
Pure error	$\displaystyle\sum_{i=1}^{m}\sum_{u=1}^{r_i} (y_{iu} - \bar{y}_i)^2$	$n - m$	MS_e
Total	$\displaystyle\sum_{i=1}^{m}\sum_{u=1}^{r_i} y_{iu}^2$	n	

Test for Lack of Fit

It is always true that

$$E(MS_e) = \sigma^2. \tag{3.11.12}$$

To evaluate the expected value of MS_L, assume that the model to be fitted is

$$y = Z\theta + \varepsilon \tag{3.11.13}$$

but the true model is

$$y = Z\theta + Z^*\theta^* + \varepsilon. \tag{3.11.14}$$

Then, with

$$A = (Z'Z)^{-1}Z'Z^* \tag{3.11.15}$$

being the alias matrix,

$$E(MS_L) = \sigma^2 + \theta^{*\prime}(Z^* - ZA)'(Z^* - ZA)\theta^*/(m - p). \tag{3.11.16}$$

Thus, provided $Z^* - ZA$ is non-null, $E(MS_L)$ will be inflated when θ^* is nonzero. If $\varepsilon \sim N(0, \sigma^2)$ and $\theta^* = 0$ it can be shown that the observed F ratio

$$F = MS_L/MS_e \tag{3.11.17}$$

follows an $F(m - p, n - m)$ distribution. Thus we can test the null hypothesis $H_0: \theta^* = 0$ versus the alternative $H_1: \theta^* \neq 0$ by comparing MS_L/MS_e to a suitable upper percentage point of the $F(m - p, n - m)$ distribution.

A large and significant value of F discredits the fitted model. In most cases, this would initiate a search for a more adequate model. The nature of the inadequacy would first be sought by analyzing the residuals, and remedial measures, perhaps involving transformation of y or of one or more of the x's or possibly the use of a radically different model, would then be taken. Occasionally, and particularly when there is a very large amount of data, investigation might show that the deficient model is, nevertheless, sufficient for the purpose at hand and therefore may be used with proper caution. (Remember that all models are wrong; the practical question is how wrong do they have to be to not be useful.)

The R^2 Statistic

The statistic

$$R^2 = \left(\hat{\theta}'Z'y - n\bar{y}^2\right)/\left(y'y - n\bar{y}^2\right) \tag{3.11.18}$$

represents the fraction of the variation about the mean that is explained by the fitted model. It is often used as an overall measure of the fit attained. Note that no model can explain pure error, so that the maximum possible value of R^2 is

$$\max R^2 = \left\{(y'y - n\bar{y}^2) - \text{pure error SS}\right\}/(y'y - n\bar{y}^2). \tag{3.11.19}$$

3.12. CONFIDENCE INTERVALS AND CONFIDENCE REGIONS

Confidence Contours

We have seen that we can obtain least-squares estimates for the parameters in the model $y = Z\theta + \varepsilon$ by evaluating $\hat{\theta} = (Z'Z)^{-1}Z'y = Ty$, say, where $T = (Z'Z)^{-1}Z'$ is the linear transformation matrix which converts the vector y into the vector of linear functions $\hat{\theta} = Ty$. Now the variance–covariance matrix of $\hat{\theta}$ is given by

$$V(\hat{\theta}) = V(Ty) = TV(y)T' = TI\sigma^2 T'$$

$$= TT'\sigma^2$$

$$= (Z'Z)^{-1}\sigma^2, \tag{3.12.1}$$

and, in fact it is true that, if the model is correct,

$$\hat{\boldsymbol{\theta}} \sim N\left(\boldsymbol{\theta}, (\mathbf{Z}'\mathbf{Z})^{-1}\sigma^2\right). \tag{3.12.2}$$

As a consequence of this result, it can be shown that the quadratic form

$$(\boldsymbol{\theta} - \hat{\boldsymbol{\theta}})'\mathbf{Z}'\mathbf{Z}(\boldsymbol{\theta} - \hat{\boldsymbol{\theta}}) \sim \sigma^2\chi^2(p) \tag{3.12.3}$$

where $\chi^2(p)$ denotes a chi-squared variable with p df and where p is, as before, the number of parameters in $\boldsymbol{\theta}$. Furthermore, if s^2 is an independent estimate of σ^2 with ν df, such that

$$\nu s^2 \sim \sigma^2\chi^2(\nu), \tag{3.12.4}$$

$$(\boldsymbol{\theta} - \hat{\boldsymbol{\theta}})'\mathbf{Z}'\mathbf{Z}(\boldsymbol{\theta} - \hat{\boldsymbol{\theta}})/(ps^2) \sim F(p, \nu) \tag{3.12.5}$$

where $F(p, \nu)$ denotes an F variable with p and ν df, because the ratio of two independent χ^2 variables each divided by their respective degrees of freedom is an F variable.

In those examples where the residual mean square with $(n - p)$ df is used to supply an estimate of error, $\nu = n - p$.

Suppose that $F_\alpha(p, \nu)$ denotes the α-significance level of the F distribution with p and $\nu = n - p$ df. Then the equation

$$(\boldsymbol{\theta} - \hat{\boldsymbol{\theta}})'\mathbf{Z}'\mathbf{Z}(\boldsymbol{\theta} - \hat{\boldsymbol{\theta}})/(ps^2) = F_\alpha(p, \nu) \tag{3.12.6}$$

defines an ellipsoidal region in the parameter space which we call a "$1 - \alpha$ confidence region for $\boldsymbol{\theta}$." This region includes the true value of $\boldsymbol{\theta}$ with probability $1 - \alpha$ in the sense that, if we imagine the model generating an infinite sequence of sets of y's for the same $\mathbf{Z}$ values, with similar calculations performed for each set, then a proportion $1 - \alpha$ of the regions so generated would actually contain the true point $\boldsymbol{\theta}$.

Individual (also called marginal) intervals for the separate θ's can be obtained as follows. The variances of $\hat{\theta}_1, \hat{\theta}_2, \ldots, \hat{\theta}_p$ are given by

$$V(\hat{\theta}_i) = c^{ii}\sigma^2 \tag{3.12.7}$$

where c^{ii} denotes the ith diagonal term of $(\mathbf{Z}'\mathbf{Z})^{-1}$. Then

$$(\hat{\theta}_i - \theta_i)/(s^2c^{ii})^{1/2} \tag{3.12.8}$$

has a t distribution with ν df and a $1 - \alpha$ confidence interval for θ_i is given by

$$\hat{\theta}_i \pm t_{\alpha/2}(\nu)(s^2 c^{ii})^{1/2} \qquad (3.12.9)$$

where $t_{\alpha/2}(\nu)$ denotes the upper $\frac{1}{2}\alpha$ percentage point of the $t(\nu)$ distribution. Confidence intervals of this type are easily obtainable and useful but they do not draw attention to the correlations between the various $\hat{\theta}_i$'s. These correlations became large if the corresponding columns of the Z matrix are highly nonorthogonal.

The situation can be understood by considering the joint estimation of the two parameters β_1 and β_2 in the first-order model

$$y = \beta_0 + \beta_1 x_1 + \beta_2 x_2 + \varepsilon. \qquad (3.12.10)$$

We shall suppose, without affecting the point at issue, that $V(\varepsilon) = \sigma^2$ is known and is equal to 1.

Design A   (An Orthogonal Design)

We shall see later that efficient estimates of the parameters in the above model can be obtained by using the 2^2 factorial design of Figure 3.8a for

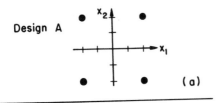

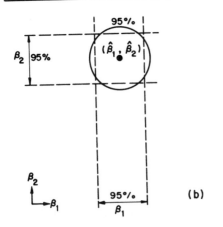

FIGURE 3.8. (a) Orthogonal design A. (b) The resulting 0.95 joint confidence region and 0.95 individual confidence intervals.

which the levels of x_1 and x_2 are

x_1	x_2
-1	-1
1	-1
-1	1
1	1

so that $\mathbf{Z'Z} = \begin{bmatrix} 4 & 0 & 0 \\ 0 & 4 & 0 \\ 0 & 0 & 4 \end{bmatrix} = 4I_3.$ (3.12.11)

With $\sigma^2 = 1$, the covariance matrix for the estimates $\hat{\beta}_0$, $\hat{\beta}_1$, $\hat{\beta}_2$ of β_0, β_1, and β_2 is then

$$(\mathbf{Z'Z})^{-1}\sigma^2 = \begin{bmatrix} \frac{1}{4} & 0 & 0 \\ 0 & \frac{1}{4} & 0 \\ 0 & 0 & \frac{1}{4} \end{bmatrix}$$ (3.12.12)

and the (circular) joint 0.95 confidence region for β_1 and β_2 is centered at the point $(\hat{\beta}_1, \hat{\beta}_2)$ and consists of all points (β_1, β_2) such that

$$4(\beta_1 - \hat{\beta}_1)^2 + 4(\beta_2 - \hat{\beta}_2)^2 \leq 5.99 = \chi^2_{0.05}(2).$$ (3.12.13)

Individual (marginal) intervals for β_1 and β_2 are given by

$$4(\beta_1 - \hat{\beta}_1)^2 = 3.84 = \chi^2_{0.05}(1) \quad \text{and} \quad 4(\beta_2 - \hat{\beta}_2)^2 = 3.84.$$ (3.12.14)

The joint and individual intervals are shown in Figure 3.8b.

Design B (A Nonorthogonal Design)

Now suppose we use a different design in x_1 and x_2 as shown in Figure 3.9a for which the levels of x_1 and x_2 are

x_1	x_2
-1.34	-1.34
0.45	-0.45
-0.45	0.45
1.34	1.34

so that $\mathbf{Z'Z} = \begin{bmatrix} 4 & 0 & 0 \\ 0 & 4 & 3.2 \\ 0 & 3.2 & 4 \end{bmatrix}.$ (3.12.15)

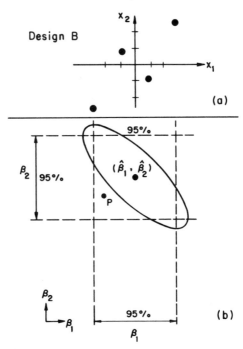

FIGURE 3.9. (a) Nonorthogonal design B. (b) The resulting 0.95 joint confidence region and 0.95 individual confidence intervals.

The levels of x_1 and x_2 have been chosen so that $\bar{x}_i = 0$ and $\sigma_{x_i} = \{\Sigma x_i^2/4\}^{1/2}$ are the same for both designs A and B. For design B, given that $\sigma^2 = 1$, the covariance matrix of $\hat{\beta}_0, \hat{\beta}_1, \hat{\beta}_2$ is

$$(\mathbf{Z'Z})^{-1}\sigma^2 = \begin{bmatrix} \frac{1}{4} & 0 & 0 \\ 0 & \frac{25}{36} & -\frac{5}{9} \\ 0 & -\frac{5}{9} & \frac{25}{36} \end{bmatrix} \qquad (3.12.16)$$

and the (elliptical) joint 0.95 confidence region for β_1 and β_2 is centered at $(\hat{\beta}_1, \hat{\beta}_2)$ and consists of all points (β_1, β_2) such that

$$4(\beta_1 - \hat{\beta}_1)^2 + 4(\beta_2 - \hat{\beta}_2)^2 + 6.4(\beta_1 - \hat{\beta}_1)(\beta_2 - \hat{\beta}_2) \le 5.99. \quad (3.12.17)$$

Individual (marginal) intervals for β_1 and β_2 are given by

$$\tfrac{25}{36}(\beta_1 - \hat{\beta}_1)^2 = 3.84 \quad \text{and} \quad \tfrac{25}{36}(\beta_1 - \hat{\beta}_1)^2 = 3.84. \quad (3.12.18)$$

The joint and individual intervals are shown in Figure 3.9b.

Notice that the orthogonal design is much more desirable than the nonorthogonal design in the senses that:

1. The area of the joint confidence region is much smaller.
2. The lengths of the individual confidence intervals are much smaller.

Correlation of the Parameter Estimates

Now consider Figures 3.8 and 3.9 together. Compare, first of all, the marginal intervals and the joint interval in Figure 3.9b for the nonorthogonal design. Consider a pair of parameter values (β_{10}, β_{20}) corresponding to the point P. We see that, although β_{10} falls within the marginal interval for β_1 and β_{20} falls within the marginal interval for β_2, the point P (β_{10}, β_{20}) itself falls outside the joint interval. The meaning to be associated with this is that, although the value β_{10} is acceptable for some values of β_2, it is not acceptable for the particular value β_{20}. In general, to understand the joint acceptability of values for a group of parameters, it is necessary to consider the joint region and these are not at all easy to visualize when we have more than two or three parameters. Figure 3.8b shows how this difficulty is considerably lessened, but not eliminated, by using an orthogonal design. Such designs lead to circular contours (for two parameters), spherical contours (for three parameters), or hyperspherical contours (for more parameters).

Greater Precision of Estimates from Orthogonal Designs

A further point which is illustrated by comparing Figures 3.8b and 3.9b is the greater precision of estimates obtained from orthogonal designs. Notice that, if such comparisons are to be made fairly, it is necessary to scale, to the same units, the designs to be compared. This has been done in this example by choosing $\Sigma x_1^2 = \Sigma x_2^2 = 4$ for both designs.

3.13. ROBUST ESTIMATION, MAXIMUM LIKELIHOOD, AND LEAST SQUARES

The method of maximum likelihood is known to produce estimates having desirable properties (see, for example, Johnson and Leone, 1977, Vol. 1, pp. 212–214). Thus least squares is often justified by the argument that the least-squares estimators of the elements of θ are the maximum likelihood estimators when the errors $\varepsilon_1, \varepsilon_2, \ldots, \varepsilon_n$ are distributed $\varepsilon \sim N(\mathbf{0}, \mathbf{I}\sigma^2)$. This can be seen as follows. Since the joint probability distribution function of the ε_u is, for $-\infty \le \varepsilon_u \le \infty$,

$$p(\boldsymbol{\varepsilon}) \propto \sigma^{-n} \exp \left\{ \frac{-\boldsymbol{\varepsilon}'\boldsymbol{\varepsilon}}{2\sigma^2} \right\}, \tag{3.13.1}$$

and since

$$\mathbf{y} = \mathbf{Z}\boldsymbol{\theta} + \boldsymbol{\varepsilon}, \tag{3.13.2}$$

the likelihood function is

$$L(\boldsymbol{\theta}|\mathbf{y}) \propto \sigma^{-n} \exp \left\{ \frac{-(\mathbf{y} - \mathbf{Z}\boldsymbol{\theta})'(\mathbf{y} - \mathbf{Z}\boldsymbol{\theta})}{2\sigma^2} \right\}.$$

$$= \sigma^{-n} \exp \left\{ \frac{-S(\boldsymbol{\theta})}{2\sigma^2} \right\}. \tag{3.13.3}$$

This is maximized with respect to $\boldsymbol{\theta}$ when the sum of squares function $S(\boldsymbol{\theta}) = (\mathbf{y} - \mathbf{Z}\boldsymbol{\theta})'(\mathbf{y} - \mathbf{Z}\boldsymbol{\theta})$ is minimized with respect to $\boldsymbol{\theta}$. Thus, the least-squares estimators are maximum likelihood estimators when $\boldsymbol{\varepsilon} \sim N(\mathbf{0}, \mathbf{I}\sigma^2)$, as stated. This, and a similar Bayesian justification, make it clear that the method of least squares is fully appropriate when the errors can be assumed (1) to be statistically independent, (2) to have constant variance σ^2, and (3) to be normally distributed. We can embrace all these assumptions by saying that, when they are true, the errors are *spherically* normally distributed. Conversely, if we know the errors to have some other distributional form, least-squares estimates of parameters would be inappropriate but we could use maximum likelihood to indicate what would then be suitable. It is instructive to consider some of the possibilities so offered.

Correlated Errors with Nonconstant Variance

Suppose we assume that, while the n errors $\varepsilon_1, \varepsilon_2, \ldots, \varepsilon_n$ are normally distributed, they have a covariance matrix $\mathbf{W}^{-1}\sigma^2$ where $\mathbf{W}^{-1}$ is known but σ^2 is not known. (Under such a setup, the previous model, in which the errors are supposed independent and with constant variance, corresponds to the special case $\mathbf{W}^{-1} = \mathbf{I}$.) Then

$$L(\boldsymbol{\theta}|\mathbf{y}) \propto \sigma^{-n} |\mathbf{W}|^{1/2} \exp \left\{ \frac{-Q(\boldsymbol{\theta})}{2\sigma^2} \right\}, \tag{3.13.4}$$

where

$$Q(\boldsymbol{\theta}) = (\mathbf{y} - \mathbf{Z}\boldsymbol{\theta})'\mathbf{W}(\mathbf{y} - \mathbf{Z}\boldsymbol{\theta}). \tag{3.13.5}$$

The likelihood is maximized with respect to θ when $Q(\theta)$ is minimized. Also, writing $\mathbf{W} = \{w_{tu}\}$, we see that the maximum likelihood estimates of θ are obtained by minimizing a *weighted* sum of *products* of discrepancies,

$$Q(\theta) = \sum_{t=1}^{n} \sum_{u=1}^{n} w_{tu}(y_t - \mathbf{z}_t'\theta)(y_u - \mathbf{z}_u'\theta) \tag{3.13.6}$$

where $\mathbf{z}_t'$ is the tth row of $\mathbf{Z}$, rather then minimizing the sum of squares

$$S(\theta) = \sum_{u=1}^{n} (y_u - \mathbf{z}_u'\theta)^2. \tag{3.13.7}$$

It is easily shown that these estimates are such that

$$\hat{\theta} = (\mathbf{Z}'\mathbf{W}\mathbf{Z})^{-1}\mathbf{Z}'\mathbf{W}\mathbf{Y} \tag{3.13.8}$$

with

$$V(\hat{\theta}) = (\mathbf{Z}'\mathbf{W}\mathbf{Z})^{-1}\sigma^2. \tag{3.13.9}$$

Also an unbiased estimate s^2 of σ^2 is provided by $Q(\hat{\theta})/(n - p)$ where p is the number of parameters.

An interesting way to view these estimates is as follows. Any symmetric positive-definite matrix $\mathbf{W}$ can be written in the form

$$\mathbf{W} = (\mathbf{W}^{1/2})'\mathbf{W}^{1/2} \tag{3.13.10}$$

say, where the choice of the matrix designated as $\mathbf{W}^{1/2}$ is not unique. Now define pseudovariables $\dot{\mathbf{Z}} = \mathbf{W}^{1/2}\mathbf{Z}$, $\dot{\mathbf{y}} = \mathbf{W}^{1/2}\mathbf{y}$ and we find that

$$\hat{\theta} = (\dot{\mathbf{Z}}'\dot{\mathbf{Z}})^{-1}\dot{\mathbf{Z}}'\dot{\mathbf{y}} \tag{3.13.11}$$

with

$$V(\hat{\theta}) = (\dot{\mathbf{Z}}'\dot{\mathbf{Z}})^{-1}\sigma^2 \tag{3.13.12}$$

which are the standard least-squares formulas for the pseudovariates $\dot{\mathbf{Z}}$ and $\dot{\mathbf{y}}$.

Autocorrelated Errors—Time Series Analysis

Frequently, business and economic data occur as time series, for example, as monthly values of the gross national product and of unemployment. In investigations aimed at discovering the relationship between such series, even though the assumption of constant error variance may be plausible, the assumption of independent errors will most often not be. It is natural for data obtained serially to be correlated serially, that is, for each error to be correlated with its near neighbors.

The use of ordinary least squares in these circumstances can lead to nonsensical results (see, e.g., Coen, Gomme, and Kendall, 1969, and Box and Newbold, 1971).

Legitimate analysis using time series analysis (see, for example, Box and Jenkins, 1976) in effect uses the data to estimate W appropriately.

Independent Errors with Nonconstant Variance—Weighted Least Squares

Another important special case occurs when it can be assumed that errors are independent but do not have constant variance so that

$$V(y_u) = \frac{\sigma^2}{w_u}. \tag{3.13.13}$$

As before, the w_u are supposed known, but σ^2 is not or, equivalently, the relative variances of the y_u's are supposed known. In this case

$$W = \mathrm{diag}(w_1, w_2, \ldots, w_n), \tag{3.13.14}$$

and the maximum likelihood estimates minimize the *weighted* sum of squares

$$Q_d(\theta) = \sum_{u=1}^{n} w_u(y_u - z'_u\theta)^2. \tag{3.13.16}$$

Pseudovariates are obtained (in this special case) by writing $\dot{z}_{iu} = w_u^{1/2} z_{iu}$ and $\dot{y}_u = w_u^{1/2} y_u$.

For simplicity, we illustrate when $p = 2$, so that the model is then

$$y_u = \theta_1 z_{1u} + \theta_2 z_{2u} + \varepsilon_u,$$

with $V(y_u) = \sigma^2/w_u$. Then writing, for example, $\Sigma wz_1 z_2$ for $\sum_{u=1}^{n} w_u z_{1u} z_{2u}$, we obtain the weighted least-squares estimates of θ_1 and θ_2 as the solutions of the normal equations

$$\hat{\theta}_1 \Sigma wz_1^2 + \hat{\theta}_2 \Sigma wz_1 z_2 = \Sigma wz_1 y,$$

$$\hat{\theta}_1 \Sigma wz_1 z_2 + \hat{\theta}_2 \Sigma wz_2^2 = \Sigma wz_2 y. \tag{3.13.17}$$

The residuals are given by

$$r_u = y_u - \hat{\theta}_1 z_{1u} - \hat{\theta}_2 z_{2u}. \qquad (3.13.18)$$

The estimate of the experimental error variance σ^2 is

$$\hat{\sigma}^2 = s^2 = \frac{\sum\limits_{u=1}^{n} w_u r_u^2}{(n-p)} \qquad (3.13.19)$$

and the covariance matrix for the estimates is given by

$$\mathbf{V}(\hat{\boldsymbol{\theta}}) = \begin{bmatrix} \Sigma w z_1^2 & \Sigma w z_1 z_2 \\ \Sigma w z_1 z_2 & \Sigma w z_2^2 \end{bmatrix}^{-1} \sigma^2, \qquad (3.13.20)$$

where $\hat{\boldsymbol{\theta}} = (\hat{\theta}_1, \hat{\theta}_2)'$.

Non-Normality

There are of course an infinite number of ways in which a distribution can be non-normal. It has been suggested in particular that error distributions in which extreme deviations occur more frequently than with the normal distribution often arise. Such distributions are said to be "heavy tailed" or "leptokurtic." For such an error distribution, the maximum likelihood estimates differ from the least-squares estimates. For example, one highly leptokurtic error distribution is the "double exponential" shown in Figure 3.10a. This has the form

$$p(\varepsilon) \propto \sigma^{-1} \exp\left\{ \left| \frac{\varepsilon}{2\sigma} \right| \right\}$$

where σ is an appropriate scale parameter. For this distribution, the maximum likelihood estimates are those obtained by minimizing the sum of absolute deviations

$$\sum_{u=1}^{n} |y_u - \mathbf{z}_u' \boldsymbol{\theta}|. \qquad (3.13.21)$$

Suppose the model is simply $y_u = \theta + \varepsilon_u$. Then the least-squares estimator of θ appropriate if the error distribution were normal would be $\hat{\theta} = \bar{y}$, the sample average. However, for the heavy tailed double exponential

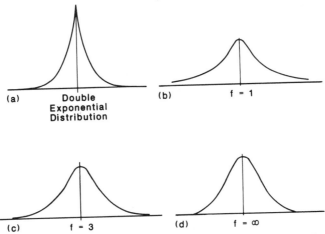

FIGURE 3.10. (*a*) The double exponential distribution. (*b*), (*c*), (*d*) *t* distributions, *f* = 1, 3, ∞ (normal).

distribution the maximum likelihood estimator that minimizes (3.13.21) is the *sample median*. Notice that this estimator is much less sensitive to the extreme deviations that occur with the double exponential distribution.

Heavy tailed error distributions may be represented somewhat more realistically by *t* distributions for which

$$p(\varepsilon) \propto \frac{1}{\sigma}\left(1 + \frac{\varepsilon^2}{f\sigma^2}\right)^{-(1/2)(f+1)},$$

where again σ is a scale parameter. Examples of *t* distributions with $f = 1$, 3, and ∞ df are shown in Figures 3.10*b*, 3.10*c*, and 3.10*d*.

The *t* distribution with $f = 1$ is extremely heavy tailed and is sometimes called the Cauchy distribution. The tendency to leptokurtosis decreases as f increases and, for $f = \infty$, the distribution is exactly the normal distribution.

Effect of Bad Values

It sometimes happens that data are afflicted with one or more bad values (also called "rogue" observations). An atypical value of this kind can arise, for example, from an unrecognized mistake in conducting an experimental run. Now least-squares estimates, which are appropriate under standard normal theory which presupposes that bad values *never* occur, can be excessively influenced by bad values. By employing alternative assumptions, which more closely model the true situation, more appropriate estimates are obtained which place less emphasis on outlying observations.

For example, the double exponential distribution of Figure 3.10a has much heavier tails than the normal and, as already observed, the maximum likelihood estimate of the population mean is the sample median rather than the sample average. Obviously the median is much less likely to be affected by outlying observations than is the sample average and so it is more robust against the possibility that the distribution has heavy tails.

Robustification Using Iteratively Reweighted Least Squares

To guard against misleading estimates caused by non-normal heavy tailed distributions and occasional bad values, the results for a number of maximum likelihood analyses may be compared using different distributional assumptions. Unfortunately, however, the direct evaluation of maximum likelihood estimates from non-normal distributions can become complicated.

An ingenious method for obtaining the required maximum likelihood estimates for a wide class of non-normal distributions employs weighted least squares iteratively. The general argument is set out in Appendix 3A. We illustrate for the important special case when it is supposed that, rather than follow the normal distribution, the errors follow a t distribution having f df. It then turns out that appropriate estimates $\hat{\theta}$ and s^2 for θ and σ^2 may be obtained by iteratively reweighted least squares with weights given by

$$w_u = \frac{(f+1)}{\left\{ f + (r_u/s)^2 \right\}}. \qquad (3.13.22)$$

In Figure 3.11, this weight function is plotted for $f = 1, 3, 5, \infty$. Recall that, when $f = \infty$, the t distribution becomes the normal distribution; the weights are then uniform and equal to unity, corresponding to the employment of ordinary least squares. As the value of f decreases (corresponding to the assumption that the error distribution is more heavy tailed than the normal) the down-weighting associated with large residuals occurs.

Iterative Reweighting

Because estimates of unknown parameters appear in the weights (3.13.22) the weighted least-squares calculations must be performed iteratively. A suitable scheme is as follows:

The iteration is started by obtaining preliminary estimates $(\hat{\theta}^{(0)}, s^{(0)})$ in some way (for example, by ordinary least squares). From these, residuals $r_u^{(0)}$ and weights $w_u^{(0)}$ may be calculated from which new weighted estimates

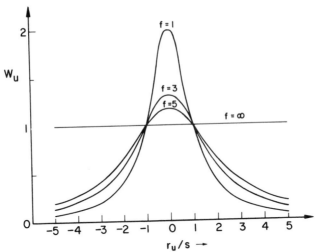

FIGURE 3.11. Weight functions appropriate when the error distribution is a t distribution with degrees of freedom $f = 1, 3, 5, \infty$.

$(\hat{\theta}^{(1)}, s^{(1)})$ are computed. These new estimates give rise to new weights and so on, until the process converges. The procedure is called iteratively reweighted least squares, or IRLS for short. (See Appendix 3A.)

Choice of f

For a given set of data we will almost certainly not know what value of f is appropriate. A useful practical procedure is to carry through the analysis for a few different values of f (for example, 1, 3, and ∞) and see how the estimates are affected.

An Example

In Figure 3.12a and Table 3.11 we show a set of data which appears to follow a straight line relationship

$$y_u = \theta_0 + \theta_1(x_u - \bar{x}) + \varepsilon.$$

Figure 3.12b shows the same data but with 13 units added to the value of y for the seventh observation making it depart markedly from the linear relationship.

The IRLS estimates, with their standard errors in parentheses and the observation weights, are shown in Table 3.12.

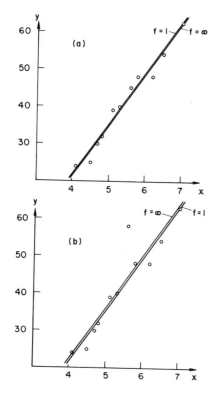

FIGURE 3.12. IRLS estimation on (*a*) original data (*b*) data with outlier.

TABLE 3.11. (a) A set of observations. (b) The same set but with 13 units added to y_7

Observation Number	(a)		(b)	
	x	y	x	y
1	4.1	24	4.1	24
2	4.5	25	4.5	25
3	4.7	30	4.7	30
4	4.8	32	4.8	32
5	5.1	39	5.1	39
6	5.3	40	5.3	40
7	5.6	45	5.6	58
8	5.8	48	5.8	48
9	6.2	48	6.2	48
10	6.5	54	6.5	54
11	7.0	63	7.0	63

TABLE 3.12. IRLS estimates, (standard errors), and observation weights for the example

	Data (a)				Data (b)			
f	∞	5	3	1	∞	5	3	1
$\hat{\theta}_0$	40.73(0.65)	40.85(0.61)	40.89(0.59)	41.00(0.54)	41.91(1.56)	41.30(1.19)	41.21(1.09)	41.13(0.90)
$\hat{\theta}_1$	13.54(0.76)	13.54(0.70)	13.55(0.68)	13.60(0.61)	13.83(1.83)	13.66(1.35)	13.63(1.23)	13.61(0.99)
Observation		Weights					Weights	
1	1	1.11	1.17	1.36	1	1.19	1.30	1.75
2	1	0.79	0.73	0.62	1	1.00	0.98	0.93
3	1	1.11	1.15	1.25	1	1.16	1.25	1.56
4	1	1.18	1.29	1.67	1	1.19	1.31	1.83
5	1	0.93	0.90	0.86	1	1.13	1.18	1.28
6	1	1.16	1.26	1.68	1	1.20	1.33	1.92
7	1	1.04	1.06	1.13	1	0.52	0.44	0.31
8	1	0.99	0.99	1.01	1	1.16	1.24	1.46
9	1	0.79	0.73	0.60	1	0.98	0.97	0.92
10	1	1.05	1.05	1.01	1	1.12	1.18	1.35
11	1	1.14	1.23	1.63	1	1.20	1.33	1.92

Two of the fitted straight lines for $f = 1$ and $f = \infty$ are shown in Figure 3.12. By considering the weights for observation 7 we can see how the influence of this observation is successively discounted as f becomes smaller. In this example it is easy to appreciate the situation by inspection of the graphs. With the more complicated models that are usually employed in this book, this is not possible. However, useful information can still be obtained by noticing the manner in which the estimates of θ and the weights change as f changes.

It is important to keep an open mind concerning the conclusions to be drawn from application of these methods. Robustification merely points out that certain observations are unusual given particular model assumptions. This does not necessarily mean that these observations are unimportant. The history of science abounds with examples where it was the *discrepant* observation that led to a new discovery. Also there are many situations where there are different plausible explanations for the same data. Robust estimation might indicate that model A was applicable if an observation y_m were discounted, but that a more complicated model B would accommodate this particular observation y_m. Distinguishing between such alternative possibilities is a matter for careful consideration and analysis and, in particular, for further experimentation in an attempt to separate these possibilities.

APPENDIX 3A. ITERATIVELY REWEIGHTED LEAST SQUARES

Consider any error distribution function $p(\varepsilon_u)$ which can be written in the form

$$p(\varepsilon_u) \propto \sigma^{-1} g\left\{\left(\frac{\varepsilon_u}{\sigma}\right)^2\right\},$$

where σ is a scale parameter, $g\{\cdot\}$ denotes a functional form, and $\varepsilon_u = y_u - \sum_i \theta_i z_{iu}$. Then the log likelihood for θ and σ^2, given a sample y of n observations, may be written

$$l(\theta, \sigma^2 | \mathbf{y}) = \text{constant} - \frac{1}{2} n \ln \sigma^2 + \sum_u \ln g\left\{\left(\frac{\varepsilon_u}{\sigma}\right)^2\right\}.$$

Now write

$$-2\left[\frac{\partial \ln g\{(\varepsilon_u/\sigma)^2\}}{\partial (\varepsilon_u/\sigma)^2}\right] = w_u(\theta, \sigma^2) = w_u,$$

where the notation $w_u(\theta, \sigma^2)$ is a reminder that w_u is a function of both θ and of σ^2. On differentiating the log likelihood, we obtain

$$\frac{\partial l}{\partial \theta_j} = \sigma^{-2} \sum_u w_u \left(y_u - \sum_i \theta_i z_{iu} \right) z_{ju}, \qquad j = 1, 2, \ldots, p,$$

$$\frac{\partial l}{\partial \sigma^2} = -\frac{1}{2} n \sigma^{-2} + \frac{1}{2} \sigma^{-4} \sum_u w_u \left(y_u - \sum_i \theta_i z_{iu} \right)^2.$$

On equating the derivatives to zero, we obtain the maximum likelihood estimates $\hat{\theta}$ and $\hat{\sigma}^2$ as the solution of the equations

$$\sum_u w_u \left(y_u - \sum_i \hat{\theta}_i z_{iu} \right) z_{ju} = 0, \qquad j = 1, 2, \ldots, p, \qquad (3A.1)$$

$$s^2 = \hat{\sigma}^2 = \frac{\sum w_u \left(y_u - \sum \hat{\theta}_i z_{iu} \right)^2}{n}, \qquad (3A.2)$$

where $w_u = w_u(\hat{\theta}, s^2)$.

Equations (3A.1) have the form of the normal equations for weighted least squares while equation (3A.2) with n replaced by $(n - p)$ has the form of the unbiased estimate of σ^2 supplied by weighted least squares [see (3.13.17), (3.13.18), (3.13.19), (3.13.20)]. The equations may be solved iteratively as described earlier. In particular, if the errors are distributed in a t distribution scaled by a parameter σ,

$$g\left\{ \left(\frac{\varepsilon_u}{\sigma} \right)^2 \right\} = \left\{ 1 + \frac{\varepsilon_u^2}{(f\sigma^2)} \right\}^{-(1/2)(f+1)}$$

and then $w_u = (f + 1)/\{f + (r_u/s)^2\}$ where r_u is the residual $y_u - \sum_i \hat{\theta}_i z_{iu}$.

APPENDIX 3B. JUSTIFICATION OF LEAST SQUARES BY THE GAUSS–MARKOV THEOREM, AND ROBUSTNESS

An alternative widely used justification for least-squares fitting of linear models employs the Gauss–Markov theorem. This states that, for the model $y = Z\theta + \varepsilon$ with the elements $\varepsilon_1, \varepsilon_2, \ldots, \varepsilon_n$ of ε pairwise uncorrelated and having equal variances σ^2, the least-squares estimators of the p parameters in θ have, individually, the smallest variances of all linear unbiased estimators of these parameters.

A linear estimator T is one which is of the form $T = l_1 y_1 + l_2 y_2 + \cdots + l_n y_n$ where the l's are selected constants. An estimator T_j of θ_j, $j = 1, 2, \ldots, p$ is unbiased if $E(T_j) = \theta_j$.

The minimum variance property is appealing and does not depend on normality. It might at first be thought that it would automatically endow ordinary least-squares estimators with robust properties. Unfortunately, this is not necessarily the case. To see why, we must consider the relationship between variance, bias, and mean square error. Suppose the true value of a parameter is θ and an estimator of it (not necessarily linear in the data) is $\tilde{\theta} = f(y_1, y_2, \ldots, y_n) = f(\mathbf{y})$. Now what we require in an estimator is that it be in some sense close to the true value. In particular, we might choose estimators with small *mean square error* (MSE), that is, ones such that

$$\text{MSE}(\tilde{\theta}) = E(\tilde{\theta} - \theta)^2$$

is small. Now write

$$\tilde{\theta} - \theta = \tilde{\theta} - E(\tilde{\theta}) + E(\tilde{\theta}) - \theta,$$

where $E(\tilde{\theta})$ is the mean value of $\tilde{\theta}$. Then

$$\text{MSE}(\tilde{\theta}) = E(\tilde{\theta} - \theta)^2$$
$$= E\{\tilde{\theta} - E(\tilde{\theta})\}^2 + \{E(\tilde{\theta}) - \theta\}^2$$
$$+ 2E\{\tilde{\theta} - E(\tilde{\theta})\}\{E(\tilde{\theta}) - \theta\}.$$

The last term is zero, $E\{\tilde{\theta} - E(\tilde{\theta})\}^2$ is the variance of $\tilde{\theta}$, and $E(\tilde{\theta}) - \theta$ is the bias of $\tilde{\theta}$, so that

$$\text{MSE}(\tilde{\theta}) = \text{variance }\{\tilde{\theta}\} + \{\text{bias }\tilde{\theta}\}^2.$$

The Gauss–Markov theorem says that *if* $\tilde{\theta} = f(\mathbf{y})$ is a *linear* function of the y's, and *if* we set the bias term equal to zero, then the variance of $\tilde{\theta}$ (which becomes, with these restrictions, the MSE) will be minimized. Now it turns out (perhaps rather unexpectedly) that, if these restrictions are relaxed and we allow nonlinear functions of the data and biased estimators to be used, then estimators with smaller MSE are possible. That is to say, by making the bias term nonzero, a more than compensating *decrease* in variance can be obtained. Robust estimators are examples of such estimators. Other examples are Stein's shrinkage estimators and Hoerl and Kennard's ridge regression estimators.

For additional information see Dempster et al. (1977, p. 19) and Andrews et al. (1972).

APPENDIX 3C.   MATRIX THEORY

Matrix, Vector, Scalar

A $p \times q$ matrix $\mathbf{M}$ is a rectangular array of numbers containing p rows and q columns written

$$\mathbf{M} = \begin{bmatrix} m_{11} & m_{12} & \cdots & m_{1q} \\ m_{21} & m_{22} & \cdots & m_{2q} \\ \cdots & & & \\ m_{p1} & m_{p2} & \cdots & m_{pq} \end{bmatrix}$$

For example,

$$
A = \begin{bmatrix} 4 & 1 & 3 & 7 \\ -1 & 0 & 2 & 2 \\ 6 & 5 & -2 & 1 \end{bmatrix}
$$

is a 3 × 4 matrix. The plural of *matrix* is *matrices*. A "matrix" with only one row is called a *row vector*: a "matrix" with only one column is called a *column vector*. For example, if

$$
a = [1, 6, 3, 2, 1], \qquad b = \begin{bmatrix} -1 \\ 0 \\ 1 \end{bmatrix}
$$

then **a** is a row vector of length five and **b** is a column vector of length three. A 1 × 1 "vector" is an ordinary number or *scalar*.

Equality

Two matrices are equal if and only if their dimensions are identical and they have exactly the same entries in the same positions. Thus a matrix equality implies as many individual equalities as there are terms in the matrices set equal.

Sum and Difference

The sum (or difference) of two matrices is the matrix each of whose elements is the sum (or difference) of the corresponding elements of the matrices added (or subtracted). For example,

$$
\begin{bmatrix} 7 & 6 & 9 \\ 4 & 2 & 1 \\ 6 & 5 & 3 \\ 2 & 1 & 4 \end{bmatrix} - \begin{bmatrix} 1 & 2 & 4 \\ -1 & 3 & -2 \\ 6 & 2 & 1 \\ 7 & 0 & 2 \end{bmatrix} = \begin{bmatrix} 6 & 4 & 5 \\ 5 & -1 & 3 \\ 0 & 3 & 2 \\ -5 & 1 & 2 \end{bmatrix}.
$$

The matrices must be of exactly the same dimensions for addition or subtraction to be carried out. Otherwise the operations are not defined.

Transpose

The transpose of a matrix **M** is a matrix **M'** whose rows are the columns of **M** and whose columns are the rows of **M** in the same original order. Thus, for **M** and **A** as defined above,

$$\mathbf{M}' = \begin{bmatrix} m_{11} & m_{21} & \cdots & m_{p1} \\ m_{12} & m_{22} & \cdots & m_{p2} \\ \cdots & & & \\ m_{1q} & m_{2q} & \cdots & m_{pq} \end{bmatrix}$$

$$\mathbf{A}' = \begin{bmatrix} 4 & -1 & 6 \\ 1 & 0 & 5 \\ 3 & 2 & -2 \\ 7 & 2 & 1 \end{bmatrix}.$$

Note that the transpose notation enables us to write, for example,

$$\mathbf{b}' = (-1, 0, 1), \quad \text{or alternatively} \quad \mathbf{b} = (-1, 0, 1)'.$$

Note: The parentheses around a matrix or vector can be square-ended or curved. Often, capital letters are used to denote matrices and lowercase letters to denote vectors.

Symmetry

A matrix $\mathbf{M}$ is said to be *symmetric* if $\mathbf{M}' = \mathbf{M}$.

Multiplication

Suppose we have two matrices, $\mathbf{A}$, which is $p \times q$, and $\mathbf{B}$, which is $r \times s$. They are *conformable* for the product $\mathbf{C} = \mathbf{AB}$ only if $q = r$. The resulting product is then a $p \times s$ matrix, the multiplication procedure being defined as follows: If

$$\mathbf{A} = \begin{bmatrix} a_{11} & a_{12} & \cdots & a_{1q} \\ a_{21} & a_{22} & \cdots & a_{2q} \\ \cdots & & & \\ a_{p1} & a_{p2} & \cdots & a_{pq} \end{bmatrix}, \quad \mathbf{B} = \begin{bmatrix} b_{11} & b_{12} & \cdots & b_{1s} \\ b_{21} & b_{22} & \cdots & b_{2s} \\ \cdots & & & \\ b_{q1} & b_{q2} & \cdots & b_{qs} \end{bmatrix},$$
$$\qquad\qquad p \times q \qquad\qquad\qquad\qquad\qquad q \times s$$

then the product

$$\mathbf{AB} = \mathbf{C} = \begin{bmatrix} c_{11} & c_{12} & \cdots & c_{1s} \\ c_{21} & c_{22} & \cdots & c_{2s} \\ \cdots & & & \\ c_{p1} & c_{p2} & \cdots & c_{ps} \end{bmatrix}$$
$$p \times s$$

is such that

$$c_{ij} = \sum_{l=1}^{q} a_{il} b_{lj},$$

that is, the entry in the ith row and jth column of $\mathbf{C}$ is the *inner product* (the element by element cross-product) of the ith row of $\mathbf{A}$ with the jth column of $\mathbf{B}$. For example,

$$
\underset{2 \times 3}{\begin{bmatrix} 1 & 2 & 1 \\ -1 & 3 & 0 \end{bmatrix}} \underset{3 \times 3}{\begin{bmatrix} 1 & 2 & 3 \\ 4 & 0 & -1 \\ -2 & 1 & 3 \end{bmatrix}}
$$

$$
= \begin{bmatrix} 1(1) + 2(4) + 1(-2) & 1(2) + 2(0) + 1(1) & 1(3) + 2(-1) + 1(3) \\ -1(1) + 3(4) + 0(-2) & -1(2) + 3(0) + 0(1) & -1(3) + 3(-1) + 0(3) \end{bmatrix}
$$

$$
= \underset{2 \times 3}{\begin{bmatrix} 7 & 3 & 4 \\ 11 & -2 & -6 \end{bmatrix}}.
$$

We say that, in the product $\mathbf{AB}$, we have *premultiplied* $\mathbf{B}$ by $\mathbf{A}$ or we have *postmultiplied* $\mathbf{A}$ by $\mathbf{B}$. Note that, in general, $\mathbf{AB}$ and $\mathbf{BA}$, even if both products are permissible (conformable), do not lead to the same result. In a matrix multiplication, the order in which the matrices are arranged is crucially important, whereas the order of the numbers in a scalar product is irrelevant.

When several matrices and/or vectors are multiplied together, the product should be carried out in the way that leads to the least work. For example, the product

$$
\underset{p \times p \ \ p \times n \ \ n \times 1}{\mathbf{W} \qquad \mathbf{Z}' \qquad \mathbf{y}}
$$

could be carried out as $(\mathbf{WZ}')\mathbf{y}$, or as $\mathbf{W}(\mathbf{Z}'\mathbf{y})$, where the parenthesized product is evaluated first. In the first case we would have to carry out pn p-length cross-products and p n-length cross-products; in the second case p p-length and p n-length, clearly a saving in effort.

Special Matrices and Vectors

We define

$$
\mathbf{I}_n = \begin{bmatrix} 1 & 0 & 0 & \cdots & 0 \\ 0 & 1 & 0 & \cdots & 0 \\ 0 & 0 & 1 & \cdots & 0 \\ & \cdots & & & \\ 0 & 0 & 0 & \cdots & 1 \end{bmatrix}
$$

a square $n \times n$ matrix with 1's on the diagonal, 0's elsewhere as the *unit matrix* or *identity matrix*. This fulfills the same role as the number 1 in ordinary arithmetic. If the size of $\mathbf{I}_n$ is clear from the context, the subscript n is often omitted. We further use $\mathbf{0}$ to denote a vector

$$\mathbf{0} = (0, 0, \ldots, 0)'$$

or a matrix

$$\mathbf{0} = \begin{bmatrix} 0 & 0 & \cdots & 0 \\ 0 & 0 & \cdots & 0 \\ \cdots & & & \\ 0 & 0 & \cdots & 0 \end{bmatrix}$$

all of whose values are zeros; the actual size of $\mathbf{0}$ is usually clear from the context. We also define

$$\mathbf{j} = (1, 1, \ldots, 1)'$$

a vector of all 1's; the size of $\mathbf{j}$ is either specified or is clear in context.

Orthogonality

A vector $\mathbf{a} = (a_1, a_2, \ldots, a_n)'$ is said to be *orthogonal* to a vector $\mathbf{b} = (b_1, b_2, \ldots, b_n)'$ if the sum of products of their elements is zero, that is, if

$$\sum_{i=1}^{n} a_i b_i = \mathbf{a'b} = \mathbf{b'a} = 0.$$

Inverse Matrix

The inverse $\mathbf{M}^{-1}$ of a square matrix $\mathbf{M}$ is the unique matrix such that

$$\mathbf{M}^{-1}\mathbf{M} = \mathbf{I} = \mathbf{M}\mathbf{M}^{-1}.$$

The columns $\mathbf{m}_1, \mathbf{m}_2, \ldots, \mathbf{m}_n$ of an $n \times n$ matrix are *linearly dependent* if there exist constants $\lambda_1, \lambda_1, \ldots, \lambda_n$, not all zero, such that

$$\lambda_1 \mathbf{m}_1 + \lambda_2 \mathbf{m}_2 + \cdots + \lambda_n \mathbf{m}_n = \mathbf{0}$$

and similarly for rows. A square matrix some of whose rows (or some of whose columns) are linearly dependent is said to be *singular* and does not possess an inverse. A square matrix that is not singular is said to be *nonsingular* and can be inverted.

If $\mathbf{M}$ is symmetric, so is $\mathbf{M}^{-1}$.

Obtaining an Inverse

The process of matrix inversion is a relatively complicated one, and is best appreciated by considering an example. Suppose we wish to obtain the inverse $\mathbf{M}^{-1}$

of the matrix

$$\mathbf{M} = \begin{bmatrix} 3 & 4 & 5 \\ 1 & 2 & 6 \\ 7 & 1 & 9 \end{bmatrix}.$$

Let

$$\mathbf{M}^{-1} = \begin{bmatrix} a & b & c \\ d & e & f \\ g & h & k \end{bmatrix}.$$

Then we must find $(a, b, c, \ldots, h, k)$ so that

$$\begin{bmatrix} a & b & c \\ d & e & f \\ g & h & k \end{bmatrix} \begin{bmatrix} 3 & 4 & 5 \\ 1 & 2 & 6 \\ 7 & 1 & 9 \end{bmatrix} = \begin{bmatrix} 1 & 0 & 0 \\ 0 & 1 & 0 \\ 0 & 0 & 1 \end{bmatrix}$$

that is, so that

$$3a + b + 7c = 1, \quad 3d + e + 7f = 0, \quad 3g + h + 7k = 0,$$

$$4a + 2b + c = 0, \quad 4d + 2e + f = 1, \quad 4g + 2h + k = 0,$$

$$5a + 6b + 9c = 0, \quad 5d + 6e + 9f = 0, \quad 5g + 6h + 9k = 1.$$

Solving these three sets of three linear simultaneous equations yields

$$\mathbf{M}^{-1} = \begin{bmatrix} \frac{12}{103} & -\frac{31}{103} & \frac{14}{103} \\ \frac{33}{103} & -\frac{8}{103} & -\frac{13}{103} \\ -\frac{13}{103} & \frac{25}{103} & \frac{2}{103} \end{bmatrix} = \frac{1}{103} \begin{bmatrix} 12 & -31 & 14 \\ 33 & -8 & -13 \\ -13 & 25 & 2 \end{bmatrix}.$$

(Note the removal of a common factor, explained below.) In general for an $n \times n$ matrix there will be n sets of n simultaneous linear equations. Accelerated methods for inverting matrices adapted specifically for use with electronic computers permit inverses to be obtained with great speed, even for large matrices.

Determinants

An important quantity associated with a square matrix is its *determinant*. Determinants occur naturally in the solution of linear simultaneous equations and in the inversion of matrices. For a 2×2 matrix

$$\mathbf{M} = \begin{bmatrix} a & b \\ c & d \end{bmatrix}$$

the determinant is defined as

$$\det \mathbf{M} = \begin{vmatrix} a & b \\ c & d \end{vmatrix} = ad - bc.$$

For a 3×3 matrix

$$\begin{bmatrix} a & b & c \\ d & e & f \\ g & h & k \end{bmatrix}$$

it is

$$a \begin{vmatrix} e & f \\ h & k \end{vmatrix} - b \begin{vmatrix} d & f \\ g & k \end{vmatrix} + c \begin{vmatrix} d & e \\ g & h \end{vmatrix} = aek - afh - bdk + bfg + cdh - ceg.$$

Notice that we expand by the first row, multiplying a by the determinant of the matrix left when we cross out the row and column containing a, multiplying b by the determinant of the matrix left when we cross out the row and column containing b, multiplying c by the determinant of the matrix left when we cross out the row and column containing c. We also attach alternate signs $+, -, +$ to these three terms, counting from the top left-hand corner element: $+$ to a, $-$ to b, $+$ to c, and so on, alternately, if there were more elements in the first row.

In fact, the determinant can be written down as an expansion of *any* row or column by the same technique. The signs to be attached are counted $+ - + -$, and so on, from the top left corner element alternating either along row or column (but *not* diagonally). In other words the signs

$$\begin{bmatrix} + & - & + \\ - & + & - \\ + & - & + \end{bmatrix}$$

are attached and any row or column is used to write down the determinant. For example, using the second row we have

$$-d \begin{vmatrix} b & c \\ h & k \end{vmatrix} + e \begin{vmatrix} a & c \\ g & k \end{vmatrix} - f \begin{vmatrix} a & b \\ g & h \end{vmatrix}$$

to obtain the same result as before.

The same principle is used to get the determinant of any matrix. Any row or column is used for the expansion and we multiply each element of the row or column by:

1. Its appropriate sign, counted as above.
2. The determinant of the submatrix obtained by deletion of the row and column in which the element of the original matrix stands.

Determinants arise in the inversion of matrices as follows. The inverse M^{-1} may be obtained by first replacing each element m_{ij} of the original matrix M by an element calculated as follows:

1. Find the determinant of the submatrix obtained by crossing out the row and column of M in which m_{ij} stands.
2. Attach a sign from the $+ - + -$ count, as above.
3. Divide by the determinant of M.

When all elements of M have been replaced, *transpose the resulting matrix.* The transpose will be M^{-1}.

The reader might like to check these rules by showing that

$$M^{-1} = \begin{bmatrix} a & b \\ c & d \end{bmatrix}^{-1} = \begin{bmatrix} d/D & -b/D \\ -c/D & a/D \end{bmatrix},$$

where $D = ad - bc$ is the determinant of M; and that

$$Q^{-1} = \begin{bmatrix} a & b & c \\ d & e & f \\ g & h & k \end{bmatrix}^{-1} = \begin{bmatrix} A & B & C \\ D & E & F \\ G & H & K \end{bmatrix},$$

where

$$A = (ek - fh)/Z, \qquad B = -(bk - ch)/Z, \qquad C = (bf - ce)/Z,$$

$$D = -(dk - fg)/Z, \qquad E = (ak - cg)/Z, \qquad F = -(af - cd)/Z,$$

$$G = (dh - eg)/Z, \qquad H = -(ah - bg)/Z, \qquad K = (ae - bd)/Z,$$

and where

$$Z = aek + bfg + cdh - afh - bdk - ceg$$

is the determinant of Q. Note that, if M is symmetric (so that $b = c$), M^{-1} is also symmetric. Also, if Q is symmetric (so that $b = d$, $c = g$, $f = h$), then Q^{-1} is also symmetric because then $B = D$, $C = G$, and $F = H$.

Common Factors

If *every* element of a matrix has a common factor, it can be taken outside the matrix. Conversely, if a matrix is multiplied by a constant c, every element of the matrix is multiplied by c. For example,

$$\begin{bmatrix} 4 & 6 & -2 \\ 8 & 6 & 2 \end{bmatrix} = 2 \begin{bmatrix} 2 & 3 & -1 \\ 4 & 3 & 1 \end{bmatrix}.$$

Note that, if a matrix is square and of size $p \times p$, and if c is a common factor, then the determinant of the matrix has a factor c^p, not just c. For example,

$$\begin{vmatrix} 4 & 6 \\ 8 & 6 \end{vmatrix} = 2^2 \begin{vmatrix} 2 & 3 \\ 4 & 3 \end{vmatrix} = 2^2 (6 - 12) = -24.$$

APPENDIX 3D.   NONLINEAR ESTIMATION

Nonlinear Least Squares

Our discussion in Chapter 3 on fitting linear models leaves open the question of how to proceed when the model we wish to fit is intrinsically nonlinear. We discuss the matter briefly here. Suppose we wish to fit the general model

$$y_u = f(\boldsymbol{\xi}_u, \boldsymbol{\theta}) + \varepsilon_u, \tag{3D.1}$$

where f is some known function, $\boldsymbol{\xi}$ is a vector of predictor variables, $\boldsymbol{\xi}_1, \boldsymbol{\xi}_2, \ldots, \boldsymbol{\xi}_k$, $\boldsymbol{\theta} = (\theta_1, \theta_2, \ldots, \theta_p)'$ is a vector of parameters to be estimated, ε is a random error from a distribution with mean zero and unknown variance σ^2, and the subscript $u = 1, 2, \ldots, n$ ranges over the n observations. We also suppose that the errors are normally distributed, so that $\varepsilon_u \sim N(0, \sigma^2)$.

To obtain least-squares estimates, we must find the value of $\boldsymbol{\theta}$ which minimizes the sum of squares

$$S(\boldsymbol{\theta}) = \sum_{u=1}^{n} \left\{ y_u - f(\boldsymbol{\xi}_u, \boldsymbol{\theta}) \right\}^2. \tag{3D.2}$$

Equating first derivatives with respect to the θ's to zero yields

$$\sum_{u=1}^{n} \left\{ y_u - f(\boldsymbol{\xi}_u, \boldsymbol{\theta}) \right\} \left\{ \frac{\partial f}{\partial \theta_i} \right\} = 0, \qquad i = 1, 2, \ldots, p. \tag{3D.3}$$

Because these normal equations are nonlinear in $\boldsymbol{\theta}$, they are not amenable to easy solution. We now describe an alternative minimizing procedure based on local linearization originally suggested by Gauss.

Linearization (or Gauss) Procedure

If the function $f(\boldsymbol{\xi}_u, \boldsymbol{\theta})$ is expanded as a Taylor series in the elements $\boldsymbol{\theta}$ to first-order terms about some guessed value $\boldsymbol{\theta}_0$, we can use linear least-squares theory to give a solution θ_1, for this *linearized* model. We can then repeat the procedure with θ_1 in place of θ_0, to give rise to θ_2, and so on. Under favorable conditions, successive iterations will converge to the least-squares estimate $\hat{\theta}$. In practice they do not always so converge. (For further technical details of this procedure, see Draper and Smith, 1981, Chapter 10.)

The Marquardt Compromise Procedure

The crude Gauss procedure outlined above has a tendency to "overshoot," that is, to go beyond points θ where smaller $S(\theta)$ values exist to points where larger $S(\theta)$ values occur. While recovery from "overshoot" sometimes occurs on the next iteration, sometimes it does not, and the iterations may continue to diverge. To avoid this problem, Levenberg (1944) suggested restraining the next iteration to points on an ellipsoid about the current best estimate. As the current iteration gets closer and closer to the minimizing value $\hat{\theta}$, overshoot becomes less likely and the dimensions of the ellipsoid can be enlarged. This is the procedure Marquardt (1963) implemented. In the appropriately scaled standard error coordinates, two constants

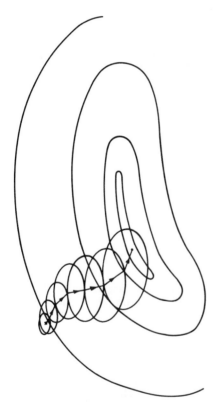

FIGURE 3D.1. Typical iterations in a Marquardt compromise procedure.

are chosen, one to set the spherical radius and one to use as an enlarging factor, as the iterations proceed; in the various programmed versions, the selection is done automatically. Figure 3D.1 shows how the iterations might proceed in a particular case.

One justification for the procedure is as follows: Initially, when steepest ascent is likely to work well, the procedure follows the path of steepest ascent orthogonal to the contours; later when linearization is likely to work well, nearer to $\hat{\theta}$, it becomes more and more like linearization. (Thus the Marquardt method *compromises* between these two approaches; hence its name.) Whether Marquardt's is actually the best procedure from this point of view has been questioned by Box and Kanemasu (1972, 1984). However, an advantage of the Marquardt compromise is that it overcomes problems that arise when there is ill-conditioning, that is, when the sum of squares surface $S(\theta)$ is approximated locally by an attenuated ridge type of surface.

Initial Values

In order to actually carry out any of the iterative procedures needed to minimize $S(\theta)$, one has to begin with a set of *initial guesses* $\theta_0 = (\theta_{10}, \theta_{20}, \ldots, \theta_{p0})'$ of the parameters. The "better" these initial guesses are, the faster will the iterative procedure converge. Thus, efforts to obtain good initial values are often worthwhile. Approximate graphical methods are frequently appropriate, and other methods based on picking p selected runs and solving for the θ's when the model is made to pass through these points exactly can sometimes be employed. However, it must also be said that even with extremely poor initial guesses, convergence can sometimes occur surprisingly fast, so that it is not *always* necessary to spend a great deal of effort selecting starting values.

Confidence Contours

By linearizing $f(\xi, \theta)$ about $\hat{\theta}$, and appealing to results that are true for the linear model case, we obtain an approximate $100(1 - \alpha)\%$ elliptical confidence region in the form

$$(\theta - \hat{\theta})'\hat{Z}'\hat{Z}(\theta - \hat{\theta}) \leq ps^2 F(p, n - p, 1 - \alpha), \qquad (3D.4)$$

where $\hat{Z}$ is an $n \times p$ matrix whose element in the uth row and ith column is

$$\left. \frac{\partial f(\xi_u, \theta)}{\partial \theta_i} \right|_{\substack{\text{evaluated} \\ \text{at } \theta = \hat{\theta}}} \qquad (3D.5)$$

and where

$$s^2 = \frac{S(\hat{\theta})}{(n - p)} . \qquad (3D.6)$$

Ways of checking the adequacy of this linear approximation have been considered, as have methods for choosing appropriate transformations of the parameters which can often greatly improve the adequacy of the linear approximation. For additional reading see, for example, Bates and Watts (1980), Beale (1960), and Box (1960b).

For wider reading on nonlinear estimation, see Draper and Smith (1981, Chapter 10) and Bates and Watts (1987).

APPENDIX 3E.   RESULTS INVOLVING $V(\hat{y})$

We have

$$\hat{y} = Z\hat{\theta} = Z(Z'Z)^{-1}Z'y$$

$$= Ry \tag{3E.1}$$

say, where $R = Z(Z'Z)^{-1}Z'$. Note that R is symmetric $(R' = R)$ and idempotent $(R^2 = RR = R)$. It follows that

$$V(\hat{y}) = R'V(y)R$$

$$= R(I\sigma^2)R$$

$$= R\sigma^2 \tag{3E.2}$$

Thus $V(\hat{y}_i)$, $i = 1, 2, \ldots, n$ is the ith diagonal term of $R\sigma^2$. Consider the sum

$$\sum_{i=1}^{n} V(\hat{y}_i) = \text{trace}(R\sigma^2) = \sigma^2 \text{ trace } R, \tag{3E.3}$$

where "trace" means "take the sum of all the diagonal elements of the square matrix indicated." Now it is true that trace $(AB) = \text{trace}(BA)$. So, taking $A = Z$ and $B = (Z'Z)^{-1}Z'$, we see that

$$\text{trace } R = \text{trace}\left(Z(Z'Z)^{-1}Z'\right) = \text{trace}\left\{(Z'Z)^{-1}Z'Z\right\}$$

$$= \text{trace}(I_p)$$

$$= p, \tag{3E.4}$$

the dimension of $Z'Z$. It follows that

$$n^{-1}\sum_{i=1}^{n} V(\hat{y}_i) = \frac{p\sigma^2}{n}. \tag{3E.5}$$

This means that the average variance of $\hat{y}$ over a set of points used for a regression calculation is $p\sigma^2/n$, and is thus fixed when p, n, and σ^2 are fixed.

EXERCISES

3.1. The data below arose in a test of certain semiconductor memory devices.

Supply voltage during "write" operation, ξ: 25.00 25.05 25.10 25.15 25.20
Retention time (h $\times 10^{-4}$), Y: 1.55   2.36   3.93   7.11   13.52

(a) Plot the data, fit the model $Y = \beta_0 + \beta_1\xi + \varepsilon$, and find the fitted values and residuals. Confirm that $\Sigma Y_i^2 = \Sigma \hat{Y}_i^2 + \Sigma(Y_i - \hat{Y}_i)^2$, within rounding error.

(b) Let ξ be coded to $x = (\xi - \xi_0)/S$. What are suitable values for ξ_0 and S?

(c) Fit the model $Y = \beta_0 + \beta_1 x + \varepsilon$ and find the fitted values and residuals.

(d) Which is preferable, model (a) or model (c)? Why?

(e) What do the residuals indicate?

(f) It is now suggested that the analysis should have been done *not* with Y but with $y = \log Y$. Fit $y = \beta_0 + \beta_1 x + \varepsilon$ and draw the fitted line on a plot of y versus x.

(g) Find the fitted values y_i and the residuals $y_i - \hat{y}_i$ in (f) and show that the vectors $\mathbf{y}$, $\hat{\mathbf{y}}$, and $\mathbf{y} - \hat{\mathbf{y}}$ form a right-angled triangle with internal angles of $2°$ (approximately), $90°$, and $88°$ (approximately).

(h) For the model in (f), test the hypothesis $(\beta_0, \beta_1) = (0.6, 0.25)$. What do you conclude?

3.2. A colleague who reads the above exercise suggests that "it might have been better to fit a quadratic $Y = \beta_0 + \beta_1 x + \beta_2 x^2 + \varepsilon$ to the original Y data." Carry out the analysis. Do you agree with her? Explain.

3.3. Consider the model $y = z\theta + \varepsilon$ where the data values of z are in $\mathbf{z}' = (1,1,1,1,1)$ and the corresponding y observations are in $\mathbf{y}' = (11, 8, 9, 10, 7)$. Estimate θ and test $H_0 : \theta = 8$ versus $H_1 : \theta \neq 8$.

3.4. The data in Table E3.4 were obtained from an experiment to determine the effect of two gasoline additives (whose percentages are x_1 and x_2) on the "cold-start ignition time" in seconds, Y, of a test vehicle. (We work with $z_1 = 10x_1$, $z_2 = 10x_2$, and $y = 100Y$ to remove decimals in what follows.)

(a) Fit the model $y = \theta_1 z_1 + \varepsilon$ by least squares. (What important assumption is being made here?)

(b) Fit $y = \theta_1 z_1 + \theta_2 z_2 + \varepsilon$.

(c) Determine $z_{2.1}$ and fit $y = \theta z_1 + \theta_2 z_{2.1} + \varepsilon$. How are $\hat{\theta}_1$, $\hat{\theta}_2$, and $\hat{\theta}$ related?

(d) Construct an analysis of variance table.

TABLE E3.4

z_1	z_2	y
3	6	176
4	4	192
5	7	262
6	3	230
7	6	308
8	4	312

(e) If model (a) is fitted but model (b) is "true," what is $E(\hat{\theta}_1)$? What relationship does this have to your answer in (c)?

(f) Construct a 95% joint confidence region for θ_1 and θ_2.

3.5. (Source: Inactivation of adrenaline and nonadrenaline by human and other mammalian liver in vitro, by W. A. Bain and J. E. Batty, *Brit. J. Pharm. Chemotherapy*, **11**, 1956, 52–57.) The data in Table E3.5 are $n = 14$ epinephrine (adrenaline) concentrations Y (erg/mL) for five "times in the lower tissues," X (min), coded to x.

(a) What is the coding?

(b) Assume for the moment that the data consist of 14 independent observations. Fit $Y = \beta_0 + \beta_1 x + \varepsilon$, but show that lack of fit exists. What alternative model would you recommend?

(c) Fit a suitable alternative model.

(d) You now look up the original paper and find that each Y column in the table is a separate experiment, in each of which samples were taken successively in time from the same tube. Might this affect your analysis? If so, how and why?

TABLE E3.5

X	x	Y			ΣY
6	−2	30.0	28.6	28.5	87.1
18	−1	8.9	8.0	10.8	27.7
30	0	4.1	—	4.7	8.8
42	1	1.8	2.6	2.2	6.6
54	2	0.8	0.6	1.0	2.4
					132.6

Factorial Designs at Two Levels

4.1. THE VALUE OF FACTORIAL DESIGNS

In the model-building process, the step of model specification or identification, as a result of which a model worthy of being entertained is put forward, is a somewhat tenuous one. It is a creative step in which the human mind, taking account of what is known about the system under study, must be allowed to interact freely with the data, making comparisons, seeking similarities, differences, trends, and so on. A class of experimental designs called "factorials" greatly facilitates this process.

We obtain a complete *factorial design* in k factors by choosing n_1 levels of factor 1, n_2 levels of factor 2,..., n_k levels of factor k, and then selecting the $n = n_1 \times n_2 \times \cdots \times n_k$ runs obtained by taking all possible combinations of the levels selected. Figure 4.1a shows a $3 \times 2 \times 2$ factorial design in the factors:

Temperature	(three levels: 180°C, 190°C, 200°C)
Pressure	(two levels: 70 psi, 90 psi)
Catalyst	(two levels: type A, type B)

Suppose the numbers shown in Figure 4.1b are the measured amounts of an impurity which are produced at the various sets of conditions. By simply looking at the figure, an observer can make many different and informative comparisons. For example, suppose we knew that the experimental error was small (that the standard deviation σ was less than 1, say). Then the "horizontal" comparisons $(7 \rightarrow 8 \rightarrow 10)$, $(6 \rightarrow 8 \rightarrow 11)$, $(3 \rightarrow 5 \rightarrow 5)$, $(3 \rightarrow 4 \rightarrow 5)$ would suggest that, over the ranges studied, increasing temperature produced increasing impurity.

Also, the "vertical" comparisons $(7 \rightarrow 6)$, $(8 \rightarrow 8)$, $(10 \rightarrow 11)$, $(3 \rightarrow 3)$, $(5 \rightarrow 4)$, $(5 \rightarrow 5)$ would suggest that there is little to choose between the two

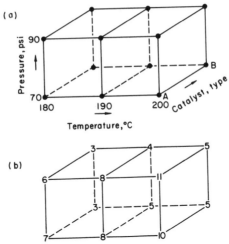

FIGURE 4.1. A $3 \times 2 \times 2$ factorial design. (*a*) The points of the design. (*b*) Impurity values observed at the 12 design point locations.

pressures examined. The six comparisons between the two types of catalysts would suggest that, over the ranges tested, *B* produces less impurity than *A*. Furthermore, the increase of impurity with increased temperature is less with *B* than with *A* (so that there is a temperature–catalyst interaction). The possibility is also seen that perhaps a fixed change in temperature produces a fixed *proportionate* response so that it is possible that the logarithm of the response is linearly and additively affected by temperature.

All of the above ideas can, of course, be subjected to more formal examination. It is the ability of the factorial design to suggest ideas to the investigator that we wish to emphasize.

In the above example, the factorial design is used to study two quantitative factors or variables (temperature and pressure) and one qualitative variable (type of catalyst). Most of this book will be concerned only with the study of quantitative variables. Factorial designs possess the following desirable properties:

1. They allow multitudes of comparisons to be made and so facilitate model creation and criticism.
2. They provide highly efficient estimates of constants (parameters), that is, estimates whose variances are as small, or nearly as small, as those that could be produced by any design occupying the same space.
3. They give rise to simple calculations.

On the mistaken supposition that simplicity of calculation is the *only* reason for their use, it is sometimes suggested that factorial designs are no longer needed because computers have made even complex calculations easy. Also it may be argued that "optimal" designs may now be computed (see Chapter 14 for details) which do not necessarily give nice patterns and yet, *if the model is assumed known*, will also give highly efficient estimates. However, this argument discounts the importance of allowing the mind to see patterns in the data at the model specification stage.

4.2. TWO-LEVEL FACTORIALS

Of particular importance for our purpose are the "two-level" factorial designs in which each variable occurs at just two levels. Such designs are especially useful at the exploratory stage of an investigation, when not very much is known about a system and the model is still to be identified. As we shall see in later examples, the initial two-level pattern can be a first building block in developing structures of many different sorts. A model-identification technique of great value that may be used in association with this type of design is the plotting of coefficients on normal probability paper (see Daniel, 1959, 1976). For additional details see Appendix 4B.

For a first illustration, we return to the 2^3 (i.e., $2 \times 2 \times 2$) factorial design introduced in Chapter 2 (see Table 2.2). This design was used to study Y = cycles to failure for combinations of three variables, each tested at two levels as follows.

Coded levels	x_i	-1	1
Length of test specimen (mm)	ξ_1	250	350
Amplitude of load cycle (mm)	ξ_2	8	10
Load (g)	ξ_3	40	50

As discussed in Chapter 2, it is convenient to code the lower and upper levels as x_1, x_2, x_3, taking the values -1 for the lower level and 1 for the upper level. Thus, for this example,

$$x_1 = \frac{(\xi_1 - 300)}{50}, \quad x_2 = \frac{(\xi_2 - 9)}{1}, \quad \text{and} \quad x_3 = \frac{(\xi_3 - 45)}{5}. \quad (4.2.1)$$

The 2^3 design is shown in Table 4.1 in both uncoded and coded form, together with the response values Y = cycles to failure and the values of $y = \log Y$. (The logarithms are taken to base 10.)

TABLE 4.1. Textile data. A 2^3 factorial design in uncoded and coded units with response Y, and $y = \log Y$ values

Uncoded			Coded			Cycles to failure, Y	$y = \log Y$
Specimen length (mm), ξ_1	Amplitude (mm), ξ_2	Load (g), ξ_3	Specimen length, x_1	Amplitude, x_2	Load, x_3		
250	8	40	−1	−1	−1	674	2.83
350	8	40	1	−1	−1	3636	3.56
250	10	40	−1	1	−1	170	2.23
350	10	40	1	1	−1	1140	3.06
250	8	50	−1	−1	1	292	2.47
350	8	50	1	−1	1	2000	3.30
250	10	50	−1	1	1	90	1.95
350	10	50	1	1	1	360	2.56

Figure 4.2 shows the eight runs set out in the space of the three variables where the design points appear as the vertices of a cube. On the vertices of the left-hand cube are shown the values of the response Y, the number of cycles to failure, and on the right-hand cube are shown the corresponding transformed values $y = \log Y$. The geometrical display again illustrates the virtues of the factorial design. It is very easy to appreciate the many comparisons that can be made. In particular, one can see the consistent increases that occur in Y with increase in length (ξ_1) and the reductions that occur with increase in amplitude (ξ_2) and with increase in load (ξ_3). Moreover, the differences in magnitudes of these changes largely disappear after log transformation, indicating the regularity that occurs in the *proportional* increases in this example.

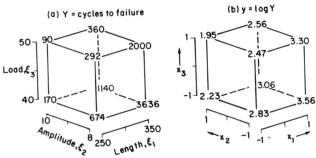

FIGURE 4.2. Textile data. A 2^3 design shown geometrically in (a) uncoded and (b) coded units, with responses (a) Y = cycles to failure and (b) $y = \log Y$.

In general, a 2^k factorial design consists of all the 2^k runs (points) with levels (coordinates)

$$(x_1, x_2, \ldots, x_k) = (\pm 1, \pm 1, \ldots, \pm 1), \qquad (4.2.2)$$

where every possible combination of $\pm$ signs is selected in turn. Geometrically the design consists of the vertices of a hypercube in k dimensions. For purposes of analysis, it is convenient to list the runs in *standard order*, not in the order (usually randomized) in which they were made. This standard order is obtained by writing alternate $-$ and $+$ signs in the column headed x_1, alternate pairs $- -$, $+ +$, in the x_2 column, alternate fours $- - - -$, $+ + + +$, in the x_3 column, and so on. For illustration, the runs are set out in standard order in Table 4.1.

Analysis of the Factorial Design

The main effect of a given variable as defined by Yates (1937) is the average difference in the level of response as one moves from the low to the high level of that variable. For illustration, consider the calculation of the main effects for $y = \log Y$. A glance at Figure 4.2b shows that the main effect indicated by 1, of variable x_1, is*

$$1 \leftarrow \tfrac{1}{4}(3.56 + 3.06 + 3.30 + 2.56) - \tfrac{1}{4}(2.83 + 2.23 + 2.47 + 1.95)$$

$$= 0.75. \qquad (4.2.3)$$

Similar calculations provide the results

$$2 \leftarrow -0.59, \qquad (4.2.4)$$

$$3 \leftarrow -0.35. \qquad (4.2.5)$$

A valuable property of the factorial design is that it makes possible not only the calculation of main effects (i.e., average effects) but also of interaction effects between variables as well.

Two variables, say 1 and 3, are said to interact (in their effect on the response) if the effect of 1 is different at the two different levels of 3. In the textile example, if we confine attention for the moment to the first four runs

---

*Here and in Chapter 5, we use the arrow pointing to the left ($\leftarrow$) to mean "is estimated by," so that Eq. (4.2.3) is read as "the main effect of variable 1 is estimated by the number 0.75." Similarly, the reverse notation "$0.75 \rightarrow 1$" is read as "0.75 is an estimate of the main effect of variable 1."

in which x_3 is at its lower level, the main (average) effect of 1 is

$$(1|x_3 = -1) \leftarrow \tfrac{1}{2}(3.56 + 3.06) - \tfrac{1}{2}(2.83 + 2.23) = 0.78. \quad (4.2.6)$$

However, for the last four runs, with x_3 at its upper level, the main effect of 1 is

$$(1|x_3 = 1) \leftarrow \tfrac{1}{2}(3.30 + 2.56) - \tfrac{1}{2}(2.47 + 1.95) = 0.72. \quad (4.2.7)$$

The interaction between variables 1 and 3 is defined as half the difference between the main effect of 1 at the upper level of x_3 and the main effect of 1 at the lower level of x_3. This interaction is denoted by the symbol 13, so that

$$13 = \tfrac{1}{2}\{(1|x_3 = 1) - (1|x_3 = -1)\} \leftarrow \tfrac{1}{2}\{0.72 - 0.78\} = -0.03. \quad (4.2.8)$$

(There is no ambiguity in this definition in the sense that interchanging the roles of variables 1 and 3 does not change the value of the interaction.)

Exercise. Evaluate $\tfrac{1}{2}\{(3|x_1 = 1) - (3|x_1 = -1)\}$ and show that it produces a result identical to the 13 value already obtained.

Via similar calculations we obtain the other two two-factor interactions as

$$12 \leftarrow -0.03, \quad\quad\quad\quad\quad\quad (4.2.9)$$

$$23 \leftarrow -0.04. \quad\quad\quad\quad\quad\quad (4.2.10)$$

Finally, the interaction 12 might be different at different levels of the variable 3. To check this we could evaluate and compare

$$(12|x_3 = -1) \leftarrow \tfrac{1}{2}\{(3.06 - 2.23) - (3.56 - 2.83)\} = 0.05,$$

$$(12|x_3 = 1) \leftarrow \tfrac{1}{2}\{(2.56 - 1.95) - (3.30 - 2.47)\} = -0.11. \quad (4.2.11)$$

Half the difference between these quantities is called the 123 interaction, that is,

$$123 = \tfrac{1}{2}\{(12|x_3 = 1) - (12|x_3 = -1)\} \leftarrow -0.08. \quad (4.2.12)$$

Exercise. Show that the same result occurs if we define the 123 interaction as half the difference between the interaction 23 at levels of $x_1 = 1$ and $x_1 = -1$, or as half the difference between the interaction 13 at levels of $x_2 = 1$ and $x_2 = -1$. Thus the three-way interaction 123 is unambiguous.

TABLE 4.2. Columns of signs and the divisors for systematically obtaining the factorial effects in a 2^3 factorial design

I	1	2	3	12	13	23	123	y
+	−	−	−	+	+	+	−	2.83
+	+	−	−	−	−	+	+	3.56
+	−	+	−	−	+	−	+	2.23
+	+	+	−	+	−	−	−	3.06
+	−	−	+	+	−	−	+	2.47
+	+	−	+	−	+	−	−	3.30
+	−	+	+	−	−	+	−	1.95
+	+	+	+	+	+	+	+	2.56
Divisor 8	4	4	4	4	4	4	4	

There is an easier systematic way of making these calculations using the columns of signs of Table 4.2. Consider the calculation of the main effect 1 as in Eq. (4.2.3). This could be written

$$1 \leftarrow \tfrac{1}{4}(-y_1 + y_2 - y_3 + y_4 - y_5 + y_6 - y_7 + y_8). \qquad (4.2.13)$$

In other words, the main effect 1 would be obtained by multiplying the column of data y by the column of − and + signs in the column labeled **1** in Table 4.2 and dividing by the divisor 4 indicated there. Note that the divisors are the numbers of + signs in the corresponding columns.
Similarly, it can readily be confirmed that

$$2 \leftarrow \tfrac{1}{4}(-y_1 - y_2 + y_3 + y_4 - y_5 - y_6 + y_7 + y_8), \qquad (4.2.14)$$

where the signs and divisor are taken from the column of Table 4.2 labeled **2**. The main effect of 3 is similarly obtained.
If the operations leading to the calculation of the two-factor interaction 13 in Eq. (4.2.8) are similarly analyzed, it will be found that

$$13 \leftarrow \tfrac{1}{4}(y_1 - y_2 + y_3 - y_4 - y_5 + y_6 - y_7 + y_8). \qquad (4.2.15)$$

This effect can be mechanically computed from the **13** column in the table; the signs in that column are simply the results of taking the products of the signs in the **1** and **3** columns, line by line. The interaction effects 12 and 23 are calculated in exactly similar fashion using their corresponding columns.

Finally, if the calculations leading to the 123 interaction are analyzed, we find that

$$123 \leftarrow \tfrac{1}{4}(-y_1 + y_2 + y_3 - y_4 + y_5 - y_6 - y_7 + y_8). \quad (4.2.16)$$

The signs are those shown in the **123** column in Table 4.2 and this column is formed by taking the triple product of signs from the **1, 2,** and **3** columns.

The table of signs is completed by the addition of a column of plus signs, shown at the left with the heading **I.** This is needed for obtaining the average of the response values, and its divisor (the number of plus signs in the column) is 8 rather than 4 as in the other columns.

A table such as Table 4.2 is very easily constructed for any 2^k factorial design as follows. Begin with a column of 1's of length 2^k. The next k columns are ± 1 signs for the design, written down in standard order. Columns **12, 13,..., 123,..., 123...k,** $(2^k - k - 1)$ in number, are then obtained by multiplying signs, row by row, in the way indicated by the headings. At the bottom are written the divisors, 2^k for the first column, and 2^{k-1} for all the others.

The table of signs is very useful for understanding the nature of the various factorial effects. However, a quicker method of obtaining the effects mechanically is available, due to Yates; this is described in Appendix 4A.

Variance and Standard Errors of Effects for 2^k Designs

For a complete 2^k design, if $V(y) = \sigma^2$,

$$V(\text{grand mean}) = \frac{\sigma^2}{2^k},$$

$$V(\text{effect}) = \frac{4\sigma^2}{2^k}. \quad (4.2.17)$$

If the responses are not "y's" but are instead "$\bar{y}$'s", the averages of (say) r y-observations, then

$$V(\text{grand mean}) = \frac{\sigma^2}{n}, \qquad V(\text{effect}) = \frac{4\sigma^2}{n}, \quad (4.2.17a)$$

where n is the total number of y-observations. Here $n = r2^k$. Formulas (4.2.17a) are also generally applicable to 2^{k-p} fractional factorial designs (see Chapter 5).

In practice, we shall need to obtain an estimate s^2 of the experimental error variance σ^2. We shall consider methods for doing this as we proceed. For the textile data, suppose that an estimate $s^2 = 0.0050$ were available. Then

$$\hat{V}(\bar{y}) = 0.000625, \qquad \hat{V}(\text{effect}) = 0.0025, \qquad (4.2.18)$$

and the corresponding standard errors are the square roots

$$s(\bar{y}) = 0.025, \qquad s(\text{effect}) = 0.05. \qquad (4.2.19)$$

The complete table of effects and their standard errors is as follows

$$I \leftarrow \bar{y} = 2.745 \pm 0.025,$$

$$1 \leftarrow 0.75 \pm 0.05,$$

$$2 \leftarrow -0.59 \pm 0.05,$$

$$3 \leftarrow -0.35 \pm 0.05,$$

$$12 \leftarrow -0.03 \pm 0.05,$$

$$13 \leftarrow -0.03 \pm 0.05,$$

$$23 \leftarrow -0.04 \pm 0.05,$$

$$123 \leftarrow -0.08 \pm 0.05. \qquad (4.2.20)$$

Effects and Regression Coefficients

If we fit a first degree polynomial to the textile data, as was done in Eq. (2.2.4), we obtain

$$\hat{y} = 2.745 + 0.375x_1 - 0.295x_2 - 0.175x_3 . \qquad (4.2.21)$$
$$\phantom{\hat{y} =} (0.025) \quad (0.025) \quad\;\; (0.025) \quad\;\; (0.025)$$

where the numbers in parentheses are the standard errors of the coefficients. Notice that the estimated regression coefficients $b_1 = 0.375$, $b_2 = -0.295$, $b_3 = -0.175$ and their standard errors are exactly half the main effects and standard errors just calculated. The factor of one-half arises because an effect was defined as the difference in response on moving from the -1 level to the $+1$ level of a given variable x_i, which corresponds to the change in y when x_i is changed by *two* units. The regression coefficient b_i is, of course, the change in y when x_i is changed by *one* unit.

TABLE 4.3. Coding of variables in the 2^6 example

Variables	ξ_i	Coded levels, x_i -1	$+1$	x_i in terms of ξ_i
Polysulfide index	ξ_1	6	7	$x_1 = (\xi_1 - 6.5)/0.5$
Reflux rate	ξ_2	150	170	$x_2 = (\xi_2 - 160)/10$
Moles polysulfide	ξ_3	1.8	2.4	$x_3 = (\xi_3 - 2.1)/0.3$
Time (min)	ξ_4	24	36	$x_4 = (\xi_4 - 30)/6$
Solvent (cm^3)	ξ_5	30	42	$x_5 = (\xi_5 - 36)/6$
Temperature (°C)	ξ_6	120	130	$x_6 = (\xi_6 - 125)/5$

TABLE 4.4. A 2^6 factorial design and resulting observations of strength, hue, and brightness

Actual run order	Polysulfide index, x_1	Reflux rate x_2	Moles polysulfide, x_3	Time, (min), x_4	Solvent (cm^3), x_5	Temperature (°C), x_6	Strength, y_1	Hue, y_2	Brightness, y_3
26	−	−	−	−	−	−	3.4	15	36
3	+	−	−	−	−	−	9.7	5	35
11	−	+	−	−	−	−	7.4	23	37
5	+	+	−	−	−	−	10.6	8	34
42	−	−	+	−	−	−	6.5	20	30
18	+	−	+	−	−	−	7.9	9	32
41	−	+	+	−	−	−	10.3	13	28
14	+	+	+	−	−	−	9.5	5	38
17	−	−	−	+	−	−	14.3	23	40
27	+	−	−	+	−	−	10.5	1	32
19	−	+	−	+	−	−	7.8	11	32
56	+	+	−	+	−	−	17.2	5	28
23	−	−	+	+	−	−	9.4	15	34
8	+	−	+	+	−	−	12.1	8	26
32	−	+	+	+	−	−	9.5	15	30
7	+	+	+	+	−	−	15.8	1	28
46	−	−	−	−	+	−	8.3	22	40
13	+	−	−	−	+	−	8.0	8	30
58	−	+	−	−	+	−	7.9	16	35
38	+	+	−	−	+	−	10.7	7	35
43	−	−	+	−	+	−	7.2	25	32
55	+	−	+	−	+	−	7.2	5	35
6	−	+	+	−	+	−	7.9	17	36
64	+	+	+	−	+	−	10.2	8	32
22	−	−	−	+	+	−	10.3	10	20
4	+	−	−	+	+	−	9.9	3	35
16	−	+	−	+	+	−	7.4	22	35
47	+	+	−	+	+	−	10.5	6	28
63	−	−	+	+	+	−	9.6	24	27
51	+	−	+	+	+	−	15.1	4	36
20	−	+	+	+	+	−	8.7	10	36
29	+	+	+	+	+	−	12.1	5	35

4.3. A 2^6 DESIGN USED IN A STUDY OF DYESTUFFS MANUFACTURE

In the manufacture of a certain dyestuff, it is important to:

1. Obtain a product of a desired hue and brightness, and of maximum strength.
2. Know what changes in the process variables should be made if the requirements of the customer (as expressed by hue and brightness of the product) should change.
3. Know what sorts of changes in the process variables might compensate for given departures from specification.

TABLE 4.4. Continued

	Design levels						Responses		
Actual run order	Polysulfide index, x_1	Reflux, rate x_2	Moles polysulfide, x_3	Time (min), x_4	Solvent (cm^3), x_5	Temperature (°C), x_6	Strength, y_1	Hue, y_2	Brightness, y_3
62	−	−	−	−	−	+	12.6	32	32
1	+	−	−	−	−	+	10.5	10	34
37	−	+	−	−	−	+	11.3	28	30
61	+	+	−	−	−	+	10.6	18	24
44	−	−	+	−	−	+	8.1	22	30
24	+	−	+	−	−	+	12.5	31	20
59	−	+	+	−	−	+	11.1	17	32
60	+	+	+	−	−	+	12.9	16	25
35	−	−	−	+	−	+	14.6	38	20
50	+	−	−	+	−	+	12.7	12	20
48	−	+	−	+	−	+	10.8	34	22
36	+	+	−	+	−	+	17.1	19	35
21	−	−	+	+	−	+	13.6	12	26
9	+	−	+	+	−	+	14.6	14	15
33	−	+	+	+	−	+	13.3	25	19
57	+	+	+	+	−	+	14.4	16	24
10	−	−	−	−	+	+	11.0	31	22
39	+	−	−	−	+	+	12.5	14	23
25	−	+	−	−	+	+	8.9	23	22
40	+	+	−	−	+	+	13.1	23	18
30	−	−	+	−	+	+	7.6	28	20
31	+	−	+	−	+	+	8.6	20	20
28	−	+	+	−	+	+	11.8	18	20
49	+	+	+	−	+	+	12.4	11	36
52	−	−	−	+	+	+	13.4	39	20
15	+	−	−	+	+	+	14.6	30	11
34	−	+	−	+	+	+	14.9	31	20
53	+	+	−	+	+	+	11.8	6	35
2	−	−	+	+	+	+	15.6	33	16
12	+	−	+	+	+	+	12.8	23	32
45	−	+	+	+	+	+	13.5	31	20
54	+	+	+	+	+	+	15.8	11	20
	64 observations Σy						711.9	1085	1810
	64 observations Σy^2						8443.41	24,421	54,340

(These problems are essentially those we have listed in Section 1.7 as "Choice of operating conditions to achieve desired specifications" and "Approximate mapping of a surface within a limited region.") The particular process under study was one whose chemical mechanism was not well understood, so that an empirical approach was mandatory.

Six variables were suggested as possibly of importance in affecting strength, hue, and brightness. They were ξ_1 = polysulfide index, ξ_2 = reflux rate, ξ_3 = moles of polysulfide, ξ_4 = reaction time, ξ_5 = amount of solvent, ξ_6 = reaction temperature. The ranges over which they were changed in the design, and the details of their coding to x's is shown in Table 4.3. The design employed was a 2^6 full factorial design.

Model Specification

Data from past manufacturing experience had indicated that effects of potential importance were not likely to be large compared with the (rather high) experimental error. Such evidence as existed suggested that, over the range of the region of interest, effects might be roughly linear and, in particular, no response was likely to be close to a maximum or other turning point. This was later checked by making additional runs both at the center of the 2^6 design and also at more extreme conditions along the six axial directions. No evidence of curved relationships was found, and we shall here use only the data from the principal part of the design, namely the complete 2^6 factorial. The results from these 64 runs (which were actually performed in random order) are shown in Table 4.4.

The preliminary statistical analysis for model specification consisted of calculating the factorial effects (see Table 4.5 and Appendix 4A) shown plotted on probability paper in Figure 4.3. For clarity, in the center of each plot only every fourth point is inserted. A brief discussion of the normal plotting of factorial effects is given in Appendix 4B. From this preliminary analysis, it appeared that the data were adequately explained in terms of linear effects in only three of the variables x_1, x_4, and x_6, that is, in polysulfide index, reaction time, and reaction temperature, respectively. The preliminary identification thus suggested that linear models might be tentatively entertained in the variables x_1, x_4, and x_6 with the residual variation ascribed principally to random error or "noise."

On this basis the data could be tentatively analyzed as if they came from a 2^3 factorial in x_1, x_4, and x_6 performed eight times over. The observations, appropriately rearranged to reflect this, are set out again in Table 4.6. The totals on the right of this table are used to facilitate the subsequent tentative analysis.

TABLE 4.5. Effects ranked in order for strength, hue, and brightness

Strength, y_1		Hue, y_2		Brightness, y_3	
Effect name	Effect value	Effect name	Effect value	Effect name	Effect value
Mean	11.12	Mean	16.95	Mean	28.28
4	2.98	1	−11.28	6	−8.87
6	2.69	6	10.84	4	−3.00
1	1.75	2456	−2.84	2346	−2.87
1245	−1.16	135	−2.78	12346	−2.81
12	0.89	2	−2.72	123456	−2.81
24	−0.86	13	2.66	234	−2.56
16	−0.82	2346	2.53	35	2.44
124	0.82	1235	2.47	1245	−2.37
1346	−0.77	46	2.34	1456	−2.37
12346	0.74	136	2.34	1356	2.19
345	0.72	256	−2.22	15	2.12
13456	0.71	12345	2.22	1246	1.94
2	0.70	1236	−2.16	56	−1.88
12345	0.70	234	2.09	126	1.75
123	−0.65	1245	−1.97	135	1.62
1235	0.65	3	−1.91	2456	−1.62
1356	−0.65	23	−1.91	125	−1.56
456	0.63	25	−1.91	1235	−1.56
23	0.60	36	−1.84	2	1.50
256	0.57	14	−1.78	1234	−1.50
146	−0.56	146	−1.72	5	−1.44
12356	0.50	23456	1.72	456	1.31
1456	−0.47	2345	−1.59	25	1.25
12456	0.47	1246	−1.47	146	1.25
5	−0.42	1356	−1.47	1256	1.25
15	−0.42	124	−1.41	134	−1.19
2346	−0.41	456	1.41	145	1.19
34	0.40	5	1.34	45	1.12
134	0.40	3456	1.34	14	1.06
234	−0.40	123	−1.22	26	1.06
2356	0.39	26	−1.16	246	1.00
236	0.35	345	1.16	3	−0.94
1345	0.34	45	1.09	12	0.94
123456	0.34	126	−1.09	16	0.94
25	−0.33	12346	1.09	136	−0.94
145	−0.33	236	1.03	156	0.94
245	−0.33	245	−1.03	346	−0.94
45	−0.32	15	−0.97	3456	−0.94
1246	−0.32	346	−0.97	235	−0.88
246	0.30	123456	−0.84	24	0.81
1236	0.29	16	0.78	2345	−0.81
3456	−0.28	34	−0.78	13	0.75
56	0.27	35	0.78	236	−0.69
125	−0.27	1256	−0.78	12345	−0.50
126	−0.25	12	0.72	12356	−0.50

TABLE 4.5. Continued.

Strength, y_1		Hue, y_2		Brightness, y_3	
Effect name	Effect value	Effect name	Effect value	Effect name	Effect value
2345	−0.24	12456	−0.66	245	0.44
356	−0.23	156	−0.53	256	−0.44
36	−0.22	1456	−0.53	1	0.38
1234	−0.22	2356	0.47	34	0.38
26	−0.18	56	0.41	345	0.38
346	0.17	4	−0.34	1236	−0.38
14	0.15	1345	0.34	23456	−0.38
13	0.14	12356	−0.34	46	−0.31
23456	0.14	1234	0.28	23	0.25
235	0.12	1346	−0.28	1346	0.25
136	0.11	246	0.22	13456	−0.25
3	0.10	356	0.22	36	0.13
156	0.10	24	0.16	123	0.06
35	0.08	125	0.16	1345	0.06
135	0.07	235	0.16	2356	−0.06
2456	0.03	13456	0.16	12456	0.06
46	0.02	134	0.03	124	0
1256	0.02	145	0.03	356	0

Model Fitting

The linear equations in x_1, x_4, and x_6 fitted by least squares to all 64 observations and rounded to two decimal places are*:

$$\text{Strength: } \hat{y}_1 = 11.12 + 0.87x_1 + 1.49x_4 + 1.35x_6 \tag{4.3.1}$$
$$\phantom{\text{Strength: } \hat{y}_1 = 11.12} (0.24) \quad (0.24) \quad (0.24) \quad (0.24)$$

$$\text{Hue: } \hat{y}_2 = 16.95 - 5.64x_1 - 0.17x_4 + 5.42x_6 \tag{4.3.2}$$
$$\phantom{\text{Hue: } \hat{y}_2 = 16.95} (0.74) \quad (0.74) \quad (0.74) \quad (0.74)$$

$$\text{Brightness: } \hat{y}_3 = 28.28 + 0.19x_1 - 1.50x_4 - 4.44x_6 \tag{4.3.3}$$
$$\phantom{\text{Brightness: } \hat{y}_3 = 28.28} (0.67) \quad (0.67) \quad (0.67) \quad (0.67)$$

*Not all the estimated regression coefficients b_1, b_4, and b_6 in Eqs. (4.3.1) to (4.3.3) are significantly different from zero, but the values shown are the best estimates available in the subspace of greatest interest. We do not substitute values of zero for the nonsignificant coefficients since these would not be the best estimates. It might be thought that this judgment is somewhat contradictory since variables x_2, x_3, and x_5 have been dropped completely. In fact, however, we can simply regard the fitted equations as the best estimates in the three-dimensional subspace of the full six-dimensional space in which x_2, x_3, and x_5 are at their average values. Our treatment implies only that, had we prepared contour diagrams for the responses as functions of x_1, x_4, and x_6 for *other* levels of x_2, x_3, and x_5 over their relevant ranges, the appearances of the various diagrams would not be appreciably different.

NORMAL PLOTS OF EFFECTS

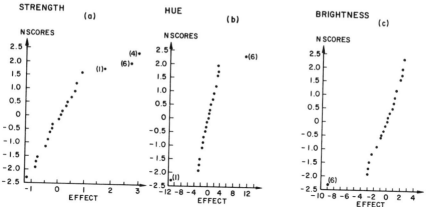

FIGURE 4.3. Normal plots for (a) strength, (b) hue, (c) brightness. (The effects plotted are those of orders 1, 2, 3, 4, 8, 12, . . . , 60, 61, 62, and 63.)

The appropriate analyses of variance are set out in Table 4.7. The estimates of the error standard deviations obtained from the residual sums of squares after fitting these equations are, respectively,

$$s_1 = 1.90, \qquad s_2 = 5.93, \qquad s_3 = 5.39. \qquad (4.3.4)$$

These estimates of the standard deviations might be biased somewhat upward because a number of small main effects and interactions have been ignored, and perhaps biased slightly downward because of the effect of selection, that is, only the large estimates were taken to be of real effects. However, these are the figures we have used in estimating the standard errors of the coefficients shown in parentheses beneath the coefficients.

4.4. DIAGNOSTIC CHECKING OF THE FITTED MODELS, 2^6 DYESTUFFS EXAMPLE

Table 4.8 shows the residuals from the models (4.3.1) to (4.3.3). The time order of the observations is also shown. Figures 4.4a–c show residuals for the fitted models plotted in various ways. There is some evidence that the variance of the response hue (y_2) is not homogeneous (see Figure 4.4b, the residuals versus $\hat{y}_2$ plot). However, in general, it appears that the tentative assumption that the changes in the responses are associated mainly with the linear effects of the three variables polysulfide index (x_1), reaction time (x_4), and reaction temperature (x_6) is borne out.

TABLE 4.6. The 64 observations on each of three responses grouped in sets of eight to reflect their dependence on the variables x_1 = polysulfide index, x_4 = time, and x_6 = temperature

x_1	x_4	x_6	- - -	+ - -	- + -	+ + -	- - +	+ - +	- + +	+ + +	Row averages	Row totals
						x_2, x_3, x_5						

Strength

x_1	x_4	x_6	- - -	+ - -	- + -	+ + -	- - +	+ - +	- + +	+ + +	Row averages	Row totals
-	-	-	3.4	7.4	6.5	10.3	8.3	7.9	7.2	7.9	7.3625	58.9
+	-	-	9.7	10.6	7.9	9.5	8.0	10.7	7.2	10.2	9.2250	73.8
-	+	-	14.3	7.8	9.4	9.5	10.3	7.4	9.6	8.7	9.6250	77.0
+	+	-	10.5	17.2	12.1	15.8	9.9	10.5	15.1	12.1	12.9000	103.2
-	-	+	12.6	11.3	8.1	11.1	11.0	8.9	7.6	11.8	10.3000	82.4
+	-	+	10.5	10.6	12.5	12.9	12.5	13.1	8.6	12.4	11.6375	93.1
-	+	+	14.6	10.8	13.6	13.3	13.4	14.9	15.6	13.5	13.7125	109.7
+	+	+	12.7	17.1	14.6	14.4	14.6	11.8	12.8	15.8	14.2250	113.8

Hue

x_1	x_4	x_6	- - -	+ - -	- + -	+ + -	- - +	+ - +	- + +	+ + +	Row averages	Row totals
-	-	-	15	23	20	13	22	16	25	17	18.875	151
+	-	-	5	8	9	5	8	7	5	8	6.875	55
-	+	-	23	11	15	15	10	22	24	10	16.250	130
+	+	-	1	5	8	1	3	6	4	5	4.125	33
-	-	+	32	28	22	17	31	23	28	18	24.875	199
+	-	+	10	18	31	16	14	23	20	11	17.875	143
-	+	+	38	34	12	25	39	31	33	31	30.375	243
+	+	+	12	19	14	16	30	6	23	11	16.375	131

Brightness

x_1	x_4	x_6	- - -	+ - -	- + -	+ + -	- - +	+ - +	- + +	+ + +	Row averages	Row totals
-	-	-	36	37	30	28	40	35	32	36	34.250	274
+	-	-	35	34	32	38	30	35	35	32	33.875	271
-	+	-	40	32	34	30	20	35	27	36	31.750	254
+	+	-	32	28	26	28	35	28	36	35	31.000	248
-	-	+	32	30	30	32	22	22	20	20	26.000	208
+	-	+	32	24	20	25	23	18	20	36	25.000	200
-	+	+	34	22	26	19	20	31	16	20	20.375	163
+	+	+	20	35	15	24	11	35	32	20	24.000	192

TABLE 4.7. Analysis of variance tables, for strength, hue, and brightness

Strength

Source of variation	SS	df	MS	F ratio
Total SS $= \Sigma y^2$	8,443.4100	64		
Correction factor, SS due to $b_0 = (\Sigma y)^2/64$	7,918.7752	1		
Corrected total SS	524.6348	63		
Due to $b_1 = b_1\Sigma x_1 y = (\Sigma x_1 y)^2/\Sigma x_1^2$	48.8252	1	48.8252	13.47*
Due to $b_4 = b_4\Sigma x_4 y = (\Sigma x_4 y)^2/\Sigma x_4^2$	142.5039	1	142.5039	39.32*
Due to $b_6 = b_6\Sigma x_6 y = (\Sigma x_6 y)^2/\Sigma x_6^2$	115.8314	1	115.8314	31.96*
Residual	217.4743	60	3.6246	
			$s_1 = 1.9038$	

$\Sigma y = 711.9; \Sigma x_1 y = 55.9; \Sigma x_4 y = 95.5; \Sigma x_6 y = 86.1$

Hue

Source of variation	SS	df	MS	F ratio
Total SS $= \Sigma y^2$	24,421.0000	64		
SS due to $b_0 = (\Sigma y)^2/64$	18,394.1406	1		
Corrected total SS	6,026.8594	63	6,026.8594	
Due to $b_1 = b_1\Sigma x_1 y = (\Sigma x_1 y)^2/\Sigma x_1^2$	2,036.2656	1	2,036.2656	57.98*
Due to $b_4 = b_4\Sigma x_4 y = (\Sigma x_4 y)^2/\Sigma x_4^2$	1.8906	1	1.8906	0.05
Due to $b_6 = b_6\Sigma x_6 y = (\Sigma x_6 y)^2/\Sigma x_6^2$	1,881.3906	1	1,881.3906	53.57*
Residual	2,107.3126	60	35.1219	
			$s_2 = 5.9264$	

$\Sigma y = 1085; \Sigma x_1 y = -361; \Sigma x_4 y = -11; \Sigma x_6 y = 347$

Brightness

Source of variation	SS	df	MS	F ratio
Total SS $= \Sigma y^2$	54,340.0000	64		
SS due to b_0	51,189.0625	1		
Corrected total SS	3,150.9375	63		
Due to b_1	2.2500	1	2.2500	0.08
Due to b_4	144.0000	1	144.0000	4.95*
Due to b_6	1260.2500	1	1260.2500	43.35*
Residual	1744.4375	60	29.0740	
			$s_3 = 5.3920$	

$\Sigma y = 1810; \Sigma x_1 y = 12; \Sigma x_4 y = -96; \Sigma x_6 y = -284$

*Significant at $\alpha = 0.05$ or smaller α-level; $P[F(1, 60) \geq 4.00] = 0.05$.

TABLE 4.8. Residuals from individual and from average observations, fitted values $\hat{y}$ obtained by using Eqs. (4.3.1) to (4.3.3), and time order of observations. (The layout is similar to that of Table 4.6 and all figures are rounded to the length of the original observations)

x_1	x_4	x_6	$---$	$+--$	$-+-$	$++-$	$--+$	$+-+$	$-++$	$+++$	From average	$\hat{y}$
							x_2, x_3, x_5					
Strength												
−	−	−	−4.0	0	−0.9	2.9	0.9	0.5	−0.2	0.5	0	7.4
+	−	−	0.5	1.4	−1.3	0.3	−1.2	1.5	−2.0	1.0	0	9.2
−	+	−	3.9	−2.6	−1.0	−0.9	−0.1	−3.0	−0.8	−1.7	−0.8	10.4
+	+	−	−1.6	5.1	0	3.7	−2.2	−1.6	3.0	0	0.8	12.1
−	−	+	2.5	1.2	−2.0	1.0	0.9	−1.2	−2.5	1.7	0.2	10.1
+	−	+	−1.4	−1.3	0.6	1.0	0.6	1.2	−3.3	0.5	−0.3	11.9
−	+	+	1.5	−2.3	0.5	0.2	0.3	1.8	2.5	0.4	0.6	13.1
+	+	+	−2.1	2.3	−0.2	−0.4	−0.2	−3.0	−2.0	1.0	−0.6	14.8
Hue												
−	−	−	−2	6	3	−4	5	−1	8	0	2	17
+	−	−	−1	2	3.	−1	2	1	−1	2	1	6
−	+	−	6	−6	−2	−2	−7	5	7	−7	−1	17
+	+	−	−5	−1	2	−5	−3	0	−2	−1	−2	6
−	−	+	4	0	−6	−11	3	−5	0	−10	−3	28
+	−	+	−7	1	14	−1	−3	6	3	−6	1	17
−	+	+	10	6	−16	−3	11	3	5	3	2	28
+	+	+	−5	2	−3	−1	13	−11	6	−6	−1	17
Brightness												
−	−	−	2	−3	−4	−6	6	1	−2	2	0	34
+	−	−	1	0	−2	4	−4	1	1	−2	0	34
−	+	−	9	1	3	−1	−11	4	−4	5	1	31
+	+	−	1	−3	−5	−3	4	−3	5	4	0	31
−	−	+	7	5	5	7	−3	−3	−5	−5	1	25
+	−	+	8	−2	−6	−1	−3	−8	−6	10	−1	26
−	+	+	−2	0	4	−3	−2	−2	−6	−2	−2	22
+	+	+	−3	12	−8	1	−12	12	9	−3	1	23
Time order												
−	−	−	26	11	42	41	46	58	43	6		
+	−	−	3	5	18	14	13	38	55	64		
−	+	−	17	19	23	32	22	16	63	20		
+	+	−	27	56	8	7	4	47	51	29		
−	−	+	62	37	44	59	10	25	30	28		
+	−	+	1	61	24	60	39	40	31	49		
−	+	+	35	48	21	33	52	34	2	45		
+	+	+	50	36	9	57	15	53	12	54		

122

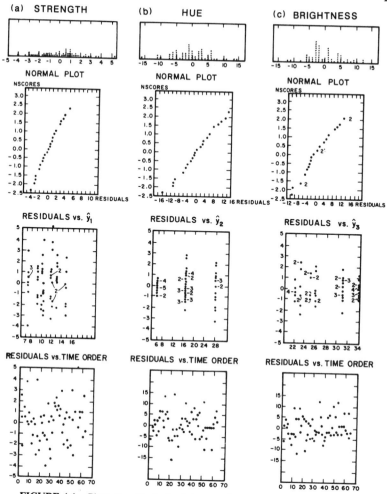

FIGURE 4.4. Plots of residuals for (*a*) strength, (*b*) hue, and (*c*) brightness.

4.5. RESPONSE SURFACE ANALYSIS OF THE 2^6 DESIGN DATA

We now consider the model's implications for the manufacturing problem, assuming that, to a sufficient approximation, the fitted models are adequate to represent the data. The immediate objective of these experiments was to determine operating conditions, if any such existed, at which a standard hue of 20 units and a standard brightness of 26 units could be obtained with maximum strength. A second objective was to understand the way in which

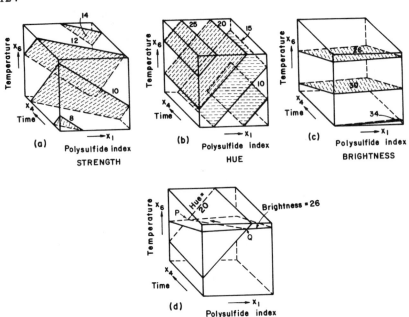

FIGURE 4.5. Contours of (*a*) strength, (*b*) hue, (*c*) brightness with (*d*) intersection of the hue = 20 and brightness = 26 planes.

the three responses were affected by the variables. Such an understanding would allow future modifications to be made in the process as necessary. Such modifications might be needed, for example, to prepare special consignments of dye whose hue and brightness had to differ from current specifications, and also to allow intelligent compensatory adjustments to be made when unknown disturbances had caused the response to depart temporarily from the standard. What was needed in this latter situation in essence was some kind of "navigation chart."

The contours of the fitted equations are shown in Figures 4.5*a–c* in cubes whose corners are the design points.

In Figure 4.5*d* are shown, together, the predicted contour planes for brightness and hue equal to their desired levels of 26 units and 20 units, respectively; the line PQ along which these two planes intersect has predicted strengths ranging from about 11.08 at Q to about 12.46 at P, as Table 4.9 indicates. Approximately, then, we expect that, anywhere along the line of intersection, satisfactory brightness and hue would be obtained and that higher strengths would result by moving in the direction indicated by the arrow in the figure. The estimated difference in strengths at the points P

TABLE 4.9. Predicted strengths at points along the line PQ in Figure 4.5d

| | Coordinates on line PQ | | | |
	x_1	x_4	x_6	Predicted strength, $\hat{y}_1$
Q	0.32	−1.00	0.87	11.08
	0.09	−0.39	0.65	11.50
	−0.05	0.00	0.51	11.77
	−0.18	0.34	0.39	12.00
P	−0.42	1.00	0.16	12.46

and Q in Figure 4.5d is given by

$$\hat{y}_P - \hat{y}_Q = b_1(x_{1P} - x_{1Q}) + b_4(x_{4P} - x_{4Q}) + b_6(x_{6P} - x_{6Q})$$

$$= 12.46 - 11.08$$

$$= 1.38 \tag{4.5.1}$$

with a variance of

$$V(\hat{y}_P - \hat{y}_Q) = \left\{(x_{1P} - x_{1Q})^2 + (x_{4P} - x_{4Q})^2 + (x_{6P} - x_{6Q})^2\right\}V(b_i)$$

$$= \overline{PQ}^2 V(b_i), \tag{4.5.2}$$

where $V(b_i) = \sigma^2/n$, and $\overline{PQ}^2$ is the squared distance between the points P and Q in the scale of the x's. Therefore the standard deviation of $(\hat{y}_P - \hat{y}_Q)$ = 1.38 is $\overline{PQ}\sigma/n^{1/2}$, which we can evaluate as

$$\left\{(-0.42 - 0.32)^2 + (1 + 1)^2 + (0.16 - 0.87)^2\right\}^{1/2}\sigma/64^{1/2} = 0.2809\sigma.$$

$$\tag{4.5.3}$$

From Table 4.7 we substitute $s_1 = 1.9038$ for σ, giving a standard error of about 0.53. The difference $\hat{y}_P - \hat{y}_Q = 1.38$ is thus about 2.6 times its standard error and we conclude that the strength is significantly higher at the position P than at the position Q.

[We note, in passing, that this design happens to be what is called a rotatable first-order design (see Chapter 14) and therefore has the interesting property that the variance of differences in estimated yields between two points is simply a function of the distance between the two points concerned. See Box and Draper, 1980.]

It is obvious that our analysis can be only an approximate one, since the data on which it is based are only approximate. However, on the assumption that the form of the prediction equation is not seriously at fault, we can readily calculate confidence limits at the points P and Q, or at any other set of values, for $E(y_1)$, $E(y_2)$, and $E(y_3)$. For any of the three responses, at a general point (x_1, x_4, x_6), we can write

$$V(\hat{y}_i) = V(b_0) + x_1^2 V(b_1) + x_4^2 V(b_4) + x_6^2 V(b_6)$$

$$= \sigma_i^2 \left[1 + x_1^2 + x_4^2 + x_6^2\right]/64, \tag{4.5.4}$$

where the σ_i^2 is the appropriate one for the response desired, and is estimated by s_i^2 from the appropriate part of Table 4.7. Thus, for example, the estimated variance of $\hat{y}_1$ at the point P with coordinates $(-0.42, 1.00, 0.16)$ is

$$V(\hat{y}_1) = \frac{3.6246(2.202)}{64} = 0.1247 \tag{4.5.5}$$

where we have substituted for σ_1^2, $s_1^2 = 3.6246$ based on 60 df from Table 4.7. Taking the square root, we then have to two decimal places that

$$\text{standard error } (\hat{y}_1) = 0.35. \tag{4.5.6}$$

Proceeding in exactly similar fashion and using $t_{0.025}(60) = 2.00$, we obtain at P and Q 95% confidence limits for $E(y_1)$, $E(y_2)$ and $E(y_3)$ as follows:

	At point P	At point Q
Yield	12.46 ± 0.70	11.08 ± 0.80
Hue	20 ± 2.20	20 ± 2.51
Brightness	26 ± 2.00	26 ± 2.28

(In Appendix 4C, methods are given for constructing confidence regions both for contour planes and for their lines of intersection.)

Confirmatory runs made in the regions indicated by this analysis verified the general conclusions given above. A three-dimensional model with contours of strength, hue, and brightness shown by different colors was constructed and proved to be of great value in improving the product and in day-to-day operation of the process.

In this investigation we were somewhat fortunate in finding that first-order expressions were sufficient to represent the three response functions. The area of application of such methods is greatly widened by allowing the possibility of nonlinear transformation in either the response or predictor variables or, sometimes, in both. Quite frequently, although the relationships may not be adequately represented by first-order expressions in the original variables, approximate first-order representation can be obtained by employing, for example, the logarithmic or inverse transformation. Methods whereby appropriate transformations may be determined are discussed later (see Chapters 8 and 13).

APPENDIX 4A.   YATES' METHOD FOR OBTAINING THE FACTORIAL EFFECTS FOR A TWO-LEVEL DESIGN

Yates' method is best understood by considering an example. Table 4A.1 shows the calculations for the textile data previously discussed in Chapter 4. Column C_1 is developed from the y_u column by these operations:

$$\text{In row 1 of } C_1 \text{ place the entry } y_2 + y_1,$$
$$2 \qquad\qquad y_4 + y_3,$$
$$3 \qquad\qquad y_6 + y_5,$$
$$4 \qquad\qquad y_8 + y_7,$$
$$5 \qquad\qquad y_2 - y_1,$$
$$6 \qquad\qquad y_4 - y_3,$$
$$7 \qquad\qquad y_6 - y_5,$$
$$\text{In row 8 of } C_1 \text{ place the entry } y_8 - y_7.$$

The identical operations performed on the C_1 column now give rise to the C_2 column, and the latter is used to calculate the C_3 column in the same manner. Each entry in the C_3 column is then divided by a divisor. The first divisor is the number of runs in the design, here 8. All the other divisors are half of this. The results of this mechanical procedure are the effects of the variables in the standard order corresponding to the plus signs in the variable settings; here, for example, the order is $I, 1, 2, 12, 3, 13, 23, 123$. Rearranging this order, we obtain the estimates already given.

Grand mean	$I \leftarrow 2.745$	$(= \bar{y})$
Main effects	$1 \leftarrow 0.75$	
	$2 \leftarrow -0.59$	
	$3 \leftarrow -0.35$	

TABLE 4A.1.   Yates' method for obtaining factorial effects

Row	Setting of variable x_1	x_2	x_3	Observations, y_u	C_1	C_2	C_3	Divisor	Effect Value	Name
1	-1	-1	-1	2.83	6.39	11.68	21.96	8	2.745	I
2	1	-1	-1	3.56	5.29	10.28	3.00	4	0.75	1
3	-1	1	-1	2.23	5.77	1.56	-2.36	4	-0.59	2
4	1	1	-1	3.06	4.51	1.44	-0.12	4	-0.03	12
5	-1	-1	1	2.47	0.73	-1.10	-1.40	4	-0.35	3
6	1	-1	1	3.30	0.83	-1.26	-0.12	4	-0.03	13
7	-1	1	1	1.95	0.83	0.10	-0.16	4	-0.04	23
8	1	1	1	2.56	0.61	-0.22	-0.32	4	-0.08	123

$$
\begin{array}{rcl}
\textit{Two-factor interactions} & 12 & \leftarrow -0.03 \\
& 13 & \leftarrow -0.03 \\
& 23 & \leftarrow -0.04 \\
\textit{Three-factor interaction} & 123 & \leftarrow -0.08
\end{array}
$$

Yates' method may be applied to any two-level 2^k factorial design by carrying out the sum and difference operations on k columns. (For its implementation on 2^{k-p} fractional factorial designs, see Section 5.5.)

APPENDIX 4B.   NORMAL PLOTS ON PROBABILITY PAPER

When two-level factorial and fractional factorial designs are run, it is frequently the case that no independent estimate of error is available. A useful technique that is often effective in distinguishing real effects from noise employs *normal probability paper*, available in technical bookstores. (Ask, for example, for Keuffel and Esser 46-8003, or Codex 3227, or Team 3211 or Dietzgen 340-PS-90.)

What Is Normal Probability Paper?

Observations y having a normal distribution with mean η and variance σ^2 have the probability density function

$$
f(y) = \frac{1}{\sigma\sqrt{2\pi}} e^{-(y-\eta)^2/(2\sigma^2)}. \tag{4B.1}
$$

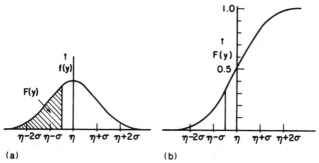

FIGURE 4B.1.   Plots of (*a*) the normal density function $f(y)$ and (*b*) the cumulative normal density function $F(y)$. Note, for example, that prob($y \le \eta - 0.5\sigma$) = 0.3085 is the area shaded in (*a*) and is the vertical height marked in (*b*).

The corresponding cumulative density function is

$$F(y) = \int_{-\infty}^{y} \frac{1}{\sigma\sqrt{2\pi}} e^{-(t-\eta)^2/(2\sigma^2)} \, dt. \qquad (4B.2)$$

Both the normal density function and its cumulative are plotted in Figure 4B.1. We see that the cumulative density function $F(y)$ is the area under the probability density function $f(y)$ from $-\infty$ to y, and that $0 \le F(y) \le 1$. The normal cumulative distribution function is an "ogive," that is, an "S" shaped curve that begins at $(-\infty, 0)$ and passes through points such as $(\eta - \sigma, 0.1587)$, $(\eta, 0.5)$, $(\eta + \sigma, 0.8413)$, and $(\infty, 1.0)$, and in general through the points $\{\eta + u\sigma, F(\eta + u\sigma)\}$ where u is called the *normal score*. Normal probability paper shown in Figure 4B.2 is arranged so that the ogive curve $F(y)$ is made into a straight line by stretching outward from the center symmetrically the vertical dimension of the paper. Shown on the vertical axis are both the probability $F(y) = F(\eta + u\sigma)$ and the normal score u.

We illustrate use of the probability paper by drawing a random sample of 20 observations from a normal distribution with mean $\eta = 7$ and variance $\sigma^2 = 0.5$. The sample is rearranged in ascending order in Table 4B.1 and plotted as a dot diagram at the bottom of Figure 4B.2. With the ith in order, $y_{(i)}$, is associated

$$F(i) = F(y_{(i)}) = \frac{(i - \frac{1}{2})}{n}, \qquad (4B.3)$$

evaluated in Table 4B.1 for $n = 20$. The points $\{y_{(i)}, F(i)\}$ are then plotted on the probability paper in Figure 4B.2.

On the assumption that the data are a sample from a normal distribution, an "eye-fitted" straight line can be used to provide rough estimates of the mean and standard deviation. The estimate of the mean η is simply the abscissa value associated with $F(y) = 0.50$. To obtain an estimate of σ, recall that prob($y \le \eta +$

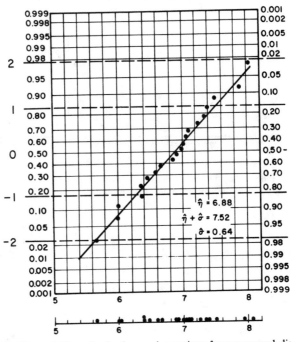

FIGURE 4B.2. Plots of 20 randomly drawn observations from a normal distribution with mean $\eta = 7.0$ and standard deviation $\sigma = 0.707$.

TABLE 4B.1. The 20 ordered observations $y_{(i)}$ and the corresponding values of $F(i) = (i - \frac{1}{2}) / 20$ for $i = 1, 2, \ldots, 20$

$y_{(i)}$	$F(i)$	$y_{(i)}$	$F(i)$
5.66	0.025	6.97	0.525
6.03	0.075	7.04	0.575
6.04	0.125	7.08	0.625
6.38	0.175	7.10	0.675
6.38	0.225	7.24	0.725
6.46	0.275	7.33	0.775
6.59	0.325	7.39	0.825
6.64	0.375	7.51	0.875
6.88	0.425	7.94	0.925
6.91	0.475	8.08	0.975

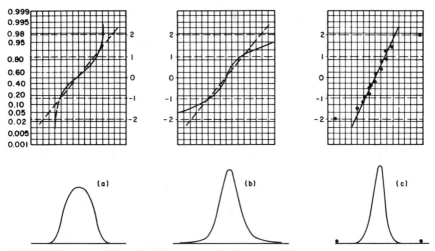

FIGURE 4B.3.   (a) Light-tailed distribution. (b) Heavy-tailed distribution. (c) Normal distribution with two outliers.

σ) = 0.8413. The abscissa values for $F(y) = 0.50$ and $F(y) = 0.8413$ are, respectively, 6.88 and 7.52; thus for this set of data, $\hat{\eta} = 6.88$ and $\hat{\sigma} = 0.64 = 7.52 - 6.88$.

If the parent distribution were not normal, the cumulative distribution would not plot as a straight line on normal probability paper. Figures 4B.3a and 4B.3b show the curved lines that result from a light-tailed distribution and a heavy-tailed distribution, respectively. The third figure 4B.3c shows a plot for a sample in which all observations are random drawings from a normal distribution except for two outliers.

Application to Estimates from 2^k Designs

If all k factors in a two-level factorial or fractional factorial design are without influence, all the $2^k - 1$ estimates of effects and interactions would be expected to be approximately normal with mean zero, and variance $\{4\sigma^2/n\}$. Thus, if a normal plot of the estimated effects approximates a straight line, there would be no reason to believe that any of the true effects were nonzero. By contrast, points that fall well off the line to the right at the top and to the left at the bottom would suggest the existence of real effects. Examples of how this works in practice are seen in Figure 4.3.

Detection of Aberrant Observations

Consider the estimates from a 2^4 factorial design in Table 4B.2. A normal plot of these is shown in Figure 4B.4. The data suggest, though not strongly, that the main

TABLE 4B.2. Ordered estimates from a 2^4 design

Rank	Effect name	Estimate	Rank	Effect name	Estimate
1	2	-4.22	9	124	0.72
2	13	-2.49	10	12	0.91
3	234	-1.58	11	4	1.01
4	24	-1.18	12	123	1.20
5	1	-0.80	13	34	1.49
6	23	-0.80	14	1234	1.52
7	14	-0.58	15	3	3.71
8	134	0.40			

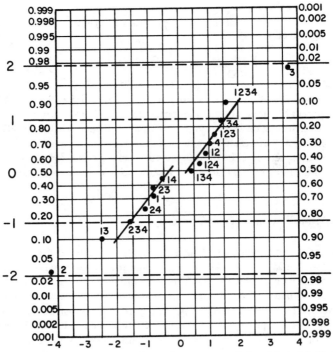

FIGURE 4B.4. Effect of an aberrant observation on a normal plot of factorial effect estimates from a 2^4 design.

132

TABLE 4B.3. Table of signs for the analysis of a 2^4 design. Matching the signs of biased effects with the run that probably causes the bias

Sign →	1	2	3	4	12	13	14	23	24	34	123	124	134	234	1234
Sign	−	−	−	+	+	−	−	−	−	+	+	+	(+)	−	+
Estimate	−0.80	−4.22	3.71	1.01	0.91	−2.49	−0.58	−0.80	−1.18	1.49	1.20	0.72	0.40	−1.58	1.52
1	−	−	−	−	+	+	+	+	+	+	−	−	−	−	+
2	+	−	−	−	−	−	−	+	+	+	+	+	+	−	−
3	−	+	−	−	−	+	+	−	−	+	+	+	−	+	−
4	+	+	−	−	+	−	−	−	−	+	−	−	+	+	+
5	−	−	+	−	+	−	+	−	+	−	+	−	+	+	−
6	+	−	+	−	−	+	−	−	+	−	−	+	−	+	+
7	−	+	+	−	−	−	+	+	−	−	−	+	+	−	+
8	+	+	+	−	+	+	−	+	−	−	+	−	−	−	−
9	−	−	−	+	+	+	−	+	−	−	−	+	+	+	−
10	+	−	−	+	−	−	+	+	−	−	+	−	−	+	+
11	−	+	−	+	−	+	−	−	+	−	+	−	+	−	+
12	+	+	−	+	+	−	+	−	+	−	−	+	−	−	−
13	−	−	+	+	+	−	−	−	−	+	+	+	−	−	+
14	+	−	+	+	−	+	+	−	−	+	−	−	+	−	−
15	−	+	+	+	−	−	−	+	+	+	−	−	−	+	−
16	+	+	+	+	+	+	+	+	+	+	+	+	+	+	+

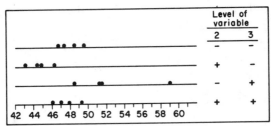

FIGURE 4B.5. Plot of replicate values when the 2^4 design is considered as a replicated 2^2 design in factors 2 and 3.

effect 2 (and possibly 3 also) is distinguishable from noise, but this plot also exhibits another striking characteristic. The inlying values, which should represent just noise, appear to be fitted not by one line, but by two parallel straight lines which "break apart" close to the zero value on the abscissa, and thus divide positive from negative effect estimates.

Now consider Table 4B.3, the body of which shows the $\pm$ signs which determine the effects, set out in the usual format. Suppose an observation were aberrant, for example, suppose observation 1 were too big by an amount δ. Then, from the table of signs, we see that effects 1, 2, 3, 4, 123, 124, 134, 234 would all be reduced by an amount $\delta/8$ and the remaining effects 12, 13, 14, 23, 24, 34, 1234 would all be increased by $\delta/8$. If we regard all the effects with the exception of the main effects 2 and 3 as possibly arising from noise, and consider the signs of the effects that remain, we see that, except for 134, these signs correspond to observation 13, suggesting that this observation is in error, and possibly too large. (Sometimes the complements of the signs make the match up, indicating a "too small" observation.)

Further, if we assume that factors x_2 and x_3 *are* the only ones of possible importance, the 2^4 design then becomes a replicated 2^2 in these factors. The replicate values are plotted in Figure 4B.5, from which it again seems likely that the thirteenth observation of 59.15 is in error.

APPENDIX 4C.   CONFIDENCE REGIONS FOR CONTOUR PLANES (see Section 4.5)

We can construct confidence regions for the contour planes as follows. Consider, for example, the equation for the contour plane of the desired level of hue, that is,

$$20 = 16.95 - 5.64x_1 - 0.17x_4 + 5.42x_6.$$

Let

$$Z_H \equiv 20 - 16.95 + 5.64x_1 + 0.17x_4 - 5.42x_6.$$

For any particular point (x_1, x_4, x_6) which lies on the "true" contour plane for a response of 20 in hue we have

$$E(Z_H) = 0.$$

Also, we know that the estimate of variance of Z_H is

$$\hat{\sigma}_{Z_H}^2 = \left(1 + x_1^2 + x_4^2 + x_6^2\right) s_2^2/64.$$

Thus a confidence region for all points (x_1, x_4, x_6) which satisfy hue = 20 is given by

$$-t_{0.025}(\nu)\hat{\sigma}_{Z_H} \leq Z_H \leq t_{0.025}(\nu)\hat{\sigma}_{Z_H}$$

where $t_{0.025}(\nu)$ is the upper $97\frac{1}{2}\%$ point of a t distribution with ν df and where $\nu = 60$ is the number of degrees of freedom on which $s_2^2 = 35.1219$ (from Table 4.7) is based.

It follows that the (x_1, x_4, x_6) boundary for the confidence region for the hue = 20 contour plane is defined by

$$(20 - 16.95 + 5.64x_1 + 0.17x_4 - 5.42x_6)^2$$

$$= \left(1 + x_1^2 + x_4^2 + x_6^2\right)2^2(35.1219)/64$$

or

$$(3.05 + 5.64x_1 + 0.17x_4 - 5.42x_6)^2 = 2.195\left(1 + x_1^2 + x_4^2 + x_6^2\right).$$

This equation represents two hyperbolic surfaces which "sandwich" the estimated hue = 20 plane and go further away from it as we travel further away from the point $(\bar{x}_1, \bar{x}_4, \bar{x}_6)$. Points (x_1, x_4, x_6) that make left-hand side (LHS) < right-hand side (RHS) lie within the confidence region. Points for which LHS > RHS lie outside it.

EXERCISES

4.1. Explain briefly, in a few sentences, Yates' method of evaluating effects in a 2^k design.

4.2. A 2^4 design gave rise to the following contrasts in standard order (1, 2, 12, 3, 13, 23, 123, 4, 14, etc.): 18, 21, 3, 6, 0, 4, -1, 24, $-8, 11, -9, 3, 1, 1, -2$. Use a normal plot technique to analyze these data.

Note: A *contrast* is a linear combination of the y's whose coefficients add to zero. Thus the factorial design estimates of main effects and of interactions are all contrasts, but $\bar{y}$ is not.

4.3.   A 2^4 factorial design yielded the following contrasts:

$$
\begin{array}{llll}
A \leftarrow 0.4   & AB \leftarrow 16.7 & ABC \leftarrow -0.1 \\
B \leftarrow -7.6   & AC \leftarrow 3.1   & ABD \leftarrow -4.7 \\
C \leftarrow 14.1   & AD \leftarrow 5.2   & ACD \leftarrow 7.7 \\
D \leftarrow 66.7   & BC \leftarrow 8.3   & BCD \leftarrow -2.3 \\
                    & BD \leftarrow -3.6 & ABCD \leftarrow 3.9 \\
                    & CD \leftarrow 14.3
\end{array}
$$

Interpret the results, employing a normal plot technique.

4.4.   (a) If a full 2^k experiment is performed and each factor has levels -1 and 1, what are the relationships between the estimates of (i) the main effects of the k factors, and (ii) the coefficients β_i in a first-order model.

      (b) How, in the same situation, are two-factor interactions represented in a regression model?

4.5.   Imagine that individual trials comprising a 2^3 factorial design are run sequentially in standard order (in complete disregard for randomization !) while, simultaneously, a fourth variable disregarded by the experimenter takes the values shown in Table E4.5.

      If the first-order model $y = \beta_0 + \beta_1 x_1 + \beta_2 x_2 + \beta_3 x_3 + \varepsilon$ is fitted by least squares, determine the bias in the estimates of b_0, b_1, b_2, and b_3 of β_0, β_1, β_2, and β_3 induced by the failure to add a term $\beta_4 x_4$ to the model.

TABLE E4.5

x_1	x_2	x_3	x_4
$-$	$-$	$-$	-7
$+$	$-$	$-$	5
$-$	$+$	$-$	7
$+$	$+$	$-$	3
$-$	$-$	$+$	-3
$+$	$-$	$+$	-7
$-$	$+$	$+$	-5
$+$	$+$	$+$	7

4.6.   Use Yates' algorithm to evaluate the effects from the observations in Table E4.6, obtained from a 2^3 factorial design.

TABLE E4.6

x_1	x_2	x_3	y
−	−	−	13
+	−	−	11
−	+	−	9
+	+	−	5
−	−	+	11
+	−	+	8
−	+	+	9
+	+	+	7

4.7. Imagine a 2^k factorial design with a center point replicated r times. What effect do these additional center observations have on the estimation of the factorial effects and on the overall mean?

4.8. The primary purpose of quenching a steel is to produce martensite. The transition from austenite to martensite depends upon the steel's composition, which affects the temperature of the martensite formation. It has long been believed that carbon and other alloying elements lower the temperature range in which martensite forms. A preliminary 2^3 factorial study to determine the effects of carbon, manganese, and nickel on "martensite start temperature" produced the data in Table E4.8.

(a) Code the predictor variables (i.e., the test conditions).

(b) Draw a diagram of the design and mark the response observations (i.e., the start temperatures) on it.

(c) Estimate all the main effects, two-factor interactions, and the three-factor interaction.

TABLE E4.8

Test conditions			Start temperature (°F)
%C	%Mn	%Ni	
0.1	0.2	0.1	940
0.9	0.2	0.1	330
0.1	1.4	0.1	820
0.9	1.4	0.1	250
0.1	0.2	0.2	840
0.9	0.2	0.2	280
0.1	1.4	0.2	760
0.9	1.4	0.2	200

(d)   Assume that the standard deviation of the response observations is $\sigma = 65$. Which estimates in (c) are large compared with their standard errors?

(e)   What are the practical conclusions of your analysis?

TABLE E4.9

X_1	X_2	X_3	y
110	2500	2	37.7
110	1000	2	25.6
60	2500	2	22.2
110	1000	7.3	36.6
110	1000	7.3	34.4
110	2500	7.3	49.1
60	1000	2	12.5
60	2500	2	21.0
110	2500	7.3	46.0
60	2500	7.3	33.5
110	1000	2	28.2
110	2500	2	39.2
60	1000	2	12.2
60	1000	7.3	19.5
60	2500	7.3	35.0
60	1000	7.3	20.8

4.9.   (Adapted from data provided by Mario Karfakis.) The data in Table E4.9 resulted from a study of the effect of three factors

$$X_1 = \text{revolutions per minute (rpm)},$$

$$X_2 = \text{downward thrust (weight), (lb)},$$

$$X_3 = \text{drilling fluid flow rate},$$

on the rate of penetration, y, of a drill. The observations are listed in the random order in which they were taken. Code the X's, and identify the type of design used. Then carry out a suitable analysis and list the practical conclusions you reach.

4.10. Four variables were examined for their effects on the yield y of a chemical process:

$$x_1 = \text{catalyst concentration},$$

$$x_2 = \text{NaOH concentration},$$

$$x_3 = \text{level of agitation},$$

$$x_4 = \text{temperature}.$$

TABLE E4.10

1	2	3	4	y	1	2	3	4	y
−	+	−	+	61	+	−	+	+	84
+	−	−	+	70	+	+	+	−	41
−	−	−	−	46	−	−	+	−	67
−	−	+	+	87	+	−	−	−	49
+	+	+	+	62	+	+	−	+	59
+	−	+	−	64	−	+	−	−	36
+	+	−	−	38	−	−	−	+	68
−	+	+	−	38	−	+	+	+	62

A 2^4 factorial design was employed. The trials were run in random order. Interpret the data, given in Table E4.10.

4.11. A manufacturer of paper bags wishes to check tear resistance, y, for which she has a long-used numerical scale. She examines three factors at each of two coded levels namely

$$x_1 = \text{type of paper,}$$

$$x_2 = \text{humidity,}$$

$$x_3 = \text{direction of tear,}$$

and obtains three replicate observations as shown in Table E4.11. (The runs have been rearranged for convenience from the random order in which they were performed.) The total sum of squares for the 24 observations is 496.87.

TABLE E4.11

x_1	x_2	x_3	Tear resistance			Row sums
−	−	−	3.8	3.1	2.2	9.1
+	−	−	6.6	8.0	6.8	21.4
−	+	−	3.4	1.7	3.8	8.9
+	+	−	6.8	8.2	6.0	21.0
−	−	+	2.3	3.1	0.7	6.1
+	−	+	4.7	3.5	4.4	12.6
−	+	+	2.1	1.1	3.6	6.8
+	+	+	4.2	4.7	2.9	11.8
						97.7

Perform a factorial analysis on the data, estimate σ^2 from the repeat runs, and formulate conclusions about the factors.

4.12. Look again at the data of Exercise 4.9. Fit the first-order model $y = \beta_0 + \beta_1 x_1 + \beta_2 x_2 + \beta_3 x_3 + \varepsilon$ by least squares, and perform the usual analyses including an overall test for lack of fit. Your conclusions should be consistent with those from the factorial analysis.

4.13. To the data Exercise 4.10 fit the model

$$y = \beta_0 + \beta_2 x_2 + \beta_3 x_3 + \beta_4 x_4 + \beta_{23} x_2 x_3 + \beta_{24} x_2 x_4 + \beta_{34} x_3 x_4 + \varepsilon.$$

Provide the associated analysis of variance table, estimate σ^2, and check the fitted coefficients for significance.

4.14. Look at the data of Exercise 4.11. We could, formally, fit a regression equation to these data but they are much more suitably analyzed by factorial methods. Why?

TABLE E4.15

Run order	x_1	x_2	x_3	y
3	-1	-1	-1	49
12	1	-1	-1	48
11	-1	1	-1	53
8	1	1	-1	54
4	-1	-1	1	57
5	1	-1	1	58
9	-1	1	1	61
7	1	1	1	55
6	0	0	0	55
2	0	0	0	54
1	0	0	0	56
10	0	0	0	54

4.15. (a) Fit the model $y = \beta_0 + \beta_1 x_1 + \beta_2 x_2 + \beta_3 x_3 + \varepsilon$ to the data given in Table E4.15. Test for lack of fit. What final model would you adopt, if any?

   (b) This design has 12 observations; exhibit the vectors c_i for 12 meaningful orthogonal contrasts which exist and which provide a complete transformation of the original data set $y_1, y_2, \ldots, y_{12}$.

4.16. (Source: Maximum data through a statistical design, by C. D. Chang, O. K. Kononenko, and R. E. Franklin, Jr., *Industrial and Engineering Chemistry*, **52**, November 1960, 939–942. Material reproduced and adapted by permis-

TABLE E4.16. A 2^4 factorial design with six center points; four responses are shown

Trial number	Factor levels				Responses			
	x_1, NH$_3$	x_2, T	x_3, H$_2$O	x_4, P	y_1, PDA	y_2, DMP	y_3, PD	y_4, R
1	-1	-1	-1	-1	1.8	58.2	24.7	84.7
2	1	-1	-1	-1	4.3	23.4	45.5	73.2
3	-1	1	-1	-1	0.4	21.9	8.6	30.9
4	1	1	-1	-1	0.7	21.8	9.1	31.6
5	-1	-1	1	-1	0.3	14.3	75.5	90.1
6	1	-1	1	-1	4.5	6.3	86.5	96.3
7	-1	1	1	-1	0.0	4.5	10.0	14.5
8	1	1	1	-1	1.6	21.8	50.1	73.5
9	-1	-1	-1	1	1.3	46.7	43.3	91.3
10	1	-1	-1	1	4.2	53.2	39.7	97.1
11	-1	1	-1	1	1.9	23.7	5.4	31.0
12	1	1	-1	1	0.7	40.3	9.7	50.7
13	-1	-1	1	1	0.0	7.5	78.8	86.3
14	1	-1	1	1	2.3	13.3	77.8	93.4
15	-1	1	1	1	0.8	49.3	21.1	71.2
16	1	1	1	1	7.3	20.1	37.8	65.2
17[a]	0	0	0	0	5.0	32.8	45.1	82.8
Pure error estimates of σ_i^2, $s_i^2 =$					1.08	9.24	15.37	8.18

[a] Average yield of six trials.

sion of the copyright holder, the American Chemical Society. The data were obtained under a grant from the Sugar Research Foundation, Inc., now the World Sugar Research Organization, Ltd.) Table E4.16 shows the results obtained from a 2^4 factorial design plus six center points. Four response variables were observed. The trials were performed in random order but are listed in standard order. Estimate all factorial main effects and all interactions for every response, and evaluate the appropriate standard errors. Which variables appear to be effective? So far, the six center points have been used only to estimate the four σ_i^2, the variances of the four responses. Calculate the values of the contrasts

$$CC_i = (\text{average response at factorial points})$$

$$- (\text{average response at center points})$$

and determine their standard errors. Which responses show evidence of pure quadratic curvature? (*Note*: In general if there are k factors, such a contrast has expectation

$$E(CC) = \beta_{11} + \beta_{22} + \cdots + \beta_{kk}.$$

Thus if the CC contrast is compared to its standard error, a test for H_0: $\Sigma \beta_{ii} = 0$ versus H_1: $\Sigma \beta_{ii} \neq 0$ may be made.) Suppose you fitted a first-order model to each response. Would the center point average response be counted as one data point or as six data points. (The answer is six; why?) A reviewer of the original article suggested that the responses might have been transformed via

$$\omega_i = \sin^{-1} \left(\frac{y_i}{100} \right)^{1/2}$$

which is a transformation often appropriate when the data are percentages, as here, or proportions (when the division by 100 would not be needed). Is such a transformation useful in the sense that it renders a simpler interpretation of the factors' effects? Transform all the response values and repeat the factorial analysis to find out. (The y_2 data will be used again in Exercise 11.16 and will be combined there with additional data, for a second-order surface fitting exercise.)

4.17. The following estimates of effects resulted from a 2^5 experiment:

$E \leftarrow$	$- 224$	$BCD \leftarrow$	$- 18$	$BE \leftarrow$	29
$C \leftarrow$	$- 153$	$ABDE \leftarrow$	$- 14$	$DE \leftarrow$	30
$ABCDE \leftarrow$	$- 77$	$D \leftarrow$	$- 9$	$ABCE \leftarrow$	31
$ACE \leftarrow$	$- 58$	$B \leftarrow$	$- 6$	$BCE \leftarrow$	39
$AD \leftarrow$	$- 54$	$CD \leftarrow$	$- 4$	$ACDE \leftarrow$	47
$BC \leftarrow$	$- 53$	$ABC \leftarrow$	0	$AC \leftarrow$	53
$ABD \leftarrow$	$- 34$	$AE \leftarrow$	2	$ABCD \leftarrow$	58
$ACD \leftarrow$	$- 33$	$BD \leftarrow$	7	$AB \leftarrow$	64
$BDE \leftarrow$	$- 28$	$CDE \leftarrow$	12	$CE \leftarrow$	83
$ABE \leftarrow$	$- 22$	$BCDE \leftarrow$	16	$A \leftarrow$	190
		$ADE \leftarrow$	21		

(a) Plot these contrasts on normal probability paper.
(b) Which effects do you think are distinguishable from noise?
(c) Replot the remaining contrasts.
(d) Estimate the standard deviation of the contrasts from plot (c). See p. 131.

[Partial answers: (b) E, C, A. (d) Roughly 40.]

Blocking and Fractionating 2^k Factorial Designs

5.1. BLOCKING THE 2^6 DESIGN

In the 2^6 design used above for illustration, the runs were made in a random order; the design was *fully randomized*. However, it might have been much better to run the arrangement in a series of *randomized blocks*. Suppose, for example, we expected trouble because of appreciable inhomogeneity of the raw material. Suppose further that a blender was available, large enough to produce homogeneous blends of raw material sufficient for making eight batches, that is, for conducting 8 of the 64 runs. Then the 64-run experiment might have been arranged using 8 blends (blocks) of homogeneous raw material within each of which 8 runs were made. The difficulty would be that differences between blends could bias the estimates of effects. R. A. Fisher found a way of circumventing this biasing problem by making the (often reasonable) assumption that the effects of blocks would be approximately additive, that is, that the differences between blends would cause the response results simply to be raised or lowered by a fixed (but unknown) amount.

Two Blocks

We introduce the method in a simpler context. Suppose first that we would like to make the 64 runs in two blocks (I and II) each with 32 runs. (Such an arrangement would be appropriate if we had a blender large enough to provide homogeneous raw material for 32 runs.) Consider the partial table of signs, Table 5.1, which shows the 2^6 design laid out in standard order. Suppose now that we allocate runs for which the product column **123456** has a minus sign to block I, and runs for which **123456** has a plus sign to block II. Then, following Fisher, we say that the block effect (denoted by B)

TABLE 5.1.   Blocking the 2^6 design into two blocks using B = 123456.

Run number	Variable number						123456	Block
	1	2	3	4	5	6		
1	−	−	−	−	−	−	+	II
2	+	−	−	−	−	−	−	I
3	−	+	−	−	−	−	−	I
4	+	+	−	−	−	−	+	II
5	−	−	+	−	−	−	−	I
6	+	−	+	−	−	−	+	II
7	−	+	+	−	−	−	+	II
8	+	+	+	−	−	−	−	I
9	−	−	−	+	−	−	−	I
10	+	−	−	+	−	−	+	II
11	−	+	−	+	−	−	+	II
12	+	+	−	+	−	−	−	I
13	−	−	+	+	−	−	+	II
14	+	−	+	+	−	−	−	I
.	.	.	.	.	.	.	.	.
.	.	.	.	.	.	.	.	.
.	.	.	.	.	.	.	.	.
63	−	+	+	+	+	+	−	I
64	+	+	+	+	+	+	+	II

is completely *confounded* with the 123456 interaction, that is,

$$\mathbf{B} = \mathbf{123456}. \tag{5.1.1}$$

Thus we lose the ability to estimate (at least with the same accuracy) the 123456 interaction. However, in most practical contexts, this very high-order interaction will probably be negligible in size and of little interest.

Now the 2^k factorial designs have an important orthogonal* property whereby a sequence of signs corresponding to a particular effect is orthogonal to every other such sequence. The important implication of this in the present context is that altering the apparent 123456 effect by superimposing on it the difference between blocks (blends) does not change the estimate of any of the other effects.

---

*Two columns of ± signs, or two columns of numbers, are said to be *orthogonal* if the sum of the cross-products of corresponding signs, or of corresponding numbers, is zero. The geometrical implication is that the two columns represent two vectors in space which are orthogonal, that is, at right angles, to each other.

Exercise. Add 3 to each of the runs for which elements of the **123456** column are negative and subtract 7 from each of the runs for which they are positive, and confirm by recalculation that the estimates of all effects except for 123456 are unchanged.

Four Blocks; Eight Blocks

Let us now get a little more ambitious and try to arrange the design in four blocks each containing 16 runs. This can be done by confounding two high-order interactions with block contrasts. Suppose, as a first shot, we associate the contrasts 123456 and 23456 with blocks. (This might seem reasonable since we shall wish to confound interactions of the highest possible order with blocks.) Thus the *blocking generators* for our design will be

$$\mathbf{B}_1 = \mathbf{123456}, \qquad \mathbf{B}_2 = \mathbf{23456}, \tag{5.1.2}$$

say. The runs would then be allocated to the four blocks according to the following scheme:

$$\mathbf{B}_1 = \mathbf{123456} \begin{cases} - \\ + \end{cases} \overbrace{\begin{array}{c|cc} & - & + \\ \hline - & I & II \\ + & III & IV \end{array}}^{\mathbf{B}_2 = \mathbf{23456}} \tag{5.1.3}$$

That is, runs would be allocated to the four blocks I, II, III, and IV as the signs associated with the columns **123456** and **23456** took the values $(--)$ $(-+)(+-)$ and $(++)$. Unfortunately, a serious difficulty occurs with this arrangement. There are, of course, 3 df (independent contrasts) among the four blocks of 16 runs each. If we associate $\mathbf{B}_1$ and $\mathbf{B}_2$ with two of these contrasts, the third must be the interaction $\mathbf{B}_1 \times \mathbf{B}_2 = \mathbf{B}_1\mathbf{B}_2$. However, then the interaction $\mathbf{B}_1\mathbf{B}_2$ will be confounded with the "interaction" between **123456** and **23456**. To see what this interaction is, consider the following table of signs in standard order. (We write them as rows to conserve space.)

123456	$+--+$	$-++-$	$-++-$	$\cdots$	$-+$
23456	$--++$	$++--$	$++--$	$\cdots$	$++$
product	$-+-+$	$-+-+$	$-+-+$	$\cdots$	$-+$
of					
above					

We see that the interaction between **123456** and **23456** is the main effect **1**. This arrangement would thus be a very poor one, because the main effect of the first variable would be confounded with block differences. To make a better choice, we need to understand how to calculate interactions between complex effects. To see how to do this, we note that if we take any set of signs for any effect (say 1) and multiply those signs by the signs of the *same* effect, we obtain a row of $+$'s which we denote by the *identity* **I**.

$$
\begin{array}{llllll}
\mathbf{1} & -+-+ & -+-+ & -+-+ & \cdots & -+ \\
\mathbf{1} & -+-+ & -+-+ & -+-+ & \cdots & -+ \\
\mathbf{1 \times 1 = 1^2 = I} & ++++ & ++++ & ++++ & \cdots & ++
\end{array}
$$

Thus, using the multiplication sign to imply the multiplication of the signs in corresponding positions in two rows (or two columns if the design is arranged vertically, as would be usual), we can write

$$\mathbf{1 \times 1 = 1^2 = I}, \quad \mathbf{2 \times 2 = 2^2 = I}, \quad \mathbf{12 \times 12 = 1^2 2^2 = I^2 = I},$$

$$(5.1.4)$$

and so on. Also, multiplication of any contrast by the identity **I** leaves the contrast unchanged. Thus

$$\mathbf{1 \times I = 1}, \quad \mathbf{23 \times I = 23}, \quad \mathbf{12345 \times I = 12345}, \quad (5.1.5)$$

and so on. Applying this rule in the case of the design above, we have that

$$\mathbf{B_1 B_2 = 123456 \times 23456 = 12^2 3^2 4^2 5^2 6^2 = 1 \times I^5 = 1}, \quad (5.1.6)$$

indicating that $B_1 B_2$ and 1 are confounded.

An arrangement in four blocks of 16 runs so that all the interactions confounded with blocks are of the highest possible order may be obtained by using as generators two four-factor interactions in which only two symbols overlap. For example, if we choose

$$\mathbf{B_1 = 1234}$$

$$\mathbf{B_2 = 3456},$$

then
$$\mathbf{B_1 B_2 = 12 56}. \quad (5.1.7)$$

We can now write out, with the design arranged in standard order, as usual, rows of signs for $\mathbf{B_1}$ and $\mathbf{B_2}$ and allocate the runs to the four blocks corresponding to the sign combinations $(\mathbf{B_1}, \mathbf{B_2}) = (-, -), (-, +), (+, -),$

$(+, +)$, as follows:

$B_1 = 1234$	+	−	−	+		−	+	+	−		−	+	+	−	$\cdots$		−	+
$B_2 = 3456$	+	+	+	+		−	−	−	−		−	−	−	−	$\cdots$		+	+
block number	IV	II	II	IV		I	III	III	I		I	III	III	I	$\cdots$		II	IV

On the assumption that the blocks contribute only additive effects, the main effects, two-factor interactions, and three-factor interactions will all remain unconfounded with any block effect.

If, as would often be acceptable, we are prepared to confound some three-factor interactions, we can split the 64 runs into eight blocks each containing eight runs. For example, choosing the generators

$$B_1 = 1234, \quad B_2 = 3456, \quad B_3 = 136, \quad (5.1.8)$$

we obtain the following confounding pattern for the seven contrasts among the eight blocks:

$$B_1 = 1234$$

$$B_2 = \quad 3456$$

$$B_1 B_2 = 12 \quad 56$$

$$B_3 = 1\ 3\quad 6$$

$$B_1 B_3 = \quad 2\ 4\ 6$$

$$B_2 B_3 = 1\quad 4\ 5$$

$$B_1 B_2 B_3 = 23\ 5. \quad (5.1.9)$$

On the assumption that blocks have only additive effects, no main effect or two-factor interaction is confounded with any block effect.

Exercise. Allocate the 64 runs of a 2^6 design to eight blocks using the generators $B_1 = 1346$, $B_2 = 1235$, and $B_3 = 156$ and write down the confounding pattern for the seven contrasts among the eight blocks. Is any main effect or two-factor interaction confounded with blocks? (No.)

A short table of useful confounding arrangements adapted from Box, Hunter, and Hunter (1978) is given in Table 5.2. Other tables will be found in Davies (1954), the National Bureau of Standards Applied Mathematics Series, No. 48 (1957), and McLean and Anderson (1984).

TABLE 5.2. Useful blocking arrangements for 2^k factorial designs

k = number of variables	Block size	Block generator	Interactions confounded with blocks
3	4	$B_1 = 123$	123
	2	$B_1 = 12, B_2 = 13$	12, 13, 23
4	8	$B_1 = 1234$	1234
	4	$B_1 = 124, B_2 = 134$	124, 134, 23
	2	$B_1 = 12, B_2 = 23, B_3 = 34$	12, 23, 34, 13, 1234, 24, 14
5	16	$B_1 = 12345$	12345
	8	$B_1 = 123, B_2 = 345$	123, 345, 1245
	4	$B_1 = 125, B_2 = 235, B_3 = 345$	125, 235, 345, 13, 1234, 24, 145
	2	$B_1 = 12, B_2 = 13, B_3 = 34, B_4 = 45$	12, 13, 34, 45, 23, 1234, 1245, 14, 1345 35, 24, 2345, 1235, 15, 25 that is, all 2fi and 4fi[a]
6	32	$B_1 = 123456$	123456
	16	$B_1 = 1236, B_2 = 3456$	1236, 3456, 1245
	8	$B_1 = 135, B_2 = 1256, B_3 = 1234$	135, 1256, 1234, 236, 245, 3456, 146
	4	$B_1 = 126, B_2 = 136, B_3 = 346, B_4 = 456$	126, 136, 346, 456, 23, 1234, 1245 14, 1345, 35, 246, 23456, 12356, 156, 25
	2	$B_1 = 12, B_2 = 23, B_3 = 34, B_4 = 45, B_5 = 56$	All 2fi, 4fi, and 6fi
7	64	$B_1 = 1234567$	1234567
	32	$B_1 = 12367, B_2 = 34567$	12367, 34567, 1245
	16	$B_1 = 123, B_2 = 456, B_3 = 167$	123, 456, 167, 123456, 2367, 1457, 23457
	8	$B_1 = 1234, B_2 = 567, B_3 = 345, B_4 = 147$	1234, 567, 345, 147, 1234567, 125, 237, 3467, 145 1357, 1267, 2356, 2457, 136, 246
	4	$B_1 = 127, B_2 = 237, B_3 = 347, B_4 = 457, B_5 = 567$	127, 237, 347, 457, 567, 13, 1234, 1245, 1256, 24, 2345, 2356, 3456, 35, 1234567, 46, 147, 13457, 135 12467, 12357, 257, 24567, 23467, 367, 15, 1456, 1346, 1236, 26, 167
	2	$B_1 = 12, B_2 = 23, B_3 = 34, B_4 = 45, B_5 = 56, B_6 = 67$	All 2fi, 4fi, and 6fi

[a] "fi" is an abbreviation for "factor interaction"; thus, for example, 2fi means two-factor interaction.

Source: Table 5.2 is reproduced and adapted with permission from the table on pp. 346–347 of *Statistics for Experimenters: A Introduction to Design, Data Analysis and Model Building,* by G. E. P. Box, W. G. Hunter, and J. S. Hunter, published by John Wiley & Sons, Inc., New York, 1978.

5.2. FRACTIONATING THE 2^6 DESIGN

To study the dyestuffs problem in Chapter 4, we used a 2^6 factorial design that required 64 runs. As it turned out, only main effects of three of the variables could be distinguished from the noise. Such a discovery naturally leads to the speculation as to whether so many runs were really needed. In this particular example, the standard deviation σ of the experimental error was rather large compared with the size of the effects of interest, so the full 64 runs *were* probably needed. However, if σ had been smaller, a design containing a smaller number of runs than the full $2^6 \doteq 64$ might have proved adequate. A class of two-level designs requiring fewer runs than the full 2^k factorials is that of the 2^{k-p} *fractional factorials*.

Half-Fractions of the 2^6 Factorial

To illustrate the nature and the utility of such designs, consider again the problem of blocking the 2^6 factorial into two blocks, each of 32 runs, with the six-factor interaction **123456** ($=$ **B**) used as the defining (or blocking) contrast. (See Section 5.1.) Let us temporarily augment our previous notation a little by using the symbol $\mathbf{I}_{32}$ to refer to a sequence of 32 plus signs. For each of the 32 runs in the first block, the signs of the six factors will multiply to give -1 and for each of the 32 runs in the second block, the signs will multiply to give $+1$. Thus,

$$\mathbf{123456} = -\mathbf{I}_{32}, \quad \text{for block I,}$$

$$\mathbf{123456} = \mathbf{I}_{32} \quad \text{for block II.} \quad (5.2.1)$$

Suppose now that, instead of performing all the 64 runs, we carried out only the 32 runs in, say, block II. These are the runs for which elements of **123456** are $+$ in Table 5.1. Let us set out a table of signs for the calculation of main effects, two-factor interactions, three-, four-, five-, and six-factor interactions, just as for a full factorial design but using only block II runs, as shown in Table 5.3.

We could certainly use the table of signs to compute all the 63 main effects and interactions. Nevertheless, something would seem to be amiss because we can scarcely expect to compute 63 independent effects from only 32 observations. The explanation lies in the duplication that occurs in the table. We see, as expected, that the signs in the **123456** column are all identical to those of $\mathbf{I}_{32}$ (because we have deliberately arranged this). Notice also, however, that the signs for the interaction column **23456** are identical to those for the column **1**, or symbolically $\mathbf{1} = \mathbf{23456}$. Additionally $\mathbf{2} = \mathbf{13456}$, and so on. Let us denote by l_1, l_2, and so on, the various linear contrasts obtained by adding together all the response values with plus signs in the **1**, **2**, and so on, column and subtracting all those with minus signs, and dividing by the appropriate column divisor. Also* let us use an arrow pointing to the right ($\rightarrow$) to mean "is an estimate of." Then, for example, l_1 provides an estimate not just of the main effect of 1, nor of the interaction 23456, but of the sum of these, that is,

$$l_1 \rightarrow 1 + 23456. \quad (5.2.2)$$

---

*As in Chapter 4, the arrow pointing to the left ($\leftarrow$) will mean "is estimated by."

TABLE 5.3. Columns of signs for calculation of effects in a half-fraction of a 2^6 design

Run number from Table 5.1	I_{32}	Main effects						Two-factor interactions				...	...	Five-factor interactions			Six-factor interaction
		1	2	3	4	5	6	12	13	14	...			12456	13456	23456	123456
1	+	−	−	−	−	−	−	+	+	+	...	...		−	−	−	+
4	+	+	+	−	−	−	−	+	−	−				−	+	+	+
6	+	+	−	+	−	−	−	−	+	−				+	−	+	+
7	+	−	+	+	+	−	−	−	−	+				−	+	−	+
10	+	+	−	−	+	−	−	−	−	+				−	−	+	+
11	+	−	+	−	+	−	−	+	+	−				−	+	−	+
13	+	−	−	+	+	−	−	+	−	−				+	−	−	+
⋮	⋮	⋮										⋮	⋮	⋮			
⋮	⋮	⋮										⋮	⋮	⋮			
64	+	+	+	+	+	+	+	+	+	+	...	...		+	+	+	+
Divisor	32	16	16	16	16	16	16	16	16	16	...			16	16	16	16

To establish this, we need only write out the formulas for the estimates of 1 and of 23456 from the full 2^6 and add the two together. It will be found that addition and cancellation leaves only the observations in block II with the signs of column 1 in Table 5.3 divided by the divisor 16. In similar fashion,

$$l_2 \rightarrow 2 + 13456 \tag{5.2.3}$$

and so on. The device that enables us quickly to identify all these *aliases* is supplied by the generator for this half-fraction of the 2^6 design which we can write as

$$I = 123456, \tag{5.2.4}$$

(where now, for simplicity, we write I for I_{32}). Equation (5.2.4) simply tells us to pick out runs from the full 64 whose product of signs **123456** is plus, same as the I column. Multiplying both sides of Eq. (5.2.4) by 1, we have 1 = **23456** which implies that $l_1 \rightarrow 1 + 23456$.

All other existing alias relationships are obtained similarly; for example, on multiplying Eq. (5.2.4) by **123** we obtain

$$\mathbf{123} = \mathbf{456}, \quad \text{implying that } l_{123} \rightarrow 123 + 456. \tag{5.2.5}$$

(Note that each l can be denoted by two equivalent names; for example, $l_{123} = l_{456}$.)

For quantitative variables, the various interactions $12, 345, \ldots$, reflect the effects of mixed derivatives of the response function η, such as

$$\frac{\partial^2 \eta}{\partial x_1 \partial x_2}, \quad \frac{\partial^3 \eta}{\partial x_3 \partial x_4 \partial x_5}, \ldots.$$

It would frequently be reasonable to assume that derivatives higher than the second could be ignored, that is, in a Taylor's series expansion of the response function, terms of third and of higher orders could be assumed to be negligible within the local region of experimentation. In these circumstances, we could, correspondingly, ignore interactions of three or more factors in the factorial analysis of the data. *Under such an assumption*, no confounding would in fact occur in the half-fraction $I = \mathbf{123456}$ of the 2^6 design above. When σ is not too large, so that fewer runs than a full factorial will provide adequate accuracy, fractional factorials are frequently very useful, particularly in experimental situations with more than three or four variables.

For our illustration above, we chose the half-fraction defined by $\mathbf{I} =$ **123456**, but we could equally well have used the other half-fraction defined by $-\mathbf{I} =$ **123456** or, equivalently, $\mathbf{I} = -$**123456**. In this case, the columns **1** and $-$**23456** will be identical and, denoting the linear contrasts defined by columns $\mathbf{1}, \mathbf{2}, \ldots$, and so on by $l'_1, l'_2, \ldots$, and so on, we shall find that

$$l'_1 \rightarrow 1 - 23456, \qquad l'_2 \rightarrow 2 - 13456, \ldots, \qquad (5.2.6)$$

and so on. Notice that, if we perform *both* halves of the 2^6 design, we can recover all the individual main effects and interactions by averaging or halving and differencing. For example,

$$\tfrac{1}{2}(l_1 + l'_1) \rightarrow \tfrac{1}{2}(1 + 23456 + 1 - 23456) = 1 \qquad (5.2.7)$$

and this would be identical to the estimate supplied by the full factorial. Also, similarly,

$$\tfrac{1}{2}(l_1 - l'_1) \rightarrow 23456, \qquad (5.2.8)$$

again identical to that supplied by the full factorial.

A one-half replicate of a full factorial in k factors, that is, $\tfrac{1}{2}$ of a 2^k, or a 2^{-1} fraction of a 2^k, is often referred to as a 2^{k-1} fractional factorial. Thus the fractions we have described above are both 2^{6-1} fractional factorial designs.

Quarter Fractions of the 2^6 Factorial

Suppose now we ran only 16 of the 64 runs of the full 2^6 design corresponding to one block of the design when it is divided into four blocks (as in Section 5.1). This *quarter replicate* would be referred to as a 2^{6-2} fractional factorial design. There are four blocks from which to choose. Suppose we use the one for which $(\mathbf{B}_1, \mathbf{B}_2) = (+, +)$. Then for all the runs in this block,

$$\mathbf{B}_1 = \mathbf{I} = \mathbf{1234}, \qquad \mathbf{B}_2 = \mathbf{I} = \mathbf{3456}, \qquad \mathbf{B}_1\mathbf{B}_2 = \mathbf{I} = \mathbf{1256}. \quad (5.2.9)$$

We can multiply through this defining relation by $\mathbf{1}, \mathbf{2}, \ldots, \mathbf{12}, \mathbf{13}, \ldots$ to give the alias pattern. For example,

$$1 = 234 = 13456 = 256,$$
$$2 = 134 = 23456 = 156,$$
$$\cdots$$
$$12 = 34 = 123456 = 56,$$
$$\cdots \qquad (5.2.10)$$

and so on. Ignoring interactions involving four or more factors we see that the design allows estimates of 16 combinations of main effects and interactions such as, for example,

$$l_1 \to 1 + (234 + 256),$$

$$l_2 \to 2 + (134 + 156),$$

$$l_3 \to 3 + (124 + 456),$$

$$l_4 \to 4 + (123 + 356),$$

$$l_5 \to 5 + (346 + 126),$$

$$l_6 \to 6 + (345 + 125),$$

$$l_{12} \to 12 + 34 + 56,$$

$$l_{13} \to 13 + 24,$$

$$l_{14} \to 14 + 23,$$

$$\cdots \qquad\qquad (5.2.11)$$

(Note that the l subscripts can be those of any of the aliased effects. For example, $l_{12} = l_{34} = l_{56}$, and so on. Typically, we choose subscripts which occur earlier in the ordering $1, 2, \ldots, 12, 13, \ldots, 123, \ldots$, eliminating aliased duplicates, but this is not essential.)

The estimation equations show that all main effects are confounded with third-order (or ignored higher-order) interactions, not with two-factor interactions or other main effects. Two-factor interactions are confounded with other two-factor interactions. The defining relation must indicate all the independent generators and their products in twos, threes, ..., all of which produce the identity **I**. For example, suppose we had used the 2^{6-2} design defined by $(\mathbf{B}_1, \mathbf{B}_2) = (-, +)$, we would have had

$$\mathbf{B}_1 = -\mathbf{I} = \mathbf{1234}, \qquad \mathbf{B}_2 = \mathbf{I} = \mathbf{3456}, \qquad \mathbf{B}_1\mathbf{B}_2 = -\mathbf{I} = \mathbf{1256}, \quad (5.2.12)$$

which leads to the defining relation

$$\mathbf{I} = -\mathbf{1234} = \mathbf{3456} = -\mathbf{1256}, \qquad (5.2.13)$$

and means that the estimates obtained would be as follows:

$$l_1' \to 1 - 234 - 256,$$

$$l_2' \to 2 - 134 - 156,$$

$$\cdots \qquad\qquad (5.2.14)$$

and so on, ignoring interactions of four or more factors.

The variance of any of the estimates in (5.2.11) or (5.2.14) is

$$V(l) = 4\sigma^2/n \qquad (5.2.15)$$

where n is the total number of individual observations (with variance σ^2) contributing to l. This formula applies to estimates throughout this chapter.

5.3. RESOLUTION OF A 2^{k-p} FACTORIAL DESIGN

The resolution R of a 2^{k-p} fractional factorial design is the length of the shortest word in the defining relation. For example, the 2^{6-1} design with defining relation **I = 123456** is of resolution VI. (Note that the resolution is expressed in Roman numerals.) The quarter fraction 2^{6-2} design with defining relation **I = 1234 = 3456 = 1256** is of resolution IV, and so on.

In this book we shall frequently use polynomial approximations of first or second degree. First-order designs to obtain data to estimate the coefficients in a first degree equation should therefore be of resolution at least III. This will ensure that no main effect is aliased with any other main effect, although main effects may (if R = III) be aliased with two-factor interactions. On the strict assumption that the response can be represented by a first degree equation, these two-factor interactions will all be zero. A resolution IV design provides added security because then no main effect would be aliased with either main effects or two-factor interactions.

We shall see later that second-order designs used to fit second degree polynomials can be formed by a suitable augmentation of two-level factorial or fractional factorial designs.

5.4. CONSTRUCTION OF 2^{k-p} DESIGNS OF RESOLUTION III AND IV

A design of resolution III can be constructed by *saturating* all the interactions of a full factorial design with additional factors. We illustrate this with an initial 2^3 design which can be saturated to produce a one-sixteenth replicate of a 2^7 design, namely a 2^{7-4} design. Consider again the columns of signs of Table 4.2 for the 2^3 factorial, now reproduced as Table 5.4.

If we use the **12** column to accommodate factor **4**, the **13** column similarly for factor **5**, the **23** column for **6**, and the **123** column for **7**, we have made the equivalances

$$\mathbf{4 = 12}, \quad \mathbf{5 = 13}, \quad \mathbf{6 = 23}, \quad \text{and} \quad \mathbf{7 = 123}. \qquad (5.4.1)$$

TABLE 5.4. A saturated, resolution III, 2^{7-4} design.

I	1	2	3	4 = 12	5 = 13	6 = 23	7 = 123
+	−	−	−	+	+	+	−
+	+	−	−	−	−	+	+
+	−	+	−	−	+	−	+
+	+	+	−	+	−	−	−
+	−	−	+	+	−	−	+
+	+	−	+	−	+	−	−
+	−	+	+	−	−	+	−
+	+	+	+	+	+	+	+

Thus the generators of the design are

$$I = 124 = 135 = 236 = 1237. \tag{5.4.2}$$

By multiplying these generators together in all possible ways, we obtain the defining relation for this 2^{7-4} design, and this leads to the alias relationships shown below. In this tabulation, we have ignored all interactions involving more than two factors for the sake of simplicity.

$$l_1 \rightarrow 1 + 24 + 35 + 67,$$

$$l_2 \rightarrow 2 + 14 + 36 + 57,$$

$$l_3 \rightarrow 3 + 15 + 26 + 47,$$

$$l_4 \rightarrow 4 + 12 + 56 + 37,$$

$$l_5 \rightarrow 5 + 13 + 46 + 27,$$

$$l_6 \rightarrow 6 + 23 + 45 + 17,$$

$$l_7 \rightarrow 7 + 34 + 25 + 16. \tag{5.4.3}$$

Eight-Run Designs for Fewer Than Seven Variables

When the number of variables examined is smaller than seven, useful designs can also be obtained by omitting one or more of the columns of the 2^{7-4} design or, more accurately, not using those columns for a variable. (The column is still used for estimating combinations of interactions but no longer specifies the levels of a variable.) The choice of which columns to drop can make use of the judgment of the investigator that only certain specific interactions are likely to occur. For example, suppose we have five factors and a particular one of these, which we designate as 1, is thought likely to interact with two others, 2 and 3, say; apart from this no other interactions are likely. Thus we would seek a design which can estimate the main effects of five factors, including 1, 2, and 3, and also the interactions 12 and 13, assuming all other interactions are zero. Looking at Table 5.4, we see that the required design is obtained by associating our five factors with the columns **1, 2, 3, 6,** and **7**. If our assumptions are correct, then the linear combinations will provide the following estimates:

$$l_1 \to 1, \qquad l_4 \to 12, \qquad l_6 \to 6,$$

$$l_2 \to 2, \qquad l_5 \to 13, \qquad l_7 \to 7.$$

$$l_3 \to 3, \tag{5.4.4}$$

For four variables, we can use variables 1, 2, 3, and 7 to obtain a design of resolution IV.

Adding Further Fractions

Sometimes, it will happen that the results from a highly fractionated design are ambiguous. In such a case we may wish to run a further fraction or fractions which will resolve the uncertainties created by the first. There are many possibilities; the following illustrations will indicate the general idea. We assume that the first fraction is the 2^{7-4} given by **I = 124 = 135 = 236 = 1237** with estimates as in Eq. (5.4.3).

Possibility 1

Suppose that, from a view of the results of the first fraction, it seems likely that a particular factor is dominating, and we decide we would like to add a further fraction and estimate the main effect of the dominating factor and all its interactions with other factors, clear of two-factor aliases. We can achieve this by choosing the second fraction whose generators are like those of the first fraction but with the sign of the variable of interest reversed everywhere. For example, if variable **1** is the designated variable we use

$$\mathbf{I} = -\mathbf{124} = -\mathbf{135} = \mathbf{236} = -\mathbf{1237} \tag{5.4.5}$$

and obtain the individual runs of the design by writing an initial 2^3 factorial in variables **1**, **2**, and **3** and then generating columns **4** $= -$ **12**, **5** $= -$ **13**, **6** $=$ **23**, and **7** $= -$ **123**. From the second fraction alone we can then estimate

$$l_1' \to 1 - 24 - 35 - 67,$$

$$l_2' \to 2 - 14 + 36 + 57,$$

$$l_3' \to 3 - 15 + 26 + 47,$$

$$l_4' \to 4 - 12 + 56 + 37,$$

$$l_5' \to 5 - 13 + 46 + 27,$$

$$l_6' \to 6 + 23 + 45 - 17,$$

$$l_7' \to 7 + 34 + 25 - 16, \tag{5.4.6}$$

which has the same basic alias structure as Eq. (5.4.3) but with certain signs reversed, as shown. It will now be seen that by combining the results from both fractions we can compute the estimates:

$$\frac{1}{2}(l_1 - l_1') \to 24 + 35 + 67, \qquad \frac{1}{2}(l_1 + l_1') \to 1,$$

$$\frac{1}{2}(l_2 - l_2') \to 14, \qquad \frac{1}{2}(l_2 + l_2') \to 2 + 36 + 57,$$

$$\frac{1}{2}(l_3 - l_3') \to 15, \qquad \frac{1}{2}(l_3 + l_3') \to 3 + 26 + 47,$$

$$\frac{1}{2}(l_4 - l_4') \to 12, \qquad \frac{1}{2}(l_4 + l_4') \to 4 + 56 + 37,$$

$$\frac{1}{2}(l_5 - l_5') \to 13, \qquad \frac{1}{2}(l_5 + l_5') \to 5 + 46 + 27,$$

$$\frac{1}{2}(l_6 - l_6') \to 17, \qquad \frac{1}{2}(l_6 + l_6') \to 6 + 23 + 45,$$

$$\frac{1}{2}(l_7 - l_7') \to 16, \qquad \frac{1}{2}(l_7 + l_7') \to 7 + 34 + 25.$$

$$\tag{5.4.7}$$

We thus see that, provided our assumption that interactions between more than two factors are negligible is true, we can isolate estimates of the main effect of factor 1 and all its interactions with other factors free from aliases.

Possibility 2

Suppose instead that we wished to run a second fraction that would provide all main effects clear of *all* two-factor interactions. This can be done by choosing a second fraction whose generators have the signs of all the variables reversed compared with the first one. For example, from Eq. (5.4.2) this procedure gives the fraction

$$I = -124 = -135 = -236 = 1237 \qquad (5.4.8)$$

whose columns are obtained by writing down a 2^3 design in the initial columns **1**, **2**, and **3** and then setting $4 = -12$, $5 = -13$, $6 = -23$, and $7 = 123$. Such a design produces the estimates

$$l_1' \rightarrow 1 - 24 - 35 - 67,$$

$$l_2' \rightarrow 2 - 14 - 36 - 57,$$

$$l_3' \rightarrow 3 - 15 - 26 - 47,$$

$$l_4' \rightarrow 4 - 12 - 56 - 37,$$

$$l_5' \rightarrow 5 - 13 - 46 - 27,$$

$$l_6' \rightarrow 6 - 23 - 45 - 17,$$

$$l_7' \rightarrow 7 - 34 - 25 - 16. \qquad (5.4.9)$$

Combining these estimates with those of Eq. (5.4.3) allows us to separate the main effects from the groups of two-factor interactions as follows:

$$\frac{1}{2}(l_1 + l_1') \rightarrow 1, \qquad \frac{1}{2}(l_1 - l_1') \rightarrow 24 + 35 + 67$$

$$\frac{1}{2}(l_2 + l_2') \rightarrow 2, \qquad \frac{1}{2}(l_2 - l_2') \rightarrow 14 + 36 + 57,$$

$$\cdots \qquad\qquad \cdots \qquad\qquad (5.4.10)$$

and so on.

Designs of Resolution IV via "Foldover"

The foregoing combination design is of resolution IV and so the method of switching all signs provides a method of building up a resolution IV design from one of resolution III. The switching can be carried out directly on the sign pattern as follows. Suppose **A** represents the block of signs shown in Table 5.4 (ignoring the **I** column for the moment) for the 2^{7-4} design $I = 124 = 135 = 236 = 1237$. Then the signs

$$\mathbf{A}$$

$$-\mathbf{A} \qquad\qquad (5.4.11)$$

provide 16 lines of signs which give a 2^{7-3} design of resolution IV. The $-\mathbf{A}$ block will be exactly the second set of runs as in "Possibility 2" above but will occur in a different order. Such a design is usually called a *foldover* design because we have "folded over" the signs, that is, switched them all. We can actually do better than this, and obtain a 2^{8-4} design of resolution IV by associating an eighth factor with the initial $\mathbf{I}_8$ column of ones. The folded design can then be written

$$\begin{array}{c} \text{runs 1--8} \\ \text{runs 9--16} \end{array} \begin{bmatrix} \mathbf{A} & \mathbf{I}_8 \\ -\mathbf{A} & -\mathbf{I}_8 \end{bmatrix}. \tag{5.4.12}$$

The design matrix of the 2^{8-4} resolution IV design obtained in this manner is shown in Table 5.5. Equation (5.4.13) shows the estimates that can be obtained from the columns whose subscripts are shown in the l's.

$$l_1 \to 1.$$
$$l_2 \to 2,$$
$$l_3 \to 3,$$
$$l_4 \to 4,$$
$$l_5 \to 5,$$
$$l_6 \to 6,$$
$$l_7 \to 7,$$
$$l_8 \to 8,$$
$$l_{12} \to 12 + 37 + 48 + 56 \qquad (B_1),$$
$$l_{13} \to 13 + 27 + 46 + 58 \qquad (B_2),$$
$$l_{14} \to 14 + 28 + 36 + 57 \qquad (B_3),$$
$$l_{15} \to 15 + 26 + 38 + 47 \qquad (B_1 B_2 B_3),$$
$$l_{16} \to 16 + 25 + 34 + 78 \qquad (B_2 B_3),$$
$$l_{17} \to 17 + 23 + 68 + 45 \qquad (B_1 B_2),$$
$$l_{18} \to 18 + 24 + 35 + 67 \qquad (B_1 B_3). \tag{5.4.13}$$

Exercise. Use the data shown in Table 5.5 and obtain the numerical estimates for the combinations of effects in Eq. (5.4.13). They are, in order, -0.7, -0.1, 5.5, -0.3, -3.8, -0.1, 0.6, 1.2, -0.6, 0.9, -0.4, 4.6, -0.3, -0.2, and -0.6. What do you conclude about the factors as a result of examining these figures? For additional details see Box, Hunter, and Hunter (1978), p. 399.

TABLE 5.5. A 2^{8-4} design of resolution IV, obtained here by the foldover technique

Run	Mold temperature	Moisture content	Holding pressure	Cavity thickness	Booster pressure	Cycle time	Gate size	Screw speed	Shrinkage
	1	2	3	4	5	6	7	8	y
1	−	−	−	+	+	+	−	+	14.0
2	+	−	−	−	−	+	+	+	16.8
3	−	+	−	−	+	−	+	+	15.0
4	+	+	−	+	−	−	−	+	15.4
5	−	−	+	+	−	−	+	+	27.6
6	+	−	+	−	+	−	−	+	24.0
7	−	+	+	−	−	+	−	+	27.4
8	+	+	+	+	+	+	+	+	22.6
9	+	+	+	−	−	−	+	−	22.3
10	−	+	+	+	+	−	−	−	17.1
11	+	−	+	+	−	+	−	−	21.5
12	−	−	+	−	+	+	+	−	17.5
13	+	+	−	−	+	+	−	−	15.9
14	−	+	−	+	−	+	+	−	21.9
15	+	−	−	+	+	−	+	−	16.7
16	−	−	−	−	−	−	−	−	20.3

Source: Table 5.5 is reproduced and adapted with permission from the table on p. 402 of *Statistics for Experimenters*: *An Introduction to Design, Data Analysis and Model Building*, by G. E. P. Box, W. G. Hunter, and J. S. Hunter, published by John Wiley & Sons, Inc., New York, 1978.

The foldover method of Eq. (5.4.12) is completely general and will provide a resolution IV design whenever **A** represents a block of signs designating a resolution III design.

Exercise.   Write down the 2^{3-1} resolution III design given by **I = 123**. Associate a fourth variable with I_4. Fold over the entire design. Confirm that the combination design is the resolution IV design **I = 1234**.

In the parentheses in Eq. (5.4.13) we show how to generate an interesting blocking arrangement which allows the 2^{8-4} design to be run in eight blocks of two runs each, without confounding main effects with blocks. The way in which this is done will be clear from Table 5.6 which shows three blocking variables associated with selected interactions. The eight possible sign combinations of the blocking variables provide the eight blocks. It is of interest to note that the two runs within any block are mirror images of each other. All the signs of any one run are reversed in its companion run.

TABLE 5.6. A 2_{IV}^{8-4} design in eight blocks of size two

2_{IV}^{8-4} design

Run	1	2	3	4	5	6	7	8	B₁ 12	B₂ 13	B₃ 14
1	−	−	−	+	+	+	−	+	+	+	−
2	+	−	−	−	+	+	+	+	−	−	−
3	−	+	+	−	−	−	+	+	−	+	+
4	+	+	+	+	−	−	−	+	+	−	+
5	−	−	+	+	+	−	+	+	+	−	−
6	+	−	+	−	−	+	+	+	−	+	−
7	−	+	−	+	+	+	−	+	−	−	+
8	+	+	−	−	+	+	+	−	+	+	+
9	+	+	+	−	−	−	+	−	+	+	−
10	−	+	+	+	+	+	−	−	−	−	−
11	+	−	−	+	+	−	−	−	−	+	+
12	−	−	−	−	+	−	+	−	+	−	+
13	+	+	−	+	−	+	−	−	+	−	−
14	−	+	−	−	+	+	+	+	−	+	−
15	+	−	+	+	−	+	−	+	−	−	+
16	−	−	+	−	−	+	+	−	+	+	+

Design rearranged in eight blocks

Block	1	2	3	4	5	6	7	8	B₁	B₂	B₃	Run
1	+	−	−	−	+	+	+	+	−	−	−	2
	−	+	+	+	+	+	−	−	−	−	−	10
2	−	−	+	+	+	−	+	+	+	−	−	5
	+	+	−	+	−	+	−	−	+	−	−	13
3	+	−	+	−	−	+	+	+	−	+	−	6
	−	+	−	−	+	+	+	+	−	+	−	14
4	−	−	−	+	+	+	−	+	+	+	−	1
	+	+	+	−	−	−	+	−	+	+	−	9
5	−	+	−	+	+	+	−	+	−	−	+	7
	+	−	+	+	−	+	−	+	−	−	+	15
6	+	+	+	+	−	−	−	+	+	−	+	4
	−	−	−	−	+	−	+	−	+	−	+	12
7	−	+	+	−	−	−	+	+	−	+	+	3
	+	−	−	+	+	−	−	−	−	+	+	11
8	+	+	−	−	+	+	+	−	+	+	+	8
	−	−	+	−	−	+	+	−	+	+	+	16

Source: Table 5.6 is reproduced and adapted with permission from the table on p. 406 of *Statistics for Experimenters: An Introduction to Design, Data Analysis and Model Building,* by G. E. P. Box, W. G. Hunter, and J. S. Hunter, published by John Wiley & Sons, Inc., New York, 1978.

161

Plackett and Burman Designs

Saturated resolution III two-level designs for examining $k = n - 1$ factors in $n = 2^q$ runs can be generated from any "core" 2^q factorial design by associating new variables with all of the interaction columns of the core design. Thus designs of this type for $n = 4, 8, 16, 32, 64, \ldots$ runs are easily generated. For intermediate values of n that are a multiple of 4 but *not* a power of 2, namely, $n = 12, 20, 24, 28, 36, \ldots$, saturated designs given by Plackett and Burman (1946) can be used. (In fact, the Plackett and Burman paper also covers $n = 2^q$ cases.) For $n = 12, 16, 20, 24$, for example, Plackett and Burman supply signs for the first row of the design matrix as follows:

```
n = 12    + + - + + + - - - + -
n = 16    + + + + - + - + + - - + - - -
n = 20    + + - - + + + + - + - + - - - - + + -
n = 24    + + + + + - + - + + - - + + - - + - + - - - -
```

To obtain the full design, cyclically permute $(n - 2)$ times the row given, and then add an nth row consisting of all minus signs.

Other constructions are different. For $n = 28$, for example, Plackett and Burman provide three 9×9 blocks of signs, A, B, and C, say, which, set side by side, form nine rows of width $k = 27$. Cyclic permutation to BCA and then CAB gives a total of 27 rows and a final row of minuses completes the $N = 28$ experimental runs. In all, Plackett and Burman (1946, 323–324) provide designs for n equal to a multiple of four for $n \leq 100$, except 92. For $n = 92$, see Baumert et al. (1962). Also see Draper (1985a).

All such designs are resolution III with complicated alias structures. Foldovers of the kinds illustrated in Eqs. (5.4.11) or (5.4.12) produce resolution IV designs. The mirror image pairs produced by foldover may be deployed in blocks of size two, if desired.

Exercise.   Write out the Plackett and Burman designs for $n = 12$ and 16. Take the first six columns of the latter and label them $1, 2, \ldots, 6$. What are the generators of this design. (Answer: Rearrange the first four columns in standard order. It is then clear that $5 = -14$ and $6 = 124$, so the generators are $I = -145 = 1246$.)

5.5.   DESIGNS OF RESOLUTION V AND OF HIGHER RESOLUTION

When five or more factors are being considered, very useful designs of resolution V or higher are provided by half-fractions, the 2^{k-1} designs. For example, the 2^{5-1} design with generator $I = 12345$ (i.e., we set $5 = 1234$

after writing down an initial 2^4 design in **1, 2, 3**, and **4**) is a 16-run design in which main effects are confounded only with four-factor interactions and two-factor interactions are confounded only with three-factor interactions. Thus, if the assumption that interactions between more than two factors are negligible is correct, the design allows unbiased estimation of all main effects and two-factor interactions. Designs of resolution five or more are of particular value as main building blocks of *composite designs* (presented later in this book) which allow estimation of all the terms in a second degree polynomial approximation. For eight factors or more, one-quarter replicates can produce designs of resolution V. Some useful resolution V designs and

TABLE 5.7. Construction and blocking of some designs of resolution V and higher so that no main effect or two-factor interaction is confounded with any other main effect or two-factor interaction

(1) Number of variables	(2) Number of runs	(3) Degree of fractionation	(4) Type of design	(5) Method of introducing "new" factors	(6) Blocking (with no main effect or 2fi confounded)	(7) Method of introducing blocks
5	16	$\frac{1}{2}$	2_V^{5-1}	$\pm 5 = 1234$	Not available	
6	32	$\frac{1}{2}$	2_{VI}^{6-1}	$\pm 6 = 12345$	Two blocks of 16 runs	$B_1 = 123$
7	64	$\frac{1}{2}$	2_{VII}^{7-1}	$\pm 7 = 123456$	Eight blocks of 8 runs	$B_1 = 1357$ $B_2 = 1256$ $B_3 = 1234$
8	64	$\frac{1}{4}$	2_V^{8-2}	$\pm 7 = 1234$ $\pm 8 = 1256$	Four blocks of 16 runs	$B_1 = 135$ $B_2 = 348$
9	128	$\frac{1}{4}$	2_{VI}^{9-2}	$\pm 8 = 13467$ $\pm 9 = 23567$	Eight blocks of 16 runs	$B_1 = 138$ $B_2 = 129$ $B_3 = 789$
10	128	$\frac{1}{8}$	2_V^{10-3}	$\pm 8 = 1237$ $\pm 9 = 2345$ $\pm \overline{10} = 1346$	Eight blocks of 16 runs	$B_1 = 149$ $B_2 = 12\overline{10}$ $B_3 = 89\overline{10}$
11	128	$\frac{1}{16}$	2_V^{11-4}	$\pm 8 = 1237$ $\pm 9 = 2345$ $\pm \overline{10} = 1346$ $\pm \overline{11} = 1234567$	Eight blocks of 16 runs	$B_1 = 149$ $B_2 = 12\overline{10}$ $B_3 = 89\overline{10}$

Source: Table 5.7 is reproduced and adapted with permission from the table on p. 408 of *Statistics for Experimenters: An Introduction to Design, Data Analysis and Model Building*, by G. E. P. Box, W. G. Hunter, and J. S. Hunter, published by John Wiley & Sons, Inc., New York, 1978.

TABLE 5.8. Two-level fractional factorial designs for k variables and N runs (numbe

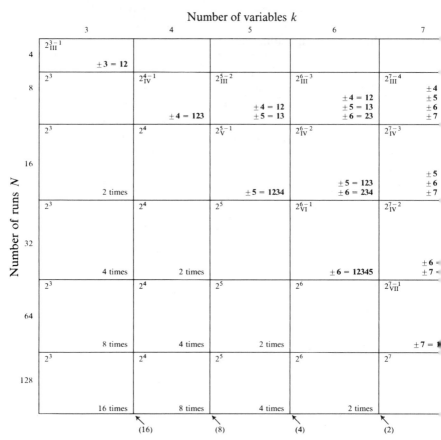

Number of variables k

Number of runs N	3	4	5	6	7
4	2_{III}^{3-1}				
	$\pm 3 = 12$				
8	2^3	2_{IV}^{4-1}	2_{III}^{5-2}	2_{III}^{6-3}	2_{III}^{7-4}
				$\pm 4 = 12$	± 4
			$\pm 4 = 12$	$\pm 5 = 13$	± 5
			$\pm 5 = 13$	$\pm 6 = 23$	± 6
		$\pm 4 = 123$			± 7
16	2^3	2^4	2_V^{5-1}	2_{IV}^{6-2}	2_{IV}^{7-3}
					± 5
				$\pm 5 = 123$	± 6
	2 times		$\pm 5 = 1234$	$\pm 6 = 234$	± 7
32	2^3	2^4	2^5	2_{VI}^{6-1}	2_{IV}^{7-2}
					$\pm 6 =$
	4 times	2 times		$\pm 6 = 12345$	$\pm 7 =$
64	2^3	2^4	2^5	2^6	2_{VII}^{7-1}
					$\pm 7 = \mathbf{1}$
	8 times	4 times	2 times		
128	2^3	2^4	2^5	2^6	2^7
	16 times	8 times	4 times	2 times	
	(16)	(8)	(4)	(2)	

Source: Table 5.8 is reproduced and adapted with permission from the table on p.
Statistics for Experimenters: *An Introduction to Design, Data analysis and Model Buila*
G. E. P. Box, W. G. Hunter, and J. S. Hunter, published by John Wiley & Sons, Inc

associated blocking arrangements are shown in Table 5.7. Note that the design resolution has been attached as a subscript, for example, 2_V^{8-2}.

A Table of 2^{k-p} Designs

Extensive tables of 2^{k-p} designs of various resolutions are available; see, for example, National Bureau of Standards (1957). A short but useful listing is given in Table 5.8.

parentheses represent replication)

Number of variables k

Number of runs N	8	9	10	11	
4					
8					
16	2_{IV}^{8-4} $\pm5 = 234$ $\pm6 = 134$ $\pm7 = 123$ $\pm8 = 124$	2_{III}^{9-5} $\pm5 = 123$ $\pm6 = 234$ $\pm7 = 134$ $\pm8 = 124$ $\pm9 = 1234$	2_{III}^{10-6} $\pm5 = 123$ $\pm6 = 234$ $\pm7 = 134$ $\pm8 = 124$ $\pm9 = 1234$ $\pm\overline{10} = 12$	2_{III}^{11-7} $\pm5 = 123$ $\pm6 = 234$ $\pm7 = 134$ $\pm8 = 124$ $\pm9 = 1234$ $\pm\overline{10} = 12$ $\pm\overline{11} = 13$	
32	2_{IV}^{8-3} $\pm6 = 123$ $\pm7 = 124$ $\pm8 = 2345$	2_{IV}^{9-4} $\pm6 = 2345$ $\pm7 = 1345$ $\pm8 = 1245$ $\pm9 = 1235$	2_{IV}^{10-5} $\pm6 = 1234$ $\pm7 = 1235$ $\pm8 = 1245$ $\pm9 = 1345$ $\pm10 = 2345$	2_{IV}^{11-6} $\pm6 = 123$ $\pm7 = 234$ $\pm8 = 345$ $\pm9 = 134$ $\pm\overline{10} = 145$ $\pm\overline{11} = 245$	$(\frac{1}{128})$
64	2_V^{8-2} $\pm7 = 1234$ $\pm8 = 1256$	2_{IV}^{9-3} $\pm7 = 1234$ $\pm8 = 1356$ $\pm9 = 3456$	2_{IV}^{10-4} $\pm7 = 2346$ $\pm8 = 1346$ $\pm9 = 1245$ $\pm10 = 1235$	2_{IV}^{11-5} $\pm7 = 345$ $\pm8 = 1234$ $\pm9 = 126$ $\pm\overline{10} = 2456$ $\pm\overline{11} = 1456$	$(\frac{1}{64})$
128	2_{VIII}^{8-1} $\pm8 = 1234567$	2_{VI}^{9-2} $\pm8 = 13467$ $\pm9 = 23567$	2_V^{10-3} $\pm8 = 1237$ $\pm9 = 2345$ $\pm10 = 1346$	2_V^{11-4} $\pm8 = 1237$ $\pm9 = 2345$ $\pm\overline{10} = 1346$ $\pm\overline{11} = 1234567$	$(\frac{1}{32})$
	(1)	$(\frac{1}{2})$	$(\frac{1}{4})$	$(\frac{1}{8})$	$(\frac{1}{16})$

York, 1978. For additional details see that book and "Minimum aberration 2^{k-p} designs," by A. Fries and W. G. Hunter *Technometrics*, **22**, 1980, 601–608, The numbers in parentheses show the replication factors in a "North-West" direction.

Calculation of Effects Using Yates' Method

In the foregoing discussion, we have shown how two-level fractional factorial designs requiring $2^{k-p} = 2^q$ runs may be generated by first writing down a complete 2^q factorial as a "core" design and then using interaction columns of the 2^q design to accommodate additional factors. To analyze the resulting fractional factorial, Yates' method may be used to compute the $2^q - 1$ contrasts *from the complete core factorial* in the usual way. The string of effects and interactions estimated by each of the $2^q - 1$ contrasts is then obtained from the alias pattern for the design. (See also Berger, 1972.)

5.6. APPLICATION OF FRACTIONAL FACTORIAL DESIGNS TO RESPONSE SURFACE METHODOLOGY

Suppose the expected response $E(y) = \eta$ is a function of k predictor variables $x_1, x_2, \ldots, x_k$, coded so that the center of the region of interest is at the origin $(0, 0, \ldots, 0)$. Consider a Taylor's series expansion

$$E(y) = \eta = \eta_0 + \sum_{i=1}^{k} \left[\frac{\partial \eta}{\partial x_i} \right]_0 x_i + \frac{1}{2} \sum_{i=1}^{k} \sum_{j=1}^{k} \left[\frac{\partial^2 \eta}{\partial x_i \, \partial x_j} \right]_0 x_i x_j + \cdots$$

(5.6.1)

where the subscript zero indicates evaluation at the origin $(0, 0, \ldots, 0)$. If we ignore terms of order higher than the first, the expansion yields the first degree (or first-order, or planar) approximation

$$\eta = \beta_0 + \sum_{i=1}^{k} \beta_i x_i.$$

(5.6.2)

If, in addition, we retain terms of second degree, we obtain the second degree (or second-order, or quadratic) approximation

$$\eta = \beta_0 + \sum_{i=1}^{k} \beta_i x_i + \sum_{i=1}^{k} \sum_{j \geq i}^{k} \beta_{ij} x_i x_j.$$

(5.6.3)

We shall use these approximations a great deal in subsequent chapters and will refer to designs suitable for collecting data to estimate the β parameters in Eqs. (5.6.2) and (5.6.3) as *first-order designs* and *second-order designs*, respectively.

Clearly, if two-level fractional factorial designs are to be used as first-order designs, they must have resolution of at least III so that, if model (5.6.2) is valid, the main effect contrasts of the design will provide estimates of all the β_i unconfounded with one another.

When a two-level design is used as a building block in second-order composite designs, it is usually* of resolution at least V. Then, if model (5.6.3) is valid, the main effect and two-factor interaction contrasts of the two-level design provide estimates of all main effect coefficients β_i and of all

*However, also see Section 15.5.

interaction coefficients β_{ij} $(i \neq j)$, unconfounded with one another. (See also Sections 9.2, 13.8, 14.3, and 15.3.)

Because we are never sure about the adequacy of an approximation of selected degree (or order), we shall need to check lack of fit as well as to estimate the model parameters. Resolution III designs that are not saturated can allow estimation of suspect interactions. Also (as we shall see in Chapter 9) the addition of center points makes possible a general curvature check. Resolution IV designs provide safer first-order designs because first-order coefficients will not then be confused with any two-factor interactions. Although the latter are ignored in a first-order fitting, they may yet exist; a resolution IV design will enable combinations of them to be estimated as a check of lack of fit, if desired.

5.7. PLOTTING EFFECTS FROM FRACTIONAL FACTORIALS ON PROBABILITY PAPER

To distinguish real effects from noise, estimates of effects from fractional factorial designs may be plotted on normal probability paper (see Appendix 4B) in just the same way as for full factorial designs.

Numerical Example, 2_{IV}^{8-4} Design

Sixteen observations were obtained from a 2_{IV}^{8-4} design given by Box, Hunter, and Hunter (1978, Table 12.11). The mean of these was $\bar{y} = 19.75$ and there were 15 estimates in addition, namely of:

1. Eight main effects: $-0.7, -0.1, 5.5, -0.3, -3.8, -0.1, 0.6, 1.2$.
2. Seven sets of four two-factor interactions: $-0.6, 0.9, -0.2, -0.4, -0.6, -0.3, 4.6$

These effects are plotted on normal probability paper in Figure 5.1. The plot suggests that two main effects (of variables 3 and 5) and one combination of two-factor interactions $(15 + 26 + 38 + 47)$ are distinguishable from the noise.

EXERCISES

5.1. Are the following statements true or false, for a 2_R^{k-p} design?
   (a) $k \geq R$
   (b) $k \geq R + p$
   Give, in a few words, reasons for your answers.

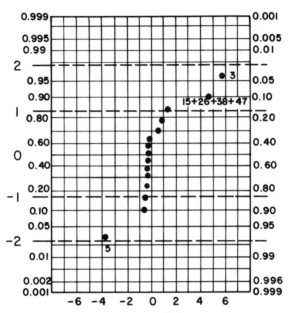

FIGURE 5.1. Plot of 15 estimated effects from a 2^{8-4} factorial on normal probability paper.

5.2. A 2_{III}^{5-2} design with generators $\mathbf{I} = \mathbf{1234} = \mathbf{135}$ is run.

(a) If it is *known* that factors 1, 2, and 3 all act independently of one another, and that factors 3, 4, and 5 all act independently of one another, what (ignoring interactions of three or more factors) can we estimate?

(b) Can you suggest a better 2_{III}^{5-2} design for these factors, or is the one given the best available?

5.3. For the construction of a one-quarter replicate of a 2^5 factorial, that is, a 2^{5-2} factorial, two generating relations are proposed:

(a) $\mathbf{I} = \mathbf{ABCDE} = \mathbf{BCDE}$

(b) $\mathbf{I} = \mathbf{ABC} = \mathbf{BCDE}$

Which design would you prefer, and why?

5.4. (a) Using the principle of foldover, show how the 2_{III}^{7-4} design generated by

$$\mathbf{I} = \mathbf{1234} = \mathbf{125} = \mathbf{146} = \mathbf{247}$$

can provide a 2_{IV}^{8-4} design.

(b) What are the generators of the new design?

(c) If this second design is again folded, but the **I** column is *not* folded with it this time, what type is the final design obtained?

5.5. Consider the 2_{IV}^{8-4} design generated by $\mathbf{I} = \mathbf{1235} = \mathbf{1246} = \mathbf{1347} = \mathbf{2348}$.

(a) By using a column that normally provides an estimate of the sum of four two-factor interactions, split the design into two blocks.

(b) Retaining two-factor interactions of all kinds show what estimates are available.

(c) If the block variable does not interact, show what estimates are available.

5.6. (a) Obtain a design for examining six variables in four blocks each containing four runs.

(b) What are the generators for the design (including blocking)?

(c) Find the defining relation (including blocking).

(d) Show what estimates can be found, assuming interactions between three or more factors are zero and that block variables do not interact with ordinary variables.

(*Hint*: Consider the 2_{IV}^{6-2} design generated by $I = 1235 = 1246$. Associate $B_1 = 12$, $B_2 = 13$, so that $B_1 B_2 = 23$. The design is then generated by $I = 1235 = 1246 = 12B_1 = 13B_2$, including blocking. In developing the estimates remember that although $B_1 B_2 5$, for example, looks like a *three*-factor interaction, it must be regarded as a *two*-factor interaction between the blocking variable $B_1 B_2$ and the ordinary variable 5.)

(e) Is the choice suggested in the hint a good one? Explain the reasons for your answer. If you decide that better choices exist, provide one.

5.7. (a) Obtain a design for examining six variables in eight blocks each containing two runs so that the estimates of main effects are not influenced by block effects.

(b) What are the generators for the design (including blocking)?

(c) Find the defining relation.

(d) Show what estimates can be found assuming interactions between three or more factors are zero and that block variables do not interact with ordinary variables.

(*Hint*: Consider the 2_{IV}^{6-2} design generated by $I = 1235 = 1246$. Associate $B_1 = 12$, $B_2 = 13$ (so that $B_1 B_2 = 23$), and $B_3 = 14$. Then $B_1 B_3 = 24$, $B_2 B_3 = 34$, and $B_1 B_2 B_3 = 1234$. The blocks are now defined. Remember that, for example, $B_1 B_2 B_3 5$ has the status of a two-factor interaction.)

5.8. An experimenter begins an investigation with a 2^{5-2} design using generators $4 = 123$ and $5 = 23$. Later he performs the eight foldover runs obtained by reversing all the signs in the five-variable design table.

(a) What is the resolution of the combined 16-run design? What generators would produce this design?

(b) What better design might the experimenter have used had he known in advance that 16 runs would be made?

5.9. Construct a 16-run two-level fractional factorial design in six factors such that all main effects and all two-factor interactions associated with factor 1 are estimable.

5.10. If the design $I = 124 = 135 = 236$ is followed by the design in which variables **1** and **2** have their signs switched, what is the alias structure of the 16-run design so formed?

5.11. An experimenter needs a 16-run two-level fractional factorial design that will estimate the main effects $1, 2, \ldots, 6$ of six variables as well as the interactions 12, 123, 13, 14, 15, 16, 23, 24 between them, assuming all interactions of three or more factors can be ignored except for 123. Does such a design exist? If so, what is its defining relation?

5.12. Verify that the 2^{8-4} resolution IV design can be used in all the following alternative ways. (Assume that all interactions between three or more factors can be neglected.)

(a) As a "main effects" design to obtain the main effects of eight factors clear of all two factor interactions.

(b) As in (a) but blocked into two, or four, or eight blocks.

(c) If there are three major variables, all of whose main effects, two-factor interactions, and three-factor interaction are needed, and eight minor variables all of whose interactions can be neglected, the design can be used to estimate all these. (*Hint*: Look at Eq. (5.4.13) and associate each $\mathbf{B}_i$ with a major variable $\mathbf{M}_i$, $i = 1, 2, 3$. Remember that all the two-factor interactions involving numbers, e.g., 12, 13, etc., are zero. Take the numbered variables as the minor variables.)

(d) As in (c) but with eight minor variables, two major variables, and in two blocks of eight runs each. (*Hint*: Replace $\mathbf{B}_1$, $\mathbf{B}_2$ and $\mathbf{B}_3$ in Eq. (5.4.13) by $\mathbf{M}_1$, $\mathbf{M}_2$, and $\mathbf{B}$.)

(e) As a design for eight minor variables and one major variable in four blocks of four runs each.

(f) As a screening design that will provide a complete factorial, twice over, in *any* set of three variables, if the other five variables have no effect. (*Hint*: If five variables are not effective we can "cross them out" in the defining relationship that leads to the alias structure. Since no words of length three or smaller appear in this—because the design is of resolution IV—all words are "deleted," no matter which five variables are chosen, implying that a full factorial design remains in the other three variables. Because the original design had 16 runs, the 2^3 is necessarily replicated.)

(g) As a screening design that will provide a complete factorial in some sets of four variables, or a 2^{4-1}, twice over, in all other sets of four variables, provided the remaining four variables *not* in the considered set have no effect. (*Hint*: If the chosen four variables form a word in the defining relationship, a replicated 2^{4-1} occurs. Otherwise we get a 2^4.)

5.13. Select a set of suitable generators for a 2_V^{10-3} design and write out the defining relation which provides the alias structure. Select a suitable blocking generator for division into two blocks. (*Hint*: See Table 5.7.)

5.14. A 2_V^{5-1} design was used to examine five factors A, B, C, D, and E. It was assumed that all interactions involving three or more variables could be neglected. From the resulting data, the following estimates (rounded to the nearest unit) were obtained:

$A \leftarrow$	0	$AB \leftarrow$	17	$BD \leftarrow$	-4
$B \leftarrow$	-8	$AC \leftarrow$	3	$BE \leftarrow$	8
$C \leftarrow$	14	$AD \leftarrow$	5	$CD \leftarrow$	14
$D \leftarrow$	67	$AE \leftarrow$	-2	$CE \leftarrow$	-5
$E \leftarrow$	4	$BC \leftarrow$	8	$DE \leftarrow$	0

Plot these results on normal probability paper and identify effects that appear to be important. Remove these, replot the rest, and estimate σ, where $V(y_i) = \sigma^2$. See Appendix 4B for additional details.

5.15. Identify the design given in Table E5.15 and perform an appropriate analysis. Assume $V(y_i) = \sigma^2 = 2$.

What factors appear to influence the response? What features need to be further elucidated? If you were allowed to perform just four more runs, what would they be? Why?

TABLE E5.15

x_1	x_2	x_3	x_4	y
1	-1	-1	-1	105
-1	1	-1	-1	107
-1	-1	1	-1	102
-1	-1	-1	1	104
1	1	1	-1	114
1	1	-1	1	111
1	-1	1	1	105
-1	1	1	1	107

5.16. What design is given in Table E5.16? Perform an appropriate factorial analysis of the data.

Would you like to have additional data to clarify what you have seen in this experiment? If four more runs were authorized what would you suggest? Why? If eight more runs were authorized what would you suggest? Why?

5.17. Consider a 2_V^{5-1} fractional factorial design defined by $I = -12345$. Suppose we wish to choose two blocking variables $\mathbf{B_1}$ and $\mathbf{B_2}$ so that the design can be divided into four blocks. Can this be done in such a way that all main effects

TABLE E5.16

x_1	x_2	x_3	x_4	x_5	y
-1	-1	1	-1	-1	18, 15
1	-1	-1	-1	1	16, 18
-1	1	-1	-1	1	18, 17
1	1	1	-1	-1	15, 16
-1	-1	1	1	1	17, 19
1	-1	-1	1	-1	31, 30
-1	1	-1	1	-1	18, 18
1	1	1	1	1	30, 27

and all two-factor interactions are *not* confounded with blocks? If your answer is yes, write out the design and show what estimates can be made.

5.18. (a) You have convinced an experimenter that she should use a 2^{7-1} design. Each run takes about 20 min, and 16 runs is the most she wishes to attempt during one day's work. Thus she wishes the design divided into four blocks of 16 runs each. Assume that the design will be generated using **I = 1234567** and blocked using blocking generators $\mathbf{B}_1$ = **1357**, $\mathbf{B}_2$ = **1256**. Write out the 16 runs in the $(\mathbf{B}_1, \mathbf{B}_2) = (+, +)$ block.

(b) After these 16 runs have been completed, the experimental unit suffers a breakdown. It will not be repaired until after the monthly meeting at which the experimenter has to make a report. Assuming that all interactions involving three or more of the seven factors are negligible, and that blocks do not interact with the factors, write out the 16 quantities that can be estimated using only the 16 runs currently available.

5.19. (Source: Toby J. Mitchell.) Ozzie Cadenza, holder of a Ph.D. in Statistics from the University of Wisconsin, and now owner and manager of Ozzie's Bar and Grill, recently decided to study the factors that influence the amount of business done at his bar.

At first, he did not know which factors were important and which were not, but he drew up the following list of six, which he decided to investigate by means of a fractional factorial experiment:

1. The amount of lighting in the bar.
2. The presence of free potato chips and chip dip at the bar.
3. The volume of the juke box.
4. The presence of Ozzie's favorite customer, a young lady by the name of Rapunzel Freeny. Miss Freeny was a real life-of-the-party type, continually chatting with the customers, passing around the potato chips, and so

on, all of which made Ozzie feel that she had a real effect on the amount of his bar business.

5. The presence of a band of roving gypsies, who had formed a musical group called the Roving Gypsy Band and who had been hired by Ozzie to play a limited engagement there.

6. The effect of the particular bartender who happened to be on duty. There were originally three bartenders, Tom, Dick, and Harry, but Harry was fired so that each factor in the experiment would have only two levels.

Plus and minus levels were assigned to these six factors as follows:

	−	+
1.	Lights are dim	Lights are bright
2.	No chips are at the bar	Chips are at the bar
3.	Juke box is playing softly	Juke box is blaring loudly
4.	Miss Freeny stays at home	Miss Freeny is there
5.	Gypsies are not there	Gypsies are there
6.	Tom is the bartender	Dick is the bartender

Ozzie decided to perform one "run" every Friday night during the cocktail hour (4:30 to 6:30 P.M.). He thought he should try a fractional factorial with as few runs as possible, since he was never quite sure just when the Roving Gypsy Band would pack up and leave. He finally decided to use a member of the family of 2_{III}^{6-3} designs with principal generating relation $I = 124 = 135 = 236$. (He had wanted to find a resolution III design which would be such that the juke box would never be blaring away while the gypsies were playing but found this requirement to be impossible.

(a) Why?

He did insist, however, that in *no* run of the experiment could variables **1**, **3**, and **5** attain their plus levels simultaneously. This restriction was made necessary by the annoying tendency of all the lights to fuse whenever the gypsies plugged in their electric zither at the same time the lights and the juke box were on full blast. Note that this restriction made it impossible for the principal member of the chosen family to be used.

(b) What members of the given family *does* this restriction allow?)

Ozzie settled on the generating relation. $I = 124 = -135 = 236$. The design matrix and the "response" (income in dollars) corresponding to each run are given in Table E5.19.

(c) Assuming third- and higher-order interactions are negligible, write down the estimates obtained from this experiment and tell what they estimate.

TABLE E5.19

1	2	3	4 = 12	5 = −13	6 = 23	y
−	−	−	+	−	+	265
+	−	−	−	+	+	155
−	+	−	−	−	−	135
+	+	−	+	+	−	205
−	−	+	+	+	−	195
+	−	+	−	−	−	205
−	+	+	−	+	+	125
+	+	+	+	−	+	315

(d) In a few sentences, tell what fraction you would perform next and why.

Being partial to Miss Freeny and encouraged by the results of the first fraction, Ozzie chose a second fraction which would give unaliased estimates of her and each of her interactions. The results of this second fraction, given with variables **1**, **2**, and **3** in standard order, were 135, 165, 285, 175, 205, 195, 295, 145.

(e) Write down the estimates obtained by combining the results of both fractions.

(f) In the light of these results, was the choice of the second fraction a wise one?

(g) Offer a brief conjecture which might explain the presence and direction of the interactions involving Miss Freeny.

5.20. (Source: Designed experiments, by K. R. Williams, *Rubber Age*, **100**, August 1968, 65–71.) Table E5.20 shows a 28-run Plackett and Burman type design for 24 variables together with the responses obtained from each run. Except for the last column, the CP row (cross-product) is the result of taking CPs of each column of signs with the y values and the EF row ("effect") is the CP row divided by 14 and rounded. In the last column, $\Sigma y = 3065$ and $\bar{y} = 109$. Confirm one or more of the effects calculations. If, a priori, it were believed that at least one-third of the factors were ineffective, but not known which factors these were, which factors would be regarded as effective ones?

5.21. A 2^3 factorial design has been partially replicated as shown in Table E5.21. Analyze the data using regression methods, with a "full factorial analysis" type model.

5.22. (Source: Sucrose-modified phenolic resins as plywood adhesives, by C. D. Chang and O. K. Kononenko, *Adhesives Age*, **5**, (7), July 1962, 36–40. Material reproduced and adapted with the permission of *Adhesives Age*. The

TABLE E5.20

Run Number	1	2	3	4	5	6	7	8	9	10	11	12	13	14	15	16	17	18	19	20	21	22	23	24	Response y
1	+	+	+	−	−	−	+	+	+	+	+	−	+	−	−	+	+	+	−	+	−	−	−	+	133
2	−	+	−	−	−	−	+	+	+	+	−	+	+	−	+	−	+	+	+	+	−	+	+	−	49
3	+	−	−	−	−	−	+	−	+	−	−	−	−	+	+	−	+	+	+	−	+	+	−	−	62
4	+	+	−	+	+	−	−	−	−	+	−	−	+	+	+	+	+	+	+	−	+	+	+	−	45
5	+	+	−	−	+	+	−	−	−	+	−	−	+	−	−	−	−	+	−	+	−	−	−	+	88
6	+	+	−	+	+	+	+	−	−	−	+	−	+	+	+	+	+	+	+	+	−	−	−	−	52
7	−	−	+	+	+	+	+	+	+	+	+	−	+	+	+	+	+	+	+	+	+	+	+	+	300
8	−	−	+	+	+	+	+	+	+	+	+	−	+	−	−	+	+	−	−	+	−	−	+	−	56
9	−	−	+	+	+	+	+	−	+	−	−	−	+	−	+	+	−	−	+	+	−	+	+	+	47
10	+	−	+	−	−	−	−	−	−	−	−	+	+	−	−	+	+	+	−	+	−	−	−	−	88
11	−	−	+	+	+	−	+	−	+	−	+	−	+	+	+	+	+	+	+	−	+	+	+	−	116
12	+	+	+	+	+	+	+	−	+	+	−	+	+	+	−	+	+	−	+	−	+	+	+	−	83
13	−	+	−	−	−	+	−	+	−	−	−	−	+	−	+	+	−	−	+	+	−	−	+	−	193
14	+	−	+	+	+	+	+	+	+	+	+	−	+	−	+	−	+	+	+	+	−	+	+	+	230
15	+	+	+	−	−	−	−	+	−	−	+	+	+	+	+	+	+	+	+	−	−	−	−	+	51
16	−	+	+	+	+	+	+	+	+	+	+	+	+	+	+	+	+	+	+	+	+	+	+	−	82
17	+	−	+	−	−	−	−	−	+	−	+	+	+	+	+	+	−	−	−	+	+	+	+	+	32
18	+	+	+	+	+	+	+	+	+	+	−	+	+	−	+	−	−	−	+	−	−	−	−	−	58
19	+	−	+	−	−	−	−	−	−	−	+	−	+	−	−	−	−	+	+	+	−	+	+	+	201
20	+	+	+	+	+	+	+	+	+	+	+	+	−	+	+	+	−	+	+	+	+	+	+	+	56
21	+	+	−	+	+	+	+	−	+	−	+	+	+	−	−	+	+	+	+	−	+	+	+	+	97
22	+	+	+	−	+	+	+	+	+	−	+	+	+	+	+	−	−	−	−	−	+	−	−	+	53
23	−	−	−	−	+	−	+	−	−	−	+	+	−	−	+	−	−	+	+	+	+	+	+	+	276
24	+	+	−	−	+	+	+	+	+	+	+	+	+	+	−	+	−	−	+	+	+	+	+	−	145
25	+	+	−	+	+	+	+	−	+	+	+	+	+	+	−	+	−	+	−	+	+	−	+	−	130
26	−	+	−	−	−	+	+	+	+	−	+	−	−	+	+	−	−	−	−	−	−	−	+	−	55
27	+	+	+	−	−	+	−	−	−	−	+	+	−	+	+	+	+	−	−	+	−	+	−	−	160
28	+	−	+	−	−	−	−	−	−	−	+	+	−	+	+	+	+	−	−	+	−	+	−	−	127
CP	−365	−281	−95	511	175	−165	−189	415	213	199	−51	−221	−321	−433	−1209	−321	−599	−191	−49	−683	−117	−451	−167	−173	3065
EF	−26	−20	−7	37	13	−12	−14	30	15	14	−4	−16	−23	−31	−86	−23	−43	−14	−4	−49	−8	−32	−12	−12	109

[a]The level of run 8, factor 20 is here shown as +. In the source reference, the lower level appears; that seems to be a typographical error.

175

TABLE E5.21

x_1	x_2	x_3	y
−	−	−	46
+	−	−	61, 57
−	+	−	37, 45
+	+	−	56
−	−	+	37, 41
+	−	+	68
−	+	+	33
+	+	+	66, 68

data were obtained under a grant from the Sugar Research Foundation, Inc., now the World Sugar Research Organization, Ltd.) The object of a research study was to improve both the dry and wet strip-shear strength in pounds per square inch (psi) of Douglas fir plywood glued with a resin modified with sugar. Here, we shall examine only the dry strength data, however.

The first part of the investigation involved six reaction variables $x_1, x_2, \ldots, x_6$, each examined at two levels coded to -1 and 1. The variables are shown in Table E5.22a, together with the coding used for each variable.

A 16-run 2_{IV}^{6-2} design generated by $I = 1235 = 2346$ (so that $I = 1456$, also) was performed, and it provided the responses in Table E5.22b. (The y values are average values of six to eight samples per run; the fact that this provides observations whose variances may vary slightly is ignored in what follows.) Five additional center-point runs at $(0, 0, \ldots, 0)$ provided a pure error estimate $s_e^2 = 95.30$ for $V(y) = \sigma^2$.

Perform a factorial analysis on these data, assuming all interactions involving three or more factors can be neglected, and so confirm the results in Table E5.22c. Note, in the left column of that table, that "other" denotes groups of interactions of order three or more. These should represent "error"

TABLE E5.22a.   **Levels of the six predictor variables examined**

Variable (units), designation	Level -1	Level 1	Coding
Sucrose (g), S	43	71	$x_1 = (S - 57)/14$
Paraform (g), P	30	42	$x_2 = (P - 36)/6$
NaOH (g), N	6	10	$x_3 = (N - 8)/2$
Water (g), W	16	20	$x_4 = (W - 18)/2$
Temperature maximum (°C), T	80	90	$x_5 = (T - 85)/5$
Time at T°C, t	25	35	$x_6 = (t - 30)/5$

TABLE E5.22b. The design and the response values

x_1	x_2	x_3	x_4	x_5	x_6	y
-1	-1	-1	-1	-1	-1	162^a
1	-1	-1	-1	1	-1	146
-1	1	-1	-1	1	1	182
1	1	-1	-1	-1	1	133
-1	-1	1	-1	1	1	228
1	-1	1	-1	-1	1	143
-1	1	1	-1	-1	-1	223
1	1	1	-1	1	-1	172
-1	-1	-1	1	-1	1	168
1	-1	-1	1	1	1	128
-1	1	-1	1	1	-1	175
1	1	-1	1	-1	-1	186
-1	-1	1	1	1	-1	197
1	-1	1	1	-1	-1	175
-1	1	1	1	-1	1	196
1	1	1	1	1	1	173

[a] Shown as 172 in the source reference; however, according to Dr. O. K. Kononenko, the 172 may be a typographical error.

if interactions between three or more factors are negligible, and so should not be statistically significant under this tentative assumption. The designation $(124 +)$ stands for $(124 + 345 + 136 + 2456)$ while $(134 +)$ represents $(134 + 245 + 126 + 3456)$. The superscript a's indicate estimated effects that exceed 2.78(s.e.) in modulus. (The standard error is based on five average readings, and thus on 4 df. Ninety-five percent of the t_4 distribution lies within ± 2.78 so that estimates exceeding 2.78 (s.e.) in modulus are those whose true values are tentatively judged to be nonzero.)

At this stage, it was decided to explore outward on a path of steepest ascent, ignoring the two significant interactions and this was successful, leading to the following tentative conclusions:

1. Variables x_5 and x_6 have shown little effect and may be dropped in future runs.

2. A good balance of dry and wet strength values was achieved around $(S, P, N, W) = (47, 36.6, 8.8, 23.8)$, and this point will be the center for a subsequent experiment.

3. A 27-run three-level rotatable design with blocking, equivalent to a rotated "cube plus star" design (see Section 15.4 and/or Box and Behnken, 1960a) with added center points, will be performed next, and a

TABLE E5.22c.   The estimated factorial effects from the 2^{6-2} experiment

Effect name	Estimate
1	-34.375^a
2	11.625
3	28.375^a
4	1.125
5	1.875
6	-10.625
12 + 35	6.375
13 + 25	-10.875
14 + 56	15.875^a
15 + 23 + 46	-6.375
16 + 45	-14.875^a
24 + 36	3.875
26 + 34	-7.375
other (124 +)	6.125
other (134 +)	6.875
s.e.	4.88
2.78 (s.e.)	13.57

^{a}Significant at $\alpha = 0.05$.

second-order surface will be fitted to the y values thus obtained. (For the continuation of this work, see Exercise 7.23. For steepest ascent, see Chapter 6.)

5.23. (Source: A first-order five variable cutting-tool temperature equation and chip equivalent, by S. M. Wu and R. N. Meyer, *J. Eng. Indus., Trans. ASME, Series B*, **87**, 1965, 395–400. Adapted with the permission of the copyright holder, The American Society of Mechanical Engineers, Copyright © 1965.) Fit a first-order model $y = \beta_0 + \beta_1 x_1 + \cdots + \beta_5 x_5 + \varepsilon$ by least squares to the data in Table E5.23. The design is a 2_V^{5-1} design generated by $x_1 x_2 x_3 x_4 x_5 = 1$ where the x's are coded predictor variables. The actual response is T (temperature in °F) but the model should be fitted in terms of $y = \ln T$. Treat the two values at each set of conditions as repeat runs for pure error purposes. Test the model for lack of fit and, if none is revealed, test if all x's are needed. Decide what model is suitable for these data, check the residuals, especially against the testing order, and interpret what you have done. (Some complications stemming from blocking have been ignored in this exercise modification; see the source reference for additional details.)

5.24. Is it possible to obtain a 2_R^{k-p} design which provides clear estimates of the main effects of nine variables in 16 runs, assuming all interactions involving three or more variables are zero? If yes, provide the design. If no, provide the best 16-run design in the circumstances and specify how many two-factor

TABLE E5.23. Coded cutting conditions and temperature results

Testing order	x_1	x_2	x_3	x_4	x_5	Temperature results T_1	T_2
4	-1	-1	-1	-1	1	462	498
16	1	-1	-1	-1	-1	743	736
9	-1	1	-1	-1	-1	714	681
8	1	1	-1	-1	1	1070	1059
11	-1	-1	1	-1	-1	474	486
5	1	-1	1	-1	1	832	810
1	-1	1	1	-1	1	764	756
15	1	1	1	-1	-1	1087	1063
14	-1	-1	-1	1	-1	522	520
6	1	-1	-1	1	1	854	828
2	-1	1	-1	1	1	773	756
10	1	1	-1	1	-1	1068	1063
7	-1	-1	1	1	1	572	574
12	1	-1	1	1	-1	831	856
13	-1	1	1	1	-1	819	813
3	1	1	1	1	1	1104	1092

interactions would need to be zero to achieve the requirement of clear estimates of main effects.

5.25. Variables **4** and **7** are dropped from the 2_{III}^{7-4} design generated by **I = 124 = 235 = 136 = 1237**. What is the resulting design?

5.26. The data in Table E5.26 were obtained from the records of a medium-sized dairy company seeking to improve sales. Explain what design is being used,

TABLE E5.26

1	2	3	4	5	Response
Daily deliveries?	Roundsman wears uniform?	Older man?	City Round?	Delivery before noon?	Sales rating
NO	NO	YES	YES	NO	4
NO	NO	YES	NO	YES	3
YES	YES	YES	YES	YES	0
NO	YES	NO	YES	NO	3
YES	NO	NO	YES	YES	0
YES	YES	YES	NO	NO	4
YES	NO	NO	NO	NO	4
NO	YES	NO	NO	YES	3

analyze the data, and suggest what sensible next step(s) should be considered by the company.

5.27. (a) A 2_{III}^{7-4} design with generators $I = -126 = 135 = -237 = -1234$ is used in a factory investigation of the time taken to pack 100 standard items. The variables investigated are given in Table E5.27. With variables **1**, **2**, and **3** in standard order, the observed packing times were, in minutes, 46.1, 55.4, 44.1, 58.7, 56.3, 18.9, 46.4, 16.4. Perform a standard analysis and state tentative conclusions.

(b) It was decided to perform a second fractional factorial, obtained from the first by reversing the signs of all variables. For these eight experiments, the packing times were (with **1**, **2** and **3** in order $+ + +, - + +, + - +, - - +, + + -, - + -, + - -, - - -,$) 41.8, 40.1, 61.5, 37.0, 22.9, 34.1, 17.7, 42.7. Perform a standard analysis.

(c) Combine the two fractions given above to obtain estimates of main effects and two factor interactions. Comment on the results. If you were the factory owner, what would you do?

TABLE E5.27

Variable	$(-1, 1)$
1: Foreman	(absent, present)
2: Sex of packer	(man, woman)
3: Time of day	(morning, afternoon)
4: Temperature	(normal, high)
5: Music	(none, piped in)
6: Age of packer	(under 25, 25 or over)
7: Factory location	(Los Angeles, New York)

TABLE E5.28

		Variable Number				
1	**2**	**3**	**4**	**5**	**6**	Response, y
+	+	−	−	−	−	29
−	+	+	−	+	−	27
+	−	−	+	+	−	42
−	+	−	+	−	+	0
−	−	−	−	+	+	30
+	+	+	+	+	+	0
−	−	+	+	−	−	39

5.28. An industrial spy is disturbed while photographing experimental records. On later developing his film, he discovers he has only seven runs of a 2_R^{k-p} design, given in Table E5.28. How can he estimate the effects from the incomplete data? Do this. The information from these tests is vital to the spy's employer who is prepared to run another 2^{k-p} block if the results from the photographed data indicate this to be a worthwhile endeavor. Should additional tests be run? If so, what would you recommend?

CHAPTER 6

The Use of Steepest Ascent
to Achieve Process Improvement

If we attempted to fit an empirical function such as a polynomial over the whole *operability region* (that is, over the whole region within which the studied system could be operated) a very complex function would usually be needed. The fitting of such a function could involve an excessive number of experiments. The exploration of the whole operability region is, however, almost never a feasible or sensible objective. First, the extent of this region is almost never known and second, the experimenter can often safely dismiss as unprofitable whole areas of the region where experiments could, theoretically, be conducted. What he usually wants to do, with good reason, is to explore a smaller subregion *of interest* which often changes as the investigation progresses. He can expect that, within this smaller subregion, a fairly simple graduating equation will be representationally adequate.

Now the mathematical procedure of fitting selects, from all possible surfaces of the degree fitted, that which approximates the responses *at the experimental points* most closely (in the least squares sense). The features of the fitted surface at points remote from the region of the experimental design need not, and usually would not, bear any resemblance to the features of the actual surface. We can therefore expect our approximation to be useful only in the immediate neighborhood of the current experimental region. If, for example, we were already close to a maximum, we might represent its main features approximately by fitting, to a suitable locally placed set of experimental points, an equation of only second degree, possibly in transformed variables. However, particularly if the system were being investigated for the first time, starting conditions would often not be very close to such a maximum. Thus, the experimenter often first needs some *preliminary procedure*, to bring him to a suitable point where the second degree equation can most usefully be employed. *One* such preliminary procedure is the one factor at a time method. An alternative which, in the authors' opinion, is usually more effective and economical in experiments (at least in the fields of application in which it has been used) is the "steepest ascent" method. This preliminary procedure, which may later be followed by the fitting of an equation of second degree (possibly in transformed variables), is the basis of a sequential experimentation method which we discuss in this chapter and which has proved extremely effective in a wide variety of applications.

6.1. WHAT STEEPEST ASCENT DOES

For illustration, consider the investigation of a chemical system whose percentage yield depends on the three predictor variables time, temperature, and concentration; a possible situation is shown in Figure 6.1. The overall objective is to find settings of the variables which will result in improved yields. Suppose we are at an early stage of investigation when considerable improvement is possible. Then the planar contours of a first degree equation can be expected to provide a fair approximation in the immediate region of a point such as P, which is far from the optimum. A direction at right angles to these contour planes is the direction of *steepest ascent*, if it points toward higher yield values. (The opposite direction is that of *steepest descent*, needed for *reducing*, and so improving, a response such as "impurity" for example.) In practice, the first degree approximating equation must be estimated via experiments. This may be done by running a design such as that indicated by the dots around the point P.

Exploratory runs performed along the path of steepest ascent and indicated by stars in Figure 6.1 will normally show increasing yield along the path. The best point found, or an interpolated estimated maximum point on the path could then be made the base for a new first-order design from which a further advance might be possible.

The direction of ascent we shall calculate is the steepest when the response surface is scaled in the units of the design as it is, for example, in Figure 6.1. The effects of scale changes are discussed in Appendix 6A.

It will usually be found that, after one or perhaps two applications of steepest ascent, first-order effects will no longer dominate and the first-order approximation will be inadequate. At this stage, further steepest ascent progress will not be possible and more sophisticated second-order methods, discussed in Chapters 7 and 9, should be applied. The steepest ascent technique is particularly effective for the initial investigation of a new process. Such an investigation typically takes place on the laboratory scale.

For an already operating process arrived at after much development work, designs performed in the vicinity of the currently known best process

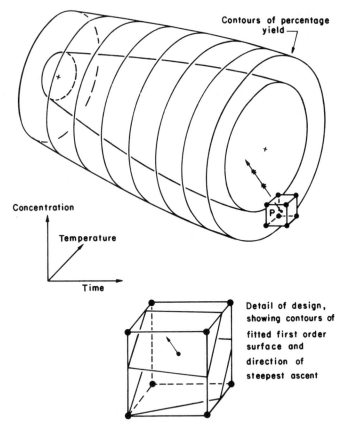

FIGURE 6.1.   A stage of the steepest ascent procedure.

may show hardly any first-order effects at all, because these have already been exploited to obtain the current conditions. In this situation then, no initial large gains may be possible using steepest ascent but, again, the more sophisticated second-order methods of Chapter 9 may still lead to substantial improvement.

6.2. AN EXAMPLE.   IMPROVEMENT OF A PROCESS USED IN DRUG MANUFACTURE

In laboratory experiments on a reaction taking place between four reagents A, B, C, and D, the amount of A was kept constant but five factors, time of reaction (ξ_1), temperature of reaction (ξ_2), amount of B (ξ_3), amount of C

TABLE 6.1. Factors ξ_i and coded levels x_i

ξ_i	Selected levels of x_i		
	Lower(-1)	Upper($+1$)	
ξ_1	6 h	10 h	so that $x_1 = (\xi_1 - 8)/2$
ξ_2	85°C	90°C	so that $x_2 = (\xi_2 - 87.5)/2.5$
ξ_3	30 cm³	60 cm³	so that $x_3 = (\xi_3 - 45)/15$
ξ_4	90 cm³	115 cm³	so that $x_4 = (\xi_4 - 102.5)/12.5$
ξ_5	40 g	50 g	so that $x_5 = (\xi_5 - 45)/5$

(ξ_4), and the amount of D (ξ_5) were varied as in Table 6.1. A 2_V^{5-1} design with $\mathbf{I} = \mathbf{12345}$ was used and the design and resulting percentage yields, y, were as shown on the left of Table 6.2.

On the assumption that the second degree polynomial

$$y = \beta_0 + \sum_{i=1}^{5} \beta_i x_i + \sum_{i=1}^{5} \beta_{ii} x_i^2 + \sum_{i<j}^{5}\sum^{5} \beta_{ij} x_i x_j + \varepsilon$$

is adequate to represent the response locally, least-squares estimates of the

TABLE 6.2. 2^{5-1} design, responses, and the Yates' analysis

Run order and number	$x_1 x_2 x_3 x_4 x_5$	y	(1)	(2)	(3)	(4)	Half-effect (5)[a]	Effect name
16	$- - - - +$	51.8	108.1	213.2	420.3	914.8	$\bar{y} = 57.1750$	0
2	$+ - - - -$	56.3	105.1	207.1	494.5	-53.6	-3.3500	1
10	$- + - - -$	56.8	112.1	242.0	-19.5	-34.6	-2.1625	2
1	$+ + - - +$	48.3	95.0	252.5	-34.1	2.2	0.1375	12
14	$- - + - -$	62.3	122.1	-4.0	-20.1	4.4	0.2750	3
8	$+ - + - +$	49.8	119.9	-15.5	-14.5	-12.0	-0.7500	13
9	$- + + - +$	49.0	132.9	-16.8	3.5	-24.2	-1.5125	23
7	$+ + + - -$	46.0	120.1	-17.3	5.7	-30.6	-1.9125	45
4	$- - - + -$	72.6	4.5	-3.0	-6.1	74.2	4.6375	4
15	$+ - - + +$	49.5	-8.5	-17.1	-1.5	-14.6	-0.9125	14
13	$- + - + +$	56.8	-12.5	-2.2	-11.5	5.6	0.3500	24
3	$+ + - + -$	63.1	-3.0	-12.3	-0.5	9.2	0.5750	35
12	$- - + + +$	64.6	-23.1	-13.0	-14.1	16.6	1.0375	34
6	$+ - + + -$	67.8	6.3	29.4	-10.1	11.0	0.6875	25
5	$- + + + -$	70.3	3.2	29.4	22.5	4.0	0.2500	15
11	$+ + + + +$	49.8	-20.5	-23.7	-53.1	-75.6	-4.7250	5

[a] Except for $\bar{y}$.

TABLE 6.3.   First- and second-order contributions to corrected total sum of squares

Source	df	SS	MS
First order	5	956.905	191.38
Second order	10	150.785	15.08
Total, corrected	15	1107.690	

five first-order coefficients β_i and the 10 two-factor interaction coefficient β_{ij} are

$$b_1 = -3.35, \qquad b_{12} = 0.14, \qquad b_{24} = 0.35,$$

$$b_2 = -2.16, \qquad b_{13} = -0.75, \qquad b_{25} = 0.69,$$

$$b_3 = 0.28, \qquad b_{14} = -0.92, \qquad b_{34} = 1.04,$$

$$b_4 = 4.64, \qquad b_{15} = 0.25, \qquad b_{35} = 0.58,$$

$$b_5 = -4.73, \qquad b_{23} = -1.51, \qquad b_{45} = -1.91.$$

It will be recalled that, because each of the x variables has been allocated the coded levels -1 and $+1$, the least-squares values of the b's will be one-half of the values obtained via Yates' method in Table 6.2. In this experiment, some of the estimated second-order effects are not particularly small compared with those of first order. On the whole, however, the first-order effects are considerably larger in magnitude than those of second order, as evidenced, for example, by the analysis of variance in Table 6.3. As an approximation, it was decided to assume tentatively that second degree effects could, for the present, be ignored and to combine the second-order effects to provide an "estimate of error" based on 10 df. From this, the standard error of each estimate is $(15.08/16)^{1/2} = 0.971$. Note that the approximation is conservative in the sense that, if second-order effects are not negligible, this calculation will tend to *inflate* the estimate of error variance. Under these assumptions we have

$$b_1 = -3.35 \pm 0.97,$$

$$b_2 = -2.16 \pm 0.97,$$

$$b_3 = 0.28 \pm 0.97,$$

$$b_4 = 4.64 \pm 0.97,$$

$$b_5 = -4.73 \pm 0.97.$$

The estimated direction of steepest ascent, when the variables are scaled in units of the design, then follows the vector of coefficient values $[-3.35, -2.16, 0.28, 4.64, -4.73]$. The length of this vector is $[(-3.35)^2 + (-2.16)^2 + \cdots + (-4.73)^2]^{1/2} = 7.74$. A vector of unit length in the direction of steepest ascent therefore has coordinates $-3.35/7.74 = -0.43$, $-2.16/7.74 = -0.28$, and so on, that is, $[-0.43, -0.28, 0.04, 0.60, -0.61]$. If we follow the steepest ascent path then, for a change of -0.43 units of the first variable we should change the second variable by -0.28 units, change the third variable by $+0.04$ units, and so on. Exploratory runs were chosen at the conditions shown in Table 6.4, obtained by multiplying the unit length vector by 2, 4, 6, and 8, respectively. These particular multiples were chosen because the first of these is close to the periphery of the experimental region. It corresponds quite closely with run number 4 which, in fact, gave the highest recorded yield of 72.6, while the spacing separating the runs E_2, E_4, E_6, and E_8 is about the same as that separating the upper and lower levels of the factorial design. Note that the decoded conditions may be obtained by inverting the formulas in Table 6.1 so that

$$\xi_1 = 2x_1 + 8,$$

$$\xi_2 = 2.5x_2 + 87.5,$$

$$\xi_3 = 15x_3 + 45,$$

$$\xi_4 = 12.5x_4 + 102.5,$$

$$\xi_5 = 5x_5 + 45.$$

Because of its closeness to the conditions of run number 4, the experiment E_2 was not actually run. The observed yields in Table 6.4 are plotted in Figure 6.2, from which it was concluded that a conditional maximum along the path of steepest ascent was located close to E_6.

TABLE 6.4.   Points on the path of steepest ascent in coded and decoded units

	Coded conditions					Decoded conditions					Observed yields
	x_1	x_2	x_3	x_4	x_5	Time	Temperature	Amount B	Amount C	Amount D	
E_2:	−0.86	−0.56	0.08	1.20	−1.22	6.3	86.1	46.2	117.5	38.9	72.6 (run 4)
E_4:	−1.72	−1.12	0.16	2.40	−2.44	4.6	84.7	47.4	132.5	32.8	85.1
E_6:	−2.58	−1.68	0.24	3.60	−3.66	2.8	83.3	48.6	147.5	26.7	82.4
E_8:	−3.44	−2.24	0.32	4.80	−4.88	1.1	81.9	49.8	162.5	20.6	80.8

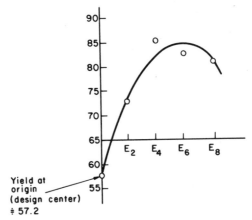

FIGURE 6.2. Yields found along the path of steepest ascent. The yield at the origin is approximated by the average 57.2 of all the 16 runs. The steepest ascent procedure has thus been very effective, moving us to a new set of conditions yielding over 80%.

6.3.   GENERAL COMMENTS ON STEEPEST ASCENT

Design

It will be remarked that in the example, the fractional factorial employed was a 2_V^{5-1}, a design of resolution V, requiring 16 runs. It might have been possible to make progress with an eight-run design but in this case the experimenters decided on a somewhat conservative course, partly because of their feeling that the first-order (i.e., main) effects might not be particularly large compared with the experimental error.

Steepest Ascent or Descent?

In this example, the *objective function* was yield and the objective was to increase it. Other objective functions that might be important in chemical investigations would be *unit cost* and *level of impurity*, both of which we would wish to *decrease* rather than increase. This would require a path of steepest *descent*, given by changing all the signs from those of the path of steepest ascent at any point. Alternatively, the same effect can be achieved by changing the sign of the objective function and following the path of steepest ascent. Thus, essentially, all problems can be set up as *ascent* problems.

When Will Ascent Methods Work?

If we are looking for, say, a maximum yield, it would be wasteful to investigate *in any detail* regions of *low* yield. Typically, the main features in such regions are the first-order (i.e., main) effects which can point a direction of ascent up a surface. Once first-order effects become small compared to those of second order, or compared to the error, or to both, it may be necessary to switch to a second-order approximation. The appropriate designs can then be chosen to cover somewhat larger regions.

Checks on the adequacy of the first-order approximation or, equivalently, on the need for second-order terms, are supplied by:

1. Examining individual interaction contrasts, which are not used for estimation of first-order effects.
2. An overall curvature check.

A useful overall curvature check is supplied by adding a number of center points to a two-level fractional factorial or factorial design. The order in which the complete design, including center points, is carried out should be random. Thus, a set of randomly replicated experimental runs from which a pure error estimate of the basic variation in the data can be calculated is obtained. The overall curvature check is supplied by the contrast

$$c = \{\text{average response in two-level factorial runs}\}$$

$$- \{\text{average response for runs at the center}\}.$$

It may be shown that, if the true response function is of second degree, c supplies an unbiased estimate of the sum of the pure quadratic effects, namely

$$c \rightarrow \sum_{i=1}^{k} \beta_{ii}.$$

For a surface that contained a minimax, the β_{ii} would be of different signs and theoretically $E(c)$ could be zero even when large curvatures β_{ii} occurred. However, minimax surfaces are rare in practice and typically the β_{ii} will be of the same sign.

There will nevertheless be intermediate situations where it is unclear whether or not the first-order approximation is good enough to allow progress via steepest ascent. In such cases, it is usually worthwhile to calculate the path of steepest ascent anyway, and to make a few trials along it to find out.

A discussion of the effect of experimental error on the estimation of the direction of steepest ascent is given in Section 6.4.

The First-Order Design as a Building Block

Suppose that, possibly after previous steepest ascent application(s), a new first-order design has just been completed. The appropriate checks, possibly augmented by tentative local exploration, indicate that no further advance by steepest ascent is likely and that we need to proceed with a second-order exploration. It will often be possible to incorporate the first-order design just completed as a building block of the second-order design now required. Consider, for instance, the chemical process example discussed above. Suppose the 2^{5-1} design had shown large second-order effects (i.e., two-factor interactions) and that these appeared to be reasonably well estimated. Then the addition of further points to the first-order design in a manner we describe in more detail in Sections 9.2 and 13.8 could convert it into a second-order design suitable for fitting a full second degree equation. The results from this second-order fit might then permit further response improvement.

Steepest Ascent as a Precursor to Further Investigation

Steepest ascent is rarely an end in itself. It is mainly of value as a preliminary procedure, to move the region of interest to a neighborhood of improved response, worthy of more thorough investigation.

Steepest Ascent Classroom Simulation

The steepest ascent procedure can be realistically simulated for teaching purposes; see, for example, Mead and Freeman (1973).

6.4. A CONFIDENCE REGION FOR THE DIRECTION OF STEEPEST ASCENT

In the example of Section 6.2 the first-order coefficients were estimated as

$$b_1 = -3.35 \pm 0.97,$$

$$b_2 = -2.16 \pm 0.97,$$

$$b_3 = 0.28 \pm 0.97,$$

$$b_4 = 4.64 \pm 0.97,$$

$$b_5 = -4.73 \pm 0.97, \tag{6.4.1}$$

where the standard error of each estimate (0.97) was based on an estimated variance having 10 df. Retaining now the convention that we consider the variables as scaled in the units of the design, the vector

$$(-3.35, -2.16, 0.28, 4.64, -4.73) \qquad (6.4.2)$$

provides the estimated best direction of advance. Assuming that a first degree response equation provides an adequate model, we may now ask how much in error due to sampling variation this direction might be. This question can be answered in a conventional way by obtaining a confidence region for the direction of steepest ascent (Box, 1954, p. 211). This confidence region will turn out to be a cone (or hypercone depending on the number of dimensions involved) whose axis is the estimated direction vector.

Suppose there are k variables and we seek a confidence region for the *true* direction of steepest ascent as defined by its direction cosines $\delta_1, \delta_2, \ldots, \delta_i, \ldots, \delta_k$. If $b_1, b_2, \ldots, b_k$ are the estimated first degree effects, the expected values of these quantities are proportional to $\delta_1, \delta_2, \ldots, \delta_k$ so that

$$E(b_i) = \gamma \delta_i, \qquad i = 1, 2, \ldots, k \qquad (6.4.3)$$

where γ is some constant. Now if we think of this relationship as a regression model in which the b_i are responses and the δ_i are the levels of a single-predictor variable, then γ is the "regression coefficient" of $b_1, \ldots, b_k$ on $\delta_1, \delta_2, \ldots, \delta_k$. The required region is supplied by those elements $\delta_1, \delta_2, \ldots, \delta_k$ which just fail to make the residual mean square significant compared with $V(b_i) = \sigma_b^2$ at some desired level α. That is, for those δ's which satisfy

$$\frac{\left\{ \sum b_i^2 - \left(\sum_{i=1}^{k} b_i \delta_i \right)^2 \bigg/ \sum \delta_i^2 \right\} \bigg/ (k-1)}{s_b^2} \leq F_\alpha(k-1, \nu_b), \qquad (6.4.4)$$

where s_b^2 is an estimate of σ_b^2 and ν_b is the number of degrees of freedom on which this estimate s_b^2 is based. For the particular case where $k = 2$ we have an application of Fieller's theorem (Fieller, 1955) of which the present development is one extension. Because all the quantities in the foregoing inequality are known except for the values of the δ's, this expression defines a set of acceptable δ's, thus a set of acceptable vectors, and hence a confidence region for the direction of steepest ascent.

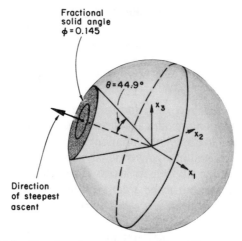

FIGURE 6.3. Direction of steepest ascent with confidence cone around it.

As an illustration we can again use the process example of Section 6.2. For simplicity and for the time being, we ignore the last two variables and treat the problem as if there had been only three variables ($k = 3$). The confidence region then takes the form of a cone in three dimensions about the estimated direction of steepest ascent as in Figure 6.3. For this reduced example, we have

$$\sum b_i^2 = (-3.35)^2 + (-2.16)^2 + (0.28)^2 = 15.9665,$$

$$\sum b_i \delta_i = -3.35\delta_1 - 2.16\delta_2 + 0.28\delta_3,$$

$$\sum \delta_i^2 = 1,$$

$$(k - 1) = 2, \qquad s_b = 0.97, \qquad F_{0.05}(2, 10) = 4.10, \qquad (6.4.5)$$

and the confidence region is thus defined by the cone with apex at the origin and such that all points $(\delta_1, \delta_2, \delta_3)$ a unit distance away from the origin satisfy

$$15.9665 - (-3.35\delta_1 - 2.16\delta_2 + 0.28\delta_3)^2 \leq 2 \times 0.97_2 \times 4.10, \quad (6.4.6)$$

that is,

$$(-3.35\delta_1 - 2.16\delta_2 + 0.28\delta_3)^2 \geq 8.2511. \qquad (6.4.7)$$

When there are more than three variables, the preparation of a diagram like that of Figure 6.3 is not possible. However, in practical applications, all we usually need to know is whether the direction of steepest ascent has been determined accurately enough for us to proceed. A good indication is supplied by the magnitude of the solid angle of the confidence cone about the estimated vector. Consider again Figure 6.3. Suppose the cap on the unit sphere centered at the origin has a surface which is a fraction ϕ of the total area of the sphere; then we shall say that the confidence cone subtends a fractional solid angle ϕ. Associated with the confidence cone we might also consider another angle, the semiplane angle θ at the vertex between a line on the surface of the cone at the origin and the axis of the cone. For a given number of dimensions, there will be a one-to-one correspondence between the fractional solid angle ϕ and the semiplane angle θ at the vertex.

This correspondence is given by tables of the t distribution. These tables are equivalent to a listing of the fractional solid angle 2ϕ (which corresponds to the double-tailed probability in the tables) in terms of the function $t = (k - 1)^{1/2}\cot\theta$. The fractional solid angle ϕ may therefore be obtained by inverse interpolation in the tables of t. Specifically, in terms of the quantities we have defined,

$$\sin\theta = \left\{ \frac{(k - 1)s_b^2 F_\alpha(k - 1, \nu_b)}{\sum_{i=1}^{k} b_i^2} \right\}^{1/2} \tag{6.4.8}$$

and

$$t_{2\phi}(k - 1) = (k - 1)^{1/2}\cot\theta$$

$$= \left\{ \frac{\sum_{i=1}^{k} b_i^2}{s_b^2 F_\alpha(k - 1, \nu_b)} - (k - 1) \right\}^{1/2} \tag{6.4.9}$$

In the simplified example above in which we have pretended that there were only three variables we have

$$\sin^2\theta = \frac{\{2F_{0.05}(2, 10)\}0.97^2}{15.9665} = \frac{2(4.10)(0.97^2)}{(15.9665)} = 0.4832, \tag{6.4.10}$$

whence $\theta = 44.0°$. Thus, the confidence region for this example consists of a cone about the direction of steepest ascent with a semiplane angle at the vertex of $\theta = 44.0°$. Applying Eq. (6.4.9) with $\theta = 44.0°$, $k = 3$, we have, since $\cot \theta = 1.034$,

$$t_{2\phi}(2) = 1.462. \tag{6.4.11}$$

Interpolation in the t tables gives $\phi = 0.141$ approximately. Thus the 95% confidence cone excludes about 85.9% of the possible directions of advance.

This kind of statement is available to us however many variables are involved. Thus, without necessarily being able to visualize the situation geometrically, we can appreciate the size of the confidence region simply by calculating ϕ.

For the original five-variable example, we would set $k = 5$, $\nu_2 = 10$ to obtain

$$t_{2\phi}(4) = \sqrt{\frac{59.8690}{0.97 \times 3.48} - 4} = 3.78. \tag{6.4.12}$$

Interpolation in the t tables yields $\phi = 0.01$ approximately. Thus, in this instance the 95% confidence cone for the direction of steepest ascent excludes about 99% of possible directions of advance. The appropriate direction is, therefore, known with some considerable accuracy.

6.5. STEEPEST ASCENT SUBJECT TO A CONSTRAINT

In some problems, we cannot proceed very far in the direction of steepest ascent before some constraint is encountered. It is then of interest to explore the modified direction of steepest ascent subject to the condition that the constraint is not violated. For example, in Figure 6.4 we show a 2^3 factorial experiment in the space of x_1, x_2, and x_3. As usual, x_1, x_2, and x_3 are standardized and relate to real variables such as temperature, ξ_1, time, ξ_2, and so on, by such equations as $x_i = (\xi_i - \xi_{i0})/S_i$. It may happen that the path of steepest ascent leads to a region of inoperability. Suppose that, at least as a local approximation, the inoperability region is bounded by the plane

$$a_0 + a_1x_1 + a_2x_2 + \cdots + a_kx_k = 0. \tag{6.5.1}$$

A vector perpendicular to this plane is then $\mathbf{a} = (a_1, a_2, \ldots, a_k)'$. Let the unmodified vector of steepest ascent be given by $\mathbf{b} = (b_1, b_2, \ldots, b_k)'$. Then the ith element of the modified vector direction of steepest ascent is

$$e_i = b_i - ca_i \qquad i = 1, 2, \ldots, k, \tag{6.5.2}$$

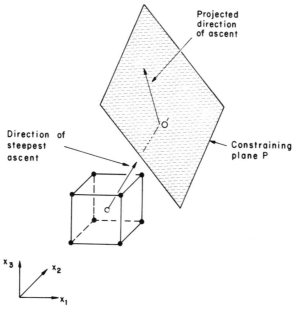

FIGURE 6.4.   Steepest ascent subject to a constraint.

where

$$c = \frac{\sum b_i a_i}{\sum a_i^2}.$$ (6.5.3)

It will be recognized that the e_i are the residuals obtained by regressing the elements of the steepest ascent vector on to the coefficients which represent the plane, the quantity c being the appropriate estimated regression coefficient. The above calculation gives the *direction* of the modified vector only. We also need to calculate the point O' where the initial vector of steepest ascent hits the constraining plane P and the new direction takes over. Clearly any point on the initial direction of steepest ascent will have coordinates

$$x_1 = \lambda b_1, \qquad x_2 = \lambda b_2, \qquad \ldots, x_k = \lambda b_k,$$ (6.5.4)

for some λ. Consequently, the required point is that which, while on the path of steepest ascent, also lies on the plane $a_0 + a_1 x_1 + a_2 x_2 + \cdots + a_k x_k = 0$. Thus, it corresponds to the value λ_0 which satisfies

$a_0 + (a_1b_1 + a_2b_2 + \cdots + a_kb_k)\lambda_0 = 0$, that is,

$$\lambda_0 = \frac{-a_0}{\sum a_i b_i}. \qquad (6.5.5)$$

An Example

The activity of a certain chemical mixture depends upon the proportion of three major ingredients A, B, and C. The mixture usually also contains an inactive diluent which is added to bring the total volume to a fixed value. A 2^3 factorial design was run to determine the effects of the ingredients on the activity y of the mixture. The lower and upper levels of the ingredients A, B, and C were as follows:

	Amount (cm^3)	
Ingredients	Lower	Upper
A	15	25
B	10	30
C	20	40

$$(6.5.6)$$

In each case sufficient diluent was added to bring up the total volume to 100 cm^3. Thus our standardized variables are

$$x_1 = \frac{(A - 20)}{5}; \qquad x_2 = \frac{(B - 20)}{10}, \qquad x_3 = \frac{(C - 30)}{10}. \qquad (6.5.7)$$

In this example, the interactions were found to be fairly small compared with the main effects and their standard errors, and the activity y of the resulting mixture was adequately represented by the linear relationship

$$\hat{y} = 12.2 + 3.8x_1 + 2.6x_2 + 1.9x_3. \qquad (6.5.8)$$

Thus, the estimated direction of steepest ascent runs from the origin $x_1 = x_2 = x_3 = 0$ along the vector $(3.8, 2.6, 1.9)$. In the·original variables, the origin corresponds to the mixture of 20 cm^3 of A, 20 cm^3 of B, and 30 cm^3 of C and, if we follow the direction of steepest ascent from this point, we shall clearly need to keep increasing the amounts of all the active ingredients. However, we are working with a system having the constraint

$$A + B + C \le 100. \tag{6.5.9}$$

Because $A = 20 + 5x_1$, $B = 20 + 10x_2$, and $C = 30 + 10x_3$, this implies that

$$20 + 5x_1 + 20 + 10x_2 + 30 + 10x_3 \le 100. \tag{6.5.10}$$

Thus, the constraining plane has the equation

$$-30 + 5x_1 + 10x_2 + 10x_3 = 0. \tag{6.5.11}$$

It follows that

$$\lambda_0 = -(-30)/\{5(3.8) + 10(2.6) + 10(1.9)\} = 0.469. \tag{6.5.12}$$

Thus the path of steepest ascent just meets the constraining plane at the point where

$$x_1 = 0.469(3.8) = 1.78, \qquad x_2 = 0.469(2.6) = 1.22,$$

$$\text{and} \quad x_3 = 0.469(1.9) = 0.89, \tag{6.5.13}$$

that is, at the point O' with coded coordinates $(1.78, 1.22, 0.89)$. To check, we note that

$$5(1.78) + 10(1.22) + 10(0.89) = 30.00 \tag{6.5.14}$$

so that this point is therefore just on the constraining plane. We now wish to calculate the modified direction of steepest ascent which is the projection of this direction in the plane $-30 + 5x_1 + 10x_2 + 10x_3 = 0$. Regressing the initial steepest ascent vector on to the coefficients of the constraining plane, we find that

$$c = \frac{3.8(5) + 2.6(10) + 1.9(10)}{5^2 + 10^2 + 10^2} = 0.284. \tag{6.5.15}$$

Thus the modified vector direction of steepest ascent from O' has elements

$$e_1 = 3.8 - 0.284(5) = 2.38,$$

$$e_2 = 2.6 - 0.284(10) = -0.24,$$

$$e_3 = 1.9 - 0.284(10) = -0.94. \tag{6.5.16}$$

Further progress in this direction of constrained steepest ascent will therefore lie along the line having coded coordinates

$$x_1 = 1.78 + \mu(2.38), \qquad x_2 = 1.22 - \mu(0.24),$$

$$\text{and} \quad x_3 = 0.89 - \mu(0.94), \tag{6.5.17}$$

where μ is some suitably chosen multiplier.

Some specimen calculations along the unmodified direction of steepest ascent and then along the modified direction of steepest ascent are shown in Table 6.5.

The first set of conditions listed in the table is for the center point. Conditions two and three lie on the steepest ascent path $x_1 = 3.8\lambda$, $x_2 = 2.6\lambda$, and $x_3 = 1.9\lambda$ where λ is 0.2 and 0.4. At the fourth set of conditions, the steepest ascent vector reaches the constraining plane ($\lambda = \lambda_0 = 0.469$). The remaining conditions listed are on the modified direction of steepest ascent defined in the foregoing display with $\mu = 0.2$ and 0.4. It will be noticed that the predicted activity continues to rise, but at a considerably reduced rate, on the modified path of ascent.

The elements comprising the vector of the path of steepest ascent are, of course, often subject to fairly large errors and, in any case, the linear approximation is probably inaccurate at points not close to the region of the design. The type of calculation given above should not therefore be regarded as an exact prediction but rather as supplying an experimental path worthy of further exploration. In some cases the constraint is not known exactly *a priori*, as in this example, but must also be estimated. It often happens that, in a given set of experimental runs, a number of responses may be measured, one of which, the principal response function, represents an objective function such as yield which it is desired to maximize and the

TABLE 6.5. Points on the original and modified directions of steepest ascent

Point number	Total (cm^3)	x_1	x_2	x_3	Predicted activity	Comments
1	70	0	0	0	12.2	Center O
2	82.8	0.76	0.52	0.38	17.2 ⎫	Unconstrained
3	95.6	1.52	1.04	0.76	22.1 ⎭	path
4	100.0	1.78	1.22	0.89	23.8	Point O'
5	100.0	2.26	1.17	0.70	25.2 ⎫	Constrained
6	99.95	2.73	1.12	0.51	26.5 ⎭	path

others are measures of physical characteristics of the product—purity, color, viscosity, odor, and so on. In those cases where the objective of the experiment is to maximize the principal response, the constraining relations may themselves be estimated from the auxiliary responses. We have seen an example of this situation earlier where we were required to maximize the strength of a dyestuff, subject to the attainment of suitable levels of hue and brightness.

To understand constraining relations between three variables, it is very worthwhile to examine appropriate sets of three-dimensional models. Computer graphics have now greatly simplified such exploration. One can say, of course, that the geometrical models do no more than express the related algebra and so consequently algebraic manipulation without the accompanying models is enough. In practice, however, it seems that there is much less likelihood of missing important aspects of the problem and of making outright mistakes, if graphical and geometrical illustrations are used whenever possible.

For an excellent pictorial representation of steepest ascent using five-factor simplex designs, see Gardiner and Cowser (1961).

APPENDIX 6A.   EFFECTS OF CHANGES IN SCALES ON STEEPEST ASCENT

In this book, we have tried to emphasize that the outcome of an investigation depends critically on a number of crucial questions which are not directly dependent on the data and which the experimenter must decide from knowledge and experience. Moreover, these decisions are typically modified as the investigation unfolds. For example, the experimenter decides *which* factors to include, the *region* of the factor space in which the experiments are to be carried out, and the *relative scales* of measurement of the factors. Such questions are quite outside the competence of the statistician, although he ought always to ask appropriately probing questions to ensure that the experimenter has properly considered all the relevant issues, and the statistician may need to point out, after the data are available, that certain suppositions of the experimenter appear to be of doubtful validity. Granted all this, it is the role of the statistician to design the best experiment within the framework currently believed to be most appropriate by the experimenter and, in particular, in the scaling of the variables the experimenter regards as most applicable at any given time. It has to be remembered that a subject as concrete and mathematically satisfying as experimental design is actually embedded in a morass of uncertainty, uncertainty due to the possibilities that the experimenter might choose wrong variables, might explore the wrong region, or might use scaling that was inappropriate. However, the gloomy view that successful experimentation is a matter of purest luck is lightened by two circumstances; first, that experimenters often *do* know a very great deal about the system they are studying and second, that, because most investigations are conducted sequentially, the experimenter does not need to guess exactly right but need only guess sufficiently right to place himself on one of the many possible paths that will lead adaptively to a satisfactory answer. In particular, the correction of any grossly unsuitable scaling of the predictor variables will normally occur as the experimental iteration proceeds.

The problem of scaling the variables in steepest ascent is one area where the general indeterminacy of experimentation does not lie conveniently hidden but makes itself manifest. In order to appreciate the problem, consider the following: Suppose the scale factors that one experimenter adopts are $s_1, s_2, \ldots, s_k$ and that the scale factors adopted by a second experimenter are $s_1' = a_1 s_1$, $s_2' = a_2 s_2, \ldots, s_k' = a_k s_k$. (For example, if in a two-level factorial or fractional factorial design, one experimenter changed temperature by $s_1 = 20°$ and another by $s_1' = 10°$, then $a_1 = s_1'/s_1 = 0.5$.) Suppose further that the direction of steepest ascent as measured in the first experimenter's scales has direction cosines proportional to the elements $1, 1, \ldots, 1$. (We can make this assumption without loss of generality because we could always rotate the axes of our x space so that the direction, whatever it may be, lies in this particular direction in new coordinates.) In this case, the direction of steepest ascent appropriate to the second experimenter would have direction cosines proportional to $a_1^2, a_2^2, \ldots, a_k^2$. Figure 6A.1 shows the situation for $k = 3$.

The fact that the direction of steepest ascent was not invariant to scale change was pointed out in the original response surface paper by Box and Wilson (1951). In the discussion of the paper, this point was taken up by N. L. Johnson who emphasized the dependence of the elements of the direction on the square of the

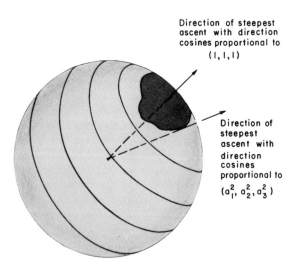

FIGURE 6A.1. Two directions of steepest ascent, one of which traces a closed curve on the unit sphere about the other (see text).

amount by which the scale factor was altered. In this same discussion it was also argued that it might be better simply to rely on the signs of the effects, which are of course scale invariant.

We can, to some extent, answer the question of how much *additional* useful information is contained in the direction of steepest ascent compared with using only the signs of the effects by examining a brief, unpublished investigation undertaken by G. E. P. Box and G. A. Coutie some years ago.

Let us choose some measure $G(|a|)$ of how widely different the second experimenter's scales are from those adopted by the first experimenter. Because multiplying all the a's by some constant factor will not change the actual situation, we need to choose, for G, some suitable homogeneous function of degree zero in the a's. One such suitable function is $G = |a|_{max}/|a|_{min}$. If we adopt this criterion, we shall regard all values of the a's that produce the same value, G_0 say, of G as representing a degree of scaling equally discordant with the original one. Such values of the a's will trace out a closed curve on the unit sphere as shown in Figure 6A.1. Thus, a measure of how much uncertainty is introduced in the direction of steepest ascent for a scaling as discordant as, or less discordant than, that represented by $G = G_0$ will be represented by the content of the closed curve bounded by $G = G_0$ on the unit sphere, taken as a proportion of the whole surface of the unit sphere. In the three-dimensional case of Figure 6A.1, this is simply proportional to the surface of the cap enclosed by the curve $G = G_0$ on the unit sphere.

Now let us write $a_i^2 = y_i$, then $G^4 = a_{max}^4/a_{min}^4 = y_{max}^2/y_{min}^2$ and the inequality $G < G_0$ is equivalent to the inequality $G^4 < G_0^4$. It can be shown that the required value is

$$\phi = \frac{1}{(2\pi)^{k/2}} \int_R \exp\left\{ -\tfrac{1}{2} \sum_{i=1}^{k} y_i^2 \right\} dy, \qquad (6A.1)$$

where the integral is taken over the region R such that $y_{max}^2/y_{min}^2 < G_0^4$. This is the same integral as would arise in a comparison of variances each having 1 df using Hartley's (1950) F_{max} criterion. H. A. David (1952) gives the upper and lower 5 and 1% points of F_{max} for a set of k variance estimates each based on ν df for $\nu \geq 2$. The test required here is for $\nu = 1$ which is not given in the tables presumably because it is usually of very little importance. The appropriate values for this integral have, however, been calculated for $\nu = 1$ to give Table 6A.1.

As an example of the use of Table 6A.1, consider the following, in which we shall leave aside the question of sampling error. Suppose we were concerned with $k = 5$ variables. Let us further suppose that experimenters may be expected to differ in their ideas about scaling by a factor G representing the ratio of the largest to the smallest modifying factor. Finally let us suppose a reference experimenter has so scaled the variables that the elements of the steepest ascent direction are proportional to $1, 1, 1, 1, 1$ in the scaling adopted. A knowledge of all the signs will reduce the uncertainty concerning the direction of advance by a factor of $1/2^5 = 1/32 = 0.03125$. If $G = 2$, the uncertainty will be reduced still further by a factor of 0.167 (obtained from Table 6A.1 with $k = 5$, $G = 2$) giving an overall uncertainty of

TABLE 6A.1.   Table of the reduction of the "percentage of uncertainty" due to changes of scale in a particular n-tant

Value of G	Values of k				
	2	3	4	5	6
$\sqrt{2}$	41.0	15.5	5.6	2.0	0.7
2	68.8	45.4	28.0	16.7	11.6
3	85.7	69.8	55.8	43.9	33.8
4	93.8	83.8	75.1	67.7	61.4

$0.167(0.03125) = 0.0052$ for $G = 2$; by a factor of 0.439 if $G = 3$ giving an overall uncertainty of $0.439(0.03125) = 0.014$ and by a factor of 0.677 for $G = 4$ giving an overall uncertainty of $0.677(0.03125) = 0.021$. Our certainty of knowledge of the direction is thus 1 minus these figures or 0.9948, 0.986, and 0.979 for $G = 2$, 3, and 4, respectively.

In summary, the present authors feel that the direction of steepest ascent correctly distills the first degree information concerning possible directions of improvement, by combining the experimenter's prior knowledge concerning the nature of the response relationship appropriately with the available data. This direction does and should change for different experimenters because different prior opinions must enter the problem.

EXERCISES

6.1. An experimenter begins a steepest ascent procedure on two variables (X_1, X_2) at the current central point (90, 20) and performs five runs with the response results provided:

X_1	80	100	80	100	90
X_2	10	10	30	30	20
y	11	0	29	6	12

Code (X_1, X_2) sensibly to variables (x_1, x_2) and fit a first-order (planar) model $\hat{y} = b_0 + b_1 x_1 + b_2 x_2$ to the data. Determine the direction of steepest ascent. The experimenter now performs six more runs:

X_1	64.5	47.5	39	30.5	43.25	34.75
X_2	38	50	56	62	53	59
y	43	58	72	62	65	68

Which of these runs lie on the path of steepest ascent you determined earlier? The experimenter decides now to combine the two runs $(X_1, X_2, y) =$ (43.25, 53, 65) and (34.75, 59, 68) with six more, namely these:

X_1	34.75	43.25	39	39	39	39
X_2	53	59	56	56	56	56
y	71	68	71	72	72	73

Fit a first-order model to those eight runs and use the repeat observations to test for lack of fit. Plot, as points on a diagram, all the runs performed so far, with their y values attached. What should the experimenter do next? Go off in a new direction of steepest ascent? Fit a second-order surface? Or what?

6.2. Two coded variables $x_1 = (X_1 - 99)/10$, $x_2 = (X_2 - 17)/20$ are examined, and the data of Table E6.2 are obtained. Is the point $(X_1, X_2) = (59, 67)$ on the path of steepest ascent?

TABLE E6.2

x_1	x_2	y
−1	−1	11
1	−1	2
−1	1	29
1	1	6
0	0	12

6.3. In a steepest ascent investigation to examine the effects of three variables, the initial design of Table E6.3 was employed.

The model $\hat{y} = 4.7 + 0.25x_1 + x_2 + 0.5x_3$ is fitted. The following restriction exists:

$$4X_1 + 5X_2 + 6X_3 \leq 371.$$

Is the point $(X_1, X_2, X_3) = (36.40, 27.10, 15.00)$ on the restricted path of steepest ascent? (Allow for the fact that round-off errors may occur unless adequate figures are retained.)

6.4. A mixture of three ingredients X_1, X_2, X_3 (plus inerts to make up to 100%) is being examined. Steepest ascent techniques are being applied to improve the response variable, percentage yield. The following fitted equation is obtained from the first set of runs made:

$$\hat{y} = 9.925 - 4.10x_1 - 9.25x_2 + 4.90x_3,$$

TABLE E6.3

Uncoded			Coded		
X_1	X_2	X_3	x_1	x_2	x_3
30	10	6	-1	-1	-1
40	10	6	1	-1	-1
30	20	6	-1	1	-1
40	20	6	1	1	-1
30	10	16	-1	-1	1
40	10	16	1	-1	1
30	20	16	-1	1	1
40	20	16	1	1	1
35	15	11	0	0	0
35	15	11	0	0	0
35	15	11	0	0	0
35	15	11	0	0	0

where

$$x_1 = X_1 - 8, \qquad x_2 = X_2 - 14, \qquad x_3 = 2(X_3 - 7).$$

If the restriction $X_1 + X_2 + X_3 \geq 26.4$ applies, is the point $X_1 = 7.11$, $X_2 = 11.29$, $X_3 = 8.00$ on the path of steepest ascent? (There will undoubtedly be a little rounding error in your calculations, remember.)

6.5. A four-dimensional steepest ascent path is obtained from the fitted coefficients $b_1 = 1$, $b_2 = 2$, $b_3 = 3$, $b_4 = 2$. If we have to keep inside or on the restricting boundary

$$-4 + x_1 + x_2 + x_3 + x_4 = 0,$$

answer this question: Is the point $(X_1, X_2, X_3, X_4) = (-5/2, 10, 45/2, 10)$ on the adjusted path of steepest ascent? Assume $x_i = (X_i - 5)/5$, $i = 1, 2, 3, 4$.

6.6. A manufacturer of circular saw blades has two empirical first-order equations for blade life in hours (η_1, observed as y_1) and for unit blade cost in cents (η_2; y_2) in terms of the manufacturing time (coded to x_1) and the hardness of the steel used (coded to x_2). These are

$$\hat{y}_1 = 20 + x_1 + x_2,$$

$$\hat{y}_2 = 50 + 4x_1 + 2x_2,$$

and they are valid only for $-2 \leq x_i \leq 2$. If the unit blade cost must be kept below 54 cents and the blade life must exceed 21.5 h, can the manufacturer produce blades? If yes, explain how he must operate. If no, explain why he cannot.

CHAPTER 7

Fitting Second-Order Models

In Chapter 6 we saw how, in appropriate situations, a fitted first degree approximation to the response function could be exploited to produce process improvement. In this chapter, we begin to consider the use of quadratic (second-order) approximating functions.

7.1. FITTING AND CHECKING SECOND DEGREE POLYNOMIAL GRADUATING FUNCTIONS

To fix ideas, we discuss in more detail an example first introduced in Chapter 2 concerning the behavior of worsted yarn under cycles of repeated loading. For simplicity in the earlier analysis, only 8 of the full set of 27 runs were utilized. These 8 runs are distinguished by superscript b's in Table 7.1, where the complete data are set out.

Reconsideration of the Textile Example

It will be recalled that, in this example, the three inputs were

$$\xi_1 = \text{the length of test specimen (mm)},$$

$$\xi_2 = \text{the amplitude of load cycle (mm)},$$

$$\xi_3 = \text{the load (g)},$$

which were conveniently coded in terms of

$$x_1 = \frac{(\xi_1 - 300)}{50}, \quad x_2 = (\xi_2 - 9), \quad x_3 = \frac{(\xi_3 - 45)}{5}. \quad (7.1.1)$$

TABLE 7.1. Textile data for 3 × 3 × 3 factorial design, from an unpublished report to the technical committee, International Wool Textile Organization by Dr. A. Barella and Dr. A. Sust

ξ_1 = length of specimen (mm) $x_1 = (\xi_1 - 300)/50$			ξ_2 = amplitude of load cycle (mm) $x_2 = (\xi_2 - 9)/1$		ξ_3 = load (g) $x_3 = (\xi_3 - 45)/5$
Run Number[a]	Length, x_1	Amplitude, x_2	Load, x_3	Y = cycles to failure	$y = (\log_{10} Y)$
1[b]	−1	−1	−1	674	2.83
2	0	−1	−1	1414	3.15
3[b]	1	−1	−1	3636	3.56
4	−1	0	−1	338	2.53
5	0	0	−1	1022	3.01
6	1	0	−1	1568	3.19
7[b]	−1	1	−1	170	2.23
8	0	1	−1	442	2.65
9[b]	1	1	−1	1140	3.06
10	−1	−1	0	370	2.57
11	0	−1	0	1198	3.08
12	1	−1	0	3184	3.50
13	−1	0	0	266	2.42
14	0	0	0	620	2.79
15	1	0	0	1070	3.03
16	−1	1	0	118	2.07
17	0	1	0	332	2.52
18	1	1	0	884	2.95
19[b]	−1	−1	1	292	2.47
20	0	−1	1	634	2.80
21[b]	1	−1	1	2000	3.30
22	−1	0	1	210	2.32
23	0	0	1	438	2.64
24	1	0	1	566	2.75
25[b]	−1	1	1	90	1.95
26	0	1	1	220	2.34
27[b]	1	1	1	360	2.56
Total sum of squares				$\Sigma Y^2 = 40,260,624,$	$\Sigma y^2 = 208.68$ (27 df

[a] Run number is used for identification only and does not indicate the order in which the runs were actually made.

[b] Runs employed in earlier analysis.

The output (or response) of interest was Y = the number of cycles to failure. However the analysis was conducted in terms of $y = \log_{10} Y$ and the first degree approximating equation fitted in Chapter 2 was

$$\hat{y} = 2.745 + 0.375x_1 - 0.295x_2 - 0.175x_3 \qquad (7.1.2)$$

which (see Table 2.3) gave fitted values agreeing very closely with the eight observations considered.

Structure of Empirical Models

Let us recapitulate certain ideas, using this example for illustration. There presumably exists some true physical relationship between the expectation $E(y) = \eta$ of the output y and the three inputs ξ_1, ξ_2, and ξ_3 via physical constants $\boldsymbol{\theta}$. The nature of this true *expectation function*

$$\eta = E(y) = f(\xi_1, \xi_2, \xi_3, \boldsymbol{\theta}) = f(\boldsymbol{\xi}, \boldsymbol{\theta}) \tag{7.1.3}$$

for this system is, however, unknown to us. In particular, we do not know the true functional form f, nor do we know the nature of the constants $\boldsymbol{\theta}$.

We therefore replace $f(\boldsymbol{\xi}, \boldsymbol{\theta})$ by a graduating function $g(\mathbf{x}, \boldsymbol{\beta})$ which we hope can approximate it locally. In the analysis of Chapter 2, it was supposed that the expected value of $y = \log Y$ could be represented by a first degree polynomial graduating function

$$g_1(\mathbf{x}, \boldsymbol{\beta}) = \beta_0 + \beta_1 x_1 + \beta_2 x_2 + \beta_3 x_3, \tag{7.1.4}$$

where each x_i was a linear coding $x_i = (\xi_i - \xi_{i0})/S_i$ of an input ξ_i. It should be remembered that, in general, the x_i's could be any *functions* $x_i = f_i(\xi_1, \xi_2, \xi_3)$, $i = 1, 2, \ldots$, of some or all of the inputs which the experimenter feels are appropriate. In some engineering applications, for instance, they might be suitably chosen dimensionless groups.

The graduating function approximates the expectation function so that

$$E(y) = \eta \simeq g(\mathbf{x}, \boldsymbol{\beta}) \tag{7.1.5}$$

which establishes an approximate link between $E(y)$ and $\boldsymbol{\xi}$ the expected output and the inputs. In practice, we do not know the mean value $E(y)$ of the output response for any particular choice of the inputs. We have only an observation y (or sometimes a number of replicated observations) subject to error $\varepsilon = y - E(y)$. Although, in any given case, the actual error ε is unknown, we suppose it to arise from a fixed distribution which we refer to as the error distribution. We shall assume that (possibly after suitable transformation of the response), the errors are to a sufficient approximation distributed normally and independently of one another with the same variance σ^2. The observed output response y is linked to the inputs in $\boldsymbol{\xi}$ in two stages. The link between y and $E(y)$ occurs via the error distribution, and the link between $E(y)$ and $\boldsymbol{\xi}$ via the graduating function:

$$y \xrightarrow{\text{error distribution}} E(y) \xrightarrow{\text{graduating function}} \boldsymbol{\xi}.$$

Thus, there are two sources of error:

1. Random error $\varepsilon = y - E(y)$,
2. Systematic error, or bias, $E(y) - g(\mathbf{x}, \boldsymbol{\beta})$, arising from the inability of the graduating function $g(\mathbf{x}, \boldsymbol{\beta})$ to exactly match the expectation function $E(y) = f(\boldsymbol{\xi}, \boldsymbol{\theta})$.

As we shall see later, it is important to keep in mind that both kinds of errors are involved. This consideration is helpful in deciding how precise a graduating function needs to be. In particular, there is little point in straining hard to reduce the level of systematic errors in the estimated response function much below that induced by random errors.

Analysis of Variance for Textile Example (Logged Data)

In our present consideration of the textile example, we now employ the whole set of data from the 27 runs set out in Table 7.1. For the time being, we continue to work in terms of $y = \log Y$, the log of the number of cycles

TABLE 7.2.   Textile data. Polynomials of zero, first, and second degree
fitted to logged data, with standard errors of coefficients
indicated by $\pm$ values

Zero Degree Polynomial

Estimated response function is $\qquad \hat{y} = \bar{y} = 2.751$
$\qquad\qquad\qquad\qquad\qquad\qquad\qquad \pm 0.016$
Regression sum of squares is $\qquad S_0 = (\Sigma y)^2/n = 204.2975$ (1 df)

First Degree Polynomial

Estimated response function is $\qquad \hat{y} = \quad 2.751 + 0.362x_1 - 0.274x_2 - 0.171x_3$
$\qquad\qquad\qquad\qquad\qquad\qquad \pm 0.016 \pm 0.020 \quad \pm 0.020 \quad \pm 0.020$
Regression sum of squares is $\qquad S_1 = 208.5292$ (4 df)

Second Degree Polynomial

Estimated response function is $\qquad \hat{y} = \quad 2.786 + 0.362x_1 - 0.274x_2 - 0.171x_3$
$\qquad\qquad\qquad\qquad\qquad\qquad \pm 0.043 \pm 0.020 \quad \pm 0.020 \quad \pm 0.020$

$\qquad\qquad\qquad\qquad\qquad\qquad -0.037x_1^2 + 0.013x_2^2 - 0.029x_3^2$
$\qquad\qquad\qquad\qquad\qquad\qquad \pm 0.034 \quad \pm 0.034 \quad \pm 0.034$

$\qquad\qquad\qquad\qquad\qquad\qquad -0.014x_1x_2 - 0.029x_1x_3 - 0.010x_2x_3$
$\qquad\qquad\qquad\qquad\qquad\qquad \pm 0.024 \quad\quad \pm 0.024 \quad\quad \pm 0.024$
Regression sum of squares is $\qquad S_2 = 208.5573$ (10 df)

TABLE 7.3. Analysis of variance for textile example (logged data)

Source	SS	df	MS
Mean (zero degree polynomial)	204.2975	1	
Added first-order terms	4.2317	3	1.4106
Added second-order terms	0.0281 ⎱ 0.1495	6 ⎱ 23	0.0047 ⎱ 0.0065
Residual	0.1214 ⎰	17 ⎰	0.0071 ⎰
Total	208.6787	27	

to failure. Later in this chapter we discuss in some detail questions concerning such data transformations.

The design used in this investigation was a $3 \times 3 \times 3$ factorial. That is, each of three levels of the three inputs were run in all combinations. The provision of three levels allows in particular a general quadratic polynomial to be fitted, and the results of doing this as well as of fitting first and zero degree polynomials to the complete set of data are set out in Table 7.2. The $\pm$ limits below the estimated coefficients are $\pm$ their standard errors, calculated from the square roots of the diagonal terms of the matrix $(\mathbf{Z'Z})^{-1} s^2$ (see Appendix 7A). We use the estimate $s^2 = 0.0071$ obtained from the residual for the second degree model in all calculations to obtain consistency between the three sets of standard errors. Also shown in Table 7.2 are the regression sums of squares and degrees of freedom associated with the fits of successively higher order. From these, the extra sums of squares $S_1 - S_0$, $S_2 - S_1$, and their associated degrees of freedom have been calculated, to yield the analysis of variance of Table 7.3.

A question of immediate interest is whether the fitted first degree equation provides an adequate representation of the response function. In this design, there was no replication and therefore no estimate of pure error with which the mean square associated with second degree terms might be compared. However, on the *assumption* that an adequate model is supplied by a polynomial of *at most* second degree the residual mean square provides an estimate of error. The F ratio

$$\frac{\text{mean square for added second degree terms}}{\text{mean square for residual}} = \frac{0.0047}{0.0071} = 0.66 \quad (7.1.6)$$

provides no reason to doubt the adequacy of the first degree polynomial representation. At this stage of our analysis, we are thus *tentatively* entertaining the fitted first degree model

$$\hat{y} = 2.751 + 0.362x_1 - 0.274x_2 - 0.171x_3. \qquad (7.1.7)$$

On this basis, the residual mean square obtained from fitting the first degree model provides an estimate of σ^2 of

$$\frac{(0.0281 + 0.1214)}{(6 + 17)} = 0.0065 \qquad (7.1.8)$$

based on 23 df. This leads to standard error values slightly smaller than those previously quoted. Note that our present fitted model based on all 27 observations agrees very well with that fitted in Chapter 2 [see Eq. (7.1.2)] which however used only 8 of the 27 observations.

Choice of Metric for the Output Response

We shall return to this example once again when we come to consider, in Chapter 8, the general question of *estimating* appropriate transformations for the output and input variables to ensure maximum simplicity in the model. For the moment, we take note of the fact that the analysis has so far been conducted in terms of the *logarithm*, $y = \log Y$, of the number of cycles to failure rather than in terms of Y itself. We have already noted that, in choosing a transformation or "metric" in terms of which the output (response) is best expressed, the implications for the *error distribution* must always be kept in mind. The estimation procedure of (unweighted) least squares is efficient only provided the standard normal theory assumptions are roughly true for the error distribution implied by the choice of model. Of major importance among these standard assumptions is the supposition that the observations have (approximately) constant variance. In Box and Cox (1964), it was pointed out that, for this particular example, the logarithmic transformation is a sensible choice on prior grounds. The number of cycles to failure has a range zero to infinity and the actual data values extend from $Y = 90$ to $Y = 3636$. It is much more likely with such data that the *percentage* errors made at different sets of experimental conditions, rather than the absolute errors, will be roughly comparable. This is equivalent to saying that, as a first guess at least, $y = \log Y$ (which has a range from minus infinity to plus infinity) might be expected to have constant variance in which case the standard least squares procedure would be efficient for a model of the form

$$\log Y = y \simeq g(\mathbf{x}, \boldsymbol{\beta}) + \varepsilon. \qquad (7.1.9)$$

Observations, fitted values, and residuals for the unlogged data Y and logged data $y = \log Y$. In each case, the first-order model form $\beta_0 + \beta_1 x_1 + \beta_2 x_2 + \beta_3 x_3 + \varepsilon$ and the second-order model form $\beta_0 + \beta_1 x_1 + \beta_2 x_2 + \beta_3 x_3 + \beta_{11} x_1^2 + \beta_{22} x_2^2 + \beta_{33} x_3^2 + \beta_{12} x_1 x_2 + \beta_{13} x_1 x_3 + \beta_{23} x_2 x_3 + \varepsilon$ have been fitted to the response data. (Note: Residuals are calculated from the fitted model *before* round-off of coefficients.)

| | | Cycles to failure | | | | | $\log_{10}$ (cycles to failure) | | | |
| | | First degree model | | Second degree model | | | First degree model | | Second degree model | |
Run number	Y	$\hat{Y}$	$Y - \hat{Y}$	$\hat{Y}$	$Y - \hat{Y}$	y	$\hat{y}$	$y - \hat{y}$	$\hat{y}$	$y - \hat{y}$
1	674	1048	−374	654	20	2.83	2.83	0.00	2.76	0.07
2	1414	1708	−294	1768	−354	3.15	3.20	−0.05	3.21	−0.06
3	3636	2368	1268	3358	278	3.56	3.56	0.00	3.57	−0.01
4	338	512	−174	156	182	2.53	2.56	−0.03	2.50	0.03
5	1022	1172	−150	813	209	3.01	2.92	0.09	2.93	0.08
6	1568	1832	−264	1947	−379	3.19	3.28	−0.09	3.28	−0.09
7	170	−24	194	210	−39	2.23	2.29	−0.06	2.26	−0.03
8	442	636	−194	409	32	2.65	2.65	0.00	2.68	−0.03
9	1140	1296	−156	1088	52	3.06	3.01	0.05	3.02	0.04
10	370	737	−367	484	−114	2.57	2.66	−0.09	2.66	−0.09
11	1198	1397	−199	1362	−164	3.08	3.02	0.06	3.07	0.01
12	3184	2057	1127	2717	467	3.50	3.39	0.11	3.41	0.09
13	266	201	65	129	137	2.42	2.39	0.03	2.39	0.03
14	620	861	−241	551	69	2.79	2.75	0.04	2.79	0.00
15	1070	1521	−451	1449	−379	3.03	3.11	−0.08	3.11	−0.08
16	118	−335	453	326	−208	2.07	2.12	−0.05	2.14	−0.07
17	332	325	7	290	42	2.52	2.48	0.04	2.53	−0.01
18	884	985	−101	733	151	2.95	2.84	0.11	2.84	0.11
19	292	426	−134	218	74	2.47	2.49	−0.02	2.50	−0.03
20	634	1086	−452	860	−226	2.80	2.85	−0.05	2.88	−0.08
21	2000	1746	254	1980	20	3.30	3.22	0.08	3.19	0.11
22	210	−109	319	6	204	2.32	2.22	0.10	2.22	0.10
23	438	551	−113	192	246	2.64	2.58	0.06	2.59	0.05
24	566	1211	−645	855	−289	2.75	2.94	−0.19	2.88	−0.13
25	90	−645	735	345	−255	1.95	1.94	0.01	1.96	−0.01
26	220	15	205	74	146	2.34	2.31	0.03	2.32	0.02
27	360	675	−315	281	79	2.56	2.67	−0.11	2.60	−0.04
Check totals (should be zero, within rounding)			3		1			−0.01		0.02

LOGGED DATA

(a)   Residuals   Overall

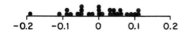

(b)   Residuals vs. ŷ.

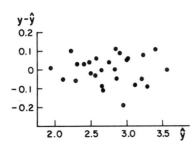

FIGURE 7.1. Plots of residuals $y_u - \hat{y}_u$ from first degree model fitted to logged data. (a) Residuals overall, (b) Residuals versus $\hat{y}_u$.

Some further checks on the adequacy of the first degree equation in log Y are provided by examination of the residuals calculated in Table 7.4. Figure 7.1a shows the residuals $y_u - \hat{y}_u$ from the fitted first degree model plotted in a dot diagram. Figure 7.1b shows the $y_u - \hat{y}_u$ plotted against $\hat{y}_u$. If the standard deviation σ_y increased as the mean value $E(y)$ increased, we should expect the range of the residuals to increase as $\hat{y}_u$ increased, giving rise to a triangular shaped scatter. The plot in Figure 7.1b shows no tendency of this sort, however. If the y's were not suitably transformed, we should expect to see a quadratic tendency in Figure 7.1b; no such tendency is evident.

To see how this quadratic tendency *can* occur, suppose in general that some model form

$$\beta_0 + g(\mathbf{x}, \boldsymbol{\beta}) \tag{7.1.10}$$

is being considered, where $g(\mathbf{x}, \boldsymbol{\beta})$ is some graduating function; for instance in the textile example, $g(\mathbf{x}, \boldsymbol{\beta}) = \beta_1 x_1 + \beta_2 x_2 + \beta_3 x_3$. Suppose that the model form would be more nearly appropriate after suitable data transformation. That is to say, a better approximation would be provided by the model form

$$\eta^{(\lambda)} = \beta_0 + g(\mathbf{x}, \boldsymbol{\beta}) \tag{7.1.11}$$

where $\eta^{(\lambda)}$ is a parametric transformation of η, such as a power transformation, for example. In these circumstances, we can employ, over moderate ranges of η, the approximation

$$\eta^{(\lambda)} = \alpha_0 + \alpha_1\eta - \alpha_2\eta^2. \tag{7.1.12}$$

Then, after equating (7.1.11) and (7.1.12), dividing through by α_1, and renaming the parameters, the model form becomes, approximately,

$$\eta = \alpha\eta^2 + \beta_0' + g(\mathbf{x}, \boldsymbol{\beta}'). \tag{7.1.13}$$

This relationship applies for each set of observations so that

$$\eta_u - \beta_0' - g(\mathbf{x}_u, \boldsymbol{\beta}') = \alpha\eta_u^2, \qquad u = 1, 2, \ldots, n. \tag{7.1.14}$$

We can obtain rough estimates of the left-hand side of (7.1.14) by fitting the model $y_u = \beta_0' + g(\mathbf{x}_u, \boldsymbol{\beta}') + \varepsilon$ by least squares, and finding the residuals $y_u - \hat{y}_u$. Also the same $\hat{y}_u$'s will provide estimates of the η_u's. Then, if $y = \log Y$ were an unsuitable transformation, we should expect a quadratic relationship to show up between the $y_u - \hat{y}_u$ and the $\hat{y}_u$ because they would reflect the relationship indicated in (7.1.14). No such tendency appears, however, as Figure 7.1b shows.

An Analysis of the Unlogged Data

To illustrate the importance of appropriate transformation and the value of the various residual checks, we consider what would have happened if the logarithmic transformation had *not* been used in this example. Table 7.5 summarizes the results of fitting polynomials of degrees 0, 1, and 2 to the original (unlogged) data Y. The corresponding analysis of variance for Y is shown in Table 7.6. If, as before, the assumption were made that an adequate model was supplied by a polynomial of *at most* second degree, then the residual mean square will provide an estimate of error. The ratio

$$\frac{\text{mean square for added second-order terms}}{\text{mean square for residual}} = 9.52 \tag{7.1.15}$$

is now much larger than for the logged data, and reference to the F table with 6 and 17 df shows it to be highly significant. Thus, for the unlogged data, second degree terms are needed to represent curvature previously taken account of by the log transformation, and the fitted equation now

TABLE 7.5.   Textile data: Polynomials of zero, first, and second degree fitted to original (unlogged) data

---

Zero Degree Polynomial

Estimated response function is $\hat{Y} = \overline{Y} = 861.3$
$\qquad\qquad\qquad\qquad\qquad\qquad\quad \pm 52.3$

Regression sum of squares is $S_0 = 20{,}031.2 \times 10^3$ (1 df)

First Degree Polynomial

Estimated response function is $\hat{Y} = 861.3 + 660.0x_1 - 535.9x_2 - 310.8x_3$
$\qquad\qquad\qquad\qquad\qquad\qquad\quad \pm 52.3 \quad \pm 64.1 \qquad \pm 64.1 \qquad \pm 64.1$

Regression sum of squares is $S_1 = 34{,}779.7 \times 10^3$ (4 df)

Second Degree Polynomial

Estimated response function is $\hat{Y} = 550.7 + 660.0x_1 - 535.9x_2 - 310.8x_3$
$\qquad\qquad\qquad\qquad\qquad\qquad\quad \pm 138.4 \quad \pm 64.1 \qquad \pm 64.1 \qquad \pm 64.1$

$\qquad\qquad\qquad\qquad\qquad\qquad\quad + 238.7x_1^2 + 275.7x_2^2 - 48.3x_3^2$
$\qquad\qquad\qquad\qquad\qquad\qquad\quad \pm 111.0 \quad\;\; \pm\; 111.0 \quad \pm 111.0$

$\qquad\qquad\qquad\qquad\qquad\qquad\quad - 456.5x_1x_2 - 235.7x_1x_3 + 143.0x_2x_3$
$\qquad\qquad\qquad\qquad\qquad\qquad\quad \pm 78.5 \qquad\;\; \pm 78.5 \qquad \pm 78.5$

Regression sum of squares is $S_2 = 39{,}004.0 \times 10^3$ (10 df)

TABLE 7.6.   Analysis of variance for unlogged textile data

Source	SS $\times 10^{-3}$	df	MS $\times 10^{-3}$
Mean (zero degree polynomial)	20,031.2	1	
Added first-order terms	14,748.5	3	4,916.2
Added second-order terms	4,224.3	6	704.1
Residual	1,256.6	17	73.9
Total	40,260.6	27	

takes the more complicated second degree form

$$\hat{Y} = 550.7 + 660.0x_1 - 535.9x_2 - 310.8x_3$$
$$+ 238.7x_1^2 + 275.7x_2^2 - 48.3x_3^2$$
$$- 456.5x_1x_2 - 235.7x_1x_3 + 143.0x_2x_3. \qquad (7.1.16)$$

Notice that this more complicated representation in terms of Y is less satisfactory in accounting for the variation in the data than the simple first-order representation in $y = \log Y$. Specifically, the F value corresponding to the ratio (regression mean square)/(residual mean square) is 7.6 times larger for the logged data employing a first-order graduating function, than for the unlogged data employing a second-order graduating function. In this example it may also be shown (Box and Fung, 1983) that the reduction factor in the variance of estimation of the response caused by employing the transformed (logged) response in combination with the simpler (first degree) model, is greater than 1 at every point in the design. At one point it is as high as 308. The values of Y, $\hat{Y}$, and $Y - \hat{Y}$ are shown in Table 7.4, both for the first degree fitted model and for the second degree fitted model. One feature immediately apparent from the fitted first degree model in the original metric is that it does not make physical sense, because four of the calculated values of $\hat{Y}$, representing the "number of cycles to failure," are negative. The need for a transformation of the original observations Y is confirmed by the quadratic tendency evident in the plots of $Y - \hat{Y}$ versus $\hat{Y}$ shown in Figures 7.2b and 7.3b for the first degree and the

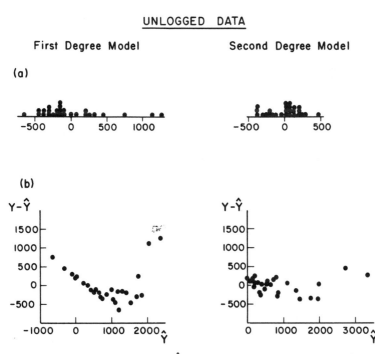

FIGURE 7.2. Plots of residuals $Y_u - \hat{Y}_u$ from first degree model fitted to unlogged data; (a) Residuals overall, (b) Residuals versus $\hat{Y}_u$.

FIGURE 7.3. Plots of residuals $Y_u - \hat{Y}_u$ from second degree model fitted to unlogged data; (a) Residuals overall, (b) Residuals versus $\hat{Y}_u$.

second degree models. Evidently the need for transformation is not totally eliminated by the introduction of second-order terms.

7.2. A COMPREHENSIVE CHECK OF THE NEED FOR TRANSFORMATION IN THE RESPONSE VARIABLE

A comprehensive check of the need for transformation is provided (Box, 1980; see also Atkinson, 1973b) by the appropriate predictive score function

$$g = \sum_{u=1}^{n} \frac{z_u r_u}{s} \tag{7.2.1}$$

where (for ln notation, see p. 269),

$$z_u = Y_u \left\{ 1 - \ln\left(\frac{Y_u}{\dot{Y}}\right) \right\}, \qquad r_u = \frac{(Y_u - \hat{Y}_u)}{s}, \qquad s^2 = \sum_{u=1}^{n} \frac{(Y_u - \hat{Y}_u)^2}{(n-p)}, \tag{7.2.2}$$

and where n is the number of observations, p is the number of parameters in the model, and

$$\ln \dot{Y} = \frac{\left(\sum_{u=1}^{n} \ln Y_u \right)}{n}, \tag{7.2.3}$$

so that $\dot{Y}$ is the geometric mean of the data.

Equivalently, a graphical check is provided by plotting (a) the residuals $z_u - \hat{z}_u$ obtained after fitting the constructed variable z_u to a model of the desired order or type against (b) the residuals obtained from fitting the same form of model to the response of current interest (Y or $y = \log_{10} Y$ as the case may be). The plots for the first degree models for unlogged and logged data are given in Figures 7.4a and 7.4b. The strong dependence shown in the "unlogged" plot clearly indicates the need for transformation. In the logged data plot, this dependence has disappeared, indicating the success of the log transformation actually made.

It is instructive to consider the relationship between the score function plot Figure 7.4a and the earlier plot of residuals overall and versus $\hat{Y}$ for the first degree model, in Figures 7.2a and 7.2b. These earlier plots are made in terms of nonstandardized quantities but we could recast them in terms of the standardized

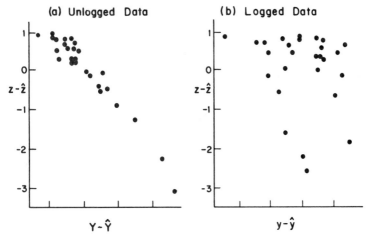

FIGURE 7.4. Plot of residuals from score function statistic versus corresponding residuals from the data, (a) unlogged, (b) logged, after fitting first degree polynomial model.

residuals $r_u = (Y_u - \hat{Y}_u)/s$ and standardized deviations of calculated values from the sample average $c_u = (\hat{Y}_u - \bar{Y})/s$ by making origin and scale changes. The shapes would not alter. In these plots the need for transformation may be evidenced by:

1. Skewness in the overall plot of the residuals r_u.
2. A tendency for the spread of the residuals r_u to become larger or smaller as c_u increases or decreases.
3. A quadratic tendency in a plot of residuals r_u versus c_u.

As was noted by Anscombe (1961) and by Anscombe and Tukey (1963), appropriate checking functions for these three characteristics are, respectively,

$$(a)\ \Sigma r_u^3 = T_{30},$$

$$(b)\ \Sigma r_u^2 c_u = T_{21},$$

$$(c)\ \Sigma r_u c_u^2 = T_{12}. \qquad (7.2.4)$$

An additional checking function that would measure skewness is

$$d = \frac{(\bar{Y} - \dot{Y})}{s}, \qquad (7.2.5)$$

the standardized difference between the arithmetic and geometric means for the original data.

Now it can be shown that z_u is closely approximated by the quadratic expression

$$z_u \doteq \dot{Y} + B(Y_u - \dot{Y})^2, \tag{7.2.6}$$

where B is a constant that can be evaluated. Thus, after writing

$$\frac{(Y_u - \dot{Y})}{s} = r_u + c_u + d \tag{7.2.7}$$

we see that

$$g \doteq Bs \sum_{u=1}^{n} (r_u + c_u + d)^2 r_u \tag{7.2.8}$$

$$\propto \{2(n-p)d + T_{30}\} + 2T_{21} + T_{12}, \tag{7.2.9}$$

after reduction, using the facts that $\Sigma r_u = 0$, for any linear model containing a β_0 term, $\Sigma r_u \hat{Y}_u = 0$ for any linear model, and $(n-p)s^2 = \Sigma(Y_u - \hat{Y}_u)^2$. Thus the score function plot appropriately brings together all the evidence for transformation.

An exact test for one type of possible discrepancy mentioned above, namely (3) a quadratic tendency in a plot of residuals r_u versus c_u, can be carried out, as shown by Tukey (1949); see also Tukey and Moore (1954) and Anscombe (1961). The test is usually referred to as "Tukey's one degree of freedom test for additivity" and is described in Appendix 7B.

7.3.   INTERIM SUMMARY

1.   Although we have used this example to illustrate the mechanism of fitting a second degree model, we have also taken the opportunity to illustrate that the mechanical application of the technique can be dangerous (as can the mechanical application of any technique).

2.   In the original analysis of these data, Barella and Sust had fitted a second degree polynomial expression to the unlogged data Y. The above analysis (see also Box and Cox, 1964) shows that a simpler and more precise representation is possible by first transforming the data via $y = \log Y$, after which a first degree polynomial adequately expresses the relationship. With the response measured in the y metric, all six of the second-order parameters $\beta_{11}, \beta_{22}, \beta_{33}, \beta_{12}, \beta_{13}, \beta_{23}$ can be dispensed with, and the representation by the first degree equation in $\log Y$ is much better than that of the second degree equation in Y itself. Indeed, a very large gain in accuracy of *estimation of the response* is achieved by using the simpler model.

3. Although, in this example, the need to fit a second degree expression is avoided by an appropriate use of a transformation, we shall later consider examples where a function of at least second degree in the inputs is essential to represent the dependence of output on inputs. This is especially so when we wish to represent a multidimensional maximum or minimum in the response.

4. The example illustrates how appropriate plotting of residuals can help to diagnose model inadequacy (and, in particular, the need for a transformation).

5. We later discuss how an appropriate transformation may be *estimated* from the data itself. When such an analysis is conducted for the textile data, it selects essentially the logarithmic transformation. However we emphasize that the choice of $\log Y$ rather than Y is a natural one * and is also suggested by physical considerations.

7.4. FURTHER EXAMINATION OF ADEQUACY OF FIT; HIGHER-ORDER TERMS

The textile example has allowed us to examine a number of general questions concerning the fitting of empirical functions and the use of transformations. We shall now employ this same example to illustrate some further general points. The analysis has so far been conducted on the assumption that, at most, an equation of second degree can represent functional dependence between the response and the inputs. It is natural to ask how the analyses would be affected if terms of third order were introduced, that is, if the "true" model were of the form

$$y = \beta_0 + \beta_1 x_1 + \beta_2 x_2 + \beta_3 x_3 + \beta_{11} x_1^2 + \beta_{22} x_2^2 + \beta_{33} x_3^2 + \beta_{12} x_1 x_2$$
$$+ \beta_{13} x_1 x_3 + \beta_{23} x_2 x_3 + \left\{ \beta_{111} x_1^3 + \beta_{222} x_2^3 + \beta_{333} x_3^3 + \beta_{122} x_1 x_2^2 \right.$$
$$+ \beta_{112} x_1^2 x_2 + \beta_{133} x_1 x_3^2 + \beta_{113} x_1^2 x_3$$
$$\left. + \beta_{233} x_2 x_3^2 + \beta_{223} x_2^2 x_3 + \beta_{123} x_1 x_2 x_3 \right\} + \varepsilon. \quad (7.4.1)$$

---

* We have sometimes been dismayed to find that the engineer newly introduced to statistical methods may put statistics and engineering in different compartments of his mind and feel that when he is doing statistics, he no longer needs to be an engineer. This is of course not so. All his engineering knowledge, hunches, and cunning concerning such matters as choice of variables and transformation of variables must still be employed in conjunction with statistical design and analysis. Statistical methods used with good engineering know how make a powerful combination, but poor engineering combined with mechanical and unimaginative use of inadequately understood statistical methods can be disastrous.

For the particular, but important, 3^3 design used in the textile experiment, we shall consider three distinct but closely related questions:

1. How would coefficient estimates, derived on the assumed adequacy of a second-order model, be biased by the existence of third-order terms?
2. How may the need for third-order terms be evidenced?
3. If real third-order terms exist, to what extent can we estimate them?

Biases in Second Degree Model Estimates Produced by Existence of Third-Order Terms

We show in Appendix 7C that, for the 3^3 design of the textile experiment, all the estimates in the second degree model except those of the first-order terms are unbiased by third-order terms. For the first-order terms, we find

$$E(b_1) = \beta_1 + \beta_{111} + \tfrac{2}{3}\{\beta_{122} + \beta_{133}\},$$

$$E(b_2) = \beta_2 + \beta_{222} + \tfrac{2}{3}\{\beta_{112} + \beta_{233}\},$$

$$E(b_3) = \beta_3 + \beta_{333} + \tfrac{2}{3}\{\beta_{113} + \beta_{223}\}. \tag{7.4.2}$$

Existence and Estimation of Third-Order Terms

We can check for overall existence of third-order terms by separating an appropriate component sum of squares in the analysis of variance table. However, a difficulty arises. For any particular regressor x_i, the 3^3 design employs only three levels which we can denote by the coded values $(-1, 0, 1)$. The added regressor x_i^3 also has the same three levels $(-1, 0, 1)$, and so the first-order and pure cubic effects (β_i and β_{iii}) cannot be separately estimated, for $i = 1, 2, 3$. The remaining seven third-order effects $\beta_{112}, \beta_{122}, \beta_{113}, \beta_{133}, \beta_{223}, \beta_{233}, \beta_{123}$ are, however, all estimable. We therefore proceed in this case by fitting the third-order expression (7.4.1) but *omitting* the regressors x_1^3, x_2^3, and x_3^3. Details will be found in Appendix 7D. An extra sum of squares based on 7 df corresponding to "added third-order terms" may be constructed and its magnitude can supply some idea of the necessity for representation by a third-order polynomial.

If this type of analysis indicates that the third-order terms we can estimate appear to be important, we might consider the possibility of appropriately transforming the response and/or predictor variables (Box and Draper, 1982a; Draper, 1983) or, failing this, of running a more elaborate design which allowed all such third-order terms to be estimated for the original response and predictors.

TABLE 7.7.   Analysis of variance of textile data showing effect of third-order terms

Source	Logged SS	Unlogged SS $\times 10^{-3}$	df	Logged data MS	Unlogged data MS $\times 10^{-3}$
Mean	204.2975	20,031.2	1	204.2975	20,031.2
Added first-order terms	4.2317	14,748.5	3	1.4106	4,916.2
Added second-order terms	0.0281	4,224.3	6	0.0047	704.1
Added third-order terms	0.0877	966.9	7	0.0125 ⎫ $F = 3.68$	138.1 ⎫ $F = 4.76$
Residual	0.0336	289.8	10	0.0034 ⎭	29.0 ⎭

The analyses of variance for both logged and unlogged data are given for the textile example in Table 7.7. Since the need for transformation of the unlogged response has been demonstrated, it is hardly surprising to find an F ratio of $138.1/29.0 = 4.76$, indicating the need for third-order terms. It will be seen, however, that a similar F ratio $0.0125/0.0034 = 3.68$ is also found for *the logged data* and this must be considered more carefully.

For the logged data the sum of squares associated with the 7 df for third-order terms is 0.0877. This sum of squares may be split into the component parts shown in Table 7.8.

The main contribution to the total arises from b_{122}, which implies that the pure quadratic contribution for predictor variable x_2 changes as the level of x_1 changes. It must be remembered that anomalous effects of this kind are sometimes produced by an individual outlying value. Examination of the residuals does not indicate that this is the case here, however. The

TABLE 7.8.   Logged data. Further analysis of the extra sum of squares
for third-order terms

Source	SS	df
b_{122}	0.0552	1
b_{112}	0.0038	1
b_{113}	0.0000	1
b_{133}	0.0156	1
b_{233}	0.0003	1
b_{223}	0.0000	1
b_{123}	0.0128	1
Total third order	0.0877	7

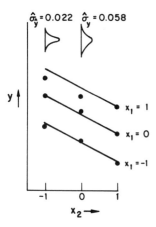

FIGURE 7.5. The b_{122} effect. Also shown are normal distributions: (i) with standard deviation $\hat{\sigma}_y = 0.058$, (ii) with standard deviation $\hat{\sigma}_{\hat{y}} = 0.022$.

effect is exhibited in Figure 7.5 by plotting, for each level of x_1, the response at each level of x_2 averaged over all levels of x_3. The straight lines in the figure are corresponding averages of values $\hat{y}$ calculated from the fitted first degree equation. Also shown in the same figure are normal distributions with standard deviations $\hat{\sigma}_y = 0.058$ and $\hat{\sigma}_{\hat{y}} = 0.022$. The numerical value $\hat{\sigma}_y^2 = 0.0034$ has been obtained from the last line of the analysis of variance of Table 7.7 and is an estimate of the experimental error variance. Also $\hat{\sigma}_{\hat{y}}^2 = 0.0034 \times \frac{4}{27}$ is an estimate of the *average* variance of the $\hat{y}$'s (see Appendix 3E). It is clear that, although the b_{122} effect is clearly visible, it is quite small in relation to the more important characteristics exhibited by the data, namely the first-order effects, which stand out very clearly. Thus, in this example, even though the existence of a significant third-order effect can be demonstrated, for most purposes the convenient first degree model in $y = \log Y$ would adequately describe the system under study over the ranges considered.

We are reminded that:

1. Just as lack of a significant effect does not demonstrate that no effect exists but only that, for the particular experiment under study, none can be demonstrated, so the presence of a significant effect does not necessarily imply an important effect that needs to be taken account of in the model.

2. If wider ranges of the variables were being studied in a subsequent trial, these high-order effects might well become important and this possibility would have to be borne in mind.

7.5. ROBUST METHODS IN FITTING GRADUATING FUNCTIONS

As we explained in Chapter 3, occasional "faulty" observations can be expected to occur in practical experimental circumstances. They could arise, for example, from copying errors, mistakes in measurement or mistakes in setting variable levels. Robust procedures can both minimize the effect of faulty values as well as help to identify them. A number of different robust estimation procedures have been proposed. These differ in their details, but

TABLE 7.9. Box–Behnken second-order design in four variables with two sets of data from (a) Bacon (1970) and (b) Box and Behnken (1960a).

Block number	Coded levels of variables				(a) Bacon data	(b) Box–Behnken data
	x_1	x_2	x_3	x_4		
1	-1	-1	0	0	84.5	84.7
	1	-1	0	0	62.9	93.3
	-1	1	0	0	90.7	84.2
	1	1	0	0	63.2	86.1
	0	0	-1	-1	70.9	85.7
	0	0	1	-1	69.2	96.4
	0	0	-1	1	80.1	88.1
	0	0	1	1	79.8	81.8
	0	0	0	0	75.1	93.8
2	-1	0	0	-1	81.8	89.4
	1	0	0	-1	61.8	88.7
	-1	0	0	1	92.2	77.8
	1	0	0	1	70.7	80.9
	0	-1	-1	0	72.4	80.9
	0	1	-1	0	76.4	79.8
	0	-1	1	0	71.9	86.8
	0	1	1	0	74.9	79.0
	0	0	0	0	74.5	87.3
3	0	-1	0	-1	69.4	86.1
	0	1	0	-1	71.2	87.9
	0	-1	0	1	77.3	85.1
	0	1	0	1	81.1	76.4
	-1	0	-1	0	88.0	79.7
	1	0	-1	0	63.0	92.5
	-1	0	1	0	86.7	89.4
	1	0	1	0	65.0	86.9
	0	0	0	0	73.0	90.7

TABLE 7.10a. Estimates of coefficients for various values of f, for Bacon (1970) data; standard errors are given in parentheses

Coefficient estimated	Value of f			
	∞	5	3	1
β_0	74.20 (0.62)	74.25 (0.58)	74.27 (0.56)	74.33 (0.50)
β_1	-11.44 (0.31)	-11.48 (0.31)	-11.49 (0.30)	-11.50 (0.30)
β_2	1.59 (0.31)	1.59 (0.29)	1.59 (0.27)	1.60 (0.24)
β_3	-0.28 (0.31)	-0.29 (0.28)	-0.29 (0.27)	-0.30 (0.23)
β_4	4.74 (0.31)	4.72 (0.29)	4.72 (0.28)	4.71 (0.25)
β_{11}	1.50 (0.47)	1.50 (0.45)	1.50 (0.44)	1.48 (0.41)
β_{22}	-0.33 (0.47)	-0.33 (0.43)	-0.33 (0.42)	-0.34 (0.38)
β_{33}	-0.03 (0.47)	-0.07 (0.43)	-0.09 (0.41)	-0.15 (0.37)
β_{44}	0.88 (0.47)	0.85 (0.44)	0.85 (0.43)	0.82 (0.40)
β_{12}	-1.48 (0.54)	-1.48 (0.53)	-1.48 (0.53)	-1.48 (0.52)
β_{13}	0.83 (0.54)	0.85 (0.50)	0.86 (0.48)	0.88 (0.45)
β_{14}	-0.38 (0.54)	-0.37 (0.57)	-0.37 (0.58)	-0.37 (0.60)
β_{23}	-0.25 (0.54)	-0.25 (0.48)	-0.25 (0.45)	-0.26 (0.38)
β_{24}	0.50 (0.54)	0.48 (0.48)	0.47 (0.46)	0.44 (0.40)
β_{34}	0.35 (0.54)	0.34 (0.48)	0.33 (0.46)	0.30 (0.40)

TABLE 7.10b. Estimates of coefficients for various values of f, for Box and Behnken (1960a) data. Standard errors are given in parentheses

Coefficient estimated	Value of f			
	∞	5	3	1
β_0	96.60 (0.84)	90.59 (0.64)	90.58 (0.59)	90.53 (0.48)
β_1	1.93 (0.42)	2.32 (0.33)	2.30 (0.30)	2.42 (0.23)
β_2	-1.96 (0.42)	-1.96 (0.31)	-1.96 (0.28)	-1.95 (0.22)
β_3	1.13 (0.42)	1.14 (0.31)	1.13 (0.28)	1.12 (0.22)
β_4	-3.68 (0.42)	-3.37 (0.33)	-3.31 (0.30)	-3.19 (0.23)
β_{11}	-1.42 (0.63)	-1.47 (0.50)	-1.49 (0.46)	-1.52 (0.39)
β_{22}	-4.33 (0.63)	-4.29 (0.48)	-4.27 (0.44)	4.22 (0.36)
β_{33}	-2.24 (0.63)	-2.21 (0.48)	-2.19 (0.44)	-2.14 (0.36)
β_{44}	-2.58 (0.63)	-2.63 (0.49)	-2.65 (0.46)	-2.69 (0.39)
β_{12}	-1.68 (0.73)	-1.67 (0.55)	-1.67 (0.49)	-1.67 (0.38)
β_{13}	-3.83 (0.73)	-3.82 (0.54)	-3.82 (0.49)	-3.82 (0.37)
β_{14}	0.95 (0.73)	-0.76 (0.70)	-0.68 (0.69)	-0.49 (0.68)
β_{23}	-1.68 (0.73)	-1.67 (0.54)	-1.66 (0.49)	-1.64 (0.40)
β_{24}	-2.63 (0.73)	-2.63 (0.54)	-2.63 (0.49)	-2.63 (0.38)
β_{34}	-4.25 (0.73)	-4.25 (0.54)	-4.25 (0.48)	-4.25 (0.36)

usually not very much in their conclusions. Here, we employ the iteratively reweighted least-squares method introduced in Section 3.13 with the weight function (3.13.22), namely,

$$w_u = \frac{(f+1)}{(f + r_u^2/s^2)},$$ (7.5.1)

where, at each iteration, the r_u are the residuals and s^2 is the residual mean square taken from the previous stage, and f is a selected constant. (Below, we choose $f = \infty, 5, 3,$ and 1.)

TABLE 7.11a. Values of weights w_u for various values of f, Bacon (1970) data

Observation number	Value of f			
	∞	5	3	1
1	1	0.99	1.00	1.03
2	1	0.85	0.80	0.70
3	1	0.96	0.96	0.99
4	1	0.87	0.83	0.74
5	1	1.19	1.32	1.90
6	1	1.02	1.03	1.03
7	1	1.12	1.18	1.38
8	1	1.20	1.33	2.00
9	1	1.00	1.01	1.08
10	1	0.75	0.68	0.57
11	1	0.85	0.81	0.72
12	1	0.82	0.77	0.67
13	1	0.78	0.72	0.62
14	1	1.20	1.33	1.98
15	1	1.11	1.17	1.36
16	1	1.09	1.14	1.27
17	1	1.20	1.32	1.95
18	1	1.19	1.31	1.92
19	1	1.00	1.00	0.98
20	1	1.17	1.28	1.81
21	1	1.20	1.33	2.00
22	1	1.12	1.19	1.41
23	1	1.18	1.28	1.65
24	1	1.05	1.08	1.17
25	1	0.91	0.87	0.78
26	1	1.10	1.13	1.20
27	1	0.90	0.86	0.78

For illustration, we consider two sets of data, each obtained when studying four predictor variables using a particular kind of three-level arrangement called a Box–Behnken design (see Chapter 15). In general, this class of second-order designs requires many fewer runs than the complete three-level factorials. As indicated in Table 7.9, the four-factor design may be run in three orthogonal blocks. Two sets of data were obtained in quite separate investigations at different times by different experimenters. One set is taken from Bacon (1970) and the other from Box and Behnken (1960a).

TABLE 7.11b. Values of weights w_u for various values of f, Box and Behnken (1960a) data

Observation number	Value of f			
	∞	5	3	1
1	1	1.06	1.11	1.39
2	1	1.16	1.28	1.85
3	1	1.09	1.16	1.50
4	1	1.14	1.25	1.78
5	1	1.16	1.21	1.67
6	1	1.09	1.16	1.57
7	1	1.16	1.28	1.90
8	1	1.18	1.31	1.95
9	1	1.01	1.00	0.93
10	1	0.43	0.34	0.22
11	1	1.20	1.32	1.88
12	1	1.66	1.26	1.62
13	1	0.51	0.43	0.32
14	1	1.17	1.28	1.72
15	1	1.15	1.22	1.39
16	1	1.08	1.09	1.07
17	1	1.20	1.33	1.94
18	1	1.05	1.07	1.15
19	1	1.08	1.15	1.48
20	1	1.20	1.33	1.72
21	1	1.20	1.33	1.78
22	1	1.07	1.13	1.41
23	1	1.17	1.30	1.96
24	1	1.09	1.15	1.49
25	1	1.10	1.17	1.56
26	1	1.16	1.29	1.92
27	1	1.20	1.33	1.95

Our purpose here is only to illustrate the estimation of the model coefficients using a robust analysis, not to discuss the resulting fitted equations. Tables 7.10a and 7.10b show the estimates of the coefficients of a second degree polynomial and (in parentheses) their standard errors using the weight function (7.5.1) with $f = \infty$, 5, 3, and 1, and with block effects eliminated, for (a) Bacon data and (b) the Box–Behnken data. The corresponding weights used are shown in Tables 7.11a and 7.11b.

Note that, for the Bacon data, the changes that occur in the estimates and their standard errors are slight. For the Box–Behnken data, they are much more pronounced, however; in particular, note the large change in the estimate of β_{14}. In Table 7.11b we see, for the Box–Behnken data, the very large down-weightings of runs 10 and 13 that occur as f is made small. This strongly suggests that observations 10 and 13 may be faulty.

Note: The following appendixes make use of various results in least-squares theory. A discussion of least squares is given in Chapter 3.

APPENDIX 7A.   FITTING OF FIRST AND SECOND DEGREE MODELS

The matrix $\mathbf{Z}_s$ of regressors for the second-order (or quadratic) model and the response data, both unlogged and logged, are shown in Table 7A.1. The matrix $\mathbf{Z}_f$ of regressors for the first-order model consists of the first four columns of $\mathbf{Z}_s$. Matrices needed for the fitting of the first- and second-order models are shown in Tables 7A.2 and 7A.3.

The matrix $\mathbf{Z}_s'\mathbf{Z}_s$ has a pattern of a special kind frequently encountered in the fitting of second degree polynomial models. Suppose in general that there are k variables $x_1, x_2, \ldots, x_k$ and that the columns of the $n \times \frac{1}{2}(k + 1)(k + 2)$ matrix $\mathbf{Z}_s$ are rearranged so that the first $k + 1$ columns are those corresponding to

$$1, x_1^2, x_2^2, \ldots, x_k^2.$$

Denote the matrix with rearranged columns by $\dot{\mathbf{Z}}$. Now suppose that $\dot{\mathbf{Z}}'\dot{\mathbf{Z}}$ is partitioned after the first $k + 1$ rows and columns. Then it is easily seen that the $\dot{\mathbf{Z}}'\dot{\mathbf{Z}}$ derived from $\mathbf{Z}_s$ in Table 7A.3 is of the form

$$\dot{\mathbf{Z}}'\dot{\mathbf{Z}} = \left[\begin{array}{c|c} \mathbf{M} & \mathbf{0} \\ \hline \mathbf{0}' & \mathbf{P} \end{array} \right] \tag{7A.1}$$

where $\mathbf{M}$ is a $k + 1$ by $k + 1$ matrix, $\mathbf{0}$ is a null matrix of size $(k + 1) \times \frac{1}{2}k(k + 1)$,

TABLE 7A.1. Matrix $\mathbf{Z}_s$ of regressors for second-order model and the data in unlogged and logged form

1	x_1	x_2	x_3	x_1^2	x_2^2	x_3^2	x_1x_2	x_1x_3	x_2x_3	Y	$y = \log Y$
1	-1	-1	-1	1	1	1	1	1	1	674	2.83
1	·	-1	-1	·	1	1	·	·	1	1414	3.15
1	1	-1	-1	1	1	1	-1	-1	1	3636	3.56
1	-1	·	-1	1	·	1	·	1	·	338	2.53
1	·	·	-1	·	·	1	·	·	·	1022	3.01
1	1	·	-1	1	·	1	·	-1	·	1568	3.19
1	-1	1	-1	1	1	1	-1	1	-1	170	2.23
1	·	1	-1	·	1	1	·	·	-1	442	2.65
1	1	1	-1	1	1	1	1	-1	-1	1140	3.06
1	-1	-1	·	1	1	·	1	·	·	370	2.57
1	·	-1	·	·	1	·	·	·	·	1198	3.08
1	1	-1	·	1	1	·	-1	·	·	3184	3.50
1	-1	·	·	1	·	·	·	·	·	266	2.42
1	·	·	·	·	·	·	·	·	·	620	2.79
1	1	·	·	1	·	·	·	·	·	1070	3.03
1	-1	1	·	1	1	·	-1	·	·	118	2.07
1	·	1	·	·	1	·	·	·	·	332	2.52
1	1	1	·	1	1	·	1	·	·	884	2.95
1	-1	-1	1	1	1	1	1	-1	-1	292	2.47
1	·	-1	1	·	1	1	·	·	-1	634	2.80
1	1	-1	1	1	1	1	-1	1	-1	2000	3.30
1	-1	·	1	1	·	1	·	-1	·	210	2.32
1	·	·	1	·	·	1	·	·	·	438	2.64
1	1	·	1	1	·	1	·	1	·	566	2.75
1	-1	1	1	1	1	1	-1	-1	1	90	1.95
1	·	1	1	·	1	1	·	·	1	220	2.34
1	1	1	1	1	1	1	1	1	1	360	2.56

and $\mathbf{P}$ is $\frac{1}{2}k(k+1) \times \frac{1}{2}k(k+1)$ and diagonal. The inverse is thus of the form

$$(\dot{\mathbf{Z}}'\dot{\mathbf{Z}})^{-1} = \left[\begin{array}{c|c} \mathbf{M}^{-1} & \mathbf{0} \\ \hline \mathbf{0}' & \mathbf{P}^{-1} \end{array}\right] \tag{7A.2}$$

and the elements of the diagonal matrix $\mathbf{P}^{-1}$ are the reciprocals of the corresponding elements in the matrix $\mathbf{P}$. Furthermore, $\mathbf{M}$ is a patterned matrix of the general form

$$\mathbf{M} = \begin{array}{c} \\ \\ \\ \\ \\ \end{array}\begin{bmatrix} d & e & e & e & \cdots & e \\ e & f & g & g & \cdots & g \\ e & g & f & g & \cdots & g \\ \vdots & & & & & \\ e & g & g & g & \cdots & f \end{bmatrix}\begin{array}{c} 0 \\ 11 \\ 22 \\ \vdots \\ kk. \end{array} \tag{7A.3}$$

with column labels $0\ \ 11\ \ 22\ \ 33\ \ \cdots\ \ kk$

TABLE 7A.2. Fitting of the first degree model

$$\mathbf{Z}_f'\mathbf{Z}_f \qquad\qquad (\mathbf{Z}_f'\mathbf{Z}_f)^{-1}$$

$$\begin{bmatrix} 27 & \cdot & \cdot & \cdot \\ \cdot & 18 & \cdot & \cdot \\ \cdot & \cdot & 18 & \cdot \\ \cdot & \cdot & \cdot & 18 \end{bmatrix} \qquad \begin{bmatrix} \frac{1}{27} & \cdot & \cdot & \cdot \\ \cdot & \frac{1}{18} & \cdot & \cdot \\ \cdot & \cdot & \frac{1}{18} & \cdot \\ \cdot & \cdot & \cdot & \frac{1}{18} \end{bmatrix}$$

$\mathbf{Z}_f'\mathbf{Y}$	$\mathbf{Z}_f'\mathbf{y}$	$\mathbf{b}_Y$	$\mathbf{b}_y$
$\begin{bmatrix} 23{,}256 \\ 11{,}880 \\ -9{,}646 \\ -5{,}594 \end{bmatrix}$	$\begin{bmatrix} 74.27 \\ 6.51 \\ -4.93 \\ -3.08 \end{bmatrix}$	$\begin{bmatrix} 861.3 \\ 660.0 \\ -535.9 \\ -310.8 \end{bmatrix}$	$\begin{bmatrix} 2.751 \\ 0.362 \\ -0.274 \\ -0.171 \end{bmatrix}$

It is readily confirmed that the inverse follows a similar pattern with elements denoted by the corresponding capital letters

$$D = H^{-1}\{f + (k-1)g\}(f - g)$$

$$E = -H^{-1}e(f - g)$$

$$F = H^{-1}\{df + (k-2)\,dg - (k-1)e^2\}$$

$$G = H^{-1}(e^2 - dg)$$

$$H = (f - g)\{df + (k-1)\,dg - ke^2\} \qquad\qquad (7A.4)$$

replacing their lowercase counterparts.

The matrices resulting from the three-level factorial design employed in the present chapter are of even more specialized form than those so far discussed. For these designs, the elements of **M** are

$$d = n, \qquad e = \frac{2n}{3}, \qquad f = \frac{2n}{3}, \qquad g = \frac{4n}{9},$$

where $n = 3^k$, whence the elements of $\mathbf{M}^{-1}$ are

$$D = \frac{(2k+1)}{n}, \qquad E = \frac{-3}{n}, \qquad F = \frac{9}{2n}, \qquad G = 0.$$

This particular result may also be obtained more directly.

TABLE 7A.3. Fitting of the second degree model

$\mathbf{Z'_\cdot Z_\cdot}$

	0	1	2	3	11	22	33	12	13	23
0	27	.	.	.	18	18	18	.	.	.
1	.	18	.	.	.	.	.	.	.	.
2	.	.	18	.	.	.	.	.	.	.
3	.	.	.	18	.	.	.	.	.	.
11	18	.	.	.	18	12	12	.	.	.
22	18	.	.	.	12	18	12	.	.	.
33	18	.	.	.	12	12	18	.	.	.
12	.	.	.	.	.	.	.	12	.	.
13	.	.	.	.	.	.	.	.	12	.
23	.	.	.	.	.	.	.	.	.	12

$(\mathbf{Z'_\cdot Z_\cdot})^{-1}$

$\dfrac{1}{108}$

28	.	.	.	-12	-12	-12	.	.	.
.	6	.	.	.	.	.	.	.	.
.	.	6	.	.	.	.	.	.	.
.	.	.	6	.	.	.	.	.	.
-12	.	.	.	18	.	.	.	.	.
-12	.	.	.	.	18	.	.	.	.
-12	.	.	.	.	.	18	.	.	.
.	.	.	.	.	.	.	9	.	.
.	.	.	.	.	.	.	.	9	.
.	.	.	.	.	.	.	.	.	9

$\mathbf{Z'_\cdot Y}$	$\mathbf{Z'_\cdot y}$	$\mathbf{b}_y$	$\mathbf{b}_\cdot$
23,256	74.27	550.6	2.786
11,880	6.51	660.0	0.362
-9,646	-4.93	-535.9	-0.274
-5,594	-3.08	-310.8	-0.171
16,936	49.29	238.7	-0.037
17,158	49.59	275.7	0.013
15,214	49.34	-48.3	-0.029
-5,478	-0.17	-456.5	-0.014
-2,828	-0.35	-235.7	-0.029
1,716	-0.12	143.0	-0.010

For the case $k = 3$ the resulting inverse is that given in Table 7A.3, after the appropriate rearrangement of rows and columns back to their original positions.

APPENDIX 7B. CHECKING FOR NONADDITIVITY WITH TUKEY'S TEST

We have seen how a graphical test for nonadditivity looks for a quadratic tendency in a plot of $y - \hat{y}$ versus $\hat{y}$. A formal exact test for such a tendency was proposed by Tukey (1949); see also Tukey and Moore (1954) and Anscombe (1961).

To perform the test, first fit the model in the ordinary way to obtain residuals $y - \hat{y}$ and calculated values $\hat{y}$. Now refit using the augmented model

$$y_u = \beta_0' + g(\mathbf{x}_u, \boldsymbol{\beta}') + \alpha q_u + \varepsilon_u \qquad (7B.1)$$

where $\beta_0' + g(\mathbf{x}_u, \boldsymbol{\beta}')$ is the original model form but with parameters renamed by the addition of primes, and where

$$q_u = \hat{y}_u^2. \qquad (7B.2)$$

[For additional detail, see the text around Eqs. (7.1.10) through (7.1.14).] The extra sum of squares for the estimate $\hat{\alpha}$ given the other parameters is associated with Tukey's "one degree of freedom for non-additivity." It may be compared with the new residual mean square, which now has 1 df fewer, to provide an exact F test.

For the textile example, the special symmetry of the original design permits the use of a more direct method of performing the calculations, as follows:

After the initial fit, $q_u = \hat{y}_u^2$ is computed and this is treated as a "pseudo-response" and is fitted to the (original) model form to give residuals $q_u - \hat{q}_u$. The required extra sum of squares corresponding to Tukey's 1 df is then

$$S_\alpha = \frac{\{\Sigma(y_u - \hat{y}_u)(q_u - \hat{q}_u)\}^2}{\Sigma(q_u - \hat{q}_u)^2}.$$

TABLE 7B.1. Adjustment of analysis of variance table to check nonadditivity

Source	SS	df	MS	F
Nonadditivity	$S_\alpha = \hat{\alpha}^2 \Sigma(q_u - \hat{q}_u)^2$	1	MS_α	$MS_\alpha/MS_{R'}$
Adjusted residual	$S_{R'} = S_R - S_\alpha$	$n - p - 1$	$MS_{R'}$	
Residual	$S_R = \Sigma(y_u - \hat{y}_u)^2$	$n - p$		

TABLE 7B.2. Checking for nonadditivity. First degree model fitted to unlogged data Y

Source	SS $\times 10^{-3}$	df	MS $\times 10^{-3}$	F
Nonadditivity	4037.0	1	4037.0	61.5
Adjusted residual	1444.0	22	65.6	
Residual	5481.0	23		

TABLE 7B.3. Checking for nonadditivity. Second degree model fitted to unlogged data Y

Source	SS $\times 10^{-3}$	df	MS $\times 10^{-3}$	F
Nonadditivity	704.3	1	704.3	20.4
Adjusted residual	552.4	16	34.5	
Residual	1256.7	17		

TABLE 7B.4. Checking for nonadditivity. First degree model fitted to logged data y

Source	SS	df	MS	F
Nonadditivity	0.00213	1	0.00213	0.32
Adjusted residual	0.14735	22	0.00670	
Residual	0.14948	23		

TABLE 7B.5. Checking for nonadditivity. Second degree model fitted to logged data y

Source	SS	df	MS	F
Nonadditivity	0.00233	1	0.00233	0.32
Adjusted residual	0.11814	16	0.00738	
Residual	0.12047	17		

The new residual sum of squares will be $S_R - S_\alpha$ and it will have 1 df fewer than S_R. The appropriate analysis of variance breakup of the residual sum of squares is shown in Table 7B.1. For the textile data we obtain the results in Tables 7B.2 through 7B.5. The analysis confirms that, while highly significant nonadditivity is found for the original data Y, use of $y = \log Y$ causes the nonadditivity to disappear.

APPENDIX 7C. THIRD-ORDER BIASES IN ESTIMATES FOR SECOND-ORDER MODEL, USING THE 3^3 DESIGN

The least-squares estimator

$$\mathbf{b}_1 = \left(\mathbf{Z}_1'\mathbf{Z}_1\right)^{-1}\mathbf{Z}_1'\mathbf{y} \qquad (7C.1)$$

TABLE 7C.1. Bias calculations for the textile example

$\mathbf{Z}_2$

	111	222	333	122	112	133	113	233	223	123
	−1	−1	−1	−1	−1	−1	−1	−1	−1	−1
	·	−1	−1	·	·	·	·	−1	−1	·
	1	−1	−1	1	−1	1	−1	−1	−1	1
	−1	·	−1	·	·	−1	−1	·	·	·
	·	·	−1	·	·	·	·	·	·	·
	1	·	−1	·	·	1	−1	·	·	·
	−1	1	−1	−1	1	−1	−1	1	−1	1
	·	1	−1	·	·	·	·	1	−1	·
	1	1	−1	1	1	1	−1	1	−1	−1
	−1	−1	·	−1	−1	·	·	·	·	·
	·	−1	·	·	·	·	·	·	·	·
	1	−1	·	1	−1	·	·	·	·	·
	−1	·	·	·	·	·	·	·	·	·
	·	·	·	·	·	·	·	·	·	·
	1	·	·	·	·	·	·	·	·	·
	−1	1	·	−1	1	·	·	·	·	·
	·	1	·	·	·	·	·	·	·	·
	1	1	·	1	1	·	·	·	·	·
	−1	−1	1	−1	−1	−1	1	−1	1	1
	·	−1	1	·	·	·	·	−1	1	·
	1	−1	1	1	−1	1	1	−1	1	−1
	−1	·	1	·	·	−1	1	·	·	·
	·	·	1	·	·	·	·	·	·	·
	1	·	1	·	·	1	1	·	·	·
	−1	1	1	−1	1	−1	1	1	1	−1
	·	1	1	·	·	·	·	1	1	·
	1	1	1	1	1	1	1	1	1	1

$\mathbf{Z}_1'\mathbf{Z}_2$

	111	222	333	122	112	133	113	233	223	123
0	·	·	·	·	·	·	·	·	·	·
1	18	·	·	12	·	12	·	·	·	·
2	·	18	·	·	12	·	·	12	·	·
3	·	·	18	·	·	·	12	·	12	·
11	·	·	·	·	·	·	·	·	·	·
22	·	·	·	·	·	·	·	·	·	·
33	·	·	·	·	·	·	·	·	·	·
12	·	·	·	·	·	·	·	·	·	·
13	·	·	·	·	·	·	·	·	·	·
23	·	·	·	·	·	·	·	·	·	·

$\mathbf{A} = (\mathbf{Z}_1'\mathbf{Z}_1)^{-1}\mathbf{Z}_1'\mathbf{Z}_2$

	111	222	333	122	112	133	113	233	223	123
0	·	·	·	·	·	·	·	·	·	·
1	1	·	·	2/3	·	2/3	·	·	·	·
2	·	1	·	·	2/3	·	·	2/3	·	·
3	·	·	1	·	·	·	2/3	·	2/3	·
11	·	·	·	·	·	·	·	·	·	·
22	·	·	·	·	·	·	·	·	·	·
33	·	·	·	·	·	·	·	·	·	·
12	·	·	·	·	·	·	·	·	·	·
13	·	·	·	·	·	·	·	·	·	·
23	·	·	·	·	·	·	·	·	·	·

associated with the postulated model

$$y = Z_1\beta_1 + \varepsilon \tag{7C.2}$$

is [with the minimal assumption $E(\varepsilon) = 0$] unbiased, so that

$$E(b_1) = \beta_1. \tag{7C.3}$$

If, however, the estimator (7C.1) appropriate to the model (7C.2) is employed, when in fact the correct model is

$$y = Z_1\beta_1 + Z_2\beta_2 + \varepsilon \tag{7C.4}$$

then, in general,

$$E(b_1) = \beta_1 + A\beta_2, \tag{7C.5}$$

that is, the estimator b_1 is now biased, and the matrix A of bias coefficients is (see Section 3.10)

$$A = (Z_1'Z_1)^{-1}Z_1'Z_2. \tag{7C.6}$$

We now illustrate this general analysis for the 3^3 design used in the textile example. Here, Z_1 is the matrix for the second degree polynomial model, called Z_s, and enumerated in Table 7A.1, whose rows are appropriate values of

$$\left\{ 1, x_1, x_2, x_3, x_1^2, x_2^2, x_3^2, x_1x_2, x_1x_3, x_2x_3 \right\}, \tag{7C.7}$$

while Z_2, given in Table 7C.1 is the matrix whose rows are appropriate values of the additional third-order terms

$$\left\{ x_1^3, x_2^3, x_3^3, x_1x_2^2, x_1^2x_2, x_1x_3^2, x_1^2x_3, x_2x_3^2, x_2^2x_3, x_1x_2x_3 \right\}. \tag{7C.8}$$

Then it is also seen that the matrix $Z_1'Z_2$ of cross-products between columns of Z_1 and Z_2 is that given in Table 7C.1 from which, after premultiplying by the inverse $(Z_1'Z_1)^{-1}$ [denoted by $(Z_s'Z_s)^{-1}$ in Table 7A.3], we obtain the bias matrix A, yielding the bias relationships of Eq. (7.4.2).

APPENDIX 7D.   ANALYSIS OF VARIANCE TO DETECT POSSIBLE THIRD-ORDER TERMS

To perform the analysis, we would first need to fit the full third degree polynomial model

$$y = Z_1\beta_1 + Z_2\beta_2 + \varepsilon \tag{7D.1}$$

in which $\mathbf{Z}_1\boldsymbol{\beta}_1$ includes terms up to second order and $\mathbf{Z}_2\boldsymbol{\beta}_2$ terms of third order, and then find the extra sum of squares due to fitting the third-order terms. An apparent difficulty for the 3^3 design is that at least four levels of a given variable are needed to estimate pure cubic coefficients $\beta_{111}, \beta_{222}, \beta_{333}$, and we have only three levels. However, any third-order coefficient of the form β_{ijj} is estimable because it measures the change in quadratic curvature associated with the jth variable as the ith variable is changed and quadratic curvature in any one input can certainly be estimated at each of the three levels of any other input.

(Note: Here and elsewhere, the generic "β_{ijj}" includes terms such as β_{113}.)

Orthogonalization of the Model

We can understand the situation more clearly if we proceed as follows. It has been noted in Chapter 3 that, by subtracting $\mathbf{Z}_1\mathbf{A}\boldsymbol{\beta}_2$ from the second term and adding it to the first in Eq. (7D.1), the model may be equivalently written so that terms of third order are orthogonal to the lower-order terms. We thus rewrite the model (7D.1) as

$$\mathbf{y} = \mathbf{Z}_1(\boldsymbol{\beta}_1 + \mathbf{A}\boldsymbol{\beta}_2) + (\mathbf{Z}_2 - \mathbf{Z}_1\mathbf{A})\boldsymbol{\beta}_2 + \boldsymbol{\varepsilon}. \qquad (7D.2)$$

When the third degree model is written in this form, it is seen that $\mathbf{y}$ is to be regressed onto $\mathbf{Z}_1$ and also onto the part, $\mathbf{Z}_2 - \mathbf{Z}_1\mathbf{A} = \mathbf{Z}_{2.1}$, of $\mathbf{Z}_2$ which is orthogonal to $\mathbf{Z}_1$. Regression of $\mathbf{y}$ onto $\mathbf{Z}_1$ (with the third-order model assumed to be true) provides the vector of estimates $\mathbf{b}_1 = (\mathbf{Z}_1'\mathbf{Z}_1)^{-1}\mathbf{Z}_1'\mathbf{y}$ of Eq. (7C.1) which, in Eq. (7D.2), are correctly shown, not as estimates of $\boldsymbol{\beta}_1$ alone, but as estimates of the biased combination of vectors $\boldsymbol{\beta}_1 + \mathbf{A}\boldsymbol{\beta}_2$; see Eq. (7C.5). Regression of $\mathbf{y}$ onto $\mathbf{Z}_{2.1} = \mathbf{Z}_2 - \mathbf{Z}_1\mathbf{A}$ produces the normal equations $\mathbf{Z}_{2.1}'\mathbf{Z}_{2.1}\mathbf{b}_2 = \mathbf{Z}_{2.1}'\mathbf{y}$. In circumstances where the $\mathbf{Z}_{2.1}'\mathbf{Z}_{2.1}$ matrix is nonsingular, this would give the unbiased estimator $\mathbf{b}_2 = (\mathbf{Z}_{2.1}'\mathbf{Z}_{2.1})^{-1}\mathbf{Z}_{2.1}'\mathbf{y}$ of $\boldsymbol{\beta}_2$. It would follow from the Gauss–Markov theorem that, provided all of the β coefficients are estimable, the unbiased least-squares estimates of the coefficients $\boldsymbol{\beta}_1$ obtained by direct fitting of the third-order model (7D.1) could be recovered from the calculation

$$\hat{\boldsymbol{\beta}}_1 = \mathbf{b}_1 - \mathbf{A}\mathbf{b}_2. \qquad (7D.3)$$

It is obvious not only that the coefficients $\beta_{111}, \beta_{222}, \beta_{333}$ cannot be estimated but also that β_{jjj} is wholly confounded with the linear coefficient β_j. This is so for any three-level factorial confined to levels $(-1, 0, 1)$, for then $x_j^3 \equiv x_j$. The fact that $\beta_{111}, \beta_{222}, \beta_{333}$ are not separately estimable shows up, as expected, in the matrix $\mathbf{Z}_{2.1}$ given in Table 7D.1, where it will be noted that the first three columns consist entirely of zeros. However, it is also clear from $\mathbf{Z}_{2.1}$ that seven estimates of β_{122}, $\beta_{112}, \beta_{133}, \beta_{113}, \beta_{233}, \beta_{223}$, and β_{123} can be obtained, and the extra sum of squares associated with these can certainly be calculated.

TABLE 7D.1. Matrix $Z_{2 \cdot 1}$ for estimation of third-order terms

$$Z_{2.1} = Z_2 - Z_1 A$$

111	222	333	122	112	133	113	233	223	123
·	·	·	$-\frac{1}{3}$	$-\frac{1}{3}$	$-\frac{1}{3}$	$-\frac{1}{3}$	$-\frac{1}{3}$	$-\frac{1}{3}$	-1
·	·	·	·	$\frac{2}{3}$	·	$\frac{2}{3}$	$-\frac{1}{3}$	$-\frac{1}{3}$	·
·	·	·	$\frac{1}{3}$	$-\frac{1}{3}$	$\frac{1}{3}$	$-\frac{1}{3}$	$-\frac{1}{3}$	$-\frac{1}{3}$	1
·	·	·	$\frac{2}{3}$	·	$-\frac{1}{3}$	$-\frac{1}{3}$	·	$\frac{2}{3}$	·
·	·	·	·	·	·	$\frac{2}{3}$	·	$\frac{2}{3}$	·
·	·	·	$-\frac{2}{3}$	·	$\frac{1}{3}$	$-\frac{1}{3}$	·	$\frac{2}{3}$	·
·	·	·	$-\frac{1}{3}$	$\frac{1}{3}$	$-\frac{1}{3}$	$-\frac{1}{3}$	$\frac{1}{3}$	$-\frac{1}{3}$	1
·	·	·	·	$-\frac{2}{3}$	·	$\frac{2}{3}$	$\frac{1}{3}$	$-\frac{1}{3}$	·
·	·	·	$\frac{1}{3}$	$\frac{1}{3}$	$\frac{1}{3}$	$-\frac{1}{3}$	$\frac{1}{3}$	$-\frac{1}{3}$	-1
·	·	·	$-\frac{1}{3}$	$-\frac{1}{3}$	$\frac{2}{3}$	·	$\frac{2}{3}$	·	·
·	·	·	·	$\frac{2}{3}$	·	·	$\frac{2}{3}$	·	·
·	·	·	$\frac{1}{3}$	$-\frac{1}{3}$	$-\frac{2}{3}$	·	$\frac{2}{3}$	·	·
·	·	·	$\frac{2}{3}$	·	$\frac{2}{3}$	·	·	·	·
·	·	·	·	·	·	·	·	·	·
·	·	·	$-\frac{2}{3}$	·	$\frac{2}{3}$	·	·	·	·
·	·	·	$-\frac{1}{3}$	$\frac{1}{3}$	$\frac{2}{3}$	·	$-\frac{2}{3}$	·	·
·	·	·	·	$-\frac{2}{3}$	·	·	$-\frac{2}{3}$	·	·
·	·	·	$\frac{1}{3}$	$\frac{1}{3}$	$-\frac{2}{3}$	·	$-\frac{2}{3}$	·	·
·	·	·	$-\frac{1}{3}$	$-\frac{1}{3}$	$-\frac{1}{3}$	$\frac{1}{3}$	$-\frac{1}{3}$	$\frac{1}{3}$	1
·	·	·	·	$\frac{2}{3}$	·	$-\frac{2}{3}$	$-\frac{1}{3}$	$\frac{1}{3}$	·
·	·	·	$\frac{1}{3}$	$-\frac{1}{3}$	$\frac{1}{3}$	$\frac{1}{3}$	$-\frac{1}{3}$	$\frac{1}{3}$	-1
·	·	·	$\frac{2}{3}$	·	$-\frac{1}{3}$	$\frac{1}{3}$	·	$-\frac{2}{3}$	·
·	·	·	·	·	·	$-\frac{2}{3}$	·	$-\frac{2}{3}$	·
·	·	·	$-\frac{2}{3}$	·	$\frac{1}{3}$	$\frac{1}{3}$	·	$-\frac{2}{3}$	·
·	·	·	$-\frac{1}{3}$	$\frac{1}{3}$	$-\frac{1}{3}$	$\frac{1}{3}$	$\frac{1}{3}$	$\frac{1}{3}$	-1
·	·	·	·	$-\frac{2}{3}$	·	$-\frac{2}{3}$	$\frac{1}{3}$	$\frac{1}{3}$	·
·	·	·	$\frac{1}{3}$	$\frac{1}{3}$	$\frac{1}{3}$	$\frac{1}{3}$	$\frac{1}{3}$	$\frac{1}{3}$	1

APPENDIX 7E. ORTHOGONALIZATION OF EFFECTS FOR THREE-LEVEL DESIGNS

Suppose a single input ξ_1 has three equally spaced levels, coded as $x_1 = -1, 0, 1$, and a quadratic model

$$y = \beta_0 x_0 + \beta_1 x_1 + \beta_{11} x_1^2 + \varepsilon \tag{7E.1}$$

is fitted, where $x_0 = 1$. If we write the model in matrix form as $y = Z\beta + \varepsilon$ the Z matrix takes the form

$$
Z = \begin{matrix} x_0 & x_1 & x_1^2 \\ \begin{bmatrix} 1 & -1 & 1 \\ 1 & \cdot & \cdot \\ 1 & 1 & 1 \end{bmatrix} \end{matrix}. \tag{7E.2}
$$

The third column is not orthogonal to the first, but we can recast the model to achieve a Z matrix with orthogonal columns by the following procedure. We regress the x_1^2 column (as "response") on to the x_0 column (as "predictor") and take residuals to give a new column. The new third column consists of elements of the form $x_{1u}^2 - \bar{x_1^2}$, that is, of the form $x_{1u}^2 - \frac{2}{3}$, and to avoid fractions we shall, in fact, use $3x_{1u}^2 - 2$. We can adjust the model to take account of this change to obtain

$$
y = \left(\beta_0 + \tfrac{2}{3}\beta_{11} \right) x_0 + \beta_1 x_1 + \tfrac{1}{3}\beta_{11}\left(3x_1^2 - 2 \right) + \varepsilon \tag{7E.3}
$$

$$
= \alpha_0 x_0 + \alpha_1 x_1 + \alpha_2 x^{(2)} + \varepsilon \tag{7E.4}
$$

say, with an obvious notation, for which the Z matrix, which we call Z_0, is shown on the left below.

$$
Z_0 = \begin{matrix} x_0 & x_1 & x_1^{(2)} = 3x_1^2 - 2 \\ \begin{bmatrix} 1 & -1 & 1 \\ 1 & \cdot & -2 \\ 1 & 1 & 1 \end{bmatrix} \end{matrix} . \, y = \begin{bmatrix} y_1 \\ y_2 \\ y_3 \end{bmatrix} = \begin{bmatrix} 1 \\ 6 \\ 5 \end{bmatrix} . \tag{7E.5}
$$

Now suppose we have a set of data like y on the right, above. Then, by regressing the columns of Z_0 onto y, we obtain the mean, linear, and quadratic orthogonal components of y, that is, the least-squares estimators a_0, a_1 and a_2 of α_0, α_1, and α_2, and we can also obtain the corresponding sums of squares. If z_i is a particular column of Z_0 then we simply calculate

$$
a_i = \frac{z_i'y}{z_i'z_i} \quad \text{and} \quad SS(a_i) = \frac{(z_i'y)^2}{z_i'z_i} . \tag{7E.6}
$$

For the data set on the right of (7E.5), we can thus evaluate

$$
a_0 = \text{mean} = \frac{1 + 6 + 5}{3} = 4,
$$

$$
a_1 = \text{linear component} = \frac{(1 \times -1) + (1 \times 5)}{2} = 2,
$$

$$
a_2 = \text{quadratic component} = \frac{(1 \times 1) + (-2 \times 6) + (1 \times 5)}{6} = -1. \tag{7E.7}
$$

Thus the fitted equation is

$$\hat{y} = 4 + 2x_1 - 1(3x_1^2 - 2) \tag{7E.8}$$

which can of course be readily rearranged into the form of the original quadratic model to give the fitted equation

$$\hat{y} = 6 + 2x_1 - 3x_1^2. \tag{7E.9}$$

For these data, the total sum of squares is $1^2 + 6^2 + 5^2 = 62$ and the component sums of squares are as follows.

Source	SS	df
Mean	$12^2/3 = 48$	1
Added linear	$4^2/2 = 8$	1
Added quadratic	$6^2/6 = 6$	1
Total	62	3

It is clearly simpler to work with the set of orthogonal variables 1, x_1, and $x_1^{(2)} = 3x_1^2 - 2$ than with the set 1, x_1 and x_1^2, and this simplicity is maintained however many inputs $x_1, x_2, \ldots, x_k$, are included within the factorial, because the balanced design ensures that the orthogonality is maintained across variables. Thus, for two inputs coded as x_1 and x_2, there are nine orthogonal components. These are indicated in Table 7E.1. The curly brackets indicate terms of different orders as indicated by the figures $0, 1, \ldots, 4$ at the heads of the columns. The appropriate

TABLE 7E.1. A 9×9 orthogonalized matrix $\mathbf{Z}_0$ for two components x_1 and x_2

Order of terms	0	1		2			3		4
	1	x_1	x_2	$x_1^{(2)}$	$x_1 x_2$	$x_2^{(2)}$	$x_1^{(2)} x_2$	$x_1 x_2^{(2)}$	$x_1^{(2)} x_2^{(2)}$
	1	-1	-1	1	1	1	-1	-1	1
	1	$\cdot$	-1	-2	$\cdot$	1	2	$\cdot$	-2
	1	1	-1	1	-1	1	-1	1	1
	1	-1	$\cdot$	1	$\cdot$	-2	$\cdot$	2	-2
$\mathbf{Z}_0 =$	1	$\cdot$	$\cdot$	-2	$\cdot$	-2	$\cdot$	$\cdot$	4
	1	1	$\cdot$	1	$\cdot$	-2	$\cdot$	-2	-2
	1	-1	1	1	-1	1	1	-1	1
	1	$\cdot$	1	-2	$\cdot$	1	-2	$\cdot$	-2
	1	1	1	1	1	1	1	1	1

adjustments must, of course, be made in the model. Because all components are orthogonal, all their individual contributions to the sums of squares are, again, given by the simple equation (7E.6).

When effects of different kinds, such as linear, pure quadratic, interaction, and so on, occur, they will, in general have different variances. To enable them to be jointly assessed using a normal probability plot, they need to be rescaled to have the same variance. A convenient choice is the rescaling that produces standardized effects A_i whose variances each equal the experimental error variance σ^2. Because the unscaled effect $a_i = \mathbf{z}_i'\mathbf{y}/\mathbf{z}_i'\mathbf{z}_i$ has variance $\sigma^2/\mathbf{z}_i'\mathbf{z}_i$, we set

$$\text{standardized effect } A_i = (\mathbf{z}_i'\mathbf{z}_i)^{1/2} a_i = \mathbf{z}_i'\mathbf{y}/(\mathbf{z}_i'\mathbf{z}_i)^{1/2}$$

Equivalently, for any effect which has 1 df, A_i is the square root of the mean square for a_i in the analysis of variance table. This is $\pm\{\mathrm{SS}(a_i)/1\}^{1/2}$ with the sign attached taken to be that of a_i.

APPENDIX 7F.   ORTHOGONAL EFFECTS ASSOCIATED WITH INDIVIDUAL DEGREES OF FREEDOM AND RELATED NORMAL PLOTS

Via details in Appendix 7E, the three-level factorial design used in the textile investigation provided an analysis of variance table in which contributions for individual degrees of freedom associated with the various effects could readily be isolated. It was thus possible to calculate standardized effects and to plot them on

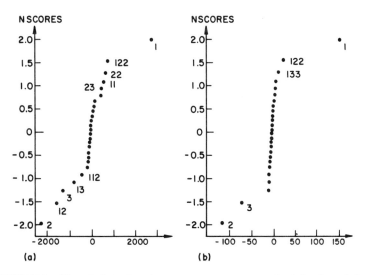

FIGURE 7F.1.   Normal plots of root mean square components for (a) unlogged data, (b) logged data ×100.

normal probability paper (see Appendix 4B) as shown in Figure 7F.1 for both unlogged and logged data. For the logged data (b), first order effects are clearly of paramount importance. For the unlogged data (a), however, the 12 and 13 interactions, for example, and various other interactions as well, are large and do not appear to be attributable to random error alone. These interactions can readily be explained therefore as *transformable* ones, that is, they arise only because of an inappropriate choice of metric.

For didactic purposes, we have introduced the analysis of individual components only after fitting some postulated models. In many investigations, particularly if identification of the model were not as clear-cut as it is with this textile example, the analysis into individual degrees of freedom, and the plotting of components on normal probability paper, is a sensible place to start, rather than to finish.

APPENDIX 7G. YATES' ALGORITHM FOR 3^k FACTORIAL DESIGNS

Yates' algorithm can be used to calculate effects and their corresponding sums of squares for 3^k designs. We illustrate this in Tables 7G.1 and 7G.2 with the 3^3 factorial textile data for unlogged and logged values.

The 27 observations are shown in columns (1) of the tables in the standard order exemplified in Table 7.1 (p. 206). There are nine sets of three observations. Let w_1, w_2 and w_3 denote the first, second, and third observations, respectively, in one such set. Then the analysis is conducted in terms of the following three operations: (0) $w_1 + w_2 + w_3$, (i) $w_3 - w_1$, (i^2) $w_1 - 2w_2 + w_3$.

The first operation, characterized by (0), corresponds to computing the mean of the three values, the second (i) to computing the linear effect, and the third (i^2) to computing the quadratic effect. (Appropriate divisors are introduced in a later stage of the calculation.)

The operation (0) is first performed on each of the nine sets and the results are entered in column (2) *as its first nine entries*. Then, the operation (i) is performed on each of the nine sets of three observations in column (1), and the nine results are set down in order in column (2), forming its second set of nine entries. Finally, the operation (i^2) is carried out on each of the nine sets in column (1), producing the third set of nine entries of column (2).

Column (3) is obtained from column (2) in exactly the same manner; then column (4) is obtained similarly from column (3). (In general, this operation is repeated as many times as there are factors.) The first entry in column (4) is now the sum of all 27 observations, and the remaining entries are sums of products $z_i' y$, corresponding to the various main effects and interactions whose identity is indicated in column (5). Thus, for example, $12^2 3$ is the interaction between the linear effects of x_1 and x_3 and the quadratic effect of x_2. The sequence of effects follows the same standard order as used for the design set out in Table 7.1 (p. 206) with $-1, 0, 1$ in the ith column replaced by $0, i, i^2$.

The (orthogonalized) effects are obtained by dividing column (4) by column (6), while the mean squares in column (8) are obtained by squaring the entries in column (4) and dividing by the corresponding divisor in column (6). The divisors of column

TABLE 7G.1.   Yates' algorithm for textile data (unlogged data)

(1)	(2)	(3)	(4)	Effect name (5)	Divisor (6)	Orthogonal effect (7)	MS (8)	$\sqrt{(8)}$
674	5724	10,404	23,256	000	27	861.33	20,031,168	
1414	2928	8,042	11,880	100	18	660.0	7,840,800	2800.1
3636	1752	4,810	4,296	1^200	54	79.56	341,771	584.6
338	4752	5,162	$-9,646$	020	18	-535.89	5,169,184	2273.6
1022	1956	4,384	$-5,478$	120	12	-456.5	2,500,707	1581.4
1568	1334	2,334	$-2,890$	1^220	36	-80.28	232,003	481.7
170	2926	1,770	4,962	02^20	54	91.89	455,953	675.2
442	1214	1,592	4,710	12^20	36	130.83	616,225	785.0
1140	670	934	4,722	1^22^20	108	43.72	206,456	454.4
370	2962	$-3,972$	$-5,594$	003	18	-310.78	1,738,491	1318.5
1198	1230	$-3,418$	$-2,828$	103	12	-235.67	666,465	816.4
3184	970	$-2,256$	-836	1^203	36	-23.22	19,414	139.3
266	2814	$-1,992$	1,716	023	12	143.0	245,388	495.4
620	804	$-2,048$	554	123	8	69.25	38,365	195.9
1070	766	$-1,438$	42	1^223	24	1.75	74	8.6
118	1708	$-1,056$	-452	02^23	36	-12.56	5,675	75.3
332	356	-820	-206	12^23	24	-8.58	1,768	42.0
884	270	$-1,014$	-950	1^22^23	72	-13.19	12,535	112.0
292	1482	1,620	-870	003^2	54	-16.11	14,017	118.4
634	-138	2,174	$-1,272$	103^2	36	-35.33	44,944	212.0
2000	426	1,168	-480	1^203^2	108	-4.44	2,133	46.2
210	1158	1,472	608	023^2	36	16.89	10,268	101.3
438	96	1,972	666	123^2	24	27.75	18,482	136.0
566	338	1,266	-430	1^223^2	72	-5.97	2,568	50.7
90	1024	2,184	$-1,560$	02^23^2	108	-14.44	22,533	150.1
220	-100	1,304	$-1,206$	12^23^2	72	-16.75	20,201	142.1
360	10	1,234	810	$1^22^23^2$	216	3.75	3,038	55.1

(6) are obtained from appropriate products of the sum of squared coefficients of the w's in the operations $(0), (i), (i^2)$. The last mentioned are:

$$\text{for } (0): \qquad 1^2 + 1^2 + 1^2 = 3,$$

$$\text{for } (i): \qquad 1^2 + 0^2 + 1^2 = 2,$$

$$\text{for } (i^2): \qquad 1^2 + 2^2 + 1^2 = 6.$$

Then, for example, the divisor for the effect $[02^23]$, that is, the interaction between the quadratic effect of x_2 and the linear effect of x_3, is $3 \times 6 \times 2 = 36$.

TABLE 7G.2. Yates' algorithm for textile data (logged data)

(1)	(2)	(3)	(4)	Effect name (5)	Divisor (6)	Orthogonal effect (7)	MS ×10^4 (8)	$\sqrt{(8)}$
2.83	9.54	26.21	74.27	000	27	2.75	2,042,975.15	
3.15	8.73	24.93	6.51	100	18	0.36	23,544.50	153.44
3.56	7.94	23.13	−0.67	1^200	54	−0.01	83.13	9.11
2.53	9.15	2.22	−4.93	020	18	−0.27	13,502.72	116.20
3.01	8.24	2.42	−0.17	120	12	−0.01	24.08	4.90
3.19	7.54	1.87	−0.37	1^220	36	−0.01	38.03	6.16
2.23	8.57	−0.22	0.23	02^20	54	0.00	9.80	3.13
2.65	7.71	−0.24	1.41	12^20	36	0.04	552.25	23.50
3.06	6.85	−0.21	1.25	1^22^20	108	0.01	144.68	12.02
2.57	0.73	−1.60	−3.08	003	18	−0.17	5,270.22	72.59
3.08	0.66	−1.61	−0.35	103	12	−0.03	102.08	10.10
3.50	0.83	−1.72	0.01	1^203	36	0.00	0.03	0.17
2.42	0.93	0.10	−0.12	023	12	−0.01	12.00	3.46
2.79	0.61	−0.05	−0.32	123	8	−0.04	128.00	11.31
3.03	0.88	−0.22	−0.24	1^223	24	−0.01	24.00	4.89
2.07	0.83	−0.10	−0.02	02^23	36	−0.00	0.11	0.33
2.52	0.43	0.07	0.34	12^23	24	0.01	48.17	6.94
2.95	0.61	−0.34	−0.26	1^22^23	72	−0.00	9.39	3.06
2.47	0.09	0.02	−0.52	003^2	54	−0.01	50.07	7.07
2.80	−0.30	0.21	−0.75	103^2	36	−0.02	156.25	12.50
3.30	−0.01	0.00	0.05	1^203^2	108	0.00	0.23	0.47
2.32	−0.09	0.24	−0.10	023^2	36	−0.00	2.78	1.66
2.64	−0.13	0.59	−0.02	123^2	24	−0.00	0.17	0.41
2.75	−0.02	0.58	−0.58	1^223^2	72	−0.01	46.72	6.83
1.95	0.17	0.68	−0.40	02^23^2	108	−0.00	14.81	3.84
2.34	−0.21	0.15	−0.36	12^23^2	72	−0.01	18.00	4.24
2.56	−0.17	0.42	0.80	$1^22^23^2$	216	0.00	29.63	5.44

To illustrate how the orthogonal effects may be employed, suppose it is assumed that a polynomial of degree 2 will provide an adequate representation. This is equivalent to supposing that all terms of the third and higher order are zero. Recalling that the appropriate orthogonal second degree contrast is $(3x_i^2 - 2)$, we may thus write down the second degree function fitted to (for example) the unlogged data as follows.

$$\hat{Y} = 861.3 + 660.0x_1 - 535.9x_2 - 310.8x_3 + 79.6[3x_1^2 - 2]$$

$$+ 91.9[3x_2^2 - 2] - 16.1[3x_3^2 - 2] - 456.5x_1x_2 - 235.7x_1x_3 + 143.0x_2x_3.$$

$$(7G.1)$$

On collecting terms, this agrees, apart from small rounding discrepancies, with the fitted second degree equation of Table 7A.3.

Normal Plots

The standardized components $z_i' y / (z_i' z_i)^{1/2}$ described in Appendix 7E are the square roots of the mean squares given in column (8) of Table 7G.2 for the logged data $y = \log Y$. Similar quantities $z' Y / (z_i' z_i)^{1/2}$ for the unlogged data Y are given in Table 7G.1. As already mentioned in Appendix 7F, these components may be plotted on normal probability paper in exactly the same manner as was done for similar quantities in the case of the 2^k factorial. This was done in Figure 7F.1. Notice that a slightly different and alternative notation is in use there, for example (122) for $(12^2 0)$.

EXERCISES

7.1. Consider the 3^2 factorial design and data shown below.

x_1	x_2	y	
-1	-1	2.57	$n = 9$
0	-1	3.08	$\Sigma y = 24.93$
1	-1	3.50	$\bar{y} = 2.770$
-1	0	2.42	$\Sigma y^2 = 70.50$
0	0	2.79	
1	0	3.03	
-1	1	2.07	
0	1	2.52	
1	1	2.95	

(a) Write down the first-order model, normal equations, fitted model, and associated analysis of variance table, and evaluate the residuals.

(b) Examine your results and make a judgment as to the adequacy of the fitted model.

7.2. Consider again the 3^2 design and data in the foregoing exercise.

(a) Write down the Z matrix associated with a *second*-order polynomial model and also the corresponding Z'Z matrix.

(b) Rewrite the second-order model so that Z'Z becomes a diagonal matrix. (*Hint*: Replace x_1^2 by $x_1^2 - \frac{2}{3}$, where $\Sigma x_{1u}^2 / n = \frac{2}{3}$. Add a term $2\beta_{11}/3$ to β_0 to compensate. Do the same for x_2^2.)

7.3. Fit the second-order model to the data given in Exercise 7.1 and construct the associated analysis of variance table. [Use either Exercise 7.2(a) or 7.2(b); your answer should be the same in the end.]

7.4. Using the data and design given in Exercise 7.1, and a first-order model, construct the 1 df test for nonadditivity. (See Appendix 7B)

7.5. Consider again the 3^2 factorial design and data given in Exercise 7.1.
   (a) Write down the standard third-order polynomial model.
   (b) Which third-order parameters can be separately estimated? Which cannot?
   (c) Develop the appropriate alias matrix given that a second-order model has been fitted, and that a third-order model might be appropriate.
   (d) Are your results in (c) consistent with what you found in (b)? (*Hint*: Do they have to be?)

7.6. (a) Using the fitted first-order model from Exercise 7.1, construct separate 95% confidence limits for each parameter in the model.
   (b) Predict the response, and determine 95% confidence limits for the true mean value of the response, at the point $x_1 = 0.7$, $x_2 = 0.5$.
   (c) Construct the joint 90% confidence region for β_1 and β_2.
   (d) Note that $(0.95)(0.95) = 0.9025 \simeq 0.90$. Compare the region defined by the two pairs of limits in (a) with the region in (c). Comment on what you see.
   (e) Show that the variance $V[y(\mathbf{x})]$ of the predicted response $\hat{y}(\mathbf{x})$ at $\mathbf{x} = (x_1, x_2)'$ is constant on circles (of radius ρ, say) centered at the point $x_1 = 0$, $x_2 = 0$.

7.7. The accompanying 3^3 factorial design and data (Table E7.7) are taken from *Design and Analysis of Industrial Experiments*, edited by O. L. Davies, Oliver and Boyd, Edinburgh, 1956, p. 333. A limp-cover revised edition was published in 1978 by the Longman Group, New York.
   (a) Fit a full second-order model to these data.
   (b) Determine the residuals and analyze them.
   (c) Make the 1 df test for nonadditivity. (See Appendix 7B.)
   (d) Using the algorithm for the three-level factorial, separately estimate all possible individual degree of freedom estimates, and plot the results on normal probability paper. (See Appendix 4B.)
   (e) Interpret your analysis.

7.8. The 3^3 factorial design of Table E7.8 arose in connection with an investigation by A. J. Feuell and R. E. Wagg, Statistical methods in detergency investigations, *Research*, 2, 1949, 334–337; see p. 335. (*Note*: *Research* later became *Research Applied in Industry*. The data were also reported again in *Statistical Analysis in Chemistry and the Chemical Industry* by C. A. Bennett and N. L. Franklin, John Wiley & Sons, New York, 1954, p. 509.) The response was a measure of washing performance (percentage reflectance in terms of white standards) and the response data were multiplied by 10 to produce the y values shown in the table. The three predictor variables were detergent level coded to x_1, sodium carbonate (x_2), and sodium carboxymethyl cellulose level (x_3). Using the 3^k algorithm, construct an appropriate model and an analysis

TABLE E7.7

x_1	x_2	x_3	y	
−1	−1	−1	159	$n = 27$
0	−1	−1	395	$\Sigma y = 7145$
1	−1	−1	149	
−1	0	−1	25	$\bar{y} = 264.63$
0	0	−1	255	
1	0	−1	251	$\Sigma y^2 = 2,328,793$
−1	1	−1	184	
0	1	−1	363	
1	1	−1	378	
−1	−1	0	260	
0	−1	0	454	
1	−1	0	112	
−1	0	0	98	
0	0	0	422	
1	0	0	270	
−1	1	0	237	
0	1	0	362	
1	1	0	363	
−1	−1	1	146	
0	−1	1	417	
1	−1	1	150	
−1	0	1	103	
0	0	1	455	
1	0	1	172	
−1	1	1	195	
0	1	1	492	
1	1	1	278	

Note: The "observations" y shown in Table E7.7 are in fact the sums of pairs of repeats, but we are ignoring this for the purposes of our question. For additional detail involving the original data, please refer to the Davies reference.

of variance table for these data, and plot the individual 1-df effects on normal probability paper.

7.9. The 3^3 factorial design, replicated three times, in Table E7.9 is from a study made to investigate the roles of three printing machine variables, speed (coded as x_1), pressure (x_2), and distance (x_3), upon the application of coloring inks onto package labels. Perform as complete an analysis as possible on these data.

TABLE E7.8

x_1	x_2	x_3	y	
-1	-1	-1	106	
0	-1	-1	198	
1	-1	-1	270	
-1	0	-1	197	
0	0	-1	329	
1	0	-1	361	
-1	1	-1	223	
0	1	-1	320	
1	1	-1	321	
-1	-1	0	149	
0	-1	0	243	
1	-1	0	315	$n = 27$
-1	0	0	255	$\Sigma y = 8068$
0	0	0	364	
1	0	0	390	
-1	1	0	294	$\bar{y} = 298.81$
0	1	0	410	
1	1	0	415	
-1	-1	1	182	$\Sigma y^2 = 2{,}609{,}558$
0	-1	1	232	
1	-1	1	340	
-1	0	1	259	
0	0	1	389	
1	0	1	406	
-1	1	1	297	
0	1	1	416	
1	1	1	387	

7.10. An investigation to examine the effects of adding sulphur (S) and asphalt (A) in the formation of sand aggregate pavements resulted in the 3^2 factorial design, replicated, shown in Table E7.10. The response y is the failure stress in pounds per square inch. Perform a complete analysis of these data.

7.11. Consider the data in Table E7.11.
   (a) Plot the data.
   (b) Fit a first-order model by least squares and do the appropriate analysis on it. Comment.
   (c) Now fit a second-order model $y = \beta_0 + \beta_1 x_1 + \beta_2 x_2 + \beta_{11} x_1^2 + \beta_{22} x_2^2 + \beta_{12} x_1 x_2 + \varepsilon$ to the data. Do the appropriate analysis and comment.
   (d) Add 37 to observations 7–12. Repeat the work of part (c) and make a list of all the differences between your original (c) results and the new ones. What is the implication of the differences and the similarities you observe?

TABLE E7.9

x_1	x_2	x_3	I	II	III	Totals
				Replicates		
-1	-1	-1	34	10	28	72
0	-1	-1	115	116	130	361
1	-1	-1	192	186	263	641
-1	0	-1	82	88	88	258
0	0	-1	44	178	188	410
1	0	-1	322	350	350	1022
-1	1	-1	141	110	86	337
0	1	-1	259	251	259	769
1	1	-1	290	280	245	815
-1	-1	0	81	81	81	243
0	-1	0	90	122	93	305
1	-1	0	319	376	376	1071
-1	0	0	180	180	154	514
0	0	0	372	372	372	1116
1	0	0	541	568	396	1505
-1	1	0	288	192	312	792
0	1	0	432	336	513	1281
1	1	0	713	725	754	2192
-1	-1	1	364	99	199	662
0	-1	1	232	221	266	719
1	-1	1	408	415	443	1266
-1	0	1	182	233	182	597
0	0	1	507	515	434	1456
1	0	1	846	535	640	2021
-1	1	1	236	126	168	530
0	1	1	660	440	403	1503
1	1	1	878	991	1161	3030
			8808	8096	8584	25488

TABLE E7.10

%A	%S	x_1	x_2	y
2	14	-1	-1	338, 344
4	14	0	-1	258, 272
6	14	1	-1	320, 334
2	16	-1	0	264, 290
4	16	0	0	242, 207
6	16	1	0	308, 310
2	18	-1	1	332, 325
4	18	0	1	258, 233
6	18	1	1	336, 350

247

TABLE E7.11

Run	x_1	x_2	y
1	-1	-1	33
2	1	-1	27
3	-1	1	28
4	1	1	52
5	0	0	45
6	0	0	41
7	$\sqrt{2}$	0	33
8	$-\sqrt{2}$	0	26
9	0	$\sqrt{2}$	37
10	0	$-\sqrt{2}$	25
11	0	0	42
12	0	0	46

7.12. Analyze the data below.

x_1	x_2	y
	Block 1	
-1	-1	73
1	-1	85
-1	1	83
1	1	79
0	0	80
0	0	81
0	0	79
		560
	Block 2	
$-\sqrt{2}$	0	67
$\sqrt{2}$	0	73
0	$-\sqrt{2}$	69
0	$\sqrt{2}$	71
0	0	71
0	0	69
0	0	70
		490

Facts:

$$\begin{bmatrix} 7 & 4 & 4 \\ 4 & 6 & 2 \\ 4 & 2 & 6 \end{bmatrix}^{-1} = \tfrac{1}{48}\begin{bmatrix} 16 & -8 & -8 \\ -8 & 13 & 1 \\ -8 & 1 & 13 \end{bmatrix}$$

$$\Sigma y = 1050 \qquad N\bar{y}^2 = 78{,}750$$

$$\Sigma y^2 = 79{,}208 \qquad \Sigma(y - \bar{y})^2 = 458$$

7.13. (Source: Statistical design of experiments for process development of MBT, by S. A. Frankel, *Rubber Age*, **89**, 1961, June, 453–458. Material reproduced and adapted by permission of the author, of his employer American Cyanamid Company, and of *Rubber Age*.) The diagram shows 25 hypothesized observations of an experimental program for investigation of two factors ξ_1 and ξ_2. Assume each ξ_i has its levels coded to $x_i = -2, -1, 0, 1, 2$. The nine encircled points are those of a composite design with four cube points at $(\pm 1, \pm 1)$, four star points at $(\pm 2, 0)$, $(0, \pm 2)$ and one center point $(0, 0)$.

   Fit a second-order model to the original 25 data points and then just to the nine composite design data points. Perform a full analysis, and predict all 25 responses for each fit. Compare the two sets of results; what do you conclude?

7.14. Fit a second-order model to the data shown in Table E7.14 and provide an analysis of variance table.

   Draw the contours of the function $V(\hat{y}(\mathbf{x}))$ within a square of side length 4 whose sides are parallel to the (x_1, x_2) axes and whose center is at $(x_1, x_2) = (0, 0)$.

7.15. (Source: Copyright © 1960. TAPPI. Reprinted and adapted from L. H. Phifer and J. B. Maginnis, Dry ashing of pulp and factors which influence it, *Tappi*, **43**(1), 1960, 38–44, with permission.) Ash values of paper pulp were widely

TABLE E7.14

x_1	x_2	y
-1	-1	130
1	-1	121
-1	1	110
1	1	125
-1.5	0	107
1.5	0	117
0	-1.5	112
0	1.5	111
0	0	149, 152. 148, 153

TABLE E7.15

Run order	x_1	x_2	y
9	-1	-1	211, 213
5	1	-1	92, 88
6	-1	1	216, 212
7	1	1	99, 83
10	-1.5	0	222, 217
3	1.5	0	48, 62
4	0	-1.5	168, 175
2	0	1.5	179, 151
1	0	0	122, 157
8	0	0	175, 146

used as a measure of the level of inorganic impurities present in the pulp. A study was carried out to examine the effects of two variables, temperature T in degrees Celsius and time t in hours, on the ash values of four specific pulps. The results for one of these, "pulp A," are shown in Table E7.15. The coded predictor variables shown are

$$x_1 = \frac{(T - 775)}{115}, \qquad x_2 = \frac{(t - 3)}{1.5},$$

and the response y is (dry ash value in %) $\times 10^3$, the factor being attached simply to remove decimals. The data were taken in the run order $1, 2, \ldots, 10$ indicated at the left of the table.

TABLE E7.16

Run number	x_1	x_2	y
1	-1	0	83.8
2	1	0	81.7
3	0	0	82.4
4	0	0	82.9
5	0	-1	84.7
6	0	1	57.9
7	0	0	81.2
8	$-\theta$	$-\theta$	81.3
9	$-\theta$	θ	83.1
10	θ	$-\theta$	85.3
11	θ	θ	72.7
12	-1	0	82.0
13	-1	1	82.1
14	1	-1	88.5
15	-1	1.3	84.4
16	-1.25	0.5	85.2
17	-1.25	1	83.2

Note: $\theta = 2^{-1/2}$ in runs 8–11.

Perform a second-order surface fitting on these data and investigate (as appropriate) lack of fit, whether or not both x_1 and x_2 should be retained, and whether or not a quadratic model is needed. Plot contours for the model you decide to retain and examine the residuals in the usual ways, including against run order. What do you conclude?

If it were difficult to control the temperature variable, how would you feel about the use of the method to compare various pulps, based on what you can see in this particular set of data?

7.16. (Source: Statistical design of experiments for process development of MBT, by S. A. Frankel, *Rubber Age*, **89**, 1961, June, 453–458. Material reproduced and adapted by permission of the author, of his employer American Cyanamid Company, and of *Rubber Age*.) In a study of the manufacture of mercaptobenzothiazole by the autoclave reaction of aniline, carbon bisulfide, and sulfur at 250°C without medium, two input variables were varied. These were ξ_1 = reaction time in hours, and ξ_2 = reaction temperature in degrees Celsius, coded to $x_1 = (\xi_1 - 12)/8$ and $x_2 = (\xi_2 - 250)/30$. The response was the percentage yield of the reaction.

First, runs 1–12 shown in Table E7.16 were performed, and a second-order surface was fitted, namely,

$$\hat{y} = 82.17 - 1.01x_1 - 8.61x_2 + 1.40x_1^2 - 8.76x_2^2 - 7.20x_1x_2.$$

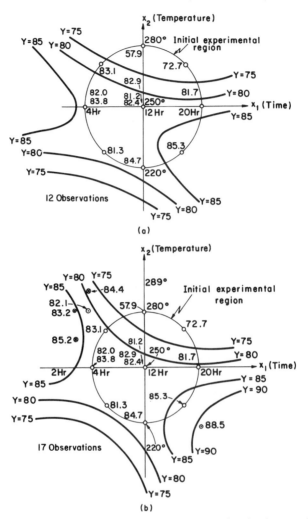

FIGURE E7.16. (*a*) Surface fitted to runs 1–12. (*b*) Surface fitted to runs 1–17.

The contours of this surface are shown in Figure E7.16a. Repeat this fitting for yourself, carrying out the usual tests for lack of fit, first- and second-order terms, and so on, and confirm the correctness of the fitted equation and of Figure E7.16a.

High yields were desirable. As a result of examining the surface above, five confirmatory runs, numbers 13–17, were performed. Are the five new response values consistent with the results that the second-order surface predicts? If so, refit the second-order surface to all 17 observations to obtain

$$\hat{y} = 81.11 - 2.09x_1 - 6.76x_2 + 1.15x_1^2 - 5.41x_2^2 - 9.49x_1x_2,$$

and carry out the usual analyses. Confirm that the surface shown in Figure E7.16b is an appropriate representation of this equation. What practical information has this investigation provided?

[*Note*: A $100(1 - \alpha)\%$ confidence prediction interval for individual observations at (x_{10}, x_{20}) is given by

$$\hat{y}_0 \pm t_{\nu, \alpha/2}\left\{1 + \mathbf{z}_0'(\mathbf{X'X})^{-1}\mathbf{z}_0\right\}^{1/2}$$

where $\hat{y}_0$ is the predicted value at (x_{10}, x_{20}), where $t_{\nu, \alpha/2}$ is the percentage point of the t distribution with ν df which leaves $\alpha/2$ in the upper tail, where $\mathbf{z}_0' = (1, x_{10}, x_{20}, x_{10}^2, x_{20}^2, x_{10}x_{20})$, and where $\mathbf{X}$ is the usual $\mathbf{X}$ matrix which occurs in the least-squares fitting of the second-order model.]

7.17. (Source: An exploratory study of Taylor's tool life equation by power transformations, by S. M. Wu, D. S. Ermer, and W. J. Hill, University of Wisconsin. Changes have been made to the original data as follows. The original x_1 is divided by 100, the original x_2 has been multiplied by 100. Responses are rounded and repeat observations appear on the same line.) Fit a second-order model to the data in Table E7.17. Test for lack of fit and, if no lack of fit is evident, test whether second-order terms are needed and whether any of the x's can be dropped. What predictive model would you use?

TABLE E7.17

x_1	x_2			y		
7	1.57	2.54	2.78	3.10	3.50	
5	1.57	8.45	9.13	9.33	10.67	
8	1.725	0	-0.51	-1.49	-2.08	
7	1.725	0	0.86	1.87	2.15	
6	1.725	3.77	4.25	4.73	5.42	
5	1.725	7.43	8.08	8.78	10.29	
4	1.725	10.99	11.33	11.47	12.06	
6	2.20	0.86	1.87	2.15	2.15	
4.5	2.20	5.80	6.38	6.80	7.22	

7.18. [Sources: (1) Maximization of potato yield under constraint, by H. P. Hermanson, *Agronomy Journal*, **57**, 1965, 210–213. Adapted by permission of the American Society of Agronomy. (2) The fertility of some Minnesota peat soils, by H. P. Hermanson, Ph.D. Thesis, University of Minnesota, 1961, 154 pp. (order No. 61-5846, Univ. Microfilms, Ann Arbor, MI)] In 1959 a peat soil fertility trial was carried out in Aitkin County, Minnesota, using Red Pontiac potatoes. Three predictor variables N = nitrogen, P = phosphorus, and K = potassium, all measured in pounds per acre were examined in a dodecahedron design for their effects upon the response Y = yield in units of 100 pounds per acre. (This soil–crop situation required appreciable potassium fertilizer but the response to nitrogen and phosphorus is strongly seasonably dependent. A frost terminated growth early that year.)

A second-order response function was fitted to the data and was slightly modified by removal of the N^2 term to provide the response equation

$$\hat{Y} = 123 - 0.476N + 4.77P + 0.917K$$

$$- 0.0936P^2 - 0.00278K^2$$

$$- 0.000670NP + 0.00100NK - 0.00192PK. \qquad (1)$$

Important in potatoes is the specific gravity G, which must lie above a certain minimum value specified by the customer to satisfy the quality specifications of chippers. The function

$$10^3(G - 1) = 62.7 - 0.0680K \qquad (2)$$

relates the specific gravity to the fertilizer rate.

The design that gave rise to the data which led to (1) was a dodecahedron type. All noncentral design points were on the surface

$$(0.0222N - c)^2 + (0.0758P - c)^2 + (0.0133K - c)^2 = c^2, \qquad (3)$$

where $c = (1 + 5^{1/2})/2 = 1.618$.

Given that we must have $G > 1.055$, determine the combination of (N, P, K) that results in the maximum response $\hat{Y}$ determined by (1).

Is this point within or on the region defined by (3)? If such a restriction were necessary, how would one determine the best fertilizer combination?

7.19. (Source: Variable shear rate viscosity of SBR-filler-plasticizer systems, by G. C. Derringer, *Rubber Chemistry and Technology*, **47**, 1974, 825–836.) Experimental studies were carried out to relate the Mooney viscosity η of various rubber compounds to three predictor variables ξ_1 = silica filler, phr, ξ_2 = oil filler, phr, and ξ_3 = shear rate, s^{-1}. (phr = parts per hundred parts of rubber.) In the accompanying tables (Tables E7.19a and E7.19b), x_1 and x_2 are the coded variables

$$x_1 = \frac{(\xi_1 - 60)}{15}, \qquad x_2 = \frac{(\xi_2 - 21)}{15},$$

TABLE E7.19a

Coded levels of		y value when ξ_3 has value			
x_1	x_2	3	75	1500	3000
−1	−1	13.71	11.01	8.32	7.71
1	−1	14.15	11.35	8.85	—
−1	1	12.87	10.40	—	7.30
1	1	13.53	10.80	8.20	7.74
−1.4	0	12.99	10.68	8.05	7.36
1.4	0	13.89	11.04	8.51	8.10
0	−1.4	14.16	11.33	8.75	8.22
0	1.4	12.90	10.57	8.00	7.45
0	0	13.75	10.96	8.36	7.86
0	0	13.66	10.94	8.38	7.80
0	0	13.86	10.95	8.40	7.80
0	0	13.63	10.88	8.35	7.80
0	0	13.74	—	8.36	—

Additional run to be added to those above: $(x_1, x_2, \xi_3, y) = (-1, 1, 1550, 7.88)$.

TABLE E7.19b

Coded levels of		y value when ξ_3 has value				
x_1	x_2	3	7.5	75	750	3000
−0.86	−0.86		12.58		8.80	
0.86	−0.86		13.12		9.09	
−0.86	0.86		11.97		8.46	
0.86	0.86		12.24		8.63	
−1.4	0			10.51		
1.4	0			10.74		
0	−1.4			11.22		
0	1.4			10.24		
0	0	13.06		10.73		7.60
0	0			10.69		
0	0			10.65		
0	0			10.67		
0	0			10.69		
0	0			10.64		

and the responses y are the observed values of $\ln \eta$. Fit, to each of these sets of data, the incomplete cubic model

$$y = \beta_0 + \beta_1 x_1 + \beta_2 x_2 + \beta_3 x_3 + \beta_{33} x_3^2 + \beta_{13} x_1 x_3 + \beta_{23} x_2 x_3$$

$$+ \beta_{133} x_1 x_3^2 + \beta_{233} x_2 x_3^2 + \varepsilon,$$

where $x_3 = \ln(\xi_3 + 1)$, and carry out appropriate tests for the validity and usefulness of the fitted surface. Note that the model can be rewritten in the form

$$y = \left(\beta_0 + \beta_3 x_3 + \beta_{33} x_3^2 \right) + \left(\beta_1 + \beta_{13} x_3 + \beta_{133} x_3^2 \right) x_1$$

$$+ \left(\beta_2 + \beta_{23} x_3 + \beta_{233} x_3^2 \right) x_2 + \varepsilon.$$

This shows how the model was conceived, namely as a "plane in x_1 and x_2" but with coefficients which are all quadratic functions of x_3.

7.20. An experimenter finds the following response surfaces for (a) yield, percent:

$$\hat{y}_1 = 80 + 4x_1 + 8x_2 - 4x_1^2 - 12x_2^2 - 12x_1 x_2,$$

and (b) unit cost per pound:

$$\hat{y}_2 = 80 + 4x_1 + 8x_2 - 2x_1^2 - 12x_2^2 - 12x_1 x_2.$$

If he must operate at yields not less than 80% and costs not more than 76 cents per pound to make a profit, what values of the predictors x_1 and x_2 are profitable ones?

7.21. Fit a second-order response surface to the coded data in Table E7.21 and provide a full analysis.

7.22. (Source: Cutting-tool temperature-predicting equation by response surface methodology, by S. M. Wu and R. N. Meyer, *J. Eng. Indus., Trans. ASME, Series B*, **86**, 1964, 150–156. Adapted with the permission of the copyright holder, The American Society of Mechanical Engineers, Copyright © 1964.) Table E7.22a shows the results obtained from a cube plus doubled star plus four center points design run on three variables, speed V, feed f, and depth of cut d, which were varied in a cutting tool investigation. The coded input variables are $x_1 = (\ln V - \ln 70)/(\ln 70 - \ln 35)$, $x_2 = (\ln f - \ln 0.010)/(\ln 0.010 - \ln 0.005)$, and $x_3 = (\ln d - \ln 0.040)/(\ln 0.040 - \ln 0.020)$. The desired response variable is $y = \ln T$ where $T°$ is the temperature in degrees Fahrenheit. Two sets of observations were obtained, so that 48 sets of (x_1, x_2, x_3, y) are available in all. The tests within each set were run in random order as denoted by the "test sequence numbers" given in the source reference and reproduced in Table E7.22a.

TABLE E7.21

x_1	x_2	x_3	y
-1	-1	-1	10.6
1	-1	-1	11.1
-1	1	-1	11.9
1	1	-1	13.0
-1	-1	1	19.7
1	-1	1	18.1
-1	1	1	18.7
1	1	1	22.1
-2	0	0	14.2, 15.3
2	0	0	17.5, 16.6
0	-2	0	12.7, 13.4
0	2	0	16.5, 17.6
0	0	-2	8.4, 9.0
0	0	2	23.8, 23.3
0	0	0	16.0, 15.5, 15.6, 16.2

Fit a second-order equation to the 48 pieces of data and provide an analysis of variance table. Test for lack of fit. (Assume that the four repeat runs at the center and the six pairs of star points provide 9 df for pure error *within each set*, and so 18 in all. Assume that there may be a block difference between the "first 24 tests" and the "second 24 tests" and take out a sum of squares for blocks.) Check the residuals, particularly against the sequence numbers of the runs.

Use your fitted equation to predict the responses for the nine confirmatory runs in Table E7.22b, and compare your predicted temperatures with those observed.

7.23. (Source: Sucrose-modified phenolic resins as plywood adhesives, by C. D. Chang and O. K. Kononenko, *Adhesives Age*, **5** (7), July 1962, 36–40. Material reproduced and adapted with the permission of *Adhesives Age*. The data were obtained under a grant from the Sugar Research Foundation, Inc., now the World Sugar Research Organization, Ltd.) This work is a continuation of that in Exercise 5.22. Four variables: (1) sucrose in grams, S, coded to $x_1 = (S - 47)/5$; (2) paraform in grams, P, coded to $x_2 = (P - 36.6)/3$; (3) NaOH in grams, N, coded to $x_3 = (N - 8.8)/2$; and (4) water in grams, W, coded to $x_4 = (W - 23.8)/7$ were examined for their effect on the dry strip-shear strength in pounds per square inch (psi) of Douglas fir plywood glued with a resin modified with sugar. Three coded levels, -1, 0, 1 were used for each factor, and 27 runs were made. These constituted a four-factor

TABLE E7.22a. Coded input values and temperatures observed

Trial number	Test sequence number		Coded cutting conditions			Response values, T (°F)	
	First 24 Tests	Second 24 Tests	x_1	x_2	x_3	First 24 tests	Second 24 tests
1	7	12	-1	-1	-1	409	398
2	1	1	1	-1	-1	711	709
3	4	3	-1	1	-1	599	604
4	9	11	1	1	-1	1025	1016
5	3	6	-1	-1	1	474	498
6	8	9	1	-1	1	856	851
7	11	8	-1	1	1	700	714
8	6	5	1	1	1	1086	1087
9	2	4	0	0	0	728	742
10	5	2	0	0	0	728	730
11	10	10	0	0	0	716	722
12	12	7	0	0	0	738	722
13	17	13	$-2^{1/2}$	0	0	498	509
14	18	14	$2^{1/2}$	0	0	1063	1059
15	14	16	0	$-2^{1/2}$	0	565	558
16	13	17	0	$2^{1/2}$	0	917	915
17	16	15	0	0	$-2^{1/2}$	633	626
18	15	18	0	0	$2^{1/2}$	785	779
19	22	22	$-2^{1/2}$	0	0	486	509
20	24	19	$2^{1/2}$	0	0	1070	1072
21	19	21	0	$-2^{1/2}$	0	559	555
22	21	23	0	$2^{1/2}$	0	917	917
23	23	20	0	0	$-2^{1/2}$	624	612
24	20	24	0	0	$2^{1/2}$	792	789

TABLE E7.22b. Confirmatory runs

x_1	x_2	x_3	Observed temperature(s) (°F)
1.28010	-0.46642	-1.23447	843, 832
-0.80735	0.76553	0.80737	711, 709
0.77761	-1.24129	-0.79837	658, 667
0.19265	0.26303	1.32194	866, 873
-1.37440	1.13750	0.0	609

Note: The x_1 and x_2 coordinates of the fifth confirmatory run have been corrected; in the source reference they do not match the V and f levels recorded.

TABLE E7.23a. A three-level, second-order, response surface design in four factors divided into three orthogonal blocks

Block	x_1	x_2	x_3	x_4	y	Σy
1	-1	-1	0	0	203	
	1	-1	0	0	189	
	-1	1	0	0	191	
	1	1	0	0	203	
	0	0	-1	-1	211	
	0	0	1	-1	206	
	0	0	-1	1	218	
	0	0	1	1	241	
	0	0	0	0	211	1873
2	-1	0	0	-1	178	
	1	0	0	-1	196	
	-1	0	0	1	210	
	1	0	0	1	209	
	0	-1	-1	0	178	
	0	1	-1	0	169	
	0	-1	1	0	211	
	0	1	1	0	160	
	0	0	0	0	184	1695
3	0	-1	0	-1	218	
	0	1	0	-1	213	
	0	-1	0	1	219	
	0	1	0	1	190	
	-1	0	-1	0	184	
	1	0	-1	0	181	
	-1	0	1	0	222	
	1	0	1	0	207	
	0	0	0	0	202	1836
						5404

cube plus star rotatable design with three center points, but one not in the conventional orientation. (For additional explanation see Section 15.4 and/or Box and Behnken, 1960a.) The design and the response observations are shown in Table E7.23a. Note that each block contains one center point and that blocks are orthogonal, that is, the blocking will not affect the estimation of a second-order model. For the conditions for this see Section 15.3. A separate set of repeat runs gave rise to a pure error estimate $s_e^2 = 50.97$.

TABLE E7.23b.   Analysis of variance for the fitted second degree equation

Source	SS	df	MS	F
Blocks	1960.5	2	980.25	5.77
First order	1944.5	4	486.13	2.86
Second order$\vert b_0$	3110.8	10	311.08	1.83
Residual	1697.6	10	169.76	
Total, corrected	8713.4	26		

Confirm that the tabled data may be fitted to a full second-order equation to provide

$$\hat{y} = 199 - 0.25x_1 - 7.667x_2 + 8.833x_3 + 5.417x_4$$

$$- 3.167x_1^2 - 6.792x_2^2 - 1.292x_3^2 + 13.833x_4^2$$

$$+ 6.5x_1x_2 - 3x_1x_3 - 4.75x_1x_4 - 10.5x_2x_3 - 6x_2x_4 + 7x_3x_4.$$

Furthermore, obtain the analysis of variance table given as Table E7.23b, and show that the lack of fit F statistic, $F(10, \nu) = 3.33$, is not significant if $\nu \leq 8$. (This number ν is not specified in the paper, but the authors remark on p. 30 that the quadratic fit is adequate. It is most likely that five runs were made, as happened in an earlier part of the study.)

We see that only blocks are significant indicating that the variation between blocks of runs is higher than any variation attributable to the factors x_1, x_2, x_3, and x_4. After three more exploratory trials to confirm this situation, the authors concluded that "it is apparent that a broad plateau has been reached" and selected the point $(S, P, N, W) = (44.6, 34, 47.1, 16.3)$ as the "optimal conditions." They reconfirmed this point with a final run.

7.24. (Source: Adapted from An agronomically useful three-factor response surface design based on dodecahedron symmetry, by H. P. Hermanson, C. E. Gates, J. W. Chapman, and R. S. Farnham, *Agronomy Journal*, **56**, 1964, 14–17, by permission of the American Society of Agronomy.) Table E7.24a shows a 30-point design divided into five blocks. Two replicate values are obtained for each treatment combination, making 60 observations in all.

(a) Confirm that the design is a rotatable dodecahedron plus 10 center points, 2 per block divided into orthogonal blocks, that is, the blocks are orthogonal to a second-order model matrix. Confirm also that "replicates" are orthogonal to the second-order model. (See Sections 14.3 and 15.3.)

(b) Obtain the fitted (by least squares) model $\hat{y} = \frac{1}{2}\{18.32 - 0.0552x_1 + 1.4063x_2 + 0.1155x_3 - 0.6225x_1^2 - 0.7075x_2^2 - 0.4850x_3^2 - 0.1667x_1x_2$

TABLE E7.24a

Block	x_1	x_2	x_3	Yield (lb/plot), y Replicate 1	Replicate 2
1	d	c	0	10.6	7.8
	0	$-d$	c	11.5	6.9
	$-c$	0	$-d$	9.7	7.9
	1	-1	-1	9.1	5.1
	0	0	0	11.0	7.4
	0	0	0	10.3	6.6
2	d	$-c$	0	9.1	5.2
	0	d	$-c$	9.8	4.7
	$-c$	0	d	8.5	7.8
	1	1	1	10.3	7.6
	0	0	0	12.1	7.0
	0	0	0	9.3	7.5
3	$-d$	c	0	10.7	7.5
	0	$-d$	$-c$	11.4	4.4
	c	0	d	8.2	5.5
	-1	-1	1	6.2	5.1
	0	0	0	10.8	8.4
	0	0	0	11.3	7.2
4	$-d$	$-c$	0	8.9	4.8
	0	d	c	10.7	7.4
	c	0	$-d$	10.3	7.8
	-1	1	-1	10.4	7.5
	0	0	0	11.3	6.4
	0	0	0	11.1	9.1
5	-1	1	1	10.4	8.7
	-1	-1	-1	11.3	6.0
	1	1	-1	11.3	7.9
	1	-1	1	9.4	6.7
	0	0	0	10.8	7.6
	0	0	0	10.7	7.3
Totals				305.5	206.8

Note: $c = \frac{1}{2}(5^{1/2} + 1) = 1.618$; $d = \frac{1}{2}(5^{1/2} - 1) = 0.618 = c^{-1}$.

TABLE E7.24b

Source	SS	df	MS	F
Replicates	165.67	1	165.67	143.69^a
First order	19.99	3	6.663	5.78^a
Second order$\mid b_0$	12.72	6	2.12	1.84
Lack of fit	47.34	39	1.214	1.32
Pure error	9.17	10	0.917	
Total, corrected	254.89	59		

^{a}Significant at $\alpha = 0.05$ or lower α.

$-0.1917x_1x_3 + 0.4167x_2x_3\} + 1.66z$. Note: (1) $z = 1$ for "replicate 1," $z = -1$ for "replicate 2." (2) The portion $\{ \ldots \}$ of the model is what is obtained if a second-order surface is fitted to totals $\{$"y for replicate 1" + "y for replicate 2"$\}$, for example, to $\{10.6 + 7.8\} = 18.4$, and so on.

(c) Confirm the analysis of variance, Table E7.24b. Note the following points, (1) Pure error is obtained only from pairs of center points within the same block *and* within the same replicate. There are 10 such pairs, for example, (11.0, 10.3), (7.4, 6.6), (12.1, 9.3), ..., (7.6, 7.3). (2) The "replicates" sum of squares is given by $(306.5)^2/30 + (206.8)^2/30 - (513.3)^2/60 = 4556.95 - 4391.28 = 165.67$. (3) Certain subtleties involving "block strips SS" (blocks) and "replicate by block interaction SS" have been ignored; see the source reference for additional details. (These details would have needed investigation if lack of fit had shown up.) (4) The pooled estimate of σ^2 given by $s^2 = (47.34 + 9.17)/(39 + 10) = 1.153$ is used to obtain all F values except that for lack of fit; the latter F value is not significant; hence the validity of the pooling.

(d) Confirm that a first-order model explains $R^2 = 0.7284$ of the variation about the mean and that the addition of the (nonsignificant) second-order terms raises this to $R^2 = 0.7812$.

(e) Check the residuals from both first- and second-order models. Would you feel happy about using the fitted first order model $\hat{y} = 8.555 + 1.66z - 0.0552x_1 + 1.4063x_2 + 0.1155x_3$? Why or why not?

7.25. (Source: Chemical process improvement by response surface methods, by P. W. Tidwell, *Industrial and Engineering Chemistry*, **52**, 1960, June, 510–512. Adapted by permission of the copyright holder, the American Chemical Society.) Fit a second-order model to each set of y data in Table E7.25. Test for lack of fit and, if no lack of fit is evident, test whether second-order terms are needed and whether any of the x's can be dropped. What predictive model would you use?

TABLE E7.25

x_1	x_2	x_3	y_1	y_2
-1	-1	-1	62.1	15.79
1	-1	-1	61.3	12.03
-1	1	-1	17.8	11.91
1	1	-1	68.8	12.37
-1	-1	1	61.3	15.77
1	-1	1	61.0	11.92
-1	1	1	16.6	11.58
1	1	1	66.5	12.27
-2	0	0	8.9	11.26
2	0	0	67.0	10.96
0	-1.65	0	62.4	13.68
0	1.65	0	45.1	12.68
0	0	-2.4	66.8	14.02
0	0	2.4	66.5	13.80
0	0	0	64.6	13.70
0	0	0	65.5	13.71
0	0	0	64.6	13.90
0	0	0	65.1	14.08

7.26. (Source: A statistical experimental design and analysis of the extraction of silica from quartz by digestion in sodium hydroxide solutions, by R. L. Stone, S. M. Wu, and T. D. Tiemann, *Trans. Soc. Mining Engineers*, **232**, June 1965, 115–124. Copyright © 1965. Adapted with the permission of the Society of Mining Engineers of AIME.) Table E7.26a shows the values of four coded variables x_1, x_2, x_3, x_4 and the response y in an experiment to determine the effects of four predictor variables, quartz size, time, temperature, and sodium hydroxide concentration on the extraction of silica from quartz. The actual predictor variables have been logged to the base e and coded, and the response is ln(extraction).

Consider runs 1–20 as one block and runs 21–28 as a second block. Is this design orthogonally blocked? Fit a second-order surface to these data, adding a blocking variable whose levels are $z = -\frac{2}{7}$ for runs 1–20 and $z = \frac{5}{7}$ for runs 21–28. Evaluate an analysis of variance table of the form indicated in Table E7.26b and also examine the usefulness of the individual x's as well as the need for quadratic terms. What model would you adopt? Examine the residuals, including a plot of residuals against test order. State your conclusions. (For orthogonal blocking, see Section 15.3.)

7.27. (Source: Use of statistics and computers in the selection of optimum food processing techniques, by A. M. Swanson, J. J. Geissler, P. J. Magnino, Jr.,

TABLE E7.26a

Test order	x_1	x_2	x_3	x_4	y
1	-1	-1	-1	-1	4.190
20	1	-1	-1	-1	5.333
11	-1	1	-1	-1	4.812
5	1	1	-1	-1	5.886
17	-1	-1	1	-1	5.333
3	1	-1	1	-1	6.225
9	-1	1	1	-1	5.905
12	1	1	1	-1	6.574
15	-1	-1	-1	1	4.533
6	1	-1	-1	1	5.642
8	-1	1	-1	1	5.199
13	1	1	-1	1	6.094
4	-1	-1	1	1	5.628
19	1	-1	1	1	6.373
14	-1	1	1	1	6.127
2	1	1	1	1	6.702
10	0	0	0	0	5.775
7	0	0	0	0	5.802
18	0	0	0	0	5.781
16	0	0	0	0	5.796
25	-2	0	0	0	4.615
21	2	0	0	0	6.690
28	0	-2	0	0	5.220
23	0	2	0	0	6.385
27	0	0	-2	0	4.796
24	0	0	2	0	6.693
26	0	0	0	-2	5.416
22	0	0	0	2	6.182

TABLE E7.26b

Source	df
b_0	1
Blocks	1
First order|blocks	4
Second order|b_0, blocks	10
Lack of fit	9
Pure error	3
Total	28

TABLE E7.27.   Data from a sterilized, concentrated, baby formula experiment

Run reference number	Levels of coded predictors					Response values (subscript = storage period in months)			
	x_1	x_2	x_3	x_4	x_5	y_0	y_3	y_6	y_9
1	−1	−1	−1	−1	−1	9.8	7.5	12.5	41.5
26	1	−1	−1	−1	1	30.2	35.0	22.5	45.0
5	−1	1	−1	−1	1	17.5	17.5	12.5	20.0
22	1	1	−1	−1	−1	12.5	10.0	7.5	12.5
3	−1	−1	1	−1	1	512.5	1950.0	2070.0	3030.0
24	1	−1	1	−1	−1	655.0	670.0	450.0	1700.0
7	−1	1	1	−1	−1	342.5	262.5	410.0	322.5
20	1	1	1	−1	1	1020.0	1050.0	970.0	1230.0
2	−1	−1	−1	1	1	82.5	145.0	162.5	145.0
25	1	−1	−1	1	−1	19.0	22.0	17.5	25.0
6	−1	1	−1	1	−1	9.3	5.8	5.0	12.5
21	1	1	−1	1	1	27.5	22.5	15.0	20.0
4	−1	−1	1	1	−1	270.0	237.5	337.5	717.5
23	1	−1	1	1	1	282.5	710.0	650.0	547.5
8	−1	1	1	1	1	172.5	237.5	210.0	190.0
19	1	1	1	1	−1	172.5	155.0	257.5	435.0
9	−1	0	0	0	0	45.8	52.5	62.5	57.5
27	1	0	0	0	0	77.5	62.5	70.0	113.8
11	0	−1	0	0	0	195.8	262.5	252.5	276.3
12	0	1	0	0	0	33.0	22.5	15.0	27.5
13	0	0	−1	0	0	20.0	15.0	17.5	17.5
14	0	0	1	0	0	337.5	117.5	105.0	177.5
15	0	0	0	−1	0	70.0	147.5	60.0	147.5
16	0	0	0	1	0	83.8	62.5	132.5	105.0
17	0	0	0	0	−2	40.0	40.0	22.5	60.0
18	0	0	0	0	2	287.5	450.0	482.5	495.0
10	0	0	0	0	0	67.5	77.5	45.0	107.5

and J. A. Swanson, *Food Technology*, **21** (11), 1967, 99–102. Adapted with the permission of the copyright holder, the Institute of Food Technologists.) The data in Table E7.27 are obtained from a larger experiment on a sterilized, concentrated, baby formula. The purpose of the experiments was to select conditions which would provide a formula with acceptable storage stability for over a year at room temperature, and which would withstand terminal sterilization once the concentrate had been diluted with water. The five factors, or predictor variables, and their ranges of interest, were as follows

1. Preheating of the milk solids for 25 min (175–205°F).
2. Addition of sodium polyphosphate (0–0.14%).

   3. Addition of sodium alginate (0–0.3%).
   4. Addition of lecithin (0–1.5%).
   5. Addition of carrageenan (0.444–0.032%).

Three levels were chosen for predictor variables 1–4, five levels for variable 5, and the coding shown in the table was adopted for these levels. The basic design is a 2_V^{5-1} fractional factorial design ($I = -12345$) plus five pairs of axial points at the extreme levels, plus one center point, that is, a cube plus star (unbalanced in x_5) plus center point. The response y_t, $t = 0, 3, 6, 9$ is the Brookfield viscosity in centipoise after t months, immediately after opening and without any agitation. To be acceptable, the product viscosity had to be less than 1000 cP. (Other responses measured, but not given in the paper, related to storage stability and the results of terminal stabilization. In fact, y_t was a subsidiary response and optimization was performed on the other responses first with a check on y_t to confirm that it was acceptable. However, for the purposes of this exercise, we regard it as the primary response.)

   Fit, to each column of response values shown, a full second-order surface in $x_1, x_2, \ldots, x_5$. Provide an analysis of variance table with as complete a breakup as possible and test for lack of fit, and for second- and first-order significant regression, as is possible.

   Evaluate the residuals and examine them. In particular, arrange the residuals in run order (shown on the left of the table) and see if any effects from that ordering are apparent.

   Consider the following question: Would it be a good idea to transform the response? Apply the Box–Cox transformation technique to find the "best" transformation in the family (see Section 8.4)

$$z = \frac{(y^\lambda - 1)}{\lambda} \qquad \lambda \neq 0$$

$$= \ln y \qquad \lambda = 0.$$

and, if a value of λ different from unity is indicated, make the transformation, and carry out all the analyses mentioned in the foregoing. What do you conclude, now?

   If the viscosity has to be kept within the range $800 \leq y_t \leq 1000$, what levels of the x_i are permissible? How about for $y_t \leq 200$?

7.28. (Source: Sequential method for estimating response surfaces, by G. C. Derringer, *Industrial and Engineering Chemistry*, **61**, 1969, 6–13. Adapted by permission of the copyright holder, the American Chemical Society.) Table E7.28 shows data from a two-factor, third-order, rotatable, orthogonally blocked, response surface design consisting of a circle of eight points, a circle of seven points, and six center points in the coded (x_1, x_2) space. Also shown are the values of six response variables $y_1, \ldots, y_6$. (Full descriptive

TABLE E7.28

Run	Block	x_1	x_2	y_1	y_2	y_3	y_4	y_5	y_6
1	1	1	1	1140	3040	3670	570	9.4	14.5
2	1	1	−1	590	1750	3430	720	13.4	20.0
3	1	−1	1	910	2390	3790	660	7.5	13.0
4	1	−1	−1	530	1500	2940	710	11.9	30.5
5	1	0	0	900	2410	3800	660	12.0	16.0
6	1	0	0	880	2410	3860	660	10.6	15.0
7	1	0	0	870	2330	3830	670	11.3	15.0
8	2	1.414	0	1040	2760	3670	600	12.7	17.5
9	2	−1.414	0	650	1830	3570	710	8.2	18.0
10	2	0	1.414	990	2560	3770	630	8.9	13.0
11	2	0	−1.414	440	1300	3040	730	20.7	41.5
12	2	0	0	840	2240	3870	690	11.2	16.0
13	2	0	0	860	2270	3690	660	11.9	16.0
14	2	0	0	860	2260	3820	680	11.9	16.5
15	3	0.667	0.836	870	2400	3440	620	10.1	14.5
16	3	−0.238	1.042	730	2000	3560	660	7.9	13.5
17	3	−0.963	0.464	570	1590	3200	700	9.4	15.0
18	3	−0.963	−0.464	490	1350	2860	710	10.5	21.0
19	3	−0.238	−1.042	460	1290	3010	740	14.5	28.0
20	3	0.667	−0.836	620	1880	3510	690	14.7	22.0
21	3	1.069	0	870	2440	3660	630	12.3	17.5

details of the experiment will be found in the source paper and are not repeated here.) The orthogonal blocking conditions are satisfied for first- and third-order terms by the nature of the design, and the second-order orthogonal blocking condition determines the radius of the seven-point circle given the radius of the eight-point circle and the number of center points used. Specifically, here, $r_2^2 = 2(7)(8)/\{14(7)\} = 8/7 = (1.069)^2$.

Fit a third-order polynomial individually to each response, perform tests to determine what order of surface is needed in each case, and plot each fitted response within the rectangle $-1.75 \le x_i \le 1.75$, $i = 1, 2$. Your model will require two blocking variables and your analysis of variance table should contain a 2-df sum of squares for blocks. Suitable values for the blocking variables are $Z_1 = -1, 0, 1$, and $Z_2 = 1, -2, 1$ for blocks 1, 2, 3, respectively, and the model should contain extra terms $\alpha_1 Z_1 + \alpha_2 Z_2$.

CHAPTER 8

Adequacy of Estimation and the Use of Transformation

Usually, the metrics in which data are recorded are chosen merely as a matter of convenience in measurement. As the textile example in the foregoing chapter illustrated, they need not be those in which the system is most simply modeled. To achieve simplicity, transformations may be applied to the response, or to the predictor variables, or both. We begin by discussing the effects of such transformations.

8.1. WHAT CAN TRANSFORMATIONS DO?

Nonlinear transformations such as the square root, log, and reciprocal of some necessarily positive response Y have the effect of expanding the scale at one part of the range and contracting it at another. This is illustrated in Table 8.1 and Figure 8.1 where, for Y covering the range 1–4, the effects of square root, log, inverse square root, and reciprocal transformations are illustrated. These are all examples* of simple power transformations characterized by $y = Y^\alpha$. Transformations such as those illustrated in which α is less than unity have the effect of contracting the range at high values and may be called contractive transformations. Power transformations with α greater than unity have the reverse effect and may be called expansive. Contractive transformations of the kind illustrated in Figure 8.1 are most often needed in practical problems. Two facts about nonlinear transformations which should be remembered are:

1. The effect of the transformation is greater the greater the ratio Y_{max}/Y_{min} considered.

---

* The log transformation can be regarded as a special case of the power transformation $y = Y^\alpha$ with $\alpha \to 0$, as explained in Section 8.4.

268

TABLE 8.1. Numerical values of contractive transformations

Y	1	1.5	2	2.5	3	3.5	4
$Y^{1/2}$	1	1.225	1.414	1.581	1.732	1.871	2
$\log_{10} Y$	0	0.176	0.301	0.398	0.477	0.544	0.602
$Y^{-1/2}$	1	0.816	0.707	0.632	0.577	0.535	0.500
Y^{-1}	1	0.667	0.500	0.400	0.333	0.286	0.250

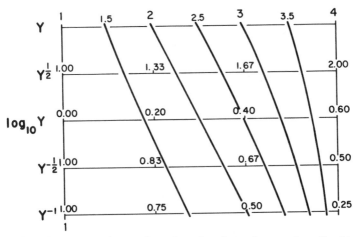

FIGURE 8.1. Some contractive transformations plotted over the range from $Y = 1$ to $Y = 4$. The scale for each transformed variable is shown on the appropriate horizontal line.

2. The effect of a power transformation $y = Y^{\alpha}$ is greater the more α differs from unity.

Note: When using the logarithmic transformation, either logarithms to the base 10 or to the base e can be used. They differ only by a constant factor, so that any subsequent analysis is unaffected except for this difference in scale. When we specifically wish to denote logarithms to the base e, we use the notation ln.

It is supposed in the above that Y is referred to a natural origin and is necessarily positive over the range considered. Thus, if we are interested in the concentration or amount of a particular ingredient in a chemical reaction, the natural origin would usually be zero. For temperature (T) measured in degrees Celsius, the functionally important origin is, most often, absolute zero ($-273°C$) and temperatures are thus usually expressed in degrees Kelvin ($T + 273$). For example, the Arrhenius equation postulates that the logarithm of the rate of a chemical reaction is a linear function of the reciprocal of the absolute temperature, namely that

$$\log(\text{rate}) = \alpha + \beta(T + 273)^{-1}.$$

If temperatures are measured in degrees Celsius, the above implies that, before transformation (to reciprocals), a constant 273 is added to the temperature. Sometimes, an *empirically chosen* constant μ is subtracted from a response Y. This produces a transformation of greater nonlinearity if μ is positive, as is usually the case. For example, for any fixed α, the effect of the transformation $(Y - \mu)^{\alpha}$ will depend on $(Y - \mu)_{\max}/(Y - \mu)_{\min}$. Thus, with the restriction $0 < \mu < Y_{\min}$, the larger μ is taken, the greater the effect induced by the transformation.

Effects of Simple Transformations on the Functional Relationships $\eta = f(\xi)$

For the moment, let us set aside the important question of the effect of transformation on distributional assumptions and consider only their effect on the functional relationship $E(Y) = \eta = f(\xi)$.

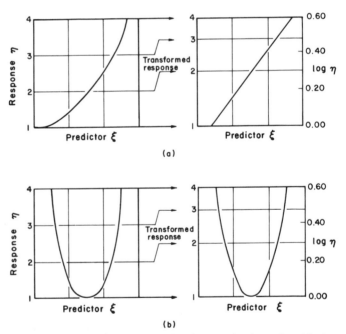

FIGURE 8.2. Transformation of a response η. (*a*) An example where a logarithmic transformation of the response variable η produces a straight line. (*b*) An example where a logarithmic transformation of the response variable η produces a quadratic curve.

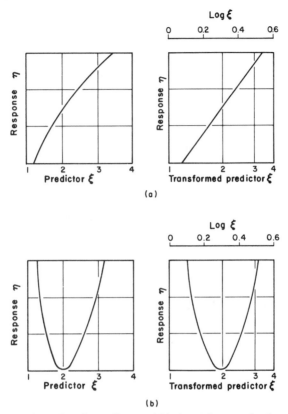

FIGURE 8.3. Transformation of a predictor variable ξ. (a) An example where a logarithmic transformation of the predictor variable ξ produces a straight line. (b) An example where a logarithmic transformation of the predictor variable ξ produces a quadratic curve.

Figure 8.2a illustrates a situation where a relationship between η and ξ is linearized by making a contractive transformation in the response η. The logarithmic transformation is actually used, so that $\log \eta = \alpha + \beta\xi$ and the original relationship is of the form $\eta = AB^\xi$. The same transformation applied in Figure 8.2b produces a quadratic relationship. Thus $\log \eta = \alpha + \beta\xi + \gamma\xi^2$ and the original relationship is of the form $Ae^{B(\xi-\xi_0)^2}$.

Figure 8.3a shows how a contractive transformation (log, again) of the *predictor variable* ξ linearizes a curved relationship. We see that a contractive transformation on the response η can linearize a "convex downward" curve (Figure 8.2a) and a contractive transformation on the predictor ξ can linearize a "convex upward" curve. A further implication which can easily be confirmed is that, where linearity can be produced by a contractive

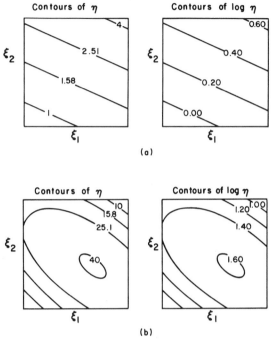

FIGURE 8.4. Transformation of a response η which depends on two predictor variables ξ_1 and ξ_2. (*a*) An example where a logarithmic transformation of the response variable η produces a planar surface. (*b*) An example where a logarithmic transformation of the response variable η produces a quadratic surface.

transformation on η, a similar result might be obtained, at least approximately, by an expansive transformation in ξ. The opposite is also true.

Figure 8.3*b* shows the same contractive transformation of the *predictor variable* ξ, producing a quadratic curve. The quadratic curves in the second parts of Figures 8.2*b* and 8.3*b* are chosen to be identical within their respective frameworks although the axial scales are different in the two cases. By examination and comparison of Figures 8.2*b* and 8.3*b* we can appreciate that a quadratic function may be used to model an unsymmetric maximum *only* if ξ is transformed. Transformation of η can produce a sharper or flatter peaked curve, but *not* an unsymmetric one.

For η depending on more than one predictor variable, Figures 8.4*a* and 8.4*b* illustrate the fact that transformation of η relabels contours without changing their shape. In the illustration of Figure 8.4*a*, log η is linear in ξ_1 and ξ_2. Notice that this implies that the contours of *any* other function of η (in particular η itself) will have parallel straight line contours. However, in

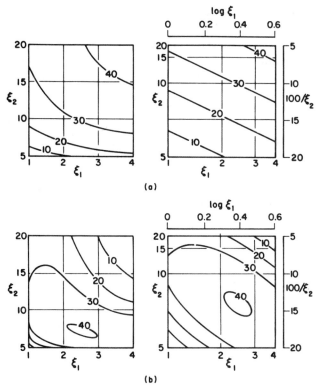

FIGURE 8.5. Transformation of two predictor variables ξ_1 and ξ_2 which affect a response η. (*a*) An example where transformation of two predictor variables ξ_1 and ξ_2 produces a planar surface. (*b*) An example where transformation of two predictor variables ξ_1 and ξ_2 produces a quadratic surface.

this particular illustration, it is only for $\log \eta$ that equal intervals in the transformed responses will be represented by equal distances in space. Thus only $\log \eta$ is *linear* in ξ_1 and ξ_2. Similarly in Figure 8.4*b*, $\log \eta$ is quadratic in ξ_1 and ξ_2 and the elliptic contours are also contours of η or any other function of η. But only $\log \eta$ is quadratic in ξ_1 and ξ_2.

When η is monotonic in ξ_1 and ξ_2 but its contours are not straight lines, it may be possible by transformations of ξ_1 and ξ_2 to produce a planar surface as in Figure 8.5*a* but transformation of η alone clearly cannot. Similarly, when a set of closed contours about a maximum are not ellipses it may, nevertheless, be possible by suitable transformations of ξ_1 and ξ_2 to produce a quadratic surface, as in Figure 8.5*b*. It is clearly not possible to do this by transformation of η alone.

Effect of Transformation of the Response on Distributional Assumptions

In practice, we do not know the expected response values (the η's); transformation must be carried out on the observed response values (the y's) which contain error. Thus, transformation of the response affects the error structure.

To illustrate how this occurs, consider an experiment designed to study how the size of the red blood cells of a rabbit were affected by two dietary substances added in amounts ξ_1 and ξ_2 to the feed. Suppose that, over the immediate region of interest $R(\xi)$, the amounts ξ_1 and ξ_2 of these two additives affected the radius y of the cells approximately linearly and that, for this measurement, the usual normal theory assumptions were approximately valid. Then y_u, the cell radius measured in the uth experiment, can be written

$$y_u = \eta_u + \varepsilon_u \tag{8.1.1}$$

with

$$\eta_u = \beta_0 + \beta_1\xi_{1u} + \beta_2\xi_{2u}.$$

Now in practice it might be most convenient to measure the area A_u of a cell or, equivalently, $A_u/\pi = Y_u$, the radius squared. But if

$$Y_u = y_u^2$$

then, using (8.1.1),

$$Y_u = \eta_u^2 + 2\eta_u\varepsilon_u + \varepsilon_u^2. \tag{8.1.2}$$

To convert (8.1.2) into a form comparable with (8.1.1), we must write

$$Y_u = \eta_u' + \varepsilon_u'$$

where

$$\eta_u' = \eta_u^2 + \sigma_\varepsilon^2 \quad \text{and} \quad \varepsilon_u' = 2\eta_u\varepsilon_u + \varepsilon_u^2 - \sigma_\varepsilon^2,$$

so that, using the normality assumption where necessary,

$$E(\varepsilon_u') = 0 \quad \text{and} \quad V(\varepsilon_u') = 2\sigma_\varepsilon^2(2\eta_u^2 + \sigma_\varepsilon^2) = 2\sigma_\varepsilon^2(2\eta_u' - \sigma_\varepsilon^2).$$

Thus, if modeling were carried out directly on the untransformed cell area, we would need a second degree equation to express the relationship between the response and the predictor variables. In addition, the errors ε'_u would have a variance which was linearly dependent on the mean level η'_u of the response, and a distribution which was non-normal. A (square root) transformation of Y however would yield a linear model with normal errors having constant variance.

From the above, it is clear that, if we regard the response Y as principally subject to error while the predictor variables are relatively error free the transformation on Y will not only affect the form of the functional relationship $\eta = f(\xi)$ but also the error structure. On the other hand, transformation of the ξ's will, on these assumptions, not affect the error structure at all. There will be examples such as the (artificial) blood cell example above and also the (real) textile example already considered, where transformation of the response can simultaneously simplify the functional relationship and improve compliance with the distributional assumptions. Other examples will occur where this is not possible. In some examples, where distribution assumptions are already adequate, appropriate simplification may be possible by transformation of the predictor variables ξ alone. In other examples, it may be necessary to transform both the response and the predictor variables to obtain simplicity of functional form and a satisfactory distribution of error.

With the possibility of simplification via transformation ever present, the task of obtaining empirically an adequate functional representation is not a trivial one; we consider various aspects of the matter in the next section. We begin by considering a measure of the adequacy of estimation of a fitted function which is related to a certain F ratio in the analysis of variance.

8.2. A MEASURE OF ADEQUACY OF FUNCTION ESTIMATION

Attempts to interpret relationships that have been inadequately estimated are likely to prove misleading and sometimes disastrous. For example, the detailed behavior of plotted contours from a fitted second degree surface ought not to be taken seriously on the basis of mere statistical significance of the "overall regression" versus "residual" F test. This section is intended to throw light on the questions: "When is a fitted function sufficiently well estimated to permit useful interpretations?" and "Should adequacy be judged on the size of F, R^2, or what else?" We follow an argument due to Box and Wetz (1973).

We can illustrate adequately the questions at issue in terms of a single-input ξ. Suppose that observations of a response y are made at n different settings of the input ξ. Figure 8.6a shows a plot of $n = 8$ observations $y_1, y_2, \ldots, y_8$ against the known levels $\xi_1, \xi_2, \ldots, \xi_8$, of the single-input ξ.

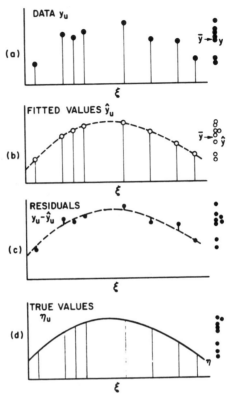

FIGURE 8.6. Data y_u, fitted values $\hat{y}_u$, residuals $y_u - \hat{y}_u$, and true values η_u for a relationship involving one input ξ. Also shown are the empirical distributions of y's, $\hat{y}$'s, $y - \hat{y}$'s, and η's, the sum of whose squares enter the analysis of variance table.

Suppose further that $\eta = g(\xi, \boldsymbol{\beta})$ is an appropriate graduating function (such as a polynomial in ξ) linear in p adjustable parameters in addition to the mean and that the curve in Figure 8.6b represents this function fitted to the data by least squares. The values that the fitted function takes at $\xi = \xi_1$, $\xi = \xi_2, \ldots, \xi = \xi_8$ are denoted by $\hat{y}_1, \hat{y}_2, \ldots, \hat{y}_8$ and are indicated in Figure 8.6b by open circles. Figure 8.6c shows the residuals $y_1 - \hat{y}_1$, $y_2 - \hat{y}_2, \ldots, y_8 - \hat{y}_8$ which measure the discrepancies between observation and fitted function at the chosen values of ξ. Figure 8.6d shows the true (expected) values, η. We know from Chapter 3 that certain identities exist between the sums of squares of these quantities. In particular, taking squares about the grand average $\bar{y}$ provides

$$\text{total sum of squares} = (\text{regression sum of squares})$$
$$+ (\text{residual sum of squares}),$$

or

$$\Sigma(y_u - \bar{y})^2 = \Sigma(\hat{y}_u - \bar{y})^2 + \Sigma(y_u - \hat{y}_u)^2. \qquad (8.2.1)$$

The corresponding identity in the degrees of freedom is

$$n - 1 = p + (n - p - 1). \qquad (8.2.2)$$

Using these identities, an appropriate analysis of variance table may be written down. For the purpose of this discussion, we shall assume for the time being that the relationship $E(y) = \eta = g(\xi, \beta)$ is representationally adequate. Then the expected values of the regression sum of squares about the mean, which is denoted by SS_R, and of the residual sum of squares, which is denoted by SS_D, are given by

$$E(SS_R) = E\{\text{regression } SS | b_0\}$$

$$= E\{\Sigma(\hat{y}_u - \bar{y})^2\} = \Sigma(\eta_u - \bar{\eta})^2 + p\sigma^2, \qquad (8.2.3)$$

$$E(SS_D) = E\{\text{residual } SS\}$$

$$= E\{\Sigma(y_u - \hat{y}_u)^2\} = (n - p - 1)\sigma^2. \qquad (8.2.4)$$

Now consider Figure 8.6d. Because of estimation errors, the estimated function $\hat{y} = g(\xi, \hat{\beta})$ differs somewhat from the true function $\eta = g(\xi, \beta)$. The values $\eta_1, \eta_2, \ldots, \eta_8$, in Figure 8.6d, provide a "sampling" of the function over the range actually observed and the mean square deviation $\Sigma(\eta_u - \bar{\eta})^2/n$ is a measure of the real change in η accounted for by the functional dependence of η on the input variable ξ. (It is the mean squared deviation of the distribution of the given η_u). We want to compare this with a measure of the uncertainty with which $\eta - \bar{\eta}$ is estimated. Such a measure is the average of the expected values of the squares of discrepancies between the estimated deviations $\hat{y}_u - \bar{y}$ and "true" deviations $\eta_u - \bar{\eta}$ measured at the n ($= 8$, here) values of ξ.

We denote this averaged mean square error of estimate by $\sigma_{\hat{y}-\bar{y}}^2$ so that

$$\sigma_{\hat{y}-\bar{y}}^2 = \left[E\{(\hat{y}_1 - \bar{y}) - (\eta_1 - \bar{\eta})\}^2 + \cdots \right.$$

$$\left. + E\{(\hat{y}_8 - \bar{y}) - (\eta_8 - \bar{\eta})\}^2 \right]/n. \qquad (8.2.5)$$

A natural measure of estimation adequacy is, therefore, the quantity

$$Q^2 = \frac{\Sigma(\eta_u - \bar{\eta})^2/n}{\sigma_{\hat{y}-\bar{y}}^2}. \qquad (8.2.6)$$

In Appendix 8A, we show that, irrespective of the experimental design, $\sigma_{\hat{y}-\bar{y}}^2 = p\sigma^2/n$. Thus, an alternative form to (8.2.6) is

$$Q^2 = \frac{\Sigma(\eta_u - \bar{\eta})^2/n}{p\sigma^2/n} = \frac{\Sigma(\eta_u - \bar{\eta})^2/p}{\sigma^2}. \qquad (8.2.7)$$

The quantity Q^2 is familiar in classical statistics as a *measure of noncentrality*, and Q may also be regarded as a measure of *signal-to-noise ratio*. Now, on the assumptions made, the residual mean square $s^2 = SS_D/(n - p - 1)$ provides an unbiased estimator of σ^2 so that, from (8.2.3), $(SS_R - ps^2)/n$ provides an unbiased estimator of $\Sigma(\eta_u - \bar{\eta})^2/n$ and ps^2/n provides an unbiased estimator of $p\sigma^2/n = \sigma_{\hat{y}-\bar{y}}^2$. Thus, an estimator having the property that the expected values of numerator and denominator are those of Q^2 is

$$\hat{Q}^2 = \frac{(SS_R - ps^2)/n}{ps^2/n} = \frac{SS_R/p}{s^2} - 1 = F - 1, \qquad (8.2.8)$$

where F is the usual ratio of regression to residual mean squares.

As an illustration of the use of the measure of estimation adequacy $\hat{Q}^2 = F - 1$, we employ it to assess the effects of transformation on the textile data already analysed in Chapter 7. The various values of $\hat{Q}^2$ are shown in Table 8.2. In the original analysis of these data, the investigators, Barella and Sust, fitted a quadratic function to the untransformed Y. This yielded a value of $\hat{Q}^2$ of 27.5. We see that, for this example, a sevenfold increase in $\hat{Q}^2$ (217.0/27.5 = 7.9) is possible by use of the simpler straight line equation fitted to $y = \log Y$. Evidently the improvement may be credited to the following facts:

1. The method of (unweighted) least squares is an efficient estimation procedure only if the data have constant variance. A change from Y to $y = \log Y$ apparently makes the constant variance assumption approximately valid.

TABLE 8.2. Values of estimation adequacy measure
$\hat{Q}^2 = F - 1$ for textile data

Response used	$y = \log Y$	Y
Linear equation	217.0[a]	65.5[b]
Quadratic equation	65.7[b]	27.5[b]

[a]Via residual from appropriate planar equation
[b]Via residual from appropriate quadratic equation

2. There are, in addition to the mean, nine parameters in the quadratic model but only three in the linear model. Thus, in the log metric, there are six redundant parameters which, of itself, should improve estimation efficiency by an additional factor of $\frac{10}{4} = 2.5$.

When Is a Response Function Adequately Estimated?

After an analysis of variance has been performed, the ratio $F =$ (regression mean square)/(error mean square) is customarily computed and compared to an F-distribution percentage point, appropriate on the assumption that the errors ε are normally distributed with constant variance. Significance of this F ratio contradicts the assertion that the noncentrality measure $Q^2 = \{\Sigma(\eta_u - \bar{\eta})^2/p\}/\sigma^2$ is *zero*, and suggests that changes of some sort are detectable in the response η. However, to permit the assertion that we have a reasonable *estimate* of the response function, much more than mere statistical significance is required.

Now consider Figures 8.6c and 8.6d. The quantity $\{\Sigma(\eta_u - \bar{\eta})^2/n\}^{1/2}$ is a root mean square measure of *change* in the function $\eta = g(\xi, \beta)$, while $\sigma_{\hat{y}-\bar{y}}$ is a root mean square measure of the average *error of estimate* of that same function. Q is the ratio of these two quantities. Before we could assert that a worthwhile *estimate* of the response function had been obtained, we would need to ensure that the change in the true function over the region of ξ considered was reasonably large compared with the error of estimate in that region. Thus we would need to be assured that Q had some minimum value Q_0.

To make assertions of this kind, we must consider the distribution of $F = \hat{Q}^2 + 1$ when the noncentrality measure is nonzero. This distribution (called the noncentral F distribution) is known and has been tabulated but, for our present purposes, it will be sufficient to use a crude approximation. Very roughly, over the range of values of interest, to assert that $Q > Q_0$ requires that the relevant significance point obtained from the F table be multiplied* by $Q_0^2 + 1$. Thus, to ensure that Q is greater than 3, we should require that the ratio (mean square regression)/(mean square residual) exceeded its normal F significance level by a factor of 10.

The general conclusion to be drawn is that adequacy of estimation of a function ought to be judged by the size of $F =$ (regression mean square)/(residual mean square). The value of F should exceed the chosen

---

*The argument involves approximation of the appropriate noncentral F by $(1 + Q_0^2)$ times a central F. This would in fact be exact if the model were a random-effects model. When, as here, the model is a fixed-effects one, the result is approximate. However, the approximation appears to be conservative; see Patnaik (1949).

tabulated significance level by a factor of $Q_0^2 + 1$, which we suggest should be about 10.

An alternative and, in spirit, equivalent procedure is to compare the range of the n fitted $\hat{y}$'s with their average standard error $(ps^2/n)^{1/2}$.

Application to the Textile Data

For the textile data, for the first degree polynomial fit, $p = 3$, $n - p - 1 = 23$ and the 5% level of $F = 3.03$. Applying the above rule, we should require, therefore, that the F value actually achieved exceeded 30. In fact, for the logged data, it is 217.

Conversely, the value $F = 217$ implies that, for the example, $Q \simeq (217 - 1)^{1/2} \simeq 14.7$, which is a very satisfactory value.

Inadequacy of Models for Unlogged Textile Data

The discussion above has been carried out as if, in all the cases considered, we could believe in the adequacy of the fitted functions and of the normal theory models implied by the least-squares analysis. In fact, as we saw in Chapter 7, if the analysis of the textile example had been conducted in terms of the unlogged data, its inadequacy would have been seen as soon as the residuals were plotted.

8.3. CHOOSING THE RESPONSE METRIC TO VALIDATE DISTRIBUTION ASSUMPTIONS

In considering the question of an appropriate choice of response metric to validate distribution assumptions, the first thing to notice is that a *linear* transformation of Y, say to $y = (Y - k_0)/k_1$ where the k's are constants, will be without effect on these assumptions since all that has happened is a rescaling and relocation of the graduating function.

[Purely for *computational* convenience it is nevertheless useful sometimes to make a linear transformation (or "coding") of the output Y just as similar linear codings are often usefully applied to the inputs. For instance, in a study of the manner in which the specific gravity Y of an aqueous solution was affected by certain input variables the range of variation in Y over the whole set of experimental conditions was from 1.0027 to 1.0086. The data invite the coding $y = 10,000(Y - 1)$. The range of variation in the coded output y's is then from 27 to 86. Such a coding reduces worries about misplaced decimals, rounding errors, and computer overflow and, when the calculations have been completed, the inverse transformation $Y = 1 + 0.0001y$ restores the calculated response to the original units.]

Although linear transformation of the response has no effect on the shape of the error distribution, *nonlinear* transformations such as $y = \log Y$, $y = \log(Y - k)$, and $y = Y^\lambda$ can, in suitable circumstances, produce marked changes in distributional shape and in variance homogeneity.

When Do Data Transformations Have an Important Effect on Distribution Assumptions?

For data that is necessarily positive, the choice of metric for Y becomes important when the ratio Y_{max}/Y_{min} is large. For if this ratio is small, moderate nonlinear transformations of Y such as $\log Y, Y^{1/2}, Y^{-1}$ are nearly linear over the relevant range and make little difference. On the other hand, when the ratio is large, transformation can bring profound changes. To see what is involved, consider Figure 8.7. Suppose that $y = \log Y$ has a normal distribution and constant variance independent of its expectation. Then the identical distributions A, B, and C on the $y = \log Y$ axis will correspond with the very different distributions A', B', and C' on the Y axis. In this illustration, the distributions A' and C' have been centered on the values $Y = 90$ and $Y = 3600$, the extreme values covered by the textile data. It is seen that normality in $y = \log Y$ will induce somewhat skewed distributions in Y because of *local* curvature in the plot of Y versus y;

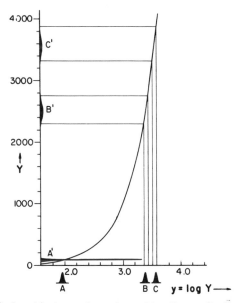

FIGURE 8.7. The logarithmic transformation and its effect on distributional assumptions.

however, such induced non-normality will be of only moderate importance. What is more important is the dramatic difference in the *standard deviation* of the distributions A' and C' because of "global" differences in gradient. In fact, in our illustration, C' has a standard deviation $3600/90 = 40$ times as large as that of A'. The variance of the distribution of C' is thus 1600 times as large as that of the distribution A'! Equivalently, the weight of an observation from A' is 1600 times as large as that of an observation from C'. This means that one observation from A' is as accurate as the mean of 1600 from C' and is so treated by the logarithmic analysis. However, if ordinary least squares is applied to Y instead of to $\log Y$, these observations of vastly different accuracy are treated as having equal weights! Notice also that the distributions B' and C', centered at 2500 and 3600, have standard deviations in proportion to $3600/2500 = 1.44$, and therefore variances in the ratio of about 2 to 1, so that, by comparison, this discrepancy is a minor one.

To sum up: Over the range that encompasses B' and C', the curve relating Y and $\log Y$ is roughly linear and the choice between Y and $\log Y$ is not very important. By contrast, the range encompassing A' and C' is highly nonlinear and, for data such as we actually have in the textile example, for which $Y_{max}/Y_{min} = 3600/90 = 40$, careful choice of transformation is essential.

For further explanation, we suppose that some nonlinear transformation $y = h(Y)$ can be approximated (see Figure 8.8) by the first two terms of its Taylor's series* expansion about some value $Y = Y_0$. Thus, approximately

$$y = h(Y) = h_0 + h_1(Y - Y_0) + \tfrac{1}{2}h_2(Y - Y_0)^2 \qquad (8.3.1)$$

where

$$h_0 = h(Y_0), \qquad h_1 = \frac{dh(Y)}{dY}\bigg|_{Y=Y_0}, \qquad h_2 = \frac{d^2h(Y)}{dY^2}\bigg|_{Y=Y_0}. \qquad (8.3.2)$$

In this formula, h_1 measures the slope at Y_0, while h_2 measures the quadratic curvature at Y_0. It is important to notice that, while the distortion of the distribution shape produced by transformation is largely due to the curvature h_2, the change in spread produced by the transformation is

---

* The formula simply says that, for a smooth function $h(Y)$, the value at some Y close to Y_0 is given by its value $h_0 = h(Y_0)$ at Y_0, plus a term to allow for the rate h_1 at which the function is changing at Y_0, plus a further term to allow for the rate h_2 at which the rate is itself changing and so on.

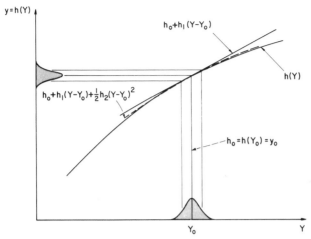

FIGURE 8.8. Effects of a nonlinear transformation $y = h(Y)$ on distribution assumptions.

associated with the change in gradient h_1, so that approximately,

$$\sigma_y \simeq h_1 \sigma_Y. \qquad (8.3.3)$$

We can thus write, for the ratio of the standard deviations of y at the extremes of the range of the data,

$$\frac{\sigma_{y_{max}}}{\sigma_{y_{min}}} \simeq \left(\frac{h_1(Y_{max})}{h_1(Y_{min})} \right) \left(\frac{\sigma_{Y_{max}}}{\sigma_{Y_{min}}} \right).$$

For the textile data, for example, if the standard deviation is roughly constant for $y = h(y) = \log Y$, then $h_1 = Y^{-1}$ and

$$\frac{\sigma_{Y_{max}}}{\sigma_{Y_{min}}} \simeq \frac{3600}{90} = 40.$$

Variance Stabilizing Transformations

We have seen that an important requirement for the efficiency and validity of unweighted least-squares estimation is that, to a rough approximation, the variance of the response Y is independent of the magnitude $E(Y) = \eta$.

Suppose this is not so and that the standard deviation σ_Y of Y is some function $F(\eta)$ of the mean value η, of Y. Then, referring to the argument

TABLE 8.3.   Appropriate variance stabilizing transformations when $\sigma_Y = F(\eta)$

Nature of dependence $\sigma_Y = F(\eta)$			Variance stabilizing transformation
$\sigma_Y \propto \eta^k$		$(Y \geq 0)$	Y^{1-k}
and in particular			
$\sigma_Y \propto \eta^{1/2}$	(Poisson)	$(Y \geq 0)$	$Y^{1/2}$
$\sigma_Y \propto \eta$		$(Y \geq 0)$	$\log Y$
$\sigma_Y \propto \eta^2$		$(Y \geq 0)$	Y^{-1}
$\sigma_Y \propto \eta^{1/2}(1 - \eta)^{1/2}$	(binomial)	$(0 \leq Y \leq 1)$	$\sin^{-1}(Y^{1/2})$
$\sigma_Y \propto (1 - \eta)^{1/2}\eta$	(negative binomial)	$(0 \leq Y \leq 1)$	$(1 - Y)^{1/2} - (1 - Y)^{3/2}/3$
$\sigma_Y \propto (1 - \eta^2)^{-2}$		$(-1 \leq Y \leq 1)$	$\log\{(1 + Y)/(1 - Y)\}$

of the foregoing section, a transformed variable or metric $h(Y)$ for which the variance will be approximately stable is such that

$$\frac{dh(Y)}{dY} \propto \frac{1}{F(Y)}. \tag{8.3.4}$$

This result leads in particular, to the important transformations shown in Table 8.3. In the textile example, an assumption that the *percentage* error was roughly constant, so that $\sigma_Y \propto \eta$, would yield the metric $y = \log Y$.

In practice, the nature of the dependence between σ_Y and η may be suggested by theoretical considerations or by preliminary empirical analysis, or by both.

Examples of Transformations Derived from Theoretical Considerations

A relationship between σ_Y and Y can sometimes be deduced by considering the theoretical distribution of Y.

Poisson Distribution

Suppose the observed response Y was the *frequency* with which an event occurred, such as a blood cell count. Then it might be expected that, under constant experimental conditions ξ, Y would vary from one count to another in a Poisson distribution

$$P(Y) = \frac{\eta^Y}{Y!}e^{-\eta} \tag{8.3.5}$$

for which

$$\sigma_Y \propto \eta^{1/2}. \tag{8.3.6}$$

Referring to Table 8.3, we see that the appropriate transformation is the square root transformation. The effect of the transformation would be such that if we now change the experimental conditions ξ, resulting in changes in $\eta = f(\xi)$, the variance of $y = Y^{1/2}$ would, nevertheless, remain approximately constant. Incidentally, the transformation would also render the distributions at various conditions ξ more nearly normal.

Binomial Distribution

As a further example suppose that Y was the proportion of "successes" out of a fixed number of trials such as the proportion of 50 tested resistors with resistances less than 1 ohm. Then Y might be expected to follow a binomial distribution. If the probability of a success is η and is constant for all resistors at a given set of experimental conditions, then

$$p(Y) \propto (\eta)^{nY}(1 - \eta)^{n(1-Y)} \tag{8.3.7}$$

for which

$$\sigma_Y \propto \eta^{1/2}(1 - \eta)^{1/2}. \tag{8.3.8}$$

From Table 8.3, we see that the appropriate transformation for which the variance would be nearly stable and independent of the true probability of success η is $\sin^{-1}(Y^{1/2})$.

Transformations Derived from Empirical Considerations

When it happens that replicate runs are made at a number of experimental conditions, a preliminary fit can be made in what is judged to be the best metric Y and then the sample variance s_ξ^2 and the mean $\bar{Y}_\xi$ may be calculated for each set of conditions ξ. The possible dependence of s_ξ^2 on the mean $\bar{Y}_\xi$ may then be judged empirically by graphical methods.

In particular, if it is assumed that σ_Y is proportional to some power of η, so that

$$\sigma_Y \propto \eta^k, \tag{8.3.9}$$

and so

$$\log \sigma_Y = c + k \log \eta, \tag{8.3.10}$$

then the slope of a plot of $\log s_\xi$ against $\log \overline{Y}_\xi$ can provide an estimate for k.

Simplicity of Response Functions and Validity of Distribution Assumptions

In the past, transformation has sometimes been carried out purely to simplify the form of the response function and insufficient consideration has been given to the vital question of how such transformations affect error assumptions, which alone determine efficient choice of the estimation procedure. For example, in physical chemistry, rate equations involving reciprocal representations, such as

$$\eta = 1/(\theta_0 + \theta_1\xi_1 + \theta_2\xi_2), \qquad (8.3.11)$$

frequently arise. The constants (the θ's) in such expressions are often estimated by making a reciprocal transformation of the response function and fitting the resulting linear expression to the reciprocals of the response data by least squares, implying a model

$$y = Y^{-1} = \theta_0 + \theta_1\xi_1 + \theta_2\xi_2 + \varepsilon. \qquad (8.3.12)$$

If, for this reciprocal scale, the normal theory constant variance assumptions were valid for ε, ordinary least squares would produce efficient estimates with *correct* standard errors, and further analysis resting on the validity of derived F and t distributions would be appropriate. However suppose, on the other hand, that the normal theory error assumptions, and particularly the assumptions about constancy of variance had been true for the model

$$Y = 1/(\theta_0 + \theta_1\xi_1 + \theta_2\xi_2) + \varepsilon'. \qquad (8.3.13)$$

Then, for Y's covering a wide range, least-squares fitting of the reciprocal function (8.3.12) could result in very inefficient estimators, and incorrect standard errors. By contrast, direct fitting of the nonlinear model (8.3.13) using *nonlinear* least squares would yield efficient estimates. (This option is discussed in Appendix 3D.) Alternatively, an appropriate weighted linear least squares analysis of the linearized form (Box and Hill, 1974) could be employed to produce a close approximation to the nonlinear fit.

Further Analysis of the Textile Data

While transformations employed merely to achieve a linear model can often seriously distort distributional assumptions, they may also occasionally make the usual assumptions more nearly applicable. For illustration, consider once more the textile data of Table 7.1, following the discussion of Box and Cox (1964). In many fields of technology, power relationships are common. For the textile example, such a relationship would postulate that cycles to failure was, apart from error, proportional to

$$\gamma_0 \xi_1^{\gamma_1} \xi_2^{\gamma_2} \xi_3^{\gamma_3} \qquad (8.3.14)$$

where $\xi_1 =$ length of test specimen (mm), $\xi_2 =$ amplitude of load cycle (mm), and $\xi_3 =$ load (g). Now we know already that the normal assumptions appear to be approximately true for $y = \log Y$. After taking logarithms, the power relationship yields the model

$$y = \log \gamma_0 + \gamma_1 \log \xi_1 + \gamma_2 \log \xi_2 + \gamma_3 \log \xi_3 + \varepsilon. \qquad (8.3.15)$$

We have already seen that the distributional assumptions appear valid for $y = \log Y$ rather than for Y itself. However, the model (8.3.15) is linear in the *logs of the input* and not in the inputs themselves which is the form we fitted in coded form (see Table 7.2) in the previous analysis. Since we know that the equation in the unlogged inputs fits very well, this might at first lead us to question whether logged inputs would be worth further consideration. However, in this particular example, the inputs (*unlike the output*), cover quite narrow ranges. The largest value of ξ_{max}/ξ_{min} is found for the first input ξ_1 (length of test specimen) for which $\xi_{1\,max}/\xi_{1\,min} = 350/250 = 1.4$. Over this small a range, the log transformation is practically linear, as we see from Figure 8.7. Thus, for these data, the fact that we get a good fit with the unlogged ξ's *guarantees* a fit which is nearly as good (or possibly slightly better) for the logged ξ's. In fact, if model (8.3.15) is estimated by least squares, an excellent fit is obtained with

$$\hat{\gamma}_1 = 4.96 \pm 0.20, \qquad \hat{\gamma}_2 = -5.27 \pm 0.30, \qquad \hat{\gamma}_3 = -3.15 \pm 0.30.$$

$$(8.3.16)$$

There is an advantage in moving as close as possible toward a model which might conceivably have more theoretical justification and it sometimes happens that interesting possibilities are suggested by the results of what began as an empirical analysis, as we now illustrate. Since in Eq. (8.3.16), $\hat{\gamma}_2 \simeq -\hat{\gamma}_1$, the quantity $\log \xi_2 - \log \xi_1 = \log(\xi_2/\xi_1)$ is suggested as being of importance. Further $\xi_2/\xi_1 = \xi_4$ say, is the fractional amplitude and is suggested by dimensional considerations. Finally, then the simple relationship

$$\text{cycles to failure} = k\xi_4^{-5}\xi_3^{-3} \qquad (8.3.17)$$

is obtained, which represents the data very well. Thus our iterative analysis leads finally to a very simple representation.

8.4. ESTIMATING RESPONSE TRANSFORMATION FROM THE DATA

The textile example illustrates how it is sometimes possible to simplify greatly the problem of empirical representation by making a suitable transformation on the response.

Consider again the general question of the choice of transformation $y = h(Y)$ for an output variable Y. If possible, such a transformation should be chosen so that the model

$$y = g(\xi, \beta) + \varepsilon \qquad (8.4.1)$$

provides an accurate and easily understood representation over the region $R(\xi)$ of immediate interest and can be readily fitted. Ideally, then, the transformation might be chosen to achieve these features:

1. $g(\xi, \beta)$ should have a simple and parsimonious form (requiring a minimum number of parameters).
2. The error variance should be approximately constant.
3. The error distributions should be approximately normal.

As has been illustrated above, scientific knowledge of the system, common sense, and the use of simple plots can greatly help in an appropriate choice. A further tool (to be used in conjunction with the above) is the likelihood function, which can help determine a suitable transformation from the data themselves.

Although the method is applicable to any class of parametric transformations we illustrate it here for the important power transformations where (supposing the data values Y are all positive) $y = Y^\lambda$. These include, for example

$\lambda = 1$	$\lambda = \frac{1}{2}$	$\lambda = 0$	$\lambda = -\frac{1}{2}$	$\lambda = -1$
$y = Y$	$y = Y^{1/2}$	$y = \log Y$	$y = 1/Y^{1/2}$	$1/Y.$
no transformation	square root	logarithm	reciprocal square root	reciprocal

Since $Y^0 = 1$ for all Y, an explanation is in order as to why the log transformation is associated with the value $\lambda = 0$. Because we can write

$$Y^\lambda = e^{\lambda \ln Y} \simeq 1 + \lambda \ln Y + \tfrac{1}{2}\lambda^2(\ln Y)^2 + \cdots, \qquad (8.4.2)$$

we have that

$$\lim_{\lambda \to 0}\left(\frac{Y^\lambda - 1}{\lambda}\right) = \ln Y. \qquad (8.4.3)$$

In essence, what this says is that, if we take a small positive or negative power of Y (say $Y^{0.01}$ or $Y^{-0.01}$), it will plot against $\log Y$ very nearly as a straight line, and linearity will be more and more nearly achieved the smaller the power we take*.

Use of the Likelihood Function in Selecting a Power Transformation

We begin by describing the mechanics of the method using the textile data for illustration. Suppose it is postulated that some power transformation $y = Y^\lambda$ exists, in terms of which the graduating function $g(\boldsymbol{\xi}, \boldsymbol{\beta})$ in (8.4.1) has some specific simple form, and in terms of which the normal theory assumptions apply. The idea is to employ the method of maximum likelihood to estimate the transformation parameter λ at the same time as the parameters $\boldsymbol{\beta}$ of the model are estimated. The necessary steps (Box and Cox, 1964) are as follows:

1. Compute the geometric mean $\dot{Y}$ of all the data from $\ln \dot{Y} = n^{-1}\Sigma \ln Y_u$.
2. For a series of suitable values of λ, compute the transformed values $Y_u^{(\lambda)}$ using the form

$$Y_u^{(\lambda)} = \begin{cases} (Y_u^\lambda - 1)/(\lambda \dot{Y}^{\lambda - 1}), & \text{if } \lambda \neq 0 \\ \dot{Y} \ln Y_u, & \text{if } \lambda = 0. \end{cases} \qquad (8.4.4)$$

3. Fit the parsimonious model $g(\boldsymbol{\xi}, \boldsymbol{\beta})$ to $Y^{(\lambda)}$ by least squares, and record the residual sum of squares $S(\lambda)$, for each chosen value of λ.
4. Plot $\ln S(\lambda)$ against λ. The value of λ which makes $\ln S(\lambda)$, and so $S(\lambda)$, smallest is the maximum likelihood value $\hat{\lambda}$.

---

*Use of logarithms to the base e (ln) is needed for Eq. (8.4.2), but logarithms to any base (e.g., 10) can be used in the transformation. The difference amounts only to a constant factor.

5. An approximate $100(1 - \alpha)\%$ confidence interval for λ is given by determining graphically the two values of λ for which

$$\ln S(\lambda) - \ln S(\hat{\lambda}) = \frac{\chi_\alpha^2(1)}{\nu_r}, \qquad (8.4.5)$$

where ν_r is the number of residual degrees of freedom, and $\chi_\alpha^2(1)$ is the upper α significance point of χ^2 with 1* df.

We illustrate the method by applying it to the untransformed textile data Y given in Table 7.1. Our object is to estimate a λ for which the errors, $y_u - g(\boldsymbol{\xi}_u, \boldsymbol{\beta})$, of the transformed observations y_u from the simple linear graduating function

$$g(\boldsymbol{\xi}, \boldsymbol{\beta}) = \beta_0 + \beta_1\xi_1 + \beta_2\xi_2 + \beta_3\xi_3, \qquad (8.4.6)$$

most nearly follow normal theory assumptions. We find that:

1. The geometric mean for the 27 observations of Table 7.1 is

$$\dot{Y} = 562.34.$$

2. The transformed data values for any given value of λ are given by

$$Y_u^{(\lambda)} = \begin{cases} \lambda^{-1}(562.34)^{1-\lambda}(Y_u^\lambda - 1), & \text{if } \lambda \neq 0, \\ (562.34)\ln Y_u, & \text{if } \lambda = 0. \end{cases}$$

3. If the linear model above is fitted to each set of transformed values $Y_u^{(\lambda)}$, $u = 1, 2, \ldots, n$ for each of the following values of λ, the residual sums of squares and their natural logarithms are as given:

λ	-1.0	-0.8	-0.6	-0.4	-0.2	0.0
$S(\lambda)$	3.9955	2.1396	1.1035	0.5478	0.2920	0.2519
$\ln S(\lambda)$	1.3852	0.7606	0.0985	-0.6018	-1.2310	-1.3787
	0.2	0.4	0.6	0.8	1.0	
	0.4115	0.8178	1.5968	2.9978	5.4810	
	-0.8879	-0.2011	0.4680	1.0979	1.7013	

*In general the transformation could involve q unknown parameters (instead of a single λ as here). In this case, the approximate expression (8.4.5) would still hold but the chi-squared table would be entered with q df.

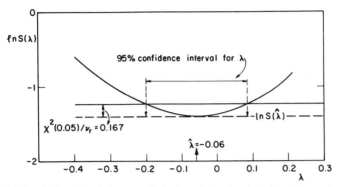

FIGURE 8.9. A plot of ln $S(\lambda)$ versus λ in the neighborhood of the minimum ln $S(\lambda)$, for the textile data.

4. The plot of ln $S(\lambda)$ against λ yields a minimum value at about $\hat{\lambda} = -0.06$. (Figure 8.9 shows the situation around the minimum.)

5. An approximate 95% confidence interval for λ is obtained by computing

$$\frac{\chi^2_{0.05}(1)}{\nu_r} = \frac{3.84}{23} = 0.167$$

and observing the two values of λ at which the horizontal line ln $S(\hat{\lambda})$ + 0.167 cuts the ln $S(\lambda)$ curve. In the present case, the 95% confidence interval extends from about $\lambda = -0.20$ to 0.08.

Our analysis confirms the conclusions previously reached on a common sense basis, and reinforces them. We see, for example, that the values $\lambda = 1$ (no transformation), $\lambda = \frac{1}{2}$ (square root transformation), and $\lambda = -\frac{1}{2}$ (reciprocal square root transformation) are all rejected by the analysis. Plausible values for λ are closely confined to a region near to $\lambda = 0$ (log transformation).

We remind ourselves once more that data transformation is of potential importance in this example because the ratio $Y_{max}/Y_{min} = 3636/90 \simeq 40$ is very large. As we have said before, when Y_{max}/Y_{min} is not large, transformations such as log, reciprocal, and square root behave locally almost like linear transformations and so do not greatly affect the analysis. If, when Y_{max}/Y_{min} is not large, the likelihood analysis is nevertheless carried through, a very wide confidence interval for λ will be obtained, indicating that it will make little difference which transformation is used.

Other Transformations

The method may be applied not only to simple power transformations but much more generally.

Suppose we consider some class of transformations y of Y which involves a vector of parameters, λ. Suppose these transformations are monotonic functions of Y for each value of λ over the admissible range. Then the analysis is made with the transformed values

$$Y_u^{(\lambda)} = \frac{y_u}{J^{1/n}}, \qquad (8.4.7)$$

where

$$J = \prod_{u=1}^{n} \left| \frac{\partial y_u}{\partial Y_u} \right|. \qquad (8.4.8)$$

For example, for the power transformations with shifted origin, $(Y + \lambda_2)^{\lambda_1}$, it is convenient to work with the form $y = \{(Y + \lambda_2)^{\lambda_1} - 1\}/\lambda_1$ which takes the value $\ln(Y + \lambda_2)$ when λ_1 is zero. The analysis is thus made in terms of

$$Y^{(\lambda)} = \frac{(Y + \lambda_2)^{\lambda_1} - 1}{\lambda_1 (Y \dot{+} \lambda_2)^{\lambda_1 - 1}}, \qquad (8.4.9)$$

where $Y \dot{+} \lambda_2$ is the geometric mean of the n values $Y_u + \lambda_2$, $u = 1, 2, \ldots, n$.

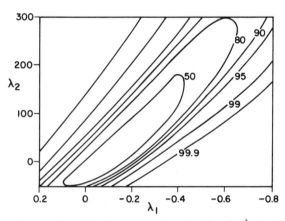

FIGURE 8.10. Checking the value of the transformation $(Y + \lambda_2)^{\lambda_1}$ for the textile data. Contours shown are those of the natural logarithm of the residual sum of squares $\ln S(\lambda_1, \lambda_2)$, labeled with their approximate confidence probability. (Reproduced from Box and Cox, 1964, by permission of the Royal Statistical Society, London.)

Box and Cox (1964) illustrated the use of this type of transformation by applying it to the textile data of Table 7.1. The function (8.4.6) was fitted by least squares to the transformed data, which this time depended on two parameters, λ_1 and λ_2. For a selected grid of points (λ_1, λ_2), the values of $\ln S(\lambda_1, \lambda_2)$, the logarithm of the residual sum of squares for the fitted model were calculated. This grid of values was used to construct Figure 8.10. The contours in that figure have been labeled to indicate the approximate probability associated with the confidence regions they encompass. There is no indication that a nonzero value of λ_2 is needed for these data, which confirms the validity of our previous analysis.

8.5. TRANSFORMATION OF THE PREDICTOR VARIABLES

So far, we have written mostly about transformation of the response variable. In some cases, as we have illustrated earlier in this chapter, needed simplification may be obtained by transforming some or all of the individual predictor variables. Occasionally, transformation of both response *and* predictor variables is necessary. We now discuss certain types of transformation of the predictor variables. Let $\xi_1^{(\alpha_1)}, \xi_2^{(\alpha_2)}, \ldots,$ denote individual parametric transformations of the predictor variables $\xi_1, \xi_2, \ldots,$ each dependent on a single parameter $\alpha_1, \alpha_2, \ldots,$ respectively, and consider a model of the form

$$y = g\left(\xi_1^{(\alpha_1)}, \xi_2^{(\alpha_2)}, \ldots, \xi_k^{(\alpha_k)}; \boldsymbol{\beta}\right) + \varepsilon. \qquad (8.5.1)$$

For illustration, Figure 8.11*a* shows contours of the function

$$\eta = 25 + 5 \ln \xi_1 + 100 \xi_2^{-1} + 6 \xi_3^{1/2}, \qquad (8.5.2)$$

over a cubical region in the $\boldsymbol{\xi}$ space, while Figure 8.11*b* shows contours of the same function plotted in terms of the transformed variables $\ln \xi_1, \xi_2^{-1}$, and $\xi_3^{1/2}$. If a polynomial were directly fitted to the ξ's, an equation of at least second degree would be needed to obtain a suitable fit, but in terms of the transformed variables the equation is first order. What we should like to do, in general, is to recognize when suitable transformation of the predictors will help us attain this sort of simplification. Transformation of the individual predictor variables should be considered whenever:

1. There is theoretical or empirical evidence to suggest that model simplification may occur as a result.

2. Transformation of some sort is clearly called for, but the error structure is satisfactory in the original response metric.

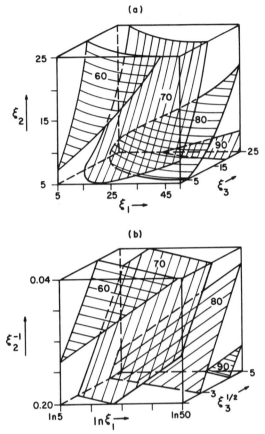

FIGURE 8.11. A three variable response function (a) before, and (b) after, transformation. (Reproduced from Box and Tidwell, 1962, by permission of the American Statistical Association.)

An appropriate transformation will frequently be suggested by theoretical considerations. The Arrhenius law, which connects the logarithm of the rate η of a chemical reaction to the absolute temperature ξ, is of the form

$$\ln \eta = \beta_0 + \beta_1 \xi^{-1}, \tag{8.5.3}$$

where β_0 and β_1 are constants. Thus if, as might well be the case, the usual normal theory distribution assumptions were expected to be true for the logarithm $y = \ln Y$ of the observed rate Y, and if also a wide temperature range were being used so that choice of transformation was important, then a linear expression in the reciprocal of the temperature ξ should be entertained, and fitted by least squares, to the y values. Again, suppose the object of interest was the time period η of a pendulum as a function of the length ξ of the string from which the plumb bob was suspended. For a "perfect" pendulum, physical theory says that

$$\eta = \beta\sqrt{\xi}, \qquad (8.5.4)$$

where β is a constant. In this instance then, if normal theory assumptions could be expected to apply to observations of the time period, then it would be natural to entertain a linear model in the square root of ξ.

Detecting the Need for, and Estimating, Power Transformations in the Predictor Variables, Using Only Linear Least-Squares Techniques

Suppose that the response η is a function of the k predictor variables in $\boldsymbol{\xi} = (\xi_1, \xi_2, \ldots, \xi_k)'$ and that it is desired to fit some graduating function g by least squares. Suppose further that, while the predictor variables have already been transformed to what is currently believed to be their most appropriate forms, yet it is possible, nevertheless, that they may require further transformation to powers $\xi_1^{\alpha_1}, \xi_2^{\alpha_2}, \ldots, \xi_k^{\alpha_k}$. Then the true response for the uth observation is a function

$$\eta_u = g(\xi_{1u}^{\alpha_1}, \xi_{2u}^{\alpha_2}, \ldots, \xi_{ku}^{\alpha_k}; \boldsymbol{\beta}^*), \qquad (8.5.5)$$

where $\boldsymbol{\beta}^* = (\beta_0^*, \beta_1^*, \beta_2^*, \ldots, \beta_p^*)'$, say, is a vector of parameters. Following Box and Tidwell (1962), we can expand η_u in a Taylor's series to first order about the values $\alpha_1 = 1, \alpha_2 = 1, \ldots, \alpha_k = 1$, to give, instead of the original model $Y_u = \eta_u + \varepsilon_u$, the approximation

$$Y_u = g(\boldsymbol{\xi}_u, \boldsymbol{\beta}^*) + \sum_{j=1}^{k} (\alpha_j - 1)\left[\frac{\partial \eta_u}{\partial \xi_j}\right]_{all\ \alpha_i = 1} (\xi_{ju}\ln \xi_{ju}) + \varepsilon_u \quad (8.5.6)$$

which can also be written as

$$Y_u - g(\boldsymbol{\xi}_u, \boldsymbol{\beta}^*) = \sum_{j=1}^{k} (\alpha_j - 1)Z_{ju} + \varepsilon_u, \qquad (8.5.7)$$

where

$$Z_{ju} = \left[\frac{\partial \eta_u}{\partial \xi_j}\right]_{all\ \alpha_i = 1} (\xi_{ju}\ln \xi_{ju}). \qquad (8.5.8)$$

Detection of Need for Transformation

Now if we fit the model

$$Y_u = g(\xi_u, \beta^*) + \varepsilon_u \qquad (8.5.9)$$

by least squares and take the residuals $e_u = Y_u - \hat{Y}_u$, we shall have estimates of the left-hand sides of (8.5.7). Moreover, the same $\hat{Y}_u$ values can be used to give us the

$$\hat{Z}_{ju} = \frac{\partial \hat{Y}_u}{\partial \xi_j} (\xi_{ju} \ln \xi_{ju}) \qquad (8.5.10)$$

which will be estimates of the Z_{ju} in (8.5.8). Thus, we can expect that, approximately, the need for transformation of the ξ_j will be indicated by a significant planar relationship apparent in plots of the e_u versus the $\hat{Z}_{ju}$, $j = 1, 2, \ldots, k$.

Estimating the Transformation

To get estimates of the α_j we can proceed as follows:

1. Fit (8.5.9) and use it to obtain the estimates $\hat{Z}_{ju}$ via (8.5.10).
2. Fit, by least squares, the model

$$Y_u = g(\xi_u, \beta^*) + \sum_{j=1}^{k} (\alpha_j - 1)\hat{Z}_{ju} + \varepsilon_u \qquad (8.5.11)$$

   which is derived from (8.5.7) by replacing Z_{ju} by $\hat{Z}_{ju}$.
3. See which $(\hat{\alpha}_j - 1)$ are large compared to their respective standard errors and use those $\hat{\alpha}_j$ to transform ξ_j to $\xi_j^{\hat{\alpha}_j}$.
4. If desired, repeat 1–3 with the transformed variables in the roles of the corresponding untransformed variables.

   This iterative process should produce improved estimates of the α's, and should converge fairly quickly in most cases.

In practice, we often work, not with the ξ_{ju}, but with scaled variables $x_{ju} = (\xi_{ju} - \xi_{j0})/S_j$ so that $\xi_{ju} = \xi_{j0} + S_j x_{ju}$. Use of the scaled variables often prevents an escalation in size of the numbers involved in the calculation as the iterations proceed. If, upon rescaling, $g(\xi_u, \beta^*)$ becomes $g(x_u, \beta)$, then we obtain, for first- and second-order polynomial models, the following.

First-Order Polynomial

$$g(\boldsymbol{\xi}_u, \boldsymbol{\beta}^*) = \beta_0^* + \sum_j \beta_j^* \xi_{ju}, \tag{8.5.12}$$

$$g(\mathbf{x}_u, \boldsymbol{\beta}) = \beta_0 + \sum_j \beta_j x_{ju}, \tag{8.5.13}$$

so that $\beta_j = S_j \beta_j^*$, and $\beta_0 = \beta_0^* + \sum_j \beta_j^* \xi_{j0}$. It follows that

$$\hat{Z}_{ju} = \hat{\beta}_j^* \xi_{ju} \ln \xi_{ju} \tag{8.5.14}$$

$$= S_j^{-1} \hat{\beta}_j (\xi_{j0} + S_j x_{ju}) \ln(\xi_{j0} + S_j x_{ju}). \tag{8.5.15}$$

Second Degree Polynomial

$$g(\boldsymbol{\xi}_u, \boldsymbol{\beta}^*) = \beta_0^* + \sum_j \beta_j^* \xi_{ju} + \sum_{i \geq j} \sum \beta_{ij}^* \xi_{iu} \xi_{ju}, \tag{8.5.16}$$

$$g(\mathbf{x}_u, \boldsymbol{\beta}) = \beta_0 + \sum_j \beta_j x_{ju} + \sum_{i \geq j} \sum \beta_{ij} x_{iu} x_{ju}, \tag{8.5.17}$$

so that

$$\beta_{ij} = S_i S_j \beta_{ij}^*, \qquad \beta_j = S_j \left(\beta_j^* + \sum_i \beta_{ij}^* \xi_{i0} \right),$$

$$\beta_0 = \beta_0^* + \sum_j \beta_j^* \xi_{j0} + \sum_{i \geq j} \sum \beta_{ij}^* \xi_{i0} \xi_{j0}. \tag{8.5.18}$$

It follows that

$$\hat{Z}_{ju} = \left\{ \hat{\beta}_j^* + 2\hat{\beta}_{jj}^* \xi_{ju} + \sum_{i \neq j} \hat{\beta}_{ij}^* \xi_{iu} \right\} \xi_{ju} \ln \xi_{ju} \tag{8.5.19}$$

$$= S_j^{-1} \left\{ \left(\hat{\beta}_j - \sum_i S_i^{-1} \hat{\beta}_{ij} \xi_{i0} + 2 S_j^{-1} \hat{\beta}_{jj} \xi_{j0} + \sum_{i \neq j} S_i^{-1} \hat{\beta}_{ij} \xi_{i0} \right) \right.$$

$$\left. + 2 \hat{\beta}_{jj} x_{ju} + \sum_{i \neq j} \hat{\beta}_{ij} x_{iu} \right\} (\xi_{j0} + S_j x_{ju}) \ln(\xi_{j0} + S_j x_{ju}). \tag{8.5.20}$$

To illustrate these ideas we consider an example with a single predictor variable ξ, taken from Box and Tidwell (1962).

Example: Yield of Polymer as a Function of Catalyst Charge

An experimenter observes the quantity of a polymer formed in 1 h, under steady-state conditions, for various (equally spaced) charges of a catalyst. The results are shown in Table 8.4 and are plotted in Figure 8.12a.

The need for a linearizing transformation is, of course, obvious in this case, and certainly one could guess a transformation for ξ which would produce near linearity. Figure 8.12b shows that $\xi^{1/2}$ *is* such a transformation. However, for illustration, we shall proceed according to the method outlined above. We want eventually to fit a straight line model

$$Y_u = \beta_0 + \beta_1 \xi_u^\alpha + \varepsilon_u \qquad (8.5.21)$$

TABLE 8.4.   Polymer formed for various catalyst charges

Grams of polymer formed, Y_u	Grams of catalyst charged to reactor, ξ_u	Values of $(13.729)\xi_u \ln \xi_u = \hat{Z}_u$
28.45	0.5	−4.76
47.78	1.5	8.34
64.35	2.5	31.45
74.67	3.5	60.20
83.65	4.5	92.92

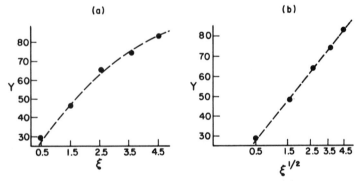

FIGURE 8.12.   Grams of polymer formed as a function of catalyst charge. (Reproduced from Box and Tidwell, 1962, by permission of the American Statistical Association.)

with the best α. We first fit (8.5.9) which, for our example, is (8.5.21) with $\alpha = 1$. This provides the fitted equation

$$\hat{Y}_u = 25.47 + 13.729\xi_u \qquad (8.5.22)$$

with derivative $\partial\hat{Y}_u/\partial\xi_u = 13.729$. We now construct the variable

$$\hat{Z}_u = 13.729\xi_u \ln \xi_u \qquad (8.5.23)$$

from (8.5.8); the values of $\hat{Z}_u$ are shown in Table 8.4. We then fit, by least squares, the model (8.5.11) which is

$$Y_u = \beta_0 + \beta_1\xi_u + (\alpha - 1)\hat{Z}_u + \varepsilon_u. \qquad (8.5.24)$$

The fitted equation is

$$\hat{Y}_u = 11.222 + 28.286\xi_u - 0.589\hat{Z}_u. \qquad (8.5.25)$$

It follows that $\hat{\alpha} = 0.41$. The standard error of $\hat{\alpha}$ is 0.08, so that transformation is clearly indicated. We could now do additional iterations to improve the estimate of α, but it is clear that a value of $\alpha = 0.5$ is both convenient and within the range of reasonable values for the parameter. Furthermore, if we use $\alpha = 0.5$, a satisfactory linearization is achieved, as Figure 8.12b has already shown. Thus the transformation analysis could sensibly be terminated at this point and the model

$$Y_u = \beta_0 + \beta_1\xi_u^{1/2} + \varepsilon_u \qquad (8.5.26)$$

could now be fitted and validated in the usual manner.

Transformations in the Predictor Variables Via Nonlinear Estimation

The procedure that we have given above is simple and useful, particularly when only one iteration is adequate for the purpose at hand. As we have intimated, the procedure could be the first step in a series of iterations which would lead to improved estimates of the α_j. When a standard nonlinear least-squares program is available, however, the above calculations can be bypassed and the nonlinear model

$$Y_u = g\left(\xi_{1u}^{\alpha_1}, \xi_{2u}^{\alpha_2}, \ldots, \xi_{ku}^{\alpha_k}; \boldsymbol{\beta}\right) + \varepsilon_u \qquad (8.5.27)$$

can be fitted directly.

APPENDIX 8A.   PROOF OF A RESULT USED IN SECTION 8.2*

We can write $\sigma^2_{\hat{y} - \bar{y}}$ as $1/n$ times S where

$$S = E\{(\hat{y} - \bar{y}j) - (\eta - \bar{\eta}j)\}'\{(\hat{y} - \bar{y}j) - (\eta - \bar{\eta}j)\}.$$

The vector $j = (1,1,\ldots,1)'$ is of length n. Now

$$\hat{y} - \bar{y}j = Ry - \frac{1}{n}jj'y = \left(R - \frac{1}{n}jj'\right)y,$$

where

$$R = X(X'X)^{-1}X'$$

and

$$\eta - \bar{\eta}j = \left(I - \frac{1}{n}jj'\right)\eta,$$

where $\eta = X\beta$. Also R, $[R - (1/n)jj']$ and $[I - (1/n)jj']$ are all idempotent, that is, $RR = R$, and so on. Thus we can write

$$S = E\left\{y'\left(R - \frac{1}{n}jj'\right)y - 2y'\left(R - \frac{1}{n}jj'\right)\left(I - \frac{1}{n}jj'\right)\eta + \eta'\left(I - \frac{1}{n}jj'\right)\eta\right\}.$$

Provided that the model has an intercept term β_0, $X'j$ is the first column of $X'X$ whereupon it follows that

$$Rj = X(X'X)^{-1}X'j = X(1,0,0,\ldots,0)' = j,$$

whereupon

$$S = E\left\{y'\left(R - \frac{1}{n}jj'\right)y - 2y'\left(R - \frac{1}{n}jj'\right)\eta + \eta'\left(I - \frac{1}{n}jj'\right)\eta\right\}$$

$$= \eta'\left(R - \frac{1}{n}jj'\right)\eta + \sigma^2 \text{trace}\left(R - \frac{1}{n}jj'\right) - 2\eta'\left(R - \frac{1}{n}jj'\right)\eta$$

$$+ \eta'\left(I - \frac{1}{n}jj'\right)\eta$$

*See Section 8.2 below Eq. (8.2.6).

[using the fact that $E(y'Ay) = \eta'A\eta + \text{trace}(\Sigma A)$ where $E(y) = \eta$, $V(y) = \Sigma$]

$$= \eta'(I - R)\eta + \sigma^2 \text{trace}\left(R - \frac{1}{n}jj'\right)$$

$$= \sigma^2 p$$

because the first term vanishes due to the fact that $(I - R)X = 0$ and the second term is σ^2 times $\text{trace}\{X(X'X)^{-1}X' - (1/n)jj'\} = \text{trace}\{(X'X)^{-1}X'X\} - (1/n)\text{trace}j'j$, using the fact that $\text{trace}(AB) = \text{trace}(BA)$, which reduces to $(p + 1) - 1 = p$, assuming there are p parameters in the model in addition to the constant term. Dividing by $1/n$ gives us $\sigma^2_{\bar{y}-\bar{y}} = p\sigma^2/n$ as required.

EXERCISES

8.1. A measure of the uncertainty associated with some event E is provided by its probability, $\text{prob}(E) = \theta$ where θ is restricted to the range $0 \le \theta \le 1$. A popular transformation of this measure of uncertainty is provided by the *odds* on an event where $\text{odds}(E) \equiv \phi = \theta/(1 - \theta)$. A further transformation is provided by the $\log(\text{odds}) = \log\{\theta/(1 - \theta)\}$. Plot, on a piece of rectangular graph paper, curves which will transform θ to ϕ to $\log \phi$.

8.2. You are given that a contour of the true response η is the following function of two variables r and θ:

$$r^2 + 4r\cos\theta + 6r\sin\theta = 12.$$

Transform this equation by $x = r\cos\theta$ and $y = r\sin\theta$, and so determine the shape of the contour in the (x, y) coordinate system.

8.3. Given the textile data of Table 7.1, fit the model

$$y \equiv \log Y = \gamma + \gamma_1\log\xi_3 + \gamma_2\log\xi_4 + \varepsilon$$

where $\xi_4 = \xi_2/\xi_1$, by least squares, and test the null hypothesis H_0: $\gamma_1 = -3$, $\gamma_2 = -5$ versus H_1: some violation of H_0. Is the remark around Eq. (8.3.17) justified, do you think? [*Hint*: On a standard regression program, it may be easier to fit $(y + 3\log\xi_3 + 5\log\xi_4) = \gamma + \delta_1\log\xi_3 + \delta_2\log\xi_4 + \varepsilon$, where $\delta_1 = \gamma_1 + 3$, $\delta_2 =: \gamma_2 + 5$, and then test H_0: $\delta_1 = \delta_2 = 0$.]

8.4. (Source: An empirical model for viscosity of filled and plasticized elastomer compounds, by G. C. Derringer, *Journal of Applied Polymer Science*, **18**, 1974, 1083–1101. Copyright © 1974, John Wiley & Sons, Inc.; adapted with their permission.) Table E8.4 shows data for two responses y_1 and y_2 at various levels of two inputs. For each response individually, choose the best value of λ in the transformation $w = (y^\lambda - 1)/\lambda$ for $\lambda \ne 0$, $w = \ln y$ for $\lambda = 0$ to

TABLE E8.4. Mooney viscosities $y_1(ML_4)$ and $y_2(MS_4)$ at 100°C for various combinations of coded variables[a] x_1 and x_2

x_1	x_2	y_1	y_2
-1	-1	58	28
-1	1	41	18
1	-1	—	93
1	1	106	54
0	$-2^{1/2}$	104	57
0	$2^{1/2}$	56	30
$-2^{1/2}$	0	38	20
$2^{1/2}$	0	—	100
0	0	97	41
0	0	74	40
0	0	78	42
0	0	83	45
0	0	90	48

[a] $x_1 = (X_1 - 40)/20$, $x_2 = (X_2 - 20)/10$, where X_1 = level of Silica A, H$_i$-Sil 233, PPG Industries, and X_2 = level of Aromatic oil, Sundex 790, Sun Oil Co., in NBR Hycar 1052, B. F. Goodrich Chemical Co.

TABLE E8.5. Mooney viscosity $y(ML_4)$ at 100°C for various combinations of coded variables[a] x_1, x_2, and x_3

x_1	x_2	x_3	y
-1	-1	1	73
-1	1	-1	78
1	-1	-1	95
1	1	1	121
1.5	0	0	114
-1.5	0	0	66
0	1.5	0	96
0	-1.5	0	80
0	0	1.5	84
0	0	-1.5	79
0	0	0	89
0	0	0	86

[a] $x_i = (X_i - 10)/5$, $i = 1, 2, 3$, where X_1 = level of Silica A, Hi-Sil 233, PPG Industries, X_2 = level of Silica B, Hi-Sil EP, PPG Industries, and X_3 = level of N774, Cabot Corp.

enable the model $w = \beta_0 + \beta_1 x_1 + \beta_2 x_2 + \varepsilon$ to be well fitted to the data by least squares. After choosing the best transformation parameter estimate $\hat{\lambda}$ in each case, carry out all the usual regression analyses including examination of the residuals. [*Note*: Once λ is chosen, the response y^λ may be used for the subsequent analysis instead of $(y^\lambda - 1)/\lambda$, if $\lambda \neq 0$. Or, if $\lambda = 0$, logarithms to any base, not necessarily e, can be used if desired. Such alternative choices do not affect the basic analysis, just the size of the numbers involved in the calculation. These remarks apply to all such transformation problems.]

8.5. (Source: An empirical model for viscosity of filled and plasticized elastomer compounds, by G. C. Derringer, *Journal of Applied Polymer Science*, **18**, 1974, 1083–1101. Copyright © 1974, John Wiley & Sons, Inc.; adapted with their permission.) Choose the best value of λ in the transformation $w = (y^\lambda - 1)/\lambda$ for $\lambda \neq 0$, $w = \ln y$ for $\lambda = 0$, to enable the model $w = \beta_0 + \beta_1 x_1 + \beta_2 x_2 + \beta_3 x_3 + \varepsilon$ to be well fitted to the data of Table E8.5. After choosing the best transformation parameter estimate $\hat{\lambda}$, carry out all the usual regression analyses including examination of the residuals. (See note attached to foregoing problem.)

Exploration of Maxima and Ridge Systems with Second-Order Response Surfaces

9.1. THE NEED FOR SECOND-ORDER SURFACES

In earlier chapters we have considered the use of polynomial graduating functions to represent locally the true expectation function $f(\xi, \theta)$. Thus it was supposed that an estimate $\hat{\beta}$ of a vector of empirical constants β could be found such that, over the region in the ξ space of immediate interest,

$$g(\xi, \hat{\beta}) \simeq f(\xi, \theta), \qquad (9.1.1)$$

where $g(\xi, \beta)$ was a polynomial of first or second degree.

In Chapter 7, an example was discussed where the observed response Y of interest was the number of cycles to failure of the tested textile specimen and where the variables ξ_1, ξ_2, and ξ_3 were length, amplitude, and load, respectively. It turned out in this example that, if we wished to represent $E(Y)$ directly by a polynomial approximation, then an expression of at least second degree in the ξ's was needed. However, if we first applied a simple transformation to the response (specifically, by use of $y = \log Y$ instead of Y), an expression of only first degree in the ξ's produced an adequate approximation. In general if, over the region of interest considered, the observed response is curved but smoothly monotonic in each of the elements of ξ, it will sometimes be possible (and where it is so, preferable) to employ a first degree expression in transformed variables, rather than a second degree expression in untransformed variables. Such a simplified representation may be possible either by transforming the output Y, as above, or by transforming some or all of the input ξ's, or by transforming both ξ's and Y simultaneously.

However, when the object is to approximate the response function in a region in which it has a turning point, we shall need a polynomial of at least second degree.* Transformed metrics may, nevertheless, make possible the use of a polynomial of lower degree than would otherwise be needed in untransformed metrics. Thus, for example, we can often avoid the need for a third degree polynomial by employing a second degree approximation in transformed variables.

As we have seen, the number of terms required by an approximating polynomial increases rapidly as the degree of the polynomial increases. Also, when N runs are used, and the polynomial model has p parameters, the average variance of estimation is

$$\overline{V}(\hat{y}) = N^{-1} \sum_{u=1}^{N} V(\hat{y}_u) = p\sigma^2/N. \qquad (9.1.2)$$

Thus, solely on the basis of consideration of the variance of the predicted values, the lower-order approximating model should always be preferred or, stated more forcefully, movement to higher-order approximating polynomial models should be resisted until all efforts at keeping to the lower-order approximations have been expended.

9.2. EXAMPLE: REPRESENTATION OF POLYMER ELASTICITY AS A QUADRATIC FUNCTION OF THREE OPERATING VARIABLES

For illustration, we describe the elucidation of the nature of a maximal region for a polymer elasticity experiment. In this example, we encounter a number of issues which will be taken up in more detail in later chapters.

A Central Composite Design

The design employed at this stage of the investigation was a second-order central *composite design*. Such a design consists of a two-level factorial or fractional factorial (chosen so as to allow the estimation of all first-order

---

*Functions other than polynomials can, of course, be used to produce curves containing maxima. One such, that is sometimes useful for a single input x, is

$$\ln Y = \alpha + \beta \ln x + \gamma x.$$

With $\beta > 0$ and $\gamma < 0$, this produces a curve containing an asymmetric maximum.

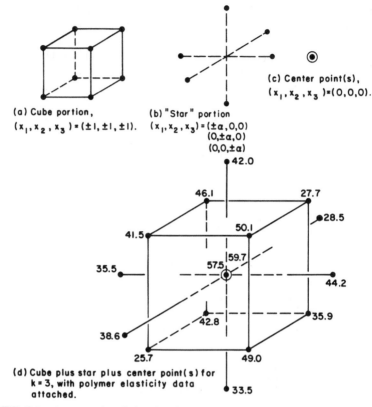

FIGURE 9.1. A composite design in three dimensions ($k = 3$). (*a*) Cube portion, $(x_1, x_2, x_3) = (\pm1, \pm1, \pm1)$. (*b*) "Star" portion, $(x_1, x_2, x_3) = (\pm\alpha, 0, 0)$, $(0, \pm\alpha, 0)$, $(0, 0, \pm\alpha)$. (*c*) Center point(s), $(x_1, x_2, x_3) = (0, 0, 0)$. (*d*) Cube plus star plus center point(s) for $k = 3$, with polymer elasticity data attached.

and two factor interaction terms) augmented with further points which allow pure quadratic effects to be estimated also. For this design, shown in Figure 9.1 for $k = 3$ inputs, the additional experimental points consist of center points plus "star" points arranged along the axes of the variables and symmetrically positioned with respect to the factorial cube. We discuss such designs in more detail in Chapter 15.

First Set of Runs

The objective of this investigation was to explore reaction conditions yielding close to maximal elasticity for a certain polymer. At this stage of

TABLE 9.1.   Coded and uncoded levels of three predictor variables

Predictor variables	Levels		Midlevel	Semirange
	−1	+1		
Percentage concentration of constituent 1, C_1	15	21	18	3
Percentage concentration of constituent 2, C_2	2.3	3.1	2.7	0.4
Temperature (°C), T	135	155	145	10

the enquiry, the variables of importance had been narrowed down to three: the percentage concentrations C_1 and C_2 of two constituents in the reaction mixture, and the reaction temperature T. Earlier groups of experiments using factorials and steepest ascent had produced conditions yielding a measured elasticity of about 55 units, and a rough estimate of the experimental error standard deviation of 2 units. A further 2^3 factorial straddling these "so far best conditions" was run using the pairs of levels shown in Table 9.1.

Thus, via the standard factorial coding

$$x_1 = \frac{(C_1 - 18)}{3}, \qquad x_2 = \frac{(C_2 - 2.7)}{0.4}, \qquad x_3 = \frac{(T - 145)}{10}. \quad (9.2.1)$$

The experimental results are given in Table 9.2. If we were to fit a

TABLE 9.2.   Results from a 2^3 factorial design

x_1	x_2	x_3	y
−1	−1	−1	25.74
1	−1	−1	48.98
−1	1	−1	42.78
1	1	−1	35.94
−1	−1	1	41.50
1	−1	1	50.10
−1	1	1	46.06
1	1	1	27.70

TABLE 9.3.   The estimates from the 2^3 design and their expected values under the assumption that the true model is the complete quadratic

Estimated coefficients $\pm$ standard errors	Expected value on the assumption that the true model is a second degree polynomial
$b_0 = 39.85 \pm 0.71$	$\beta_0 + \beta_{11} + \beta_{22} + \beta_{33}$
$b_1 = 0.83 \pm 0.71$	β_1
$b_2 = -1.73 \pm 0.71$	β_2
$b_3 = 1.49 \pm 0.71$	β_3
$b_{12} = -7.13 \pm 0.71$	β_{12}
$b_{13} = -3.27 \pm 0.71$	β_{13}
$b_{23} = -2.73 \pm 0.71$	β_{23}

polynomial of first degree to these data we would obtain

$$\hat{y} = 39.85 \quad + 0.83x_1 \quad - 1.73x_2 \quad + 1.49x_3 \,, \qquad (9.2.2)$$
$$\pm 0.71 \quad\;\; \pm 0.71 \quad\;\; \pm 0.71 \quad\;\; \pm 0.71$$

where the standard errors of the coefficients are evaluated on the assumption that $\sigma = 2$. If we attempt to fit a second degree polynomial to the present data, the pure quadratic effects are not of course separately estimable, because the design contains only two levels of each of the predictor variables. In fact, the quadratic effects appear as aliases of the constant term, as indicated in Table 9.3. It is now seen that those estimates of second-order coefficients which are available, namely, b_{12}, b_{13}, b_{23}, are large not only compared with their own standard errors, but also compared with the estimates b_1, b_2, b_3 of terms of first order. From this analysis, it is evident therefore that we have moved into a region where the first degree polynomial no longer provides a good approximation. Further experimentation to allow estimation of all second-order effects (pure quadratic terms as well as interaction terms) is needed to clarify the situation.

Second Set of Runs

The three-dimensional star of experimental points and two center points were therefore symmetrically added to the cube at this stage. The two sets of experiments together form the composite design of Figure 9.1*d*. The experimental results from this second group of experiments are shown in Table 9.4. The least-squares calculations for the combined set of 16 results are shown in Table 9.5. In this table, the rows of the 16 by 10 matrix $\mathbf{Z}_1$ contain the elements 1, x_1, x_2, x_3, x_1^2, x_2^2, x_3^2, x_1x_2, x_1x_3, x_2x_3 corre-

TABLE 9.4. Results from the added star and center point runs

		Uncoded levels		Coded levels			Response
	$C_1\%$	$C_2\%$	$T°C$	x_1	x_2	x_3	y
Center points	18	2.7	145	0	0	0	57.52
	18	2.7	145	0	0	0	59.68
	12	2.7	145	−2	0	0	35.50
	24	2.7	145	2	0	0	44.18
Star points	18	1.9	145	0	−2	0	38.58
	18	3.5	145	0	2	0	28.46
	18	2.7	125	0	0	−2	33.50
	18	2.7	165	0	0	2	42.02

sponding (row by row) to the 16 sets of experimental conditions. The 10 by 1 vector of estimates $\mathbf{b}_1$ with elements b_0, b_1, b_2, b_3, b_{11}, b_{22}, b_{33}, b_{12}, b_{13}, b_{23} is given by

$$\mathbf{b}_1 = (\mathbf{Z}_1'\mathbf{Z}_1)^{-1}\mathbf{Z}_1'\mathbf{y}, \tag{9.2.3}$$

and the standard errors of the various estimates are the square roots of the corresponding diagonal elements of $(\mathbf{Z}_1'\mathbf{Z}_1)^{-1}\sigma^2$. Carrying through these calculations, we obtain the fitted second degree polynomial as

$$\begin{aligned}
\hat{y} = 57.3 + \underset{\pm 0.5}{1.5x_1} - \underset{\pm 0.5}{2.1x_2} + \underset{\pm 0.5}{1.8x_3} - \underset{\pm 0.5}{4.7x_1^2} - \underset{\pm 0.5}{6.3x_2^2} \\
- \underset{\pm 0.5}{5.2x_3^2} - \underset{\pm 0.7}{7.1x_1x_2} - \underset{\pm 0.7}{3.3x_1x_3} - \underset{\pm 0.7}{2.7x_2x_3},
\end{aligned} \tag{9.2.4}$$

where, once again, the standard errors of the coefficients (shown below the corresponding coefficients) are computed on the assumption that $\sigma = 2$. An analysis of variance table is shown in Table 9.6.

Removal of the Block Difference

In this particular investigation, more than a week elapsed between the performance of the first set of eight runs (the 2^3 factorial "cube") and of the second set of eight runs (the "star" plus two center points). There were, therefore, opportunities for systematic differences to occur between the first and second sets of eight runs, possibly associated with slight changes in the

TABLE 9.5. Least-squares calculations for polymer elasticity example

					$\mathbf{Z}_1$				
1	x_1	x_2	x_3	x_1^2	x_2^2	x_3^2	x_1x_2	x_1x_3	x_2x_3
1	−1	−1	−1	1	1	1	1	1	1
1	1	−1	−1	1	1	1	−1	−1	1
1	−1	1	−1	1	1	1	−1	1	−1
1	1	1	−1	1	1	1	1	−1	−1
1	−1	−1	1	1	1	1	1	−1	−1
1	1	−1	1	1	1	1	−1	1	−1
1	−1	1	1	1	1	1	−1	−1	1
1	1	1	1	1	1	1	1	1	1
1	·	·	·	·	·	·	·	·	·
1	·	·	·	·	·	·	·	·	·
1	−2	·	·	4	·	·	·	·	·
1	2	·	·	4	·	·	·	·	·
1	·	−2	·	·	4	·	·	·	·
1	·	2	·	·	4	·	·	·	·
1	·	·	−2	·	·	4	·	·	·
1	·	·	2	·	·	4	·	·	·

		$\mathbf{Z}_2^*$			$\mathbf{Z}_2^0$	y
$x_1^{(3)}$	$x_2^{(3)}$	$x_3^{(3)}$	$x_1x_2x_3$	x_B	x_E	
1	1	1	−1	−1	·	25.74
−1	1	1	1	−1	·	48.98
1	−1	1	1	−1	·	42.78
−1	−1	1	−1	−1	·	35.94
1	1	−1	1	−1	·	41.50
−1	1	−1	−1	−1	·	50.10
1	−1	−1	−1	−1	·	46.06
−1	−1	−1	1	−1	·	27.70
·	·	·	·	1	−1	57.52
·	·	·	·	1	1	59.68
−2	·	·	·	1	·	35.50
2	·	·	·	1	·	44.18
·	−2	·	·	1	·	35.58
·	2	·	·	1	·	28.46
·	·	−2	·	1	·	33.50
·	·	2	·	1	·	42.02

TABLE 9.5. Continued

$Z_1'Z_1$ and $Z_1'y$:

	0	1	2	3	11	22	33	12	13	23	$Z_1'y$
0	16	.	.	.	16	16	16	.	.	.	658.24
1	.	16	.	.	.	.	.	.	.	.	24.00
2	.	.	16	.	.	.	.	.	.	.	−34.08
3	.	.	.	16	.	.	.	.	.	.	28.96
11	16	.	.	.	40	8	8	.	.	.	637.52
22	16	.	.	.	8	40	8	.	.	.	586.96
33	16	.	.	.	8	8	40	.	.	.	620.88
12	.	.	.	.	.	.	.	8	.	.	−57.04
13	.	.	.	.	.	.	.	.	8	.	−26.16
23	.	.	.	.	.	.	.	.	.	8	−21.84

$32(Z_1'Z_1)^{-1}$ and b_1:

	0	1	2	3	11	22	33	12	13	23	b_1
0	14	.	.	.	−4	−4	−4	.	.	.	57.31
1	.	2	.	.	.	.	.	.	.	.	1.50
2	.	.	2	.	.	.	.	.	.	.	−2.13
3	.	.	.	2	.	.	.	.	.	.	1.81
11	−4	.	.	.	2	1	1	.	.	.	−4.69
22	−4	.	.	.	1	2	1	.	.	.	−6.27
33	−4	.	.	.	1	1	2	.	.	.	−5.21
12	.	.	.	.	.	.	.	4	.	.	−7.13
13	.	.	.	.	.	.	.	.	4	.	−3.27
23	.	.	.	.	.	.	.	.	.	4	−2.73

raw materials employed, in the standard solutions used in the analyses, or in the setting of the instruments. Following the terminology introduced in Chapter 5, we shall say that the experiment was run in two *blocks* of eight runs.

A systematic difference between two blocks may be estimated and eliminated as follows. If all 16 runs had been performed under the conditions of the first block, suppose the constant term in the second degree polynomial would have been $\beta_0 - \delta$, while if all runs had been performed under the conditions of the second block, suppose the constant term would have been $\beta_0 + \delta$. Then the true mean difference between blocks is 2δ and the model becomes

$$y = g(\mathbf{x}, \boldsymbol{\beta}) + x_B \delta + \varepsilon \tag{9.2.5}$$

TABLE 9.6. Analysis of variance for polymer elasticity example

Source	SS	df	MS
Mean (zero degree polynomial)	27,078.0	1	
Added first-order terms (given b_0)	161.0 ⎫	3 ⎫	53.7 ⎫
Added second-order terms (given zero and first-order terms)	1,290.6 ⎬ 1,451.6	6 ⎬ 9	215.1 ⎬ 161.3
Residual	41.6	6	6.9
Total	28,571.2	16	

where the indicator variable x_B (the *blocking variable* shown in Table 9.5 in the first column of $\mathbf{Z}_2^0$) is -1 in the first block and $+1$ in the second block. A rather remarkable circumstance will be noted for this particular design. The indicator variable x_B is orthogonal to each of the columns of $\mathbf{Z}_1$. It follows (see Section 3.9) that

$$\hat{\delta} = \frac{\left(\sum\limits_{j=1}^{16} x_{Bj} y_j\right)}{\sum\limits_{j=1}^{16} x_{Bj}^2} = 1.29. \tag{9.2.6}$$

Moreover, we can separate out from the residual sum of squares, a "blocks" contribution with 1 df given by (see, also, p. 513)

$$16\hat{\delta}^2 = 26.63. \tag{9.2.7}$$

TABLE 9.7. Further analysis of the residual sum of squares in the polymer elasticity example

Source	SS	df	MS	
Blocks	26.6	1	26.6	$F = 8.9$
Residual, after removal of blocks	15.0	5	3.0	
Residual, before removal of blocks	41.6	6		

Thus, the 6 df labeled "residual" in the analysis of variance of Table 9.6 can be further analyzed as indicated in Table 9.7. The contribution of blocks in this example is clearly quite substantial. The estimated standard deviation computed from the residual mean square is now $s = 1.73$, close to the value $\sigma = 2$ assumed previously.

Orthogonal Blocking and Sequential Design

Because the blocking is orthogonal in this example, the removal of the contribution associated with blocks does not change the estimated coefficients in the fitted function (9.2.4). However, that portion of the original residual sum of squares accounted for by the systematic block difference is removed, and the accuracy of the experiment is correspondingly increased.

The Importance of Blocking in the Sequential Assembly of Designs

Blocking, and where possible, orthogonal blocking, is of great importance in sequential experimentation. When the 2^3 factorial corresponding to the first half of this design was planned, several possibilities were open for the succeeding step. If the first-order effects had been fairly large (a) compared with their standard errors *and* (b) compared with the estimated interaction terms, then an application of steepest ascent would have been appropriate and this might have moved the center of interest to a new location possibly yielding higher elasticity where a further design would have been run. In the event, as we have seen, an inspection of the set of eight factorial runs made it seem likely that a second degree exploration in the immediate neighborhood was necessary. The second part of the design corresponding to the star with center points was added with the knowledge that the second degree polynomial equation could now be conveniently estimated, even though a change in level could have occurred between the two blocks of runs. This possibility, of *sequential assembly* of different kinds of design building blocks as the need for them is demonstrated, is a very valuable one and we discuss it in more detail in Section 15.3.

9.3. EXAMINATION OF THE FITTED SURFACE

The residual variance of 3.0 obtained after removing blocks is based on 5 df and agrees quite well with previous expectation. We shall show later how it is possible to analyze this residual variance further and to isolate measures of adequacy of fit. Here, these checks do not show any abnormalities, and we proceed on the basis that the fit is adequate. If the residual variance of

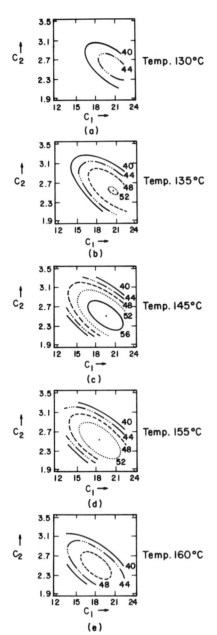

FIGURE 9.2. The fitted surface: Contours show polymer elasticity as a function of concentrations C_1 and C_2 for various values of temperature T. (a) Temperature 130°C ($x_3 = -1.5$); (b) temperature 135°C ($x_3 = -1.0$); (c) temperature 145°C ($x_3 = 0$); (d) temperature 155°C ($x_3 = 1.0$); (e) temperature 160°C ($x_3 = 1.5$).

3.0 is used as an estimate of error, the overall F value associated with the complete second degree regression equation is $161.3/3.0 = 53.8$. The tabled 5% value of F for 9 and 5 df is 4.8. The observed value of F exceeds by more than 10 times its 5% significance level, and application of the rough guideline introduced in Section 8.2 indicates that, in this example, we would probably have something worthwhile to interpret from the fitted surface. For this example, where there are just three predictor variables, we can easily visualize what the second degree fitted equation (9.2.4) is telling us, by mapping elasticity contours in the space of two of the predictor variables for a series of chosen values of the third. In Figures 9.2a to 9.2e, the fitted contours of elasticity are drawn in the space of C_1 and C_2 (x_1 and x_2) for five values of the temperature T (x_3). The fitted surface shows a maximum, with some factor dependence between the variables. This dependence is such that a higher concentration C_1 is, to some extent, compensated by a lower concentration C_2, and higher temperatures T are, to some extent, compensated by lower levels of C_1.

Location of the Maximum of the Fitted Surface

The maximum of the fitted surface may be located by differentiating the fitted second degree polynomial with respect to each predictor variable and equating derivatives to zero. For instance, for $k = 3$, differentiation with respect to x_1, x_2, and x_3 of

$$\hat{y} = b_0 + b_1 x_1 + b_2 x_2 + b_3 x_3 + b_{11} x_1^2 + b_{22} x_2^2 + b_{33} x_3^2$$

$$+ b_{12} x_1 x_2 + b_{13} x_1 x_3 + b_{23} x_2 x_3, \tag{9.3.1}$$

yields the equations

$$2b_{11} x_1 + b_{12} x_2 + b_{13} x_3 = -b_1,$$

$$b_{12} x_1 + 2b_{22} x_2 + b_{23} x_3 = -b_2,$$

$$b_{13} x_1 + b_{23} x_2 + 2b_{33} x_3 = -b_3, \tag{9.3.2}$$

from which the form for general k is obvious. The main diagonal coefficients on the left are twice the estimated pure quadratic effects, while the off-diagonal coefficients are the interaction effects. The right-hand sides are the estimated linear effects with signs reversed. For the present example, after changing signs, we obtain

$$9.38x_1 + 7.13x_2 + 3.27x_3 = 1.50,$$

$$7.13x_1 + 12.54x_2 + 2.73x_3 = -2.13,$$

$$3.27x_1 + 2.73x_2 + 10.42x_3 = 1.81. \qquad (9.3.3)$$

Solving these equations yields

$$x_1^* = 0.460, \qquad x_2^* = -0.465, \qquad x_3^* = 0.151, \qquad (9.3.4)$$

for the coordinates of the maximum of the fitted surface, or in terms of the original predictor variables,

$$C_1^* = 19.4\%, \qquad C_2^* = 2.5\%, \qquad T^* = 146.5°C. \qquad (9.3.5)$$

Also, it is easily shown that $\hat{y}^*$, the value given by the fitted equation at this maximum point is

$$\hat{y}^* = b_0 + \tfrac{1}{2}(b_1 x_1^* + b_2 x_2^* + b_3 x_3^*). \qquad (9.3.6)$$

Thus, for the present example,

$$\hat{y}^* = 58.29.$$

On the assumption of the adequacy of the second degree equation, a confidence region for the maximum point may be computed. For details, see Box and Hunter (1954).

9.4. INVESTIGATION OF ADEQUACY OF FIT: ISOLATION OF RESIDUAL DEGREES OF FREEDOM FOR THE COMPOSITE DESIGN

Before accepting that the fitted second degree equation provides a satisfactory approximation to the true surface, one must of course consider possible evidence for lack of fit. In this example, there are only five residual degrees of freedom but nevertheless they can provide some information about lack of fit.

In general, suppose N observations are fitted to a linear model containing p parameters. Then the fitting process itself will introduce p linear relationships among the N residuals $y - \hat{y}$. If N is large with respect to p, then the effect of this induced dependence among the residuals will be slight, and the plotting techniques employed to examine residuals will be useful in revealing any inadequacies in the model. As p becomes larger and, in particular, as it approaches N, patterns caused by the induced dependencies become dominant, and can mask those due to model inadequacies. However, as we saw for the factorial design in Section 7.4, it may be possible to obtain information on the adequacy of fit by isolating, and identifying, individual residual degrees of freedom associated with feared model inadequacies. In particular, when considering the fitting of a polynomial of a given degree, it is natural to be concerned with the possibility that, in fact, a polynomial of higher degree may be needed. This immediately focuses attention on the characteristics of the estimates when the feared alternative model applies, but the simpler assumed model has been fitted. In these circumstances, contemplation of the fitted model embedded in a more complex one makes it possible to answer two kinds of questions.

1. To what extent are the original estimates of the coefficients biased if the more complex model is true?
2. What are appropriate checking functions to warn of the possible need for a more complex model?

Both questions are critically affected by the particular choice of design.

For illustration, we shall consider the possibility that a third degree polynomial might be needed in the elasticity example.

Bias Characteristics of the Design

We may write the extended (third-order) polynomial model in the form

$$y = \mathbf{Z}_1\boldsymbol{\beta}_1 + \mathbf{Z}_2\boldsymbol{\beta}_2 + \boldsymbol{\varepsilon},$$

or, arguing as in Section 3.9, we can write it in the orthogonalized form

$$y = \mathbf{Z}_1(\boldsymbol{\beta}_1 + \mathbf{A}\boldsymbol{\beta}_2) + (\mathbf{Z}_2 - \mathbf{Z}_1\mathbf{A})\boldsymbol{\beta}_2 + \boldsymbol{\varepsilon} \qquad (9.4.1)$$

where $\mathbf{Z}_1\boldsymbol{\beta}_1$ includes all terms up to and including second order, and $\mathbf{Z}_2\boldsymbol{\beta}_2$ includes all terms of third order.

The alias (or bias) matrix $\mathbf{A}$ (Table 9.8) shows that only the estimates of first-order terms are biased by third-order terms. In fact, if a third-order model is appropriate, and if b_1, b_2, and b_3 are our previous least-squares estimates (assuming a second-order model has been fitted), then

$$E(b_1) = \beta_1 + 2.5\beta_{111} + 0.5\beta_{122} + 0.5\beta_{133},$$

$$E(b_2) = \beta_2 + 2.5\beta_{222} + 0.5\beta_{112} + 0.5\beta_{233},$$

$$E(b_3) = \beta_3 + 2.5\beta_{333} + 0.5\beta_{113} + 0.5\beta_{223}. \qquad (9.4.2)$$

TABLE 9.8. Calculations leading to the alias matrix for the polymer elasticity example

$$\mathbf{Z}_2$$

111	222	333	122	112	133	113	233	223	123
−1	−1	−1	−1	−1	−1	−1	−1	−1	−1
1	−1	−1	1	−1	1	−1	−1	−1	1
−1	1	−1	−1	1	−1	−1	1	−1	1
1	1	−1	1	1	1	−1	1	−1	−1
−1	−1	1	−1	−1	−1	1	−1	1	1
1	−1	1	1	−1	1	1	−1	1	−1
−1	1	1	−1	1	−1	1	1	1	−1
1	1	1	1	1	1	1	1	1	1
.	.	.	.	.	.	.	.	.	.
.	.	.	.	.	.	.	.	.	.
−8	.	.	.	.	.	.	.	.	.
8	.	.	.	.	.	.	.	.	.
.	−8	.	.	.	.	.	.	.	.
.	8	.	.	.	.	.	.	.	.
.	.	−8	.	.	.	.	.	.	.
.	.	8	.	.	.	.	.	.	.

$$\mathbf{Z}_1'\mathbf{Z}_2$$

	111	222	333	122	112	133	113	233	223	123
0	.	.	.	.	.	.	.	.	.	.
1	40	.	.	8	.	8	.	.	.	.
2	.	40	.	.	8	.	.	8	.	.
3	.	.	40	.	.	.	8	.	8	.
11	.	.	.	.	.	.	.	.	.	.
22	.	.	.	.	.	.	.	.	.	.
33	.	.	.	.	.	.	.	.	.	.
12	.	.	.	.	.	.	.	.	.	.
13	.	.	.	.	.	.	.	.	.	.
23	.	.	.	.	.	.	.	.	.	.

TABLE 9.8.   Continued

$$\mathbf{Z}_{2\cdot 1} = \mathbf{Z}_2 - \mathbf{Z}_1 \mathbf{A}$$

111	222	333	122	112	133	113	233	223	123
1.5	1.5	1.5	−0.5	−0.5	−0.5	−0.5	−0.5	−0.5	−1
−1.5	1.5	1.5	0.5	−0.5	0.5	−0.5	−0.5	−0.5	1
1.5	−1.5	1.5	−0.5	0.5	−0.5	−0.5	0.5	−0.5	1
−1.5	−1.5	1.5	0.5	0.5	0.5	−0.5	0.5	−0.5	−1
1.5	1.5	−1.5	−0.5	−0.5	−0.5	0.5	−0.5	0.5	1
−1.5	1.5	−1.5	0.5	−0.5	0.5	0.5	−0.5	0.5	−1
1.5	−1.5	−1.5	−0.5	0.5	−0.5	0.5	0.5	0.5	−1
−1.5	−1.5	−1.5	0.5	0.5	0.5	0.5	0.5	0.5	1
.	.	.	.	.	.	.	.	.	.
.	.	.	.	.	.	.	.	.	.
−3	.	.	1	.	1	.	.	.	.
3	.	.	−1	.	−1	.	.	.	.
.	−3	.	.	1	.	.	1	.	.
.	3	.	.	−1	.	.	−1	.	.
.	.	−3	.	.	.	1	.	1	.
.	.	3	.	.	.	−1	.	−1	.

$$\mathbf{A} = (\mathbf{Z}_1'\mathbf{Z}_1)^{-1}\mathbf{Z}_1'\mathbf{Z}_2$$

	111	222	333	122	112	133	113	233	223	123
0	.	.	.	.	.	.	.	.	.	.
1	2.5	.	.	0.5	.	0.5	.	.	.	.
2	.	2.5	.	.	0.5	.	.	0.5	.	.
3	.	.	2.5	.	.	.	0.5	.	0.5	.
11	.	.	.	.	.	.	.	.	.	.
22	.	.	.	.	.	.	.	.	.	.
33	.	.	.	.	.	.	.	.	.	.
12	.	.	.	.	.	.	.	.	.	.
13	.	.	.	.	.	.	.	.	.	.
23	.	.	.	.	.	.	.	.	.	.

Checking Functions for the Design

Examination of the matrix $\mathbf{Z}_{2\cdot 1} = \mathbf{Z}_2 - \mathbf{Z}_1\mathbf{A}$ in Table 9.8 reveals a rather remarkable circumstance. Although the matrix has 10 columns, only 4 of these are independent and these 4 are all simple multiples of the 4 columns of $\mathbf{Z}_2^*$ in Table 9.5. Consider, for example, the first column (labeled 111) of the matrix $\mathbf{Z}_{2\cdot 1}$. We can write this as

$$1.5x_1^{(3)}, \tag{9.4.3}$$

where

$$\mathbf{x}_1^{(3)} = (1, -1, 1, -1, 1, -1, 1, -1, 0, 0, -2, 2, 0, 0, 0, 0)' \quad (9.4.4)$$

is the first column of $\mathbf{Z}_2^*$. This vector is, at first sight, rather like the $\mathbf{x}_1$ vector which is the second column in the matrix $\mathbf{Z}_1$. Notice however that the signs of the first eight elements of $\mathbf{x}_1^{(3)}$ are the reverse of those of $\mathbf{x}_1$, while the signs associated with the elements $(-2, 2)$ are the same as those of $\mathbf{x}_1$. In spite of their similar form, $\mathbf{x}_1$ and $\mathbf{x}_1^{(3)}$ are, of course, orthogonal. Further inspection shows that the elements of the columns labeled 122 and 133 in $\mathbf{Z}_{2 \cdot 1}$ are those of $-0.5\mathbf{x}_1^{(3)}$ in each case.

We now define a normalized linear combination l_{111} of the observations which uses the elements of $\mathbf{x}_1^{(3)}$ as coefficients. This is

$$l_{111} = \frac{\mathbf{x}_1^{(3)'}\mathbf{y}}{\mathbf{x}_1^{(3)'}\mathbf{x}_1^{(3)}} = \frac{\mathbf{x}_1^{(3)'}\mathbf{y}}{16}. \quad (9.4.5)$$

Then

$$E(l_{111}) = 1.5\beta_{111} - 0.5\beta_{122} - 0.5\beta_{133}. \quad (9.4.6)$$

Similarly, we can define estimates l_{222}, l_{333}, and obtain

$$E(l_{222}) = 1.5\beta_{222} - 0.5\beta_{112} - 0.5\beta_{233},$$

$$E(l_{333}) = 1.5\beta_{333} - 0.5\beta_{113} - 0.5\beta_{223}. \quad (9.4.7)$$

A parallel definition of

$$l_{123} = \Sigma x_1 x_2 x_3 y / 8, \quad (9.4.8)$$

will give us

$$E(l_{123}) = \beta_{123}. \quad (9.4.9)$$

Thus, although we cannot obtain estimates of each of the third-order effects individually, with this design, we can isolate certain linear combinations of them (that is, certain alias groups). The size of these combinations can indicate particular directions in which there may be lack of fit. In particular, suppose that l_{jjj} were excessively large. This could indicate, for example, that a transformation of x_j might be needed to obtain an adequate representation using a second degree equation. This transformation aspect is discussed in more detail in Section 13.8.

Complete Breakup of the Residual Sum of Squares in the Elasticity Example.

There are 16 runs in the composite design of the elasticity example. Ten degrees of freedom are used in the estimation of the second degree polynomial. Of the remaining 6 df, one is used for blocking, and one is a pure error comparison in which the two center points are compared. These are associated with the two vectors in $\mathbf{Z}_2^0$ of Table 9.5. Four degrees of freedom remain, and these can be associated with possible lack of fit from neglected third-order terms (or alternatively with the need for transformation of the variables). We notice that the columns of $\mathbf{Z}_2^*$ and $\mathbf{Z}_2^0$ are all orthogonal to one another and to the columns of $\mathbf{Z}_1$ so that, for the elasticity example, we can work out each contribution separately and easily. For example,

$$l_{111} = \tfrac{1}{16}\{+25.74 - 48.98 + \cdots + (2 \times 44.18)\} = 0.67, \quad (9.4.10)$$

and the corresponding contribution to the residual sum of squares is

$$16 l_{111}^2 = 7.18. \tag{9.4.11}$$

For all four, we obtain

$$l_{111} = 0.67, \qquad l_{222} = -0.40, \qquad l_{333} = -0.32, \qquad l_{123} = 0.39, \quad (9.4.12)$$

and a detailed analysis of the residual sum of squares is shown in Table 9.9. (Note that 8, not 16, is the multiplier for the SS calculation for 123.)

TABLE 9.9.   **Analysis of the residual sum of squares into individual degrees of freedom for the elasticity example**

Source	SS	df	MS
Blocks	26.6	1	26.6
111	7.2	1	7.2
222	2.6	1	2.6
333	1.6	1	1.6
123	1.2	1	1.2
Pure error	2.3	1	2.3
Residual sum of squares	41.6[a]	6	

[a] The slight discrepancy in addition is due to round-off error.

None of the mean squares is excessively large compared with the others, nor do they contradict the earlier supposition that $\sigma \doteq 2$, that is, $\sigma^2 \doteq 4$. Thus we have no reason to suspect lack of fit.

Conclusions

What does this investigation allow us to conclude? We can say that:

1. In the immediate neighborhood of its maximum value, elasticity can be represented approximately by a second degree polynomial in the three predictor variables, concentration 1, concentration 2, and temperature, given by (9.2.4).
2. Examination of individual residual degrees of freedom gives no indication that the fit is inadequate.
3. The overall F value associated with the complete second degree equation is more than 10 times its 5% significance level implying that the fitted surface is worthy of interpretation.
4. The maximum point is estimated to be at $C_1^* = 19.4\%$, $C_2^* = 2.5\%$, $T^* = 146.5°C$, and the estimated elasticity there is $\hat{y}^* = 58.3$.
5. Contour plots indicate some factor dependence between the variables.

CHAPTER 10

Occurrence and Elucidation of Ridge Systems, I

10.1. INTRODUCTION

In the polynomial elasticity example discussed in Chapter 9, the nature of the fitted surface was explored by plotting contours of estimated elasticity $\hat{y}$ in the space of the three predictor variables, or factors, C_1, C_2 and T. The plots revealed some factor dependence. For example, study of Figure 9.2 shows that the concentration C_1 giving the maximum elasticity is different depending on the levels of the second concentration C_2 and the temperature T. Examples frequently occur which show factor dependence of a much stronger kind. Also there may be more than two or three factors involved. In this chapter we consider the nature of likely factor dependencies and show, in an elementary way, how they may be elucidated.

Response Functions Near Maxima, Minima, and Stationary Values

Consider, to begin with, an example where there is a single predictor variable "residence time" and the response of interest is the "percentage yield." Thus, in our usual notation η = mean yield (%), ξ_1 = residence time (min) and $\eta = f(\xi_1, \boldsymbol{\theta})$. As usual, $\eta = f(\xi_1, \boldsymbol{\theta})$ denotes the true, but we shall presume unknown, relationship between mean yield η and residence time ξ_1. We shall assume that this relationship contains a maximum and that, in the neighborhood of this maximum,* the relationship can be represented by a graph like that in Figure 10.1.

---

*We talk in terms of a maximum but, when the response η passes through a maximum, the response $c - \eta$, where c is any constant, passes through a corresponding minimum. Thus discussion of a maximum includes that of a minimum also, throughout what follows.

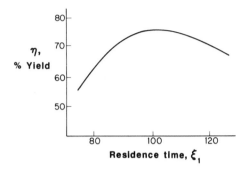

FIGURE 10.1. A relationship between η and ξ_1 in the region of a maximum.

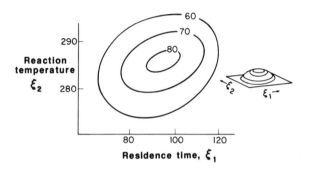

FIGURE 10.2. Yield contours for a roughly symmetric maximum.

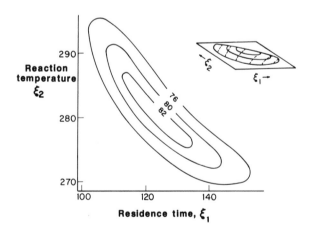

FIGURE 10.3. Yield contours for an attenuated maximum.

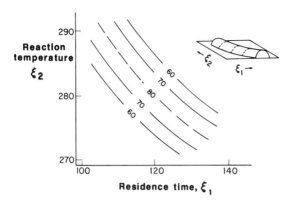

FIGURE 10.4.   Yield contours for a stationary ridge.

However, suppose now that two variables are studied, say ξ_1 = residence time, and ξ_2 = reaction temperature. It is tempting to assume that the response function can be represented by a surface like a more or less symmetric mound, the contour representation of which might be like that shown in Figure 10.2.

As soon as experiments began to be carried out in which actual response surfaces could be roughly plotted, it was found that such a generalization was frequently inadequate. It became clear not only that surfaces were often attenuated in the neighborhood of maxima to form an oblique ridge like

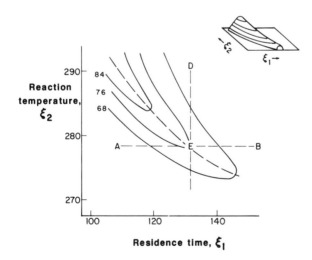

FIGURE 10.5.   Yield contours for a rising ridge.

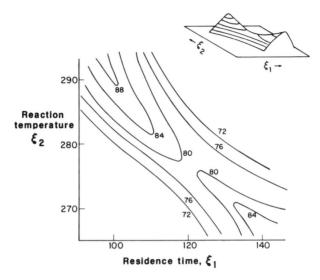

FIGURE 10.6. Yield contours for a "minimax," "saddle," or "col."

that in Figure 10.3, but also that oblique stationary and rising ridge systems like those of Figures 10.4 and 10.5 were of quite common occurrence as well.

Another possibility, which however seems to occur much less frequently in practice, is illustrated in Figure 10.6. This surface contains a ridgy *saddle* system, also often called a *col* or *minimax*.

It will be noted, in Figures 10.2 to 10.5, that any section of the response surfaces taken parallel to either the ξ_1 or the ξ_2 axis will yield a curve like that in Figure 10.1, so that all of these surfaces are entirely built up from such curves.

10.2. FACTOR DEPENDENCE

The reason for the occurrence of ridge systems of the kind displayed in Figures 10.3 to 10.5 can be seen when it is remembered that factors like temperature, time, pressure, concentration, and so on are regarded as "natural" variables only because they happen to be quantities that can be conveniently manipulated and measured. Individual fundamental variables not directly controlled or measured, but in terms of which the behavior of

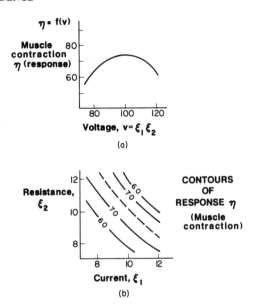

FIGURE 10.7. Generation of a ridge system. (a) Response function in η versus $v = \xi_1\xi_2$. (b) Response function in terms of ξ_1 and ξ_2.

the system could be described more economically (e.g., frequency of a particularly type of molecular collision), will often be a function of two or more natural variables.

For this reason, many different combinations of levels of the natural variables may correspond to the best level of a fundamental variable. A similar rationale may be found for many biological and economic systems which display maxima. As an example of the former, consider an investigation of the muscular contraction of a frog's leg resulting from the passage of an electric current. Suppose that the response η = muscle contraction was a function $\eta = f(v)$ only of the voltage v applied and that initially, when v was increased, the contraction increased but, beyond a certain voltage, the muscle was paralyzed so that further increases in v resulted in reduced contraction. The η plotted versus v could give a curve like that of Figure 10.7a, with a maximum muscle contraction of about 74 occurring when a voltage of $v = 100$ was applied. Suppose, however, that this relationship was unknown to the investigator, whose experimental apparatus was set up in such a way that the variables he could actually manipulate were the current ξ_1 and the resistance ξ_2 in the circuit. Suppose finally that the investigator was unaware of the existence of Ohm's law which would have told him that the voltage applied, $v = \xi_1 \xi_2$, was the product of current and resistance in the circuit. By carrying out experiments in which current ξ_1 and resistance ξ_2 were varied separately, he would be exploring a system for which the response surface was a stationary ridge like that shown* in Figure 10.7b.

Other functions $v = f(\xi_1, \xi_2)$ can produce ridges with sections that contain a maximum. It will be found, for example, that if b and c are positive constants, the functions $v = a + b\xi_1 + c\xi_2$, $v = a\xi_1^b \xi_2^c$, and $v = a\xi_1^b \exp\{-c/\xi_2\}$ will all produce diagonal ridge systems running from the upper left to the lower right of the diagram, as does the one shown in Figure 10.7b, while the functions $v = a + b\xi_1 - c\xi_2$, $v = a\xi_1^b/\xi_2^c$, and $v = a\xi_1^b \exp\{-c\xi_2\}$ all produce ridges running from lower left to upper right. These types of ridge systems are, of course, associated with "interaction" between the predictor variables ξ_1 and ξ_2, the former with a negative two-factor interaction between ξ_1 and ξ_2, the latter with a positive two-factor interaction.

The simplest type of ridge system is produced when v is a linear function of ξ_1 and ξ_2, that is, $v = a + b\xi_1 + c\xi_2$ with any choice of signs. The ridge systems so generated will have parallel straightline contours running in a direction determined by the relative magnitudes of b and c. A section at right angles to these contours will reproduce the original graph of η on v.

We say, in the example quoted above, that there is a "redundancy" of one variable, or that the maximum "possesses 1df." This is because the apparently two-variables system can in fact be expressed in terms of a single fundamental "compound" variable v. The physical and biological sciences abound with examples where, over suitable ranges of the variables, relationships exist similar to those given above.

Frequently the surface which describes the system, while not of the form shown in Figure 10.4, nevertheless contains a marked component of this type together with an additional component resulting in an elongated maximum like that shown in Figure 10.3, or a system like that shown in Figure 10.5, where the ridge is steadily rising to higher yields. In the latter "rising ridge" case, the best practical combination of levels to use may be the most extreme point on the ridge that can be attained by the experimenter. Thus, in Figure 10.5, the best conditions could be those using the highest possible temperature (with the appropriate residence time for that temperature).

---

*Figure 10.7b was constructed by drawing contour lines through those points giving a constant product $\xi_1 \xi_2$, the appropriate yield being read off from Figure 10.7a.

The examples of Figures 10.3 to 10.6 may be said to show *factor dependence** in the sense that the response function for one factor is not independent of the levels of the other factors. The investigation of factor dependence is of considerable practical importance in response surface studies, for the following reasons.

1. *Alternative optima.* Where the surface is like that in Figure 10.4, not one , but a whole range of alternative optimum processes, corresponding to points along the crest of the ridge, will be available from which to choose. In practice, some of these processes may be far less costly, or far more convenient to operate, than others. The factors in this system may be said to be compensating in the sense that departure from the maximum response due to change in one variable can be compensated by a suitable change in another variable. The direction of the ridge indicates how much one factor must be changed to compensate for a given change in the other.

2. *Optimizing, or improving a second response.* If we imagine the contours for some auxiliary response η_2 superimposed on those for the major response η_1 we see that (provided the contours of the two systems are not exactly parallel) we could select, for our optimum process, that point near the crest of the ridge for the major response that yielded the most satisfactory value of the auxiliary response.

3. *Directions of insensitivity.* When the surface is like that in Figure 10.3, the direction of attenuation of the surface indicates those directions in which departures can be made from the optimum process with only small losses in response.

4. *Yield improvement by simultaneous changes in several variables.* The detection of a rising ridge in the surface like that in Figure 10.5 supplies the knowledge that, if the variables are changed together in the direction of the axis of the ridge, then yield improvement is possible even though no improvement may be possible by changing any single variable.

---

*The idea of factor dependence is analogous to that of stochastic dependence, and diagrams like Figure 10.3 call to mind the familiar contour representation of a bivariate probability surface. Again, there exists the same analogy between the "fundamental variables" we have discussed, and the "factors" in factor analysis and principal component analysis.

These analogies are helpful, but it must be emphasized that the two types of dependence are quite distinct and care should be taken to differentiate between them. We are concerned *here* with the deterministic relationship between the mean response η and the levels ξ of a set of predictor variables that can be varied at our choice such as temperature, time, and so on. We do not need any ideas of probability to define *this* relationship. In the analogous stochastic situation, "probability" takes the place of the response, and random variables such as "test scores" take the place of the predictor variables such as temperature and time.

5. *Suggestion of mechanism or "natural law."* Of by no means least importance is the fact that discovery of factor dependence of a particular type may, in conjunction with the experimenter's theoretical knowledge, lead to a better understanding of the basic mechanism of the reaction. Thus, if experimentation with the variables ξ_1 and ξ_2 produced a surface like that of Figure 10.7*b* we might suspect that some more fundamental variable of the type $v = \xi_1 \xi_2$ existed. (In our frog's leg example, a biologist estimating muscle response might accidentally discover Ohm's law!)

More Than Two Predictor Variables

So far we have illustrated our discussion with examples in which there were only two predictor variables ξ_1 and ξ_2. The situations that can occur as we add more predictor variables become progressively more complicated. Figure 10.8 shows a contour representation for three predictor variables in the neighborhood of a point maximum (analogous to that in two dimensions shown in Figure 10.2). The contours enclose the maximum point like the skins of an onion enclose its center. Insensitivity of the response to changes in conditions when predictor variables were changed *together* in a certain way would correspond to attenuation of the contours in the direction of the compensatory changes. As an extreme form of such an attenuation, we could imagine a line maximum in the space (i.e., a line such that all sets of conditions on it gave the same maximum yield) surrounded by cylindrical

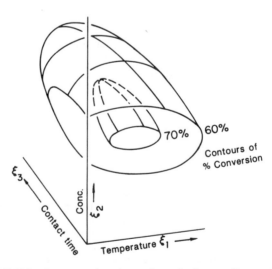

FIGURE 10.8. Representation of a maximum in three predictor variables.

contours of falling yield. We would call this a *line stationary ridge* because it is analogous to the two-dimensional case illustrated in Figure 10.4. Many other possibilities occur, but we shall leave their fuller discussion until later.

It should not be thought that factor dependence is to be expected only in connection with well-defined physical and chemical phenomena, nor that "single variable" redundancy is all that need concern us. Suppose, for example, the problem concerned the making of a cake, the object being to bring some desirable property such as "texture" (which is our response η and which, we suppose, can be measured on some suitable numerical scale) to the "best" level. The natural variables, k in number, whose levels we denote by $\xi_1, \xi_2, \ldots, \xi_k$ might be numerous and might involve, among other things, the amounts of such substances as baking powder (ξ_1), flour (ξ_2), egg white (ξ_3), and citric acid (ξ_4).

Suppose it had happened that the texture η depended in reality on only two fundamental variables, the consistency of the mix ω_1 and its acidity ω_2, and that optimum texture was attained whenever ω_1 and ω_2 were at their optimum levels ω_{10}, ω_{20}, say. For simplicity, we shall make the assumption that ω_1 and ω_2, measured in some suitable way, could be locally approximated by linear functions of the amounts $(\xi_1, \xi_2, \ldots, \xi_k)$ of the various ingredients, that is,

$$\text{consistency} \quad \omega_1 = a_1 + b_1\xi_1 + c_1\xi_2 + d_1\xi_3 + \cdots + p_1\xi_k \quad (10.2.1)$$

$$\text{acidity} \quad \omega_2 = a_2 + b_2\xi_1 + c_2\xi_2 + d_2\xi_3 + \cdots + p_2\xi_k. \quad (10.2.2)$$

Each of the coefficients $b_1, c_1, \ldots, p_1$ in the first equation would measure the change in consistency due to adding a unit amount of the corresponding ingredient. Thus the coefficient might be positive for solid substances like flour, and negative for liquid substances like water. The optimum consistency would thus be obtained on some $k - 1$ dimensional planar surface in the k-dimensional space of the factors defined by $\omega = \omega_{10}$, that is, such that

$$a_1 + b_1\xi_1 + c_1\xi_2 + d_1\xi_3 + \cdots + p_1\xi_k = \omega_{10}. \quad (10.2.3)$$

Similarly, each of the coefficients $b_2, c_2, \ldots, p_2$ in the second equation would measure the change in acidity due to the addition of a unit of the corresponding ingredient, and would be positive for acid substances and negative for alkaline substances. The optimum acidity would thus be obtained on some second $k - 1$ dimensional planar surface given by $\omega_2 = \omega_{20}$, that is, such that

$$a_2 + b_2\xi_1 + c_2\xi_2 + d_2\xi_3 + \cdots + p_2\xi_k = \omega_{20}. \quad (10.2.4)$$

The intersections of the two planes would give all the ξ levels for which both acidity and consistency were at their best levels, and hence for which optimum texture was attained. Optimum texture would thus be attained on a subspace of $k - 2$ dimensions. That is to say, there would be $k - 2$ directions at right angles to each other in which we could move while still maintaining optimum texture for the cake. We can express the k variable system in terms of two fundamental "compound" variables ω_1 and ω_2; there is thus a redundancy of $k - 2$ variables corresponding to the $k - 2$ dimensions in which the maximum is attained. Thus we could say that the maximum "had $k - 2$ df."

The surface for this example would again be a generalization of the stationary ridge of Figure 10.4. Rising ridges, attenuated maxima, and combinations of these phenomena, all have multidimensional generalizations to which we return later.

10.3. ELUCIDATION OF STATIONARY REGIONS, MAXIMA, AND MINIMA BY CANONICAL ANALYSIS

Let us suppose that a representationally adequate second degree equation has been fitted* in a near stationary region in which there appears to be some kind of maximum, so that, near the center of the last experimental design, the gradients, as measured by the b_j, $j = 1, 2, \ldots, k$, are not large compared with second-order effects. It is now important to determine the nature of the local surface which we tentatively believe contains a maximum. Some necessary questions are: (a) Is it, in fact, a maximum? (b) If yes, is this maximum most helpfully approximated by a point, a line, or a space (or, approximately speaking, are there 0 , 1, 2 or more, df in the maximum)? (c) How can we define the point, line, or space, on which a maximal response is obtained? *Canonical analysis* of the fitted second degree equation can illuminate these questions.

Canonical Analysis

Canonical analysis is a method of rewriting a fitted second degree equation in a form in which it can be more readily understood. This is achieved by a rotation of axes which removes all cross-product terms. We call this simplification the *A canonical form*. If desired, this may be accompanied by a

---

*Typically, this would be after suitable preliminary experimentation involving (perhaps) steepest ascent to come close to a region of interest, and (possibly) theoretical and empirical knowledge to arrive at suitable transformations of the response and/or the predictor variables.

change of origin to remove first-order terms as well. We call the result the *B canonical form*.

We first give a brief technical description of the analysis. This is then explained and illustrated.

The *A* Canonical Form

Consider a fitted second degree model

$$\hat{y} = b_0 + \sum_{j=1}^{k} b_j x_j + \sum\sum_{i \geq j} b_{ij} x_i x_j. \qquad (10.3.1)$$

If we write*

$$\mathbf{x} = \begin{bmatrix} x_1 \\ x_2 \\ \cdots \\ x_k \end{bmatrix}; \quad \mathbf{b} = \begin{bmatrix} b_1 \\ b_2 \\ \cdots \\ b_k \end{bmatrix}; \quad \mathbf{B} = \begin{bmatrix} b_{11} & \frac{1}{2}b_{12} & \cdots & \frac{1}{2}b_{1k} \\ \frac{1}{2}b_{12} & b_{22} & \cdots & \frac{1}{2}b_{2k} \\ \cdots & & & \\ \frac{1}{2}b_{1k} & \frac{1}{2}b_{2k} & \cdots & b_{kk} \end{bmatrix}$$

$$(10.3.2)$$

the fitted equation is

$$\hat{y} = b_0 + \mathbf{x'b} + \mathbf{x'Bx}. \qquad (10.3.3)$$

Let $\lambda_1, \lambda_2, \ldots, \lambda_i, \ldots, \lambda_k$ be the eigenvalues[†] of the symmetric matrix $\mathbf{B}$, and $\mathbf{m}_1, \mathbf{m}_2, \ldots, \mathbf{m}_i, \ldots, \mathbf{m}_k$ the corresponding eigenvectors[‡] so that, by definition,

$$\mathbf{Bm}_i = \mathbf{m}_i \lambda_i, \qquad i = 1, 2, \ldots, k. \qquad (10.3.4)$$

If we standardize each eigenvector so that $\mathbf{m}'_i \mathbf{m}_i = 1$ and if the $k \times k$ matrix $\mathbf{M}$ has $\mathbf{m}_i$ for its ith column, then $\mathbf{M}$ is an orthonormal matrix and the k equations (10.3.4) may be written simultaneously as

$$\mathbf{BM} = \mathbf{M\Lambda} \qquad (10.3.5)$$

---

*Note that, in this discussion, we shall write **b** for a vector of first-order estimated parameters. However, in Chapter 3, **b** was used to denote the full vector of estimates. The reader must be careful not to confuse the two notations.
[†]Also called characteristic roots, or latent roots.
[‡]Also called characteristic vectors, or latent vectors.

where Λ is a diagonal matrix whose ith diagonal element is λ_i. Premultiplying by $\mathbf{M}'(=\mathbf{M}^{-1})$ gives

$$\mathbf{M}'\mathbf{BM} = \Lambda. \tag{10.3.6}$$

Now by making use of the fact that $\mathbf{MM}' = \mathbf{I}$, we can write Eq. (10.3.3) as

$$\hat{y} = b_0 + (\mathbf{x}'\mathbf{M})(\mathbf{M}'\mathbf{b}) + (\mathbf{x}'\mathbf{M})\mathbf{M}'\mathbf{BM}(\mathbf{M}'\mathbf{x}). \tag{10.3.7}$$

If we now write $\mathbf{X} = \mathbf{M}'\mathbf{x}$ and $\boldsymbol{\theta} = \mathbf{M}'\mathbf{b}$, (or, equivalently, $\mathbf{x} = \mathbf{MX}$ and $\mathbf{b} = \mathbf{M\theta}$) this can be rewritten as

$$\hat{y} = b_0 + \mathbf{X}'\boldsymbol{\theta} + \mathbf{X}'\Lambda\mathbf{X}, \tag{10.3.8}$$

that is, as

$$\hat{y} = b_0 + \theta_1 X_1 + \cdots + \theta_k X_k + \lambda_1 X_1^2 + \cdots + \lambda_k X_k^2. \tag{10.3.9}$$

This constitutes the A canonical form which eliminates cross-product terms by rotating the axes. By differentiating (10.3.9) with respect to $X_1, \ldots, X_k$ we find that the coordinates of the stationary point $(X_{1S}, \ldots, X_{kS})$ in relation to the rotated axes are such that

$$X_{iS} = \frac{-\theta_i}{2\lambda_i}. \tag{10.3.10}$$

The A canonical form has a useful story to tell. The sizes and signs of the λ_i determine the type of second-order surface we have fitted. The θ_i measure the slopes of the surface at the original origin $\mathbf{x} = \mathbf{0}$, in the directions of the (rotated) coordinate axes $X_1, \ldots, X_k$. The values of the X_{iS}'s tell us how far along the various canonical axes is the stationary point S. We illustrate this with a simple constructed example for $k = 2$ variables.

An Example of Canonical Analysis for $k = 2$ Variables

Suppose the data obtained from a design indicated by dots in Figure 10.9 had led to the fitted second degree equation

$$\hat{y} = 78.8988 + 2.272x_1 + 3.496x_2 - 2.08x_1^2 - 2.92x_2^2 - 2.88x_1x_2,$$

$$\tag{10.3.11}$$

and that the usual checks indicated that this equation was representationally

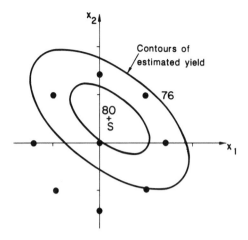

FIGURE 10.9. Contours $\hat{y} = 76$, and 79 of the fitted equation

$$\hat{y} = 78.8988 + 2.272x_1 + 3.496x_2$$

$$-2.08x_1^2 - 2.92x_2^2 - 2.88x_1x_2.$$

adequate. The elliptical contours plotted in Figure 10.9 show that the fitted equation describes a surface with a maximum at a stationary point S which is in the neighborhood of the center of the design. Furthermore, we see that the maximal region is somewhat attenuated in the NW–SE direction. Canonical analysis of the equation makes it possible to deduce this information without graphical representation. For this example,

$$\mathbf{x} = \begin{bmatrix} x_1 \\ x_2 \end{bmatrix}; \quad \mathbf{b} = \begin{bmatrix} 2.272 \\ 3.496 \end{bmatrix}; \quad \mathbf{B} = \begin{bmatrix} -2.08 & -1.44 \\ -1.44 & -2.92 \end{bmatrix}. \quad (10.3.12)$$

The fitted equation in matrix form is therefore

$$\hat{y} = 78.8988 + [x_1, x_2]\begin{bmatrix} 2.272 \\ 3.496 \end{bmatrix} + [x_1, x_2]\begin{bmatrix} -2.08 & -1.44 \\ -1.44 & -2.92 \end{bmatrix}\begin{bmatrix} x_1 \\ x_2 \end{bmatrix}.$$

$$(10.3.13)$$

The eigenvalues of $\mathbf{B}$ and the corresponding orthonormal eigenvectors are

$$\lambda_1 = -4, \quad \lambda_2 = -1, \quad (10.3.14)$$

$$\mathbf{m}_1 = \begin{bmatrix} 0.6 \\ 0.8 \end{bmatrix}, \quad \mathbf{m}_2 = \begin{bmatrix} -0.8 \\ 0.6 \end{bmatrix}. \quad (10.3.15)$$

Thus

$$\mathbf{M} = \begin{bmatrix} 0.6 & -0.8 \\ 0.8 & 0.6 \end{bmatrix}, \qquad \Lambda = \begin{bmatrix} -4 & 0 \\ 0 & -1 \end{bmatrix}, \qquad (10.3.16)$$

and

$$\boldsymbol{\theta} = \mathbf{M}'\mathbf{b} = \begin{bmatrix} 0.6 & 0.8 \\ -0.8 & 0.6 \end{bmatrix} \begin{bmatrix} 2.272 \\ 3.496 \end{bmatrix} = \begin{bmatrix} 4.16 \\ 0.28 \end{bmatrix}, \qquad (10.3.17)$$

whereupon

$$X_{1S} = \frac{4.16}{(2 \times 4)} = 0.52,$$

$$X_{2S} = \frac{0.28}{(2 \times 1)} = 0.14. \qquad (10.3.18)$$

The A canonical form is thus

$$\hat{y} = 78.8988 + 4.16X_1 + 0.28X_2 - 4X_1^2 - 1X_2^2, \qquad (10.3.19)$$

where

$$X_1 = 0.6x_1 + 0.8x_2,$$

$$X_2 = -0.8x_1 + 0.6x_2. \qquad (10.3.20)$$

As illustrated in Figure 10.10, the A canonical reduction refers the equation to the rotated axes (X_1, X_2) which are chosen to lie parallel to the principal axes of the system.

Important quantities to consider in this canonical form are:

1. The signs and relative magnitudes of the eigenvalues λ_i.
2. The size of the coordinates X_{iS} of the stationary point.
3. The nature of the transformation $\mathbf{X} = \mathbf{M}'\mathbf{x}$.

We now discuss these points in detail for our example, in which $k = 2$.

1. For our example, the eigenvalues, which are the coefficients of the quadratic terms X_1^2 and X_2^2 are $\lambda_1 = -4$, $\lambda_2 = -1$. (a) We know, because both signs are negative, that the curvature is negative in both canonical directions and so we must be dealing with a maximum. If λ_1 and λ_2 had both been positive, we would have a minimum, and if they were of different

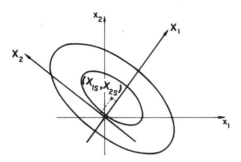

FIGURE 10.10. A rotation of axes produces canonical form A with fitted equation

$$\hat{y} = 78.8988 + 4.16X_1 + 0.28X_2 - 4X_1^2 - X_2^2.$$

signs, a saddle point. (b) The lengths of the principal axes of the ellipses are proportional to $|\lambda_1|^{-\frac{1}{2}}$ and $|\lambda_2|^{-\frac{1}{2}}$, respectively. Thus, in our example, we know that the fitted surface is somewhat attenuated along the X_2 axis.

2. In the agreed scaling of the response function, distance is preserved by the transformation $\mathbf{X} = \mathbf{M}'\mathbf{x}$, so that we know, from the small magnitudes of X_{1S} and X_{2S}, that the center of the system is close to the center of the design.

3. The transformation $\mathbf{X} = \mathbf{M}'\mathbf{x}$, or equivalently, $\mathbf{x} = \mathbf{M}\mathbf{X}$ allows us to determine the directions of the canonical axes. For the example, the coordinate X_2 along which the contours are attenuated is defined by $X_2 = -0.8x_1 + 0.6x_2$. The perpendicular axis shown is of course $X_1 = 0.6x_1 + 0.8x_2$.

The B Canonical Form

By equating to zero derivatives with respect to $\mathbf{x}$ in (10.3.3) and $\mathbf{X}$ in (10.3.8) we find the coordinates of the stationary point S to be the solution of the equations

$$-2\mathbf{B}\mathbf{x}_S = \mathbf{b}, \quad \text{or of} \quad -2\mathbf{\Lambda}\mathbf{X}_S = \mathbf{\theta}, \quad (10.3.21)$$

and the fitted response at this point is

$$\hat{y}_S = b_0 + \tfrac{1}{2}\mathbf{x}_S'\mathbf{b} \quad \text{or, equally,} \quad \hat{y}_S = b_0 + \tfrac{1}{2}\mathbf{X}_S'\mathbf{\theta}. \quad (10.3.22)$$

We now adopt the notation $\tilde{x}_i = x_i - x_{iS}$, $\tilde{X}_i = X_i - X_{iS}$, and the corresponding vector forms $\tilde{\mathbf{x}}$ and $\tilde{\mathbf{X}}$, so that the $\tilde{x}_i$'s and $\tilde{X}_i$'s refer to coordi-

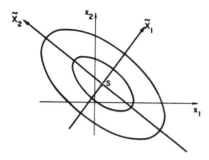

FIGURE 10.11. A rotation of axes and a shift of origin produces canonical form B with fitted equation

$$\hat{y} = 80 - 4\tilde{X}_1^2 - \tilde{X}_2^2.$$

nates measured from the stationary point S. Then, if in addition to rotating the axes, we also shift the origin to the stationary point S, we obtain the form

$$\hat{y} = \hat{y}_S + \tilde{\mathbf{X}}'\mathbf{\Lambda}\tilde{\mathbf{X}}, \qquad (10.3.23)$$

that is

$$\hat{y} = \hat{y}_S + \lambda_1 \tilde{X}_1^2 + \cdots + \lambda_k \tilde{X}_k^2. \qquad (10.3.24)$$

We call this the B canonical form; see Figure 10.11.

For our constructed example with $k = 2$, the equations $-2\mathbf{B}\mathbf{x}_S = \mathbf{b}$ become

$$2\begin{bmatrix} 2.08 & 1.44 \\ 1.44 & 2.92 \end{bmatrix}\begin{bmatrix} x_{1S} \\ x_{2S} \end{bmatrix} = \begin{bmatrix} 2.272 \\ 3.496 \end{bmatrix} \qquad (10.3.25)$$

and these have the solution $x_{1S} = 0.2$, $x_{2S} = 0.5$. Also, the estimated response at S, given by the formula $\hat{y}_S = b_0 + \frac{1}{2}\mathbf{x}_S'\mathbf{b}$ is

$$\hat{y}_S = 78.8988 + \frac{1}{2}(0.2, 0.5)\begin{bmatrix} 2.272 \\ 3.496 \end{bmatrix} = 80.0. \qquad (10.3.26)$$

Thus, finally, moving the origin to S, we express the canonical form $\hat{y} = \hat{y}_S + \tilde{\mathbf{X}}'\mathbf{\Lambda}\tilde{\mathbf{X}}$ as

$$\hat{y} = 80 - 4\tilde{X}_1^2 - 1\tilde{X}_2^2 \qquad (10.3.27)$$

with $\tilde{\mathbf{X}} = \mathbf{M}'\tilde{\mathbf{x}} = \mathbf{M}'(\mathbf{x} - \mathbf{x}_S)$, namely,

$$\tilde{X}_1 = 0.6(x_1 - 0.2) + 0.8(x_2 - 0.5),$$

$$\tilde{X}_2 = -0.8(x_1 - 0.2) + 0.6(x_2 - 0.5). \qquad (10.3.28)$$

As we see from Figure 10.11, the B canonical form refers the fitted equation to new axes ($\tilde{X}_1$, $\tilde{X}_2$) which are rotated as before but, in addition, a shift of origin to the stationary point S is made.

Application of Canonical Forms A and B

For two or three predictor variables ($k = 2$ or 3) it is fairly easy to appreciate the nature of various second degree systems, but for $k > 3$ it becomes more difficult. Canonical analysis makes it possible to understand the system even when k is greater than 3.

In design units, the distance of the center of the system (the stationary point) S from the design center O is given by

$$D = OS = \left\{ \sum_{i=1}^{k} x_{iS}^2 \right\}^{1/2} = \left\{ \sum_{j=1}^{k} X_{jS}^2 \right\}^{1/2} \tag{10.3.29}$$

When S is close to O (say, D not much greater than 1), the simpler B canonical form is adequate for understanding the fitted surface. However, particularly with ridge systems, the center S of the fitted system may be remote from the center O of the design. In this case, the analysis is better carried out in terms of the A canonical form. We first illustrate canonical analysis for the simple system fitted to the elasticity data (see Sections 9.2 and 9.3) using the B canonical form. In Chapter 11, we consider more complicated examples which require use of the A canonical form.

EXAMPLE 2.   *Canonical Analysis for the Polymer Elasticity Data*   We now re-examine a three variable fitted surface previously discussed in Sections 9.2 and 9.3. It will be recalled that, in that example, the response y, measured elasticity, was considered as a function of three predictor variables, percentage concentrations C_1 and C_2, and reaction temperature T. The variables were coded as

$$x_1 = \frac{(C_1 - 18)}{3}, \qquad x_2 = \frac{(C_2 - 2.7)}{0.4}, \qquad x_3 = \frac{(T - 145)}{10}. \tag{10.3.30}$$

The fitted second degree equation in a near stationary region was found to be

$$\hat{y} = 57.31 + 1.50x_1 - 2.13x_2 + 1.81x_3$$
$$- 4.69x_1^2 - 6.27x_2^2 - 5.21x_3^2 - 7.13x_1x_2 - 3.27x_1x_3 - 2.73x_2x_3.$$

$$\tag{10.3.31}$$

The fitted surface was studied in Chapter 9 by plotting contours of $\hat{y}$ (Figure 9.2) in the space of x_1 and x_2 for various fixed values of x_3. The contour plots showed that the stationary point S was at a maximum of the fitted surface. We now look at the corresponding canonical analysis.

Canonical Analysis

We have

$$b_0 = 57.31, \quad \mathbf{b} = \begin{bmatrix} 1.50 \\ -2.13 \\ 1.81 \end{bmatrix}, \quad \mathbf{B} = -\begin{bmatrix} 4.690 & 3.565 & 1.635 \\ 3.565 & 6.270 & 1.365 \\ 1.635 & 1.365 & 5.210 \end{bmatrix}.$$

$$(10.3.32)$$

We thus obtain the coordinates of the stationary point

$$\mathbf{x}_S = -\tfrac{1}{2}\mathbf{B}^{-1}\mathbf{b} = (0.4603, -0.4645, 0.1509)', \quad (10.3.33)$$

at which point the estimated response is

$$\hat{y}_S = b_0 + \tfrac{1}{2}\mathbf{x}_S'\mathbf{b} = 58.29. \quad (10.3.34)$$

The distance from the stationary point S to the center of the design is thus

$$D = \{\mathbf{x}_S'\mathbf{x}_S\}^{1/2} = \left\{ (0.4603)^2 + (-0.4645)^2 + (0.1509)^2 \right\}^{1/2} = 0.6711. \quad (10.3.35)$$

Thus the point S is close to the center of the experimental region and so we shall proceed directly to the B canonical form. The eigenvalues of $\mathbf{B}$ in order of absolute magnitude, and their associated orthonormal eigenvectors are displayed below.

λ_j	-10.03	-4.37	-1.77
Corresponding	0.5899	0.1312	0.7967
$\mathbf{m}_j$	0.7024	0.4034	-0.5865
vector	0.3983	-0.9056	-0.1458

The 3 by 3 lower right portion of this display comprises the matrix $\mathbf{M}$ featured in Eq. (10.3.5) and elsewhere. We obtain $\mathbf{M}^{-1} = \mathbf{M}'$ by transposing **M**. This, by use of the fact that $\tilde{x} = x - x_S$, leads to the equations

$$\tilde{X}_1 = 0.5899x_1 + 0.7024x_2 + 0.3983x_3 - 0.00537,$$

$$\tilde{X}_2 = 0.1312x_1 + 0.4034x_2 - 0.9056x_3 + 0.2636,$$

$$\tilde{X}_3 = 0.7967x_1 - 0.5865x_2 - 0.1458x_3 - 0.6171. \qquad (10.3.36)$$

The *B* canonical form is

$$\hat{y} = 58.29 - 10.03\tilde{X}_1^2 - 4.37\tilde{X}_2^2 - 1.77\tilde{X}_3^2. \qquad (10.3.37)$$

This canonical analysis tells us the following important facts, without reference to contour plots:

1. Equation (10.3.35) shows that the center of the fitted system lies close to the center $(0, 0, 0)$ of the experimental region.

2. At the center, the fitted response is $\hat{y} = 58.29$.

3. The negative sign attached to each of the eigenvalues tells us that the fitted response value 58.29 at the center of the fitted surface is a maximum.

4. Because (a) we established in Chapter 9 that the fitted quadratic surface seems to provide a reasonably good explanation of the variation in the data and (b) the appropriate standard errors for the canonical coefficients are roughly of the same size as those of the quadratic coefficients (± 0.5), it seems likely that the *underlying true surface* also contains a maximum. (A more accurate picture of what is known of the location of the true maximum at this stage of experimentation can be gained by the confidence region calculation described in Box and Hunter, 1954.)

5. The elements of the eigenvectors, which are also the coefficients in Equations (10.3.36) can provide important information about the relationships of the predictors to the response. This information may be put to purely empirical use, or may possibly aid theoretical conjectures. For example, the coefficients of the eigenvector associated with the smallest (in absolute value) eigenvalue $\lambda_3 = -1.77$ are

$$(0.7967, -0.5865, -0.1458).$$

This suggests that, in the $\tilde{X}_3$ direction, which is the direction of greatest elongation of the ellipsoid, the concentrations C_1 and C_2 are, to some extent, compensatory. A reduction in one concentration can be, to some extent, compensated by a corresponding increase in the other.

The last point brings us to the most important practical use of canonical analysis, the detection, exploration, and exploitation of ridge systems. We discuss this further in Chapter 11.

APPENDIX 10A.   A SIMPLE EXPLANATION OF CANONICAL ANALYSIS

We now provide a simple geometrical explanation of the mathematics of canonical analysis. We illustrate for $k = 3$ dimensions, and give a numerical example for $k = 2$. Suppose we have a fitted second-order equation

$$\hat{y} = b_0 + \mathbf{x'b} + \mathbf{x'Bx}, \tag{10A.1}$$

with a stationary point at $\mathbf{x}_S$. We wish to make a rotation to a coordinate system whose axes lie along the principal axes of the ellipsoidal contours. To do this, we imagine a sphere *of radius unity* centered at S. (Actually any radius choice will serve, but we might as well take it as unity.) The equation of such a sphere is

$$\tilde{\mathbf{x}}'\tilde{\mathbf{x}} = 1, \tag{10A.2}$$

or equivalently,

$$(\mathbf{x} - \mathbf{x}_S)'(\mathbf{x} - \mathbf{x}_S) = 1. \tag{10A.3}$$

Now consider the response $\hat{y}$ *on the surface of this sphere.* (See Figure 10A.1). In general the response will be stationary at $2k$ points $P_1, P_1'; \ P_2, P_2'; \ldots; P_k, P_k'$, and the straight lines $P_i S P_i'$, for $i = 1, 2, \ldots, k$, will lie along the k principal axes of the second degree system. We adopt the lines $P_i S P_i'$ as our new coordinate axes. S, of course, is the new center.

The $2k$ stationary points P_i, P_i' may be obtained by using Lagrange's method of undetermined multipliers. Our objective function to be differentiated is

$$F(\mathbf{x}, \lambda) = b_0 + \mathbf{x'b} + \mathbf{x'Bx} - \lambda\{(\mathbf{x} - \mathbf{x}_S)'(\mathbf{x} - \mathbf{x}_S) - 1\}. \tag{10A.4}$$

Equating to zero the derivatives with respect to $\mathbf{x}'$ provides the k simultaneous equations

$$\mathbf{b} + 2\mathbf{Bx} - 2\lambda(\mathbf{x} - \mathbf{x}_S) = \mathbf{0}. \tag{10A.5}$$

Substitution for $\mathbf{b}$ from (10.3.21) provides

$$\mathbf{B}(\mathbf{x} - \mathbf{x}_S) - \lambda(\mathbf{x} - \mathbf{x}_S) = 0 \tag{10A.6}$$

or,

$$(\mathbf{B} - \lambda\mathbf{I})\tilde{\mathbf{x}} = 0, \tag{10A.7}$$

where

$$\tilde{\mathbf{x}} = \mathbf{x} - \mathbf{x}_S. \tag{10A.8}$$

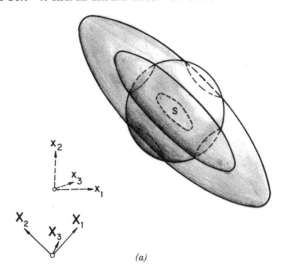

(a)

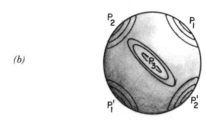

(b)

FIGURE 10A.1. (*a*) $k = 3$. A sphere centered at S cuts through contours of the fitted second degree response system. (*b*) A birds-eye view from the $\tilde{X}_3$ axis of the response contours on the sphere, showing the stationary points.

For this set of homogeneous equations to have a solution, the determinantal equation

$$|\mathbf{B} - \lambda \mathbf{I}| = 0 \qquad (10A.9)$$

must hold. The solutions λ_i, $i = 1, 2, \ldots, k$ of this determinantal equation are called the eigenvalues of the matrix $\mathbf{B}$. We need consider here only the case where there are k distinct eigenvalues $\lambda_1, \lambda_2, \ldots, \lambda_k$. Theoretical complications which we consider later, can occur when some of the λ_i are equal to zero or when certain λ_i are equal to one another. However, since we are dealing with experimental data subject to error, neither of these circumstances will ever occur exactly in practice.

TABLE 10A.1.   Coordinates of points in original and rotated systems

| Point | Coordinates in the $\tilde{x}$ system | | | | Coordinates in the $\tilde{X}$ system | | | |
	$\tilde{x}_1$	$\tilde{x}_2$	$\cdots$	$\tilde{x}_k$	$\tilde{X}_1$	$\tilde{X}_2$	$\cdots$	$\tilde{X}_k$
P_1	m_{11}	m_{12}	$\cdots$	m_{1k}	1	0	$\cdots$	0
P_2	m_{21}	m_{22}	$\cdots$	m_{2k}	0	1	$\cdots$	0
$\cdots$	$\cdots$				$\cdots$			
P_k	m_{k1}	m_{k2}	$\cdots$	m_{kk}	0	0	$\cdots$	1
S	0	0	$\cdots$	0	0	0	$\cdots$	0
P_1'	$-m_{11}$	$-m_{12}$	$\cdots$	$-m_{1k}$	-1	0	$\cdots$	0
P_2'	$-m_{21}$	$-m_{22}$	$\cdots$	$-m_{2k}$	0	-1	$\cdots$	0
$\cdots$	$\cdots$							
P_k'	$-m_{k1}$	$-m_{k2}$	$\cdots$	$-m_{kk}$	0	0	$\cdots$	-1

Now for each λ_i we solve the set of equations

$$(\mathbf{B} - \lambda_i\mathbf{I})\tilde{\mathbf{x}} = 0 \tag{10A.10}$$

which yields the solution $\tilde{\mathbf{x}} = \mathbf{m}_i$. [These equations are dependent. To solve them we set one element of $\tilde{\mathbf{x}}$ to any nonzero value, e.g., 1, or some other convenient value, solve any $k - 1$ of the k equations (10A.10), and then normalize this intermediate solution so that the sum of squares of the elements of the solution is 1. The result is $\mathbf{m}_i$ and $\mathbf{m}_i'\mathbf{m}_i = 1$.]

If we take the coordinates of the points $P_1, P_2, \ldots, P_k$ to be, respectively,

$$\tilde{\mathbf{x}} = \mathbf{m}_1, \mathbf{m}_2, \ldots, \mathbf{m}_k \tag{10A.11}$$

then the coordinates of $P_1', P_2', \ldots, P_k'$ will be

$$\tilde{\mathbf{x}} = -\mathbf{m}_1, -\mathbf{m}_2, \ldots, -\mathbf{m}_k. \tag{10A.12}$$

If $SP_1, SP_2, \ldots, SP_k$ are designated as the coordinate axes of the new canonical variables, the correspondence in Table 10A.1 can be set out, where $\mathbf{m}_i = (m_{i1}, m_{i2}, \ldots, m_{ik})'$. The elements of the vector $\mathbf{m}_i = (m_{i1}, m_{i2}, \ldots, m_{ik})'$ are, in fact, the *direction cosines* of the $\tilde{X}_i$ axis in the $\tilde{x}$ space. Furthermore, if we construct a matrix $\mathbf{M}$ whose ith column is $\mathbf{m}_i$, the transformation from old ($\tilde{x}$) to new ($\tilde{X}$) coordinates is given by

$$\tilde{\mathbf{x}} = \mathbf{M}\tilde{\mathbf{X}}. \tag{10A.13}$$

Equivalently, for the respectively parallel coordinate systems x and X, measured from the natural origin, we have the transformation

$$\mathbf{x} = \mathbf{M}\mathbf{X}. \tag{10A.14}$$

Because the transformation is an orthogonal one and $\mathbf{m}'_i \mathbf{m}_j = 0$, $\mathbf{m}'_i \mathbf{m}_i = 1$, $i, j = 1, 2, \ldots, k$, it follows that $\mathbf{M}'\mathbf{M} = \mathbf{M}\mathbf{M}' = \mathbf{I}$, that is, $\mathbf{M}^{-1} = \mathbf{M}'$. Thus the inverse transformations are

$$\tilde{\mathbf{X}} = \mathbf{M}'\tilde{\mathbf{x}} \quad \text{and} \quad \mathbf{X} = \mathbf{M}'\mathbf{x}. \tag{10A.15}$$

From these forms, we see that the jth *row* of $\mathbf{M}$, with elements $m_{1j}, m_{2j}, \ldots, m_{kj}$ provides the direction cosines of the $\tilde{x}_j$ axis in the $\tilde{X}$ space (or of the x_j axis in the X space).

Efficient computer programs are available for the actual determination of eigenvalues and eigenvectors. However, to illustrate the above theory, we carry through the calculation for the constructed ($k = 2$) example of Eq. (10.3.11). Using Eq. (10A.9), we obtain the eigenvalues from the solution of the determinantal equation

$$\begin{vmatrix} -2.08 - \lambda & -1.44 \\ -1.44 & -2.92 - \lambda \end{vmatrix} = 0, \tag{10A.16}$$

namely $(-2.08 - \lambda)(-2.92 - \lambda) - (-1.44)^2 = 0$. This equation has two roots $\lambda_1 = -4$, $\lambda_2 = -1$, and these are the eigenvalues. The corresponding eigenvectors are

$$\mathbf{m}_1 = \begin{bmatrix} 0.6 \\ 0.8 \end{bmatrix}, \qquad \mathbf{m}_2 = \begin{bmatrix} -0.8 \\ 0.6 \end{bmatrix}. \tag{10A.17}$$

These are obtained by successively substituting in Eqs. (10A.7), namely in $(\mathbf{B} - \lambda \mathbf{I})\tilde{\mathbf{x}} = 0$. For example, putting $\lambda_1 = -4$, Eqs. (10A.7) are

$$(-2.08 + 4)\tilde{x}_1 - 1.44\tilde{x}_2 = 0$$

$$-1.44\tilde{x}_1 + (-2.92 + 4)\tilde{x}_2 = 0. \tag{10A.18}$$

These are two dependent equations and so their solution is not unique. If we set $\tilde{x}_1 = 1$, however, we find $\tilde{x}_2 = \frac{4}{3}$ from either equation. This solution $(1, \frac{4}{3})$ is now normalized (uniquely apart from sign) by making the sum of squares of the elements equal to 1 so that

$$\mathbf{m}_1 = \begin{bmatrix} 0.6 \\ 0.8 \end{bmatrix} \tag{10A.19}$$

as above. Similarly $\lambda_2 = -1$ leads to a solution $\mathbf{m}_2$ as above, whereupon

$$\mathbf{M} = \begin{bmatrix} 0.6 & -0.8 \\ 0.8 & 0.6 \end{bmatrix}. \tag{10A.20}$$

CHAPTER 11

Occurrence and Elucidation of Ridge Systems, II

11.1. RIDGE SYSTEMS APPROXIMATED BY LIMITING CASES

As we pointed out in Chapter 10, a maximum* in k variables is often most revealingly approximated not by a point but by a line, plane, or hyperplane. Furthermore, the detection, description, and exploitation of such ridge systems is one of the most important uses of response surface techniques.

If we can employ a second degree polynomial to represent the response in the neighborhood of interest, theoretical systems that approximate ridges occurring in practice can be thought of as limiting cases of the canonical forms considered earlier. We first consider the possibilities for $k = 2$ variables to illustrate the basic ideas.

Stationary ridge, $k = 2$

Consider the canonical form B for a fitted second degree equation

$$\hat{y} = \hat{y}_S + \lambda_1 \tilde{X}_1^2 + \lambda_2 \tilde{X}_2^2. \tag{11.1.1}$$

Suppose that λ_1 and λ_2 are both negative, corresponding to a surface like that in Figure 11.1a. Now suppose that λ_1 is kept fixed, while λ_2 is made smaller and smaller in absolute value. Then the contours of the resulting system become more and more drawn out along the $\tilde{X}_2$ axis until, when λ_2 reaches zero, the stationary ridge system of Figure 11.1c is obtained. If we now allow λ_2 to *pass through* zero, and to become positive, we shall have an attenuated minimax as in Figure 11.1b. Thus Figure 11.1c can equally well be regarded as a limiting case of Figure 11.1b with the positive λ_2 tending

*As before, we shall discuss only maxima but parallel considerations apply to minima, also.

346

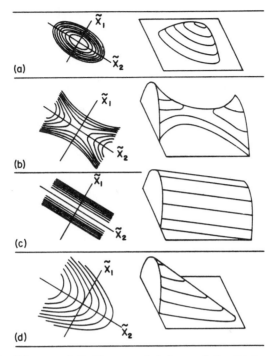

FIGURE 11.1. Primary and limiting canonical forms of the fitted quadratic equation $\hat{y} = b_0 + b_1 x_1 + b_2 x_2 + b_{11} x_1^2 + b_{22} x_2^2 + b_{12} x_1 x_2$.

to zero. In practice, a system that is *exactly* of the stationary ridge form is unlikely; experimental error and mild lack of fit can blur the picture somewhat. Thus, for an estimated surface, we would not expect to find a coefficient λ_2 *exactly* zero, rather we might find such a coefficient that was small compared with λ_1 and of about the same size as its standard error. This would imply an attenuation of the surface in the $\tilde{X}_2$ direction. Such an occurrence puts the experimenter on notice that it would be wise to explore the possibility that locally, and approximately, there is a line maximum rather than a point maximum.

Rising Ridge, $k = 2$

A further practical possibility is approximated by a canonical representation of the form

$$\hat{y} = \hat{y}_S + \lambda_1 \tilde{X}_1^2 + \theta_2 \tilde{X}_2 \tag{11.1.2}$$

corresponding to the rising ridge of Figure 11.1d. The system is similar to that of Figure 11.1c but the crest of the ridge is tipped and has a gradient θ_2 in the $\tilde{X}_2$ direction instead of being level. The contours are parabolas. The system can be imagined as a limiting case either of Figure 11.1a or of Figure 11.1b. It occurs if λ_1 is kept fixed and λ_2 is allowed to become vanishingly small *at the same time as the center of the system is moved to infinity*. In practice, the limiting forms corresponding to Figures 11.1c and 11.1d do not occur exactly, as we have already mentioned; however these forms may be regarded as reference marks. When we say that an empirical surface is like a limiting case, we mean that it is best thought of as approximated by that case. Approximation by a limiting form of some kind is suggested whenever one of the canonical coefficients is close to zero and small in absolute

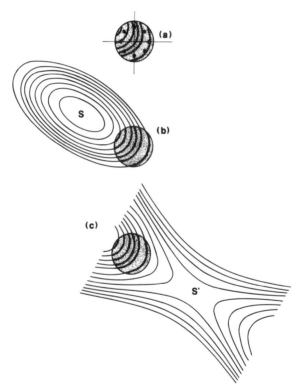

FIGURE 11.2. When data are taken at experimental points on a true surface such as (a), the fitted second degree equation may be part of a system of ellipses centered at S [see (b)] or of a system of hyperbolas centered at S' [see (c)]. In either case, an understanding of what is happening in the (shaded) region of interest is best gained by using canonical form A.

magnitude compared with the other. When this happens, it may also happen that the stationary point for the fitted surface will be remote from the center of the design. The reason for this is that, when the system contains a ridge, steepest ascent will bring the experimental region *close to the ridge*, but the system that represents that ridge may have a remote center.

To see how this can happen, consider Figure 11.2. Suppose that, in the immediate neighborhood of the current experimental region, indicated by shading in Figure 11.2a, the true surface is the rising ridge indicated by the contours. The second degree equation that is being fitted is, as we have seen, capable of taking on a number of different shapes. Our method of surface fitting will choose that system which most closely represents (in the least-squares sense) the experimental data. It might do this by using part of a system of elliptical contours with a center at *S*, as shown in Figure 11.2b. Alternatively, it might use part of a system of hyperbolic contours from a minimax system with center at *S'* as indicated in Figure 11.2c. A small change in the pattern of experimental error could result in a switch from one of these forms to the other. Now obviously we are interested only in approximating the response in the immediate experimental neighborhood, that is, the shaded region in the figure, and either system can do this. By using canonical analysis in the *A* form, we will see how we can be led to approximately the same canonical form whichever of the apparently quite different fitted representations arise. (Also, see Figure 11.4.)

11.2. CANONICAL ANALYSIS IN THE *A* FORM TO CHARACTERIZE RIDGE PHENOMENA

In view of the considerations discussed above, when the center *S* of the fitted system is remote from the region of experimentation, it is most useful to present the reduced equation as canonical form *A*, namely,

$$\hat{y} = b_0 + \theta_1 X_1 + \theta_2 X_2 + \cdots + \theta_k X_k + \lambda_1 X_1^2 + \lambda_2 X_2^2 + \cdots + \lambda_k X_k^2.$$

$$(11.2.1)$$

For illustration, and further to illuminate the discussion that follows, we show in Figure 11.3 a number of possibilities for *theoretical* second-order surfaces in $k = 3$ dimensions. Table 11.1 shows the values of the θ's and λ's associated with each of these theoretical possibilities and also the characteristics to be expected of a corresponding surface *fitted to data subject to error*. (It is assumed that steepest ascent has been applied, so that a near stationary region has been attained.)

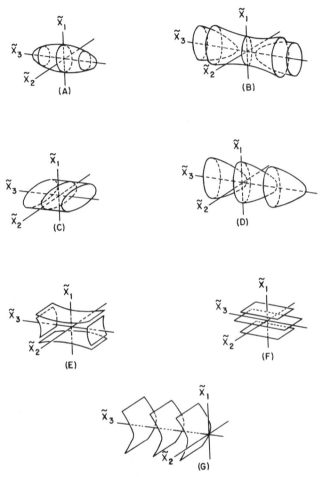

FIGURE 11.3. Contour systems for some second degree equations in three predictor variables, including some important limiting forms.

In practice, we may have to deal with complicated ridge systems in many dimensions. Certain quantities calculated from the canonical form A can be of great help in characterizing such systems. It is important to remember that, since we are dealing with data subject to error, we never have to deal with the limiting cases ($\lambda_i = 0$, $\theta_j = \infty$) exactly. What happens in practice is that, within the region of interest, the surface is approximated by a piece of some system in which some of the λ's are small and some of the θ's large.

The A canonical form of Eq. (11.2.1) uses the center of the design as its origin; we shall call this point the *design origin O*. Obviously, to identify

TABLE 11.1. Characteristics of the theoretical second degree equations whose contours are shown in Figure 11.3, and of the corresponding equations obtained from data subject to error, after steepest ascent has been applied

Labels of system in Figure 11.3	Values of $\theta_1\;\theta_2\;\theta_3$ / Signs^a of $\lambda_1\;\lambda_2\;\lambda_3$	Conditions implied by the diagram as drawn	$\tilde{X}_1\tilde{X}_2$	$\tilde{X}_1\tilde{X}_3$	$\tilde{X}_2\tilde{X}_3$	System center in relation to fitted center	Special features found in practice
	Characteristics of the theoretical canonical forms; in practice these characteristics are only approximately attained by the fitted surface →		*Type of contours when a cross-section is taken in the variables shown*^{b} (Figures 11.1a–d) →			*Characteristics of canonically reduced fitted surfaces approximating the theoretical forms* →	
(A)	$0\;\;-\;\;0$ / $-\;\;-\;\;-$	$\lvert\lambda_2\rvert > \lvert\lambda_1\rvert > \lvert\lambda_3\rvert$	(a)	(a)	(a)	Close	—
(B)	$0\;\;-\;\;0$ / $-\;\;-\;\;+$	$\lvert\lambda_2\rvert > \lvert\lambda_1\rvert > \lvert\lambda_3\rvert$	(a)	(b)	(b)	Close	When λ_3 is small, true surface may be like (C)
(C)	$0\;\;-\;\;0$ / $-\;\;-\;\;0$	Line of centers on the $\tilde{X}_3$ axis	(a)	(c)	(c)	Indeterminate	λ_3 will have a small value (negative or positive)
(D)	$0\;\;-\;\;+$ / $-\;\;-\;\;0$	Center at infinity on the $\tilde{X}_3$ axis	(a)	(d)	(d)	Remote	λ_3 will have a small value (negative or positive)
(E)	$0\;\;-\;\;0$ / $-\;\;-\;\;+$	Line of centers on the $\tilde{X}_2$ axis	(c)	(b)	(c)	Indeterminate	When λ_3 is small, true surface may be like (F)
(F)	$0\;\;-\;\;0$ / $-\;\;-\;\;0$	Plane of centers given by $\tilde{X}_1 = 0$	(c)	(c)	Constant response	Indeterminate	λ_2 and λ_3 will have small values (negative or positive)
(G)	$0\;\;-\;\;+$ / $-\;\;-\;\;0$	Center at infinity on the $\tilde{X}_3$ axis	(c)	(c)	Parallel straight lines	Remote	λ_2 and λ_3 will have small values (negative or positive)

^{a}A complete switch of all signs does not alter the contour forms, but reverses the directions in which response increases. "0" indicates that this value is exactly zero in the theoretical canonical representation. For practical cases, see the last column.

^{b}The heading $\tilde{X}_1\tilde{X}_2$ indicates the form of the contours in $\tilde{X}_1$ and $\tilde{X}_2$ when a cross-section ($\tilde{X}_3$ = constant) is taken.

stationary or rising ridges, it is important to know how close the ridge system is to the design origin. We can find this out as follows. As we have seen in Eq. (10.3.10) the coordinates of the center of the system are given by

$$X_{iS} = -\frac{\theta_i}{2\lambda_i}, \qquad i = 1, 2, \ldots, k. \tag{11.2.2}$$

These coordinates provide important clues to the nature of the system and should be examined together with the θ's and the λ's.

Distance to the Ridge

Consider now a theoretical system in k variables in which there is a p-dimensional ridge evidenced by p zero eigenvalues. Then let the coordinates of the point R on the ridge which is nearest to the design center O have $(k - p)$ coordinates $\{X_{iS}\}$, with the omitted p coordinates corresponding to the zero eigenvalues. The shortest distance OR to the ridge will thus be

$$OR = \left\{ \Sigma X_{iS}^2 \right\}^{1/2} \tag{11.2.3}$$

where the summation is taken only over those coordinates i corresponding to nonzero eigenvalues. For fitted surfaces, eigenvalues will not be exactly zero and a corresponding summation may be taken omitting subscripts whose eigenvalues are small.

For illustration, consider the situation of Figure 11.4. The rising ridge system depicted by the contours would be approximated either by a system of ellipsoids with contours like those of Figure 11.3A centered on a fairly remote point on the $\tilde{X}_3$ axis such as S, or by a system such as that in Figure 11.3B centered remotely at S'. Thus, the approximating system would have λ_1 and λ_2 both comparatively large and negative while λ_3 would be close to zero and with a negative [for case (A)] or positive [for case (B)] sign. In either case, the point R on the ridge closest to the design origin O has coordinates $(X_{1S}, X_{2S}, 0)$ and the distance OR is given by $OR = (X_{1S}^2 + X_{2S}^2 + 0^2)^{1/2}$.

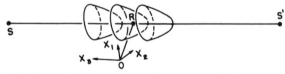

FIGURE 11.4. A rising ridge may be approximated by ellipsoids centered at S or by hyperboloids centered at S'.

Stationarity on the Ridge

For a suspected ridge system with p small eigenvalues $\lambda_{k-p-1}, \ldots, \lambda_k$, a point R having coordinates $(X_{1S}, \ldots, X_{k-p, S}, 0, \ldots, 0)$ will be chosen as being "closest to the ridge". If we now move a distance of one unit in either direction away from R along the ith of the p "small eigenvalue" axes to points R_{i1} and R_{i2}, we can write the three calculated responses in the form

$$
\begin{array}{ll}
\hat{y}_R & \text{at } R, \\
\hat{y}_R + \theta_i + \lambda_i & \text{at } R_{i1}, \\
\hat{y}_R - \theta_i + \lambda_i & \text{at } R_{i2}.
\end{array}
\tag{11.2.4}
$$

The average of these three responses is $\hat{y}_R + \frac{2}{3}\lambda_i$. The sample variance of these three responses is $\theta_i^2 + \frac{1}{3}\lambda_i^2$. Now, if we have a sample of three observations from a normal distribution, the standard deviation* of the distribution may be approximated by (sample range)/$3^{1/2}$, so that, for the three responses above, the sample range is approximated by

$$
r_i = \left\{ 3\theta_i^2 + \lambda_i^2 \right\}^{1/2}
\tag{11.2.5}
$$

and this will provide a measure of how much change occurs, in the calculated response, along "near zero eigenvalue" axes. [The reader may wonder why we use (11.2.5) instead of obtaining the actual range of the three responses. This is because we can then avoid working out each of the numbers (11.2.4) to see which is the smallest and largest for each i, in favor of a routine calculation r_i which is of adequate accuracy in practice.]

Summary of Results

It is now possible to summarize the results of the canonical analysis in a table as follows.

Slopes	θ_1	θ_2	$\cdots$	θ_k
Curvatures	λ_1	λ_2	$\cdots$	λ_k
Distances from O of system center	X_{1S}	X_{2S}	$\cdots$	X_{kS}
Changes in $\hat{y}$	r_1	r_2	$\cdots$	r_k

*For a sample of size two, the formula is (sample range)/$2^{1/2}$ and is *exact*, whether the distribution is normal or not. For a sample of size three, the rule is not very sensitive to the normality assumption.

Approximate Standard Errors for θ_i and λ_i

In the rotated axes coordinate system $(X_1, X_2, \ldots, X_k)$, the θ_i represent linear effects and the λ_i are quadratic effects. The second-order designs we use typically have the property (discussed in Section 14.3) of either exact or approximate rotatability. Under these circumstances, $V(\theta_i) = V(b_i)$, and $V(\lambda_i) = V(b_{ii})$, approximately. We have used these results to obtain the "standard errors" for θ_i and λ_i in the examples discussed below.

EXAMPLE 11.1. Constructed data, $k = 2$. The constructed data analyzed in Section 10.3 gave rise to the fitted equation $\hat{y} = 78.8988 + 2.272x_1 + 3.496x_2 - 2.08x_1^2 - 2.92x_2^2 - 2.88x_1x_2$. We thus find the following.

| | | Subscript | | "Standard |
Description		1	2	error"
Slopes	θ	4.2	0.3	0.3
Curvatures	λ	−4.0	−1.0	0.4
Distances	X_S	0.5	0.1	
Changes, $\hat{y}$	r	8.3	1.1	

The λ's differ significantly from zero, and their sizes show there is more response change in the X_1 (or $\tilde{X}_1$) direction than in X_2 (or $\tilde{X}_2$). We have a point maximum at the center of a set of elliptical contours and we are close to the center of the system. We can take $R = S$ and $OR = OS = [0.5^2 + (-0.1)^2]^{1/2} \doteq 0.5$.

EXAMPLE 11.2. Polymer elasticity data. (See Section 10.3.) The fitted equation was $\hat{y} = 57.31 + 1.50x_1 - 2.13x_2 + 1.81x_3 - 4.69x_1^2 - 6.27x_2^2 - 5.21x_3^2 - 7.13x_1x_2 - 3.27x_1x_3 - 2.73x_2x_3$.

Description		1	2	3	"Standard error"
Slopes	θ	0.1	−2.3	2.2	0.5
Curvatures	λ	−10.0	−4.4	−1.8	0.5
Distances	X_S	0.0	−0.3	0.6	
Changes, $\hat{y}$	r	10.0	5.9	4.2	

Although the λ's are considerably different in size, they all apparently differ significantly from zero, so changes in $\hat{y}$ are substantial in all three directions X_1, X_2, X_3. We have a point maximum like Figure 11.3(A), and we are close to the center of the system S. For this situation $R = S$, and $OR = OS = (0^2 + (-0.3^2) + 0.6^2)^{1/2} \doteq 0.7$.

We next consider in detail some other examples which provide interesting ridge systems.

11.3. EXAMPLE 11.3: A CONSECUTIVE CHEMICAL SYSTEM YIELDING A NEAR STATIONARY PLANAR RIDGE MAXIMUM

Table 11.2 shows data from an investigation of a consecutive chemical system. The products from a batch reaction conducted in an autoclave were analyzed at various conditions of temperature (T), concentration of catalyst (C), and reaction time (t). The initial planning and analysis were conducted in the coded variables

$$\dot{x}_1 = \frac{(T - 167)}{5}, \qquad \dot{x}_2 = \frac{(C - 27.5)}{2.5}, \qquad \dot{x}_3 = \frac{(t - 6.5)}{1.5}, \quad (11.3.1)$$

whose values will be found in the table under the heading "first coding."

It will be seen that the first 15 runs recorded in the table were made in accordance with a central composite design in the "first coding" units. The

TABLE 11.2. Experimental data from 19 runs on a consecutive chemical system

Run number	Levels of predictor variables			First coding			Second coding			Observed percentage yield, y
	$T(°C)$	$C(\%)$	$t(h)$	$\dot{x}_1$	$\dot{x}_2$	$\dot{x}_3$	x_1	x_2	x_3	
1	162	25	5	-1	-1	-1	-1	-1	-1	45.9
2	162	25	8	-1	-1	1	-1	-1	1	53.3
3	162	30	5	-1	1	-1	-1	1	-1	57.5
4	162	30	8	-1	1	1	-1	1	1	58.8
5	172	25	5	1	-1	-1	1	-1	-1	60.6
6	172	25	8	1	-1	1	1	-1	1	58.0
7	172	30	5	1	1	-1	1	1	-1	58.6
8	172	30	8	1	1	1	1	1	1	52.4
9	167	27.5	6.5	0	0	0	0.01	0.05	0.12	56.9
10	177	27.5	6.5	2	0	0	1.97	0.05	0.12	55.4
11	157	27.5	6.5	-2	0	0	-2.03	0.05	0.12	46.9
12	167	32.5	6.5	0	2	0	0.01	1.87	0.12	57.5
13	167	22.5	6.5	0	-2	0	0.01	-2.16	0.12	55.0
14	167	27.5	9.5	0	0	2	0.01	0.05	1.73	58.9
15	167	27.5	3.5	0	0	-2	0.01	0.05	-2.52	50.3
16	177	20	6.5	2	-3	0	1.97	-3.47	0.12	61.1
17	177	20	6.5	2	-3	0	1.97	-3.47	0.12	62.9
18	160	34	7.5	-1.4	2.6	0.7	-1.41	2.36	0.72	60.0
19	160	34	7.5	-1.4	2.6	0.7	-1.41	2.36	0.72	60.6

four additional runs were then made for confirmation purposes after a preliminary analysis of the initial findings. Subsequently, consideration of the *likely* mechanism underlying the process indicated (Box and Youle, 1955) that it would be more advantageous to work with other predictor variables, namely the reciprocal of the absolute temperature $(T + 273)$, the logarithm of the reaction time t, and the function $f(C)$ of concentration, where

$$f(C) = \ln\left[\frac{C - 4}{\ln(2C - 4) - \ln C}\right]. \qquad (11.3.2)$$

If, for our convenience, we recode these variables in such a way that the upper and lower levels of the factorial design remain at -1 and $+1$, we obtain the coding

$$x_1 = 88 - \frac{38,715}{(T + 273)},$$

$$x_2 = -38.2056 + 10.5124 f(C),$$

$$x_3 = -7.84868 + 4.25532 \ln t. \qquad (11.3.3)$$

The fitted second degree equation is then

$$\hat{y} = 59.16 + 2.01x_1 + 1.00x_2 + 0.67x_3 - 2.01x_1^2 - 0.72x_2^2$$

$$- 1.01x_3^2 - 2.79x_1x_2 - 2.18x_1x_3 - 1.16x_2x_3. \qquad (11.3.4)$$

Canonical Analysis

For this example, we find that $b_0 = 59.16$,

$$\mathbf{b} = \begin{bmatrix} 2.01 \\ 1.00 \\ 0.67 \end{bmatrix}, \qquad \mathbf{B} = -\begin{bmatrix} 2.01 & 1.39 & 1.09 \\ 1.39 & 0.72 & 0.58 \\ 1.09 & 0.58 & 1.01 \end{bmatrix} \qquad (11.3.5)$$

and solution of the equations $2\mathbf{B}\mathbf{x}_S = -\mathbf{b}$ gives

$$\mathbf{x}_S = (-0.05, 0.90, -0.13)' \qquad (11.3.6)$$

which is close to the center $(0, 0, 0)$ of the experimental region. At $\mathbf{x}_S$, the fitted value is

$$\hat{y}_S = b_0 + \tfrac{1}{2}\mathbf{x}'_S\mathbf{b} = 59.52. \tag{11.3.7}$$

The eigenvalues in order of absolute magnitude, and their associated eigenvectors are as follows:

$$\lambda_1 = -3.51, \quad \mathbf{m}_1 = (0.762, 0.473, 0.443)',$$

$$\lambda_2 = -0.42, \quad \mathbf{m}_2 = (0.291, 0.361, -0.886)',$$

$$\lambda_3 = 0.18, \quad \mathbf{m}_3 = (0.579, -0.804, -0.138)'. \tag{11.3.8}$$

Thus the canonical form is

$$\hat{y} = 59.52 - 3.51\tilde{X}_1^2 - 0.42\,\tilde{X}_2^2 + 0.18\tilde{X}_3^2, \tag{11.3.9}$$

where

$$\tilde{X}_1 = 0.762x_1 + 0.473x_2 + 0.443x_3 - 0.330,$$

$$\tilde{X}_2 = 0.291x_1 + 0.361x_2 - 0.886x_3 - 0.426,$$

$$\tilde{X}_3 = 0.579x_1 - 0.804x_2 - 0.138x_3 + 0.731. \tag{11.3.10}$$

We see that the eigenvalues λ_2 and λ_3 are both small in absolute magnitude compared with λ_1. Moreover, the standard errors of the λ's will be, very roughly, of the same magnitude as those of the quadratic coefficients in the fitted equation. These latter are about 0.4. Thus the fitted surface can be realistically approximated by the limiting system

$$\hat{y} = \hat{y}_S + \lambda_1\tilde{X}_1^2, \tag{11.3.11}$$

where, approximately, $\hat{y}_S = 60$, $\lambda_1 = -3.51$, and $\tilde{X}_1 = 0.76x_1 + 0.47x_2 + 0.44x_3 - 0.33$. The contours for such a stationary plane ridge system are shown in Figure 11.5.

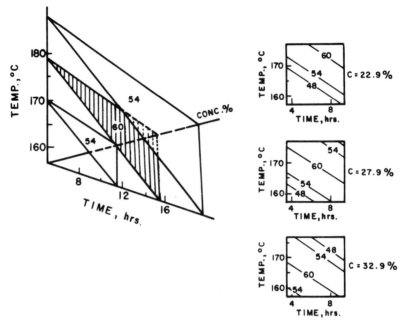

FIGURE 11.5. Contours of empirical yield surface, with sections at three levels of concentration.

The associated tabular display is as follows.

Description		1	2	3	"Standard error"
Slopes	θ	2.3	0.3	0.3	0.3
Curvatures	λ	-3.5	-0.4	0.2	0.4
Distances	X_S	0.3	0.4	-0.7	
Changes, $\hat{y}$	r	5.3	0.7	0.6	

Only λ_1 is substantially and significantly different from zero. The values of θ and λ, taken together with those of r, indicate that very little is happening in any direction except X_1. We have a ($p = 2$) two-dimensional ridge like Figure 11.3(F). The distance OR is quite short, namely, $OR = (0.3^2)^{1/2} = 0.3$. See Eq. (11.2.3) on page 352.

Direct Fitting of the Canonical Form

When originally performed (Box, 1954), this analysis led to the realization that kinetic theory could readily produce a surface which would be approximated by a stationary ridge of this type. We discuss this development later. For the moment, it is sufficient to say that some definite scientific interest is associated with a model of the form of Equation (11.3.11). As a result, this type of canonical form was fitted directly by maximum likelihood. The model, which is now nonlinear in the parameters, can be written

$$y = \gamma_0 - \lambda(\alpha_{11}x_1 + \alpha_{12}x_2 + \alpha_{13}x_3 + \alpha_{10})^2 + \varepsilon \quad (11.3.12a)$$

with the restriction $\alpha_{11}^2 + \alpha_{12}^2 + \alpha_{13}^2 = 1$. For fitting purposes it is preferable to cast the model in unrestricted form by writing $\gamma_{1j} = \alpha_{1j}\lambda^{1/2}$, $j = 0, 1, 2, 3$, whereupon the model becomes

$$y = \gamma_0 - (\gamma_{11}x_1 + \gamma_{12}x_2 + \gamma_{13}x_3 + \gamma_{10})^2 + \varepsilon \quad (11.3.12b)$$

and has five independent parameters γ_0, γ_{11}, γ_{12}, γ_{13}, and γ_{10}. On the assumption that the errors ε are normally and independently distributed with constant variance, maximum likelihood estimates of the parameters which appear nonlinearly in Eq. (11.3.12b) are obtained by application of iterative nonlinear least squares, briefly described in Appendix 3D. The resulting fitted equation, recast in the form of (11.3.11), is

$$\hat{y} = 59.93 - 3.76X^2,$$

where

$$X = 0.780x_1 + 0.426x_2 + 0.458x_3 - 0.320. \quad (11.3.13)$$

The constants in this equation are all readily interpreted; $\hat{y}_S = 59.93$ is the estimated response on the fitted maximal plane, while the constants 0.780, 0.426, and 0.458 are the direction cosines of a line perpendicular to this plane. The value 0.320 measures the distance in design units of the nearest point on the fitted plane to the design origin. The quantity $\lambda_1 = -3.76$ measures the quadratic fall-off in a direction perpendicular to the fitted maximal plane. The residual sum of squares after fitting the canonical form (11.3.13) is 45.2. There are five adjustable coefficients in the new equation and thus an approximate analysis of variance (approximate because the canonical model is nonlinear in the parameters) may now be obtained as in Table 11.3.

TABLE 11.3. **Approximate analysis of variance table for the fitted canonical form compared with the full second degree fit, Section 11.3**

Sources		SS		df		MS		F
Full second degree equation after eliminating mean	Canonical form after eliminating mean	380.2 {	355.7	9 {	4	42.24	88.92	27.53
	Remainder Residual from nonlinear model		24.5		5		4.90	
Residual from full second degree model		20.7	45.2	9	14	2.30		3.23
Total after eliminating mean		400.9		18				

The analysis shows that the canonical form, with only five adjustable constants, explains the data almost as well as does the full quadratic equation with 10 adjustable constants. (Remember that the table entries for df are one fewer in each case, due to elimination of the mean).

Exploitation of the Canonical Form

When this investigation was completed, great practical interest was associated with the fact that, for this system, there was not a single point maximum but, locally at least, a plane of near-maxima. This allowed considerable choice of operating conditions (Box, 1954). In particular (see Figure 11.6) it allowed conditions to be chosen so that two other criteria (the level of a certain impurity, and cost) attained their best levels.

11.4. EXAMPLE 11.4: A SMALL REACTOR STUDY YIELDING A RISING RIDGE SURFACE

Table 11.4 shows data for an experiment performed sequentially in four blocks. (The blocking sequence has some interesting design aspects which are discussed more fully in Section 15.3.) The series of runs was part of a pilot plant study on a small continuous reactor in which the predictor variables were

$$r = \text{flow rate in liters per hour,}$$

$$c = \text{concentration of catalyst,}$$

$$T = \text{temperature,}$$

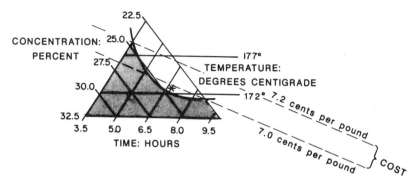

FIGURE 11.6. Alternative processes giving near-maximum yield, with cost contours superimposed. In the shaded region, the biproduct is unacceptably high. The asterisk indicates the process chosen for future operation.

and the response variable was y = concentration of product. The runs were randomized within blocks, but are shown in a standard order in Table 11.4.

The stated objective was to find, and explore, conditions giving higher concentrations of product. (This part of the investigation was undertaken after application of the steepest ascent method had resulted in some progress toward higher values of y, but had also indicated the need for a more detailed examination of the response surface.) Flow rate and catalyst concentration were varied on a log scale, and temperature on a linear scale in this design; the actual coding was as follows:

$$x_1 = \frac{(\ln r - \ln 3)}{\ln 2}, \qquad x_2 = \frac{(\ln c - \ln 2)}{\ln 2}, \qquad x_3 = \frac{(T - 80)}{10}. \quad (11.4.1)$$

The choice of relative ranges for the variables was based on the chemist's guess that a halving of the flow rate, a doubling of the concentration of catalyst, and a 10-degree increase in temperature would each individually about double the reaction rate.

The second degree equation fitted to the 24 runs by least squares was

$$\hat{y} = 51.796 + 0.745x_1 + 4.813x_2 + 8.013x_3 - 3.833x_1^2$$

$$+ 1.217x_2^2 - 6.258x_3^2 + 0.375x_1x_2 + 10.350x_1x_3 - 2.825x_2x_3.$$

$$(11.4.2)$$

TABLE 11.4.   Data for small reactor study (Example 11.4)

Block	Run	x_1	x_2	x_3	Concentration of product, y	Block total, T_i
1	1	-1	-1	1	40.0	
	2	1	-1	-1	18.6	
	3	-1	1	-1	53.8	
	4	1	1	1	64.2	
	5	0	0	0	53.5	
	6	0	0	0	52.7	282.8
2	7	-1	-1	-1	39.5	
	8	1	-1	1	59.7	
	9	-1	1	1	42.2	
	10	1	1	-1	33.6	
	11	0	0	0	54.1	
	12	0	0	0	51.0	280.1
3	13	$-\sqrt{2}$	0	0	43.0	
	14	$\sqrt{2}$	0	0	43.9	
	15	0	$-\sqrt{2}$	0	47.0	
	16	0	$\sqrt{2}$	0	62.8	
	17	0	0	$-\sqrt{2}$	25.6	
	18	0	0	$\sqrt{2}$	49.7	272.0
4	19	$-\sqrt{2}$	0	0	39.2	
	20	$\sqrt{2}$	0	0	46.3	
	21	0	$-\sqrt{2}$	0	44.9	
	22	0	$\sqrt{2}$	0	58.1	
	23	0	0	$-\sqrt{2}$	27.0	
	24	0	0	$\sqrt{2}$	50.7	266.2
						1101.1
					$(\Sigma y_i)^2/n = 50{,}517.55$	

The resulting analysis of variance is given in Table 11.5. For additional detail, see Appendix 11A. For a plot of the surface, see Figure 11.7. See also Figure 12.1a.

No evidence of lack of fit appears. Furthermore, the observed F ratio of 94.03 with 9 and 11 df is over 32 times the percentage point (2.90) needed for significance at the $\alpha = 0.05$ test level. Thus, on the basis of the Box–Wetz criterion discussed in Section 8.2, the equation merits further interpretation.

TABLE 11.5. Analysis of variance table for second degree fit with blocks, Example 11.4

Source	SS		df		MS	F
First degree equation	1,406.8		3			
Extra from second degree equation $\|b_0$	1,597.4	3004.2	6	9	333.8	94.03
Blocks	28.8		3			
Lack of fit	11.8		3		3.93	
Error	27.2	39.0	8	11	3.40	3.55
Total (after eliminating mean)	3,072.0		23			
Mean (b_0)	50,517.6		1			
Total	50,889.6		24			

Canonical Analysis

For this example, $b_0 = 51.796$,

$$\mathbf{b} = \begin{bmatrix} 0.7446 \\ 4.8133 \\ 8.0125 \end{bmatrix}, \quad \mathbf{B} = \begin{bmatrix} -3.8333 & 0.1875 & 5.1750 \\ 0.1875 & 1.2167 & -1.4125 \\ 5.1750 & -1.4125 & -6.2583 \end{bmatrix}, \quad (11.4.3)$$

and the solutions of the equations $2\mathbf{B}\mathbf{x}_S = -\mathbf{b}$; $\hat{y}_S = b_0 + \frac{1}{2}\mathbf{x}'_S\mathbf{b}$ give the coordinates of the stationary point and the corresponding response as

$$\mathbf{x}_S = (25.78, 15.48, 18.46)'; \qquad \hat{y}_S = 172.61. \qquad (11.4.4)$$

Obviously the fitted equation can have no relevance at this very remote stationary point. Nevertheless its location is of importance as a "construction point" in the analysis that follows. The eigenvalues in order of absolute magnitude, and the associated eigenvectors are as follows:

$$\lambda_1 = -10.4893, \quad \mathbf{m}_1 = (-0.6123, 0.1044, 0.7837)',$$

$$\lambda_2 = 1.7109, \quad \mathbf{m}_2 = (-0.2969, 0.8884, -0.3502)',$$

$$\lambda_3 = -0.0965, \quad \mathbf{m}_3 = (0.7328, 0.4471, 0.5129)'. \qquad (11.4.5)$$

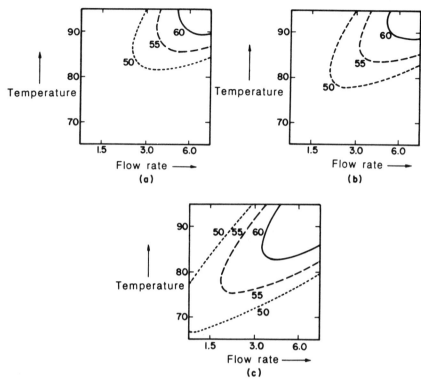

FIGURE 11.7. Small reactor study. Sections of fitted response surface. (a) Slice of fitted surface at $x_2 = -1$ (i.e., catalyst concentration 1 mole percent). (b) Slice of fitted surface at $x_2 = 0$ (i.e., catalyst concentration 2 mole percent). (c) Slice of fitted surface at $x_2 = 1$ (i.e., catalyst concentration 4 mole percent).

The B canonical form is thus

$$\hat{y} = 172.61 - 10.4893\tilde{X}_1^2 + 1.7109\tilde{X}_2^2 - 0.0965\tilde{X}_3^2, \quad (11.4.6)$$

where, from equations $\tilde{\mathbf{X}} = \mathbf{M}'\tilde{\mathbf{x}} = \mathbf{M}'(\mathbf{x} - \mathbf{x}_S)$, we have

$$\tilde{X}_1 = -0.6123x_1 + 0.1044x_2 + 0.7837x_3 - 0.3015,$$

$$\tilde{X}_2 = -0.2969x_1 + 0.8884x_2 - 0.3502x_3 + 0.3649,$$

$$\tilde{X}_3 = 0.7328x_1 + 0.4471x_2 + 0.5129x_3 - 35.2796. \quad (11.4.7)$$

This canonical form is (as we have seen) centered in a region very remote from the one in which experiments were actually conducted, and of course the response equation has no relevance in that remote region. What we now must do is to determine what the canonical equation tells us about the estimated response *in the region in which it is relevant*, that is, primarily, the region where the experiments themselves were conducted centered at $(x_1, x_2, x_3) = (0, 0, 0)$. We do this by referring the canonical form to a new origin S' situated as close as possible to the design origin. Setting x_1, x_2, and $x_3 = 0$ in Eqs. (11.4.7) we find that the design origin has $\tilde{X}$ coordinates

$$\tilde{X}_1 = -0.3015, \qquad \tilde{X}_2 = 0.3649, \qquad \tilde{X}_3 = -35.2796. \quad (11.4.8)$$

[Note that this implies that $(X_{1S}, X_{2S}, X_{3S}) = (0.3015, -0.3649, 35.2796)$.] Consider the plane $\tilde{X}_1 = 0$ associated with the first eigenvalue $\lambda_1 = -10.49$. The point S' on this plane and closest to the design origin is that for which $\tilde{X}_1 = 0$, $\tilde{X}_2 = 0.3649$, $\tilde{X}_3 = -35.2796$. If coordinates referred to this new local region are denoted by dotted X's, then $\tilde{X}_2 = \dot{X}_2 + 0.3649$, $\tilde{X}_3 = \dot{X}_3 - 35.2796$. [Because $\dot{X}_1 = \tilde{X}_1$ we do not use $\dot{X}_1$.] Substitution in the canonical equation gives

$$\hat{y} = 172.61 - 10.4893\tilde{X}_1^2 + 1.7109\left(\dot{X}_2 + 0.3649 \right)^2 - 0.0965\left(\dot{X}_3 - 35.2796 \right)^2.$$

$$(11.4.9)$$

After expansion, and dropping the squared terms in $\dot{X}_2$ and $\dot{X}_3$ on the grounds that they have relatively small local influence compared with the squared term in $\tilde{X}_1$, we obtain the approximating form which roughly describes the behavior of the fitted function in the neighborhood where it is relevant as

$$\hat{y} = 57.71 - 10.49\tilde{X}_1^2 + 1.25\dot{X}_2 + 6.53\dot{X}_3. \quad (11.4.10)$$

In this approximating canonical form, the quadratic effect is entirely confined to $\tilde{X}_1$ and the coefficients of $\dot{X}_2$ and $\dot{X}_3$ indicate the orientation of the straight line contours of the rising ridge on the plane $\tilde{X}_1 = 0$.

Now $\dot{X}_2 = \mathbf{m}'_2\mathbf{x}$ and $\dot{X}_3 = \mathbf{m}'_3\mathbf{x}$. Thus the first-order part of the model is

$$1.25\dot{X}_2 + 6.53\dot{X}_3 = 4.41x_1 + 4.03x_2 + 2.91x_3$$

$$= 6.645(0.664x_1 + 0.606x_2 + 0.438x_3)$$

$$= 6.645X, \quad (11.4.11)$$

where, two lines up, the coefficients of x_1, x_2, and x_3 have been normalized so that their sum of squares is unity. Equation (11.4.10) is therefore equivalent to

$$\hat{y} = 57.71 - 10.49\tilde{X}_1^2 + 6.645X. \qquad (11.4.12)$$

In tabular form, we can display:

Description		1	2	3	"Standard error"
Slopes	θ	6.3	1.3	6.8	0.47
Curvatures	λ	-10.5	1.7	-0.1	0.55
Distances	X_S	0.3	-0.4	35.3	
Changes, $\hat{y}$	r	15.1	2.8	11.8	

The eigenvalue λ_1 is dominant and there are considerable response changes in the X_1 direction (where the surface is mostly quadratic) and in the X_3 direction (mostly linear); not much is happening in the X_2 direction. We thus have a rising ridge of the type shown in Figure 11.3(G).

Direct Fitting of the Canonical Form

In this example, as in the previous one, it is possible to derive a simple kinetic relationship which would produce a rising ridge of the sort found here. We discuss in Appendix 3D how the nonlinear mechanistic model may be fitted directly to the data. For the moment, we consider the result of fitting, by maximum likelihood, a reduced model of the form of (11.4.12). The postulated model is

$$y = \gamma_0 - \lambda_1(\alpha_{11}x_1 + \alpha_{12}x_2 + \alpha_{13}x_3 + \alpha_{10})^2$$
$$+ \lambda_2(\alpha_{21}x_1 + \alpha_{22}x_2 + \alpha_{23}x_3) + \varepsilon \qquad (11.4.13)$$

where $\alpha_{11}^2 + \alpha_{12}^2 + \alpha_{13}^2 = 1 = \alpha_{21}^2 + \alpha_{22}^2 + \alpha_{23}^2$. Alternatively, we can write this as a seven-parameter model

$$y = \hat{y}_S - (L_{11}x_1 + L_{12}x_2 + L_{13}x_3)^2 + c_1x_1 + c_2x_2 + c_3x_3 + \varepsilon \quad (11.4.14)$$

and drop the restrictions. If we fit the seven-parameter form and then convert it to the form (11.4.13), we obtain

$$\hat{y} = 54.07 - 10.812\{-0.612x_1 + 0.104x_2 + 0.784x_3 - 0.301\}^2$$
$$+ 6.923(0.683x_1 + 0.597x_2 + 0.420x_3). \qquad (11.4.15)$$

Interpreting this equation, we see that $\hat{y}_{S'} = 54.07$ is the estimated response at S', the point nearest to the origin on the rising planar ridge. The

TABLE 11.6.   Approximate analysis of variance table for second degree and canonical form fitted equations

Source		SS	df	MS	F
Full second degree equation after allowance for mean	⎧ Canonical form, after allowance for mean	⎧ 2965.4	⎧ 6	⎧ 494.23	
		3004.2	9	333.8	
	⎩ Remainder	38.8	⎩ 3	⎩ 12.93	3.64
Blocks		28.8	3		
Residual		39.0	11	3.55	
Total, corrected for mean		3072.0	23		

quantities $-0.612, 0.104, 0.784$ are the direction cosines of lines perpendicular to this plane. The size of the constant -0.301 measures the shortest distance of the planar ridge from the design origin. The constant -10.812 measures the rate of quadratic fall-off as we move away from the plane of the ridge. The values $0.683, 0.597, 0.420$ are the direction cosines of the line of steepest ascent up the planar ridge. Finally, 6.923 is the linear rate of increase of yield as we move up the ridge. The analysis of variance table is shown as Table 11.6.

For an approximate test of the remainder sum of squares, we compare the ratio (remainder mean square)/(residual mean square) $= 3.64$ with the upper 5% point of an $F(3, 11)$ distribution, namely 3.59. We see that it is just significant, which implies that some variation is not explained by the fitted canonical form compared with the full second-order model. However, the fitted canonical form model does pick up a proportion $2965.4/3072 = 0.9653$ of the variation about the mean using only six additional parameters, whereas the full second-order model picks up a proportion $3004.2/3072 = 0.9779$, using nine additional parameters, so that, from a practical point of view, the canonical form does provide a worthwhile and parsimonious representation.

Interpretation

In this particular investigation, the maximum temperature attainable with the apparatus used was 100°C, and the highest flow rate possible was 12 L/h. Further experimentation and study showed, as might be expected from this analysis, that a product concentration of about 75 was attainable with the temperature just below 100°C, with a high catalyst concentration of about 5%, and with the maximum permissible flow rate. The most important fact revealed by this response surface study was that even higher concentrations of product were likely to be produced at temperatures higher than 100°C. Further experimentation was therefore conducted in which the reaction was carried out under pressure in an autoclave where temperatures above 100°C were possible.

11.5. EXAMPLE 11.5: A STATIONARY RIDGE IN FIVE VARIABLES

To assist geometric understanding, we have so far illustrated canonical analysis for examples with only two or three predictor variables. The technique becomes even more valuable when there are more than three predictors, for it enables the basic situation to be understood, even though it cannot be easily visualized geometrically. In this next example (Box, 1954), we shall see a two-dimensional ridge system embedded in the five-dimensional space of five predictor variables.

The chemical process under study had two stages, and the five factors considered were the temperatures (T_1, T_2), and the times (t_1, t_2), of reaction at the two stages, and the concentration (C) of one of the reactants at the first stage. Evidence was available which suggested that the times of reaction were best varied on a logarithmic scale, and it was known that the second reaction time needed to be varied over a wide range. A preliminary application of the steepest ascent procedure had brought the experimenter close to a near-stationary region, and a second-order model was now to be fitted and examined. The coding used in the experiments was

$$x_1 = \frac{(T_1 - 122.5)}{7.5},$$

$$x_2 = \left\{ \frac{2(\log t_1 - \log 5)}{\log 2} \right\} + 1,$$

$$x_3 = \frac{(C - 70)}{10},$$

$$x_4 = \frac{(T_2 - 32.5)}{7.5},$$

$$x_5 = \left\{ \frac{2(\log t_2 - \log 1.25)}{\log 5} \right\} + 1. \tag{11.5.1}$$

Thirty-two observations of percentage yield were available in all, and these are given in Table 11.7. A second degree equation fitted to the 32 observed responses produced the estimated coefficients shown (rounded to three places of decimals):

$$b_0 = 68.717, \qquad \mathbf{b} = \begin{bmatrix} 3.258 \\ 1.582 \\ 1.161 \\ 3.474 \\ 1.488 \end{bmatrix},$$

$$\mathbf{B} = \begin{bmatrix} -1.610 & -0.952 & 1.051 & -0.176 & -0.022 \\ & -1.354 & 0.299 & -0.084 & -0.549 \\ & & -2.584 & -1.775 & -0.376 \\ & & & -2.335 & 0.201 \\ \text{symmetric} & & & & -1.421 \end{bmatrix}. \quad (11.5.2)$$

The standard computations led to the canonical form

$$\hat{y} = 72.51 - 4.46\tilde{X}_1^2 - 2.62\,\tilde{X}_2^2 - 1.78\tilde{X}_3^2 - 0.40\,\tilde{X}_4^2 - 0.04\,\tilde{X}_5^2, \quad (11.5.3)$$

where $\mathbf{x}_S = (2.52, -1.10, 1.27, -0.32, 0.53)'$, and

$$\tilde{\mathbf{X}} = \begin{bmatrix} -0.284 & -0.138 & 0.743 & 0.589 & 0.026 \\ -0.639 & -0.599 & -0.003 & -0.434 & -0.213 \\ -0.254 & 0.255 & 0.225 & -0.383 & 0.820 \\ -0.369 & 0.730 & 0.188 & -0.222 & -0.496 \\ 0.558 & -0.157 & 0.601 & -0.518 & -0.185 \end{bmatrix} \mathbf{x} + \begin{bmatrix} -0.203 \\ -0.925 \\ 0.077 \\ -1.684 \\ 2.408 \end{bmatrix}$$

$$(11.5.4)$$

(or $\tilde{\mathbf{X}} = \mathbf{M}'\tilde{\mathbf{x}}$ where $\mathbf{M}'$ is the 5×5 matrix shown). The center of the system lies within the immediate region of experimentation. Also, the canonical variables $\tilde{X}_4$ and $\tilde{X}_5$ play a comparatively minor role in describing the function, their coefficients λ_4 and λ_5 being of the same order of magnitude as their standard errors. At least locally, most of the change in response can be well described by the canonical variables $\tilde{X}_1, \tilde{X}_2, \tilde{X}_3$. Thus, the system in five variables has, approximately, a two-dimensional maximum. The sets of conditions which yield, approximately, the same maximum response result of about 72.51 are those in the plane defined by the equations $\tilde{X}_1 = 0$, $\tilde{X}_2 = 0$, $\tilde{X}_3 = 0$. These three equations are readily transformed to three equations in the five unknowns x_1, x_2, x_3, x_4, and x_5 and thence into three equations in the five natural (uncoded) variables T_1, T_2, t_1,

TABLE 11.7. Thirty-two observations for Example 11.5

Run number	x_1	x_2	x_3	x_4	x_5	Response, y
1	−1	−1	−1	−1	1	49.8
2	1	−1	−1	−1	−1	51.2
3	−1	1	−1	−1	−1	50.4
4	1	1	−1	−1	1	52.4
5	−1	−1	1	−1	−1	49.2
6	1	−1	1	−1	1	67.1
7	−1	1	1	−1	1	59.6
8	1	1	1	−1	−1	67.9
9	−1	−1	−1	1	−1	59.3
10	1	−1	−1	1	1	70.4
11	−1	1	−1	1	1	69.6
12	1	1	−1	1	−1	64.0
13	−1	−1	1	1	1	53.1
14	1	−1	1	1	−1	63.2
15	−1	1	1	1	−1	58.4
16	1	1	1	1	1	64.3
17	3	−1	−1	1	1	63.0
18	1	−3	−1	1	1	63.8
19	1	−1	−3	1	1	53.5
20	1	−1	−1	3	1	66.8
21	1	−1	−1	1	3	67.4
22	1.23	−0.56	−0.03	0.69	0.70	72.3
23	0.77	−0.82	1.48	1.88	0.77	57.1
24	1.69	−0.30	−1.55	−0.50	0.62	53.4
25	2.53	0.64	−0.10	1.51	1.12	62.3
26	−0.08	−1.75	0.04	−0.13	0.27	61.3
27	0.78	−0.06	0.47	−0.12	2.32	64.8
28	1.68	−1.06	−0.54	1.50	−0.93	63.4
29	2.08	−2.05	−0.32	1.00	1.63	72.5
30	0.38	0.93	0.25	0.38	−0.24	72.0
31	0.15	−0.38	−1.20	1.76	1.24	70.4
32	2.30	−0.74	1.13	−0.38	0.15	71.8

t_2, and C. We have 2 df in the choice of conditions in the sense that, within the local region considered, if we choose values for any two of the predictor variables, we obtain three equations in three unknowns which can be solved to give appropriate levels for the remaining three predictor variables. To demonstrate the practical implications to the experimenter of these findings, a table of approximate alternatives, which covered the ranges of interest,

TABLE 11.8. Sets of conditions that produce, approximately, the same maximum response values of about 72.51, for Example 11.5. All these sets of conditions lie on the plane $\tilde{X}_1 = \tilde{X}_2 = \tilde{X}_3 = 0$ in the five dimensional space of the predictor variables. Rough contours of $t_1 + t_2$, the total reaction time, are sketched over the figures.

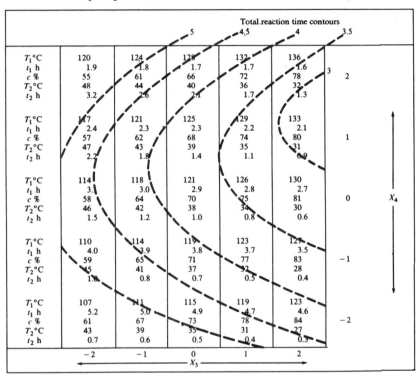

Total reaction time contours

T_1 °C	120	124	128	132	136		3	
t_1 h	1.9	1.8	1.7	1.7	1.6			
c %	55	61	66	72	78		2	
T_2 °C	48	44	40	36	32			
t_2 h	3.2	2.6	2.1	1.7	1.3			
T_1 °C	117	121	125	129	133			
t_1 h	2.4	2.3	2.3	2.2	2.1			
c %	57	62	68	74	80		1	
T_2 °C	47	43	39	35	31			
t_2 h	2.2	1.8	1.4	1.1	0.9			
T_1 °C	114	118	121	126	130			
t_1 h	3.1	3.0	2.9	2.8	2.7			
c %	58	64	70	75	81		0	X_4
T_2 °C	46	42	38	34	30			
t_2 h	1.5	1.2	1.0	0.8	0.6			
T_1 °C	110	114	119	123	127			
t_1 h	4.0	3.9	3.8	3.7	3.5			
c %	59	65	71	77	83		−1	
T_2 °C	45	41	37	32	28			
t_2 h	1.0	0.8	0.7	0.5	0.4			
T_1 °C	107	111	115	119	123			
t_1 h	5.2	5.0	4.9	4.7	4.6			
c %	61	67	73	78	84		−2	
T_2 °C	43	39	35	31	27			
t_2 h	0.7	0.6	0.5	0.4	0.3			
	−2	−1	0	1	2			

X_5

was calculated as shown in Table 11.8. Sketched over the table are rough contours of total reaction time, $t_1 + t_2$.

It must be understood that any analysis of this kind is, almost invariably, a tentative one, subject to practical confirmation. Estimation errors, and biases arising from the inability of the second degree equation adequately to represent the system will almost guarantee that, in practice, the findings are not very precise. If the tentative results look at all useful and interesting, therefore, confirmatory runs are essential. In particular, these should be made near conditions likely to be of economic importance. In the present example, the most intriguing possibility was that of achieving the maximum yield of about 73% *at high throughput*. The conditions at the top right-hand corner of Table 11.8 suggest that, with suitable levels for the concentration and the temperatures, a near maximum yield can be obtained with a total reaction time of only $1.6 + 1.3 = 2.9$ h. This would allow almost double the throughput possible with the conditions at the bottom left-hand corner of Table 11.8 where the total reaction time is $5.2 + 0.7 = 5.9$ h. Subsequent experimental work, therefore, was directed to investigating the possibilities of the high throughput process.

11.6. THE ECONOMIC IMPORTANCE OF THE DIMENSIONALITY OF MAXIMA AND MINIMA

The example in the foregoing section illustrates the possible great economic value of the study of the "degrees of freedom" in maxima and minima. Three general facts of importance to be kept in mind are:

1. The natural parsimony of practical systems guarantees that multidimensional maxima and minima will be often best approximated by subspaces (e.g., lines, planes, or hyperplanes) rather than by points. When we consider chemical or biological systems, for example, it is difficult if not impossible to think of experimental variables which behave independently of one another and so yield symmetrical maxima. It is much more likely that the system is controlled essentially by a few (and possibly even only one) functions of the predictor variables, in the neighborhood of the maximum.

2. Classical systems of experimentation such as the one factor at a time method will normally lead the experimenter close to only one set of "best" conditions on a stationary or rising ridge, and will, further, mislead him into believing these conditions are:

i.   The *only* maximal conditions (which they are not when the ridge is stationary).
ii. Necessarily maximal (which they are not when the ridge is rising).

3. When there are degrees of freedom in the maximum, they can almost always be profitably exploited (e.g., to give an equally high yield at less cost and/or greater purity).

These three facts have, in turn, these implications:

a.   Even processes that have been thoroughly explored by the one factor at a time method can often be dramatically improved by the application of response surface methods.

b. An understanding of the geometrical redundancies in the fitted response function can enable the predictor variables to be adjusted to maintain optimal response levels as prices, specifications, and requirements change.

c. Knowledge of the geometry of maxima can sometimes help the experimenter to postulate possible theoretical mechanisms which would describe the system under study more fundamentally. (We discuss this aspect more fully in Chapter 12.)

11.7. A METHOD FOR OBTAINING A DESIRABLE COMBINATION OF SEVERAL RESPONSES

When several response variables $y_1, y_2, \ldots, y_m$ have been represented by fitted equations $\hat{y}_1, \hat{y}_2, \ldots, \hat{y}_m$ based on the same set of coded input variables $x_1, x_2, \ldots, x_k$, the question often arises of where in the x space the best overall set of response values might be obtained.

When there are only two or three important input variables, it is possible to solve this problem by looking at the response contours for the different fitted responses $\hat{y}_1, \hat{y}_2, \ldots, \hat{y}_m$ simultaneously, for example, by overlays of various contour diagrams. When the problem can be expressed as one of maximizing or minimizing one response subject to constraints in the others, linear programming methods can sometimes be used advantageously. A different approach which is sometimes useful if carefully applied, is to introduce an overall criterion of desirability. One interesting variant on this method is due to G. C. Derringer and R. Such, and is described in "Simultaneous optimization of several response variables," *J. Quality Technology*, **12**, 1980, 214–219. Suppose we choose, for each response, levels $A \leq B \leq C$ such that the product is unacceptable if $y < A$ or $y > C$, and so that the desirability d of the product increases between A and B and decreases between B and C. Define

$$d = \begin{cases} \left\{ \dfrac{(\hat{y} - A)}{(B - A)} \right\}^s, & \text{for } A \leq \hat{y} \leq B, \\[2em] \left\{ \dfrac{(\hat{y} - C)}{(B - C)} \right\}^t, & B \leq \hat{y} \leq C, \end{cases}$$

and $d = 0$ outside (A, C). By choosing the powers s and t in various ways, we can attribute various levels of desirability to various values of $\hat{y}$ as shown in Figure 11.8. This is a two-sided choice with B the most desirable

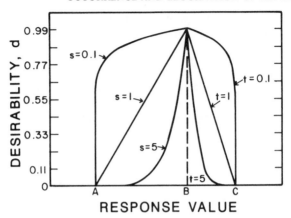

FIGURE 11.8.   Two-sided desirability functions.

level of $\hat{y}$. A one-sided choice can be made in either of two ways:

1. Let $A = B$ and define $d = 1$ for $\hat{y} \leq B$.
2. Let $B = C$ and define $d = 1$ for $\hat{y} \geq B$. This second choice is shown in Figure 11.9.

If we write $D = (d_1 d_2 d_3 \dots d_m)^{1/m}$, where d_i is the desirability function for the ith response, $i = 1, 2, \dots, m$, we have defined an overall desirability function whose value is specified at each point of the x space. The maximum of D can be sought to give the most desirable point in the x space

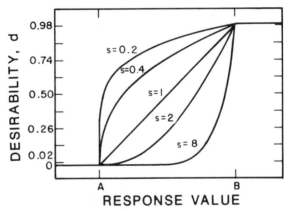

FIGURE 11.9.   One-sided desirability functions.

for all the responses simultaneously. Of course any selected combination of the d_i can be defined as the desirability function; D is simply a specific choice suggested by Derringer and Suich. For examples of this technique, see the quoted reference and also "A statistical methodology for designing elastomer formulations to meet performance specifications," by G. C. Derringer, *Kautschuk + Gummi · Kunststoffe*, **36**, 1983, 349–352.

Great caution is needed in the application of these methods. If used mechanically, they can sometimes lead to unanticipated conditions that are not practically desirable.

APPENDIX 11A. CALCULATIONS FOR THE ANALYSIS OF VARIANCE, TABLE 11.5

The "error" sum of squares is comprised of the sums of squares arising from various comparisons, all of which have a zero expectation if the true model is of cubic, or smaller, order. These 8 df arise as follows.

1. There are six comparisons between pairs of runs $(13, 19)$, $(14, 20)$, ..., $(18, 24)$. The associated sum of squares is $\frac{1}{2}(43.0 - 39.2)^2 + \frac{1}{2}(43.9 - 46.3)^2 + \cdots + \frac{1}{2}(49.7 - 50.7)^2 = 24.83$. Adjusting this for the sum of squares for the difference between blocks 3 and 4, we obtain an error sum of squares with 5 df of $24.83 - 2.803 = 22.027$.

2. In each of blocks 1 and 2 there is a comparison between the mean of the 2^{3-1} points and the mean of the center points. The difference between these two comparisons $\mathbf{c}'\mathbf{y}$, say, has a sum of squares $(\mathbf{c}'\mathbf{y})^2/\mathbf{c}'\mathbf{c}$, with 1 df. Now $\mathbf{c}' = (\frac{1}{4}, \frac{1}{4}, \frac{1}{4}, \frac{1}{4}, -\frac{1}{2}, -\frac{1}{2}; -\frac{1}{4}, -\frac{1}{4}, -\frac{1}{4}, -\frac{1}{4}, \frac{1}{2}, \frac{1}{2}, 0; 0)$, $\mathbf{c}'\mathbf{y} = -0.15$, $\mathbf{c}'\mathbf{c} = 1.5$, so SS $= (-0.15)^2/1.5 = 0.015$.

3. Runs $(5, 6)$ and runs $(11, 12)$ provide two error contrasts, yielding a sum of squares with 2 df of $\frac{1}{2}(53.5 - 52.7)^2 + \frac{1}{2}(54.1 - 51.0)^2 = 5.125$.

The three portions sum to 27.167, rounded to 27.2 in Table 11.5.

APPENDIX 11B. "RIDGE ANALYSIS" OF RESPONSE SURFACES

The technique of ridge analysis was suggested by A. E. Hoerl (1959) and later urged by R. W. Hoerl (1985). In one sense, it is an alternative to canonical reduction. More accurately, it is the method of steepest ascent applied to *second-order* surfaces (rather than first-order) and has its origin in Section 2.2.1 of Box and Wilson (1951). While we ourselves do not favor use of this technique, we nevertheless provide a brief description for those who might. Much of what follows is adapted from Draper (1963); see also Hoerl (1959).

Derivation of the Technique

Consider the second-order response surface in k variables $x_1, x_2, \ldots, x_k$, given by

$$\hat{y} = b_0 + b_1 x_1 + b_2 x_2 + \cdots + b_k x_k + b_{11} x_1^2 + b_{22} x_2^2 + \cdots + b_{kk} x_k^2$$

$$+ b_{12} x_1 x_2 + \cdots + b_{k-1, k} x_{k-1} x_k. \qquad (11B.1)$$

The point $(0, 0, \ldots, 0)$ is the origin of measurement of the variables $x_1, x_2, \ldots, x_k$. If the data used to obtain (11B.1) resulted from a designed experiment, it would usually be the center of the design also. Suppose now we imagine a sphere, center at the origin $(0, 0, \ldots, 0)$ and of radius R, drawn in the x space. Then somewhere on the sphere there will be a maximum $\hat{y}$ and elsewhere a minimum $\hat{y}$, and possibly also [depending on the type of quadratic surface (11B.1) and the value of R] values of $\hat{y}$ which are local maxima or minima, that is, maxima or minima for all nearby points on the sphere, but not absolute maxima and minima when all points of the sphere are taken into consideration. If we investigate the stationary values of the function $\hat{y}$ on the sphere, that is, the stationary values subject to the restriction

$$g(x_1, x_2, \ldots, x_k) \equiv x_1^2 + x_2^2 + \cdots + x_k^2 - R^2 = 0, \qquad (11B.2)$$

we shall be able to find all these local and absolute maxima and minima. We can then plot against R as abscissa the following $(k + 1)$ ordinates: $x_1, x_2, \ldots, x_k, \hat{y}$, for, say, the absolute maximum of $\hat{y}$ found on the sphere radius R.

If we change R slightly, the appropriate values of $x_1, x_2, \ldots, x_k$ and $\hat{y}$ for the absolute maximum will also change slightly and so, by varying R, we can construct $(k + 1)$ curves showing how the position and magnitude of the absolute maximum $\hat{y}$ change as R changes. We can thus find, for any selected R, the place of maximum yield on the response surface. Such a plot can also be made for the absolute minimum or for intermediate stationary values, as desired. Mathematically, then, we wish to find the stationary values of $\hat{y} = f(x_1, x_2, \ldots, x_k)$, from Eq. (11B.1), subject to the restriction $g(x_1, x_2, \ldots, x_k) = 0$ as in Eq. (11B.2).

Application of the method of Lagrange multipliers (see Draper, 1963, for omitted details) leads to the equations, which must be solved for $\mathbf{x} = (x_1, x_2, \ldots, x_k)'$:

$$(\mathbf{B} - \lambda \mathbf{I})\mathbf{x} = -\tfrac{1}{2}\mathbf{b} \qquad (11B.3)$$

where

$$\mathbf{B} = \begin{bmatrix} b_{11} & \tfrac{1}{2}b_{12} & \cdots & \tfrac{1}{2}b_{1k} \\ \tfrac{1}{2}b_{12} & b_{22} & \cdots & \tfrac{1}{2}b_{2k} \\ \cdots & & & \\ \tfrac{1}{2}b_{1k} & \tfrac{1}{2}b_{2k} & \cdots & b_{kk} \end{bmatrix}, \quad \mathbf{b} = \begin{bmatrix} b_1 \\ b_2 \\ \vdots \\ b_k \end{bmatrix} \qquad (11B.4)$$

and $\mathbf{I}$ is the k by k unit matrix. (Note: λ is *not* an eigenvalue; eigenvalues are denoted by $\mu_1, \mu_2, \ldots, \mu_k$ in this section.)

Now, theoretically, the $(k + 1)$ equations (11B.3) and (11B.2) can be solved for sets of $x_1, x_2, \ldots, x_k$, and λ corresponding to the various stationary values of $\hat{y}$ on the sphere radius R. Since the solution in this form leads to messy calculations, a simpler and equivalent method of solution may be used as follows.

1. Regard R as variable, but fix λ instead.
2. Insert the selected value of λ in Eqs. (11B.3) and solve them for $x_1, x_2, \ldots, x_k$. The solution is used in steps 3 and 4.
3. Compute $R = (x_1^2 + x_2^2 + \cdots + x_k^2)^{1/2} = (\mathbf{x}'\mathbf{x})^{1/2}$, where $\mathbf{x} = (x_1, x_2, \ldots, x_k)'$.
4. Evaluate $\hat{y}$.

We now have a set of numbers $(\lambda, x_1, x_2, \ldots, x_k, R, \hat{y})$ and know that on the sphere radius R, center the origin, there is a stationary value of $\hat{y}$, value determined, at the point $(x_1, x_2, \ldots, x_k)$. Several different values of λ will give rise to several stationary points which lie on the same sphere radius R. Whether a particular stationary value is the absolute maximum, absolute minimum, a local maximum or a local minimum is determined, as we shall see, by the value of λ.

Properties of the Stationary Values

Let the eigenvalues or latent roots of the matrix $\mathbf{B}$ be denoted by $\mu_i (i = 1, 2, \ldots, k)$. Then the μ_i are such that

$$\mathbf{Bx} = \mu\mathbf{x}, \qquad (11B.5)$$

or

$$(\mathbf{B} - \mu\mathbf{I})\mathbf{x} = 0. \qquad (11B.6)$$

Hence

$$\det(\mathbf{B} - \mu\mathbf{I}) = 0, \qquad (11B.7)$$

where "det" denotes "the determinant of," provides a kth degree equation with roots $\mu_1, \mu_2, \ldots, \mu_k$, say. Note that when a standard canonical reduction is made of Eq. (11B.1), $\mu_1, \mu_2, \ldots, \mu_k$ are the latent roots needed to reduce $\hat{y}$ to the form

$$\hat{y} = \hat{y}_S + \mu_1 X_1^2 + \mu_2 X_2^2 + \cdots + \mu_k X_k^2.$$

By comparing the value λ, which corresponds to any particular stationary value of $\hat{y}$ on a sphere of radius R, with the latent roots μ_i we shall be able to determine what sort of stationary value has been obtained.

Suppose $\lambda = \lambda_1$ and $\lambda = \lambda_2$ are substituted in Eq. (11B.3) and the solutions $\mathbf{x}_1 = (a_1, a_2, \ldots, a_k)'$ and $\mathbf{x}_2 = (c_1, c_2, \ldots, c_k)'$ result, thus providing two stationary values $\hat{y}_1$ and $\hat{y}_2$ of $\hat{y}$ on the spheres $\mathbf{x}'\mathbf{x} = R_1^2$ and $\mathbf{x}'\mathbf{x} = R_2^2$, respectively.

Then the following results are true.

Result 1: If $R_1 = R_2$ and $\lambda_1 > \lambda_2$, then $\hat{y}_1 > \hat{y}_2$.

Result 2: If $R_1 = R_2$, $\mathbf{M}(\mathbf{x}_1)$ is positive definite, and $\mathbf{M}(\mathbf{x}_2)$ is indefinite, then $\hat{y}_1 < \hat{y}_2$. [$\mathbf{M}(\mathbf{x})$ is a matrix of second-order derivatives; see Draper, 1963.]

Result 3: If $\lambda_1 > \mu_i$ (all i), then $\mathbf{x}_1$ is a point at which $\hat{y}$ attains a local maximum on the sphere radius R_1; if $\lambda_1 < \mu_i$ (all i), then $\mathbf{x}_1$ is a point at which $\hat{y}$ attains a local minimum on the sphere radius R_1. (As will be seen later, we obtain the absolute maximum and minimum in this way, not only the local maximum and minimum.)

Result 4: Suppose, as R increases, we trace a locus of stationary points (the absolute maximum, absolute minimum, or a local maximum or minimum) and examine the changing values of $\hat{y}$. Then, as R increases, $\hat{y}$ changes in one of the following ways (when the response surface is quadratic):

a. Decreases monotonically.

b. Increases monotonically.

c. Passes through a maximum and then decreases monotonically.

d. Passes through a minimum and then increases monotonically.

If c or d happens, it is because the locus has passed through the center of the quadratic system.

The four results have the following implication. If we wish to follow a locus of the absolute maximum $\hat{y}$ for increasing R, we should substitute in Eq. (11B.3) only values of λ *greater* than all the latent roots of $\mathbf{B}$. This will ensure that $\hat{y}$ is a local maximum for every solution $\mathbf{x}$. (It is in fact an absolute maximum as we shall soon see.) No value of λ less than the greatest latent root should be considered in such a case because, while values of λ between eigenvalues may provide a local maximum or minimum, they cannot provide an absolute maximum or minimum.

In fact the total range of λ, namely $-\infty$ to ∞, is divided into sections by the latent roots $\mu_1, \mu_2, \ldots, \mu_k$. Suppose $\mu_1 < \cdots < \mu_k$. Then we have $(k + 1)$ intervals $(-\infty, \mu_1), (\mu_1, \mu_2), \ldots, (\mu_{k-1}, \mu_k)$ and (μ_k, ∞).

As $\lambda \to \mu_i (i = 1, 2, \ldots, k)$, the resulting solution $\mathbf{x} \to \pm \infty$ so that $R \to \infty$. As $\lambda \to \pm \infty$, $\mathbf{x} \to 0$ and so $R \to 0$. Furthermore the value of $\partial^2 R / \partial \lambda^2$ is positive except when $R = 0$, when it takes the value zero.

From the above, we see that the graph of R, plotted as ordinate against λ as abscissa, acts as follows.

At $\lambda = -\infty$, $R = 0$ and R increases steadily to infinity at $\lambda = \mu_1$; between pairs of latent roots, R passes down from infinity at μ_i through a stationary value and up to infinity again at μ_{i+1}. Finally R passes from infinity at μ_k to zero at $\lambda = \infty$. (See Figure 11B.1.)

Suppose we consider what happens for various values of R. Each value of R can give rise to, at most, $2k$ corresponding values of λ. The number will be less if some of the loops in Figure 11B.1 have their lowest point above the value of R being

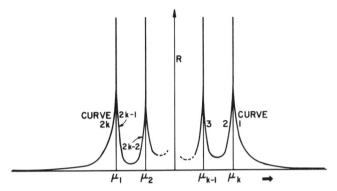

FIGURE 11B.1.   The dependence of R on λ.

considered. It is clear too, that if we wish to find the locus of the absolute minimum of $\hat{y}$ as R varies we can substitute any value of λ less than the smallest latent root μ_1 into (11B.3) and obtain a point on the locus, since there is only one such locus and thus there can be no ambiguity. A similar remark, but with $\lambda > \mu_k$, is true for the locus of the absolute maximum $\hat{y}$ as R varies. When we choose values of λ between latent roots, however, we may be on either of two loci of stationary values, depending on whether we are to the right or left of the value of λ for which R is stationary.

As indicated above, not all of the loci appear for every value of R, but as R increases, more and more appear. Since the fitted model can be considered accurate only within the region of the experimental design, loci which do not appear except for large R are usually of little interest.

To summarize the main practical feature of this work: Suppose we wish to follow the absolute maximum predicted value of $\hat{y}$ on a sphere of radius R, as R increases. Find the latent roots of $\mathbf{B}$, choose values of λ greater than all of these roots, and substitute them into (11B.3). Solve for x, evaluate $R^2 = \mathbf{x}'\mathbf{x}$ and $\hat{y}$, and plot $\hat{y}, x_1, x_2, \ldots, x_k$ against R. (Similar work, choosing values of λ less than all of the latent roots of B, can be carried out for an investigation of the absolute minimum value of $\hat{y}$ on spheres of radius R.)

Example

This example was used by Hoerl (1959). Consider the response surface in two factors

$$\hat{y} = 80 + 0.1x_1 + 0.2x_2 + 0.2x_1^2 + 0.1x_2^2 + x_1x_2.$$

Thus

$$\mathbf{B} = \begin{bmatrix} 0.2 & 0.5 \\ 0.5 & 0.1 \end{bmatrix}, \qquad \mathbf{b} = \begin{bmatrix} 0.1 \\ 0.2 \end{bmatrix}$$

Equations (11B.3) become

$$(0.2 - \lambda) x_1 + 0.5x_2 = -0.05,$$

$$0.5x_1 + (0.1 - \lambda) x_2 = -0.10,$$

with solution

$$x_1 = \frac{(9 + 10\lambda)}{2D},$$

$$x_2 = \frac{(1 + 20\lambda)}{2D}, \tag{11B.8}$$

where

$$D = 100 \det(\mathbf{B} - \lambda \mathbf{I}) = 100\lambda^2 - 30\lambda - 23. \tag{11B.9}$$

The eigenvalues or latent roots of $\mathbf{B}$ are given by $D = 0$, whence

$$\lambda = 0.652 \quad \text{or} \quad -0.352. \tag{11B.10}$$

If now we wish to look for the locus of the absolute minimum (or maximum) of $\hat{y}$ on circles $x_1^2 + x_2^2 = R^2$ of radius R, we should insert in Eqs. (11B.8) values of λ less (or greater) than both eigenvalues (11B.10), that is, $\lambda < -0.352$ (or $\lambda > 0.652$).

Suppose we select a value $\lambda = 0.2$. Then Eqs. (11B.8) and (11B.9) yield solution $(x_1, x_2) = (-0.22, -0.10)$. Then $R = 0.242$, so that on the circle $x_1^2 + x_2^2 = 0.242$, $\hat{y}$ is stationary at the point $(-0.22, -0.10)$ but, because $-0.352 < \lambda = 0.2 < 0.652$, this stationary value $\hat{y} = 79.99$ is neither an absolute maximum or minimum.

Continued substitution of values of λ into Eqs. (11B.8) and (11B.9) will yield four loci of stationary values as R increases and these, as evaluated by Hoerl (1959), are shown in Figure 11B.2.

The loci of absolute maximum and absolute minimum, curves 1 and 4, begin at $R = 0$ and correspond to values of λ beginning at $\lambda = \infty$ and $\lambda = -\infty$, respectively. The two loci of intermediate stationary values do not begin until $R = 0.195$ and correspond to $\lambda = -0.003$, when $\partial R / \partial \lambda = 0$, that is, we are at the bottom of the loop of R, plotted against λ, which lies between the latent roots $\mu_1 = -0.352$ and $\mu_2 = 0.652$. Because of the scale of the diagram, the difference in starting points cannot be distinguished.

The response surface given is in fact a saddle, rising in the first and third quadrants of the (x_1, x_2) plane, falling in the second and fourth quadrants, with ridges oriented approximately 45° to the axes and with center slightly off the origin at $(-\frac{9}{46}, -\frac{1}{46})$. Thus the locus of absolute maxima in Figure 11B.2 passes from the origin out the first quadrant of the (x_1, x_2) plane, the locus of absolute minima passes out the fourth quadrant, and the other two loci of stationary points, which are loci of neither the absolute maximum nor the absolute minimum, pass out the second and third quadrants.

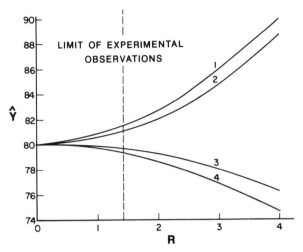

FIGURE 11B.2.   Loci of stationary values as R varies.

EXERCISES

11.1.   (Source: The influence of temperature on the scouring of raw wool, by C. C. Grove, *American Dyestuffs Reporter*, **54**, 1965, 13–16.) A two-factor four-level design was used to investigate the response $y = $ "consumption of detergent in lb/100 lb of raw wool" in a wool-scouring experiment. Four bowls were used. In each of the first two (15 ft long, 1 ft wide, capacity 170 gal each) four levels of temperature, 50, 55, 60, 65°C, were investigated. The third bowl (wash, 10 ft by 1 ft, 100 gal) was maintained at 50°C, and the fourth (rinse, 10 ft by 1 ft, 100 gal) at 40°C. The rate of detergent addition was determined by frequent checks on the variable $z = $ "the residual grease content of the wool" which "was kept at approximately 0.6%." The table shows the 16 design points and the results. The temperatures have been coded to $-3, -1, 1, 3$ via $x_i = $ (temperature$_i$ $- 57.5)/2.5$, for bowls $i = 1, 2$, the y's are in the body of Table E11.1, and the z's are below the y's in parentheses.

Fit a second-order response surface to these data and check whether use of second-order terms is necessary. Check whether your chosen surface is worthy of further interpretation, and plot the contours, if so. If a second-order surface is retained, also perform the canonical reduction. If low levels of detergent consumption are desirable, what combination of first and second bowl temperatures would you recommend?

In your checks of residuals, plot residuals versus z values. Does the presence of the one atypical z value (0.81) alter your conclusions above?

11.2.   Perform a canonical reduction on the fitted second-order equations:

(a) $\hat{y} = 60 + 0.35x_1 + 0.22x_2 + 0.12x_1^2 - 0.63x_2^2 - 3.67x_1x_2.$

(b) $\hat{y} = \frac{1}{4}\{(12 + 2\sqrt{3}) - (10 + 2\sqrt{3})x_1 - (14 + 2\sqrt{3})x_2 + 5x_1^2 + 7x_2^2 + 2\sqrt{3}x_1x_2\}.$

TABLE E11.1

x_2	x_1			
	-3	-1	1	3
-3	2.038	0.890	0.475	0.528
	(0.81)	(0.56)	(0.56)	(0.60)
-1	1.192	0.689	0.424	0.369
	(0.66)	(0.58)	(0.56)	(0.58)
1	0.592	0.561	0.333	0.413
	(0.59)	(0.60)	(0.60)	(0.61)
3	0.523	0.517	0.318	0.393
	(0.59)	(0.54)	(0.57)	(0.53)

11.3. Perform a canonical analysis of the following second-order fitted surfaces:
   (a) $\hat{y} = 60 - 44x_1 - 30x_2 + 8x_1^2 + 3x_2^2 + 12x_1x_2$.
   (b) $\hat{y} = 23 - 404x_1 + 396x_2 + 101x_1^2 + 101x_2^2 - 198x_1x_2$.
   (c) $\hat{y} = 75 + 2.06x_1 + 0.85x_2 - 4x_1x_2$.

11.4. Perform a canonical reduction on the fitted equation

$$\hat{y} = b_0 + b_1x_1 + b_2x_2 + b_{11}x_1^2 + b_{22}x_2^2 + b_{12}x_1x_2$$

for each of the sets of b's given in Table E11.4. In each case find and plot the canonical axes, state the type of surface found, and draw rough contours. (Accurate plots can also be made if desired.)

TABLE E11.4

Problem number	1	2	3	4
b_0	75.6	26.7	76.4	29.8
b_1	-7.6	-3.1	-1.7	-3.8
b_2	1.5	-10.8	-1.7	-0.2
b_{11}	2.5	0.3	1.3	-0.4
b_{22}	-0.5	3.8	1.7	-0.6
b_{12}	-2.6	2.2	-0.9	1.0

11.5. (Source: Simultaneous optimization of several response variables, by G. Derringer and R. Suich, *J. of Quality Technology*, **12**, October 1980, 214–219. Adapted with the permission of the copyright holder, the American Society

TABLE E11.5

Compound number	x_1	x_2	x_3	Y_1	Y_2	Y_3	Y_4
1	−1	−1	1	102	900	470	67.5
2	1	−1	−1	120	860	410	65
3	−1	1	−1	117	800	570	77.5
4	1	1	1	198	2294	240	74.5
5	−1	−1	−1	103	490	640	62.5
6	1	−1	1	132	1289	270	67
7	−1	1	1	132	1270	410	78
8	1	1	−1	139	1090	380	70
9	−1.633	0	0	102	770	590	76
10	1.633	0	0	154	1690	260	70
11	0	−1.633	0	96	700	520	63
12	0	1.633	0	163	1540	380	75
13	0	0	−1.633	116	2184	520	65
14	0	0	1.633	153	1784	290	71
15	0	0	0	133	1300	380	70
16	0	0	0	133	1300	380	68.5
17	0	0	0	140	1145	430	68
18	0	0	0	142	1090	430	68
19	0	0	0	145	1260	390	69
20	0	0	0	142	1344	390	70

$x_1 = (\text{phr silica} - 1.2)/0.5.$
$x_2 = (\text{phr silane} - 50)/10.$
$x_3 = (\text{phr sulfur} - 2.3)/0.5.$

for Quality Control, Inc.) Table E11.5 shows four columns of response data obtained in the development of a tire tread compound on the four responses: PICO Abrasion Index, Y_1; 200% modulus, Y_2; elongation at break, Y_3; hardness, Y_4. Each column was taken at the 20 sets of conditions shown, where x_1, x_2, x_3 are coded levels of the variables X_1 = hydrated silica level, X_2 = silane coupling agent level, and X_3 = sulfur. (The coding is given below the table; phr = parts per hundred.) Fit a full second-order model to each set of data individually. From a practical viewpoint, it is desirable that $Y_1 > 120$, $Y_2 > 1000$, $600 > Y_3 > 400$, and $75 > Y_4 > 60$. What values of (x_1, x_2, x_3) achieve this, based on the fitted surfaces? If this question seems too difficult to handle, consider this simpler question: Which (x_1, x_2, x_3) points in the actual data satisfy all the restrictions? Reduce to canonical form those fitted second-order surfaces that have significant curvature.

11.6. (Source: Kinetics of Catalytic Isomerisation of n-Pentane, by N. L. Carr, *Industrial and Engineering Chemistry*, **52**, 1960, 391–396. The responses are

TABLE E11.6a

x_1	x_2	x_3	y_1	y_2
-1	-1	-1	19.2	4.39
1	-1	-1	29.5	5.43
-1	1	-1	21.6	4.65
1	1	-1	33.8	5.82
-1	-1	1	36.1	6.01
1	-1	1	86.6	9.30
-1	1	1	37.4	6.11
1	1	1	44.8	6.70
-1.68	0	0	27.3	5.22
1.68	0	0	45.0	6.71
0	-1.68	0	20.3	4.51
0	1.68	0	30.0	5.47
0	0	-1.68	23.9	4.89
0	0	1.68	42.3	6.50
0	0	0	33.7	5.81
0	0	0	26.8	5.17
0	0	0	28.7	5.35
0	0	0	26.7	5.17

TABLE E11.6b

x_1	x_2	x_3	y_1	y_2
0	-1.31	-0.38	33.3	5.77
-1.93	1.75	-1.57	16.3	4.03
1.17	-1.29	-1.07	31.8	5.64
-0.49	3.56	-1.53	28.0	5.29
-0.50	0.18	-1.61	17.9	4.23
-1.55	3.68	-1.52	22.5	4.74

rounded. Adapted by permission of the copyright holder, the American Chemical Society.) Fit a second-order model to the data in Table E11.6a. Test for lack of fit. If no lack of fit is shown, test whether or not second-order terms are needed, and also whether or not all the x's should be retained. Make your choice of the fitted model to use (if any) on the basis of these tests. Then, use your selected model to predict responses at the additional points given in Table E11.6b, and check the values obtained against the responses actually recorded.

If canonical reduction is justified for either or both responses, perform the additional analysis.

11.7. (Source: The effects of various levels of sodium citrate, glycerol, and equilibrium time on survival of bovine spermatazoa after storage at $-79°C$, by R. G. Cragle, R. M. Myers, R. K. Waugh, J. S. Hunter, and R. L. Anderson, *Journal of Dairy Science*, **38** (5), 1955, 508–514.) A procedure for storing bovine semen involved the addition of $\xi_1\%$ sodium citrate and $\xi_2\%$ glycerol, and maintainance at 5°C for an equilibration time of ξ_3 h before freezing. It was desired to know how these three factors affected the percentage survival rate of motile spermatozoa. The coding

$$x_1 = \frac{(\xi_1 - 3)}{0.7}, \qquad x_2 = \frac{(\xi_2 - 8)}{3}, \qquad x_3 = \frac{(\xi_3 - 16)}{6},$$

was used and a three-factor cube plus star ($\alpha = 2$) plus one center point design was used to provide the data shown in Table E11.7. Fit a second-order response surface to these data and provide all the usual regression analyses. Reduce the selected surface to canonical form, interpret the results, and form conclusions about the best levels of (ξ_1, ξ_2, ξ_3).

TABLE E11.7. Percentage survival rates of motile spermatozoa (y) from a designed experiment

x_1	x_2	x_3	y
-1	-1	-1	57
1	-1	-1	40
-1	1	-1	19
1	1	-1	40
-1	-1	1	54
1	-1	1	41
-1	1	1	21
1	1	1	43
0	0	0	63
-2	0	0	28
2	0	0	11
0	-2	0	2
0	2	0	18
0	0	-2	56
0	0	2	46

11.8. (Source: Composite designs in agricultural research, by P. Robinson and K. F. Nielsen, *Canadian J. of Soil Science*, **40**, 1960, 168–176.) Tomato plants in half-gallon glazed pots, grown in North Gower clay loam, received various

TABLE E11.8

	Coded levels			Yields of tops (gm. dm./pot)	
	x_1	x_2	x_3	Replicate 1	Replicate 2
	−1.5	0	0	5.00	5.14
	1.5	0	0	12.71	14.29
	0	−1.5	0	5.52	4.20
	0	1.5	0	12.59	10.43
	0	0	−1.5	9.96	12.24
	0	0	1.5	10.23	10.38
	0	0	0	9.78	9.40
	−1	−1	−1	6.82	7.83
	−1	−1	1	7.08	6.21
	−1	1	−1	8.85	8.58
	−1	1	1	6.97	6.07
	1	−1	−1	9.49	7.80
	1	−1	1	7.80	8.85
	1	1	−1	16.25	10.83
	1	1	1	15.48	14.56
	−1	−1	0	5.38	6.78
	−1	0	−1	5.90	7.42
	−1	0	0	6.44	6.80
	−1	0	1	5.68	5.22
	−1	1	0	7.03	7.70
	0	−1	−1	9.23	8.50
	0	−1	0	6.52	9.82
	0	−1	1	8.08	7.31
	0	0	−1	8.86	11.56
	0	0	1	12.62	11.63
	0	1	−1	13.68	10.58
	0	1	0	11.90	11.50
	0	1	1	13.00	14.03
	1	−1	0	8.38	9.48
	1	0	−1	13.89	12.58
	1	0	0	12.60	14.79
	1	0	1	13.23	13.51
	1	1	0	14.40	15.58
SS (composite)	12.5	12.5	12.5		
SS (3^3)	18.	18.	18.		

combinations of three nutrients, nitrogen (N), phosphorus (P_2O_5), and potassium (K_2O). The 33 runs exhibited in Table E11.8 show the levels of the coded variables $x_1 = (N - 150)/100$, $x_2 = (P_2O_5 - 240)/160$, and $x_3 = (K_2O - 210)/140$, the integers being in units of pounds per acre. Replicate response values of yields of plant tops were measured, as shown.

The first 15 runs form a composite design. The remaining 18 runs are the additional runs necessary to provide a complete 3^3 factorial "enclosed within" the composite set of treatments. (Another experiment was performed using Castor silt loam in which the composite design was "enclosed within" the factorial. This is described in a subsequent exercise.) One purpose of this experiment was to compare the relative efficacies of the two designs, composite versus 3^3 factorial. (Note that the comparison is somewhat biased in favor of the 3^3 factorial in that its 27 points are generally "more spread out" than the 15 points of the composite design, as measured by $\sum_u x_{iu}^2$; these values are shown at the bottom of the table. The factorial points are distant $3^{1/2} = 1.732$ or $2^{1/2} = 1.414$ from the origin, while the axial points are 1.5 units distant, as far as individual points are concerned.) Perform the following calculations.

(a) Fit a second-order response surface to the 30 observations from the replicated composite design.

(b) Fit a second-order response surface to the 54 observations from the replicated 3^3 factorial design.

(c) Fit a second-order response surface to all 66 observations.

In (a), (b), and (c) perform all the usual response surface analyses and tests, and also answer this question:

(d) Do we need to retain variable x_3 in the model? (*Hint*: Get the extra sum of squares for all b's with a 3 subscript.)

(e) For a *reduced* second-order model in x_1 and x_2 only, repeat (a), (b), and (c).

(f) Plot the three surfaces you obtain in (e) in the square $-1.5 \le x_i \le 1.5$, $i = 1, 2$.

(g) State your conclusions about the form of the response surfaces and the comparative efficacies of the two designs, composite and 3^3.

11.9. (Source: Same as the foregoing exercise.) Tables E11.9a and E11.9b are for tomatoes grown in Castor silt loam, and give response values for a replicated composite design and for a 3^3 design which "encloses" it, and has wider spread, as measured by second moments of the design. Notice also that the 3^3 observations designated by superscript a's are merely values repeated from replicate 1 of the doubled composite design, and not new observations. Thus, when the designs are combined, these runs must be included *only once*.

(a) Fit a second-order response surface to the 30 observations from the replicated composite design.

(b) Fit a second-order response surface to the 27 observations from the 3^3 factorial design.

TABLE E11.9a. A replicated composite design (30 runs)

Coded levels				Yields of tops (gm. dm./pot)	
x_1	x_2	x_3		Replicate 1	Replicate 2
-1	-1	-1		3.70	3.80
-1	-1	1		2.94	2.11
-1	1	-1		4.94	3.92
-1	1	1		2.31	2.03
1	-1	-1		5.68	5.29
1	-1	1		3.31	3.09
1	1	-1		10.57	11.83
1	1	1		16.82	9.84
0	0	0		5.86	7.24
-1.5	0	0		2.69	2.03
1.5	0	0		13.20	10.09
0	-1.5	0		0.91	0.82
0	1.5	0		6.50	7.09
0	0	-1.5		6.85	7.28
0	0	1.5		6.01	5.53
12.5	12.5	12.5	SS[a]		

[a] For two replicates, SS = 25. This compares with 40.5 for the corresponding 3^3 design which is thus more "widely spread."

(c) Fit a second-order response surface to all 50 observations. (*Reminder*: Observations with superscript *a*'s attached should be omitted for this.)
In (a), (b), and (c) perform all the usual response surface analyses.

(d) Plot the three surfaces you obtain in (a), (b), and (c), for slices at $x_3 = -1.5, 0, 1.5$ and for $-1.5 \leq x_i \leq 1.5$, $i = 1, 2$.

(e) State your conclusions about the form of the response surfaces and the comparative efficacies of the two designs, composite and 3^3.

(f) Perform a canonical analysis on all three surfaces and compare the three canonical forms.

11.10. (Source: The influence of fertilizer placement and rate of nitrogen on fertilizer phosphorus utilization by oats as studied using a central composite design, by M. H. Miller and G. C. Ashton, *Canadian J. of Soil Science*, **40**, 1960, 157–167.)

Three predictor variables X_1 = horizontal distance from seed, inches, X_2 = depth below seed, inches, and X_3 = rate of nitrogen applied, pounds per acre, were examined. The variables were coded as $x_1 = X_1 - 2$, $x_2 = X_2 - 1$, and $x_3 = (X_3 - 10)/5$ and a 38-point design consisting of a doubled

TABLE E11.9b. A 3^3 factorial design (27 runs)

Coded levels			Yields of tops (gm. dm./pot)
x_1	x_2	x_3	Replicate 1
-1.5	-1.5	-1.5	0.80
-1.5	-1.5	0	0.87
-1.5	-1.5	1.5	0.92
-1.5	0	-1.5	2.53
-1.5	0	0	2.69^a
-1.5	0	1.5	1.66
-1.5	1.5	-1.5	2.78
-1.5	1.5	0	2.02
-1.5	1.5	1.5	1.61
0	-1.5	-1.5	0.71
0	-1.5	0	0.91^a
0	-1.5	1.5	0.52
0	0	-1.5	6.85^a
0	0	0	5.86^a
0	0	1.5	6.01^a
0	1.5	-1.5	6.00
0	1.5	0	6.50^a
0	1.5	1.5	6.19
1.5	-1.5	-1.5	0.33
1.5	-1.5	0	0.42
1.5	-1.5	1.5	0.05
1.5	0	-1.5	6.91
1.5	0	0	13.20^a
1.5	0	1.5	10.88
1.5	1.5	-1.5	7.70
1.5	1.5	0	11.19
1.5	1.5	1.5	12.43
40.5	40.5	40.5	SS

^{a}Response values are those of replicate 1 in the composite design.

(replicated) cube ($\pm 1, \pm 1, \pm 1$), a doubled star ($\pm 2, 0, 0$), ($0, \pm 2, 0$), ($0, 0, \pm 2$), plus 10 center points ($0, 0, 0$), was performed. The response data for y_t = fertilizer phosphorus absorption by oats in milligrams per pot (taken at $t = 14, 28, 42, 58$ days after planting) are not given in the paper, but the estimated regression coefficients for the second order surfaces fitted are given, and they are reproduced in Table E11.10. The authors state that these

TABLE E11.10

Coefficient	Time, days after planting			
	14	28	42	58
b_0	0.3021^b	2.3507^b	3.0303^b	7.4846^b
b_1	-0.3587^b	-0.3802^b	-0.0893	-0.5334^a
b_2	0.0378	0.3087^b	0.2757^b	-0.2484
b_3	0.0099	0.7634^b	0.7188^b	0.9903^b
b_{11}	0.1178^b	-0.1622^a	-0.1424	-0.4302^a
b_{22}	-0.0367	0.0509	-0.0003	-0.0277
b_{33}	-0.0103	-0.0189	0.0948	-0.0027
b_{12}	-0.0294	-0.0162	0.0218	0.3681
b_{13}	0.0404	0.1610	0.2549	-0.0544
b_{23}	0.0057	0.0616	0.2300	0.0681
R^2	0.825	0.815	0.743	0.517

^{a}Significant at 0.05 probability level.
^{b}Significant at 0.01 probability level.
R is the multiple correlation coefficient.

coefficients are determined from the equations

$$b_0 = 0.0945945(0y) - 0.027027\Sigma(iiy),$$

$$b_i = 0.03125(iy),$$

$$b_{ii} = -0.027027(0y) + 0.015625(iiy) + 0.0054899\Sigma(iiy),$$

$$b_{ij} = 0.0625(ijy),$$

where, for example, $(ijy) = \Sigma_{u=1}^n x_{iu}x_{ju}y_u$, and $\Sigma(iiy) = (11y) + (22y) + (33y)$.

(a) Confirm that these equations are correct.

(b) Use the information in the equations and in the table to fit reduced equations of the following forms via least squares:

   i. $y_{14} = \beta_0 + \beta_1 x_1 + \beta_{11} x_1^2 + \varepsilon.$

   ii. $y_{28} = \beta_0 + \beta_1 x_1 + \beta_2 x_2 + \beta_3 x_3 + \beta_{11} x_1^2 + \beta_{12} x_1 x_2 + \beta_{13} x_1 x_3 + \varepsilon.$

   iii. $y_{42} = \beta_0 + \beta_1 x_1 + \beta_2 x_2 + \beta_3 x_3 + \varepsilon.$

   iv. $y_{58} = \beta_0 + \beta_1 x_1 + \beta_2 x_2 + \beta_3 x_3 + \beta_{11} x_1^2 + \beta_{12} x_1 x_2 + \beta_{13} x_1 x_3 + \epsilon.$

(c) Plot the above surfaces for $x_3 = -2, -1, 0, 1, 2$ and comment upon what these show.

(d) Suppose we code time as a fourth predictor $x_4 = (t - 35)/7$ so that the levels of x_4 become $-3, -1, 1, 3.2857$. Do we have adequate information to fit a full second-order model in x_1, x_2, x_3, and x_4? Fit such a model if so, and determine which coefficients are "significant" at the 0.05 and 0.01 levels.

(e) Fit a reduced model retaining all "significant" coefficients plus any "nonsignificant" coefficients relating to terms of equal or lesser order in x's whose estimated coefficients were significant. Examples: (i) If b_{11} were significant, retain $\beta_{11}, \beta_{12}, \beta_{13}, \beta_{14}, \beta_1$; (ii) If b_{12} were significant, retain $\beta_{11}, \beta_{12}, \beta_{13}, \beta_{14}, \beta_{22}, \beta_{23}, \beta_{24}, \beta_1, \beta_2$.

(f) Perform a canonical reduction on your reduced model and interpret the results.

11.11. (Source: A statistical approach to catalyst development, part I: the effect of process variables in the vapour phase oxidation of napthalene, by N. L. Franklin, P. H. Pinchbeck, and F. Popper, *Transactions of the Institution of Chemical Engineers*, **34**, 1956, 280–293. See also **36**, 1958, 259–269.) Table E11.11 provides 80 observations on the variables

$\xi_1 =$ air to naphthalene ratio (L/pg),

$\xi_2 =$ contact time (s),

$\xi_3 =$ bed temperature $(^\circ C)$

$Y_P =$ percentage mole conversion of naphthalene to phthalic anhydride,

$Y_N =$ percentage mole conversion of naphthalene to naphthoquinone,

$Y_M =$ percentage mole conversion of naphthalene to maleic anhydride,

$Y_C =$ percentage mole conversion of naphthalene

by complete oxidation to CO and CO_2,

which arose in an investigation of the oxidation of napthalene to phthalic anhydride. For a full description, the reader should consult the references given above. Fit, to the response variable Y, by least squares, a full quadratic model in the predictor variables

$$X_1 = \log \xi_1, \qquad X_2 = \log(10\xi_2), \qquad X_3 = 0.01(\xi_3 - 330),$$

where log denotes logarithm to the base 10. Carry out the usual analysis of variance and checks of residuals. Apply the methods of canonical reduction

TABLE E11.11

ξ_1	ξ_2	ξ_3	Y_P	Y_N	Y_M	Y_C
7	2.35	404	63.3	2.2	4.7	27.5
5	2.35	406	61.6	2.5	4.2	26.3
7	2.75	403	60.9	1.9	4.7	28.3
9	1.88	403	67.0	1.1	5	24.5
11	1.38	400	73.0	1.9	3.8	14.8
13	0.84	400	75.8	3.4	3	18
15.2	0.31	402	72.4	8.2	3.6	2
13	0.60	405	75.3	3	5	19.9
15	0.84	400	75.2	3.2	3.1	19
15	0.60	399	74.9	2.7	3.8	17.5
10	0.50	407	77.5	2.4	4.8	4
8	0.38	428	80.0	1.2	4	0
7.2	0.30	428	75.8	1.9	5.2	0
10	0.85	402	72.5	1.7	4.3	11.8
16.5	0.40	400	76.4	4	2	6
18	0.80	395	77.5	3.8	2.8	9
20	1.00	397	73.7	3.1	3.3	11.4
10	0.60	405	75.5	2.4	4.8	4
12.3	0.59	405	78.0	3	5	0
10	0.86	404	72.5	1.7	4.3	11.8
12.9	0.84	400	75.8	3.4	3	18
10.4	0.62	385	67.8	12.6	4.2	2
12.3	0.60	385	69.5	12	3.9	1
10	0.88	386	71.2	6.4	4.1	0
12.5	0.89	383	77.0	5.3	4.4	4.3
9.05	0.46	421	80.0	2.5	4.1	6.6
7.9	0.40	413	79.6	3.7	4	0
21	0.50	408	83.2	2	4.4	1.3
40.2	0.20	406	78.0	6.8	2.5	7
10.7	0.50	405	75.0	4.6	4.3	4.3
7	0.59	405	74.5	5.7	5.4	0
29	0.29	404	84.2	3.7	2.4	7.3
25.3	0.79	404	75.8	1.4	5.7	9
25	0.39	404	83.8	1.5	3	7
9.45	1.63	404	65.2	1.3	4.4	25.8
9	1.42	404	70.5	1.9	4.8	20
25	0.60	403	82.4	1.6	2.9	10.7
10.2	0.30	403	62.5	14.5	3.7	1
7.25	0.90	403	74.2	4.7	5.3	5.3
5.75	0.435	403	66.0	11.7	6.9	0
40.2	0.43	401	81.4	3.7	4.1	8.3
12.5	1.07	401	73.0	2.2	1.8	22

TABLE E11.11. (*Continued*)

ξ_1	ξ_2	ξ_3	Y_P	Y_N	Y_M	Y_C
10.7	1.70	401	71.0	2.2	4.1	22
16.2	0.80	400	74.7	1.8	2.8	10.5
12.1	1.46	400	69.0	2.1	3	18.6
10	0.61	400	74.0	6.8	5.4	0
11.5	0.73	398	79.0	4.4	5.3	1
10.25	0.87	397	77.5	4.1	3.3	0.5[a]
28.5	0.50	396	79.2	3.1	5.9	2.6
12.15	0.87	395	76.4	2.9	3.3	4.7
25.7	0.45	392	75.0	2.5	3.4	7.1
12.4	1.20	392	79.0	2.3	5.2	9.6
12.15	0.59	392	72.0	8.9	3.4	1.4
25.3	0.66	391	83.5	1.6	2.3	8.3
19.4	1.05	391	82.5	1.4	3.3	7.2
24.7	1.6	390	75.0	0.5[a]	3.5	0
16.7	0.87	390	82.5	2.8	5.3	6.3
15.5	1.19	390	77.3	1.6	4.5	9.8
40.2	0.605	389	78.5	2.7	2.9	16.5
31.8	1.21	389	78.5	1.3	2.5	14.7
25.4	0.905	389	84.0	1.8	2.5	10
30.4	0.355	383	75.4	7.6	2.3	7.8
22.7	0.51	383	80.5	4	3.9	4.2
71.5	0.50	382	80.4	4.1	2.6	11.4
39.5	0.50	382	83.0	7.3	3.4	8.8
12.4	1.19	382	72.1	2.9	2.6	7
58.5	0.60	381	87.0	1.9	2.4	10.4
16	1.20	381	78.8	2.9	3.8	8.6
81	0.70	380	84.1	1.3	5.2	9
29.6	0.71	380	82.7	3.6	3.9	7.1
25.3	1.01	380	81.0	2.2	3.7	8
25	1.63	380	77.5	0.5[a]	4	15.7
16	0.88	380	80.0	4	3.7	7.6
39.5	1.21	379	81.5	1.8	2.7	15
19.7	1.08	379	78.5	1.8	3.3	2.2
38	0.81	378	82.5	2.1	3.75	8.8
30.9	1.23	373	77.5	2	2.8	11.6
76.5	0.80	365	86.7	3.1	3.8	0
42	2.35	365	74.2	0.5[a]	3.4	24
103	1.25	334	87.5	4.6	3.5	10.5

[a]Actually recorded as " < 1.0;" the value 0.5 was arbitrarily inserted.

to elucidate the type of surface represented by the fitted model, and come to appropriate conclusions. (This exercise can be performed for any or all of the four response variables shown.)

11.12. Reduce, to canonical form, the second-order fitted equation $\hat{y} = 67.711 + 1.944x_1 + 0.906x_2 + 1.069x_3 - 1.539x_1^2 - 0.264x_2^2 - 0.676x_3^2 - 3.088x_1x_2 - 2.188x_1x_3 - 1.212x_2x_3$.

11.13. Perform a canonical analysis on the fitted equation

$$\hat{y} = b_0 + b_1x_1 + b_2x_2 + b_3x_3 + b_{11}x_1^2 + b_{22}x_2^2 + b_{33}x_3^2$$

$$+ b_{12}x_1x_2 + b_{13}x_1x_3 + b_{23}x_2x_3$$

for each of the sets of b's given in Table E11.13. In each case find and plot the canonical axes, state the type of surface found, and draw rough contours.

TABLE E11.13

Problem number	1	2	3	4
b_0	52	43	79	71
b_1	2.01	−1.67	−0.39	1.73
b_2	1.60	−0.69	0.27	0.90
b_3	−0.46	−0.21	0.69	0.41
b_{11}	0.38	2.85	0.99	−0.25
b_{22}	−1.04	1.58	0.11	−1.90
b_{33}	−4.25	1.57	0.02	−1.85
b_{12}	−0.82	−0.18	0.22	−3.28
b_{13}	−3.18	−0.92	−0.10	−1.71
b_{23}	−4.99	−0.97	0.01	0.37

11.14. (Source: Response surface for dry modulus of rupture and drying shrinkage, by H. Hackney and P. R. Jones, *Amer. Ceramic Soc. Bulletin*, **46**, 1967, 745–749.) Carry out a canonical reduction on the fitted surfaces:
  (a) $\hat{y} = 6.88 + 0.0325x_1 + 0.2588x_2 - 0.1363x_3$
      $- 0.1466x_1^2 - 0.0053x_2^2 + 0.1359x_3^2$
      $+ 0.1875x_1x_2 + 0.2050x_1x_3 - 0.1450x_2x_3$.
  (b) $\hat{y} = 2.714 - 0.0138x_1 - 0.1300x_2 - 0.1525x_3$
      $+ 0.0333x_1^2 - 0.0342x_2^2 + 0.4833x_3^2$
      $+ 0.175x_1x_2 + 0.28x_1x_3 + 0.5625x_2x_3$.

11.15. (Source: Subjective responses in process investigation, by H. Smith and A. Rose, *Indus. and Eng. Chem.*, **55** (7), 1963, 25–28. Adapted by permission of the copyright holder, the American Chemical Society.) Three predictor variables, the amounts of flour, water, and shortening were varied in a pie crust recipe. Three responses, flakiness, gumminess (toughness), and specific volume in cm^3/g were measured. The fitted second-order surface obtained for flakiness in terms of coded predictors was

$$\hat{y} = 6.89462 + 0.06323x_1 - 0.12318x_2 + 0.15162x_3$$
$$- 0.11544x_1^2 - 0.03997x_2^2 - 0.11544x_3^2$$
$$+ 0.09375x_1x_2 - 0.34375x_1x_3 - 0.03125x_2x_3.$$

Reduce this equation to canonical form. If flakiness should exceed 7, in what region of (x_1, x_2, x_3) should the crust be made? [*Note*: The original paper is concerned also with limits on the other two responses. Moreover, because the observed F for (regression given b_0) for flakiness is only 1.1 times the $\alpha = 0.05$ percentage point, there is some question as to how well established the fitted surface is, and thus whether canonical reduction is justified. In fact, the original authors did not use canonical reduction, but simply plotted response contours and looked for desirable regions. Properties observed could then be checked with confirmatory runs.]

11.16. (Source: Maximum data through a statistical design, by C. D. Chang, O. K. Kononenko, and R. E. Franklin, Jr., *Industrial and Engineering Chemistry*, **52**, November 1960, 939–942. Material reproduced and adapted by permission of the copyright holder, the American Chemical Society. The data were obtained under a grant from the Sugar Research Foundation, Inc., now the World Sugar Research Organization, Ltd.) Consider the data for y_2(DMP) in Exercise 4.16. Remember to count "Trial 17" as six individual trials, each with response 32.8, for the surface fitting and to count, as pure error sum of squares, the quantity $(6 - 1)s_2^2 = 5(9.24) = 46.2$ in the analysis of variance table. Combine the previous data with those of trials 18–25 given in Table E11.16. Fit a full second-order surface to all 30 data points and check for

TABLE E11.16. **Extra trials completing a central composite design**

Trial number	Factor levels				Response y_2, DMP
	x_1, NH_3	x_2, T	x_3, H_2O	x_4, P	
18	−1.4	0	0	0	31.1
19	1.4	0	0	0	28.1
20	0	−1.4	0	0	17.5
21	0	1.4	0	0	49.7
22	0	0	−1.4	0	49.9
23	0	0	1.4	0	34.2
24	0	0	0	−1.4	31.1
25	0	0	0	1.4	43.1

lack of fit. If no lack of fit exists, is the regression significant and if so, what part(s), first order, second order, both? Is it worthwhile to canonically reduce the fitted surface? Repeat the entire exercise on the response

$$w_i = \sin^{-1}\left(\frac{y_i}{100}\right)^{1/2},$$

and compare and contrast the two analyses. What would you report to the experimenter?

11.17. (Source: Evaluation of variables in the pressure-kier bleaching of cotton, by J. J. Gaido and H. D. Terhune, *American Dyestuffs Reporter*, **50**, October 16, 1961, 23–26 and 32). Four predictor variables, coded here as

$$x_1 = (\text{temperature } °F - 235)/10$$

$$x_2 = (\text{time in minutes} - 60)/20$$

$$x_3 = (\text{albone hydrogen peroxide concentration } \% - 3)/0.2$$

$$x_4 = (\text{NaOH concentration } \% - 0.3)/0.1$$

were examined in a cotton-bleaching experiment. Three responses, coded here as

$$y_1 = \text{whiteness } \% - 80$$

$$y_2 = \text{absorbency (s)}$$

$$y_3 = \text{fluidity (rhes)}$$

were recorded. The results from 35 experimental runs appear in Table E11.17.

To each response, fit a second-order surface in the x's. Test for lack of fit, and if no lack of fit is found, test for the need for quadratic terms and/or the need for specific x variables. Note especially the peculiarity of the response data in run 16. The authors remark that this sample is "obviously under-bleached" and has an exceptionally high absorbency; thus you should treat this observation, certainly in its y_1 and y_2 values, at least, as a potential outlier.

Whatever is your final choice of fitted surface for each response, examine whether or not it is worth further interpretation and, if it is, plot or canonically reduce it for that interpretation. What are your overall conclusions?

11.18. (Source: Effect of raw-material ratios on absorption of whiteware compositions, by W. C. Hackler, W. W. Kriegel, and R. J. Hader, *J. Amer. Ceramic*

TABLE E11.17

Run number	Coded predictor levels				Response values		
	x_1	x_2	x_3	x_4	y_1	y_2	y_3
1	1.3	−2	2	2	8.2	1.2	1.8
2	1.3	−1	3	−1	8.3	1.8	1.5
3	1.3	−1	−3	1	8.0	1.6	1.5
4	1.3	−2	−2	−2	4.0	5.6	1.2
5	0.3	−1.5	2	2	8.4	2.0	1.8
6	0.3	−2.25	3	−1	4.6	4.7	1.2
7	0.3	−2.25	−3	1	5.2	1.5	1.4
8	0.3	−1.5	−2	−2	4.3	3.0	0.9
9	−0.3	3	2	2	9.4	1.2	2.2
10	−0.3	0	3	−1	6.9	3.6	1.5
11	−0.3	0	−3	1	7.7	1.7	1.5
12	−0.3	3	−2	−2	7.0	1.7	1.5
13	−1.3	−0.75	2	2	7.9	2.2	1.6
14	−1.3	1.5	3	−1	7.9	2.5	1.5
15	−1.3	1.5	−3	1	7.7	1.8	1.4
16	−1.3	−0.75	−2	−2	2.2	27.0	0.9
17	1.5	−2	0	0	6.9	1.0	1.4
18	−1.5	6	0	0	8.5	1.1	1.5
19	1.0	1	0	0	8.8	1.0	1.2
20	−1.0	−0.75	0	0	6.3	1.4	1.2
21	0	0	5	1	9.0	0.7	1.6
22	0	0	−5	−1	6.1	1.2	1.3
23	0	0	−1	3	8.4	0.6	1.8
24	0	0	1	−3	5.8	1.5	1.1
25	0	−1.5	0	0	5.7	2.3	1.3
26	0	−1.5	0	0	6.7	2.0	1.5
27	0	0	0	0	8.4	1.8	1.5
28	0	0	0	0	8.3	1.9	1.7
29	0	0	0	0	8.3	0.8	1.6
30	0	0	0	0	7.9	1.3	1.6
31	0	−1.5	0	0	6.4	1.4	1.4
32	1.3	−2	2	5	7.4	2.4	2.0
33	1.3	−1	2	2	7.8	2.4	2.1
34	1.3	−1	2	2	8.7	1.5	2.3
35	1.3	1	2	2	9.6	1.1	2.6
Sums	5.2	−9.25	8	11	250.9	90.5	54.1

TABLE E11.18

Coded levels of the variables			Value of $100y$ when $x_4 =$		
x_1	x_2	x_3	-1	0	1
-1	-1	-1	1484	881	636
-1	-1	1	1354	726	392
-1	1	-1	1364	746	376
-1	1	1	1286	684	380
1	-1	-1	1171	642	383
1	-1	1	1331	778	550
1	1	-1	1133	597	358
1	1	1	1136	604	362
0	0	0	1304	756	419
-2	0	0	1577	882	455
2	0	0	1070	534	342
0	-2	0	1312	796	488
0	2	0	1278	711	411
0	0	-2	1281	706	418
0	0	2	1238	682	359

Soc., **39** (1), 1956, 20–25.) A study of four whiteware raw material components was carried out in the following way. First, three ratios of the four ingredients were formed. This avoided the problem arising from the fact that the components necessarily totaled 100% in each run. The three ratios were coded to variables x_1, x_2, and x_3. A fourth predictor variable, firing temperature, was coded and named x_4. The 45 responses obtained are shown in Table E11.18. The response variable y is the "fired absorption" of the whiteware "adjusted for location in the kiln and averaged for the two batches and two firings at each temperature." (For details of the adjustments, see pp. 21–22 of the source reference.)

Fit a second-order model to the data. Use the extra sum of squares principle to investigate if second-order terms are needed, if x_1 is needed, if x_2 is needed, and so on. When you have decided what model to adopt, check whether canonical reduction is worthwhile. If so, perform the canonical reduction and comment on the shape of the surface. If reduced absorption were desirable, what settings of the x's would you recommend for future work?

11.19. (Source: Starch vinylation, by J. W. Berry, H. Tucker, and A. J. Deutshman; *Indus. Eng. Proc., Des. Devel.*, **2** (4), 1963, 318–322. Also see correspondence by I. Klein and D. I. Marshall, **3** (3), 1964, 287–288. Adapted by permission of the copyright holder, the American Chemical Society.) A study of starch vinylation involved five input variables and two responses (only one of which is presented here). Table E11.19 shows a (coded) 2^{5-1} fractional

TABLE E11.19

		Coded values			
x_1	x_2	x_3	x_4	x_5	y
-1	1	-1	1	-1	0.94
-1	1	1	1	1	0.84
1	1	-1	-1	-1	0.75
0	0	0	0	0	0.84
-1	-1	-1	1	1	0.46
1	1	1	1	-1	0.66
-1	-1	1	-1	1	0.24
0	0	0	0	0	0.82
1	1	1	-1	1	0.60
1	-1	1	1	1	0.74
1	-1	-1	-1	1	0.26
1	-1	-1	1	-1	0.10
1	-1	1	-1	-1	0.80
-1	1	1	-1	-1	0.80
0	0	0	0	0	0.84
1	1	-1	1	1	0.78
-1	1	-1	-1	1	0.41
-1	-1	-1	-1	-1	0.25
-1	-1	1	1	-1	0.75
0	0	0	0	0	0.83
-2	0	0	0	0	0.73
0	2	0	0	0	0.72
0	0	-2	0	0	0.73
0	0	0	0	2	0.43
0	-2	0	0	0	0.30
2	0	0	0	0	0.82
0	0	0	-2	0	0.36
0	0	2	0	0	0.89
0	0	0	0	0	0.84
0	0	0	2	0	0.88
0	0	0	0	-2	0.95
0	0	0	0	0	0.84

factorial ($x_1 x_2 x_3 x_4 x_5 = -1$) plus 10 axial points distant ± 2 units from the origin, plus 6 center points, and the 32 responses that were observed.

Fit a second-order surface to these data and check for lack of fit and for the needed presence of second-order terms or of any of the x's.

Perform a canonical reduction on the fitted surface you decide to adopt, and interpret the results.

TABLE E11.20

Run number	x_1	x_2	x_3	x_4	x_5	y
1	-2	0	0	0	0	67
2	2	0	0	0	0	238
3	0	-2	0	0	0	197
4	0	2	0	0	0	174
5	0	0	-2	0	0	221
6	0	0	2	0	0	152
7	0	0	0	-2	0	195
8	0	0	0	2	0	154
9	0	0	0	0	-2	213
10	0	0	0	0	2	136
11	0	0	0	0	0	221
12	0	0	0	0	0	221
13	-1	-1	-1	-1	1	196
14	1	1	-1	-1	1	198
15	1	1	1	-1	-1	192
16	-1	-1	1	-1	-1	203
17	-1	-1	-1	1	-1	193
18	1	1	-1	1	-1	211
19	-1	-1	1	1	1	76
20	1	1	1	1	1	218
21	0	0	0	0	0	229
22	0	0	0	0	0	225
23	0	0	0	0	0	211
24	1	-1	-1	-1	-1	210
25	-1	1	-1	-1	-1	166
26	1	-1	1	-1	1	251
27	-1	1	1	-1	1	92
28	1	-1	-1	1	1	221
29	-1	1	-1	1	1	81
30	1	-1	1	1	-1	238
31	-1	1	1	1	-1	114
32	0	0	0	0	0	226
33	0	0	0	0	0	234
34	0	0	0	0	0	235

11.20. (Source: Statistically designed experiments, by G. C. Derringer, *Rubber Age*, **101**, 1969, November, 66–76.) Five major compounding ingredients in a silica filled rubber compound, coded to $x_1, \ldots, x_5$, were studied via a five-factor central composite design. The design was split into three blocks as indicated in Table E11.20.

Block 1: Ten axial points plus two center points.
Block 2: Half of a 2^{5-1} design ($x_1 x_2 x_3 x_4 x_5 = 1$), such that $x_1 x_2 = 1$, plus three center points.
Block 3: Half of a 2^{5-1} design ($x_1 x_2 x_3 x_4 x_5 = 1$), such that $x_1 x_2 = -1$, plus three center points.

The 2^{5-1} design was blocked on the signs of $x_1 x_2$ because it was known from previous work that the interaction between factors 1 and 2 was insignificant. The observed response shown is $y =$ (tensile strength of the compound, psi)/10.

Fit to the data a full second-order model in $x_1, \ldots, x_5$ with two blocking variables z_1 and z_2. Many choices of z_1 and z_2 levels are possible, but values $z_1 = (0, 1, 0)$ and $z_2 = (0, 0, 1)$ for blocks (1, 2, 3) respectively, are suggested. Check for lack of fit and whether or not a reduced model is possible. If appropriate, perform a canonical reduction of the fitted equation you finally select. (*Note*: Blocks are not orthogonal to the model. Various orders of terms in the analysis of variance table are possible. We suggest the sequence: b_0; blocks given b_0; first-order given b_0 and blocks; second-order given preceding items; residual, split up into lack of fit and pure error. Pure error is calculated within blocks only and has thus $1 + 2 + 2 = 5$ df.)

11.21. (Source: A statistical approach to catalyst development, part II: The integration of process and catalyst variables in the vapour phase oxidation of napthalene, by N. L. Franklin, P. H. Pinchbeck, and F. Popper, *Transactions of the Institution of Chemical Engineers*, **36**, 1958, 259–269.) Two five-predictor-variable fitted response surfaces for yield and purity were obtained as follows.

$$\text{yield} = -110.2 + 88.1 X_1 + 119.8 X_2$$

$$+ 152.1 X_3 - 10.7 X_4 + 27.6 X_5$$

$$- 15.8 X_1^2 - 10.8 X_2^2 - 32.9 X_3^2$$

$$- 6.5 X_4^2 - 7.9 X_5^2 - 14.1 X_1 X_2$$

$$- 28.9 X_1 X_3 + 0.2 X_1 X_4 - 6.3 X_1 X_5$$

$$- 52.2 X_2 X_3 - 1.9 X_2 X_4 - 30.8 X_2 X_5$$

$$+ 5.2 X_3 X_4 + 13.2 X_3 X_5 + 19.4 X_4 X_5.$$

$$\text{purity} = -292.90 + 137.36\,X_1 + 221.37 X_2$$

$$+ 247.75\,X_3 + 20.44\,X_4 + 46.14\,X_5$$

$$- 17.91\,X_1^2 - 48.35\,X_2^2 - 38.65\,X_3^2$$

$$- 9.17 X_4^2 - 14.09\,X_5^2$$

$$- 13.23\,X_1 X_2 - 47.90\,X_1 X_3$$

$$- 6.04\,X_1 X_4 - 10.90\,X_1 X_5$$

$$- 81.24\,X_2 X_3 - 11.63\,X_2 X_4$$

$$- 18.74\,X_2 X_5 - 12.66\,X_3 X_4$$

$$- 15.10\,X_3 X_5 + 27.95\,X_4 X_5 .$$

Reduce these surfaces to canonical form and indicate their main features. Does it appear possible to improve yield and purity at the same time (to the extent that the contours can be relied upon as accurate approximations to the underlying true surfaces)? Sketch the contours in the (X_4, X_5) plane for the following three sets of conditions (where log = logarithm to the base 10):

(a) $X_1 = \log 321$, $X_2 = \log 7$, $X_3 = 0.5$.
(b) $X_1 = \log 1001$, $X_2 = \log 7$, $X_3 = 0.7$.
(c) $X_1 = \log 321$, $X_2 = \log 7$, $X_3 = 0.7$.

CHAPTER 12

Links Between Empirical and Theoretical Models

12.1. INTRODUCTION

The underlying process mechanism determines the geometrical characteristics of a response surface, particularly the nature of ridge forms. Conversely, such characteristics of the surface, determined approximately by fitting purely empirical models, can often provide clues to the nature of the underlying mechanism.

We use, for illustration, the data of Example 11.4 in Section 11.4 which consist of the chemical conversions achieved at various conditions of flow rate (r), catalyst concentration (c), and temperature (T) in a continuous stirred reactor. Our earlier empirical analysis of these data showed the existence of a rising ridge surface in the variables r, c, and T.

12.2. A MECHANISTIC MODEL

The Form of the Mechanistic Model

Consider again the data of Section 11.4, set out again in Table 12.1. As we shall show subsequently, a mechanistic model which adequately accounts for the system under study, and is suggested purely by physical considerations, yields the following relationship between mean conversion η, and flow rate r, catalyst concentration c, and temperature T.

$$\eta = f(r, c, T | \theta_0, \theta_0', \theta_1, \theta_1', \theta_2, \theta_2')$$

$$= \frac{rc^{\theta_1}\theta_0\exp\{-\theta_2(x - \bar{x})\}}{[r + c^{\theta_1}\theta_0\exp\{-\theta_2(x - \bar{x})\}][r + c^{\theta_1'}\theta_0'\exp\{-\theta_2'(x - \bar{x})\}]}, \quad (12.2.1)$$

TABLE 12.1.   Observed percentage conversion y of reactant to desired product, at various conditions of flow rate (r) catalyst concentration (c), and temperature (T). The fitted values and residuals shown are for the fitted mechanistic model.

Block	Observation number	Flow rate, r	Catalyst concentration, c	Temperature, T	Percentage conversion		
					Observed, y	Fitted, $\hat{y}$	Residual, $y - \hat{y}$
1	1	1.5	1	90	40.0	38.9	1.1
	2	6.0	1	70	18.6	19.8	−1.2
	3	1.5	4	70	53.8	49.6	4.2
	4	6.0	4	90	64.2	65.2	−1.0
	5	3.0	2	80	53.5	53.7	−0.2
	6	3.0	2	80	52.7	53.7	−1.0
2	7	1.5	1	70	39.5	37.3	2.2
	8	6.0	1	90	59.7	57.2	2.5
	9	1.5	4	90	42.2	40.8	1.4
	10	6.0	4	70	33.6	33.4	0.2
	11	3.0	2	80	54.1	53.7	0.4
	12	3.0	2	80	51.0	53.7	−2.7
3	13	1.13	2	80	43.0	43.7	−0.7
	14	8.0	2	80	43.9	45.5	−1.6
	15	3.0	0.75	80	47.0	45.6	1.4
	16	3.0	5.33	80	62.8	60.1	2.7
	17	3.0	2	65.9	25.6	27.4	−1.8
	18	3.0	2	94.1	49.7	50.2	−0.5
4	19	1.13	2	80	39.2	43.7	−4.5
	20	8.0	2	80	46.3	45.5	0.8
	21	3.0	0.75	80	44.9	45.6	−0.7
	22	3.0	5.33	80	58.1	60.1	−2.0
	23	3.0	2	65.9	27.0	27.4	−0.4
	24	3.0	2	94.1	50.7	50.2	0.5

where

$$x = \frac{1}{(T + 273)}$$

is the reciprocal of the absolute temperature. The model contains six constants (parameters to be estimated), θ_0, θ_0'; θ_1, θ_1'; θ_2, θ_2' whose physical meanings will become clear later. Fitting this model to the data in Table 12.1 by nonlinear least squares yields the estimates

$$\hat{\theta}_0 = 5.90 \ (\pm 0.43),$$

$$\hat{\theta}_0' = 1.15 \ (\pm 0.09),$$

$$\hat{\theta}_1 = 0.53 \ (\pm 0.08),$$

$$\hat{\theta}_1' = -0.01 \ (\pm 0.07),$$

$$\hat{\theta}_2 = 15,475 \ (\pm 625),$$

$$\hat{\theta}_2' = 7489 \ (\pm 598). \tag{12.2.2}$$

TABLE 12.2. Approximate analysis of variance for the fitted mechanistic model (12.2.1)

Source	SS	df	MS
Nonlinear model	53,504.3	6	8917.4
Residual	$S(\hat{\theta}) = 85.3$	18	$s^2 = 4.7$
Total	53,589.6	24	

The parenthetical figures are ($\pm$ one standard error), obtained on the usual linearizing approximation assumption. (See Appendix 3D.) The residual sum of squares which remains unexplained by this model is 85.3. The resulting fitted values $\hat{y}_u$, and the residuals $y_u - \hat{y}_u$, are shown in Table 12.1 along with the original data. (For the coded form of the predictors, see Table 11.4.) The basic approximate analysis of variance table (obtained from the division $y'y = S(\hat{\theta}) + SS$ (model), with 6 df ascribed to the model —see Appendix 3D) is shown in Table 12.2. A comparative analysis for the various models is given in Appendix 12A. This shows that the six-parameter mechanistic model accounts for the data at least as well as the empirical quadratic model with 10 parameters.

The plot of residuals versus r exhibited a hint (but no more) of possible inhomogeneity of variance. This arose from the fact that the two most extreme residuals were at adjacent small r values. Other standard plots of residuals showed no abnormalities, however, and it was concluded that the theoretical model (12.2.1) provided an adequate representation of the data. Figure 12.1 shows diagrams of the contour surfaces produced by:

(a) The fitted second degree polynomial (11.4.2).
(b) The directly fitted canonical form (11.4.15).
(c) The nonlinear mechanistic model, (12.2.1) with (12.2.2).

The agreement between the representations, particularly that between (b) and (c), is very striking.

Derivation of the Mechanistic Model (12.2.1)

We now consider (a) how the theoretical model was obtained, (b) what physical meanings are associated with the parameters, and, most importantly, (c) how the form of the model induces the rising ridge behavior already seen in our earlier analysis.

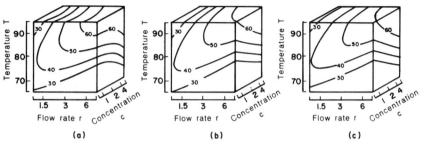

FIGURE 12.1. Contour diagrams of fitted surfaces. (a) Fitted second degree polynomial.
$\hat{y} = 51.8 + 0.74x_1 + 4.81x_2 + 8.01x_3 - 3.83x_1^2 + 1.22x_2^2 - 6.26x_3^2 + 0.38x_1x_2 + 10.35x_1x_3$
$- 2.83x_2x_3$. (b) Fitted canonical form, $\hat{y} = 53.1 - (-2.01x_1 + 0.34x_2 + 2.58x_3)^2 + 0.74x_1$
$+ 4.81x_2 + 8.01x_3$. (c) Fitted mechanistic model, $\hat{y} = f(r, c, T|\theta_0, \theta_0'; \theta_1, \theta_1'; \theta_2, \theta_2')$; see Eqs.
(12.2.1) and (12.2.2).

Point (c) is most important because it can lead to an understanding of
how empirical exploration can supply clues to mechanistic behavior. Specifically, the nature of the fitted empirical polynomial response surface can help
to suggest possible theoretical mechanisms, and thus may well lead to better
scientific understanding.

A diagrammatic representation of the chemical reaction in the continuous stirred tank reactor is shown in Figure 12.2. A solution of a chemical X
is flowing into a reactor at a rate of r L/min. Inside the reactor, it is
thoroughly mixed and comes into contact, at some fixed temperature T,
with a catalyst whose concentration is fixed at a level c. A reaction occurs in
which a proportion of X decomposes to form the desired product Y which,
in turn, decomposes to form an *un*desired biproduct Z. Suppose that, at a
particular flow rate, and after a suitable lapse of time to achieve equilibrium, the output stream from the process is found to contain $100\xi\%$ of X,
$100\eta\%$ of Y, and $100\zeta\%$ of Z. (The catalyst remains in the reactor and is
not part of the output stream.) Note that $\xi + \eta + \zeta = 1$.

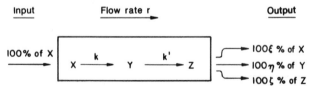

FIGURE 12.2. Diagrammatical representation of a stirred tank reactor showing input and
output.

Taking Account of Flow Rates

We might expect that the composition of the output and, in particular, the proportion of the desired product Y would depend on the flow rate r. A simple model that takes account of flow rate mechanistically may be derived on the basis of "collision" theory and the supposition of perfect mixing. Under these assumptions, the probability of a molecule of X reacting to form Y is proportional to the probability of a molecule of X colliding* with a catalyst molecule. Since the concentration of the catalyst is held constant,

$$\text{rate of chemical utilization of } X \text{ in the reactor} = k\xi. \quad (12.2.3)$$

Similarly, it follows that

$$\text{rate of chemical formation of } Y \text{ in the reactor} = k\xi - k'\eta. \quad (12.2.4)$$

To build the model, we take note of two further facts:

$$\text{rate of supply of } X \text{ to the reactor} = r - \xi r, \quad (12.2.5)$$

and

$$\text{rate of loss of } Y \text{ from the reactor} = r\eta. \quad (12.2.6)$$

Now if, as we suppose, the process has been allowed to come to equilibrium at the given flow rate, so that the concentrations (ξ, η, ζ) are constant, then the right-hand sides of Eqs. (12.2.3) and (12.2.5) must be equal, so that

$$r - \xi r = k\xi$$

or

$$\xi = \frac{r}{(r + k)}. \quad (12.2.7)$$

Similarly, equating the right-hand sides of (12.2.4) and (12.2.6), we have that

$$k\xi - k'\eta = r\eta$$

or

$$\eta = \frac{k\xi}{(r + k')}. \quad (12.2.8)$$

*This version of the theory is, of course, a vast oversimplification of what actually happens but can yield useful results.

It follows, from (12.2.7) and (12.2.8), that

$$\eta = \frac{kr}{(r+k)(r+k')}$$

(12.2.9)

and, furthermore, because $\xi + \eta + \zeta = 1$,

$$\zeta = \frac{kk'}{(r+k)(r+k')}$$

(12.2.10)

The Effect of Catalyst Concentration c

For the purposes of this elementary discussion, we define a catalyst as an agent which speeds up a chemical reaction without directly taking part in it. Its effect on k and k' could often be represented, locally at least, by power laws as follows:

$$k \propto c^{\theta_1} \quad \text{and} \quad k' \propto c^{\theta_1'},$$

(12.2.11)

where θ_1 and θ_1' would be referred to as the "orders," or "pseudo-orders," of the reactions with respect to catalyst concentration c. In particular, orders of 0, $\frac{1}{2}$, 1, or 2 are commonly found.

The Effect of Temperature T

Study of a variety of chemical reactions has shown that the relationship between the rate of a reaction and temperature T can often be represented quite well (over a moderate range of temperature) by an equation due to Arrhenius. This postulates that the natural logarithm of the rate constant k is a linear function of the reciprocal of the absolute temperature. [Absolute temperature is measured in degrees Kelvin and is numerically equal to $(T + 273)$ where T is the temperature measured in degrees Celsius.] Thus, for a fixed catalyst concentration c,

$$k \propto \exp\{-\theta_2 x\} \quad \text{and} \quad k' \propto \exp\{-\theta_2' x\},$$

where we define $x = 1/(T + 273)$. If we write

$$\bar{x} = \sum_{u=1}^{n} (T_u + 273)^{-1} n^{-1}$$

for the average x value observed in the data, this being a constant, we can express k and k' equivalently as

$$k \propto \exp\{-\theta_2(x - \bar{x})\} \quad \text{and} \quad k' \propto \exp\{-\theta_2'(x - \bar{x})\}. \quad (12.2.12)$$

[It is often better to center data values about their observed mean like this, or some other local convenient origin, in model formulation, because it tends to lead to more stable estimation of the parameters; see Box (1964) and Draper and Smith (1981, pp. 488–489).] The parameters θ_2 and θ_2' are constant multiples of the so called "activation energies" E and E' for the two reactions. Combining (12.2.11) and (12.2.12), we have

$$k = \theta_0 c^{\theta_1} \exp\{-\theta_2(x - \bar{x})\},$$

$$k' = \theta_0' c^{\theta_1'} \exp\{-\theta_2'(x - \bar{x})\}, \quad (12.2.13)$$

where θ_0 and θ_0' are two additional constants. Substitution for k and k' in (12.2.13) into (12.2.9) now gives the mechanistic model (12.2.1) in which η is expressed as a function of r, c, and $T(\equiv x^{-1} - 273)$.

We have seen already that this model fits the data very well. Assuming that it reflects the main characteristics of the mechanism, we now consider what insights it provides.

12.3. INSIGHT PROVIDED BY THE FITTED MECHANISTIC MODEL

Changing the Flow Rate

Equations (12.2.7), (12.2.9), and (12.2.10) give expressions for the proportions ξ, η, ζ of the feed reactant X, the desired intermediate product Y, and the final undesired by-product Z, as follows:

$$\xi = \frac{r}{(r + k)}, \quad (12.3.1)$$

$$\eta = \frac{kr}{(r + k)(r + k')}, \quad (12.3.2)$$

$$\zeta = \frac{kk'}{(r + k)(r + k')}. \quad (12.3.3)$$

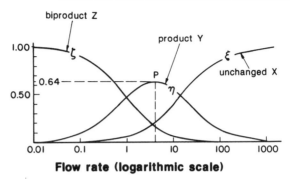

FIGURE 12.3. Theoretical compositions of reactor output for various flow rates when $k = 16$, $k' = 1$.

For illustration, we take the case $k = 16$, $k' = 1$, and draw the corresponding response curves in Figure 12.3, which shows the composition of the output stream of the reactor at various flow rates.

Figure 12.3 may be interpreted as follows. If the flow rate r is very high so that there is a very short residence time in the reactor, the reactions will not have a chance to get going, and most of the output stream will be unchanged X. On the other hand, if the flow rate r is very low so that the residence time is very long, most of the desired product Y will have been formed by the first reaction but will have been destroyed by the second before it can emerge. Between these two extremes is a flow rate ($r = 4$ in our example) which provides the greatest conversion to the desired product Y.

Choosing Flow Rate to Maximize Conversion

We refer to η, the proportion of the feed that is converted to Y as the *conversion*. From the figure, we see that the maximum conversion is attained at a turning point of η. Differentiating (12.3.2),

$$\frac{\partial \eta}{\partial r} = \frac{k(k'k - r^2)}{\{(r + k)(r + k')\}^2}$$

which vanishes when the flow rate is $r = r_m$ where

$$r_m = (kk')^{1/2}, \tag{12.3.4}$$

because r, k, and k' are essentially positive from physical considerations.

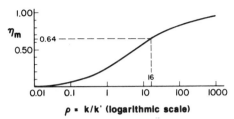

FIGURE 12.4. Maximum conversion, η_m, plotted against $\rho = k/k'$ for the theoretical model.

(It is easy to confirm that, at this value of r, the second derivative $\partial^2\eta/\partial r^2$ is negative, implying that η *is* maximized at $r = r_m$.) Note that, in our example of Figure 12.3, $k = 16$, and $k' = 1$, so that $r_m = (16 \times 1)^{1/2} = 4$.

The maximum value η_m of the conversion η is found by substituting the value of r given by (12.3.4) into (12.3.2) to give

$$\eta_m = \left(1 + \rho^{-1/2}\right)^{-2}, \tag{12.3.5}$$

where

$$\rho = \frac{k}{k'} \tag{12.3.6}$$

is the ratio of the rate constants. Thus we see that the maximum conversion η_m is a monotonically increasing function of the ratio of the rate constants. A plot of η_m versus ρ is shown in Figure 12.4.

For our example, $k = 16$, $k' = 1$, $\rho = 16$, and $\eta_m = (1 + \frac{1}{4})^{-2} = 0.64$.

The relationship shown in Figure 12.4 merits study. It takes a form to be expected on commonsense grounds, namely, the greater the speed of the first reaction (responsible for forming the desired product Y) relative to the speed of the second reaction (responsible for destroying the desired product Y), the greater is the maximum conversion achievable when the flow rate is suitably adjusted to its best level.

Another point to be noted is that, while we can always select a flow rate of r_m so as to maximize η, the maximum conversion η_m so achieved is itself an increasing function only of the ratio, $\rho = k/k'$, of the rate constants. To raise the value of η_m, then, some variable must be introduced which increases the rate k of the first reaction relative to the rate k' of the second reaction. Two variables that might do this (and which were, clearly, introduced by the investigators for that reason) are catalyst concentration c and temperature T.

Changing Catalyst Concentration c

We now return to (12.2.11). If we substitute in the estimates for θ_1 and θ_1' given in (12.2.2) we have

$$k \propto c^{0.53}, \qquad k' \propto c^{-0.01}. \tag{12.3.7}$$

Thus the speed of the first reaction increases approximately as the square root of the catalyst concentration c, whereas that of the second reaction is essentially unaffected by c. It follows that, over the ranges of the predictor variables being considered, increasing the catalyst concentration favors higher maximum conversions η_m. It is assumed that the flow rate r is appropriately adjusted to $r_m = (kk')^{1/2}$ as c is increased.

Changing Temperature

Substituting the estimates for θ_2 and θ_2' given in (12.2.2) into (12.2.12) provides

$$k \propto \exp\{-15{,}475(x - \bar{x})\}, \qquad k' \propto \exp\{-7489(x - \bar{x})\}, \tag{12.3.8}$$

where $x = (T + 273)^{-1}$ is the inverse of the absolute temperature. As we have said, the constants estimated by 15,475 and 7489 are proportional to the respective activation energies of the two reactions. To see more clearly what this implies, consider the effect of increasing the reaction temperature from 80 to 90°C, that is, from 353 to 363°K. Denoting the rate constants at the two temperatures by appropriate subscripts, we find from (12.3.8) that

$$\frac{k_{90}}{k_{80}} = \exp\{-15{,}475(363^{-1} - 353^{-1})\} = 3.34,$$

$$\frac{k_{90}'}{k_{80}'} = \exp\{-7489(363^{-1} - 353^{-1})\} = 1.79.$$

We see that this 10°C increase in temperature T speeds up the first reaction about twice as much as it does the second. Over the ranges of predictor variables in which the model applies, higher temperatures favor higher maximum conversions.

We already knew, from our empirical analysis in Chapter 11, that high conversion can be attained by using high catalyst concentration c and high temperature T with a suitably adjusted flow rate r. The present discussion allows us to better understand *why* this is so.

We now consider in more detail the link between the geometry of the surface and the characteristics of the underlying mechanism.

12.4. RIDGE PROPERTIES AND THE MECHANISM

We have previously emphasized that the geometry of response surfaces, and particularly that of ridge systems, is related to mechanism. We now illustrate this point in more detail for the present example.

The Concentration–Flow Rate Ridge

For the moment. we consider the temperature fixed at some value $T = T^*$, say. Now consider the conversion surface plotted as a function of $\ln r$ and $\ln c$. This will have the general appearance shown in Figure 12.5. The readers attention is directed particularly to the tangents of the angles ϕ and ψ. We shall say that $\tan \phi$ determines the "slope" of the concentration ridge (with respect to the $\ln c$ axis) while $\tan \psi$ determines the "tilt" of the ridge (also with respect to the $\ln c$ axis.)

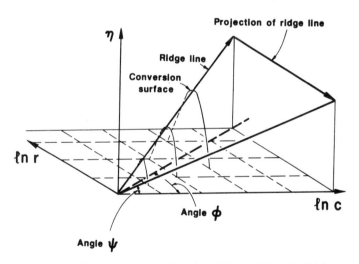

FIGURE 12.5. Conversion η plotted as a function of $\ln r$ and $\ln c$ for fixed temperature $T = T^*$. The angle ϕ measures the *slope* of the ridge with respect to the $\ln c$ axis. The angle ψ measures the tilt of the ridge (when projected onto a plane of constant $\ln r$) with respect to the $\ln c$ axis; $\tan \phi = \frac{1}{2}(\theta_1 + \theta_1')$, and $\tan \psi \approx (\eta_m^* - \eta_m^{*3/2})(\theta_1 - \theta_1')$.

The Slope of the Concentration–Flow Rate Ridge Is Determined by the Average Reaction Order

We recall from (12.3.4) that, for any given value of the concentration c, the maximizing value r_m of the flow rate is $r_m = (kk')^{1/2}$ or, taking natural logarithms,

$$\ln r_m = \tfrac{1}{2}(\ln k + \ln k'). \tag{12.4.1}$$

Substituting for $k \propto c^{\theta_1}$, $k' \propto c^{\theta_1'}$ from (12.2.11) gives

$$\ln r_m = \text{constant} + \tfrac{1}{2}(\theta_1 + \theta_1')\ln c, \tag{12.4.2}$$

where (we recall) θ_1 and θ_1' are the reaction orders with respect to the catalyst concentration. Equivalently, the average order of the two reactions can be considered as

$$\tfrac{1}{2}(\theta_1 + \theta_1') = \frac{\partial(\ln r_m)}{\partial(\ln c)} = \tan\phi, \tag{12.4.3}$$

say, obtained by differentiating (12.4.2).

The Tilt of the Concentration–Flow Rate Ridge Is Determined by the Difference of the Reaction Orders

From (12.3.5) we can obtain

$$-\ln\!\left(\eta_m^{-1/2} - 1\right) = \tfrac{1}{2}\ln\rho. \tag{12.4.4}$$

We denote the left-hand side of this equation by $F(\eta_m)$, and substitute first for ρ from (12.3.6), and then for k and k' from (12.2.11) to give

$$F(\eta_m) \equiv -\ln\!\left(\eta_m^{-1/2} - 1\right) = \text{constant} + \tfrac{1}{2}(\theta_1 - \theta_1')\ln c. \tag{12.4.5}$$

This shows that the slope of $F(\eta_m)$ with respect to $\ln c$ is exactly $\tfrac{1}{2}(\theta_1 - \theta_1')$ and so is dependent on the difference between the reaction rates. Figure 12.6 which provides a plot of $F(\eta_m)$ versus η_m indicates that their relationship is roughly linear over moderate ranges of η_m. Thus, we may conclude that,

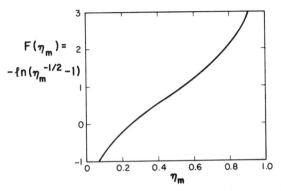

FIGURE 12.6. Plot of $F(\eta_m)$ versus η_m, exhibiting a roughly straight line relationship over moderate ranges of η_m (and particularly in the central range).

approximately, the tilt of the ridge $\partial \eta_m / \partial(\ln c)$ is directly dependent on the value $\theta_1 - \theta_1'$. More accurately we can differentiate (12.4.5) with respect to $\ln c$ to give

$$\tfrac{1}{2}\left(\eta_m - \eta_m^{3/2}\right)^{-1}\left\{\frac{\partial \eta_m}{\partial(\ln c)}\right\} = \tfrac{1}{2}(\theta_1 - \theta_1') \qquad (12.4.6)$$

and it follows that, at a specific concentration $c = c^*$ (at which $\eta_m = \eta_m^*$), the tilt of the ridge is given by

$$\tan \psi = \left[\frac{\partial \eta_m}{\partial(\ln c)}\right]_{c=c^*} = \left(\eta_m^* - \eta_m^{*3/2}\right)(\theta_1 - \theta_1'). \qquad (12.4.7)$$

We see that the tilt of the ridge at $c = c^*$ is directly proportional to the difference $\theta_1 - \theta_1'$ between the two reaction rates.

Some General Conclusions About the Geometry of the Concentration–Flow Rate Ridge

For this system, physical considerations suggested that the reactions, if they took place at all, could occur only in a forward direction. Under these assumptions then, $k > 0$, $k' > 0$, and for *any* system of the kind discussed here:

1. If both reactions are of zero order with respect to catalyst concentration c, there will be a stationary ridge *in the direction of the concentration*

axis. In other words, catalyst concentration will be a redundant variable.

2. Otherwise, there will always be an obliquely sloping concentration–flow rate ridge of some kind.

3. This ridge will be a stationary ridge (with zero tilt) if the two reaction orders are identical, so that $\theta_1 - \theta_1' = 0$. Obviously, in this case, although we can speed up the system, we cannot favor one or other of the reactions by altering the concentration.

Notice that, for *any* system of this kind, if catalyst concentration has any effect at all, the conversion surface in the coordinates of catalyst concentration and flow rate will *always* take the form of an obliquely oriented ridge of some kind. In particular, this implies that there will always be an "interaction" between concentration and flow rate.

The Temperature–Flow Rate Ridge

Very similar conclusions result if we suppose the catalyst concentration fixed at some value $c = c^*$ and consider the conversion surface plotted as a function of $\ln r$ and $x = (T + 273)^{-1}$. This surface will, again, have the general appearance of Figure 12.5 with the coordinate x replacing $\ln c$.

The Slope of the Temperature–Flow Rate Ridge is Determined by the Average Activation Energy

Proceeding as before and substituting (12.2.12) into (12.4.1) we find

$$\ln r_m = \text{constant} - \tfrac{1}{2}(\theta_2 + \theta_2')(x - \bar{x}), \qquad (12.4.8)$$

where $x = (T + 273)^{-1}$. The slope of this line is obviously $-\tfrac{1}{2}(\theta_2 + \theta_2')$, that is, minus the average activation energy.

The Tilt of the Temperature–Flow Rate Ridge is Determined by the Difference in Activation Energies

By arguments similar to those already given we have that

$$-\left[\frac{\partial \eta_m}{\partial x}\right]_{T=T^*} = \left(\eta_m^* - \eta_m^{*3/2}\right)(\theta_2 - \theta_2'), \qquad (12.4.9)$$

where $T = T^*$ is some selected temperature at which $\eta_m = \eta_m^*$.

Some General Conclusions About the Geometry of the Temperature–Flow Rate Ridge

General conclusions can be drawn similar to those for the concentration–flow rate ridge:

1. If (as would be most unusual) both reactions were unaffected by temperature, there would be a stationary ridge in the direction of the temperature axis and the predictor variable temperature would be redundant.
2. Otherwise (as is almost always the case), there would be an obliquely oriented temperature–flow rate ridge of some kind (and consequently an interaction between temperature and flow rate).
3. The ridge will be stationary (that is, have zero tilt) if the two activation energies θ_2, θ_2' are equal. In this case, although we can speed up the whole system, we cannot favor one or other of the reactions by increasing temperature.

As an exercise, the reader should sketch the appearance of three-dimensional systems of the type we have described in these special cases:

a. The orders of the two reactions are both zero, and the activation energies are equal.
b. The orders of the two reactions are equal but greater than zero, and the activation energies are equal.
c. The orders of the two reactions are different and the activation energies are equal.

12.5. RELATIONSHIPS OF EMPIRICAL COEFFICIENTS TO MECHANISTIC PARAMETERS: THEIR USE IN THE IDENTIFICATION OF POSSIBLE THEORETICAL MODELS

In Chapter 11, for Example 11.4, (see Section 11.4 and Appendix 11A) we fitted, on a purely empirical basis, a canonical form

$$\hat{y} = b_0 - (m_1 x_1 + m_2 x_2 + m_3 x_3)^2 + b_1 x_1 + b_2 x_2 + b_3 x_3. \quad (12.5.1)$$

The estimated coefficients we obtained, subsumed in Eq. (11.4.15), were

$$b_0 = 53.0865, \qquad m_1 = -2.0132, \qquad b_1 = 0.7446,$$
$$m_2 = 0.3432, \qquad b_2 = 4.8136,$$
$$m_3 = 2.5772, \qquad b_3 = 8.0130, \quad (12.5.2)$$

and the coded predictors are related to the original variables by the relationships

$$x_1 = \frac{(\ln r - \ln 3)}{\ln 2}, \qquad x_2 = \frac{(\ln c - \ln 2)}{\ln 2}, \qquad x_3 = \frac{(T - 80)}{10}. \quad (12.5.3)$$

Obviously, if the physical system can be represented by some consecutive mechanism of the type we have described, the coefficients in the empirical fitted equation must be related to the fundamental constants (parameters) of the system. It is instructive to consider the nature of these relationships. In particular, such consideration is helpful in identifying, and assessing the plausibility of, a postulated or contemplated mechanism. If we differentiate (12.5.1) with respect to x_1 and equate the result to zero, we have

$$m_1 x_1 + m_2 x_2 + m_3 x_3 = \frac{b_1}{2m_1}. \quad (12.5.4)$$

This is the equation of a plane and defines, for specified values of x_2 and x_3, the value x_{1m} at which the maximum conversion $\hat{y}_m$ is attained (for the given x_2, x_3 values). In fact

$$x_{1m} = \frac{b_1}{2m_1^2} - \left(\frac{m_2}{m_1}\right) x_2 - \left(\frac{m_3}{m_1}\right) x_3. \quad (12.5.5)$$

Differentiating (12.5.5) with respect to x_2 and x_3, and substituting the numerical values of (12.5.2) into the results yields

$$\frac{\partial x_{1m}}{\partial x_2} = \frac{-m_2}{m_1} \approx 0.1705,$$

$$\frac{\partial x_{1m}}{\partial x_3} = \frac{-m_3}{m_1} \approx 1.2802. \quad (12.5.6)$$

From (12.5.1), (12.5.4), and (12.5.5) we now obtain

$$\hat{y}_m = b_0 + \frac{1}{4}\left(\frac{b_1}{m_1}\right)^2 + \left(b_2 - \frac{b_1 m_2}{m_1}\right) x_2 + \left(b_3 - \frac{b_1 m_3}{m_1}\right) x_3. \quad (12.5.7)$$

It follows that

$$\frac{\partial \hat{y}_m}{\partial x_2} = b_2 - \frac{b_1 m_2}{m_1} = 4.9406,$$

$$\frac{\partial \hat{y}_m}{\partial x_3} = b_3 - \frac{b_1 m_3}{m_1} = 8.9662. \tag{12.5.8}$$

Now because x_1, x_2, and x_3 are related to r, c, and T by (12.5.3) and $x = 1/(T + 273)$,

$$\frac{\partial (\ln r_m)}{\partial (\ln c)} = \frac{\partial x_{1m}}{\partial x_2} \approx 0.1705, \tag{12.5.9}$$

$$\left[\frac{\partial (\ln r_m)}{\partial x} \right]_{T=80} = \left[\frac{\partial (\ln r_m)}{\partial x_{1m}} \frac{\partial x_{1m}}{\partial x_3} \frac{\partial x_3}{\partial x} \right]_{T=80}$$

$$\approx 0.6931 \times 1.2802 \times (-12,460.9)$$

$$= -11,057. \tag{12.5.10}$$

Also,

$$\frac{\partial \hat{y}_m}{\partial (\ln c)} = \frac{\partial \hat{y}_m}{\partial x_2} \frac{\partial x_2}{\partial (\ln c)} \approx \frac{4.9406}{(\ln 2)} = 7.1278,$$

$$\frac{\partial \hat{y}_m}{\partial x} = \frac{\partial \hat{y}_m}{\partial x_3} \frac{\partial x_3}{\partial x} \approx -12,460.9 \times 8.9662 = -111,726.9. \tag{12.5.11}$$

We now apply results given in Section 12.4. From (12.4.3) and (12.5.9) we see that, roughly,

$$\tfrac{1}{2}(\theta_1 + \theta_1') = 0.1705. \tag{12.5.12}$$

From (12.4.7) and (12.5.11) and noting that η_m^* is approximated by $b_0 = 53.0865$ in percentage units when $(r, c, T) = (3, 2, 80)$ we have, roughly,

$$\theta_1 - \theta_1' = \frac{0.01(7.1278)}{\left\{ (0.01b_0) - (0.01b_0)^{3/2} \right\}}$$

$$= 0.4947, \tag{12.5.13}$$

the 0.01 being a "percentage to proportion" scale factor. From (12.5.12) and (12.5.13) it follows that, roughly, $\theta_1 = 0.42$, $\theta_1' = -0.08$. (In practice, because we are assuming here that the reactions are, if anything *forward* reactions, i.e., $\theta_1' \geq 0$, we would replace -0.08 by 0.) Similarly from (12.4.8) and (12.5.10), approximately,

$$\frac{1}{2}(\theta_2 + \theta_2') = -\frac{\partial(\ln r_m)}{\partial x} = 11{,}057, \qquad (12.5.14)$$

while (12.4.9) and (12.5.11) provide, roughly,

$$\theta_2 - \theta_2' = \frac{-0.01(-111{,}726.9)}{\left\{(0.01b_0) - (0.01b_0)^{3/2}\right\}},$$

$$= 7755, \qquad (12.5.15)$$

the 0.01 converting percentages to proportions. We see from (12.5.14) and (12.5.15) that, roughly, $\theta_2 = 14{,}935$, $\theta_2' = 7180$.

The above rough values for θ_1, θ_1', θ_2, and θ_2' can be compared with the corresponding least squares values in (12.2.2):

Parameter	Rough value	Least-squares values
θ_1	0.42	0.53 (± 0.08)
θ_1'	-0.08	-0.01 (± 0.07)
θ_2	14,935	15,475 (± 625)
θ_2'	7180	7489 (± 598)

We see that the agreement is good. If the rough values were obtained as part of a preliminary analysis based on the fitted empirical surface, they would encourage the experimenter to proceed with the theoretical analysis. For they indicate that the first reaction is about half-order with respect to catalyst concentration while the second reaction is virtually unaffected by this variable, a result that is theoretically sensible. Moreover, the rough values of θ_2 and θ_2' yield reasonable values for the activation energies.

APPENDIX 12A. A MORE DETAILED ANALYSIS OF THE FITTED MECHANISTIC MODEL

The Mechanistic Model

The mechanistic model (12.2.1) does not contain an adjustable mean, that is, a β_0 term, as do the empirical models. Also, while for the quadratic model, the blocking variables are orthogonal to the predictor variables, permitting the clean separation of a sum of squares for blocks, this is not true for the nonlinear mechanistic model. An approximate analysis of variance for the model (12.2.1), fitted by least squares as in (12.2.2), which will detect and eliminate block differences, is given in Table 12A.1.

In the table, the "fitted model" sum of squares was obtained by fitting the nonlinear model $f(r, c, T|\theta)$ of (12.2.1) directly to the data, and calculating $\Sigma y_i^2 - S(\hat{\theta})$ where $S(\hat{\theta})$ is the minimum value 85.3 of the sum of squares function $S(\theta)$. The "extra for blocks" sum of squares was the extra sum of squares associated with the estimates of the blocking parameters β_1, β_2, β_3, and β_4 in the model

$$y = f(r, c, T|\theta) + \beta_1 z_1 + \beta_2 z_2 + \beta_3 z_3 + \beta_4 z_4 + \varepsilon$$

where the dummy variables z_i are such that

$$z_i = \begin{cases} 1 & \text{for observations in the } i\text{th block,} \\ 0 & \text{otherwise.} \end{cases}$$

The "error" sum of squares was obtained as in Section 11.4 and Appendix 11B (see Table 11.5), and the "lack of fit" by difference.

It is clear from the table that the nonlinear model has taken up the greater part of the total variation in the observations and there is no evidence that block effects or lack of fit exist. As a result we pool blocks, lack of fit, and pure error into a residual which yields an estimate $\hat{\sigma} = (4.74)^{1/2} = 2.18$ for the experimental error standard deviation.

TABLE 12A.1. Approximate analysis of variance table for the fitted nonlinear mechanistic model

Source	SS		df		MS	
Fitted model	53,504.3		6		8,917.38	
Extra for blocks	21.5 ⎫		4 ⎫		5.38 ⎫	
Lack of fit	36.6 ⎬	(85.3)	6 ⎬	(18)	6.10 ⎬	(4.74)
Error	27.2 ⎭		8 ⎭		3.39 ⎭	
Total	53,589.6		24			

TABLE 12A.2. Comparison of extra sums of squares taken up by various models

Model	Source	Extra SS	df	MS
Second	Model	3004.2	9	333.8
degree	Residual	67.8	14	4.8
Canonical	Model	2965.4	6	494.2
form	Residual	106.6	17	6.3
Mechanism	Model	2986.7	5	597.3
	Residual	85.3	18	4.7

Comparison of Various Models

Some measure of the comparative representational ability of the various models we have fitted to the conversion data can be had by comparing the sums of squares explained by the models, and their corresponding degrees of freedom. It is convenient to examine the *extra* sums of squares beyond that accounted for by fitting only a mean, namely, $y = \beta_0 + \varepsilon$. (It must be remembered that the mechanistic model does not have a β_0 term in it, so that the "extra" sum of squares in this case is not one that can be attributed to specific "extra" parameters.) These are shown in Table 12A.2, the degrees of freedom in each case being the total number of parameters minus one for b_0. The mechanistic model appears to possess remarkably good representational properties, for comparatively few parameters.

CHAPTER 13

Design Aspects of Variance, Bias, and Lack of Fit

13.1. THE USE OF APPROXIMATING FUNCTIONS

Typically, in a scientific investigation, some response y is measured whose mean value $E(y) = \eta$ is believed to depend on a set of variables $\boldsymbol{\xi} = (\xi_1, \xi_2, \ldots, \xi_k)'$. The exact functional relationship between them,

$$E(y) = \eta = \eta(\boldsymbol{\xi}) \tag{13.1.1}$$

is usually unknown and possibly unknowable. We often represent the function $\eta(\boldsymbol{\xi})$ as the solution to some set of time- and space-dependent ordinary or partial differential equations. However, we have only to think of the flight of a bird, the fall of a leaf, or the flow of water through a valve, to realize that, even using such equations, we are likely to be able to approximate only the main features of such a relationship. In this book, we employ even cruder approximations using polynomials that exploit local smoothness properties. Approximations of this type are, however, often perfectly adequate locally. Over a short distance, the flight of the bird might be approximated by a straight line function of time; over somewhat longer distances, a quadratic function might be used. Over narrow ranges, the flow of water through the valve might be similarly approximated by a straight line function of the valve opening.

Mathematically, this idea may be expressed by saying that, over a limited range of interest $R(\boldsymbol{\xi})$, the main characteristics of a smooth function may be represented by the low-order terms of a Taylor series approximation. It should be understood that the *region of interest* $R(\boldsymbol{\xi})$ will lie within a larger *region of operability* $O(\boldsymbol{\xi})$, defined as the region over which experiments *could* be conducted if desired, using the available apparatus. For example, suppose we are anxious to discover the water temperature ξ which will extract the maximum amount of caffeine η when hot water is added to a particular kind of tea leaves. Suppose, further, that we have a pressure cooker at our disposal. The operability region $O(\xi)$ would define the temperatures at which the cooker could be operated; these might extend from 30 to 120°C (from room temperature to the temperature attainable when maximum pressure was applied). However, the region of interest $R(\xi)$ typically would be much more limited; initially, at least, $R(\xi)$ would likely be in the immediate vicinity of the boiling point of water—from 95 to 100°C, say. Depending on what the experiments conducted in this smaller region show, the region of interest might then be changed to cover some other more promising region $R'(\xi)$ within $O(\xi)$, and so on.

It should be carefully noted in this formulation that, in order best to explore some region $R(\xi)$, of current interest, we do not need to perform all the experimental runs within $R(\xi)$. Some runs might be inside and some outside, or all might possibly be outside.

Now, writing $f(\xi)$ for the polynomial approximation, we have

$$E(y) = \eta(\xi) \tag{13.1.2}$$

for all ξ, and

$$\eta(\xi) \doteq f(\xi) \tag{13.1.3}$$

over some limited region of interest $R(\xi)$: The fact that the polynomial is an approximation does not necessarily detract from its usefulness because all models are approximations. Essentially, all models are wrong, but some are useful. However, the approximate nature of the model must always be borne in mind. For, if $\boldsymbol{\varepsilon} = (\varepsilon_1, \varepsilon_2, \ldots, \varepsilon_n)'$ is a vector of random errors having zero vector mean, and if $\mathbf{y} = (y_1, y_2, \ldots, y_n)'$ and $\mathbf{f}(\boldsymbol{\xi}) = \{f(\boldsymbol{\xi}_1), f(\boldsymbol{\xi}_2), \ldots, f(\boldsymbol{\xi}_n)\}$ where $\boldsymbol{\xi}_1, \boldsymbol{\xi}_2, \ldots, \boldsymbol{\xi}_n$ are n observations on $\boldsymbol{\xi}$, the true model is not

$$\mathbf{y} = \mathbf{f}(\boldsymbol{\xi}) + \boldsymbol{\varepsilon} \tag{13.1.4}$$

but

$$\mathbf{y} = \boldsymbol{\eta}(\boldsymbol{\xi}) + \boldsymbol{\varepsilon}, \tag{13.1.5}$$

that is

$$\mathbf{y} = \mathbf{f}(\boldsymbol{\xi}) + \boldsymbol{\delta}(\boldsymbol{\xi}) + \boldsymbol{\varepsilon}, \tag{13.1.6}$$

where

$$\boldsymbol{\delta}(\boldsymbol{\xi}) = \boldsymbol{\eta}(\boldsymbol{\xi}) - \mathbf{f}(\boldsymbol{\xi}) \tag{13.1.7}$$

is the vector discrepancy, which we would like to be small over $R(\xi)$, between actual and approximate models. In other words, there are *two* types of errors which must be taken into account:

1. Systematic, or bias, errors $\delta(\xi) = \eta(\xi) - f(\xi)$, the difference between the expected value of the response, $E(y) = \eta(\xi)$ and the approximating function $f(\xi)$.
2. Random errors ϵ.

Although the above implies that systematic errors $\delta(\xi)$ are always to be expected, there has been, since the time of Gauss (1777–1855), an unfortunate tendency to ignore them, and to concentrate only on the random errors ϵ. This seems to have been because nice mathematical results are possible when this is done. In choosing an experimental design, the ignoring of systematic error is *not* an innocuous approximation and misleading results may well be obtained, as we shall see in due course.

13.2. THE COMPETING EFFECTS OF BIAS AND VARIANCE

Consider first the elementary case where there is only one important predictor variable ξ. Figure 13.1 represents a typical situation in which the

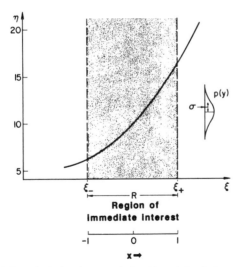

FIGURE 13.1.   A hypothetical unknown relationship existing between $E(y) = \eta$ and ξ. At the right is shown the error distribution $p(y)$ which has standard deviation σ.

investigator might desire to approximate a relationship

$$E(y) = \eta(\xi) \tag{13.2.1}$$

over the region (here, interval) of interest R defined by $\xi_- \le \xi \le \xi_+$, by a straight line

$$f(\xi) = \alpha + \beta\xi. \tag{13.2.2}$$

In practice, of course, the true functional relationship $\eta(\xi)$ indicated by the curve drawn in the figure would be unknown.

Suppose that the errors ε that occur in the observations have a variance σ^2, also unknown in practice, as indicated by the distribution on the right of the figure. By applying the coding transformation

$$x = \frac{\xi - \frac{1}{2}(\xi_+ + \xi_-)}{(\xi_+ - \xi_-)} \tag{13.2.3}$$

we can always convert the interval of interest, (ξ_-, ξ_+) in the variable ξ, to the interval $(-1, 1)$ in the standardized variable x. Suppose a particular set of data provides the least-squares fitted equation

$$\hat{y}_x = a + bx. \tag{13.2.4}$$

The mean squared error associated with estimating η_x by $\hat{y}_x$, standardized for the number, N, of design points and the error variance σ^2 is

$$\frac{NE(\hat{y}_x - \eta_x)^2}{\sigma^2} = \frac{NE\{\hat{y}_x - E(\hat{y}_x) + E(\hat{y}_x) - \eta_x\}^2}{\sigma^2}$$

$$= \frac{NV(\hat{y}_x)}{\sigma^2} + \frac{N\{E(\hat{y}_x) - \eta_x\}^2}{\sigma^2}, \tag{13.2.5}$$

after reduction. We can represent this symbolically as

$$M_x = V_x + B_x. \tag{13.2.6}$$

In words, the standardized mean squared error at x is equal to the variance V_x at x plus the squared bias B_x at x.

For the special case of a straight line fitted to observations made at locations $x_1, x_2, \ldots, x_u, \ldots, x_N$, and such that $\bar{x} = 0$, we find that

$$V_x = 1 + \frac{x^2}{\sigma_x^2}, \tag{13.2.7}$$

where

$$\sigma_x^2 = \sum_{u=1}^{N} \frac{(x_u - \bar{x})^2}{N}, \qquad \bar{x} = \sum_{u=1}^{N} \frac{x_u}{N}. \qquad (13.2.8)$$

Also, it is easily shown that

$$E(\hat{y}_x) = E(a) + E(b)x \qquad (13.2.9)$$

which, theoretically, could be obtained by fitting a straight line to the *true values* η_u at the design points.

An Illustration Using Symmetric Three Point Designs

Consider the special case of designs consisting of just three points at $(-x_0, 0, x_0)$, so that $\sigma_x^2 = 2x_0^2/3$, or, conversely, $x_0 = 1.225\sigma_x$. Clearly

$$V_x = 1 + \frac{1.5x^2}{x_0^2}. \qquad (13.2.10)$$

In particular, consider the two specific designs shown in Figure 13.2a and 13.2b with $x_0 = \frac{2}{3}$ and $x_0 = \frac{4}{3}$, respectively. Thus design 1 consists of the

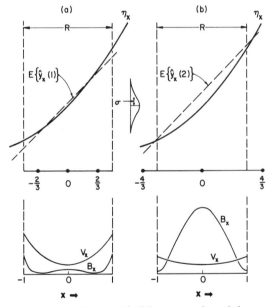

FIGURE 13.2. Two three-point designs with different spreads, and the corresponding values of V_x and B_x over the region of interest $-1 \le x \le 1$.

points $(-\frac{2}{3}, 0, \frac{2}{3})$, and design 2 of the points $(-\frac{4}{3}, 0, \frac{4}{3})$. The expected straight lines

$$E(\hat{y}_x) = E(a) + E(b)x \qquad (13.2.11)$$

are drawn in the figures and denoted by $E\{\hat{y}_x(1)\}$ and $E\{\hat{y}_x(2)\}$, where $\hat{y}_x(1)$ and $\hat{y}_x(2)$ are the corresponding fitted lines. For design 1, we have

$$V_x(1) = 1 + \frac{27x^2}{8}, \qquad B_x(1) = \left[E\{\hat{y}_x(1)\} - \eta_x\right]^2, \quad (13.2.12)$$

and for design 2,

$$V_x(2) = 1 + \frac{27x^2}{32}, \qquad B_x(2) = \left[E\{\hat{y}_x(2)\} - \eta_x\right]^2. \quad (13.2.13)$$

The functions V_x and B_x are plotted below their corresponding designs. From Figure 13.2a we see that the variance error $V_x(1)$ is quite high compared to $B_x(1)$ and to $V_x(2)$. Now, when $\bar{x} = 0$, $V_x = 1 + x^2/\sigma_x^2$ and, for the particular type of design shown, $V_x = 1 + 1.5x^2/x_0^2$. Thus we make V_x smaller (except when $x = 0$) by making σ_x^2 larger, that is, by taking x_0 to be larger, and spreading the design points more widely, as is done in design 2. Unfortunately, as we increase x_0, the representational ability of the straight line is strained more severely and thus, as we see from the figure, the squared bias B_x is greatly increased.

13.3. INTEGRATED MEAN SQUARED ERROR

Overall measures of variance and squared bias over any specified region of interest R may be obtained by averaging V_x and B_x over R to provide the quantities

$$V = \frac{\int_R V_x \, dx}{\int_R dx}, \qquad B = \frac{\int_R B_x \, dx}{\int_R dx}. \qquad (13.3.1)$$

Denoting the mean squared error integrated over R by M, we can then write

$$M = V + B. \qquad (13.3.2)$$

For the specific situations illustrated in Figures 13.2a and b, these equations become:

$$2.58 = 2.12 + 0.46 \qquad \text{design 1}$$

$$5.41 = 1.28 + 4.13 \qquad \text{design 2}. \qquad (13.3.3)$$

More generally, if we run a design at levels $(-x_0, 0, x_0)$ and fit a straight line when the true function and error distribution are those shown in Figure 13.1, then the overall integrated mean squared error M, and the contributions to M from V and B are as shown in Figure 13.3.

This diagram repays careful study. Consider the individual components V and B. The averaged (over R) variance V becomes very large if the spread of the design is made very small (that is, as x_0 approaches zero, V approaches infinity). Also, if the design is made very large, V slowly approaches its minimum value of unity (that is, as x_0 approaches infinity, V approaches 1). The average squared bias B, on the other hand, has a minimum value when x_0 is about 0.7, and increases for larger or smaller designs. The reason for this behavior is not difficult to see. If x_0 is

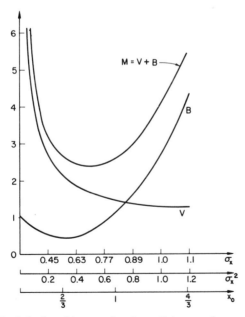

FIGURE 13.3. The behavior of integrated variance V, integrated squared bias B, and their total $M = V + B$, for three-point designs $(-x_0, 0, x_0)$ as the value of the spread $\sigma_x^2 = 2x_0^2/3$ varies.

close to 0.7, bias over the whole interval R can be made small. However, if x_0 is less than about 0.7, so that the design is made smaller, the straight line will be close to the true curve only in the middle of the interval. On the other hand, if x_0 exceeds about 0.7, so that the design is made larger, the straight line will be close to the true curve only nearer the extremes of the interval. The curve M shows the overall effect on averaged mean squared error $M = V + B$. A rather flat minimum for M will be observed near $x_0 = 0.79$, or $\sigma_x = 0.65$. Notice that the design which minimizes averaged mean squared error M in this case is not very different from the design which minimizes averaged squared bias B but is very different from that which minimizes averaged variance V.

The Relative Magnitudes of V and B

In practice, of course, the true relationship $E(y_x) = \eta_x$ would be unknown. To make further progress, we can proceed as follows:

1. Given that we are going to fit a polynomial of degree d_1 (say) to represent the function over some interval R, we can suppose that the true function $E(y_x) = \eta_x$ is a polynomial of degree d_2, greater than d_1. Thus, for example, in Figure 13.1, given that the investigator will use a straight line to represent the function in the interval R, we might assume that the true function was a second degree polynomial. In this case, then, $d_1 = 1$ and $d_2 = 2$.

2. We need also to say something about the *relative* magnitudes of systematic (bias) and random (variance) errors that we could expect to meet in practical cases. An investigator might typically employ a fitted approximating function such as a straight line, if he believed that the average departure from the truth induced by the approximating function were no worse than that induced by the process of fitting. We shall suppose this to be so, and will assume, therefore, that the experimenter will tend to choose the size of his region R, and the degree of his approximating function in such a way that the integrated random error and the integrated systematic error are about equal. Thus we shall suppose that the situation of typical interest is that where B is roughly equal to V.

13.4. REGION OF INTEREST AND REGION OF OPERABILITY IN k DIMENSIONS

So far we have illustrated the argument for the case where there is only one input variable ($k = 1$). We now consider the extension of these ideas to the general k case.

Readers of this book are already aware of the context in which response surface designs are customarily used. Iterative experimentation results in a sequence of designs which move about in the predictor space and which tend to become more complex as the investigation progresses. Also, from one phase to the next, scales, transformations, and the identities of the variables considered will all tend to change.

At a given stage of the investigation then, we suppose there is a region of current interest $R(\xi)$ within which it is hoped that the currently entertained approximating function $f(\xi)$ will provide adequate local representation. When the investigation proceeds to the next stage, both $R(\xi)$ and $f(\xi)$ can change. This region $R(\xi)$ is, as we have said, contained within a much larger region of operability $O(\xi)$, often only vaguely known,* within which experiments *could* be performed using the current experimental apparatus. The approximating function $f(\xi)$ would usually provide a totally inadequate representation over the whole operability region $O(\xi)$.

In routine use of standard designs, both the choice of approximating function $f(\xi)$ and of the selected neighborhood $R(\xi)$ in which $f(\xi)$ is expected to apply, are implicit as soon as the experimenter decides on the type of design he intends to employ, the variables he wishes to investigate, the levels of those variables, and the transformations of them he will use. Suppose, for example, that a chemist decides to run a 2^2 factorial with a center point in which he varies the amounts ξ_1 and ξ_2 of two catalysts A and B. Suppose he currently believes that a ratio ξ_1/ξ_2 of $10:1$ by weight is "about right" when a total weight of $\xi_1 + \xi_2 = 5$ g of catalyst is to be used. He further decides to vary the ratio from $9:1$ to $11:1$ and to vary the total catalyst weight from 4 to 6 g. Figure 13.4a shows the design in the coordinate system

$$x_1 = \frac{\xi_1}{\xi_2} - 10, \qquad x_2 = \xi_1 + \xi_2 - 5 \qquad (13.4.1)$$

which the experimenter has implicitly chosen to consider. As we have seen in Chapter 4, the design he has chosen is suitable for estimating and checking the applicability of an approximating plane in the immediate neighborhood of the experimental points. More specifically, if a polynomial

---

*Occasionally, delineation of certain features of $O(\xi)$ is itself the objective. Thus the experimenter might conceivably want to know: "What is the lowest temperature at which this apparatus will still function?" Sometimes questions of this kind are best put into the context of problems of specification; see Section 1.7. For example, the question "How high a dose of this drug could be given?" might best be studied by jointly considering two separate response surfaces measuring therapeutic value on the one hand, and undesirable side effects on the other.

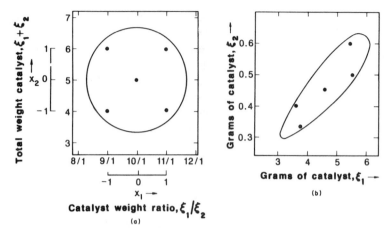

FIGURE 13.4. The effect of transformation of predictor variables on the region and the design. (*a*) A 2^2 factorial plus center point design set in a circular region of interest in the variables $x_1 = \xi_1/\xi_2 - 10$ and $x_2 = \xi_1 + \xi_2 - 5$. (*b*) The same design and region described in the variables ξ_1 and ξ_2.

of degree $d_1 = 1$,

$$\eta = \beta_0 + \beta_1 x_1 + \beta_2 x_2 \qquad (13.4.2)$$

were adequate, one could estimate its β coefficients and, also, check the adequacy of the approximation by obtaining estimates of β_{12} and $(\beta_{11} + \beta_{22})$ in the polynomial of degree $d_2 = 2$ given by

$$\eta = \beta_0 + \beta_1 x_1 + \beta_2 x_2 + \beta_{11} x_1^2 + \beta_{22} x_2^2 + \beta_{12} x_1 x_2. \qquad (13.4.3)$$

We could express this more formally by saying that the experimenter's choice implies that:

1. For a region of interest $R(\xi)$, approximately represented by the circle in Figure 13.4*a*, the experimenter hopes that a polynomial of degree $d_1 = 1$ in the variables $\xi_1/\xi_2 - 10$ and $\xi_1 + \xi_2 - 5$ will provide adequate approximation to the true but unknown response function.

2. The experimenter feels that, if the first-order polynomial is not adequate, a polynomial of degree $d_2 = 2$ should fit, and he will check the adequacy of the first degree model by looking at estimates of the representative combinations of second-order effects β_{12} and $(\beta_{11} + \beta_{22})$.

Notice that the choices of design levels and of transformations of the original predictor variables ξ_1 and ξ_2 determine the main characteristics of the region $R(\xi)$. Notice also that the circular region of interest in the (x_1, x_2) space of Figure 13.4a is equivalent to the oval region in the original (ξ_1, ξ_2) space, underlining the facts that the type of design to be used, the forms of the transformations of the variables, and the choice of levels of those transformed variables—or, almost equivalently, the choice of d_1, d_2, and $R(\xi)$—are matters of judgment. At any given stage of an investigation, this judgment will vary from one investigator to another. For example, a second investigator might employ the same design and the same transformations but, instead of varying the total catalyst level 1 g either way about an average of 5 g, might vary it 2 g either way about an average of 7 g. The second investigator would be likely also to choose the levels for catalyst ratio differently. A third investigator might not have set out the design in terms of the ratio ξ_1/ξ_2 of the concentrations and their sum $\xi_1 + \xi_2$, but might have treated ξ_1 and ξ_2 themselves as predictor variables. A fourth might have considered different, or additional, variables at this stage of the investigation.

Such differences would not necessarily have any adverse effect on the *end result* of the investigation. The iterative strategy we have proposed for the exploration of response surfaces is designed to be adaptive and self-correcting. For example, an inappropriate choice of scales or of a transformation can be corrected as the iteration proceeds. However, for a given experimental design, a change of scale can have a major influence on:

1. The variances of estimated effects.
2. The sizes of systematic errors.

The purpose of this discussion is to remind the reader that optimal design properties are critically dependent on this matter of choice of region. Because this choice contains so many arbitrary elements, it is somewhat dubious how far we should push the finer points of optimal theory, which we now discuss.

13.5. AN EXTENSION OF THE IDEA OF A REGION OF INTEREST: THE WEIGHT FUNCTION

Our talk of a *"region of interest R"* thus far has implied that accuracy of prediction is of uniform importance throughout the entire region R and of no interest at all outside R. In some instances, this might not be the case. Sometimes, for example, we might need high accuracy at some point P in the predictor variable space and could tolerate reduced accuracy as we

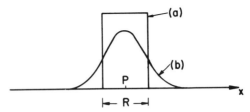

FIGURE 13.5. Two possible weight functions for $k = 1$. (a) "Uniform over R" type indicating uniform interest over R, no interest outside R. (b) Normal distribution shape, giving greater weight to points nearer P. See Exercise 13.4.

moved away from P. We can express such ideas by introducing a weight function $w(\mathbf{x})$ and minimizing a weighted mean squared error integrated over the whole of the operability region O. With this more general formulation, we have

$$M = V + B \tag{13.5.1}$$

with

$$M = \left(\frac{N}{\sigma^2}\right)\int_O w(\mathbf{x})E\{\hat{y}(\mathbf{x}) - \eta(\mathbf{x})\}^2 d\mathbf{x}, \tag{13.5.2}$$

$$V = \left(\frac{N}{\sigma^2}\right)\int_O w(\mathbf{x})E\{\hat{y}(\mathbf{x}) - E\hat{y}(\mathbf{x})\}^2 d\mathbf{x}, \tag{13.5.3}$$

$$B = \left(\frac{N}{\sigma^2}\right)\int_O w(\mathbf{x})\{E\hat{y}(\mathbf{x}) - \eta(\mathbf{x})\}^2 d\mathbf{x}. \tag{13.5.4}$$

Two possible weight functions for the $k = 1$ case are shown in Figure 13.5. Note that the vertical scale would be chosen so that $w(\mathbf{x})$ was normalized, that is,

$$\int_O w(\mathbf{x}) = 1.$$

The Importance of "All-Bias" Designs Illustrated for $k = 1$

We now return to the case where there is only one input variable ξ (so that $k = 1$) and where $d_1 = 1, d_2 = 2$. We generalize the idea of a region of interest by supposing that we have a symmetric weight function $w(x)$

centered at $x = 0$. The fitted equation is

$$\hat{y}_x = b_0 + b_1 x \tag{13.5.5}$$

while the true model, valid throughout an extensive region O, is

$$\eta_x = \beta_0 + \beta_1 x + \beta_{11} x^2. \tag{13.5.6}$$

Suppose that N runs are to be made at levels $x_1, x_2, \ldots, x_N$, and that $\Sigma x_i = N\bar{x} = 0$. Define

$$m_p = \frac{\Sigma x_u^p}{N}, \qquad \mu_p = \int_O w(x) x^p \, dx, \tag{13.5.7}$$

and note that $m_1 = 0$ and, because of the assumed symmetry, $\mu_p = 0$ for all odd p. Then it is readily shown (see Appendix 13A) that the integrated mean squared error for any design is

$$M \doteq V + B = \left\{ 1 + \frac{\mu_2}{m_2} \right\} + \alpha^2 \left\{ (m_2 - \mu_2)^2 + \left(\mu_4 - \mu_2^2 \right) + \frac{m_3^2 \mu_2}{m_2^2} \right\}, \tag{13.5.8}$$

where

$$\alpha = \frac{N^{1/2} \beta_{11}}{\sigma}. \tag{13.5.9}$$

If M is to be minimized, clearly we should choose $m_3 = 0$ whatever values are taken by α and m_2. One way to do this is to make the design symmetric about its center. The only design characteristic remaining in the reduced expression is m_2, which measures the spread of the x values. Consider, first, the two extreme situations:

The "All-Variance" Case

Here, the bias term B is entirely ignored. (This extreme is the case most often considered by theoretical statisticians.) It is easy to see that the optimal value of m_2 is infinity. The practical interpretation of this is that we are told to spread the design as widely as possible in the operability region O.

The "All-Bias" Case

Here, the variance term V is entirely ignored and it is assumed that fitting discrepancies arise only from inadequacy of the model. M (with $m_3 = 0$) is minimized when $m_2 = \mu_2$, that is, the design points are restricted to mimic the weight function in location (because $m_1 = \mu_1 = 0$) and spread ($m_2 = \mu_2$) and in third moment ($m_3 = \mu_3 = 0$).

For cases intermediate between these extremes, the minimizing choice of m_2 will depend on the value held by α. Arguing as in Section 13.3, however, we can minimize M (with $m_3 = 0$) with respect to m_2 subject to the constraint that the fraction $V/M = V/(V + B)$ of the mean squared error M accounted for by V has some specific value. As we have seen, special interest attaches to a value of $V/(V + B)$ close to $\frac{1}{2}$ when the contributions from variance and bias are about equal. Table 13.1 shows the optimum spreads of such designs, as measured by the ratio $(m_2/\mu_2)^{1/2}$, for various selected values of the fraction $V/(V + B)$, and for both a uniform weight function and a normal distribution weight function. The relative shapes of these two weight functions are shown in Figure 13.5.

It is very evident that the optimal value of $(m_2/\mu_2)^{1/2}$ for $V = B$, that is, $V/(V + B) = \frac{1}{2}$, is close to that for the all-bias designs $V/(V + B) = 0$, and is dramatically different from that for the all-variance designs, $V/(V + B) = 1$. Furthermore, the function relating $(m_2/\mu_2)^{1/2}$ to $V/(V + B)$ remains extremely flat except when $V/(V + B)$ is close to 1, namely, except when we approach the all-variance case. This suggests that, if a simplification is to be made in the design problem, it might be better to ignore the effects of sampling variation rather than those of bias.

TABLE 13.1. Optimum spread of design points as measured by $(m_2 / \mu_2)^{1/2}$ for various $V / (V + B)$ ratios and two types of weight functions

		Spread of design points as measured by $(m_2/\mu_2)^{1/2}$	
Design	$V/(V + B)$	For uniform weight function	For normal weight function
All-bias	0	1.00	1.00
	0.2	1.02	1.04
	0.333	1.04	1.09
$V = B$	0.5	1.08	1.16
	0.667	1.14	1.26
	0.8	1.25	1.41
All-variance	1.0	∞	∞

13.6. DESIGNS THAT MINIMIZE SQUARED BIAS

The foregoing section indicates that designs that minimize squared bias are of practical importance. We therefore consider next the properties of such designs, using an important general result. Suppose a polynomial model of degree d_1

$$\hat{y}(\mathbf{x}) = \mathbf{x}_1' \mathbf{b}_1 \tag{13.6.1}$$

is fitted to the data, while the true model is a polynomial of degree d_2,

$$\eta(\mathbf{x}) = \mathbf{x}_1' \boldsymbol{\beta}_1 + \mathbf{x}_2' \boldsymbol{\beta}_2. \tag{13.6.2}$$

Thus, for the complete set of N data points

$$\hat{y}(\mathbf{x}) = \mathbf{X}_1 \mathbf{b}_1,$$

$$\eta(\mathbf{x}) = \mathbf{X}_1 \boldsymbol{\beta}_1 + \mathbf{X}_2 \boldsymbol{\beta}_2. \tag{13.6.3}$$

Let us now write

$$\mathbf{M}_{11} = N^{-1} \mathbf{X}_1' \mathbf{X}_1, \qquad \mathbf{M}_{12} = N^{-1} \mathbf{X}_1' \mathbf{X}_2,$$

$$\boldsymbol{\mu}_{11} = \int_O w(\mathbf{x}) \mathbf{x}_1 \mathbf{x}_1' \, d\mathbf{x}, \qquad \boldsymbol{\mu}_{12} = \int_O w(\mathbf{x}) \mathbf{x}_1 \mathbf{x}_2' \, d\mathbf{x}. \tag{13.6.4}$$

It can now be shown (see Appendix 13B) that, whatever the values of $\boldsymbol{\beta}_1$ and $\boldsymbol{\beta}_2$, a necessary and sufficient condition for the squared bias B to be minimized is that

$$\mathbf{M}_{11}^{-1} \mathbf{M}_{12} = \boldsymbol{\mu}_{11}^{-1} \boldsymbol{\mu}_{12}. \tag{13.6.5}$$

A sufficient (but not necessary) condition for B to be minimized is that

$$\mathbf{M}_{11} = \boldsymbol{\mu}_{11} \quad \text{and} \quad \mathbf{M}_{12} = \boldsymbol{\mu}_{12}. \tag{13.6.6}$$

Now the elements of $\boldsymbol{\mu}_{11}$ and $\boldsymbol{\mu}_{12}$ are of the form

$$\int_O w(\mathbf{x}) x_1^{\alpha_1} x_2^{\alpha_2} \cdots x_k^{\alpha_k} \, d\mathbf{x} \tag{13.6.7}$$

and the elements of $\mathbf{M}_{11}$ and $\mathbf{M}_{12}$ are of the form

$$N^{-1} \sum_{u=1}^{N} x_{1u}^{\alpha_1} x_{2u}^{\alpha_2} \cdots x_{ku}^{\alpha_k}. \tag{13.6.8}$$

These typical elements are, respectively, moments of the weight function and moments of the design points of order

$$\alpha = \alpha_1 + \alpha_2 + \cdots + \alpha_k. \tag{13.6.9}$$

Thus, the sufficient condition above states that, up to and including order $d_1 + d_2$, all the moments of the design are equal to all the moments of the weight function.

We have already seen an elementary example in which this result applied when, in the foregoing section, we considered the case $k = 1$, $d_1 = 1$, $d_2 = 2$. There, the all-bias design was obtained by setting $m_2 = \mu_2$ and $m_3 = \mu_3$, where $\mu_3 = 0$, because the weight function was assumed to be symmetric about 0.

We now consider two examples where $k > 1$, and we shall suppose, for illustration, that the weight function is uniform over a sphere of radius 1.

EXAMPLE 13.1.   First-order minimum bias design, $d_1 = 1$, $d_2 = 2$. In this example, we are interested in fitting a plane

$$\hat{y} = b_0 + \sum_{i=1}^{k} b_i x_i \tag{13.6.10}$$

to describe the response function within a spherical region of interest R of radius 1. The true function over the whole operability region O can be represented by the second degree polynomial

$$\eta = \beta_0 + \sum_{i=1}^{k} \beta_i x_i + \sum_{i \geq j}^{k} \sum_{}^{k} \beta_{ij} x_i x_j. \tag{13.6.11}$$

To minimize squared bias, the moments of the design points up to and including order $d_1 + d_2 = 3$ are chosen to match those of a uniform density taken over a k-dimensional sphere. Thus,

$$w(\mathbf{x}) = \begin{cases} c, & \text{a constant,* within the spherical region } R, \\ 0, & \text{elsewhere.} \end{cases} \tag{13.6.12}$$

*The value of

$$c = \frac{\left\{ \Gamma\!\left(\frac{1}{2}\right) \right\}^{k}}{\Gamma(k/2 + 1)},$$

the inverse of the volume of a sphere of radius 1.

If we take the origin to be at the center of the spherical region, it is at once obvious from symmetry that all moments of the region of orders 1, 2, and 3 are zero except for the pure second moments. For the ith of these moments we have

$$c\int_R x_i^2 \, d\mathbf{x} = \frac{1}{(k+2)}. \qquad (13.6.13)$$

Thus, the moments of the design will be equal to the moments of the region if

$$N^{-1}\Sigma x_{iu} = 0,$$

$$N^{-1}\Sigma x_{iu}^2 = \frac{1}{(k+2)},$$

$$N^{-1}\Sigma x_{iu}x_{ju} = 0,$$

$$N^{-1}\Sigma x_{iu}^3 = 0,$$

$$N^{-1}\Sigma x_{iu}^2 x_{ju} = 0,$$

$$N^{-1}\Sigma x_{iu}x_{ju}x_{hu} = 0, \qquad (13.6.14)$$

where all summations are over $u = 1, 2, \ldots, N$. These conditions may be met in a variety of ways and, in particular, are satisfied by any two-level fractional factorial design of resolution IV, appropriately scaled.

We may illustrate the actual placing of the points by using the 2^3 factorial $(\pm a, \pm a, \pm a)$ for the case $k = 3$ as shown in Figure 13.6. When $k = 3$, the condition on the pure second moments becomes

$$0.2 = N^{-1} \sum_{u=1}^{N} x_{iu}^2 = a^2 \qquad (13.6.15)$$

so that $a = 0.447$. The radial distance of each point from the center is thus $(3a^2)^{1/2} = 0.7746$. Thus the design points are the vertices of a cube of half-side $a = 0.447$; the vertices lie on a sphere of radius 0.7746 embedded in a spherical region of interest of radius 1. Because the desirable properties of this arrangement are retained when the design and the region are (together) subjected to any linear transformation, we can obtain minimum bias designs for any ellipsoidal region as well.

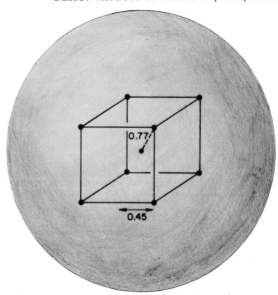

FIGURE 13.6. A first-order (two-level factorial) design in three factors which minimizes squared bias from second-order terms when the region of interest is a sphere of unit radius.

EXAMPLE 13.2.   Second-order minimum bias designs, $d_1 = 2, d_2 = 3$. We again assume a spherical region of interest R of unit radius, and consider the choice of a second-order design given that, over the wider operability region O, a third degree polynomial model exactly represents the functional relationship. A design that minimizes squared bias may be obtained by choosing the moments of the design, up to and including order $d_1 + d_2 = 5$, to match those of the k-dimensional sphere. This implies that all moments of orders 1, 2, 3, 4, and 5 are zero, except for the following, which are evaluated under the assumption of Eq. (13.6.12):

$$\int_O x_i^2 w(\mathbf{x})\, d\mathbf{x} = \frac{1}{k+2}; \quad \int_O x_i^4 w(\mathbf{x})\, d\mathbf{x} = \frac{3}{(k+2)(k+4)};$$

$$\int_O x_i^2 x_j^2 w(\mathbf{x})\, d\mathbf{x} = \frac{1}{(k+2)(k+4)} \tag{13.6.16}$$

The designs so specified are second-order rotatable.* It follows, in particular, that one class of suitable designs consists of the "cube plus star plus center points" composite design class, in which the "cube" portion is

*A design is rotatable if $V\{\hat{y}(\mathbf{x})\}$ is a function of $r = (x_1^2 + x_2^2 + \cdots + x_k^2)^{1/2}$ only, that is, if the accuracy of prediction of the response depends only on the distance from the center of the design. See Section 14.3.

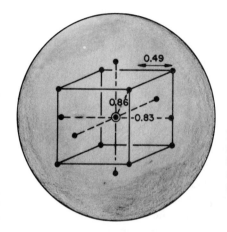

FIGURE 13.7. A second-order composite rotatable design which minimizes squared bias from third-order terms when a weight function uniform over a spherical region of interest of radius 1 is assumed.

taken as a two-level factorial, or fractional factorial of resolution V. Figure 13.7 shows such a design for $k = 3$. The design consists of a cube of side $2a$, six axial points at a distance αa from the origin, and n_0 center points. Any such design has the required zero moments, and we obtain appropriate values of a, α, and n_0 from the following equations, obtained by equating design moments to region moments.

$$\sum x_{iu}^2 = 8a^2 + 2\alpha^2 a^2 = \frac{1}{5}N,$$

$$\sum x_{iu}^4 = 8a^4 + 2\alpha^4 a^4 = \frac{3}{35}N,$$

$$\sum x_{iu}^2 x_{ju}^2 = 8a^4 = \frac{1}{35}N,$$

$$8 + 6 + n_0 = N. \tag{13.6.17}$$

These equations have the solution $\alpha = 2^{3/4} = 1.682$, $a = 0.4938$, $n_0 = 2.653$. Thus, the design of Figure 13.7 will (very nearly) give minimum average mean squared bias if we set $a = 0.49$, $\alpha a = 0.83$, and use three center points. The points of the cube will lie a distance $(3a^2)^{1/2} = 0.86$ from the center.

Other Weight Functions, Other Regions of Interest

If more importance were attached to the central portion of the region than to its extremes and a weight function with spherical contours were employed, then the all-bias moment requirements would still call for the use of a second-order rotatable design with scale suitably adjusted. Interest has been shown by some authors in cuboidal regions of interest or, more generally, in regions that can be reduced to cubes by linear transformations. More generally, in that context, we can consider weight functions whose contours are concentric cubes. It is easy to show that the moment requirements for the case $d_1 = 2$, $d_2 = 3$ are again met by suitably chosen (but not rotatable) composite designs. (See Draper and Lawrence, 1965c.)

13.7. ENSURING GOOD DETECTABILITY OF LACK OF FIT

Our aims in choosing a design are necessarily many faceted. In particular, on the one hand when, as inevitably happens, we fit a somewhat inexact model, we do not want our estimate $\hat{y}$ of η to be too far wrong—we want our design to allow *good estimation* of η. On the other hand, when large discrepancies occur which are likely to invalidate the model, we want to know about them so that we can modify the model and, in some cases, augment the data appropriately—we want our design to allow *detection of model inadequacy*. In their general discussion, Box and Draper (1959, 1963) consider a strategy of experimental design in which first a class of designs that minimize mean squared error to satisfy the first requirement is found; then, a further choice is made of a subset of designs that also satisfy the second requirement. It is this second aspect of the choice of design—to give good detection of model inadequacy—that we now consider.

Consider the mechanics of making a test of goodness of fit using the analysis of variance. Suppose we are estimating p parameters, observations are made at $p + f$ distinct points, and repeated observations are made at certain of these points to provide e pure error degrees of freedom, so that the total number of observations is $N = p + f + e$. The resulting analysis of variance table is of the type shown in Table 13.2.

TABLE 13.2.   Skeleton analysis of variance table

Source	df	$E(MS)$
Parameter estimates	p	
Lack of fit	f	$\sigma^2 + \Lambda^2/f$
Pure error	e	σ^2
Total	N	

Under the usual assumptions, the expectation of the unbiased pure error mean square is σ^2, the experimental error variance, and the expected value of the lack of fit mean square equals $\sigma^2 + \Lambda^2/f$ where Λ^2 is a noncentrality parameter. The test for goodness of fit is now made by comparing the mean square for lack of fit against the mean square for error, via an $F(f, e)$ test.

In general, the noncentrality parameter takes the form

$$\Lambda^2 = \sum_{u=1}^{N} \{ E(\hat{y}_u) - \eta_u \}^2 = E(S_L) - f\sigma^2, \qquad (13.7.1)$$

where S_L is the lack of fit sum of squares. Thus, good detectability of general lack of fit can be obtained by choosing a design that makes Λ^2 large. It turns out that this requirement of good detection of model inadequacy can, like the earlier requirement of good estimation, be achieved by certain conditions on the design moments. Thus, under certain sensible assumptions, it can be shown that a dth order design would provide high detectability for terms of order $(d + 1)$ if (1) all odd design moments of order $(2d + 1)$ or less are zero, and (2) the ratio

$$\frac{\displaystyle\sum_{u=1}^{N} r_u^{2(d+1)}}{\left\{ \displaystyle\sum_{u=1}^{N} r_u^2 \right\}^{d+1}} \qquad (13.7.2)$$

is large, where

$$r_u^2 = x_{1u}^2 + x_{2u}^2 + \cdots + x_{ku}^2.$$

In particular, this would require that, for a first-order design $(d = 1)$ the ratio $\Sigma r_u^4/\{\Sigma r_u^2\}^2$ should be large to provide high detectability of quadratic lack of fit; for a second-order design $(d = 2)$, the ratio $\Sigma r_{iu}^6/\{\Sigma r_{iu}^2\}^3$ should be large to provide high detectability of cubic lack of fit. While these general results are of interest, careful attention must be given in practice to the detailed nature of Λ^2, as is now illustrated.

The Nature of the Noncentrality Parameter Λ^2

Suppose that we fit the model

$$y = X\beta + \varepsilon \qquad (13.7.3)$$

but that the true model is

$$y = X\beta + X_2\beta_2 + \varepsilon = \eta + \varepsilon. \tag{13.7.4}$$

Then (Section 3.10) the estimate $b = (X'X)^{-1}X'y$ of β is biased and

$$E(b) = \beta + A\beta_2, \tag{13.7.5}$$

where the bias matrix A is

$$A = (X'X)^{-1}X'X_2. \tag{13.7.6}$$

Comparing the formulas for b and A, we readily see that A can be regarded as the matrix of regression coefficients of X_2 on X. Now with $\hat{y} = Xb$,

$$\Lambda^2 = \{ E(\hat{y}) - \eta \}'\{ E(\hat{y}) - \eta \} = \beta_2'X_2'(I - R)'(I - R)X_2\beta_2 \tag{13.7.7}$$

where* $R = X(X'X)^{-1}X'$, and using $E(\hat{y}) = XE(b)$. Also,

$$(I - R)X_2 = X_2 - XA$$

$$= X_2 - \hat{X}_2$$

$$= X_{2\cdot1}, \tag{13.7.8}$$

say, where the elements of $\hat{X}_2$ are the "calculated values" obtained by regressing X_2 on X, just as the elements of $\hat{y}$ are the calculated values obtained via the regression of y on X. It follows that

$$\Lambda^2 = \beta_2'X_{2\cdot1}'X_{2\cdot1}\beta_2. \tag{13.7.9}$$

Notice that, if the jth column of X_2 was a linear combination of columns of X, then the jth column of $X_{2\cdot1} = X_2 - \hat{X}_2$ would consist of zeros and no term involving β_{2j} would occur in Λ^2. In such a case, however large the parameter β_{2j} might be, its effect could not be detected by the lack of fit test. Conversely, if the jth column of X_2 were orthogonal to all the columns of X, and to all the other columns of X_2, then the term involving β_{2j} in Λ^2 would be $\beta_{2j}^2\Sigma x_{2j}^2$. In this case, by making the value of Σx_{2j}^2 large, the lack of fit test could be made very sensitive to discrepancies arising from a nonzero value of β_{2j}. We now illustrate these points via an example.

---

*Both R and $I - R$ are symmetric (e.g., $R' = R$) and idempotent (e.g., $RR = R^2 = R$).

TABLE 13.3. Two first-order designs with associated matrices and skeleton
analyses of variance

$$\textit{Design A} \qquad\qquad\qquad \textit{Design B}$$

$$
\begin{array}{c}
\begin{array}{ccc} 0 & 1 & 2 \end{array} \\
\mathbf{X} = \begin{bmatrix} 1 & -1 & -1 \\ 1 & 1 & -1 \\ 1 & -1 & 1 \\ 1 & 1 & 1 \end{bmatrix}
\end{array},\;
\begin{array}{c}
\begin{array}{ccc} 11 & 22 & 12 \end{array} \\
\mathbf{X}_2 = \begin{bmatrix} 1 & 1 & 1 \\ 1 & 1 & -1 \\ 1 & 1 & -1 \\ 1 & 1 & 1 \end{bmatrix}
\end{array}
\qquad
\begin{array}{c}
\begin{array}{ccc} 0 & 1 & 2 \end{array} \\
\mathbf{X} = \begin{bmatrix} 1 & -2^{1/2} & 0 \\ 1 & 0 & -2^{1/2} \\ 1 & 0 & 2^{1/2} \\ 1 & 2^{1/2} & 0 \end{bmatrix}
\end{array},\;
\begin{array}{c}
\begin{array}{ccc} 11 & 22 & 12 \end{array} \\
\mathbf{X}_2 = \begin{bmatrix} 2 & 0 & 0 \\ 0 & 2 & 0 \\ 0 & 2 & 0 \\ 2 & 0 & 0 \end{bmatrix}
\end{array}
$$

$$
\mathbf{X}_{2 \cdot 1} = \begin{bmatrix} 0 & 0 & 1 \\ 0 & 0 & -1 \\ 0 & 0 & -1 \\ 0 & 0 & 1 \end{bmatrix}
\qquad\qquad
\mathbf{X}_{2 \cdot 1} = \begin{bmatrix} 1 & -1 & 0 \\ -1 & 1 & 0 \\ -1 & 1 & 0 \\ 1 & -1 & 0 \end{bmatrix}
$$

$$
\mathbf{X}'_{2 \cdot 1}\mathbf{X}_{2 \cdot 1} = \begin{bmatrix} 0 & 0 & 0 \\ 0 & 0 & 0 \\ 0 & 0 & 4 \end{bmatrix}
\qquad\qquad
\mathbf{X}'_{2 \cdot 1}\mathbf{X}_{2 \cdot 1} = \begin{bmatrix} 4 & -4 & 0 \\ -4 & 4 & 0 \\ 0 & 0 & 0 \end{bmatrix}
$$

Source	df	E(MS)	Source	df	E(MS)
Parameter estimates	3		Parameter estimates	3	
Lack of fit	1	$\sigma^2 + 4\beta_{12}^2$	Lack of fit	1	$\sigma^2 + 4(\beta_{22} - \beta_{11})^2$

EXAMPLE 13.3. Suppose $k = 2$, $d_1 = 1$, and $d_2 = 2$ so that initially we fit
the model

$$y = \beta_0 + \beta_1 x_1 + \beta_2 x_2 + \varepsilon, \qquad\qquad (13.7.10)$$

but we bear in mind the possibility that a second degree approximation

$$y = \beta_0 + \beta_1 x_1 + \beta_2 x_2 + \beta_{11} x_1^2 + \beta_{22} x_2^2 + \beta_{12} x_1 x_2 + \varepsilon \quad (13.7.11)$$

might be needed. Consider the two specific designs whose characteristics are
displayed in Table 13.3. See also Figure 13.8. Design A is a 2^2 factorial
design with points $(\pm 1, \pm 1)$, and design B is the same design turned
through an angle of $45°$, so that the axial distances of the points from the
origin are all $2^{1/2}$.

Design A is sensitive to lack of fit produced by the *interaction* term in the
second-order model, but is completely insensitive to lack of fit produced by

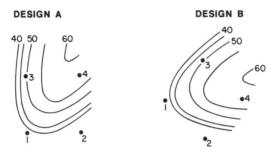

FIGURE 13.8. Two alternative designs A and B in relation to contours of a possible response function.

the pure *quadratic* terms, while design B has just the opposite sensitivities. The role played by the vectors of X_2, and their relationships with the vectors of X is clearly seen. In design A, the first two columns of X_2 are identical with the first column of X and therefore are annihilated when we take residuals while, in design B, the elements of the final column of X_2 are all zero initially and therefore their residuals are necessarily zero also. These results are to be expected because, with design A, we would have estimated the interaction term coefficient β_{12} but not the quadratic coefficients β_{11} and β_{22}, while with design B, we could have estimated the difference $\beta_{11} - \beta_{22}$ of the quadratic terms, but not the interaction term.

Interim Discussion

With this general theory of tests of lack of fit, and with this specific illustration in mind, we are in a position now to consider some of the more subtle aspects of the problem of checking fit. First of all, we must consider the following dilemma. Suppose we have a design suitable for fitting a polynomial of a certain degree d_1, which we hope provides adequate representation, but which we fear may not. Then we could check the fit by running a larger design of order $d_1 + 1$. In the above example, for instance, where $d_1 = 1$, if instead of running either design A or design B we added center points to separate the quadratic effects, and combined these two designs, we would obtain a second-order ($d_1 + 1 = 2$) composite design and could then estimate *all* of the coefficients in the second degree equation. The adequacy of fit test would then consist of checking if the additional second-order terms were negligible. Usually, however, economy in experimentation is important and if, for example, we believed that a first degree approximation should be adequate, we should not want to expend all the additional runs needed to fit an approximation of second degree. We thus need to seek some principle or principles on which we can choose an intermediate design which supplies some information on the "most feared" discrepancies from the model of order d_1, but is not as comprehensive as a design of order $d_1 + 1$.

If we think in terms of our example, we are thus led to the question "Is it more important to be able to check if β_{12} is small, in which case design A should be used, or is it more important to check if $(\beta_{22} - \beta_{11})$ is small, in which case design B will be appropriate?" However, a little thought reveals this question as somewhat naive. Consider the first-order design A in relation to the response surface whose contours are shown in Figure 13.8a. Obviously, in the situation depicted, serious second-order lack of fit would occur and we would wish to detect it. Now consider design B in relation to the same surface rotated through 45°; this is shown in Figure 13.8b. Since design B is a rotation through 45° of design A, the whole of Figure 13.8b, surface *and* design, is now simply a 45° rotation of Figure 13.8a. Any lack of fit detected by design A for the first surface must therefore be equally detectable by design B for the rotated surface. However suppose, as might well be the case, that at the time the experiment was being planned, a surface oriented like that in Figure 13.8a was regarded equally as likely to occur as a surface oriented like that in Figure 13.8b. Then there would clearly be nothing to choose between the designs so far as their ability to detect lack of fit was concerned.

To explore the same idea algebraically, consider the second-order terms

$$\beta_{11}x_1^2 + \beta_{22}x_2^2 + \beta_{12}x_1x_2 \qquad (13.7.12)$$

in an equation of second degree for two predictor variables x_1 and x_2. A rotation of axes through 45° corresponds to the transformation to $\dot{x}_1$ and $\dot{x}_2$ via

$$x_1 = \frac{(\dot{x}_1 - \dot{x}_2)}{2^{1/2}}, \qquad x_2 = \frac{(\dot{x}_1 + \dot{x}_2)}{2^{1/2}}. \qquad (13.7.13)$$

In the new coordinate system the second-order terms become

$$\dot{\beta}_{11}\dot{x}_1^2 + \dot{\beta}_{22}\dot{x}_2^2 + \dot{\beta}_{12}\dot{x}_1\dot{x}_2 \qquad (13.7.14)$$

with

$$\dot{\beta}_{11} = \tfrac{1}{2}(\beta_{11} + \beta_{22} - \beta_{12}), \qquad (13.7.15)$$

$$\dot{\beta}_{22} = \tfrac{1}{2}(\beta_{11} + \beta_{22} + \beta_{12}), \qquad (13.7.16)$$

and

$$\dot{\beta}_{12} = \beta_{22} - \beta_{11}, \qquad (13.7.17)$$

the last mentioned being of special relevance to the present discussion. Equivalently we can write

$$\beta_{12} = \dot{\beta}_{22} - \dot{\beta}_{11}. \qquad (13.7.18)$$

Thus the contents of the expected mean squares, $E(MS)$ in Table 13.3, namely $\sigma^2 + 4\beta_{12}^2$ for design A and $\sigma^2 + 4(\beta_{22} - \beta_{11})^2$ for design B, refer to the same quadratic lack of fit measured in two different orientations that are unlikely to be distinguishable a priori.

Yet another way to understand the comparison between the designs is to consider, for data y for design A, the contrast

$$\tfrac{1}{2}(y_1 - y_2 - y_3 + y_4). \qquad (13.7.19)$$

If the true model is quadratic, this contrast provides, for design A, an unbiased estimate of β_{12}, that is, of $\dot{\beta}_{22} - \dot{\beta}_{11}$. For data $\dot{y}$ from design B, the corresponding contrast $\tfrac{1}{2}(\dot{y}_1 - \dot{y}_2 - \dot{y}_3 + \dot{y}_4)$ provides an unbiased estimate of $\beta_{22} - \beta_{11}$, that is, of $\dot{\beta}_{12}$.

This raises the question of whether there are measures of quadratic lack of fit which are invariant under rotation of axes. One measure of this kind is the sum of the quadratic effects

$$\beta_{11} + \beta_{22} + \cdots + \beta_{kk}, \qquad (13.7.20)$$

which becomes $\dot{\beta}_{11} + \dot{\beta}_{22} + \cdots + \dot{\beta}_{kk}$ after rotation. [For the case $k = 2$, this is obvious from Eqs. (13.7.15) and (13.7.16).] This measure is worth practical consideration because we shall often know from the nature of the problem that, while a maximum* showing some degree of ridginess is likely, a minimax is not. This is equivalent to saying that all of the β_{ii}'s are likely to be negative or zero. Consequently, their sum will be a good overall measure of quadratic lack of fit.

Any first-order design with added center points will allow the estimation of $\Sigma\beta_{ii}$ and so will be sensitive to this kind of quadratic lack of fit. For illustration, suppose that under study are eight input variables whose levels, after suitable transformation and scaling, are denoted by $x_1, x_2, \ldots, x_8$ and that it is expected that, at this stage of experimentation, first-order effects

---

*As before, when the same kind of knowledge is available about a minimum, a parallel discussion arises.

TABLE 13.4. A first-order design for $k = 8$ variables which checks for certain types of quadratic bias

x_1	x_2	x_3	x_4	x_5	x_6	x_7	x_8	
1	−1	−1	−1	1	1	1	−1	
1	1	−1	−1	−1	−1	1	1	
1	−1	1	−1	−1	1	−1	1	
1	1	1	−1	1	−1	−1	−1	16-point
1	−1	−1	1	1	−1	−1	1	2^{8-4}_{IV}
1	1	−1	1	−1	1	−1	−1	design
1	−1	1	1	−1	−1	1	−1	with
1	1	1	1	1	1	1	1	generating
−1	1	1	1	−1	−1	−1	1	relation
−1	−1	1	1	1	1	−1	−1	I = 1235
−1	1	−1	1	1	−1	1	−1	= 1246
−1	−1	−1	1	−1	1	1	1	= 1347
−1	1	1	−1	−1	1	1	−1	= 2348
−1	−1	1	−1	1	−1	1	1	
−1	1	−1	−1	1	1	−1	1	
−1	−1	−1	−1	−1	−1	−1	−1	
0	0	0	0	0	0	0	0	
0	0	0	0	0	0	0	0	n_0
...								center
0	0	0	0	0	0	0	0	points

will be dominant. We have seen that first-order arrangements which minimize squared bias are provided by two-level fractional factorial designs of resolution IV or of higher resolution. [See under Eq.(13.6.14).]

For example, a good design for $d_1 = 1$, $d_2 = 2$, $k = 8$ is that shown in Table 13.4. This consists of a resolution IV fractional factorial design containing $n = 16$ runs, with n_0 added center points for a total of $N = 16 + n_0$ points. On the assumption that a polynomial model of degree $d_2 = 2$ is needed, the design allows estimation of the measure of convexity ($\beta_{11} + \beta_{22} + \cdots + \beta_{88}$) and of seven combinations of two-factor interactions, namely, ($\beta_{12} + \beta_{35} + \beta_{46} + \beta_{78}$), ($\beta_{13} + \beta_{25} + \beta_{47} + \beta_{68}$), ($\beta_{14} + \beta_{26} + \beta_{37} + \beta_{58}$), ($\beta_{15} + \beta_{23} + \beta_{48} + \beta_{67}$), ($\beta_{16} + \beta_{24} + \beta_{38} + \beta_{57}$), ($\beta_{17} + \beta_{28} + \beta_{34} + \beta_{56}$), and ($\beta_{18} + \beta_{27} + \beta_{36} + \beta_{45}$). Thus the residual sum of squares can be broken up as shown in Table 13.5. In this table,

$$c = \frac{\sum_{u=1}^{n} x_{iu}^2}{N}, \qquad i = 1, 2, \ldots, k$$

$$= \frac{16}{(16 + n_0)}. \qquad (13.7.21)$$

TABLE 13.5. Breakup of the residual sum of squares and its expected value for the first-order design of Table 13.4

	SS	df	Expected value of mean square
Lack of fit, S_L	$n_0 n(\bar{y}_f - \bar{y}_0)^2/N$	1	$\sigma^2 + c^2 n_0 N \left(\sum_{i=1}^{8} \beta_{ii} \right)^2 / n$
	$n\left(\sum_{u=1}^{N} x_{1u}x_{2u}y_u \right)^2 /(c^2 N^2)$	1	$\sigma^2 + c^2 N^2 (\beta_{12} + \beta_{35} + \beta_{46} + \beta_{78})^2 / n$
	$n\left(\sum_{u=1}^{N} x_{1u}x_{3u}y_u \right)^2 /(c^2 N^2)$	$8 \left\{ \begin{array}{c} 1 \end{array} \right.$ $\quad 8\sigma^2 + \Lambda^2$	$\sigma^2 + c^2 N^2 (\beta_{13} + \beta_{25} + \beta_{47} + \beta_{58})^2 / n$
	$\cdots$	$\cdots$	$\cdots$
	$n\left(\sum_{u=1}^{N} x_{1u}x_{8u}y_u \right)^2 /(c^2 N^2)$	1	$\sigma^2 + c^2 N^2 (\beta_{18} + \beta_{27} + \beta_{35} + \beta_{45})^2 / n$
Pure error, S_e	$\sum_{u=1}^{n_0} (y_{0u} - \bar{y}_0)^2$	$n_0 - 1$	σ^2

Also $\bar{y}_f$ and $\bar{y}_0$ are the averages of the factorial point responses y_u, $u = 1, 2, \ldots, n$, and of the center points responses $y_{0u}, u = 1, 2, \ldots, n_0$, respectively.

13.8. CHECKS BASED ON THE POSSIBILITY OF OBTAINING A SIMPLE MODEL BY TRANSFORMATION OF THE PREDICTORS

According to the argument outlined above, an efficient design of order d should (a) achieve small mean square error in estimating a polynomial of degree d and (b) allow the adequacy of the dth degree polynomial to be checked in a parsimonious manner. Parsimony in checking must rely on *a priori* knowledge of what kinds of discrepancies are likely to occur. Knowledge of this kind is utilized when, for example, knowing that a maximum* is more likely to arise than a saddle point, we provide a check of the overall curvature measure $\Sigma \beta_{ii}$ by adding center points to a first-order design.

It is also possible to build in checks for investigating the feasibility of transformations on the predictor variables. Both theoretical knowledge and practical experience frequently indicate that simpler models can be obtained by suitably transforming experimental variables $\xi_1, \xi_2, \ldots, \xi_k$, such as temperature, concentration, pressure, feed rate, and reaction time, and then

---

*Or minimum.

fitting a polynomial to the transformed variables. We consider first the application of this idea to first-order designs.

Possibility of a First-Order Model in Transformed Predictor Variables

When a curvilinear response relationship exists which is monotonic in the predictor variables over the current region of interest, it may be possible to use a first-order model in which power transformations $\xi_1^{\lambda_1}, \xi_2^{\lambda_2}, \ldots, \xi_k^{\lambda_k}$ are applied to the ξ's.

Assume that, *at worst*, the response may be represented by a second degree polynomial in transformed variables, namely by

$$F(\boldsymbol{\xi}, \boldsymbol{\lambda}) = \dot{\beta}_0 + \sum_{i=1}^{k} \dot{\beta}_i \xi_i^{\lambda_i} + \sum_{i=1}^{k} \sum_{j \geq i}^{k} \dot{\beta}_{ij} \xi_i^{\lambda_i} \xi_j^{\lambda_j}. \qquad (13.8.1)$$

Then a first degree polynomial model will be appropriate if the λ_i may be chosen so that $\dot{\beta}_{ij} = 0$ for all i and j. In Appendix 13E, we show that this requires that

$$\eta_{ij} = 0, \qquad i \neq j, \qquad (13.8.2)$$

$$\eta_{ii} + (S_i/\xi_i)(1 - \lambda_i)\eta_i = 0, \qquad i = 1, 2, \ldots, k, \qquad (13.8.3)$$

where

$$\eta_i = \frac{\partial F}{\partial x_i}, \qquad \eta_{ij} = \frac{\partial^2 F}{\partial x_i \partial x_j}, \qquad (13.8.4)$$

where

$$x_i = \frac{(\xi_i - \xi_{i0})}{S_i}, \qquad i = 1, 2, \ldots, k.$$

(To obtain Eqs. (13.8.2) and (13.8.3), set Eqs. (13E.2a) and (13E.2b) equal to zero.) We define

$$\delta_i = \frac{S_i}{\xi_{i0}}. \qquad (13.8.5)$$

Now suppose that a full second-order model

$$\hat{y} = b_0 + \Sigma b_i x_i + \sum \sum_{i \leq j} b_{ij} x_i x_j$$

has been fitted to the data from an appropriate design. Then we could approximate the derivatives of Eq. (13.8.4) by the corresponding derivatives of $\hat{y}$, evaluated at $\mathbf{x} = \mathbf{0}$ where necessary, namely by

$$\hat{\eta}_i = b_i, \qquad \hat{\eta}_{ij} = b_{ij}, \qquad i \neq j, \qquad \hat{\eta}_{ii} = 2b_{ii}. \qquad (13.8.6)$$

Also, we approximate S_i/ξ_i by δ_i in Eq. (13.8.5). Then (a) the possibility of a first-order representation in transformed variables $\xi_i^{\lambda_i}$ is contraindicated [see Eq. (13.8.2)] if any interaction estimate b_{ij}, $i \neq j$, is significantly different from zero, and (b) supposing such a transformation to be possible, the appropriate transformation parameters are roughly* estimated by

$$\hat{\lambda}_i = 1 + \frac{2b_{ii}}{\delta_i b_i}, \qquad i = 1, 2, \ldots, k. \qquad (13.8.7)$$

As always with experimental designs, we can obtain only those features for which we are prepared to pay. However, the above analysis indicates what these features might be:

1. Suppose we are prepared to assume that, if the function is not adequately approximated by a first-order model in $\xi_1, \xi_2, \ldots, \xi_k$, it will be by one in $\xi_1^{\lambda_1}, \xi_2^{\lambda_2}, \ldots, \xi_k^{\lambda_k}$. Then designs that provide appropriate checks will allow separate estimation of the β_{ii}. One source of such designs which provide maximum economy in the number of runs to be made are the three-level fractional factorials proposed by Finney. For example, for $k = 4$, and for standardized variables $x_i = (\xi_i - \xi_{i0})/S_i$ chosen at the three levels ± 1 and 0, we might use the fractional factorial design requiring nine runs shown in Table 13.6(a).

This design is readily derived by associating the three levels of x_3 and x_4, respectively, with the three Roman (A, B, C) and Greek (α, β, γ) letters in the Graeco–Latin square shown in Table 13.6(b). The levels of x_1 and x_2 are shown bordering the square. Regarded as a design to estimate the β_i and β_{ii}, this is a fully saturated arrangement, and must be used with care because ambiguities in interpretation will arise if terms other than those specifically allowed for in the model occur.

2. If we wish to check that a first-order representation is possible in $\xi_1^{\lambda_1}, \xi_2^{\lambda_2}, \ldots, \xi_k^{\lambda_k}$, then we need a design that permits estimation of at least some of the two-factor interactions.

---

*More precise estimates can be found by application of standard nonlinear least squares, fitting the model of Eq. (13.8.1) with $\hat{\beta}_{ij} = 0$ directly to the data.

TABLE 13.6. A design in four variables suitable for estimating power transformations

(a) Design

x_1	x_2	x_3	x_4
−1	−1	−1	1
0	−1	0	0
1	−1	1	−1
−1	0	1	0
0	0	−1	−1
1	0	0	1
−1	1	0	−1
0	1	1	1
1	1	−1	0

(b) Association scheme

		x_1		
		−1	0	1
	−1	$A\gamma$	$B\beta$	$C\alpha$
x_2	0	$C\beta$	$A\alpha$	$B\gamma$
	1	$B\alpha$	$C\gamma$	$A\beta$

$$x_3 = \begin{array}{ccc} -1 & 0 & 1 \\ A & B & C \end{array} \qquad x_4 = \begin{array}{ccc} -1 & 0 & 1 \\ \alpha & \beta & \gamma \end{array}$$

Possibility of a Second-Order Model in Transformed Predictor Variables

We now consider second-order designs that have built in checks sensitive to the possibility that a second degree approximation in transformed metrics, rather than the original metrics, is needed. For example, the true underlying function may possess an asymmetrical maximum which, *after suitable transformation of the* ξ_i, may be represented well enough by a quadratic function. (A quadratic function necessarily has a symmetric maximum.) Figure 13.9 illustrates this idea for a single predictor variable ξ, so that $k = 1$. The "suitable transformation" here is $\zeta = \xi^{1/2}$.

It turns out that a parsimonious class of designs of this type consists of the central composite arrangements in which a "cube," consisting of a two-level factorial with coded points $(\pm 1, \pm 1, \dots, \pm 1)$ or a fraction of

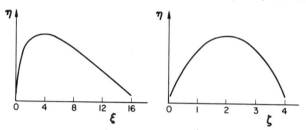

FIGURE 13.9. An asymmetrical maximum in ξ, on the left, is converted to a symmetrical maximum in ζ, on the right, by the transformation $\zeta = \xi^{1/2}$.

resolution $R \geq 5$ is augmented by an added "star," with axial points at coded distance α, and by n_0 added center points at $(0, 0, \ldots, 0)$. More generally, both the cube and the star might be replicated. A simple example of a design of this type for $k = 2$ is defined by the columns headed x_1 and x_2 in Table 13.7. (We use n_c and n_s for the number of cube and star points and n_{c0} and n_{s0} for the number of center points associated with them, respectively. Thus $n_0 = n_{c0} + n_{s0}$.) Also shown are some manufactured data generated by adding random error to response values read from Figure 8.5b, left diagram. The fitted least-squares second degree equation is

$$\hat{y} = \underset{\pm 0.25}{30.59} \quad \underset{\pm 0.17}{-4.22x_1} \quad \underset{\pm 0.17}{-5.91x_2} \quad \underset{\pm 0.14}{-1.66x_1^2} \quad \underset{\pm 0.14}{-1.44x_2^2} \quad \underset{\pm 0.24}{-3.41x_1x_2}, \quad (13.8.8)$$

where $\pm$ limits beneath each estimated coefficient indicate standard errors, calculated by using the pure error estimate $s_e^2 = 0.457$ to estimate σ^2. An associated analysis of variance table is shown as Table 13.8.

Before accepting the utility of the fitted equation we would need to be reassured on two questions: (a) Is there evidence from the data of serious lack of fit? If not, (b) is the change in $\hat{y}$, over the experimental region explored by the design, large enough compared with the standard error of $\hat{y}$ to indicate that the response surface is adequately estimated?

The analysis of variance of Table 13.8 sheds light on both these questions. Its use to throw light on the second was discussed in Section 8.2.

Clearly, for this example, it is the marked lack of fit of the second-order model which immediately concerns us. In particular, a need for transformation would be associated with the appearance of third-order terms. Associated with the design of Table 13.7 are four possible third-order columns, namely, those formed by creating entries of the form

$$\left(x_1^3, x_1x_2^2 \right); \quad \left(x_2^3, x_2x_1^2 \right). \quad (13.8.9)$$

These form two sets of two items, as indicated by the parentheses.

ABLE 13.7. A composite design for $k = 2$ predictor variables and its associated estimator columns; $n_c = 8, n_{c0} = 1, n_s = 4, n_{s0} = 5, \alpha = 2$

1	x_1	x_2	x_1^2	x_2^2	$x_1 x_2$	$x_{111}/1.5$	$x_{222}/1.5$	$8CC^a$	Blocks	y
1	-1	-1	1	1	1	1	1	1	-1	37.5
1	-1	-1	1	1	1	1	1	1	-1	38.3
1	1	-1	1	1	-1	-1	1	1	-1	34.7
1	1	-1	1	1	-1	-1	1	1	-1	35.1
1	-1	1	1	1	-1	1	-1	1	-1	27.7
1	-1	1	1	1	-1	1	-1	1	-1	29.2
1	1	1	1	1	1	-1	-1	1	-1	12.2
1	1	1	1	1	1	-1	-1	1	-1	11.4
1	.	.	.	.	.	.		-8	-1	30.1
1	-2	.	4	.	.	-2	.	-1	1	30.2
1	2	.	4	.	.	2	.	-1	1	16.1
1	.	-2	.	4	.	.	-2	-1	1	31.4
1	.	2	.	4	.	.	2	-1	1	16.7
1	.	.	.	.	.	.	.	0.8	1	30.5
1	.	.	.	.	.	.	.	0.8	1	29.9
1	.	.	.	.	.	.	.	0.8	1	29.9
1	.	.	.	.	.	.	.	0.8	1	29.8
1	.	.	.	.	.	.	.	0.8	1	30.2

The $8CC$ column provides a "curvature contrast." If the assumptions made about he model being quadratic are true, this contrast has expectation zero. See Box and Draper (1982a, Section 3.4), or Eq. (13.8.27).

ABLE 13.8. Analysis of variance associated with the second-order model and its checks, for the data of Table 13.7

Source		df	SS	MS	F
Mean		1	13,938.934		
Blocks		1	7.347	7.347	
First order extra		2	842.907	421.453	
Second order extra		3	188.142	62.714	
Lack	b_{111}		$\Big\lbrace$ 7.701	$\Big\lbrace$ 7.701	$\Big\lbrace$ 16.74
of	b_{222}	3 $\Big\lbrace$ 1	88.923 $\Big\lbrace$ 79.656	29.691 $\Big\lbrace$ 79.656	64.86 $\Big\lbrace$ 173.17
fit	CC		$\Big\lbrace$ 1.567	$\Big\lbrace$ 1.567	$\Big\lbrace$ 3.14
Pure error		8	3.657	0.457	
Total		18	15,069.910		

Now suppose these third-order columns are orthogonalized with respect to the lower-order $\mathbf{X}$ vectors. This may be accomplished by regressing them against the first six columns and taking residuals to yield columns x_{111} (from x_1^3), x_{122} (from $x_1 x_2^2$), and so on. Then, in vector notation,

$$\mathbf{x}_{iii} = -3\mathbf{x}_{ijj}, \qquad i \neq j \tag{13.8.10}$$

and the residual vectors are confounded in two sets of two. Furthermore, the columns $\mathbf{x}_{111}$ and $\mathbf{x}_{222}$ are orthogonal to each other. These vectors, reduced by a convenient factor of 1.5 to show their somewhat remarkable basic form, are given in Table 13.7.

Consider now the column $\mathbf{x}_{111}$ in relation to Figure 13.10, which shows the projection of the points of the composite design onto the x_1 axis. Denoting the average of the responses at $x_1 = -\alpha, -1, 1, \alpha$ by $\bar{y}_{-\alpha}$, $\bar{y}_{-1}$, $\bar{y}_1$, and $\bar{y}_\alpha$, respectively, we see that a contrast c_{31} associated with $\mathbf{x}_{111}$ is

$$c_{31} = \frac{1}{36} \mathbf{x}_{111}' \mathbf{y} = \frac{1}{3} \left\{ \frac{\bar{y}_\alpha - \bar{y}_{-\alpha}}{2\alpha} - \frac{\bar{y}_1 - \bar{y}_{-1}}{2} \right\}, \tag{13.8.11}$$

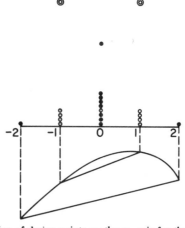

FIGURE 13.10. Projection of design points on the x_1 axis for the composite design in two factors given in Table 13.7. The contrast c_{31} is an estimate of the difference between the slopes of the two chords.

where, for our example, $\alpha = 2$. The expression in the parentheses is an estimate of the difference in slopes of the two chords joining points equi-distant from the design center. For a quadratic response curve this difference is zero. Thus c_{31} is a natural measure of overall nonquadraticity in the x_1 direction. A corresponding measure in the x_2 direction is, of course, given by $c_{32} = \mathbf{x}'_{222}\mathbf{y}/36$.

The corresponding sums of squares for these contrasts, given in Table 13.8, indicate a highly significant lack of fit. Corresponding plots of the residuals against x_1 and against x_2 show a characteristic pattern. A line joining residuals for observations at $x_i = \alpha$ and $x_i = -\alpha$ slopes up, while the tendency of the remaining residuals is down as x_i is increased. We return to discuss these data later.

General Formulas

In general, a composite design contains:

1. A "cube," consisting of a 2^k factorial, or a 2^{k-p} fractional factorial, made up of points of the type $(\pm 1, \pm 1, \ldots, \pm 1)$, of resolution $R \geq 5$ (Box and Hunter, 1961a, b) replicated r_c (≥ 1) times. There are thus $n_c = r_c 2^{k-p}$ such points (where p may be zero).

2. A "star," that is, $2k$ points $(\pm\alpha, 0, 0, \ldots, 0)$, $(0, \pm\alpha, 0, \ldots, 0), \ldots$, $(0, 0, 0, \ldots, \pm\alpha)$ on the predictor variable axes, replicated r_s times, so that there are $n_s = 2kr_s$ points in all. We assume $\alpha \neq 1$; see Draper (1983).

3. Center points $(0, 0, \ldots, 0)$, n_0 in number, of which n_{c0} are in cube blocks and n_{s0} in star blocks.

It is shown in Appendix 13F that, for any such design, k sets of columns can be isolated with the ith set containing the k columns $x_i x_j^2$, $j = 1, 2, \ldots, k$. This ith set is associated with a single vector $\mathbf{x}_{iii}$ which is orthogonal to the $(k + 1)(k + 2)/2$ columns required for fitting the second degree equation and is also orthogonal to the $(k - 1)$ similarly constructed vectors $\mathbf{x}_{jjj}$, $j \neq i$.

The elements of these vectors are such that:

For the cube points, $x_{iii} = \phi x_i$, with $\phi = 2r_s\alpha^2(1 - \alpha^2)/(n_c + 2r_s\alpha^2)$.
For the star points, $x_{iii} = \gamma x_i$, with $\gamma = n_c(\alpha^2 - 1)/(n_c + 2r_s\alpha^2)$.
For the center points $x_{iii} = 0 = x_i$.

Thus, the k estimates of third-order lack of fit, $c_{31}, c_{32}, \ldots, c_{3k}$ are

$$c_{3i} = \frac{\mathbf{x}'_{iii}\mathbf{y}}{\mathbf{x}'_{iii}\mathbf{x}_{iii}} = \frac{1}{\alpha^2 - 1}\left\{\frac{\bar{y}_{\alpha i} - \bar{y}_{-\alpha i}}{2\alpha} - \frac{\bar{y}_{1i} - \bar{y}_{-1i}}{2}\right\} \qquad (13.8.12)$$

with standard deviation

$$\sigma_{c_3} = \frac{1}{\alpha^2 - 1} \left\{ \frac{1}{n_c} + \frac{1}{2r_s\alpha^2} \right\}^{1/2} \sigma. \qquad (13.8.13)$$

Also

$$E(c_{3i}) = \beta_{iii} + (1 - \alpha^2)^{-1} \sum_{j \neq i}^{k} \beta_{ijj}. \qquad (13.8.14)$$

and the contribution to the lack of fit sum of squares is

$$\text{SS}(c_{3i}) = \frac{(\alpha^2 - 1)^2 c_{3i}^2}{\{1/n_c + 1/(2r_s\alpha^2)\}}. \qquad (13.8.15)$$

Note that even if $E(c_{3i}) = 0$, this does not necessarily mean there are no cubic coefficients. A combination of *nonzero* β_{iii} and β_{ijj} could occur for which $\beta_{iii} + (1 - \alpha^2)^{-1}\Sigma\beta_{ijj} = 0$. It is, of course, impossible to guard against every such possibility unless the full cubic model is fitted.

We now consider more explicitly what relation these lack of fit measures have to the possibility of transformation of the predictor variables. Suppose that, at worst, the response function may be represented by a third-order model in the *transformed* predictor variables. In Appendix 13E we show that the conditions that must then apply if all third-order coefficients of the transformed $\xi_i^{\lambda_i}$ are to be zero are [see Eqs. (13E.5), (13E.3), and (13E.4)]

$$\eta_{ijl} = 0, \qquad \text{all } i \neq j \neq l = 1, 2, \ldots, k; \quad (13.8.16)$$

$$\eta_{ijj} + \delta_i(1 - \lambda_j)\eta_{ij} = 0, \qquad i \neq j = 1, 2, \ldots, k; \qquad (13.8.17)$$

$$\eta_{iii} + 3\delta_i(1 - \lambda_i)\eta_{ii} + \delta_i^2(1 - \lambda_i)(1 - 2\lambda_i)\eta_i = 0, \qquad i = 1, 2, \ldots, k.$$

$$(13.8.18)$$

Thus the possibility of second-order representation in the transformed variables is contraindicated if any interaction estimate b_{ijl}, $i \neq j \neq l$, is nonzero.

In practice, the estimation of the transformation (when not contraindicated) is best done using nonlinear least squares directly on the model of Eq. (13.8.1); however some interesting light on how the curvature measures c_{3i} relate to these transformations is obtained by considering how Eqs. (13.8.17) and (13.8.18) could be used to obtain estimates of the λ_i.

A composite design does not permit all third-order terms to be separately estimated. Suppose, however, that a second-order model augmented with only cubic terms

$$\beta_{111}x_1^3 + \beta_{222}x_2^3 + \cdots + \beta_{kkk}x_k^3, \tag{13.8.19}$$

was fitted. If the response η could be represented by a full third-order model, the estimates b_i and b_{iii} obtained from the composite design would have expectations

$$E(b_i) = \eta_i - \frac{1}{2}\alpha^2(1 - \alpha^2)^{-1}\sum_{j\neq i}^{k}\eta_{ijj}, \tag{13.8.20}$$

$$E(b_{iii}) = \frac{1}{6}\eta_{iii} + \frac{1}{2}(1 - \alpha^2)^{-1}\sum_{j\neq i}^{k}\eta_{ijj}. \tag{13.8.21}$$

If now b_i and b_{iii} are used as estimates of the quantities shown as their expectations then, after appropriate substitutions have been made in Eqs. (13.8.16) to (13.8.18), we obtain the following k equations for the λ_i. (In these equations, $b_{iii} = c_{3i}$.)

$$b_{iii} + \delta_i(1 - \lambda_i)b_{ii} + \frac{1}{6}\delta_i^2(1 - \lambda_i)(1 - 2\lambda_i)b_i$$

$$+ \frac{1}{2}(1 - \alpha^2)^{-1}\left\{1 - \frac{1}{6}\alpha^2\delta_i^2(1 - 2\lambda_i)(1 - \lambda_i)\sum_{j\neq i}^{k}\delta_j(1 - \lambda_j)b_{ij}\right\} = 0,$$

$$i = 1, 2, \ldots, k. \tag{13.8.22}$$

These equations can be solved iteratively. Guessed values for the λ_i are first substituted in the grouping $(1 - \lambda_i)(1 - 2\lambda_i)$ wherever it occurs and the resulting linear equations solved to provide improved estimates for a second iteration, and so on.

For the example data, this procedure converges to the values $\hat{\lambda}_1 = -0.23$, $\hat{\lambda}_2 = -0.93$. These may be compared with the values $\hat{\lambda}_1 = 0.09$, $\hat{\lambda}_2 = -0.82$ provided by nonlinear least squares (these are maximum likelihood estimates under the standard normal error assumptions) and with $\lambda_1 = 0$, $\lambda_2 = -1$, the values used to generate the data; see Figure 8.5b.

An analysis of variance for the transformed data is shown in Table 13.9 where, as anticipated, no lack of fit appears.

A Curvature Contrast

We have already considered, indirectly, the curvature contrast

$$c_2 = \bar{y}_c - \bar{y}_{c0}, \tag{13.8.23}$$

TABLE 13.9. Analysis of variance for second-order model in predictor variables ln ξ_1 and ξ_2^{-1}

Source		df		SS		MS		F
Mean		1		13,938.934				
Blocks		1		7.347		7.347		
First-order extra		2		552.713		276.357		535.57
Second-order extra		3		565.238		188.413		365.14
Lack of fit $\left\{\begin{array}{l} CC \\ \text{Third order} \end{array}\right.$	$3\left\{\begin{array}{l} 1 \\ 2 \end{array}\right.$		$2.021\left\{\begin{array}{l} \\ \end{array}\right.$		$\left\{\begin{array}{l} 1.396 \\ 0.625 \end{array}\right.$	$0.674\left\{\begin{array}{l} \\ \end{array}\right.$	$1.47\left\{\begin{array}{l} 1.396 \\ 0.313 \end{array}\right.$	$\left\{\begin{array}{l} 3.05 \\ 0.68 \end{array}\right.$
Pure error		8		3.657		0.457		
Total		18		15,069.910				

which compares the average response at the factorial ("cube") points of a two-level factorial design with the average response at the center of the design. The expected value is

$$E(c_2) = \beta_{11} + \beta_{22} + \cdots + \beta_{kk}, \tag{13.8.24}$$

and this contrast contributes, to the analysis of variance table for a first-order design, a sum of squares

$$\frac{n_0 n_c (\bar{y}_0 - \bar{y}_{c0})^2}{(n_0 + n_c)} \tag{13.8.25}$$

where n_c, n_0 are the number of factorial and center points, respectively. (See Table 13.5 with slightly different notation.) Now when, as in the design of Table 13.7, center points are available both in the factorial block(s) *and* in the star block(s), several (two for our example) such measures are available. Consider, specifically, the two block case for a moment. If the average response at the center of the star is $\bar{y}_{s0}$ and the average over all the star points is $\bar{y}_s$, then the contrast

$$c_2' = \frac{k}{\alpha^2}(\bar{y}_s - \bar{y}_{s0}) \tag{13.8.26}$$

also has expectation $\Sigma \beta_{ii}$. Thus the statistic

$$c_2 - c_2' = \bar{y}_c - \bar{y}_{c0} - \frac{k}{\alpha^2}(\bar{y}_s - \bar{y}_{s0}), \tag{13.8.27}$$

which is the difference of the two measures of overall curvature, should be zero if the assumptions made about the model being quadratic are true.

From Figure 13.10, we see that the curvature measure c_2 associated with the cube (open circles) is contrasted with c_2' associated with the star (black dots). In general the distance from the center of the design to the cube points is $k^{1/2}$ and that for the star points is α. When, as in our example, $k^{1/2}$ and α are different, a significant value of $c_2 - c_2'$ could indicate (for example) a *symmetric* departure from quadratic fall-off on each side of the maximum, such as we see (for example) in a normal distribution curve.

In general, for two blocks, the standard deviation for $c_2 - c_2'$ is given by

$$\sigma_{c_2 - c_2'} = \left\{ \frac{1}{n_c} + \frac{1}{n_{c0}} + \frac{k^2}{\alpha^4} \left[\frac{1}{2kr_s} + \frac{1}{n_{s0}} \right] \right\}^{1/2} \sigma \qquad (13.8.28)$$

and the associated sum of squares for the analysis of variance table entry of Table 13.8 is obtained from

$$\mathrm{SS}(c_2 - c_2') = \frac{(c_2 - c_2')^2}{\left\{ 1/n_c + 1/n_{c0} + k/(2r_s\alpha^4) + k^2/(\alpha^4 n_{s0}) \right\}}. \qquad (13.8.29)$$

For our example we find

$$c_2 - c_2' = 1.3925 \pm 0.7520 \qquad (13.8.30)$$

with associated sum of squares 1.567 as shown in Table 13.8. There is clearly no evidence of this sort of lack of fit. When transformed predictors are used, again no lack of fit of this kind is evident, as we see from Table 13.9. No reduction in the degrees of freedom is made for the estimates of λ_1 and λ_2 (Box and Cox 1964, p. 240) because these degrees of freedom are identical to those attributed to third order.

Interaction With Blocks?

When composite designs are run in blocks, and if we allow the possibility that effects from the predictor variables could interact with blocks, then the various measures of lack of fit would be confounded with block-effect interactions. Although such contingencies must always be borne in mind, it should be remembered that these particular block-effect interactions are no more likely than any others.

APPENDIX 13A.   DERIVATION OF EQ. (13.5.8)

From Eqs. (13.5.3), and (13.5.5), we have

$$V = \left(\frac{N}{\sigma^2}\right)\int_O w(x)\,E\{(b_0 - Eb_0) + (b_1 - Eb_1)x\}^2\,dx$$

$$= \left(\frac{N}{\sigma^2}\right)\int_O w(x)\Big\{ E(b_0 - Eb_0)^2$$

$$+ 2xE[(b_0 - Eb_0)(b_1 - Eb_1)] + x^2 E(b_1 - Eb_1)^2\Big\}\,dx$$

$$= \frac{N\{V(b_0) + \mu_2 V(b_1)\}}{\sigma^2}, \qquad\qquad [\text{because } \mu_1 = 0],$$

$$= 1 + \frac{\mu_2}{m_2}, \qquad\qquad [\text{because } \bar{x} = 0].$$

Applying the work of Section 3.10 to Eqs. (13.5.5) and (13.5.6), we find that

$$E(b_0) = \beta_0 + m_2\beta_{11}, \qquad E(b_1) = \beta_1 + \frac{m_3\beta_2}{m_2}.$$

Thus,

$$B = \alpha^2\int_O w(x)\left(m_2 - x^2 + \frac{m_3 x}{m_2}\right)^2 dx, \qquad \left(\text{where } \alpha^2 = \frac{N\beta_{11}^2}{\sigma^2}\right)$$

$$= \alpha^2\int_O w(x)\left(m_2^2 - 2m_2 x^2 + x^4 + \frac{m_3^2 x^2}{m_2^2} + 2m_3 x - \frac{m_3 x^3}{m_2}\right) dx$$

$$= \alpha^2\left(m_2^2 - 2\mu_2 m_2 + \mu_4 + \frac{m_3^2\mu_2}{m_2^2}\right), \qquad [\text{because } \mu_p = 0, \text{ for } p \text{ odd}],$$

$$= \alpha^2\left\{(m_2 - \mu_2)^2 + (\mu_4 - \mu_2^2) + \frac{m_3^2\mu_2}{m_2^2}\right\}.$$

Combination of V and B gives the desired result.

APPENDIX 13B.   NECESSARY AND SUFFICIENT DESIGN CONDITIONS TO MINIMIZE BIAS

From (13.5.4), (13.6.3) and (13.6.4), we have

$$B = \left(\frac{N}{\sigma^2}\right) \int_O w(\mathbf{x})\, \boldsymbol{\beta}_2'(\mathbf{x}_1'\mathbf{A} - \mathbf{x}_2')'(\mathbf{x}_1'\mathbf{A} - \mathbf{x}_2')\boldsymbol{\beta}_2\, d\mathbf{x}$$

$$= \boldsymbol{\alpha}_2' \boldsymbol{\Delta} \boldsymbol{\alpha}_2,$$

where

$$\boldsymbol{\alpha}_2 = N\boldsymbol{\beta}_2/\sigma^2$$

$$\mathbf{A} = (\mathbf{X}_1'\mathbf{X}_1)^{-1}\mathbf{X}_1'\mathbf{X}_2 = \mathbf{M}_{11}^{-1}\mathbf{M}_{12},$$

$$\boldsymbol{\Delta} = \mathbf{A}'\boldsymbol{\mu}_{11}\mathbf{A} - \boldsymbol{\mu}_{12}'\mathbf{A} - \mathbf{A}'\boldsymbol{\mu}_{12} + \boldsymbol{\mu}_{22}$$

with $\boldsymbol{\mu}_{11}$ and $\boldsymbol{\mu}_{12}$ defined as in Eqs. (13.6.4) and $\boldsymbol{\mu}_{22}$ being similarly defined with appropriate changes in subscripts. Now $\boldsymbol{\Delta}$ can be rewritten as $\boldsymbol{\Delta}_1 + \boldsymbol{\Delta}_2$ where

$$\boldsymbol{\Delta}_1 = \boldsymbol{\mu}_{22} - \boldsymbol{\mu}_{12}'\boldsymbol{\mu}_{11}^{-1}\boldsymbol{\mu}_{12}$$

and

$$\boldsymbol{\Delta}_2 = \left(\mathbf{M}_{11}^{-1}\mathbf{M}_{12} - \boldsymbol{\mu}_{11}^{-1}\boldsymbol{\mu}_{12}\right)' \boldsymbol{\mu}_{11} \left(\mathbf{M}_{11}^{-1}\mathbf{M}_{12} - \boldsymbol{\mu}_{11}^{-1}\boldsymbol{\mu}_{12}\right).$$

It is shown below that both $\boldsymbol{\mu}_{11}$ and $\boldsymbol{\Delta}_1$ are positive semidefinite. It follows that, whatever the value of $\boldsymbol{\beta}_2$, B is minimized if and only if $\mathbf{M}_{11}^{-1}\mathbf{M}_{12} = \boldsymbol{\mu}_{11}^{-1}\boldsymbol{\mu}_{12}$. (If $\mathbf{M}_{11} = \boldsymbol{\mu}_{11}$ and $\mathbf{M}_{12} = \boldsymbol{\mu}_{12}$, than B is also minimized but these conditions are sufficient and not necessary.)

To Show Both $\boldsymbol{\mu}_{11}$ and $\boldsymbol{\Delta}_1$ Are Positive Semidefinite

It is obvious that

$$\int_O w(\mathbf{x})[(\mathbf{x}_1', \mathbf{x}_2')\boldsymbol{\beta}]^2\, d\mathbf{x} \geq 0$$

where $\boldsymbol{\beta} = (\boldsymbol{\beta}_1', \boldsymbol{\beta}_2')'$. Writing this as a quadratic form in $\boldsymbol{\beta}$ and performing the integration in the matrix of the quadratic form gives

$$\boldsymbol{\beta}' \begin{bmatrix} \boldsymbol{\mu}_{11} & \boldsymbol{\mu}_{12} \\ \boldsymbol{\mu}_{12}' & \boldsymbol{\mu}_{22} \end{bmatrix} \boldsymbol{\beta} \geq 0,$$

so that both μ_{11} and the matrix shown are positive semidefinite; but if we define the nonsingular matrix T as

$$T = \begin{bmatrix} I_{11} & -\mu_{11}^{-1}\mu_{12} \\ 0 & I_{22} \end{bmatrix}$$

then

$$\begin{bmatrix} \mu_{11} & 0 \\ 0 & \Delta_1 \end{bmatrix} = \begin{bmatrix} \mu_{11} & 0 \\ 0 & \mu_{22} - \mu_{12}'\mu_{11}^{-1}\mu_{12} \end{bmatrix} = T'\begin{bmatrix} \mu_{11} & \mu_{12} \\ \mu_{12}' & \mu_{22} \end{bmatrix}T.$$

The right-hand matrix is clearly positive semidefinite. It follows that Δ_1 must be positive semidefinite also.

APPENDIX 13C.　MINIMUM BIAS ESTIMATION

In the work we have described, in which designs are sought which minimize integrated variance plus bias, the estimation procedure used was that of least squares. An extra dimension can be created in this work by allowing other, perhaps biased, estimators and treating the least-squares method as a special case of a more general estimation class. Specifically, the use of *minimum bias estimation* was suggested by Karson, Manson, and Hader (1969). In this procedure, the estimator for β in the regression model is chosen to minimize the integrated bias B. Provided that $(X'X)$ is nonsingular*, the estimator that achieves this takes the form

$$b_1(MB) = \left\{I, \mu_{11}^{-1}\mu_{12}\right\}(X'X)^{-1}X'y,$$

where

$$X = (X_1, X_2).$$

This estimator can be written in the alternative form

$$b_1(MB) = \left\{I, \mu_{11}^{-1}\mu_{12}\right\}\begin{bmatrix} b_1^* \\ b_2^* \end{bmatrix}$$

$$= b_1^* + \mu_{11}^{-1}\mu_{12}b_2^*,$$

where b_1^*, b_2^* are the least-squares estimators of β_1 and β_2 *when the latter are estimated together by least squares*. The minimum value of B arising from the

*If it is not, the estimator may still exist. The requirement is then that $(I, \mu_{11}^{-1}\mu_{12})(\beta_1', \beta_2')'$ be estimable, and the estimator is of the form $T'Y$ where T' must satisfy the condition $T'(X_1, X_2) = (I, \mu_{11}^{-1}\mu_{12})$.

estimator b_1(MB) is

$$\alpha_2' \left(\mu_{22} - \mu_{12}' \mu_{11}^{-1} \mu_{12} \right) \alpha_2$$

and this is achieved for any experimental design for which

$$\{ I, \mu_{11}^{-1}\mu_{12} \} \begin{pmatrix} \beta_1 \\ \beta_2 \end{pmatrix}$$

is estimable. Note that *if* a design is an "all-bias design" (see Eq. (13.6.5) or Appendix 13B), then

$$\mu_{11}^{-1}\mu_{12} = (X_1'X_1)^{-1}X_1'X_2 = M_{11}^{-1}M_{12} = A,$$

where A is the usual bias or alias matrix and then

$$b_1(MB) = b_1^* + Ab_2^*$$

$$= (X_1'X_1)^{-1}X_1'y$$

the usual least squares estimator for β_1 estimated alone. The reduction to this arises from the fact that

$$b^* = \begin{bmatrix} b_1^* \\ b_2^* \end{bmatrix} = \begin{bmatrix} X_1'X_1 & X_1'X_2 \\ X_2'X_1 & X_2'X_2 \end{bmatrix}^{-1} \begin{bmatrix} X_1'y \\ X_2'y \end{bmatrix} = \begin{bmatrix} M_{11} & M_{12} \\ M_{12}' & M_{22} \end{bmatrix} \begin{bmatrix} X_1'y \\ X_2'y \end{bmatrix}$$

$$= \begin{bmatrix} C_{11}^{-1} & -M_{11}^{-1}M_{12}C_{22}^{-1} \\ -M_{22}^{-1}M_{21}C_{11}^{-1} & C_{22}^{-1} \end{bmatrix} \begin{bmatrix} X_1'y \\ X_2'y \end{bmatrix},$$

where

$$C_{ii} = M_{ii} - M_{ij}M_{jj}^{-1}M_{ji}, \quad i, j = 1, 2, \quad i \neq j, \quad \text{and} \quad M_{21} = M_{12}'.$$

Premultiplication of b^* by $(I, M_{11}^{-1}M_{12})$ produces the stated result, after some reduction.

We see immediately some advantages and some drawbacks in minimum bias estimation. Advantages are that it reduces the integrated bias B to its minimum value, and we can then choose a design to minimize the integrated variance V, thus achieving a smaller integrated variance plus bias than would be possible when using least squares; for details see Karson et al. (1969). A drawback is that, for nonsingular* $X'X$, we need to be able to estimate *both* β_1 and β_2 from the data via least squares to achieve minimum bias. (In the least-squares case, only β_1 need be

*See previous footnote.

estimable.) Moreover, methods of proceeding statistically after the minimum bias estimates have been obtained are unknown. For example, there are no minimum bias estimation equivalents to the analysis of variance table, nor to t or F tests. Our opinion is that minimum bias estimation is an ingenious idea which needs further development before it can be usefully employed.

APPENDIX 13D.   A SHRINKAGE ESTIMATION PROCEDURE

Another type of biased estimation procedure was proposed by Kupper and Meydrech (1973, 1974). They suggested using an estimator for β_1 of form

$$\mathbf{b}_1(KM) = \mathbf{K}\mathbf{b}_1$$

where $\mathbf{K}$ is a square matrix of the same order as the size of $\mathbf{b}_1$ and $\mathbf{b}_1 = (\mathbf{X}_1'\mathbf{X}_1)^{-1}\mathbf{X}_1'\mathbf{y}$ is the usual least-squares estimator of β_1. The basic idea is to choose $\mathbf{K}$ in an advantageous manner, to minimize the integrated variance plus bias. In fact, it can be shown that, if at least one of the elements in $\alpha_1 = N\beta_1/\sigma^2$ can be bounded, however loosely, a $\mathbf{K}$ can be found which will produce smaller integrated variance plus bias than would the least-squares choice $\mathbf{K} = \mathbf{I}$. Consider, for example, the case of a cuboidal region of interest in k dimensions, and a fitted polynomial of first order, with a true model of second order. Then, if $\alpha_i^2 \leq M_i$ where M_i is some given positive constant, the corresponding diagonal element of $\mathbf{K}$ is chosen as

$$k_i = \frac{m_2 M_i}{(1 + m_2 M_i)}$$

where

$$m_2 = N^{-1} \sum_{u=1}^{N} x_{iu}^2, \qquad i = 1, 2, \ldots, k.$$

Overall, $\mathbf{K}$ is chosen as the diagonal matrix whose elements are either k_i as given above for finite M_i, or 1 for $M_i = \infty$. In fact, the integrated variance plus bias will be reduced even if α_i^2 is as large as $2M_i$. For details, see Kupper and Meydrech (1973, 1974).

This procedure produces estimates that are shrunken versions (because $k_i \leq 1$) of the least-squares estimators. It is related in spirit to ridge regression methods and is perfectly appropriate in situations where values of M_i can be realistically settled upon. A disadvantage the method shares with minimum bias estimation is that methods of proceeding after estimation are not known. Again, our opinion is that this is an ingenious idea which needs further development to be of practical value.

APPENDIX 13E.   CONDITIONS FOR EFFICACY OF TRANSFORMATIONS ON THE PREDICTOR VARIABLES

Suppose that an experimental design is run in the k coded variables

$$x_i = \frac{(\xi_i - \xi_{i0})}{S_i}, \qquad i = 1, 2, \ldots, k, \tag{13E.1a}$$

in a situation where the underlying response function can be approximated by a second degree polynomial $F\{(\xi_i^{\lambda_i})\}$ in the transformed *original* variables $\xi_i^{\lambda_i}$. Thus

$$F(\xi, \lambda) = \dot{\beta}_0 + \sum_{i=1}^{k} \dot{\beta}_i \xi_i^{\lambda_i} + \sum_{i=1}^{k} \sum_{j \geq i}^{k} \dot{\beta}_{ij} \xi_i^{\lambda_i} \xi_j^{\lambda_j}. \tag{13E.1b}$$

Assume all $\lambda_i \neq 0$. [The case when any $\lambda_i = 0$ can be handled as a limiting case using the fact (see, e.g., Box and Cox, 1964) that $(\xi^\lambda - 1)/\lambda$ tends to $\ln \lambda$ as λ tends to zero.]

If Eq. (13E.1b) is to be a suitable representation, then all third derivatives with respect to the $\xi_i^{\lambda_i}$ must vanish identically. Note that

$$\frac{\partial F}{\partial \xi_i^{\lambda_i}} = \frac{\partial F}{\partial \xi_i} \frac{\partial \xi_i}{\partial \xi_i^{\lambda_i}} = F_i \left(\frac{\xi_i^{1-\lambda_i}}{\lambda_i} \right)$$

say, where $F_i = \partial F / \partial \xi_i$. Moreover, because of Eq. (13E.1a),

$$\eta_i = \frac{\partial F}{\partial x_i} = \frac{\partial F}{\partial \xi_i} \frac{\partial \xi_i}{\partial x_i} = F_i S_i$$

so that

$$\frac{\partial F}{\partial \xi_i^{\lambda_i}} = \eta_i S_i^{-1} \left(\frac{\xi_i^{1-\lambda_i}}{\lambda_i} \right).$$

The obvious extensions of these results also follow for the higher derivatives which involve terms of the form

$$\eta_{ij} = \frac{\partial^2 F}{\partial x_i \partial x_j} \quad \text{and} \quad \eta_{ijl} = \frac{\partial^3 F}{\partial x_i \partial x_j \partial x_l}.$$

If we now carry out the appropriate differentiations, we obtain

$$\frac{\partial F}{\partial \xi_i^{\lambda_i}} = \frac{\xi_i^{1-\lambda_i}}{S_i \lambda_i} \eta_i,$$

$$\frac{\partial^2 F}{\partial \xi_i^{\lambda_i} \partial \xi_j^{\lambda_j}} = \frac{\xi_i^{1-\lambda_i} \xi_j^{1-\lambda_j}}{S_i S_j \lambda_i \lambda_j} \eta_{ij}, \qquad (13E.2a)$$

$$\frac{\partial^2 F}{\partial^2 (\xi_i^{\lambda_i})} = \frac{\xi_i^{1-2\lambda_i}}{S_i \lambda_i^2} \left[(1 - \lambda_i) \eta_i + \frac{\xi_i}{S_i} \eta_{ii} \right], \qquad (13E.2b)$$

$$\frac{\partial^3 F}{\partial \xi_i^{\lambda_i} \partial^2 (\xi_j^{\lambda_j})} = \frac{\xi_i^{1-\lambda_i} \xi_j^{1-2\lambda_j}}{S_i S_j \lambda_i \lambda_j^2} \left[\frac{\xi_j}{S_j} \eta_{ijj} + (1 - \lambda_j) \eta_{ij} \right] = 0, \qquad (13E.3)$$

$$\frac{\partial^3 F}{\partial^3 (\xi_i^{\lambda_i})} = \frac{\xi_i^{1-3\lambda_i}}{S_i \lambda_i^3} \left[\left(\frac{\xi_i}{S_i} \right)^2 \eta_{iii} + \frac{3\xi_i}{S_i} (1 - \lambda_i) \eta_{ii} + (1 - \lambda_i)(1 - 2\lambda_i) \eta_i \right] = 0,$$

$$(13E.4)$$

$$\frac{\partial^3 F}{\partial \xi_i^{\lambda_i} \partial \xi_j^{\lambda_j} \partial \xi_l^{\lambda_l}} = \frac{\xi_i^{1-\lambda_i} \xi_j^{1-\lambda_j} \xi_l^{1-\lambda_l}}{S_i S_j S_l \lambda_i \lambda_j \lambda_l} \eta_{ijl} = 0. \qquad (13E.5)$$

Equations (13E.3)–(13E.5) are exact conditions on the λ's. The η_i, η_{ij}, and η_{ijl} also involve the λ's, and cannot be specifically evaluated if the λ's are unknown. If *estimates* of the η_i, η_{ij}, and η_{ijl} are substituted, however, the three equations can be solved to provide estimates of the λ's.

We now assume that the response surface can be approximately represented by a *cubic* polynomial in the coded predictor variables, namely, by

$$\hat{\eta} = b_0 + b_1 x_1 + \cdots + b_k x_k$$

$$+ b_{11} x_1^2 + \cdots + b_{kk} x_k^2$$

$$+ b_{12} x_1 x_2 + \cdots + b_{k-1, k} x_{k-1} x_k$$

$$+ b_{111} x_1^3 + b_{122} x_1 x_2^2 + \cdots$$

$$+ b_{222} x_2^3 + b_{112} x_1^2 x_2 + \cdots$$

$$+ b_{kkk} x_k^3 + b_{11k} x_1^2 x_k + \cdots$$

$$+ b_{123} x_1 x_2 x_3 + \cdots .$$

We can estimate the η_i, η_{ij}, and η_{ijl} by the corresponding derivatives of $\hat{\eta}$ evaluated at the center of the design, that is, at $\mathbf{x} = \mathbf{0}$. In general then,

$$\hat{\eta}_i = b_i, \qquad \hat{\eta}_{ii} = 2b_{ii}, \qquad \hat{\eta}_{iii} = 6b_{iii},$$

$$\hat{\eta}_{ij} = b_{ij}, \qquad \hat{\eta}_{ijj} = 2b_{ijj}, \qquad \hat{\eta}_{ijl} = b_{ijl}, \qquad (13E.6)$$

and we would substitute these in Eqs. (13E.3)–(13E.5), at the same time setting $\xi_i = \xi_{i0}$, its value when $x_i = 0$. Note that, from Eq. (13E.5), we require that $\eta_{ijl} = 0$, which implies that the assumed transformations $\xi_i^{\lambda_i}$ are not suitable representations unless all three-factor interactions are zero. In practice, then, we would want b_{ijl} to be small and nonsignificant when $i \neq j \neq l$ or else we have a clear indication that the $\xi_i^{\lambda_i}$ will *not* provide a satisfactory second-order representation in (13E.1b). If this aspect is satisfied, however, we now solve the $\frac{1}{2}k(k+1) + k = \frac{1}{2}k(k+3)$ simultaneous equations:

$$\hat{\eta}_{ijj} + \delta_i(1 - \lambda_j)\hat{\eta}_{ij} = 0, \qquad i \neq j = 1, 2, \ldots, k,$$

$$\hat{\eta}_{iii} + 3\delta_i(1 - \lambda_i)\hat{\eta}_{ii} + \delta_i^2(1 - \lambda_i)(1 - 2\lambda_i)\hat{\eta}_i = 0, \qquad i = 1, 2, \ldots, k \quad (13E.7)$$

where $\delta_i = S_i/\xi_{i0}$, and we have divided through Eqs. (13E.3) and (13E.4) by factors assumed to be non-zero, namely S_i, ξ_{i0}, λ_i, using the $\hat{\eta}$ values from Eq. (13E.6).

Composite Designs

An additional complication arises with composite designs. For such designs, we cannot estimate all the third-order coefficients individually; see Appendix 13F. This means that, while Eqs. (13E.7) are still valid, the values in Eq. (13E.6) cannot be used. For a composite design, the column vectors with elements x_i, x_i^3, and $x_i x_j^2$ are linearly dependent via

$$x_i x_j^2 = \frac{\left(-\alpha^2 x_i + x_i^3\right)}{(1 - \alpha^2)}$$

Thus, in the general cubic model, we cannot estimate all $k + k + k(k - 1) = k(k + 1)$ coefficients in the terms

$$\beta_i x_i + \beta_{iii} x_i^3 + \sum_{j \neq i}^{k} \beta_{ijj} x_i x_j^2$$

but only the $2k$ coefficients of

$$\left\{ \beta_i - \frac{\alpha^2}{1 - \alpha^2} \sum_{j \neq i}^{k} \beta_{ijj} \right\} x_i + \left\{ \beta_{iii} + \frac{1}{1 - \alpha^2} \sum_{j \neq i}^{k} \beta_{ijj} \right\} x_i^3. \qquad (13E.8)$$

It follows that, in the model so reduced,

$$
b_i \text{ estimates } \eta_i - \frac{\alpha^2}{1 - \alpha^2} \frac{1}{2} \sum_{j \neq i}^{k} \eta_{ijj}
$$

while

$$
b_{iii} \text{ estimates } \frac{1}{6} \eta_{iii} + \frac{1}{1 - \alpha^2} \frac{1}{2} \sum_{j \neq i}^{k} \eta_{ijj}. \tag{13E.9}
$$

[By examining Eqs. (13.8.14), (13E.6), and the second portion of Eq. (13E.8), we infer that $b_{iii} = c_{3i}$.] Alternatively, if the model is fitted using x_i and the "orthogonalized x_i^3", namely $x_{iii} = x_i^3 - \psi x_i$, the terms of the model are

$$
\left\{ \beta_i + \psi \beta_{iii} + \theta \sum_{j \neq i}^{k} \beta_{ijj} \right\} x_i + \left\{ \beta_{iii} + \frac{1}{1 - \alpha^2} \sum_{j \neq i}^{k} \beta_{ijj} \right\} (x_i^3 - \psi x_i),
$$

where

$$
\theta = \frac{n_c}{\left(n_c + 2 r_s \alpha^2 \right)}, \qquad \psi = \frac{\left(n_c + 2 r_s \alpha^4 \right)}{\left(n_c + 2 r_s \alpha^2 \right)}.
$$

In this form, we have that

$$
b_i^* \text{ estimates } \eta_i + \frac{1}{6} \psi \eta_{iii} + \frac{1}{2} \theta \sum_{j \neq i}^{k} \eta_{ijj}
$$

and

$$
b_{iii}^* \text{ estimates } \frac{1}{6} \eta_{iii} + \frac{1}{1 - \alpha^2} \frac{1}{2} \sum_{j \neq i}^{k} \eta_{ijj},
$$

where b_i^* is the estimated coefficient of x_i and b_{iii}^* is that for $(x_i^3 - \psi x_i)$. (Note that $b_{iii}^* = b_{iii} = c_{3i}$, but that $b_i^* = b_i + \psi b_{iii}$.) We now describe how this affects the estimation of the λ_i. From Eqs. (13E.3) and (13E.4), and setting $\xi_i = \xi_{i0}$ we have

$$
\eta_{ijj} + \delta_j (1 - \lambda_j) \eta_{ij} = 0 \tag{13E.10}
$$

and

$$
\eta_{iii} + 3 \delta_i (1 - \lambda_i) \eta_{ii} + \delta_i^2 (1 - \lambda_i)(1 - 2\lambda_i) \eta_i = 0. \tag{13E.11}
$$

We now combine $\frac{1}{6}$ times Eq. (13E.11) with

$$\left\{ \frac{1}{2(1 - \alpha^2)} - \frac{\alpha^2}{12(1 - \alpha^2)} \delta_i^2(1 - 2\lambda_i)(1 - \lambda_i) \right\}$$

times Eq. (13E.10) summed over $j \neq i$ to give, for $i = 1, 2, \ldots, k$,

$$\left\{ \frac{1}{6}\eta_{iii} + \frac{1}{2}\frac{1}{1 - \alpha^2} \sum_{j \neq i}^{k} \eta_{ijj} \right\}$$

$$+ \frac{1}{2}\delta_i(1 - \lambda_i)\eta_{ii}$$

$$+ \frac{1}{6}\delta_i^2(1 - \lambda_i)(1 - 2\lambda_i)\left\{ \eta_i - \frac{\alpha^2}{1 - \alpha^2}\frac{1}{2}\sum_{j \neq i}^{k}\eta_{ijj} \right\}$$

$$+ \frac{1}{2}\frac{1}{1 - \alpha^2}\left\{ 1 - \frac{1}{6}\alpha^2\delta_i^2(1 - 2\lambda_i)(1 - \lambda_i) \right\} \sum_{j \neq i}^{k} \delta_j(1 - \lambda_j)\eta_{ij} = 0.$$

Thus, if the third-order model is fitted in terms of x_i and x_i^3 [rather than x_i and $(x_i^3 - \psi x_i)$ as described above] we can substitute appropriate estimates to give the k simultaneous equations for $i = 1, 2, \ldots, k$:

$$b_{iii} + \delta_i(1 - \lambda_i)b_{ii} + \frac{1}{6}\delta_i^2(1 - \lambda_i)(1 - 2\lambda_i)b_i$$

$$+ \frac{1}{2}\frac{1}{1 - \alpha^2}\left\{ 1 - \frac{1}{6}\alpha^2\delta_i^2(1 - 2\lambda_i)(1 - \lambda_i) \right\} \sum_{j \neq i}^{k} \delta_j(1 - \lambda_j)b_{ij} = 0 \quad (13E.12)$$

which can be solved for $\lambda_1, \lambda_2, \ldots, \lambda_k$. These awkward equations can be difficult to solve unless some care is applied. We suggest an iterative procedure in which rough estimates of λ_i are used in the grouping $Q_i \equiv (1 - \lambda_i)(1 - 2\lambda_i)$ which occurs in two positions in Eq. (13E.12). The resulting linear equations in $\theta_i = 1 - \lambda_i$ are straightforward to solve, and the results are used in the grouping $(1 - \lambda_i)(1 - 2\lambda_i)$ for a second iteration and so on, until convergence is achieved. To aid convergence, each new iteration can be started from the midpoint of the old and new values, if desired.

Alternative Cubic Case

If the cubic is fitted with estimated terms $b_i^* x_i + b_{iii}^*(x_i^3 - \psi x_i)$ as described above, we obtain b_i and b_{iii} from

$$b_i = b_i^* - \psi b_{iii}^* \quad \text{and} \quad b_{iii} = b_{iii}^*.$$

Related Work

The expression in Eq. (13E.9) can be written alternatively in the form

$$\beta_{iii} + (1 - \alpha^2)^{-1}\Sigma\beta_{iij}.$$

Draper and Herzberg (1971, p. 236) show that the sum of squares of these quantities for $i = 1, 2, \ldots, k$ occurs in the expected value of a general measure of lack of fit L_1. Thus, the c_{3i} contrasts essentially provide a split-up of L_1 which permits a more detailed and sensitive analysis. The remaining degrees of freedom pertaining to L_1 can be attributed to other contrasts as already described.

APPENDIX 13F.   THE THIRD-ORDER COLUMNS OF THE X MATRIX FOR A COMPOSITE DESIGN

The form of the X matrix in the full cubic regression model $y = X\beta + \varepsilon$ when the design consists of r_c "cubes" plus r_s "stars" plus n_0 center points is as shown in Table 13F.1. We can denote columns by placing square brackets around the column head; for example $[x_1]$ will denote the x_1 column, and so on. We write $n_c = r_c 2^{k-p}$ for the number of cube points.

All of the cubic columns are orthogonal to all of the other columns with the following exceptions: $[x_i^3]$ is not orthogonal to $[x_i]$, nor to $[x_i x_j^2]$; $[x_i x_j^2]$ is not orthogonal to $[x_i]$, nor to $[x_i^3]$, nor to $[x_i x_l^2]$. The first step is to regress the $[x_i^3]$ and $[x_i x_j^2]$ vectors on the $[x_i]$ and take residuals. Because the columns involved are orthogonal to $[x_0]$, no adjustment for means is needed. We denote the "cube portion" of the $[x_i]$ and $[x_i x_j^2]$ vectors by c_i, as indicated in the table. These two sets of residuals are, where the prime denotes transpose,

$$[x_{iii}] = [x_i^3] - \left\{\frac{[x_i]'[x_i^3]}{[x_i]'[x_i]}\right\}[x_i]$$

and

$$[x_{ijj}] = [x_i x_j^2] - \left\{\frac{[x_i]'[x_i x_j^2]}{[x_i]'[x_i]}\right\}[x_i]$$

both of which reduce to multiples $(1 - \alpha^2)m$ and m, respectively, where $m = 2r_s\alpha^2/(n_c + 2r_s\alpha^2)$, of the same vector. For example,

$$[x_{111}]' = [c_1', d, -d, d, -d, \ldots, d, -d, 0, 0, \ldots, 0](1 - \alpha^2)m,$$

where

$$d = \frac{n_c}{(2r_s\alpha)},$$

TABLE 13F.1. The X matrix for a cubic model in k predictors $x_1, x_2, \ldots, x_k$

x_0	x_1	x_2	$\cdots$	x_k	x_1^2	x_2^2	$\cdots$	x_k^2	x_1x_2	$\cdots$	$x_{k-1}x_k$	x_1^3	$x_1x_2^2$	$\cdots$	$x_1x_k^2$	$\cdots$	$x_1x_2x_3$	$\cdots$
	$\mathbf{c}_1$	$\mathbf{c}_2$	$\cdots$	$\mathbf{c}_k$														
1																		
1			±1's				1's			±1's			±1's			±1's		
1			2^{k-p} design							interaction patterns			each column same pattern as in x_1 column					
⋮																		
1																		
1	$-\alpha$	0	$\cdots$	0	α^2	0	$\cdots$	0	0	0	0	$-\alpha^3$	0	0	0	0	0	0
1	α	0	$\cdots$	0	α^2	0	$\cdots$	0				α^3						
⋮			r_s sets					$\cdots$				$\cdots$						
1	0	0	permuted	0	0	0		0				$-\alpha^3$						
1	0	0	down	0	0	0		0				α^3						
⋮	$\cdots$		columns	$\cdots$	$\cdots$			$\cdots$				$\cdots$						
1	0	0		$-\alpha$	0	0		α^2	0	0	0	0	0	0	0	0	0	0
1	0	0		α	0	0		α^2										
⋮	$\cdot$			$-\alpha$	$\cdot$			α^2										
1	0			α	0			α^2										
1	0	0		0	0	0		0	0	0	0	0	0	0	0	0	0	0
1																		
1			0				0			0			0			0		
⋮																		
1																		

473

and where there are r_s sets of $(d, -d)$'s in the vector. In general, for $[x_{iii}]'$, $\mathbf{c}'_1$ will be replaced by $\mathbf{c}'_i$ and the position of the $\pm d$'s will correspond to those of the $\mp \alpha$'s in the corresponding $[x_i]'$ vector. Note that, because $\mathbf{c}'_i \mathbf{c}_j = 0$, $i \neq j$, it is obvious that $[x_{iii}]$ and $[x_{jjj}]$ are orthogonal.

It follows that the k cubic coefficients β_{iii}, β_{ijj} ($j \neq i$, $j = 1, 2, \ldots, k$, otherwise) cannot be estimated individually but only in linear combination, and that an appropriate normalized estimating constrast for this is

$$l_{iii} = \frac{[x_{iii}]' y}{[x_{iii}]'[x_{iii}]}$$

$$= \frac{\{\mathbf{c}_i y_1 + d(-r_s \bar{y}_{\alpha i} + r_s \bar{y}_{-\alpha i})\}}{\{n_c(1 - \alpha^2)\}},$$

where y_1 is the portion of y corresponding to the cube part of the design, and $\bar{y}_{\alpha i}$, $\bar{y}_{-\alpha i}$ are, respectively, the averages of observations taken at the α and $-\alpha$ axial points on the x_i axis. If we similarly denote by $\bar{y}_{1i}$ and $\bar{y}_{-1i}$ the averages of the $n_c/2$ observations in y_1 corresponding to 1 and -1 in $\mathbf{c}_i$, respectively, it follows quickly that $l_{iii} = c_{3i}$ where c_{3i} is given in Eq. (13.8.12). The expected value is

$$E(c_{3i}) = \frac{[x_{iii}]' X \beta}{[x_{iii}]'[x_{iii}]},$$

where X is as in Table 13F.1 and the coefficients of β correspond to the columns in the obvious manner. Because $[x_{iii}]$ is orthogonal to all columns of X except the $[x_i^3]$ and $[x_i x_j^2]$ columns, Eq. (13.8.14) emerges almost immediately.

EXERCISES

13.1. Suppose we wish to choose a design to fit a straight line $y = \beta_0 + \beta_1 x + \varepsilon$ over the region of interest $-1 \leq x \leq 1$ and also guard against the possibility that the true model might be a general polynomial of degree $d_2 > 1$. If there were little or no variance error (so that V could effectively be ignored) show that an appropriate "all-bias" design $(x_1, x_2, \ldots, x_N)$ would have moments

$$N^{-1} \sum_{u=1}^{N} x_u^i = \begin{cases} 0, & \text{for } i \text{ odd,} \\ (i+1)^{-1}, & \text{for } i \text{ even,} \end{cases} \quad i \leq d_2 + 1.$$

If $d_2 = 3$, $N = 6$, find a specific design which satisfies these moment conditions.

13.2. Suppose we wish to choose a design $x_1, x_2, \ldots, x_N$ to fit the model $y = \beta_0 + \beta_1 x + \beta_{11} x^2 + \varepsilon$ over the coded range of interest $-1 \leq x \leq 1$, but also

TABLE E13.2

Number of points at					Value of	
$-b$	$-a$	0	a	b	a	b
1	4	0	4	1	0.478	1.009
1	3	2	3	1	0.567	0.984
1	2	4	2	1	0.790	0.828
2	1	4	1	2	0.777	0.815
2	3	0	3	2	0.358	0.880
2	2	2	2	2	0.452	0.873

wish to guard against the possibility that the true model contains an additional term $\beta_{111}x^3$. We furthermore specify that $N = 10$ points should be allocated in some manner to the five levels $x = (-b, -a, 0, a, b)$ where a and b are to be determined. It can be shown (see The choice of a second order rotatable design, by G. E. P. Box and N. R. Draper, *Biometrika*, **50**, 1963, 335–352) that if variance and bias errors are to be roughly the same size we should choose (approximately) $\Sigma x_u^2 = 3.866$, $\Sigma x_u^4 = 2.494$. Table E13.2 shows six designs that approximately achieve those figures. Evaluate the detectability ratio defined in Section 13.7; this is $\Sigma x_u^6/\{\Sigma x_u^2\}^3$ but, since Σx_u^2 is approximately constant, evaluate Σx_u^6 for each design instead. Which design would you use? Why?

13.3. Suppose we wish to choose an "all-bias" design to fit a plane $y = \beta_0 + \beta_1 x_1 + \beta_2 x_2 + \cdots \beta_k x_k + \varepsilon$ while guarding against the possibility that the true model might be a second-order polynomial. Show that, if $\mathbf{M}$ is any $k \times k$ orthogonal matrix with a first column of all $+1$'s, then the design matrix

$$\mathbf{D} = \begin{bmatrix} \mathbf{M} \\ -\mathbf{M} \end{bmatrix}$$

provides a suitable design.

13.4. (Source: Sequential designs for spherical weight functions, by N. R. Draper and W. E. Lawrence, *Technometrics*, **4**, 1967, 517–529.) Consider the density function in k dimensions

$$W(x) = C\exp\left\{-\tfrac{1}{2}\left[\left(x_1^2 + x_2^2 + \cdots + x_k^2\right)/\sigma^2\right]^{1/(1+\beta)}\right\}, (-1 < \beta \leq 1)$$

where the normalizing constant C is given by

$$C = 2^{-(1/2)k(1+\beta)}\sigma^{-k}(1+\beta)^{-1}\pi^{-k/2}\frac{\Gamma(k/2)}{\Gamma\{\tfrac{1}{2}k(1+\beta)\}}.$$

Show that, if $k = 1$, the cases (a) $\beta \to -1$, and (b) $\beta = 0$, give rise to the two weight functions shown in Figure 13.5. In general, $W(x)$ provides a versatile weight function for k dimensions.

13.5.   Consider the design and observations y_u given in Table E13.5. Will a second-order model in the x's provide a satisfactory fit to these data? If not, and given that the codings from x's to ξ's are as given below, will a second-order model in $\xi_i^{\lambda_i}$, $i = 1, 2, 3$ provide a satisfactory fit? (In both parts of the question provide all the usual analyses.)

$$\text{Codings: } \xi_1 = 0.85 + 0.09x_1,$$

$$\xi_2 = 50 + 24x_2,$$

$$\xi_3 = 3.5 + 0.9x_3.$$

TABLE E13.5

x_1	x_2	x_3	y	x_1	x_2	x_3	y
-1	-1	-1	353	0	-1.68	0	98
1	-1	-1	361	0	1.68	0	3958
-1	1	-1	2643	0	0	-1.68	1223
1	1	-1	2665	0	0	1.68	1207
-1	-1	1	350				
1	-1	1	354	0	0	0	1216
-1	1	1	2639	0	0	0	1212
1	1	1	2652	0	0	0	1216
				0	0	0	1215
-1.68	0	0	1204	0	0	0	1213
1.68	0	0	1223	0	0	0	1218

CHAPTER 14

Variance-Optimal Designs

14.1. INTRODUCTION

We said, in Section 13.7, that our aims in selecting an experimental design must necessarily be multifaceted. The problem of selecting a suitable design is thus a formidable one. Some properties (given by Box and Draper, 1975) of a response surface design, any, all, or some of which might be important, depending on the experimental circumstances, are as follows. The design should:

1. Generate a satisfactory distribution of information throughout the region of interest, R.
2. Ensure that the fitted value at $\mathbf{x}$, $\hat{y}(\mathbf{x})$, be as close as possible to the true value at $\mathbf{x}$, $\eta(\mathbf{x})$.
3. Give good detectability of lack of fit.
4. Allow transformations to be estimated.
5. Allow experiments to be performed in blocks.
6. Allow designs of increasing order to be built up sequentially.
7. Provide an internal estimate of error.
8. Be insensitive to wild observations and to violation of the usual normal theory assumptions.
9. Require a minimum number of experimental runs.
10. Provide simple data patterns that allow ready visual appreciation.
11. Ensure simplicity of calculation.
12. Behave well when errors occur in the settings of the predictor variables, the x's.
13. Not require an impractically large number of levels of the predictor variables.
14. Provide a check on the "constancy of variance" assumption.

These requirements are sometimes conflicting, sometimes confluent and, in any specific situation, judgment is required to select the design that achieves the best compromise. This judgment is aided by knowledge of the behavior of these design properties. In particular, as we pointed out in Chapter 13, the trade-off between variance and bias is of critical practical importance. Because polynomials, or any other type of empirical function, are approximations whose adequacy depends on the size of the region covered by the design, we must balance off the reduced variance achieved by increasing the size of a design against the extra bias so induced.

Much theoretical work has been done without considering the effect of model inadequacy. The conclusions obtained from such studies can, nevertheless, be of value in providing information for design choice, as long as the possibility of model inadequacy is kept in mind.

14.2. ORTHOGONAL DESIGNS

The work of R. A. Fisher and F. Yates pointed to orthogonality as an important design principle long ago. We first explain and discuss *orthogonality* and then introduce *rotatability* as a logical extension of it.

For simplicity, consider a simple linear regression setup with only two experimental variables ξ_1 and ξ_2, n observations, $u = 1, 2, 3, \ldots, n$, and model

$$y_u = \beta_0 + \beta_1 \tilde{\xi}_{1u} + \beta_2 \tilde{\xi}_{2u} + \varepsilon, \tag{14.2.1}$$

where $\tilde{\xi}_{iu} = \xi_{iu} - \bar{\xi}_i$. If we use least squares to obtain estimates $b_0 = \bar{y}$, b_1, b_2 of β_0, β_1, β_2, respectively, then it is easy to show that the variances and covariances of these estimates are

$$V(b_0) = \frac{\sigma^2}{n}, \tag{14.2.2}$$

$$V(b_1) = \left\{ \frac{1}{(1 - \cos^2\theta)S_1^2} \right\} \frac{\sigma^2}{n}, \qquad V(b_2) = \left\{ \frac{1}{(1 - \cos^2\theta)S_2^2} \right\} \frac{\sigma^2}{n}, \tag{14.2.3}$$

$$\text{cov}(b_0, b_1) = 0,$$

$$\text{cov}(b_0, b_2) = 0, \qquad \text{cov}(b_1, b_2) = \left\{ \frac{-\cos\theta}{(1 - \cos^2\theta)S_1 S_2} \right\} \frac{\sigma^2}{n}, \tag{14.2.4}$$

where

$$S_1^2 = n^{-1}\Sigma\tilde{\xi}_{1u}^2, \qquad S_2^2 = n^{-1}\Sigma\tilde{\xi}_{2u}^2, \qquad \cos\theta = \frac{\Sigma\tilde{\xi}_{1u}\tilde{\xi}_{2u}}{\left\{(\Sigma\tilde{\xi}_{1u}^2)(\Sigma\tilde{\xi}_{2u}^2)\right\}^{1/2}},$$

$$(14.2.5)$$

and all summations are over $u = 1, 2, \ldots, n$. Given that we have decided to make a fixed number of runs, n, and that the experimental error variance $V(\varepsilon_u) = \sigma^2$ is fixed, how should we design an experiment, that is, how should we choose the levels (ξ_{1u}, ξ_{2u}), $u = 1, 2, \ldots, n$? Obviously the choice will depend on the criterion of excellence applied. Suppose we require that the β's be estimated with smallest possible variances. Clearly $V(b_0)$ will be unaffected by choice of design because σ^2/n is fixed by definition of our problem. Now S_1 and S_2 are measures of spread of the design points in the ξ_1 and ξ_2 directions and θ is the angle between the design vectors $\tilde{\xi}_1 = (\tilde{\xi}_{11}, \tilde{\xi}_{12}, \ldots, \tilde{\xi}_{1n})'$ and $\tilde{\xi}_2 = (\tilde{\xi}_{21}, \tilde{\xi}_{22}, \ldots, \tilde{\xi}_{2n})'$, so that $-1 \le \cos\theta \le 1$. Thus we can make $V(b_1)$ and $V(b_2)$ as small as possible by choosing (a) S_1^2 and S_2^2 as large as possible, or (b) $\cos^2\theta$ as small as possible. In practice, we shall need to choose the scale factors S_1 and S_2 together with the location measures $\bar{\xi}_1$ and $\bar{\xi}_2$ so that the design spans the region of interest in the predictor variable space. If this is done and if, in addition, the design is selected so that $\cos\theta = 0$, then for given S_1 and S_2, $V(b_1)$ and $V(b_2)$ will be as small as possible. The condition $\cos\theta = 0$ is the requirement that the n-dimensional design vectors $\tilde{\xi}_1$ and $\tilde{\xi}_2$ are at right angles, that is, *orthogonal* to each other. Since S_1 and S_2 are, by assumption, greater than zero, the condition will be met by making $\Sigma\tilde{\xi}_{1u}\tilde{\xi}_{2u} = 0$. Notice that $\cos\theta$ is a measure of the correlation between ξ_1 and ξ_2 so that orthogonality implies that the regressors ξ_1 and ξ_2 are uncorrelated. Furthermore, $-\cos\theta$ is the correlation between the two estimates b_1 and b_2; thus, choice of an orthogonal design ensures that these estimates are uncorrelated.

It is possible to show that the models employed in the analysis of randomized blocks, Latin squares, and factorials can all be written in the general linear regression form of Eq. (3.9.1) using indicator or dummy variables. These designs are said to be orthogonal because the dummy X's that carry, for example, the treatment and block contrasts define subspaces that are all orthogonal to one another. This ensures that, for example, the treatment effects from the randomized block design may be estimated independently of the block effects.

Further illustration of the property of orthogonality and, in particular, how it relates to the size of the confidence region for the parameters will be found in Chapter 3; see Figure 3.9 in particular.

Useful arrangements such as factorials may be employed in many different experimental contexts. Therefore it is convenient to write such designs in a standard form in which the predictor variables are located and scaled in a standard way. We often work in terms of the *coded* design variables

$$x_{iu} = \frac{(\xi_{iu} - \bar{\xi}_i)}{S_i}, \qquad (14.2.6)$$

where we can choose

$$S_i^2 = n^{-1}\Sigma(\xi_{iu} - \bar{\xi}_i)^2. \qquad (14.2.7)$$

Note that this means that

$$n^{-1}\Sigma x_{iu}^2 = 1, \qquad (14.2.8)$$

where, again, all summations are over $u = 1, 2, \ldots, n$. Although the discussion to this point has involved just two regressors, for simplicity, the arguments would apply generally to k regressors, $k = 1, 2, \ldots$. Consider the model

$$y_u = \beta_0 + \sum_{i=1}^{k} \beta_i x_{iu} + \varepsilon_u. \qquad (14.2.9)$$

Then the variances of the estimates b_i of β_i, $i = 1, 2, \ldots, k$ would again attain their smallest values if

$$\Sigma x_{iu} x_{ju} = 0, \qquad \text{for all } i \neq j. \qquad (14.2.10)$$

The first-order response surface designs we have discussed have all been orthogonal. In particular, the two-level factorial and fractional factorials discussed in Chapters 4 and 5 are of this type. To illustrate the coding we have adopted, we give below a 2^2 factorial design both in original and coded predictor variables formats.

Temperature, $\xi_1(°C)$	Concentration, $\xi_2(\%)$	x_1	x_2
160	20	-1	-1
180	20	1	-1
160	50	-1	1
180	50	1	1

The appropriate coding here requires $\bar{\xi}_1 = 170$, $\bar{\xi}_2 = 35$, $S_1 = 10$, $S_2 = 15$, so that

$$x_1 = \frac{(\xi_1 - 170)}{10}, \qquad x_2 = \frac{(\xi_2 - 35)}{15}.$$

First-order orthogonal designs which are not factorials are occasionally of value. In particular, they may be used for the elimination of time trends; see, for example, Box (1952), Box and Hay (1953), and Hill (1960).

14.3. THE VARIANCE FUNCTION

If it were true that bias could be ignored so that the postulated model function were capable of exactly representing reality, then at $\mathbf{x} = (x_1, x_2, \ldots, x_k)'$

$$E(y_{\mathbf{x}}) = \eta_{\mathbf{x}} \tag{14.3.1}$$

and the standardized mean squared error associated with estimating it by $\hat{y}_{\mathbf{x}}$ would be

$$V_{\mathbf{x}} = \left(\frac{n}{\sigma^2}\right) V(\hat{y}_{\mathbf{x}}), \tag{14.3.2}$$

where $V_{\mathbf{x}}$ will be called the *variance function*. Equivalently, we may consider the *information function* of the design defined by

$$I_{\mathbf{x}} = V_{\mathbf{x}}^{-1}. \tag{14.3.3}$$

For example, for the factorial design illustration in the foregoing section,

$$I_{\mathbf{x}} = \frac{1}{\left(1 + x_1^2 + x_2^2\right)}. \tag{14.3.4}$$

This is plotted in Figure 14.1a. Obviously, if we were prepared to make the unrealistic assumption of no bias, the information function tells us all we can know about the design's ability to estimate the response η. For example, inspection of Figure 14.1a shows that a large amount of information on $\eta_{\mathbf{x}}$ is generated at the design origin and the information falls off rapidly as we move outward from the center. We also notice that, for this particular design, setting $\rho = (x_1^2 + x_2^2)^{1/2}$,

$$I_{\mathbf{x}} = I_{\rho} = \frac{1}{\left(1 + \rho^2\right)} \tag{14.3.5}$$

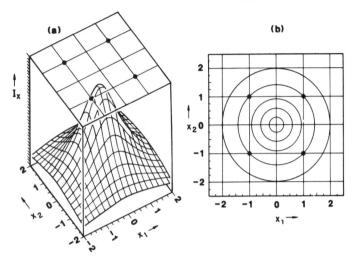

FIGURE 14.1. (a) Information surface for a 2^2 factorial used as a first-order design. (b) The corresponding information contours.

so that, in units of (x_1, x_2), the function is constant on circles centered at the origin. Thus the information surface is one of rotation. We say that the design is *rotatable*. As far as $V(\hat{y})$ is concerned, any rotation of the original 2^2 factorial design will generate exactly the same information function.

Evidently, rotatable designs are appropriate if we suppose that the experimenter would try to choose the scales and transformations of his predictor variables so that the desired distribution of information is, in his chosen metrics, spherically distributed about the design origin. This seems to be a reasonable goal. When an experimenter chooses a wide scale for one of his variables, it implies that his prior knowledge about the behavior of that variable suggests that a wide range is necessary to make the expected effect of that variable roughly equivalent to that of other variables. Similarly, suppose an experimenter were particularly interested in exploring a diagonal region in which, say, X_1/X_2 varied by only a small percentage, but $X_1 X_2$ varied by a large percentage. This would suggest the use *not* of the original variables X_1 and X_2, but of the transformed variables $\xi_1 = X_1/X_2$ and $\xi_2 = X_1 X_2$. After the design had been appropriately scaled by $x_1 = (\xi_1 - \bar{\xi}_1)/S_1$, $x_2 = (\xi_2 - \bar{\xi}_2)/S_2$, the desired distribution of information could be attained by making the design rotatable in x_1 and x_2. The process, which we must always face, of choosing appropriate scales and transformations for our variables is thus clearly reduced to that of taking account of what is known or suspected about the variables, *a priori*. The design is then superimposed on the chosen system of metrics (in terms of which a spherical distribution of information makes most sense).

So far we have seen that, for a first-order arrangement, orthogonality and rotatability are both desirable. In fact, *for the particular case of first-order designs*, they are equivalent. When we consider designs of higher order, however, this equivalence does not hold. For illustration, consider a 3^2 factorial used as a second-order design for $k = 2$ variables. With the postulated model in the form $y = X\beta + \varepsilon$, the X, $X'X$, and $(X'X)^{-1}$ matrices are as follows.

$$
X =
\begin{array}{c}
\begin{array}{cccccc}
0 & 1 & 2 & 11 & 22 & 12
\end{array} \\
\begin{bmatrix}
1 & -1 & -1 & 1 & 1 & 1 \\
1 & 0 & -1 & 0 & 1 & 0 \\
1 & 1 & -1 & 1 & 1 & -1 \\
1 & -1 & 0 & 1 & 0 & 0 \\
1 & 0 & 0 & 0 & 0 & 0 \\
1 & 1 & 0 & 1 & 0 & 0 \\
1 & -1 & 1 & 1 & 1 & -1 \\
1 & 0 & 1 & 0 & 1 & 0 \\
1 & 1 & 1 & 1 & 1 & 1
\end{bmatrix}
\end{array}
,
\qquad (14.3.6)
$$

$$
X'X =
\begin{array}{c}
\begin{array}{cccccc}
0 & 1 & 2 & 11 & 22 & 12
\end{array} \\
\begin{array}{c}
0 \\ 1 \\ 2 \\ 11 \\ 22 \\ 12
\end{array}
\begin{bmatrix}
9 & \cdot & \cdot & 6 & 6 & \cdot \\
\cdot & 6 & \cdot & \cdot & \cdot & \cdot \\
\cdot & \cdot & 6 & \cdot & \cdot & \cdot \\
6 & \cdot & \cdot & 6 & 4 & \cdot \\
6 & \cdot & \cdot & 4 & 6 & \cdot \\
\cdot & \cdot & \cdot & \cdot & \cdot & 4
\end{bmatrix}
\end{array}
,
\qquad (14.3.7)
$$

$$
(X'X)^{-1} = \tfrac{1}{9}
\begin{array}{c}
\begin{array}{cccccc}
0 & 1 & 2 & 11 & 22 & 12
\end{array} \\
\begin{bmatrix}
5 & \cdot & \cdot & -3 & -3 & \cdot \\
\cdot & 1.5 & \cdot & \cdot & \cdot & \cdot \\
\cdot & \cdot & 1.5 & \cdot & \cdot & \cdot \\
-3 & \cdot & \cdot & 4.5 & \cdot & \cdot \\
-3 & \cdot & \cdot & \cdot & 4.5 & \cdot \\
\cdot & \cdot & \cdot & \cdot & \cdot & 2.25
\end{bmatrix}
\end{array}
.
\qquad (14.3.8)
$$

The 3^2 factorial is an orthogonal design in the sense that no two of the estimates for first- and second-order effects are correlated, as is clear from the form of $(X'X)^{-1}$. (The estimate b_0 is, however, inevitably correlated with b_{11} and with b_{22}.) Suppose we are to use the 3^2 factorial to explore a response surface which, it is believed, can be represented by a second degree expression, but about which little else is known. Now imagine the second degree surface referred to its principal axes; see Figure 11.1. One thing in particular that will usually not be known will be the orientation of those principal axes, and frequently it will be true that one orientation is as likely as another. Instead of considering rotations of the surface relative to the design, we may equivalently consider rotations of the design relative to the surface.

Figure 14.2 shows how the variances of the second-order terms and the correlations between them change as the 3^2 design is rotated through an angle θ. (For simplicity, we set $\sigma^2 = 1$ here.) Thus, although this 3^2 second-order design is orthogonal (in the sense described) in its initial orientation, it loses this property on rotation, unlike the first-order design. In view of this, it would seem most sensible to choose, if possible, a second-order design for which the variances and covariances of the estimates stay constant as the design is rotated.

Consider the information function for the 3^2 factorial design. Writing $\mathbf{z} = (1, x_1, x_2, x_1^2, x_2^2, x_1 x_2)'$, this is

$$I_x = \left\{ n\mathbf{z}'(\mathbf{X}'\mathbf{X})^{-1}\mathbf{z} \right\}^{-1} \tag{14.3.9}$$

$$= \left\{ 5 - 4.5\left(x_1^2 + x_2^2\right) + 4.5\left(x_1^2 + x_2^2\right)^2 - 6.75 x_1^2 x_2^2 \right\}^{-1} \tag{14.3.10}$$

which is graphed in Figure 14.3a and has its contours plotted in Figure 14.3b. It will be seen that the design generates four pockets of high information which seemingly have little to do with the needs of an experimenter. Rather than attempt to generalize the property of orthogonality to second order, we shall instead generalize the property of rotatability. We shall see that it is possible to choose designs of second and higher orders for which the information contours are spherical. Equivalently, these rotatable designs have the property that the variances and covariances of the effects remain unaffected by rotation. One such rotatable design for two factors, consisting of eight points evenly distributed on a circle with four added center points is shown in Figure 14.4a together with its information function. The corresponding information contours appear in Figure 14.4b.

A k-dimensional design will be rotatable if the information function I_x depends only on $\rho = (x_1^2 + x_2^2 + \cdots + x_k^2)^{1/2}$, the distance from the origin. Such a design has an $\mathbf{X}'\mathbf{X}$ matrix of special form. If we scale the variables so that

$$\sum_{u=1}^{n} x_{iu}^2 = n, \qquad i = 1, 2, \ldots, k, \tag{14.3.11}$$

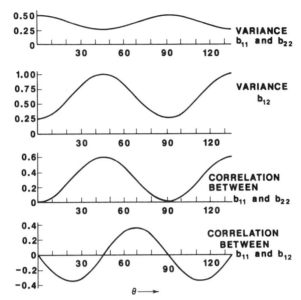

FIGURE 14.2. Variances of, and correlations between, second-order coefficients estimated from a 3^2 factorial design rotated through an angle θ.

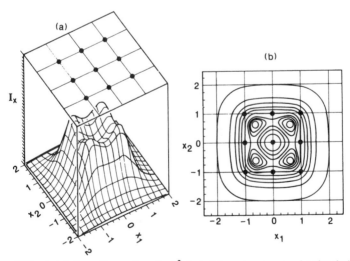

FIGURE 14.3. (*a*) Information surface for 3^2 factorial used as a second-order design. (*b*) The corresponding information contours.

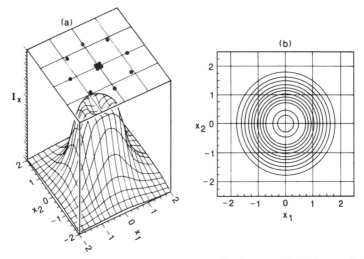

FIGURE 14.4. (*a*) Information function for a second-order rotatable design consisting of eight points on a circle plus four center points. (*b*) The corresponding information contours.

then, for example, for $k = 2$,

$$
n^{-1}\mathbf{X}'\mathbf{X} =
\begin{array}{c}
 \\
0 \\
1 \\
2 \\
11 \\
22 \\
12
\end{array}
\begin{array}{c}
\begin{array}{cccccc}
0 & 1 & 2 & 11 & 22 & 12
\end{array} \\
\left[
\begin{array}{cccccc}
1 & \cdot & \cdot & 1 & 1 & \cdot \\
\cdot & 1 & \cdot & \cdot & \cdot & \cdot \\
\cdot & \cdot & 1 & \cdot & \cdot & \cdot \\
1 & \cdot & \cdot & 3\lambda & \lambda & \cdot \\
1 & \cdot & \cdot & \lambda & 3\lambda & \cdot \\
\cdot & \cdot & \cdot & \cdot & \cdot & \lambda
\end{array}
\right]
\end{array}.
\qquad (14.3.12)
$$

One adjustable parameter λ is at our disposal. This determines the level of the information function as we move out from the center of the design along any radius vector in the x space. Figure 14.5 shows a plot of I_ρ versus ρ for the indicated values of λ for $k = 2$. The function plotted is

$$
I_\rho = \frac{\{4\lambda(2\lambda - 1)\}}{\{8\lambda^2 + 8\lambda(\lambda - 1)\rho^2 + (3\lambda - 1)\rho^4\}}. \qquad (14.3.13)
$$

[See Box and Hunter, 1957, p. 213, for details, especially Eq. (48).] The value attained by λ for a rotatable design is determined by the arrangement of design points and especially by the number n_0 of points placed at the center of the design, as we now explain.

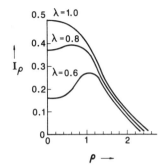

FIGURE 14.5. Plots of information function versus $\rho = (x_1^2 + x_2^2)^{1/2}$ for various values of λ for a central composite design in $k = 2$ variables.

Central Composite Designs

One convenient class of rotatable designs (available for all values of k) is that of the suitably dimensioned central composite designs already discussed in Section 9.2, for $k = 3$, and in Section 13.8. Figure 14.6 shows such a design for $k = 3$. If we scale the eight cube points so that their coordinates are $(\pm 1, \pm 1, \pm 1)$ then, to attain rotatability, the six axial points $(\pm \alpha, 0, 0)$, $(0, \pm \alpha, 0)$, $(0, 0, \pm \alpha)$ must be located at distances $\alpha = 8^{1/4} = 1.682$ from

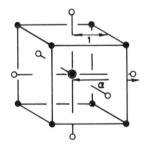

● Factorial (or "cube") points, with
   coordinates $(\pm 1, \pm 1, \pm 1)$

O Axial (or "star") points, distance α from origin

◉ Center points. If the design is blocked into cube
   and star portions, some center points would be
   in the cube block, some in the star block.
   See Chapter 15.

FIGURE 14.6. Central composite design for $k = 3$ variables.

the origin. In general, the cube may be replicated r_c times and the axial points replicated r_s times. In that event, the rotatable value of α would be given by the more general formula $\alpha = (8r_c/r_s)^{1/4}$. For k factors, when a 2^{k-p} fractional factorial, r_c times over, is combined with $2k$ axial points taken r_s times, the appropriate formula is

$$\alpha = \left(\frac{2^{k-p}r_c}{r_s} \right)^{1/4}, \tag{14.3.14}$$

the foregoing special case being that of $k = 3$, $p = 0$.

To ensure that all coefficients in the second-order model are estimable, points are needed at the center of the composite design whenever all the noncentral points fall on a sphere in k dimensions (i.e., on a circle when $k = 2$). When the design is rotatable, their number n_0 then determines λ according to the formula

$$\lambda = \frac{\left(r_c 2^{k-p} + 2kr_s + n_0 \right)}{\left\{ \left(r_c 2^{k-p} \right)^{1/2} + 2r_s^{1/2} \right\}^2}. \tag{14.3.15}$$

(Thus diagrams such as Figure 14.5 could also be constructed and marked using values of n_0 instead of values of λ.) As we have seen, the choice of λ will, in turn, affect the information profile. The question of choosing λ suitably will be discussed in Chapter 15.

Rotatability assures a spherically uniform distribution of information which, in the circumstances we have outlined, seems sensible. However, approximate sphericity is all that will be needed in practice. Whenever the criterion of rotatability is in conflict with some other important consideration (such as, for example, the need to split the design into orthogonal blocks) a moderate departure from exact rotatability may be made to achieve this.

General Conditions for Design Rotatability of First, Second, and Third Order

Suppose there are n design points $(x_{1u}, x_{2u}, \ldots, x_{ku})$, $u = 1, 2, \ldots, n$, in a k factor design. The quantity

$$n^{-1}\Sigma x_{1u}^p x_{2u}^q \cdots x_{ku}^r,$$

where this and subsequent summations are over $u = 1, 2, \ldots, n$, is a design moment of order $p + q + \cdots + r$. If multiplied by n, it is called a "sum of powers" (if only one of $p, q, \ldots, r$ is non-zero) or a "sum of products" (if two or more of $p, q, \ldots, r$ are non-zero) of the same order.

A third order rotatable design must be such that

$$\Sigma x_{iu}^2 = n\lambda_2, \tag{14.3.16}$$

$$\Sigma x_{iu}^4 = 3\Sigma x_{iu}^2 x_{ju}^2 = 3n\lambda_4, \tag{14.3.17}$$

$$\Sigma x_{iu}^6 = 5\Sigma x_{iu}^2 x_{ju}^4 = 15\Sigma x_{iu}^2 x_{ju}^2 x_{lu}^2 = 15n\lambda_6, \tag{14.3.18}$$

where $i \neq j \neq l$, and all other sums of powers and products up to and including order $2d$ (where here $d = 3$) are zero. (See Box and Hunter, 1957, p. 209.) These equations serve to define $\lambda_2, \lambda_4, \lambda_6$. For second order rotatability we need only (14.3.16), (14.3.17), and $2d = 4$; for first order rotatability, which is simply orthogonality with equal scaling on the axes, we need only (14.3.16) and $2d = 2$. These various assumptions determine appropriate forms for $X'X$. For a second order design to be nonsingular (i.e., to provide a nonsingular $X'X$ matrix), we need

$$\frac{\lambda_4}{\lambda_2^2} > \frac{k}{k+2}. \tag{14.3.19}$$

The (minimum) value $k/(k + 2)$ occurs when all the design points lie on a sphere centered at the origin; the addition of center points will then always satisfy (14.3.19). For a third order design to be nonsingular, both (14.3.19) and

$$\frac{\lambda_6 \lambda_2}{\lambda_4^2} > \frac{k+2}{k+4} \tag{14.3.20}$$

must hold. Design points must lie on at least *two* spheres of *non-zero* radii to achieve (14.3.20); see Draper (1960b). If this is so, then (14.3.19) is automatically satisfied. (See, for example, Exercise 7.28.)

14.4. THE ALPHABETIC OPTIMALITY APPROACH

A somewhat different approach, which has been very thoroughly explored mathematically, concerns various individual optimal characteristics of the matrix $X'X$ which occurs in the least-squares fitting of the response function. Such studies can be revealing and are of considerable interest, but they are concerned with optimality of a very narrow kind and their limitations must be clearly understood if the practitioner is not to be misled. To aid this understanding, a brief discussion of some of the main ideas of alphabetic optimality, and a critical appraisal of their application in a response surface context is given here. (Also see Box, 1982.)

Aspects of Alphabetic Optimal Design Theory

Consider a response η which is supposed to be an exactly known function $\eta = z'\beta$ linear in p coefficients β, where $z = \{f_1(\chi), \ldots, f_p(\chi)\}'$ is a vector of p functions of k experimental variables $\chi = (\chi_1, \chi_2, \ldots, \chi_k)'$. Suppose a design is to be run defining n sets of k experimental conditions given by the $n \times k$ design matrix whose uth row is $\chi'_u = (\chi_{1u}, \chi_{2u}, \ldots, \chi_{ku})$, and yielding n observations $\{y_u\}$, so that

$$\eta_u = z'_u\beta, \qquad (u = 1, 2, \ldots, n), \tag{14.4.1}$$

where $y_u - \eta_u = \varepsilon_u$ is distributed $N(0, \sigma^2)$ and the $n \times p$ matrix $X = \{z'_u\}$ consists of n rows with $z'_u = \{f_1(\chi_u), f_2(\chi_u), \ldots, f_p(\chi_u)\}$ in the uth row. The ε_u are assumed to be independently distributed.

The elements of $\{c_{ij}\} = (X'X)^{-1}$ are proportional to the variances and covariances of the least-squares estimates b. Within this specification, the problem of experimental design is that of choosing the design $\{\chi'_u\}$ so that the elements c_{ij} are to our liking. Because there are $\frac{1}{2}p(p + 1)$ of these, simplification is desirable.

One motivation for simplification is provided by considering the confidence region* for β

$$(\beta - b)'X'X(\beta - b) = \text{constant} \tag{14.4.2}$$

which defines an ellipsoid in p parameters. The eigenvalues $\lambda_1, \lambda_2, \ldots, \lambda_p$ of $(X'X)^{-1}$ are proportional to the squared lengths of the p principal axes of this ellipsoid. Suppose their maximum, arithmetic mean, and geometric mean are indicated by $\lambda_{\max}$, $\bar{\lambda}$, and $\tilde{\lambda}$. Then it is illuminating to consider the transformation of the $\frac{1}{2}p(p + 1)$ elements c_{ij} to an identical number of criteria measuring volume, nonsphericity, and orientation, the meanings of which are easier to comprehend. These are:

1. $D = |X'X| = \tilde{\lambda}^{-p}$ (so that $D^{-1/2} = \tilde{\lambda}^{p/2}$ is proportional to the volume of the confidence ellipsoid).

---

*There are also parallel fiducial and Bayesian rationalizations.

2. $H_1, H_2, \ldots, H_{p-1}$, a set of $p - 1$ homogeneous independent functions of order zero in the λ's, which measure the *nonsphericity* or state of ill-conditioning of the ellipsoid. In particular we might choose, for two of these, $H_1 = \bar{\lambda}/\tilde{\lambda}$ and $H_2 = \lambda_{\max}/\tilde{\lambda}$, both of which would take the value unity for a spherical region.

3. $\frac{1}{2}p(p-1)$ independent direction cosines which determine the *orientation* of the orthogonal axes of the ellipsoid.

It is traditionally assumed that the $\frac{1}{2}p(p-1)$ elements concerned with orientation of the ellipsoid are of no interest, and attention has been concentrated on particular design criteria that measure in some way or another the sizes of the eigenvalues. All such criteria thus measure some combination of size and sphericity of the confidence ellipsoid. Among these criteria are

$$D = |\mathbf{X}'\mathbf{X}| = \prod \lambda_i^{-1} = \tilde{\lambda}^{-p},$$

$$A = \sum \lambda_i = \mathrm{tr}(\mathbf{X}'\mathbf{X})^{-1} = \sigma^{-2} \sum \mathrm{var}(b_i) = p\bar{\lambda}H_1,$$

$$E = \max\{\lambda_i\} = \tilde{\lambda}H_2. \tag{14.4.3}$$

The D criterion is a function of $\tilde{\lambda}$ only and is thus a measure of region size alone. The A and E criteria both contain the same size measure $\tilde{\lambda}$ multiplied by a measure of nonsphericity.

The desirability of a design, as measured by the D, A, and E criteria, increases as $\tilde{\lambda}$, $\tilde{\lambda}H_1$, and $\tilde{\lambda}H_2$ respectively, are decreased. However, in practical situations, because $\tilde{\lambda}$ is common to all three, each of these criteria will take smaller and hence more desirable values as the ranges for the experimental variables χ_u are taken larger and larger. To cope with this problem it is usually assumed that the experimental variables χ_u may vary only within some exactly known region in the space of χ, but not outside it. We call this permissible region RO, a notation that (purposely) combines the R and O notations of Chapter 13.

Another characteristic of the problem which makes its study mathematically difficult is the necessary discreteness of the number of runs that can be made at any given location. In a technically brilliant paper, Kiefer and Wolfowitz (1960) dealt with this obstacle by introducing a continuous design measure which determines the *proportion* of runs that should ideally be made at each of a number of points in the χ space. Realizable designs that closely approximated the optimal *measure design(s)* distribution could then be used in practice.

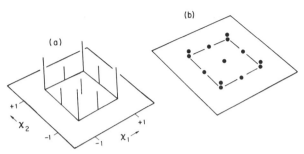

FIGURE 14.7. (*a*) Fedorov's *D/G* optimal second-order measure design for the square inner region. The vertical scale indicates the proportion of runs to be made at the various points. (*b*) A realizable design with 13 runs roughly approximating Fedorov's *D/G* optimal design.

A further important result of Kiefer and Wolfowitz linked the problem of estimating β with that of estimating the response η via the property of "*G* optimality." *G* optimal designs were defined as those that minimized the maximum value of $V(\hat{y}_x)$ *within RO*. The authors were then able to show, for their measure designs, the equivalence of *G* and *D* optimality. Furthermore, they showed that, for such a design, within the region *RO*, the maximum value of $nV(\hat{y}_x)/\sigma^2$ was p, and this value was actually attained at each of the design points.

For illustration we consider a second-order measure design in two variables; that is, a design appropriate for the fitting of the second degree polynomial. Such a design which is both *D* and *G* optimal for a square region *RO* with vertices $(\pm 1, \pm 1)$ was given by Fedorov (1972) (see also Herzberg, 1979). The design places 14.6% of the measure at each of the four vertices, 8.0% at each of the midpoints of the edges, and 9.6% at the origin. The distribution measure for this design is shown in Figure 14.7*a*. A design realizable in practice which would roughly approximate the measure design is shown in Figure 14.7*b*. This would place two points at each corner of the square region and one point each at the centers of the edges and at the design center.

As we have already seen, a function that represents the overall informational characteristic of the design, and not merely an isolated part of it, is the information function. A plot of this information function for Fedorov's second-order *D/G* optimal design over the square permissible *RO* region $(\pm 1, \pm 1)$ referred to earlier, is shown in Figure 14.8*a*. The corresponding contour plot and the measure values are shown in Figure 14.8*b*. To set this design into the kind of scientific context in which it would have to be used,

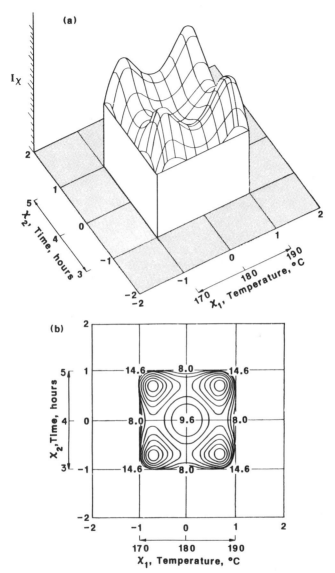

FIGURE 14.8. (*a*) Information function for a second-order D/G optimal design within the *RO* region $170 \leq \chi_1 \leq 190$, $3 \leq \chi_2 \leq 5$. (*b*) The corresponding contour plots and percentage distribution of the design measure.

we have supposed that the variables to be studied are χ_1 = temperature (°C) and χ_2 = time (h), and that they have been coded via $x_1 = (\chi_1 - 180)/10$, $x_2 = \chi_2 - 4$. This corresponds to a choice of the RO region in which experimentation is allowed within the limits $\chi_1 = 170$ to $190°C$ and $\chi_2 = 3$ to 5 h, but not outside these limits. The extensions of the information function and contour plot over a wider region are shown for comparison in Figure 14.9a and 14.9b. We return to these on pages 497–500.

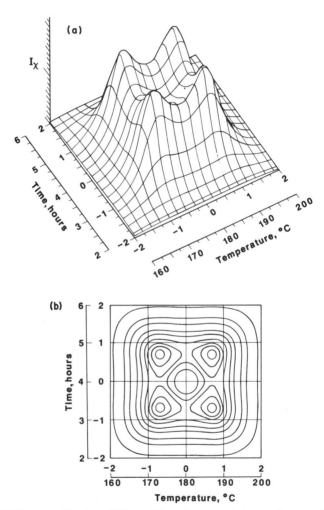

FIGURE 14.9. The plots of Figure 14.8 are here extended over a larger region.

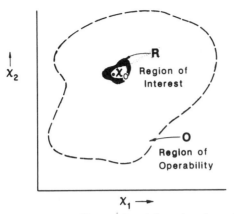

FIGURE 14.10.   The current region of interest R and the region of operability O in the space of two continuous experimental variables χ_1 and χ_2.

14.5. ALPHABETIC OPTIMALITY IN THE RESPONSE SURFACE CONTEXT

In the response surface context, a number of questions arise concerning the appropriateness of the specification set out above for alphabetic optimality. These concern:

1. Formulation in terms of the RO region.
2. Distribution of information over a wider region.
3. Sensitivity of criteria to size and shape of the RO region.
4. Ignoring of bias.

As an example, suppose it is desired to study some chemical system, with the object of obtaining a higher value for a response η such as yield which is initially believed to be some function $\eta = g(\chi)$ of k continuous input variables $\chi = (\chi_1, \chi_2, \ldots, \chi_k)'$ such as reaction time, temperature, or concentration. As explained in Chapter 13 and again illustrated in Figure 14.10, it is usually known initially that the system can be operated at some point χ_0 in the space of χ and is expected to be capable of operating over some much more extensive region O called the *operability region*, which*, however, is usually unknown or poorly known. Response surface methods of the

*A subsidiary objective of the investigation may be to find out more about the operability region O.

kind discussed in this book are employed when the nature of the true response function $\eta = g(\chi)$ is also unknown* or is inaccessible.

Suppose then that, over some immediate *region of interest* R in the neighborhood of χ_0, it is guessed that a graduating function, such as a dth degree polynomial in x,

$$\eta_\chi \doteq z'\beta \qquad (14.5.1)$$

might provide a *locally adequate approximation* to the true function $\eta_\chi = g(\chi)$ where, as before, z is a p-dimensional vector of suitably transformed input variables $z' = \{ f_1(\chi), f_2(\chi), \ldots, f_p(\chi) \}$, and β is a vector of coefficients occurring linearly which may be adjusted to approximate the unknown true response function $\eta_\chi = g(\chi)$. Then progress may be achieved by using a sequence of such approximations. For example, when a first degree polynomial approximation could be employed it might, via the method of steepest ascent, be used to find a new region of interest R_1 where, say, the yield was higher. Also a maximum in many variables is often represented by some rather complicated ridge system† and a second degree polynomial approximation, when suitably analyzed, might be used to elucidate, describe, and exploit such a system.

Design Information Function

In general, to throw light on the desirable characteristics (1) and (2) of Section 14.1, we again consider the variance function

$$V_\chi = \frac{n \, \text{var}(\hat{y}_\chi)}{\sigma^2} = n z'(X'X)^{-1} z \qquad (14.5.2)$$

or equivalently the *information function*

$$I_\chi = V_\chi^{-1}. \qquad (14.5.3)$$

---

*Occasionally the true functional form $\eta = g(\chi)$ may be known, or at least conjectured, from knowledge of physical mechanisms. Typically, $g(\chi)$ will then appear as a solution of a set of differential equations that are nonlinear in a number of parameters that may represent physical constants. Problems of nonlinear experimental design then arise which are of considerable interest, although they have received comparatively little attention; see, for example, Box and Lucas (1959), Cochran (1973), and Fisher (1935; see 8th ed., 1966, pp. 218–223). We do not discuss these problems further here, but see Draper and Smith (1981, Chapter 10).

†Empirical evidence suggests this. Also, integration of sets of differential equations which describe the kinetics of chemical systems almost invariably leads to ridge systems; see Box (1954), Box and Youle (1955), Franklin, Pinchbeck, and Popper (1956), and Pinchbeck (1957).

If we were to make the unrealistic assumption (made in alphabetic optimality) that the graduating function $\eta = z'\beta$ is capable of *exactly* representing the true function $g(\chi)$, then the information function would tell us all we could know about the design's ability to estimate η. For example, we have already seen information functions and associated information contours for a 2^2 factorial used as a first-order design (Figure 14.1) and for a 3^2 factorial used as a second-order design (Figure 14.3).

Formulation in Terms of the RO Region

As has been pointed out, in response surface studies it is typically true that, at any given stage of an investigation, the current region of interest R is much smaller than the region of operability O which is, in any case, usually unknown. In particular, it is obvious that this must be so for any investigation in which we allow the possibility that results of one design may allow progress to a *different* unexplored region. Consequently we believe that formulation in terms of an RO region which assumes that the region of interest R and the region of operability O are identical is artificial and limiting. In particular, to obtain a good approximation *within R* it might be sensible to put some experimental points *outside R* and, as long as they are within O, there is no practical reason not to do so. Also, because typically R is only vaguely known, we will want to consider the information function over a wider region, as is done, for example, in Figure 14.9 for Fedorov's second-order D optimal design. We shall soon discuss how the information function for this design compares, over this wider region, with that for the 3^2 factorial, already shown in Figure 14.3.

Distribution of Information over a Wider Region

In the response surface context, the coefficients β of a graduating function $\eta_x \doteq z'\beta$, acting as they do merely as adjustments to a kind of mathematical French curve, are not usually of individual interest except insofar as they affect η, in which case only the G optimality criterion among those considered is of direct interest. For response surface studies however, it is far from clear how desirable is the property of G optimality itself. We illustrate by comparing the information surface for the 3^2 factorial design (Figure 14.3) with that of Fedorov's optimal design (Figure 14.9).

The profiles of Figure 14.11, which were made by taking sections of the surfaces of Figure 14.3 and Figure 14.9, show that neither the D/G optimal design, nor the 3^2 design, is universally superior. In some subregions one design is slightly better, and in others the other design is slightly better. Both information functions, and particularly that of the D/G optimal

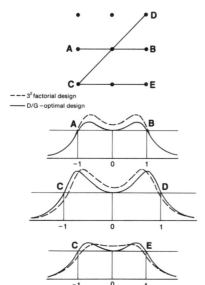

FIGURE 14.11. Profiles of I_x for the second-order D/G optimal design and for the 3^2 factorial design.

design, show a tendency to sag in the middle. This happens for the D/G optimal design because the G optimality characteristic guarantees that (maximized) minima for I_x, each equal to $1/p$, where p is the number of parameters in the model, occur at every design point, *which must include the center point*. However, this sagging information pattern of the second-order design is not of course a characteristic of the first-order design of Figure 14.1 which is also D/G optimal but contains no center points. Because the information function for a design is a basic characteristic measuring how well we can estimate the response at any given point in the space of the variables, it seems unsatisfactory that the shape of that function should depend so very much on the order of the design under consideration. Indeed, it follows from the Kiefer–Wolfowitz theorem that a second-order design for the $(\pm 1, \pm 1)$ region whose information function did not sag in the middle would necessarily not be D optimal. As we saw earlier, D optimality is only one of many single-valued criteria that might be used in attempts to describe some important characteristic of the $X'X$ matrix. Others for example would be A optimality and E optimality, and these would yield different information profiles. However, because the information function *itself* is the most direct measure of desirability so far as the single issue of variance properties is concerned, our best course would seem to be to choose the design directly by picking a suitable information function, and not indirectly by finding some extremum for A, E, D, or some other arbitrary criterion.

Sensitivity of Criteria to Size and Shape

In the process of scientific investigation, the investigator and the statistician must do a great deal of guesswork. In matching the region of interest R and the degree of complexity of the approximating function, they must try to take into account, for example, that the more flexible second degree approximating polynomial can be expected to be adequate over a larger region R than the first degree approximation. Obviously different experimenters would have different ideas of appropriate locations and ranges for experimental variables. In particular, ranges could easily differ from one experimenter to another by a factor of 2 or more.* In view of this, extreme sensitivity of design criteria to scaling would be disturbing.[†] Such sensitivity can be studied as follows. Suppose each dimension of a dth order experimental design is increased[‡] by a factor c. Then the D criterion is increased by a factor of c^q where

$$q = \frac{2k(k+d)!}{(k+1)!(d-1)!}.$$ (14.5.4)

Equivalently a confidence region of the same volume as that for a D optimal design can be achieved for a design of given D value by increasing the scale for each variable by a factor of $c = (D_{opt}/D)^{1/q}$, thus increasing the volume occupied by the design in the χ space, by a factor $c^k = (D_{opt}/D)^{k/q}$. For example, the D value for the 3^2 factorial design of Figure 14.3 is 0.98×10^{-2} as compared with a D value of 1.14×10^{-2} for the D optimal design. For $k = 2$, $d = 2$, we find $q = 16$, and $c = (1.14/0.98)^{1/16}$

*Over a sequence of designs, initial bad choices of scale and location would tend to be corrected, of course.

[†] In particular, designs can only be fairly compared if they are first scaled to be of the "same size." But how is size to be measured? It was suggested by Box and Wilson (1951) that designs should be judged as being of the same size when their marginal second moments $\Sigma(x_{iu} - \bar{x}_i)^2/n$ were identical. This convention is not entirely satisfactory, but will of course give very different results from those which assume design points to be all included in the same region RO. It is important to be aware that the apparent superiority of one design over another will often disappear if the method of scaling the design is changed. In particular this applies to comparisons such as those made by Nalimov et al. (1970) and Lucas (1976).

[‡] A measure of efficiency of a design criterion (see, for example, Atwood, 1969, and Box, 1968) is motivated by considering the ratio of the number of runs necessary to achieve the optimal design to the number of runs required for the suboptimal design to obtain the same value of the criterion (supposing fractional numbers of runs to be allowed). In particular for the D criterion, this measure of D efficiency is $(D/D_{opt})^{1/k}$. Equivalently here, to illustrate scale sensitivity, we concentrate attention on the factor c by which each scale would need to be inflated to achieve the same value of the criterion.

= 1.009. Thus the same value of D (the same volume of a confidence region for the β's) as is obtained for the D optimal design would be obtained from a 3^2 design if *each side of the square region were increased by less than 1%!* Equivalently, the area of the region would be increased by less than 2%. Using the scaling that was used in Figure 14.8 for illustration, we would have to change the temperature by 20.18°C instead of 20°C, and the time by 2 h and 1 min instead of 2 h, for the 3^2 factorial to give the same D value as the D/G optimal design. Obviously, no experimenter can guess to anything approaching this accuracy what are suitable ranges over which to vary these factors. Thus if we limit consideration to alphabetic optimality, the apparent difference between the 3^2 factorial and the D-optimal design is seen, after all, to be entirely trivial, and any practical choice between these designs should be based on other criteria.

Choice of region and choice of information function are, in fact, closely interlinked. For example, *any* set of $N = k + 1$ points in k space which have no coplanarities is obviously a D optimal first-order design for *some**** ellipsoidal region. Furthermore the information function for a design of order d is a smooth function whose harmonic average over the n experimental points (which can presumably be regarded as representative of the region of interest) is always $1/p$ wherever we place the points. Thus the problem of design is not so much a question of choosing the design to *increase* total information as of *spreading* the total information around in the manner desired.

Scaleless D Optimality

One way to eliminate this massive dependence on scale is to employ the design coding

$$x_{iu} = \frac{(x_{iu} - \bar{x}_i)}{S_i}, \quad \text{where} \quad S_i^2 = n^{-1} \sum_{u=1}^{n} (x_{iu} - \bar{x}_i)^2, \quad i = 1, 2, \ldots, k.$$

$$(14.5.5)$$

As an example, consider the choice of a first-order design in $k = 2$ variables. In the selected scaling, the design's D criterion takes the form

$$D = n^3(1 - r^2),$$ $$(14.5.6)$$

where $nr = \Sigma x_{1u} x_{2u}$. This is maximized when $r = 0$ and the design is

---

* Namely, for that region enclosed within the information contour $I_{\mathbf{x}} = 1/p$ which must pass through all the $k + 1$ experimental points.

orthogonal. In general, for a first-order design in k variables it can be shown that, after scale factors are removed in accordance with Eq. (14.3.11), D optimality requires simply that the design should be orthogonal. This line of argument thus leads to the conclusion that, if the criterion of *scaleless D* optimality is applied to the choice of a first-order design, an orthogonal arrangement is best. In practice, the orthogonal scaleless pattern would be related by the experimenter to the experimental variables using the scales and locations that he deemed appropriate. (This is, in fact, traditionally done with factorial and other standard designs.)

Ignoring of Bias

The whole of the above theory rests on the assumption that the graduating polynomial *is* the response function. As we have argued in detail in Chapter 13, such a polynomial must in fact always be regarded as a mathematical French curve which graduates locally the true but unknown response function. Thus, as we have said, there are two sources of error, variance error and bias error, and both should be taken into account in choosing a design. In particular, we have seen that designs which take account of bias tend *not* to place points at the extremes of the region of interest where the credibility of the approximating function is most strained.

Acknowledgment

The diagrams in this chapter were adapted from originals computer-generated by Conrad Fung.

CHAPTER 15

Practical Choice of a Response Surface Design

15.1. INTRODUCTION

We saw in Chapter 14 that the choice of an experimental design requires multifaceted consideration. Like a sailboat, an experimental design has many characteristics whose relative importance differs in different circumstances. In choosing a sailboat, the relative importance of characteristics such as size, speed, sea-worthiness, and comfort will depend greatly on whether we plan to sail on the local pond, undertake a trans-Atlantic voyage, or compete in the America's Cup contest. Some of the characteristics of experimental designs and some of the circumstances that influence their importance are as follows.

Characteristics of Design	Some Relevant Experimental Circumstances
Allows check of fit	Size of the experimental region
	Smoothness of the response function
	Complexity of the model
Allows estimation of transformations	Lack of fit that could be corrected by transformation
Permits sequential assembly	Ability to perform runs sequentially
	Ability to move in the space of the variables
Can be run in blocks	Homogeneity of experimental materials
	State of control of the process
Provides an independent estimate of error	Number of runs permissible
	Possibility of large experimental error
	Existence of reliable prior estimate of error

Characteristics of Design	**Some Relevant Experimental Circumstances**
Robustness of distribution of design points	Possibility of occasional aberrant runs and or observations
	Nature of the error function
Number of runs required	Cost of making runs
Simplicity of data pattern	Need to visualize data to motivate model

In choosing a design, we have to take account of the circumstances listed on the right to decide on the relative importance of the design characteristics listed on the left. When we are ignorant about the experimental circumstances, then we need to insure against our fears.

We saw in Section 1.7 that response surface methods are used for various kinds of applications. We mentioned the approximate mapping of a surface, the achievement of a desired specification, and the attainment of best (or at least, good) operating conditions.

Sequential Assembly

In the applications discussed, we have employed first- and second-order designs, and sometimes made suitable transformations of the response and/or the input variables. The designs have often been used sequentially. For example, in investigating a new process, one or more phases of steepest ascent using first-order designs might be necessary before a sufficiently promising region was found within which it would be worth fitting a second-order surface. Because we do not know how such an investigation will proceed until we actually begin to carry it out, a good deal of flexibility in our choice of designs is desirable.

For illustration, Figure 15.1 shows the sequential assembly of a design arranged in three orthogonal blocks, each of six runs, labeled I, II, and III. We first suppose that only the six runs of block I have been performed. Note that these six runs comprise an orthogonal first-order design which also provides a check for overall curvature, obtained by contrasting the average response of the center points with the average response on the cube. Moreover, a single contrast in the center responses is available as a gross check on previous information about experimental error. Suppose now we found that first-order effects were very large compared with their standard errors, and that no serious curvature was exhibited. Then it would make sense immediately to explore the indicated direction of steepest ascent. Alternatively, if after analyzing the results from block I there were doubts about the adequacy of a first degree polynomial model, block II could be performed. It uses the complementary simplex, and while the two parts

I   I + II   I + II + III

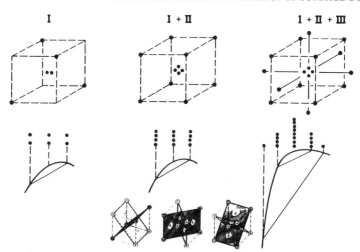

FIGURE 15.1.   An example of sequential assembly, showing checks of linearity and quadraticity. Block I: Regular simplex (see p. 507) plus two center points. Block II: Complementary simplex plus two center points. Block III: Six axial points (star).

together (I + II) form a design of only first order, this combination has much greater ability to detect lack of fit due to the addition of contrasts which permit estimation of two-factor interactions. Again, at this point, if first-order terms dominate, there is an opportunity to move to a more promising region. On the other hand, if it is found that the first-order effects are small and the two-factor interactions dominate and/or there is strong evidence of curvature, then block III can be added to produce a composite design (I + II + III) which allows a full second degree approximating equation to be fitted. The complete design also provides orthogonal checking contrasts for second-order adequacy in each of the three directions. These contrasts can also be regarded as checking the need for transformation in each of the x's. Finally, if it were decided that more information about experimental error was desirable, the replication of the star in a further block IV could furnish this, and also provide some increase in the robustness of the design to wild observations

Robustness

Approaches to the robust design of experiments have been reviewed by Herzberg (1982); see also Herzberg and Andrews (1976). In particular, Box and Draper (1975) suggested that the effects of wild observations could be minimized by making $r = \Sigma r_{uu}^2$ small, where $\mathbf{R} = \{r_{tu}\} = \mathbf{X}(\mathbf{X}'\mathbf{X})^{-1}\mathbf{X}'$.

This is equivalent to minimizing $\Sigma r_{uu}^2 - p^2/N = N\sigma^{-4}\,\text{var}\{V(\hat{y})\}$ which takes the value zero when $V(\hat{y}_u) = p/N$ $(u = 1, 2, \ldots, N)$, p being the number of parameters in the model and N the total number of observations. Thus, G optimal designs are optimally robust in this sense.

Number of Runs in an Experimental Design

A good experimental design is one which focuses experimental effort on what is judged important in the particular current experimental context. Suppose that, in addition to estimating the p parameters of the assumed model form, it is concluded that $f \geq 0$ contrasts are needed to check adequacy of fit, $b \geq 0$ further contrasts are needed for blocking, and that an estimate of experimental error is needed having $e \geq 0$ df. To obtain independent estimates of all items of interest we then require a design containing at least $p + f + b + e$ runs. However, the relative importance of checking fit, blocking, and obtaining an independent estimate of error will differ in different circumstances, and the minimum value of N will thus correspondingly differ. But this minimum design will in any case only be adequate if σ^2 is below some critical value. When σ^2 is larger, designs larger than the minimum design will be needed to obtain estimates of sufficient precision. In this circumstance, rather than merely replicating the minimum design, opportunity may be taken to employ a higher-order design allowing the fitting of a more elaborate approximating function which can then cover a wider experimental region. Notice that even when σ^2 is small, designs for which N is larger than p are not necessarily wasteful. This depends on whether the additional degrees of freedom are genuinely used to achieve the experimenter's objectives.

Simple Data Patterns

It has sometimes been argued that we may as well choose points randomly to cover the "design region" or employ some algorithm that distributes them evenly, even though this does not result in a simple data pattern such as is achieved by factorials and composite response surface designs. In favor of this idea, it has been urged that the fitting of a function by least squares to a haphazard set of points is a routine matter, given modern computational devices. This is true, but overlooks an important attribute of designs which form simple patterns. The statistician's task as a member of a scientific team is a dual one, involving inductive criticism and deductive estimation. The latter involves deducing in the light of the data the consequences of given assumptions (estimating the fitted function), and this can certainly be done with haphazard designs. However, the former involves the questions (a) of what function should be fitted in the first place, and (b) of how to examine residuals from the fitted function in an attempt to understand deviations from the initial model, in particular in relation to the predictor variables, in order to be led to appropriate model modification.

Designs such as factorials and composite response surface designs employ patterns of experimental points that allow many such comparisons to be made, both for the original observations and for the residuals from any fitted function. For example, consider a 3^2 factorial design used to elucidate the effects of temperature and concentration on some response such as yield. Intelligent inductive criticism is greatly enhanced by the possibility of being able to plot the original data and residuals against temperature for each individual level of concentration, and against concentration for each individual level of temperature.

We now present, in relation to the criteria discussed above, some of the characteristics of various useful response surface designs.

15.2. FIRST-ORDER DESIGNS

First-order designs of great value in response surface methodology are the two-level factorial and fractional factorials discussed in Chapters 4 and 5. In particular, we have discussed in Chapter 6 their use in estimating the direction of steepest ascent. It will be recalled that a convenient curvature check is obtained by adding center points to the factorial points; see Section 6.3. These first-order designs can also play the role of initial building blocks in the construction of second-order designs.

In some situations, Plackett and Burman designs are useful first-order designs; see Section 5.4. They can be used as initial building blocks for some small second-order designs, as described in Section 15.5.

For extreme economy of experimentation, the Koshal type first-order design can be used; see Section 15.5. Another economical alternative would be to use a regular simplex design, described below in material adapted from Box (1952).

Regular Simplex Designs

We first note that if $\mathbf{O}$ is *any* $N \times N$ *orthogonal* matrix, then $\mathbf{O'O} = \mathbf{OO'} = \mathbf{I}_N$. Suppose now that $\mathbf{Q}$ is an orthogonal $N \times N$ matrix whose first column has all elements equal to $N^{-1/2}$. Consider the $N \times N$ matrix $N^{1/2}\mathbf{Q}$. This has a first column of ones and the sum of squares of any row, or of any column, is N. Pairs of rows or columns are orthogonal. By deleting any $N - k - 1$ columns *except* the first, we can obtain the $N \times (k + 1)$ **X**-

matrix of a first-order orthogonal design for k variables in N runs, where $k \le N - 1$. The columns of $\mathbf{X}$, other than the first, define the design matrix $\mathbf{D}$ and the coded variable scales are such that

$$\sum_{u=1}^{N} x_{iu}^2 = N, \qquad i = 1, 2, \ldots, k. \qquad (15.2.1)$$

When $k = N - 1$, we obtain the maximum number of factors that can be accommodated and

$$\mathbf{X} = N^{1/2}\mathbf{Q}, \qquad (15.2.2)$$

no deletion being necessary. It can be shown (see Box, 1952) that the $k + 1$ design points must be distributed over a k-dimensional sphere, centered at the origin, radius $(N - 1)^{1/2}$, in the x-space, in such a way that the angle subtended at the origin by any pair of points is the same and has cosine $-1/(N - 1)$. Thus the design consists of the vertices of a *regular k-dimensional simplex*.

(Note: A *simplex* in k dimensions is the figure formed by joining any $k + 1 = N$ points that do not lie in a $(k - 1)$-dimensional space. For example, any three points not all on the same straight line are the vertices of a simplex (a triangle) in two dimensions. Any four points that do not all lie in the same plane form a simplex (a tetrahedron) in three dimensions. A simplex in k dimensions has $N = k + 1$ points and $k(k + 1)/2$ edges. A regular simplex is one that has all its edges equal. When $k = 2$, it is an equilateral triangle; when $k = 3$, an equilateral tetrahedron, and so on.)

A regular simplex design can be oriented in any manner in the x-space without affecting its orthogonal properties. Thus such a design is first order rotatable (see Section 14.3). For certain values of k, and in certain orientations, the regular simplex becomes a familiar design. For $k = 3$, it can take the form of a 2^{3-1} design defined by either $x_1 x_2 x_3 = 1$ or $x_1 x_2 x_3 = -1$. For $k = 7$, it can take the form of a 2^{7-4} design; see Table 5.4.

One particular type of orientation allows us to write down a regular simplex design for any value of k. We illustrate for $k = 3$. Suppose one face is taken parallel to the plane defined by the x_1 and x_2 axes and one of the sides of that face is taken parallel to the x_1 axis. The design points can then be allocated coordinates as follows, the scaling being chosen so that $\sum_{u=1}^{4} x_{iu}^2 = N = 4$, for $i = 1, 2, 3$:

u	x_{1u}	x_{2u}	x_{3u}
1	$-2^{1/2}$	$-(2/3)^{1/2}$	$-1/3^{1/2}$
2	$2^{1/2}$	$-(2/3)^{1/2}$	$-1/3^{1/2}$
3	0	$2(2/3)^{1/2}$	$-1/3^{1/2}$
4	0	0	$3^{1/2}$

Removing a factor from each column, we obtain:

u	x_{1u}	x_{2u}	x_{3u}
1	-1	-1	-1
2	1	-1	-1
3	0	2	-1
4	0	0	3
Scale factor	$2^{1/2}$	$(2/3)^{1/2}$	$1/3^{1/2}$

Extensions to higher values of k are now straightforward. For example, if a fourth factor had to be introduced, we would add an x_{4u} column containing $-1, -1, -1, -1, 4$, a row $(5, 0, 0, 0)$ under the $(u, x_{1u}, x_{2u}, x_{3u})$ columns and either select a scaling factor of $1/5^{1/2}$ for x_{4u} so that $\sum_{u=1}^{5} x_{4u}^2 = 4$ or else rescale all the columns so that $\sum_{u=1}^{5} x_{iu}^2 = N = 5$ ($i = 1, 2, 3, 4$), or any other selected value. Because the actual experimental levels will have to be coded to design levels in any case, there is no necessity to take $\sum_{u=1}^{N} x_{iu}^2 = N$, $i = 1, 2, \ldots, k$. If desired, the scaling *could* be different for each x-variable, but then the design would no longer be rotatable. For the general extension of this method, see Exercise 15.9.

15.3. SECOND-ORDER COMPOSITE DESIGNS

We next consider the features of some of the more useful second-order designs. Consider a composite design consisting of:

1. $n_c = 2^{k-p} r_c$ points with coordinates of the type $(\pm 1, \pm 1, \ldots, \pm 1)$ forming the "cube" part of the design, of resolution at least V.
2. $n_s = 2k r_s$ axial points with coordinates $(\pm \alpha, 0, \ldots, 0)$, $(0, \pm \alpha, \ldots, 0), \ldots, (0, 0, \ldots, \pm \alpha)$ forming the "star" part of the design.
3. n_0 center points at $(0, 0, \ldots, 0)$.

Thus $r_c = n_c / 2^{k-p}$ measures the degree of replication of the chosen cube and $r_s = n_s / (2k)$ is the number of times the star is replicated.

Rotatability

It is shown in Section 14.3 that the design will be rotatable if

$$\alpha = \left(\frac{n_c}{r_s} \right)^{1/4} = \left(\frac{2k n_c}{n_s} \right)^{1/4} \tag{15.3.1}$$

Orthogonal Blocking

A second-order design is said to be orthogonally blocked if it is divided into blocks in such a manner that block effects do not affect the usual estimates of the parameters of the second-order model. Box and Hunter (1957, pp. 226–234) show that two conditions must be fulfilled to achieve this. Suppose there are n_w runs in the wth block, then:

1. Each block must itself be a first-order orthogonal design, that is,

$$\sum_{u}^{n_w} x_{iu} x_{ju} = 0, \qquad i \neq j = 0, 1, 2, \ldots, k \quad \text{all } w. \qquad (15.3.2)$$

   (Note: $x_{0u} = 1$ for all u.)

2. The fraction of the total sum of squares of each variable contributed by every block must be equal to the fraction of the total observations allotted to the block, that is,

$$\frac{\displaystyle\sum_{u}^{n_w} x_{iu}^2}{\displaystyle\sum_{u=1}^{N} x_{iu}^2} = \frac{n_w}{N}, \qquad i = 1, 2, \ldots, k, \quad \text{all } w. \qquad (15.3.3)$$

Consider first the division of the design into just two blocks, one consisting of the n_c runs from the cube portion plus n_{c0} center points, and the other of the n_s runs from the star portion plus n_{s0} center points. Such blocks obviously satisfy the first condition for orthogonality. The second is satisfied if

$$\frac{n_c}{2r_s\alpha^2} = \frac{n_c + n_{c0}}{n_s + n_{s0}} \qquad (15.3.4)$$

that is, if

$$\alpha^2 = \frac{kn_c}{n_s} \left\{ \frac{n_s + n_{s0}}{n_c + n_{c0}} \right\} = k \left\{ \frac{1 + n_{s0}/n_s}{1 + n_{c0}/n_c} \right\} \qquad (15.3.5)$$

or

$$\alpha^2 = k \left\{ \frac{1 + p_s}{1 + p_c} \right\}, \qquad (15.3.6)$$

where $p_s = n_{s0}/n_s$ and $p_c = n_{c0}/n_c$ are the proportions of center points relative to the noncentral points in the star and cube, respectively. Now p_s and p_c are typically small fractions of about the same magnitude, so that the factor $(1 + p_s)/(1 + p_c)$ is close to unity and hence, for an orthogonally blocked design, α would be close to $k^{1/2}$. In other words, the star points would be roughly the same distance from the design center as the cube points.

Simultaneous Rotatability and Orthogonal Blocking

For a composite design to be both rotatable and orthogonally blocked, Eqs. (15.3.1) and (15.3.6) must be satisfied simultaneously, implying that

$$\frac{r_s}{r_c} = \frac{2^{k-p}}{k^2} \left(\frac{1 + p_c}{1 + p_s} \right)^2 \qquad (15.3.7)$$

Because n_c, n_s, n_{c0}, and n_{s0} must all be integers in practice, it will not always be possible to meet the *exact* requirements for both rotatability and orthogonal blocking. However, neither of these properties is critical. In particular, rotatability introduced to ensure a spherical distribution of information can be relaxed somewhat without appreciable loss, and the number of center points can be adjusted to achieve orthogonal blocking if desired, using Eq. (15.3.5).

Relative Replications of Star and Cube When $p_c = p_s$

If the factor $(1 + p_c)/(1 + p_s) = 1$, then

$$\frac{r_s}{r_c} = \frac{2^{k-p}}{k^2}$$

provides the desirable relative replications of star and cube when $p_c = p_s$. When $p = 0$, substitution of various values of k provides the details shown in Table 15.1.

For example, the entry $r_s/r_c = 4$ for $k = 8$ implies that, if a complete 2^8 factorial were run with proportion of center points for cube and star the same, then four replicates of the star would yield a design which was rotatable and orthogonally blockable. Conversely, if we look at the ratio r_c/r_s, we find the ideal fractional replicate of the cube to be associated with one complete star. So, for $k = 8$, a single star would require a 2^{8-2} design cube portion. When $k = 3, 5, 6,$ or 7, exact achievement of rotatability and

TABLE 15.1. **Relative proportions of r_c and r_s needed to give both rotatability and orthogonal blocking when $p_c = p_s$ for a composite design which uses a full 2^k factorial**

$k =$	2	3	4	5	6	7	8
$\dfrac{r_s}{r_c} = \dfrac{2^k}{k^2} =$	1	0.89	1	1.28	1.78	2.61	4
$\dfrac{r_c}{r_s} = \dfrac{k^2}{2^k} =$	1	1.13	1	0.78	0.56	0.38	0.25
Smallest[a] available cube fractionation	1	1	1	$\frac{1}{2}$	$\frac{1}{2}$	$\frac{1}{2}$	$\frac{1}{4}$

[a] To retain a resolution V cube portion.

orthogonal blocking is not possible for $p_c = p_s$, but the table entries show the smallest fractional cube of resolution V roughly appropriate for each star.

Center Points in Composite Designs

Various criteria have been suggested for deciding the number of center points $n_0 = n_{c0} + n_{s0}$ that should be used in a composite design. For a discussion, see Draper (1982). The criteria include choosing n_0 to:

1. Produce a good profile for the information function.
2. Minimize integrated mean squared error.
3. Give good detectability of third-order lack of fit.
4. Give a robust design insensitive to bad values.

The values of n_0 that satisfy these various criteria do not differ very much and, in the table of useful designs that follows (Table 15.2), convenient compromise values have been chosen. It should be understood that, if for some reason (to obtain more information about σ^2, for example, or to achieve equal block sizes, or orthogonal blocking) we wish to introduce more centerpoints than the numbers indicated, nothing will be lost by this except the cost of performing the additional runs.

TABLE 15.2.   Some useful second-order composite designs and their characteristics

k	2	3	4	5	5	6	7	8
Cube fraction, 2^{-p}	1	1	1	1	$\frac{1}{2}$	$\frac{1}{2}$	$\frac{1}{2}$	$\frac{1}{4}$
r_s	1	1	1	1	1	1	1	1
α for rotatability $(n_c/r_s)^{1/4}$	1.41	1.68	2	2.38	2.00	2.38	2.83	2.83
$k^{1/2}$	1.41	1.73	2	2.24	2.24	2.45	2.65	2.83
n_c	4	8	16	32	16	32	64	64
$n_s = 2k$	4	6	8	10	10	12	14	16
Minimal $\begin{cases} n_0 \\ n_{c0} \\ n_{s0} \end{cases}$	2–4 1(2) 1(2)	2–4 2 1	2–4 2 1	2–4 2 1	1–4 2 2	2–4 2 1	2–4 2 1	2–4 4 1
Blocks in cube	—	2 × 4	2 × 8	4 × 8	—	2 × 16	4 × 16 8 × 8	4 × 16
Blocking generators	—	B = 123	B = 1234	B_1 = 123 B_2 = 2345	—	B = 123	B_1 = 1357 B_2 = 1256 B_3 = 1234	B_1 = 135 B_2 = 348
Fractional design generators	—	—	—	—	I = 12345	I = 123456	I = 1234567	I = 12347 I = 12568

Smaller Blocks

So far, we have considered the splitting of the composite design into only two blocks, one associated with the cube and one with the star. Smaller blocks can be used if the factorial or fractional factorial part of the design can be further split into blocks that are first-order orthogonal designs. By splitting the n_{c0} center points equally among these smaller blocks, the requirements for orthogonal blocking will remain satisfied. (Naturally, n_{c0} must be a multiple of the number of smaller cube blocks here.) Also, if the star is replicated, then each replication, with appropriate equal allocation of the n_{s0} center points, may be made one of the set of orthogonal blocks.

Useful Second-Order Composite Designs

A short table of useful designs is shown as Table 15.2. We now offer some suggestions for using this table.

If the design is *not* to be run in blocks, the α may as well be chosen to achieve exact rotatability from the row $\alpha = (n_c/r_s)^{1/4}$. If the design is to be blocked then, if possible, the number of center points should be chosen as indicated, first of all. For example, if $k = 4$, then place one center point in each of the two blocks into which the 2^4 cube is divided and one center point in the star portion. Suppose, however, that $k = 7$ and that the 2^{7-1} cube is to be run in eight blocks of eight runs each. To achieve balance, we could put one center point in each cube block and one in the star block. This would, of course, require nine center points rather than the recommended 2–4 if the design were unblocked. As we have already noted, however, nothing is lost by increasing the number of center points except the cost of the additional runs. In any case, having decided on the allocation of points to the various parts of the blocked design, the best choice of α is now the one that achieves orthogonal blocking, namely

$$\alpha^2 = k\frac{(1 + n_{s0}/n_s)}{(1 + n_{c0}/n_c)}.$$

In general, this will not achieve exact rotatability, although in some instances it happens to, but in all cases it should provide variance contours which are adequately close to the spherical ones which exact rotatability would provide.

Design Examples

Tables 15.3, 15.5, and 15.6 contain examples of useful orthogonally blocked second-order designs for $k = 4$, 5, and 6 factors. For the case $k = 4$ of Table 15.3, orthogonal blocking can be achieved by various combinations of center points, using an appropriate value of α; for examples, see Table 15.4. A similar flexibility is, of course, also achievable for other values of k; however, Tables 15.5 and 15.6 contain a specific example design only.

Note that, for $k = 5$ (Table 15.5), blocks I and III (or II and III) can be used as an orthogonally blocked second-order design with the same value of α. This provides a 30-run design for five factors in two blocks consisting of a 2^{5-1} design plus three center points and a star plus one center point. The design is again not rotatable since $\alpha \neq 2^{(k-1)/4} = 2$.

An excellent 24-run, three-factor, second-order, orthogonally blocked design is given in Table 11.4 of Section 11.4.

For other examples of designs that have been used in various applications, see the exercises in Chapters 7 and 11.

Other Effects of Orthogonal Blocking

1. Pure error. Runs that would be repeat runs in an unblocked design are often divided among the blocks. In such a case, these runs are no longer repeat runs *unless they occur in the same block*, and the pure error must be calculated on that basis.

2. Analysis of variance. When a design is blocked, the analysis of variance table must contain a sum of squares for blocks. For blocks orthogonal to the model, the appropriate sum of squares for blocks is

TABLE 15.3. Orthogonally blocked second-order designs for four factors; with $\frac{1}{2}n_{c0}$ center points in each of blocks I and II and n_{s0} center points in block III, $\alpha^2 = 4(1 + n_{s0}/8)/(1 + n_{c0}/16)$

Block I $(2_{IV}^{4-1}$ with $x_1x_2x_3x_4 = 1$, plus center points)				Block II $(2_{IV}^{4-1}$ with $x_1x_2x_3x_4 = -1$, plus center points)				Block III (star, axial distance α, plus center points)			
x_1	x_2	x_3	x_4	x_1	x_2	x_3	x_4	x_1	x_2	x_3	x_4
-1	-1	-1	-1	-1	-1	-1	1	$-\alpha$	0	0	0
-1	-1	1	1	-1	-1	1	-1	α	0	0	0
-1	1	-1	1	-1	1	-1	-1	0	$-\alpha$	0	0
-1	1	1	-1	-1	1	1	1	0	α	0	0
1	-1	-1	1	1	-1	-1	-1	0	0	$-\alpha$	0
1	-1	1	-1	1	-1	1	1	0	0	α	0
1	1	-1	-1	1	1	-1	1	0	0	0	$-\alpha$
1	1	1	1	1	1	1	-1	0	0	0	α
0	0	0	0	0	0	0	0	0	0	0	0
$\vdots$	$\vdots$	$\vdots$	$\vdots$	$\vdots$	$\vdots$	$\vdots$	$\vdots$	$\vdots$	$\vdots$	$\vdots$	$\vdots$
0	0	0	0	0	0	0	0	0	0	0	0

514

TABLE 15.4. Values of α which achieve orthogonal blocking for various selected center point choices for the four-factor design of Table 15.3

| n_{c0} | n_{s0} | Center points in block | | | α |
		I	II	III	
4	2	2	2	2	2.000[a]
4	3	2	2	3	2.098
4	4	2	2	4	2.191
8	2	4	4	2	1.826
8	3	4	4	3	1.915
8	4	4	4	4	2.000[a]

[a] Note that when α = 2, the design is also rotatable.

usually

$$SS(\text{blocks}) = \sum_{w=1}^{m} \frac{B_w^2}{n_w} - \frac{G^2}{N}, \quad \text{with } (m-1)\,\text{df},$$

where B_w is the total of the n_w observations in the wth block (there are m blocks in all) and G is the grand total of all the observations in all of the m blocks. When blocks are not orthogonal to the model, the extra sum of squares principle applies. (It can, of course, be applied in all cases, orthogonally blocked or not, and produces the answer given above in the former case. See page 60.)

15.4. SECOND-ORDER DESIGNS REQUIRING ONLY THREE LEVELS

Second-order composite designs usually require five levels coded $-\alpha, -1, 0, 1, \alpha$, for each of the variables. Circumstances occur, however, where second-order arrangements are required which, for ease of performance, must employ the smallest number of different levels, namely three. One useful class of such designs, which are economical in the number of runs required, is due to Box and Behnken (1960a).

The designs are formed by combining two-level factorial designs with incomplete block designs in a particular manner. This is best illustrated by an example. Table 15.7(a) shows a balanced incomplete block design for testing $k = 4$ varieties in $b = 6$ blocks of size $s = 2$. If this design were

TABLE 15.5.   An orthogonally blocked second-order design for five factors

Block I (2^{5-1}, with $x_1x_2x_3x_4x_5 = 1$ plus three center points)					Block II (2^{5-1}, with $x_1x_2x_3x_4x_5 = -1$ plus three center points)					Block III (star, with $\alpha = (88/19)^{1/2} = 2.152110$* plus one center point)				
x_1	x_2	x_3	x_4	x_5	x_1	x_2	x_3	x_4	x_5	x_1	x_2	x_3	x_4	x_5
-1	-1	-1	-1	1	-1	-1	-1	-1	-1	$-\alpha$	0	0	0	0
1	-1	-1	-1	-1	1	-1	-1	-1	1	α	0	0	0	0
-1	1	-1	-1	-1	-1	1	-1	-1	1	0	$-\alpha$	0	0	0
1	1	-1	-1	1	1	1	-1	-1	-1	0	α	0	0	0
-1	-1	1	-1	-1	-1	-1	1	-1	1	0	0	$-\alpha$	0	0
1	-1	1	-1	1	1	-1	1	-1	-1	0	0	α	0	0
-1	1	1	-1	1	-1	1	1	-1	-1	0	0	0	$-\alpha$	0
1	1	1	-1	-1	1	1	1	-1	1	0	0	0	α	0
-1	-1	-1	1	-1	-1	-1	-1	1	1	0	0	0	0	$-\alpha$
1	-1	-1	1	1	1	-1	-1	1	-1	0	0	0	0	α
-1	1	-1	1	1	-1	1	-1	1	-1	0	0	0	0	0
1	1	-1	1	-1	1	1	-1	1	1					
-1	-1	1	1	1	-1	-1	1	1	-1					
1	-1	1	1	-1	1	-1	1	1	1					
-1	1	1	1	-1	-1	1	1	1	1	*The design is not rotatable				
1	1	1	1	1	1	1	1	1	-1	because $\alpha \neq 2.378$.				
0	0	0	0	0	0	0	0	0	0					
0	0	0	0	0	0	0	0	0	0					
0	0	0	0	0	0	0	0	0	0					

employed in the usual way, varieties 1 and 2 denoted by x_1 and x_2 would be tested in the first block, varieties 3 and 4 in the second, and so on.

A basis for a three-level design in four variables is obtained by combining this incomplete block design with the 2^2 factorial of Table 15.7(b). The two asterisks in every row of the incomplete block design are replaced by the $s = 2$ columns of the two-level 2^2 design. Wherever an asterisk does not appear, a column of zeros is inserted. The design is completed by the addition of a number of center points $(0, 0, 0, 0)$, about three being desirable with this arrangement. The resulting design is shown in Table 15.8.

This design is a rotatable second-order design[†] suitable for studying four variables in 27 trials and is capable of being orthogonally blocked in three sets of nine trials, separated by dashed lines in Table 15.8. (Also see Table 15.9 and Exercise 7.23.)

Table 15.9 provides useful designs of this kind for $k = 3, 4, 5, 6, 7$ factors. Note that the design for $k = 6$ is not symmetrical in all factors, because it is

---

[†] This particular design is, in fact, a rotation of the four variable central composite design with three center points. However, it is not generally true that the present class of designs can be generated from the central composite designs by rotation.

TABLE 15.6. An orthogonally blocked second-order design for six factors

Block I (2^{6-2}, with $x_5 = -x_1x_2$, $x_6 = -x_3x_4$, plus four center points)						Block II (2^{6-2}, with $x_5 = x_1x_2$, $x_6 = x_3x_4$, plus four center points)						Block III (star, with $\alpha = 2(7/5)^{1/2}$ = 2.366432* plus two center points)					
x_1	x_2	x_3	x_4	x_5	x_6	x_1	x_2	x_3	x_4	x_5	x_6	x_1	x_2	x_3	x_4	x_5	x_6
-1	-1	-1	-1	-1	-1	-1	-1	-1	-1	1	1	$-\alpha$	0	0	0	0	0
1	-1	-1	-1	1	-1	1	-1	-1	-1	-1	1	α	0	0	0	0	0
-1	1	-1	-1	1	-1	-1	1	-1	-1	-1	1	0	$-\alpha$	0	0	0	0
1	1	-1	-1	-1	-1	1	1	-1	-1	1	1	0	α	0	0	0	0
-1	-1	1	-1	-1	1	-1	-1	1	-1	1	-1	0	0	$-\alpha$	0	0	0
1	-1	1	-1	1	1	1	-1	1	-1	-1	-1	0	0	α	0	0	0
-1	1	1	-1	1	1	-1	1	1	-1	-1	-1	0	0	0	$-\alpha$	0	0
1	1	1	-1	-1	1	1	1	1	-1	1	-1	0	0	0	α	0	0
-1	-1	-1	1	-1	1	-1	-1	-1	1	1	-1	0	0	0	0	$-\alpha$	0
1	-1	-1	1	1	1	1	-1	-1	1	-1	-1	0	0	0	0	α	0
-1	1	-1	1	1	1	-1	1	-1	1	-1	-1	0	0	0	0	0	$-\alpha$
1	1	-1	1	-1	1	1	1	-1	1	1	-1	0	0	0	0	0	α
-1	-1	1	1	-1	-1	-1	-1	1	1	1	1	0	0	0	0	0	0
1	-1	1	1	1	-1	1	-1	1	1	-1	1	0	0	0	0	0	0
-1	1	1	1	1	-1	-1	1	1	1	-1	1						
1	1	1	1	-1	-1	1	1	1	1	1	1						
0	0	0	0	0	0	0	0	0	0	0	0	*The design is not rotatable					
0	0	0	0	0	0	0	0	0	0	0	0	because $\alpha \neq 2.378$, but is					
0	0	0	0	0	0	0	0	0	0	0	0	nearly so.					
0	0	0	0	0	0	0	0	0	0	0	0						

TABLE 15.7. Constructing a three-level second-order design for four variables

(a) A balanced incomplete block design for four varieties in six blocks

(b) A 2^2 factorial design

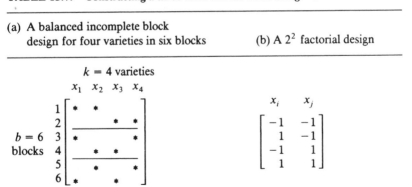

TABLE 15.8. An incomplete 3^4 factorial in three blocks of nine experimental runs

x_1	x_2	x_3	x_4	
-1	-1	0	0	
1	-1	0	0	
-1	1	0	0	
1	1	0	0	
0	0	-1	-1	Block 1
0	0	1	-1	
0	0	-1	1	
0	0	1	1	
0	0	0	0	
-1	0	0	-1	
1	0	0	-1	
-1	0	0	1	
1	0	0	1	
0	-1	-1	0	Block 2
0	1	-1	0	
0	-1	1	0	
0	1	1	0	
0	0	0	0	
0	-1	0	-1	
0	1	0	-1	
0	-1	0	1	
0	1	0	1	
-1	0	-1	0	Block 3
1	0	-1	0	
-1	0	1	0	
1	0	1	0	
0	0	0	0	

based on a *partially* balanced incomplete block design with first associates $(1, 4)$, $(2, 5)$, and $(3, 6)$. The designs for $k = 4$ and 7 are both rotatable. The requirements for orthogonal blocking may be met as follows.

1. Where "replicate sets" can be found in the generating incomplete block design, these provide a basis for orthogonal blocking. These replicate sets are subgroups within which each variety is tested the same number of times.

TABLE 15.9. Some useful three-level second-order response surface designs

Number of factors (k)	Design matrix	Number of points
3	$\begin{bmatrix} \pm1 & \pm1 & 0 \\ \pm1 & 0 & \pm1 \\ 0 & \pm1 & \pm1 \\ 0 & 0 & 0 \end{bmatrix}$	$\left.\begin{array}{c} \\ \end{array}\right\}12$ $\dfrac{3}{N=15}$

| 4 | $\left[\begin{array}{cccc} \pm1 & \pm1 & 0 & 0 \\ 0 & 0 & \pm1 & \pm1 \\ 0 & 0 & 0 & 0 \\ \hline \pm1 & 0 & 0 & \pm1 \\ 0 & \pm1 & \pm1 & 0 \\ 0 & 0 & 0 & 0 \\ \hline \pm1 & 0 & \pm1 & 0 \\ 0 & \pm1 & 0 & \pm1 \\ 0 & 0 & 0 & 0 \end{array}\right]$ | $\left.\begin{array}{c}\\\end{array}\right\}8$
 1
 $\left.\begin{array}{c}\\\end{array}\right\}8$
 1
 $\left.\begin{array}{c}\\\end{array}\right\}8$
 1
 $\dfrac{}{N=27}$ |

| 5 | $\left[\begin{array}{ccccc} \pm1 & \pm1 & 0 & 0 & 0 \\ 0 & 0 & \pm1 & \pm1 & 0 \\ 0 & \pm1 & 0 & 0 & \pm1 \\ \pm1 & 0 & \pm1 & 0 & 0 \\ 0 & 0 & 0 & \pm1 & \pm1 \\ 0 & 0 & 0 & 0 & 0 \\ \hline 0 & \pm1 & \pm1 & 0 & 0 \\ \pm1 & 0 & 0 & \pm1 & 0 \\ 0 & 0 & \pm1 & 0 & \pm1 \\ \pm1 & 0 & 0 & 0 & \pm1 \\ 0 & \pm1 & 0 & \pm1 & 0 \\ 0 & 0 & 0 & 0 & 0 \end{array}\right]$ | $\left.\begin{array}{c}\\\\\end{array}\right\}20$
 3
 $\left.\begin{array}{c}\\\\\end{array}\right\}20$
 3
 $\dfrac{}{N=46}$ |

| 6 | $\left[\begin{array}{cccccc} \pm1 & \pm1 & 0 & \pm1 & 0 & 0 \\ 0 & \pm1 & \pm1 & 0 & \pm1 & 0 \\ 0 & 0 & \pm1 & \pm1 & 0 & \pm1 \\ \pm1 & 0 & 0 & \pm1 & \pm1 & 0 \\ 0 & \pm1 & 0 & 0 & \pm1 & \pm1 \\ \pm1 & 0 & \pm1 & 0 & 0 & \pm1 \\ 0 & 0 & 0 & 0 & 0 & 0 \end{array}\right]$ | $\left.\begin{array}{c}\\\\\end{array}\right\}48$
 6
 $\dfrac{}{N=54}$ |

| 7 | $\left[\begin{array}{ccccccc} 0 & 0 & 0 & \pm1 & \pm1 & \pm1 & 0 \\ \pm1 & 0 & 0 & 0 & 0 & \pm1 & \pm1 \\ 0 & \pm1 & 0 & 0 & \pm1 & 0 & \pm1 \\ \pm1 & \pm1 & 0 & \pm1 & 0 & 0 & 0 \\ 0 & 0 & \pm1 & \pm1 & 0 & 0 & \pm1 \\ \pm1 & 0 & \pm1 & 0 & \pm1 & 0 & 0 \\ 0 & \pm1 & \pm1 & 0 & 0 & \pm1 & 0 \\ 0 & 0 & 0 & 0 & 0 & 0 & 0 \end{array}\right]$ | $\left.\begin{array}{c}\\\\\end{array}\right\}56$
 6
 $\dfrac{}{N=62}$ |

2. Where the component factorial designs can be divided into blocks which confound only interactions of more than two factors, these can provide a basis for orthogonal blocking.

An illustration of the first method of blocking has already been given in Table 15.8. In Table 15.9, for $k = 4$ and 5, dotted lines indicate the appropriate divisions into replicate sets. Using these divisions, the design for $k = 4$ can be split into three blocks, the design for $k = 5$ into two blocks. The center points *must* be distributed equally among blocks to retain orthogonality.

The second method may be illustrated with the design for $k = 6$, for which the first method cannot be employed. The basis for the design consists of 48 trials generated from six 2^3 factorial designs. If we were running a single 2^3 factorial design, it could be performed in two sets of four trials, confounding the three-factor interaction with blocks. Trials with levels $(1, 1, 1)$, $(1, -1, -1)$, $(-1, -1, 1)$, $(-1, 1, -1)$ would be included in one set (called the positive set) and trials with levels $(-1, -1, -1)$, $(-1, 1, 1)$, $(1, -1, 1)$, $(1, 1, -1)$ in the other (called the negative set). The complete group of 48 trials can be split into two orthogonal blocks of 24 by allocating one set (either positive or negative) from each of the 2^3 factorial designs to one block, and the remainder to the other. This method can also be used for the design for $k = 7$. Three level designs for $k > 7$ and their blocking arrangements are given in the original reference, Box and Behnken (1960a).

15.5. DESIGNS REQUIRING ONLY A SMALL NUMBER OF RUNS

We have seen that, if a response function containing p adjustable parameters is to be fitted, we need at least p runs. It sometimes happens that designs containing close to this minimal number of runs are of interest. This could be so for a variety of reasons:

1. Runs might be extremely expensive.
2. The checking of assumptions, the need for an internal error estimate, the need to check fit and, in particular, the need to consider possible transformations of the predictor variables, might not be regarded as important in a particular application.
3. The objective might be to approximate locally, by a polynomial, a function that can be computed *exactly* at any given combination of the input variables; that is, there is no experimental error.

We now discuss some of the possibilities available for such situations.

TABLE 15.10. Points needed by some small composite designs

Factors, k	2	3	4	5	6	7	8	9
Coefficients $\frac{1}{2}(k+1)(k+2)$	6	10	15	21	28	36	45	55
Points in Box–Hunter (1957) designs	8	14	24	26	44	78	80	146
Hartley's number of points	6	10	16	26	28	46	80	82
Westlake's number of points	—	—	—	22	—	40	—	62

Small Composite Designs

The composite designs recommended in Section 15.3 for fitting second-order models in k factors all contain cube portions of resolution at least V, plus axial points, plus center points. Of course there must be at least $\frac{1}{2}(k+1)$ $(k+2)$ points in the design, this being the number of coefficients to estimate. Hartley (1959) pointed out that the cube portion of the composite design need not be of resolution V. It could, in fact, be of resolution as low as III, provided that two-factor interactions were not aliased with two-factor interactions. (Two-factor interactions could be aliased with main effects, because the star portion provides additional information on the main effects.) This idea permitted much smaller cubes to be used. Westlake (1965) took this idea further by finding even smaller cubes for the $k = 5$, 7, and 9 cases. Table 15.10 shows the numbers of points in the various designs suggested, for $2 \le k \le 9$.

Westlake (1965) provided three examples of 22-run designs for $k = 5$, one example of a 40-run design for $k = 7$, and one example of a 62-run design for $k = 9$. He noted that for $k = 7$ or 9, "systematic generation of all possible designs... appears to be almost out of the question" (p. 332).

Subsequent work by Draper (1985b) showed that designs even smaller than Westlake's are possible. Moreover, for $k = 5$ and 9 it is possible to *equal* Westlake's number of runs in a simple manner, and for $k = 7$, simple designs are available with only 42 runs, two more than Westlake's 40. The overall advantage of these suggested designs is that none of the ingenuity shown by Westlake (1965) is needed, thanks to Plackett and Burman (1946), and yet an apparently large selection of possible designs is immediately available. (As we shall see later, the selection is not as large as first appears!)

The basic method can be simply stated: (a) Use, for the cube portion of the design, k columns of a Plackett and Burman (1946) design. (b) Where repeat runs exist, remove one of each duplicate pair to reduce the number of runs. (c) If $\alpha = k^{1/2}$, center points are mandatory to prevent the design from being singular.

For $k = 5$, for example, we use 5 (of the 11) columns of a 12-run Plackett and Burman (1946) design. There are $\binom{11}{5}$ that is, 462 possible choices, all of which produce nonsingular designs. These require 22 runs, the same number as Westlake's. A detailed examination of the cube portions for the designs shows that there are two basic types. Sixty-six choices of five columns produce a design with a duplicate run that can then be deleted to provide a 21-run design (which beats Westlake's designs by one run). The remaining 396 designs have one pair of runs that are "opposite,"—that is, are mirror images, or foldover runs. An example of a duplicate-run design is the one formed by columns $(1, 2, 3, 9, 11)$ of the Plackett and Burman design; runs 3 and 11 are duplicates. Columns $(1, 2, 3, 7, 11)$ produce a design in which runs 2 and 9 are mirror images. [The column numbers are those obtained if the design matrix is written down by following the formula suggested by Plackett and Burman (1946), and cyclically permuting rows by moving the signs to the left. The consequent column numbering is different from that used by Box, Hunter, and Hunter (1978, p. 398), for which the column transformation $(1, 2, 3, \dots, 11) \rightarrow (1, 11, 10, \dots, 2)$ should be applied. This notational difference arises because Box, Hunter, and Hunter use the Plackett and Burman first row as a column, and they cycle down rather than up.]

There are, as we have said, 462 choices of five columns from the 12-run Plackett and Burman design. How many intrinsically different designs are there? An investigation reveals that there are only two essentially different cube portions, shown in Table 15.11. All others are derivable from these by switching signs in one or more columns and then rearranging rows and columns.

For additional details and for the $k = 7$ and 9 cases, see Draper (1985b).

Koshal Designs

A polynomial of degree d in k variables contains $p = \{(k + d)!\}/(d!k!)$ coefficients. An example of a design that contains exactly this number of runs for the case $d = 2$ was given by Koshal (1933). He wished to determine the maximum of a likelihood function which could be calculated numerically at any given value, but was difficult to handle analytically. Such designs are occasionally of use in response surface work, particularly in applications like Koshal's; see, for example, Kanemasu (1979). Koshal's design is readily extended to any order d and to any number of predictor variables k.

TABLE 15.11.   Two essentially different choices of five columns from a 12-run Plackett and Burman design

(a) with a pair of repeat runs					(b) with a mirror-image pair of runs				
−	−	−	−	−	−	−	−	−	−
−	−	−	−	−	+	+	+	+	+
−	−	+	+	+	−	−	+	+	+
−	+	−	+	+	−	+	+	−	+
−	+	+	−	+	+	−	−	+	+
−	+	+	+	−	+	+	−	+	−
+	−	−	+	+	+	+	+	−	−
+	−	+	−	+	+	−	+	−	−
+	−	+	+	−	+	−	−	−	+
+	+	−	−	+	−	+	−	+	−
+	+	−	+	−	−	+	−	−	+
+	+	+	−	−	−	−	+	+	−

Note: All other choices are equivalent to one of these, subject to changes in signs throughout one or more columns, renaming of variables and reordering runs.

First-Order Koshal Designs (d = 1)

These designs are the well-known "one factor at a time" designs with $N = k + 1$ runs. Thus for $k = 2$, the three constants in a first-order model can be determined using the design

x_1	x_2
0	0
1	0
0	1

$d = 1$
$k = 2$
$N = p = 3$

where 0, 1 denote the two levels selected for each variable. The generalization to any number of variables k is immediate. For example, the $k = 4$ Koshal first-order design is

x_1	x_2	x_3	x_4
0	0	0	0
1	0	0	0
0	1	0	0
0	0	1	0
0	0	0	1

$d = 1$
$k = 4$
$N = p = 5$

TABLE 15.12.   Second-order ($d = 2$) Koshal designs for $k = 3$ predictor variables

(a)			(b)			
x_1	x_2	x_3	x_1	x_2	x_3	
0	0	0	0	0	0	
1	0	0	1	0	0	
0	1	0	0	1	0	
0	0	1	0	0	1	$d = 2$
2	0	0	-1	0	0	$k = 3$
0	2	0	0	-1	0	$N = p = 10$
0	0	2	0	0	-1	
1	1	0	1	1	0	
1	0	1	1	0	1	
0	1	1	0	1	1	

Second-Order Koshal designs ($d = 2$)

Consider, for $k = 3$ variables, the alternative designs of Table 15.12. Such designs are obviously readily generalized for any k. For designs of type (a), the numbers included in the rows of the design are essentially those of the subscripts of the coefficients of a polynomial of second degree or, equivalently, relate to the partial derivatives entering a Taylor expansion of second order.

It is easy to produce modifications of Koshal designs for various purposes. Design (b) has the possible advantage of providing a less asymmetric pattern than (a). Other variations include the switching of signs in the "interaction" rows.

TABLE 15.13.   Third-order ($d = 3$) Koshal design for $k = 3$ predictor variables

x_1	x_2	x_3	x_1	x_2	x_3	
0	0	0	3	0	0	
1	0	0	0	3	0	
0	1	0	0	0	3	
0	0	1	1	2	0	$d = 3$
2	0	0	2	1	0	$k = 3$
0	2	0	1	0	2	$N = p = 20$
0	0	2	2	0	1	
1	1	0	0	1	2	
1	0	1	0	2	1	
0	1	1	1	1	1	

Higher-Order Koshal Designs

The generalization to higher-order designs is also straightforward. For illustration, consider a third-order design in three variables with $N = p = 20$ runs, shown in Table 15.13. Again, obvious substitutions and sign switchings are possible (for example, to achieve greater symmetry) but we will not pursue these modifications further here.

EXERCISES

15.1. Consider, in $k = 2$ dimensions, the following design: "cube" $(\pm 1, \pm 1)$, plus "star" $(\pm \alpha, 0)$, $(0, \pm \alpha)$, plus four center points. Find the variance function appropriate for the second-order model $y = \beta_0 + \beta_1 x_1 + \beta_2 x_2 + \beta_{11} x_1^2 + \beta_{22} x_2^2 + \beta_{12} x_1 x_2 + \varepsilon$, and hence derive the value of α for which the design is second-order rotatable. Show that, in this case, the variance two units from the origin is $9\sigma^2/4$, where σ^2 is the variance of the individual observations.

15.2. (Source: Agnes M. Herzberg.) Consider the design in $k = 2$ variables which consists of the nine points $(x_1, x_2) = (0, \pm a)$, $(\pm b, \pm c)$, $(0, 0)$ three times, where $a = 3^{1/2}$, $b = 1.5$, and $c = 3^{1/2}/2$.
   (a) Is it possible to estimate a full second-order model in x_1 and x_2 using this design? If so, is the design second-order rotatable?
   (b) Evaluate and plot the variance function $NV\{\hat{y}(\mathbf{x})\}/\sigma^2$.
   (c) If the true response function were a third degree polynomial, what biases would be induced in the expected values of the second-order estimated coefficients?

15.3. Consider a rotatable design for $k = 3$ consisting of cube, plus star, plus n_0 center points. Find $V(r) = V[\hat{y}(\mathbf{x})]$ as a function of $r = (x_1^2 + \cdots + x_k^2)^{1/2}$, and plot $V(r)$ against r for $n_0 = 2(2)10$. How many center points are needed to achieve $V(0) = V(1)$, approximately? [*Note:* In working this question, keep the cube points fixed at $(\pm 1, \pm 1, \pm 1)$ and *do not* scale so that $\sum_{u=1}^{N} x_{iu}^2 = N$.] How many center points will achieve $V(0) = V(\lambda_2^{1/2})$, approximately, where $\lambda_2 N = \sum_{u=1}^{N} x_{iu}^2$? How many center points will achieve $V(0) = V(k^{1/2})$, approximately?

15.4. Consider the four points of a 2^{3-1} design given by $(1, 1, 1)$, $(1, -1, -1)$, $(-1, 1, -1)$, $(-1, -1, 1)$. If we "add the points in pairs" and attach a multiplier $\frac{1}{2}\alpha$, we get six points: $(\alpha, 0, 0)$ from $\alpha\{(1, 1, 1) + (1, -1, -1)\}$, $(0, \alpha, 0)$, $(0, 0, \alpha)$, $(0, 0, -\alpha)$, $(0, -\alpha, 0)$, $(-\alpha, 0, 0)$, namely the six points of a "star." If we "add the points in threes" with a multiplier of unity we get the other 2^{3-1} points $(-1, -1, -1)$, $(-1, 1, 1)$, $(1, -1, 1)$, $(1, 1, -1)$. Addition of all four points with any multiplier gives the center point $(0, 0, 0)$. Thus we obtain the point locations of a "cube plus star plus center points" composite design for $k = 3$ factors. This is the essence of the "simplex-sum" method of design construction due to G. E. P. Box and D. W. Behnken and described in "Simplex-sum designs: a class of second order rotatable designs derivable from those of first order," *Annals of Math. Statist.*, **31**, 1960, 838–864.

Apply this same method to the eight points of a 2^{7-4} design and show that it leads to the 56-point rotatable design in seven factors given by G. E. P. Box and D. W. Behnken in "Some new three level designs for the study of quantitative variables," *Technometrics*, **2**, 1960, 455–475, and shown in Table 15.9. [*Note*: In the example above, all the point sets generated were used. Here you will need only the pairs, the 6's, and the sum of all eight, which provides the center points. Note also the mirror image folded property of the r's and the $(N - r)$'s.]

15.5. (Source: Mark Thornquist.) Consider the 12-point design below. Confirm that the points of this design form an icosahedron, a regular figure in three dimensions. Confirm also that all the points lie on a sphere of radius unity, so that center points would be needed to enable inversion of the $\mathbf{X}'\mathbf{X}$ matrix, if such a design were used.

$$(1,0,0)$$
$$(a,b,0)$$
$$(a, b\cos\theta, b\sin\theta)$$
$$(a, b\cos 2\theta, b\sin 2\theta)$$
$$(a, b\cos 3\theta, b\sin 3\theta)$$
$$(a, b\cos 4\theta, b\sin 4\theta)$$
$$(-a, -b, 0)$$
$$(-a, -b\cos\theta, -b\sin\theta)$$
$$(-a, -b\cos 2\theta, -b\sin 2\theta)$$
$$(-a, -b\cos 3\theta, -b\sin 3\theta)$$
$$(-a, -b\cos 4\theta, -b\sin 4\theta)$$
$$(-1,0,0)$$

$$\theta = \frac{2\pi}{5}$$

$$a = \frac{\cos\theta}{(1 + \cos\theta)}$$

$$b = (1 - a^2)^{1/2}$$

[Note that the design consists of a single point at $(x_1, x_2, x_3) = (-1,0,0)$, a circle of five equally spaced points at $x_1 = -a$, a similar circle but rotated an angle of $\frac{1}{2}\theta$ at $x_1 = a$, and a single point at $(1,0,0)$. These points are scaled and rotated forms of the usual representation $(\pm p, \pm q, 0)$, $(0, \pm p, \pm q)$, $(\pm q, 0, \pm p)$ where $p^2 - pq - q^2 = 0$.]

15.6. Consider, for $k = 3$, the cube plus star plus center points composite response surface design. Suppose that the design is split into three blocks.

$$2^{3-1} \ (\mathbf{I} = \mathbf{123}) \text{ plus } \tfrac{1}{2}c_0 \text{ center points,}$$

$$2^{3-1} \ (\mathbf{I} = -\mathbf{123}) \text{ plus } \tfrac{1}{2}c_0 \text{ center points,}$$

Star plus s_0 center points.

Confirm that, if $c_0 = 4$, $s_0 = 2$, and $\alpha^2 = \frac{8}{3}$, the design is orthogonally blocked. Find the formula for values of α^2 that will give an orthogonally blocked design for other values of c_0 and s_0, and tabulate the numerical values of α^2 for $0 \le c_0 \le 8$, $0 \le s_0 \le 5$. Is it possible for the design to be rotatable and orthogonally blocked?

15.7. Consider, for $k = 4$, the cube plus star plus center points composite response surface design. Suppose that the design is split into three blocks

$$2^{4-1} \ (\mathbf{I} = \mathbf{1234}) \text{ plus } \tfrac{1}{2}c_0 \text{ center points,}$$

$$2^{4-1} \ (\mathbf{I} = -\mathbf{1234}) \text{ plus } \tfrac{1}{2}c_0 \text{ center points,}$$

$$\text{Star plus } s_0 \text{ center points.}$$

Confirm that if $c_0 = 4$, $s_0 = 2$, and $\alpha = 2$, the design is orthogonally blocked *and* rotatable. Find the formula for values of α^2 that will give an orthogonally blocked design for other values of c_0 and s_0, and tabulate the numerical values of α^2 for $0 \le c_0 \le 8$, $0 \le s_0 \le 5$.

15.8. Consider a composite design for $2 \le k \le 8$ consisting of a cube (2^{k-p} points with $p = 0$ for $k = 2, 3$, and 4; $p = 1$ for $k = 5, 6, 7$, and $p = 2$ for $k = 8$), a replicated star (each of the two with axial distance α), and center points. Divide the cube up into b blocks where $b = 1$ for $k = 2$ and $k = 5$ ($p = 1$), $b = 2$ for $k = 3, 4, 6$ ($p = 1$), $b = 8$ for $k = 7$ ($p = 1$), and $b = 4$ for $k = 8$ ($p = 2$). Each star is itself another block. Consider the cases where there are $A = n_{c0}/b = 0$ or 1 or 2 center points in *each* of the b cube blocks (i.e., no more than 2). Of these 18 situations, which allow a choice of α for which the design can be *both* rotatable *and* orthogonally blocked? Tabulate the α and $B = \frac{1}{2}n_{s0}$ (= number of center points in each star block) values for each case in which it can be achieved.

15.9. Using the method described in Section 15.2, write down regular simplex designs for $k = 4$, 5, and 6, scaling the variables so that $\sum_{u=1}^{k+1} x_{iu}^2 = N = k + 1$, $i = 1, 2, \ldots, k$. Show that the length of an edge is $\sqrt{10}$, $\sqrt{12}$, and $\sqrt{14}$ units respectively in the three cases. Show more generally that, for the general regular simplex design for k variables in $k + 1 = N$ runs obtained by this method, the length of an edge is $(2Nc)^{1/2}$ where

$$\sum_{u=1}^{N} x_{iu}^2 = Nc, \qquad i = 1, 2, \ldots, k.$$

Answers to Exercises

CHAPTER 3

3.1. (a) $\hat{Y} = -1434.544 + 57.38\xi$. Note that the b_0 value is attained at $\xi = 0$, well away from the data, and is not practically meaningful there.

 (b) $\xi_0 = 25.10$, $S = 0.05$; $x = -2, -1, 0, 1, 2$, then.

 (c) $\hat{Y} = 5.694 + 2.869x$. The fitted values and residuals are identical to those in (a).

 (d) The models are identical via substitution of $x = (\xi - 25.1)/0.05$.

 (e) The residuals, $1.594, -0.465, -1.764, -1.453, 2.088$, which correctly sum to zero, indicate curvature around the fitted line, obvious from the plot. Either transformation of Y or use of a quadratic in x might be sensible.

 (f) $\hat{y} = 0.6281 + 0.2360x$.

x	y		$\hat{y}$		$e = y - \hat{y}$	
-2	0.1903		0.1561		0.0342	
-1	0.3729		0.3921		-0.0192	
0	0.5944		0.6281		-0.0337	
1	0.8519		0.8641		-0.0122	
2	1.1310		1.1001		0.0309	
0	3.1405		3.1405		0.0	Sum
10	2.533474	$=$	2.529508	$+$	0.003778	SS

(The SS equality has a rounding error of 0.000188.) Taking logs does not remove the curvature pattern.

 (g) Apply formula (3.4.4).

(h) In addition to the columns in (f) we need the following, where $\eta_0 = 0.6 + 0.25x$.

η_0	$y - \eta_0$	$\hat{y} - \eta_0$	
0.10	0.0903	0.0561	
0.35	0.0229	0.0421	
0.60	−0.0056	0.0281	
0.85	0.0019	0.0141	
1.10	0.0310	0.0001	
3.00	0.1405	0.1405	Sum
	96.74	59.08	SS × 10⁴

Source	SS × 10⁴	df	MS	F
Model	$\Sigma(\hat{y} - \eta_0)^2 = 59.08$	2	29.54	2.346
Residual	$\Sigma(y - \hat{y})^2 = 37.78$	3	12.59	
Total	$\Sigma(y - \eta_0)^2 = 96.74$	5		

We do not reject H_0: $\beta_0 = 0.6$, $\beta_1 = 0.25$ at the $\alpha = 0.10$ level. (Note round-off error effect in SS column.)

3.2.

$$\mathbf{b} = \begin{bmatrix} 5 & 0 & 10 \\ 0 & 10 & 0 \\ 10 & 0 & 34 \end{bmatrix}^{-1} \begin{bmatrix} 28.47 \\ 28.69 \\ 69.75 \end{bmatrix} = \begin{bmatrix} 34/70 & 0 & -10/70 \\ 0 & 0.1 & 0 \\ -10/70 & 0 & 5/70 \end{bmatrix} \begin{bmatrix} 28.47 \\ 28.69 \\ 69.75 \end{bmatrix}$$

$\hat{Y} = 3.864 + 2.869x + 0.915x^2$.

Y	$\hat{Y}$	$e = Y - \hat{Y}$
1.55	1.786	−0.236
2.36	1.910	0.450
3.93	3.864	0.066
7.11	7.648	−0.538
13.52	13.262	0.258

The quadratic requires an extra parameter but the residuals are less systematic. (Of course in practice one can not make much out of just five observations! The data do, however, illustrate the calculations and considerations involved in this sort of situation.)

3.3.

$$\hat{\theta} = (\Sigma z^2)^{-1}(\Sigma zy) = 5^{-1}(45) = 9. \ \eta_0 = 8.$$

$$\Sigma(\hat{y} - \eta_0)^2 = 5, \ \Sigma(y - \hat{y})^2 = 10. \ F = \frac{(5/1)}{(10/4)} = 2.0.$$

Prob($F(1,4) > 2.0$) exceeds 0.10, so do not reject H_0.

3.4. (a) $\hat{y} = 43.40704z_1$. It is being assumed that the model passes through the $(z_1, y) = (0,0)$ origin, that is, that with no additives the vehicle starts immediately.

(b) $\hat{y} = 31.79864z_1 + 14.34827z_2$.

(c) $z_{2 \cdot 1} = z_2 - (\Sigma z_1 z_2 / \Sigma z_1^2)z_1 = z_2 - 0.809045z_1$.
The $z_{2 \cdot 1}$ entries are $z'_{2 \cdot 1} = (3.57286, 0.76382, 2.95477, -1.85427, 0.33668, -2.47236)$ and the vector is orthogonal to z_1 of course. (You may wish to check this.)
$\hat{y} = 43.40704z_1 + 14.34828z_{2 \cdot 1}$.
$\hat{\theta} = \hat{\theta}_1 + 0.809045\hat{\theta}_2$.

(d)

Source	SS	df	MS
z_1 only	$\hat{\theta}^2\Sigma z_1^2 = 374950$	1	374950
Extra for z_2	$\hat{\theta}_2^2\Sigma z_{2 \cdot 1}^2 = \ \ 6535$	1	6535
Residual	102	4	026
Total	3815.92	6	

(Note rounding error in table.)
Clearly both variables play a significant role.

(e) Alias "matrix" is $(z'_1 z_1)^{-1}z'_1 z_2 = 161/199 = 0.809045$, so that $E(\hat{\theta}_1) = \theta_1 + 0.809045\theta_2$. This should be identical to the relationship in (c) and it is.

(f) Region is defined by the interior of the ellipse

$$[(\theta_1 - 31.80), (\theta_2 - 14.35)]\begin{bmatrix} 199 & 161 \\ 161 & 162 \end{bmatrix}\begin{bmatrix} \theta_1 - 31.80 \\ \theta_2 - 14.35 \end{bmatrix}$$

$$= 2(0.26)6.94 = 3.61.$$

3.5. (a) $x = (X - 30)/12$.
     (b) $\hat{Y} = 9.471 - 6.3500x$.

Source	SS	df	MS	F
b_0	1255.91	1		
$b_1 \vert b_0$	1209.67	1		
Lack of fit	374.15	3	124.72	186.1
Pure error	6.07	9	0.67	
Total	2845.80	14		

Clear lack of fit is shown. The residuals indicate curvature. Possible alternatives are $Y = \beta_0 + \beta_1 x + \beta_{11} x^2 + \varepsilon$ and $y \equiv \ln Y = \beta_0 + \beta_1 x + \varepsilon$.

(c) $\hat{Y} = 3.326 - 6.3500x + 2.8678x^2$.
     $\hat{y} = 1.5219 - 0.86634x$.
     The residuals from the first equation produce (in order of increasing x) three pluses, three minuses, five pluses, and three minuses, a disturbing pattern which indicates possible *systematic* departure from the fitted model. The second fitted equation looks preferable.

(d) This discovery raises the possibility of serial correlation between the response values in a column, requiring estimation of the correlation(s) and use of generalized least squares or an alternative analysis. For some pertinent remarks see, for example, "Problems in the analysis of growth and wear curves," by G. E. P. Box, *Biometrics*, **6**, 1950, 362–389.

CHAPTER 4

4.1. See Appendix 4A.

4.2. Real effects (in descending order) appear to be 4 (estimate 24), 2 (21) and 1 (18). Replotting the remaining contrasts, we obtain an estimate of 5.25 for the standard deviation of the effects, that is, $\hat{\sigma} = \frac{1}{2}(5.25)(16)^{1/2}$ is the estimated standard deviation of individual observations.

4.3. The only real effect appears to be D with a value of 66.7. Replotting the remaining contrasts gives a standard error (effect) = 8, that is, $\hat{\sigma} = \frac{1}{2}(8)(16)^{1/2} = 16$ for the estimated standard deviation of individual observations.

4.4. (a) (i) $ME_i = (\Sigma y\text{'s at upper level of factor } i - \Sigma y\text{'s at lower level})/2^{k-1}$.
         (ii) $b_i = (\text{same})/2^k$.
         Thus $ME_i = 2b_i$.

(b) (i) (Two-factor interaction)$_{ij}$ = (TFI)$_{ij}$ = {(Σy's for which factor i level times factor j level is $+1$) − (Σy's for ... is -1)}/2^{k-1}.
   (ii) b_{ij} = (same)/2^k.
   Thus (TFI)$_{ij}$ = $2b_{ij}$.

4.5. We first add a column of 1's and define the four columns of $\mathbf{X}$ as the columns of $(1, x_1, x_2, x_3)$. The single column of $\mathbf{X}_2$ is defined as the values of x_4 shown. Applying the formula $E(\mathbf{b}) = \boldsymbol{\beta} + (\mathbf{X}'\mathbf{X})^{-1}\mathbf{X}'\mathbf{X}_2\boldsymbol{\beta}_2$, where $\boldsymbol{\beta} = (\beta_0, \beta_1, \beta_2, \beta_3)'$, $\boldsymbol{\beta}_2 = \beta_4$, and $\mathbf{b}$ estimates $\boldsymbol{\beta}$ in the model, we find $(\mathbf{X}'\mathbf{X})^{-1}\mathbf{X}'\mathbf{X}_2 = (0, 2, 3, -2)'$ so that b_0 is unbiased, while

$$E(b_1) = \beta_1 + 2\beta_4,$$

$$E(b_2) = \beta_2 + 3\beta_4,$$

$$E(b_3) = \beta_3 - 2\beta_4.$$

4.6. The effects are

$$I \leftarrow 9.125, \quad 1 \leftarrow -2.75, \quad 2 \leftarrow -3.25 \quad 3 \leftarrow -0.75,$$

$$12 \leftarrow -0.25, \quad 13 \leftarrow 0.25, \quad 23 \leftarrow 1.75, \quad 123 \leftarrow 0.75.$$

4.7. Center point observations added to a 2^k design have no effect whatsoever on the factorial effects; they must be included in the mean calculation, however. Also, a comparison can be made between the factorial and center point averages to give an idea of whether curvature is present. (See Section 6.3.)

4.8. (a) $x_1 = (\%C - 0.5)/0.4$,
   $x_2 = (\%Mn - 0.8)/0.6$,
   $x_3 = (\%Ni - 0.15)/0.05$.
   (c) $\bar{y} = 552.5$ $1 \leftarrow -575$ $12 \leftarrow 10$ $123 \leftarrow -10$
   $\phantom{(c) \bar{y} = 552.5}$ $2 \leftarrow -90$ $13 \leftarrow 15$
   $\phantom{(c) \bar{y} = 552.5}$ $3 \leftarrow -65$ $23 \leftarrow 10$.
   (d) $V(\text{effect}) = 4\sigma^2/n = 4(65)^2/8 = (45.962)^2$, so that standard error (effect) = 46, approximately. Only the 1-effect is larger than $\pm 2(46) = \pm 92$.
   (e) The 1-effect is negative, so increasing the %C level decreases the start temperature.

4.9. With the coding $x_1 = (X_1 - 85)/25$, $x_2 = (X_2 - 1750)/750$, $x_3 = (X_3 - 4.65)/2.65$, the design is recognizable as a replicated 2^3 experiment with runs $(\pm 1, \pm 1, \pm 1)$. The mean is $\bar{y} = 29.59375$ and the factorial effects, rounded to two decimal places are

$$1 \leftarrow 15.01 \quad 12 \leftarrow 0.06 \quad 123 \leftarrow 1.09$$

$$2 \leftarrow 11.73 \quad 13 \leftarrow -0.69$$

$$3 \leftarrow 9.54 \quad 23 \leftarrow 1.34$$

each with variance $4\sigma^2/16 = \sigma^2/4$. We estimate σ^2 from the repeat runs; each pair [e.g., 12.2, 12.5 from the $(-,-,-)$ runs, etc.] provides a 1-df estimate of σ^2 via the special formula for a pair of repeat runs, for example, $\frac{1}{2}(12.2 - 12.5)^2 = 0.045$. The eight such estimates can be averaged to give $s^2 = 1.8081$ based on 8 df. [Remember that, in general, the weighted average $(v_1 s_1^2 + v_2 s_2^2 + \cdots + v_r s_r^2)/(v_1 + v_2 + \cdots + v_r)$ is needed, but here all the individual degrees of freedom are $v_i = 1$ so that the general formula implies simple averaging in such a case.] The standard error (effect) = $\{1.8081/4\}^{1/2}$ = 0.672 and only the three main effects are significant. All are positive, so that increases in X_1, X_2, and X_3 lead to increased penetration rates. In drilling, however, there are other responses to consider such as the wear on the drill, and the consequent overall cost of a succession of drilling jobs. Thus, increasing the levels of the predictors will improve the response, but it is not necessarily the best overall practical policy. Additional information is needed to determine this. See, for example, "Tool life testing by response surface methodology," by S. M. Wu, *J. Engineering Industry, Trans. ASME*, **B-86**, 1964, 105–116. See also a paper by S. M. Wu and R. N. Meyer in the November 1965 issue.

4.10.

$\bar{y} =$ 58.25		
1 ← 0.25	12 ← 0.5	123 ← 1.75
2 ← −17.25	13 ← −1.	124 ← −0.75
3 ← 9.75	14 ← −1.	134 ← 0.25
4 ← 21.75	23 ← −7.5	234 ← 0.25
	24 ← 1.	1234 ← 0.00
	34 ← −0.5	

We have no estimate for σ, but it is clear from the sizes of the main effects that factors 2, 3, 4 are important, and that the 23 interaction is probably operating, also. It is tempting to drop variable 1 entirely. If we did this, we would have replicated 2^3 design in variables 2, 3, and 4, and could use the ignored effects to estimate σ^2.

For a regression treatment that provides an estimate of σ^2, see Exercise 4.13.

4.11. The results are $\bar{y} = 4.071$,

1 ← 2.99	12 ← −0.14	123 ← −0.11
2 ← −0.06	13 ← −1.08	
3 ← −1.93	23 ← 0.04	

$s_e^2 = 16.06/16 = 1.004$. Standard error (effect) = $(4s_e^2/24)^{1/2} = 0.41$, so the 1, 3, and 13 effects all exceed 2 standard error in size. We rearrange the data in a two-way table of averages of six observations in factors 1 and 3:

	2.15	4.07	
↑			
(3)	3	7.07	(1) →

Increased tear resistance is clearly attained by the " + " paper but tearing is greatly affected by the direction of tear.

4.12.

$$\hat{y} = 29.594 + 7.506x_1 + 5.869x_2 + 4.769x_3.$$

Note that $b_0 = \bar{y}$ and the b_i are twice the factorial effects.

Source	SS	df	MS	F
b_0	14,012.64	1	—	
b_1	901.50	1	901.50	382.80
b_2	551.08	1	551.08	234.00
b_3	363.85	1	363.85	154.50
Lack of fit	13.79	4	3.448	1.906
Pure error	14.47	8	1.809	
Total	15,857.33	16		

There is no lack of fit so $s^2 = (13.79 + 14.47)/(4 + 8) = 2.355$ estimates σ^2. The regression coefficients b_1, b_2, b_3 are all highly significant.

4.13. $\hat{y} = 58.25 - 8.625x_2 + 4.875x_3 + 10.875x_4$
$\qquad - 3.75x_2x_3 + 0.5x_2x_4 - 0.25x_3x_4.$

Source	SS	df	MS	F
b_0	54,289.00	1	(same as	
b_2	1,190.25	1	SS's)	393.83
b_3	380.25	1		125.50
b_4	1,892.25	1		624.50
b_{23}	225.00	1		74.26
b_{24}	4.00	1		1.32
b_{34}	1.00	1		0.33
Residual	24.25	9	$s^2 = 2.69$	
Total	58,006.00	16		

The b_2, b_3, b_4, and b_{23} coefficients are significant.

4.14. Two of the factors, $x_1 =$ type of paper, and $x_3 =$ direction of tear, are qualitative and would usually not be suitably extrapolated or interpolated. The only quantitative factor, $x_2 =$ humidity, has little effect on the response.

4.15. (a) The fitted equation is $\hat{y} = 54.5 - 0.625x_1 + 1.375x_2 + 3.375x_3$ with analysis of variance table as below.

Source	SS	df	MS	F
First order	109.375	3	36.485	9.85
Lack of fit	26.875	5	5.375	5.86
Pure error	2.750	3	0.917	
Total, corrected	139.000	11		

Lack of fit is not significant; $s^2 = (26.875 + 2.75)/(5 + 3) = 3.703$. First-order terms are significant. A splitup of the first-order sum of squares into its three component pieces of $3.125 + 15.125 + 91.125$ for b_1, b_2, and b_3, respectively, shows that only b_3 is significant. A model $\hat{y} = 54.5 + 3.375x_3$ explains $R^2 = 0.6556$ of the variation about the mean. (Because pure error cannot ever be explained, the maximum possible value of R^2 is 0.9802.)

   (b) A set of 12 meaningful orthogonal contrasts consists of the following columns: I(twelve 1's); x_1, x_2, x_3 (as shown); x_1x_2, x_1x_3, x_2x_3, $x_1x_2x_3$; a column generated from $x_i^2 - \overline{x_i^2}$, formed from any x_i^2 column, then made orthogonal to the I column, and consisting of eight $\frac{1}{3}$'s followed by four $-\frac{2}{3}$'s; three pure error contrast columns which have zeros in the first eight positions and any orthogonal pattern such as, for example,

$$
\begin{array}{rrr}
-1 & -1 & 1 \\
1 & 1 & -1 \\
-1 & 1 & -1 \\
1 & 1 & 1
\end{array}
$$

for their last four entries.

4.16. Table A4.16a shows the factorial effects and the CC contrasts, with their respective standard errors. The asterisks indicate contrasts which exceed, in modulus, 2.57 times their standard error. The value 2.57 is the upper $2\frac{1}{2}\%$ point of the t distribution with 5 df. A confused pattern is shown and curvature, as measured by the estimate of $(\beta_{11} + \beta_{22} + \beta_{33} + \beta_{44})$, appears for all responses. The standard errors are $s_i/2$ for the factorial effects and $(11s_i^2/48)^{1/2}$ for the CC contrasts which are the differences between two averages of 16 and 6 observations and so have variance $\sigma^2/16 + \sigma^2/6 = 11\sigma^2/48$.

Transformation to $\arcsin(y/100)^{1/2}$ gives the effects in Table A4.16b. We cannot work out the standard errors here using appropriate pure error mean squares because we cannot transform the individual center point observa-

TABLE A4.16a.   Factorial effects and *CC* contrasts for the original data

Main effect or interaction	Response number			
	1	2	3	4
1	2.39*	−3.24	11.10*	10.13*
2	−0.66	−2.44	−40.00*	−42.98*
3	0.19	−19.01*	31.45*	12.50*
4	0.61	10.24*	0.45	11.43*
12	−0.59	4.39*	4.30	8.23*
13	1.26	−0.29	5.60*	6.45*
14	0.24	3.16	−7.00*	−3.48
23	1.31	16.01*	−9.90*	7.55*
24	1.39*	5.61*	−1.40	5.48*
34	0.39	0.59	−2.10	−1.00
123	0.99	−6.81*	7.40*	1.70
124	0.61	−10.61*	2.10	−8.03*
134	0.51	−11.34*	−1.85	−12.55*
234	0.86	5.11*	2.45	8.30*
1234	1.09	−4.46*	−4.95	−8.45*
Standard error (effects)	0.52	1.52	1.96	1.43
CC	−2.99*	−6.16*	−6.13*	−15.24*
Standard error (*CC*)	0.50	1.46	1.88	1.37

tions, which we do not know. As a very crude approximation, we use $q_i = \arcsin(s_i/100)^{1/2}$ and then evaluate $q_i/2$ and $q_i(11/48)^{1/2}$. We see that a much simpler picture emerges after transformation. See page 538.

CHAPTER 5

5.1.   (a)   True. The largest value of R is given by the half-fractions $I = \pm\mathbf{12}\ldots\mathbf{k}$, so that R cannot exceed k. For example, the 2_{III}^{3-1} with generator $I = \mathbf{123}$ has $k = R = 3$.

   (b)   False. For example, the 2_{III}^{3-1} generated by $I = \mathbf{123}$ has $k = 3$, $R = 3$, $p = 1$ so that $3 = k < R + p = 4$.

TABLE A4.16b. Factorial effects and *CC* contrasts for the transformed data, after multiplication by 1000

Main effect or interaction	Response number			
	1	2	3	4
1	94	−25	132	127
2	−28	−15	−460*	−508*
3	−15	−235*	357*	132
4	22	122	5	134
12	−25	59	53	69
13	61	5	63	78
14	−1	32	−77	−31
23	46	178	−83	66
24	56	70	−11	46
34	8	17	−16	−34
123	25	−60	77	13
124	−4	−133	26	97
134	13	−121	31	147
234	32	64	37	116
1234	20	−71	57	−78
Standard error (effects)	51	88	100	85
CC	−101	−88	−81	−147
Standard error (*CC*)	49	84	95	81

5.2. (a) Assume all interactions (i) between 1, 2, and 3 are zero; (ii) between 3, 4, and 5 are zero. The defining relation is **I** = **1234** = **135** = **245**. Ignoring interactions of three or more factors, we can estimate

Mean	(12) + (34)
1 + (35)	(13) + 24 + 5
2 + (45)	(23) + 14
3 + 15	(123) + 4 + 25

Two main effects are clear, three are confounded. (The interactions in parentheses are assumed to be zero, but are listed for information.)

(b)   A better design is given by $I = 123 = 345$ ($= 1245$) in which we can estimate

Mean $4 + (35)$
$1 + (23)$ $5 + (34)$
$2 + (13)$ $14 + 25$
$3 + (12) + (45)$ $24 + 15$

Here, because of the assumptions, all five main effects are clear and the remaining four two-factor interactions are grouped in two pairs.

5.3.   Design (a) has $I = A$, while design (b) is of resolution III. So (b) is *much* better.

5.4.   (a)   First fraction $I = 8 = 1234 = 125 = 146 = 247$.
            Folded fraction $I = -8 = 1234 = -125 = -146 = -247$.
      (b) $I = 1234 = 1258 = 1468 = 2478$, a 2_{IV}^{8-4}.
      (c)   Suppose design (b) has matrix

$$\mathbf{B} = \begin{bmatrix} \mathbf{A} \\ -\mathbf{A} \end{bmatrix}.$$

The new design will have matrix

$$\begin{bmatrix} \mathbf{B} \\ -\mathbf{B} \end{bmatrix} = \begin{bmatrix} \mathbf{A} \\ -\mathbf{A} \\ -\mathbf{A} \\ \mathbf{A} \end{bmatrix}$$

which is a 2_{IV}^{8-4} design, replicated.

5.5.   The 2_{IV}^{8-4} design, $I = 1235 = 1246 = 1347 = 2348$, in two blocks of eight runs each.
      Set $\mathbf{B} = 12$ so that $I = 12B$ generates the blocking. Then we can obtain estimates of

$$I, (1 + 2B), (2 + 1B), (3 + 5B), (4 + 6B),$$

$$(5 + 3B), (6 + 4B), (7 + 8B), (8 + 7B),$$

$$(B + 12 + 35 + 46 + 78), (13 + 25 + 47 + 68),$$

$$(14 + 26 + 37 + 58),$$

$$(23 + 15 + 48 + 67), (24 + 16 + 38 + 57),$$

$$(34 + 17 + 28 + 56), (45 + 36 + 27 + 18).$$

The main effects are clear if the block variable does not interact.

5.6.   The defining relation for the design given in the hint is

$$I = 1235 = 1246 = 12B_1 = 13B_2 = 3456$$

$$= 35B_1 = 25B_2 = 46B_1 = 2346B_2$$

$$= 23B_1 B_2 = 123456B_1 = 1456B_2$$

$$= 15B_1 B_2 = 1346B_1 B_2 = 2456B_1 B_2.$$

The 16 estimates available are those of the mean I, the six main effects, $(B_1 + 12 + 35 + 46)$, $(B_2 + 13 + 25)$, $(B_1 B_2 + 15 + 23)$, $(14 + 26)$, $(16 + 24)$, $(34 + 56)$, (higher-order interactions), (higher-order interactions), and $(36 + 45)$, if block variables do not interact.

It might be slightly better to block the design by $B_1 = 134$, and $B_2 = 234$. This will provide estimates of the mean I, the six main effects, $(B_1 B_2 + 12 + 35 + 46)$, $(13 + 25)$, $(15 + 23)$, $(14 + 26)$, $(16 + 24)$, $(34 + 56)$, B_1, B_2, $(36 + 45)$, if block variables do not interact. Two of the block effects have been disassociated from the sets of two-factor interactions. However, the first design will provide two estimates of higher-order interactions which can be used to estimate σ^2, which the second design does not provide.

5.7.   Generators (including blocking) are

$$I = 1235 = 1246 = 12B_1 = 13B_2 = 14B_3.$$

Defining relation:

$$I = 1235 = 1246 = 12B_1 = 13B_2 = 14B_3 = 3456 = 35B_1 = 25B_2$$

$$= 2345B_3 = 46B_1 = 2346B_2 = 26B_3 = 23B_1 B_2 = 24B_1 B_3 = 34B_2 B_3$$

$$= 123456B_1 = 1456B_2 = 1356B_3 = 15B_1 B_2 = 1345B_1 B_3 = 1245B_2 B_3$$

$$= 1346B_1 B_2 = 16B_1 B_3 = 1236B_2 B_3 = 1234B_1 B_2 B_3 = 2456B_1 B_2$$

$$= 2356B_1 B_3 = 56B_2 B_3 = 45B_1 B_2 B_3 = 36B_1 B_2 B_3 = 1256B_1 B_2 B_3.$$

If block variables do not interact, and interactions between three or more factors are ignored, the following estimates are available: $I, 1, 2, 3, 4, 5, 6$, $(B_1 + 12 + 35 + 46)$, $(B_2 + 13 + 25)$, $(B_3 + 14 + 26)$, $(B_1 B_2 + 15 + 23)$, $(B_1 B_3 + 16 + 24)$, $(B_2 B_3 + 34 + 56)$, (third-order interactions), (third-order interactions), $(B_1 B_2 B_3 + 36 + 45)$.

When there are eight blocks it is not possible to improve on this estimation setup. For example, if we set $B_1 = 134$, $B_2 = 234$, then $B_1 B_2 = 12$. The allocation of B_3 to any of the remaining columns aliases a main effect with a block effect. Thus, the three blocking variables must be allocated to columns that usually estimate two-factor interaction combinations. (Note that, as in a previous example, if we want *four* blocks, we can set $\mathbf{B}_1 = 134$ and $\mathbf{B}_2 = 234$ without trouble.)

5.8. (a)   The defining relation for the initial eight runs is $\mathbf{I} = 1234 = 235 = 145$. When a design is folded over, the signs of each variable are reversed, so that the defining relation for the second eight-run fraction is $\mathbf{I} = 1234 = -235 = -145$. Combining the two designs (symbolically, "averaging" the two defining relations) gives $\mathbf{I} = 1234$ and thus the design is a 2^{5-1} of resolution IV.

(b)   A 2_V^{5-1} design $\mathbf{I} = 12345$ is usually better because all main effects and two-factor interactions can be separately estimated.

5.9.   First write down an eight-run resolution III design in six factors, for example, $\mathbf{I} = 124 = 135 = 236$. Add a second fraction in which only the sign of variable 1 is switched, namely, $\mathbf{I} = -124 = -135 = 236$. The defining relations of the two designs are, respectively,

$$\mathbf{I} = 124 = 135 = 236 = 2345 = 1346 = 1256 = 456,$$

$$\mathbf{I} = -124 = -135 = 236 = 2345 = -1346 = -1256 = 456.$$

The joint design thus has $\mathbf{I} = 236 = 2345 = 456$, and any two of the three indicated words will generate it. The main effect of 1 is associated with four- or five-factor interactions and all two-factor interactions involving factor 1 are associated with three- or four-factor interactions, achieving the desired aim.

5.10.   The two defining relationships are

$$\mathbf{I} = 124 = 135 = 236 = 2345 = 1346 = 1256 = 456,$$

$$\mathbf{I} = 124 = -135 = -236 = -2345 = -1346 = 1256 = 456,$$

with common parts

$$\mathbf{I} = 124 = 1256 = 456.$$

If we ignore interactions involving three or more factors, the alias structure is thus

I	13
1 = 24	15 = 26
2 = 14	16 = 25
3	23
4 = 12 = 56	34
5 = 46	35
6 = 45	36

We see that variable 3 and all its two-factor interactions are estimable individually. This happens because changing the signs of **1** and **2** produces exactly the same effect here as changing the sign of **3**.

5.11. We first set down a 2^4 factorial design in variables **1, 2, 3**, and **4**. Clearly the columns **12, 13, 14, 23, 24, 123** *cannot* be used to accommodate new variables or these interactions would not be individually estimable. Five choices remain, namely **34, 124, 134, 234**, and **1234**. If, for example, we choose **5 = 34** and **6 = 1234** the defining relation is **I = 345 = 12346 = 1256**. However, this aliases **5** with **34** and **15** with **26**, and so is no good. No other pairs of choices work, either; such a design does not exist.

(*Note*: Some readers interpret this question to mean that two factor interactions not specifically mentioned are zero. If this *were* true, setting **5 = 34** and **6 = 1234** *would* work.)

5.12–5.14. Solutions are implicit in the questions.

5.15. The design is a 2_{IV}^{4-1} generated by $I = -1234$. We can estimate:

$$1 - 234 \leftarrow \quad 3.75, \qquad\qquad I \leftarrow 106.875,$$

$$2 - 134 \leftarrow \quad 5.75, \qquad 12 - 34 \leftarrow \quad 1.75,$$

$$3 - 124 \leftarrow \quad 0.26, \qquad 13 - 24 \leftarrow \quad 1.25$$

$$4 - 123 \leftarrow -0.25, \qquad 14 - 23 \leftarrow -1.25.$$

Standard error (effect) $= 4\sigma^2/8 = 1$. Factors 1 and 2 appear to be the only influential ones. It is, of course, possible that a pair of interactions may be canceling each other out. One possibility for four extra runs is to do a 2^2 factorial in factors 1 and 2, with factors 3 and 4 held fixed. This would provide a recheck on the main effects of 1 and 2 and give an estimate of the interaction 12 (and so of 34 when combined with the results above).

5.16. This 2_{III}^{5-2} design has defining relation $I = 123 = 345 = 1245$. Ignoring interactions of three or more factors, we can estimate the following:

$$1 + 23 \qquad \leftarrow \quad 5.375, \qquad 5 + 34 \leftarrow \quad 0.125,$$

$$2 + 13 \qquad \leftarrow \quad 0.625, \qquad 14 + 25 \leftarrow \quad 6.125,$$

$$3 + 12 + 45 \leftarrow -1.125, \qquad 15 + 24 \leftarrow -0.375,$$

$$4 + 35 \qquad \leftarrow \quad 7.125, \qquad\qquad I \leftarrow \quad 20.1875.$$

Pure error $s^2 = 1.8125$; standard error(effects) $= (s^2/4)^{1/2} = 0.673$. The important effects are $(1 + 23)$, $(4 + 35)$, $(14 + 25)$. A likely possibility is that factors 1 and 4 and the interaction 14 are large. A 2^2 factorial in 1 and 4 would be a useful four-run followup to check this. Other factors would be held constant.

If eight new runs were permissible, the fold-over design $\mathbf{I} = -\mathbf{123} = -\mathbf{345} = \mathbf{1245}$, when combined with the original design would give clear estimates of all main effects and two-factor interactions except for $(12 + 45)$, $(14 + 25)$ and $(15 + 24)$, assuming all interactions of three or more factors are zero.

5.17.   This cannot be done. $\mathbf{B}_1$ and $\mathbf{B}_2$ must be set equal to interactions with at least three letters, or blocks are confounded with two-factor interactions immediately. However, if we choose, say, $\mathbf{B}_1 = \mathbf{123}$, then also $\mathbf{B}_1 = \mathbf{123} \, (-\mathbf{12345}) = -\mathbf{45}$. Similarly, a choice of $\mathbf{B}_1 = \mathbf{1234}$ implies $\mathbf{B}_1 = -\mathbf{5}$.

5.18.   (a)   The $(\mathbf{B}_1, \mathbf{B}_2) = (+, +)$ block is defined by

$$\mathbf{I} = \mathbf{1234567} = \mathbf{1357} = \mathbf{1256}$$

$$(= \mathbf{246} = \mathbf{347} = \mathbf{2367} = \mathbf{145}).$$

In other words, we have a 2_{III}^{7-3} design which we can generate from initial columns $\mathbf{1}$, $\mathbf{2}$, $\mathbf{3}$, and $\mathbf{4}$ (the basic 2^4 design) by choosing $\mathbf{5} = \mathbf{14}$, $\mathbf{6} = \mathbf{24}$, and $\mathbf{7} = \mathbf{34}$.

(b)   Estimates available consist of the average response $\bar{y} \rightarrow I$, and those for

$1 + 45$,	$12 + 56$,
$2 + 46$,	$13 + 57$,
$3 + 47$,	$16 + 25$,
$4 + 26 \ + 37 + 15$,	$17 + 35$,
$5 + 14$,	$23 + 67$,
$6 + 24$,	$27 + 36$.
$7 + 34$,	

5.19.   (a)   Ozzie wanted to find a resolution III design which would be such that variables $\mathbf{3}$ and $\mathbf{5}$ would never attain their $+$ levels simultaneously. However, every resolution III design gives a 2^2 factorial (possibly replicated) in any two variables. Hence any eight run resolution III design will give a (replicated) 2^2 factorial in variables $\mathbf{3}$ and $\mathbf{5}$, which means that there will be at least one run for which variables $\mathbf{3}$ and $\mathbf{5}$ attain their $+$ levels simultaneously.

(b)   The principal generating relation is $\mathbf{I} = \mathbf{124} = \mathbf{135} = \mathbf{236}$. If we do not rename the variables, the only members of this family which can be used are $\mathbf{I} = \pm \mathbf{124} = -\mathbf{135} = \pm \mathbf{236}$. There are four such fractions. If we are allowed to rename the variables, the question becomes: for what members of the family can we find three variables that do not attain their $+$ levels simultaneously? We see, for example, that in the fraction with generating relation $\mathbf{I} = -\mathbf{124} = \mathbf{135} = \mathbf{236}$ our requirement will be satisfied for the three variables $\mathbf{1}$, $\mathbf{2}$, and $\mathbf{4}$. In fact, if we put a minus sign in front of *any* word in the principal generating relation, our requirement will be satisfied for the three variables in that word. Therefore, the *only* member of the family which *cannot* be used is that associated with the principal generating relation.

(c) Defining relation: $I = 124 = -135 = 236 = -2345 = 1346 = -1256 = -456$

$$1 + 24 - 35 \leftarrow 40,$$
$$2 + 14 + 36 \leftarrow -10,$$
$$3 - 15 + 26 \leftarrow 20,$$
$$4 + 12 - 56 \leftarrow 90,$$
$$5 - 13 - 46 \leftarrow -60,$$
$$6 + 23 - 45 \leftarrow 30,$$
$$34 - 25 + 16 \leftarrow 0.$$

(d) One possibility is to switch all the signs and separate out all the main effects from combinations of two-factor interactions. Ozzie was more interested in learning about Miss Freeny's interactions, however, than he was in learning about all the main effects. The fact that the estimates that have the largest magnitude involve either the main effect of Miss Freeny or her interactions gives justification to his decision to switch the sign of variable **4** in the second fraction. (Note that a similar argument could be made for switching the sign of only variable **5** in the second fraction. However, we have to face the fact that Ozzie was far more interested in Miss Freeny than in the gypsy band.)

The defining relation for the second fraction is

$$I = -124 = -135 = 236 = 2345 = -1346 = -1256 = 456.$$

The calculation matrix for the second fraction is

I	1	2	3	4	5	6	123	y
+	−	−	−	−	−	+	−	135
+	+	−	−	+	+	+	+	165
+	−	+	−	+	−	−	+	285
+	+	+	−	−	+	−	−	175
+	−	−	+	−	+	−	+	205
+	+	−	+	+	−	−	−	195
+	−	+	+	+	+	+	−	295
+	+	+	+	−	−	+	+	145

Design matrix

The estimates obtained from the second fraction are

$$1 - 24 - 35 \leftarrow -60,$$
$$2 - 14 + 36 \leftarrow 50,$$
$$3 - 15 + 26 \leftarrow 20,$$
$$4 - 12 + 56 \leftarrow 70,$$
$$5 - 13 + 46 \leftarrow 20,$$
$$6 + 23 + 45 \leftarrow -30,$$
$$-34 - 25 + 16 \leftarrow 0.$$

(e) Combining the two pieces gives the following estimates:

$1 - 35 \leftarrow -10,$	$24 \leftarrow 50,$
$2 + 36 \leftarrow 20,$	$14 \leftarrow -30,$
$3 - 15 + 26 \leftarrow 20,$	$\text{error} \leftarrow 0,$
$4 \leftarrow 80,$	$12 - 56 \leftarrow 10,$
$5 - 13 \leftarrow -20,$	$46 \leftarrow 40,$
$6 + 23 \leftarrow 0,$	$45 \leftarrow -30,$
$-25 + 16 \leftarrow 0,$	$34 \leftarrow 0.$

(f) The second fraction was a wise choice. The main effect 4 of Miss Freeny is large and so are four of her five interactions 24, 14, 46, and 45.

(g) Miss Freeny's presence clearly has a positive effect on bar business, an effect that is enhanced when there are chips available for her to pass around (effect $24 \leftarrow 50$), the lights are low (effect $14 \leftarrow -30$), the band is not playing (effect $45 \leftarrow -30$), and Dick is the bartender (effect $46 \leftarrow 40$). Obviously, the customers were much more attentive to Miss Freeny when the lights were low and they were not distracted by the band; they consumed more potato chips, became thirstier, and drank faster. Unfortunately, Tom the bartender, who was generally more alert than Dick, became so enamored of Miss Freeny that he seldom noticed the empty glasses, hence the negative 46 interaction. Note that this particular explanation is just one of several that are consistent with the data. To appreciate that it was probably the right one, you had to be there.

5.20. The sum of squares of a linear combination $c'y$ is defined as $(c'y)^2/c'c$. Because all columns of signs are orthogonal, this calculation can be applied to each column individually, with $c'y = CP$ value and $c'c = 28$. The sums of squares are, in order, 4758.04, 2820.04, 322.32, 9325.75, 1093.75, 972.32, 1275.75, 6150.89, 1620.32, 1414.32, 92.89, 1744.32,

3680.04, 6696.04, 52,202.89, 3680.04, 12,814.32, 1302.89, 85.75, 16,660.32, 488.89, 7264.32, 996.04, 1068.89 totaling 138,531.14 with 24 df. The analysis of variance table is then as follows.

Source	SS	df	MS
Mean	335,508.04	1	
Effects	138,531.14	24	
Residual	4,273.82	3	$1424.61 = s^2$
Total	478,313.00	28	

On the basis of our *a priori* assumption, we shall transfer, to the residual, the smallest eight sums of squares.

Source	SS	df	MS
16 larger SS	132,896.58	16	
Residual	9,908.38	11	900.76
Total, corrected	142,804.96	27	

If we use the upper 5% point of the $F(1, 11)$ distribution as a point of comparison, an individual 1 df sum of squares would be declared "large" if

$$\frac{(SS/1)}{900.76} > 4.84,$$

namely if SS > 4360. By this criterion, factors 1, 4, 8, 14, 15, 17, 20, and 22 are designated effective. [This conclusion is the one reached by the original author, who added that seven of these factors were confirmed as effective in subsequent work. The eighth (17) "was not studied further."]

5.21. We note that the replicated runs are the 2^{3-1} design for which $x_1 x_2 x_3 = 1$, so that pairs like (b_1, b_{23}) will not be independently estimated. If we fit the full factorial model, the **Z** matrix has column heads 1, x_1, x_2, x_3, $x_1 x_2$, $x_1 x_3$, $x_2 x_3$, $x_1 x_2 x_3$ and is a 12×8 matrix. The solution

$\mathbf{b} = (\mathbf{Z}'\mathbf{Z})^{-1}\mathbf{Z}'\mathbf{y}$ is then

$$
\begin{bmatrix}
12 & & & & & & & & 4 \\
 & 12 & & & & & & 4 & \\
 & & 12 & & & & 4 & & \\
 & & & 12 & 4 & & & & \\
 & & & 4 & 12 & & & & \\
 & & 4 & & & 12 & & & \\
 & 4 & & & & & 12 & & \\
4 & & & & & & & & 12
\end{bmatrix}^{-1}
\begin{bmatrix}
615 \\
137 \\
-5 \\
11 \\
13 \\
45 \\
47 \\
209
\end{bmatrix}
=
\begin{bmatrix}
51.125 \\
11.375 \\
-1.875 \\
0.625 \\
0.875 \\
4.375 \\
0.125 \\
0.375
\end{bmatrix}
$$

Source	SS	df	MS	F
b_0	31,518.75	1	(Same	—
b_1	1,564.08	1	as SS's)	125.13
b_2	2.08	1		0.17
b_3	10.08	1		0.81
$b_{12}\vert b_3$	8.17	1		0.65
$b_{13}\vert b_2$	204.17	1		16.33
$b_{23}\vert b_1$	0.17	1		0.01
$b_{123}\vert b_0$	1.50	1		0.12
Residual (pure error)	50.00	4	$s_e^2 = 12.50$	
Total	33,359.00	12		

Note that the residual here consists only of the pure error. For all the b's, $\hat{V}(b) = 0.09375(12.50) = 1.172$, so that standard error $(b) = 1.08$.

5.22. Solution is implicit in the questions asked.

5.23. (Wu and Meyer, 1965).

$$\hat{y} = 6.637 + 0.202x_1 + 0.169x_2 + 0.025x_3 + 0.038x_4 + 0.015x_5.$$

Source	SS	df	MS	F
b_1	1.3069	1		
b_2	0.9147	1		
b_3	0.0194	1		
b_4	0.0451	1		
b_5	0.0070	1		
Lack of fit	0.0443	10	0.00443	11.25
Pure error	0.0063	16	0.00039	
Total, corrected	2.3437	31		

Lack of fit is indicated. $R^2 = 0.9784$. Thus, most of the variation in the data is explained by the model, but the variation that remains unexplained is still large compared to the pure error variation shown between pairs of runs at the same experimental conditions.

5.24. It is not possible, because no 2_{IV}^{9-5} exists. However suppose we use, for example, the 2_{III}^{9-5} given by $I = 1235 = 1246 = 1347 = 2348 = 129$. The full defining relation contains 26 more "words" including the following (only) of length three: **789**. Thus all main effects are estimated clear except for those in the combinations:

$$1 + 29$$

$$2 + 19$$

$$7 + 89$$

$$8 + 79$$

$$9 + 12 + 78.$$

So we would need six two-factor interactions to be zero before we could obtain clear main effects.

5.25. We first obtain the defining relation of this eight-run design by multiplying together all possible combinations of generators. Rearranging this gives $I = 235 = 136 = 1256$ $[= 124 = 1237 = 1345 = 2346 = 347 = 157 = 267 = 1234567 = 3567 = 1467 = 2457 = 456]$. All the "words" in the square brackets contain either a 4 or a 7 which we drop. Moreover **1256** is the product of **235** and **136**. Thus the reduced design is generated by $I = 235 = 136$ so that $5 = 23$ and $6 = 13$ columns can be obtained from an initial 2^3 design in variables **1**, **2**, and **3**. The design is a 2_{III}^{5-2} in variables **1, 2, 3, 5,** and **6**.

5.26. Inspection shows the design to be a 2_{III}^{5-2} generated by $I = 123 = 145$ $(= 2345)$. If we ignore interactions involving three or more factors, we obtain estimates as follows.

$$
\begin{aligned}
1 + 23 + 45 &\leftarrow -1.25^*, \\
2 + 13 &\leftarrow -0.25, \\
3 + 12 &\leftarrow \ \ 0.25, \\
4 + 15 &\leftarrow -1.75^*, \\
5 + 14 &\leftarrow -2.25^*, \\
24 + 35 &\leftarrow -0.25, \\
25 + 34 &\leftarrow \ \ 0.25.
\end{aligned}
$$

The larger estimates (marked with asterisks) involve factors 1, 4, and 5 and/or their interactions. A sensible next step would be to run a second design in which the signs of all variables were switched, namely $I =$

$-123 = -145 \; (= 2345)$. This would enable the main effects to be split off separately.

5.27. (a) The full defining relation is

$$I = -126 = 135 = -237 = -1234 = -2356 = 1367 = 346$$

$$= -1257 = -245 = 147 = 567 = 1456 = -2467 = 3457$$

$$= -1234567.$$

I	1	2	3	$4 = -123$	$5 = 13$	$6 = -12$	$7 = -23$	y
+	−	−	−	+	+	−	−	46.1
+	+	−	−	−	−	+	−	55.4
+	−	+	−	−	+	+	+	44.1
+	+	+	−	+	−	−	+	58.7
+	−	−	+	−	−	−	+	56.3
+	+	−	+	+	+	+	+	18.9
+	−	+	+	+	−	+	−	46.4
+	+	+	+	−	+	−	−	16.4

Ignoring all interactions involving three or more factors, we obtain

$$l_I = \;\;\; 42.8 \to \text{average}, \qquad l_4 = \;\; -0.5 \to 4 + 36 - 25 + 17,$$
$$l_1 = -10.9 \to 1 - 26 + 35 + 47, \qquad l_5 = -22.8 \to 5 + 13 - 24 + 67,$$
$$l_2 = \;\; -2.8 \to 2 - 16 - 37 - 45, \qquad l_6 = \;\; -3.2 \to 6 - 12 + 34 + 57,$$
$$l_3 = -16.6 \to 3 + 15 - 27 + 46, \qquad l_7 = \;\;\;\; 3.4 \to 7 - 23 + 14 + 56,$$

Conclude: Relatively large effects are

$$1 - 26 + 35 + 47,$$

$$3 + 15 - 27 + 46,$$

$$5 + 13 - 24 + 67.$$

All decrease packing time. This can be explained in several ways, any of which *might* be correct:

1, 3, 5 are all important main effects, each causing a decrease in packing time.
Or 1 and 3 are the only important main effects, and they interact.
Or 1 and 5 are the only important main effects, and they interact.
Or 3 and 5 are the only important main effects, and they interact.
Or 1 is the only important main effect, but it interacts with 3 and 5.

Or 3 is the only important main effect, but it interacts with 1 and 5.

Or 5 is the only important main effect, but it interacts with 1 and 3.

Other conclusions might be drawn, but these seem most plausible. Since this experiment leads us to at least seven fairly reasonable but differing conclusions, the advantage of performing the second fractional factorial becomes apparent.

(b) For the second fraction

$$I = 126 = -135 = 237 = -1234 = -2356 = 1367 = -346 = -1257$$

$$= 245 = -147 = -567 = 1456 = -2467 = 3457 = 1234567.$$

I	1	2	3	4 = -123	5 = -13	6 = 12	7 = 23	y
+	+	+	+	−	−	+	+	41.8
+	−	+	+	+	+	−	+	40.1
+	+	−	+	+	−	−	−	61.5
+	−	−	+	−	+	+	−	37.0
+	+	+	−	+	+	+	−	22.9
+	−	+	−	−	−	−	−	34.1
+	+	−	−	−	+	−	+	17.7
+	−	−	−	+	−	+	+	42.7

$l_1' = 37.2 \to$ average,

$l_1' = -2.5 \to 1 + 26 - 35 - 47$,

$l_2' = -5.0 \to 2 + 16 + 37 + 45$,

$l_3' = 15.8 \to 3 - 15 + 27 - 46$,

$l_4' = 9.2 \to 4 - 36 + 25 - 17$,

$l_5' = -15.6 \to 5 - 13 + 24 - 67$,

$l_6' = -2.3 \to 6 + 12 - 34 - 57$,

$l_7' = -3.3 \to 7 + 23 - 14 - 56$.

(c) Combining (a) and (b) we obtain estimates as follows

$\frac{1}{2}(l_1 + l_1') = 40.0 \to$ average,

$\frac{1}{2}(l_1 + l_1') = -6.7 \to 1$,

$\frac{1}{2}(l_2 + l_2') = -3.9 \to 2$,

$\frac{1}{2}(l_3 + l_3') = -0.4 \to 3$,

$\frac{1}{2}(l_4 + l_4') = 4.4 \to 4$,

$\frac{1}{2}(l_5 + l_5') = -19.2 \to 5$,

$\frac{1}{2}(l_6 + l_6') = -2.8 \to 6$,

$\frac{1}{2}(l_7 + l_7') = 0.1 \to 7$,

$l_1 - l_1' = 5.6 \to$ block effect,

$\frac{1}{2}(l_1 - l_1') = -4.2 \to -26 + 35 + 47$

$\frac{1}{2}(l_2 - l_2') = 1.1 \to -16 - 37 - 45$

$\frac{1}{2}(l_3 - l_3') = -16.2 \to 15 - 27 + 46$

$\frac{1}{2}(l_4 - l_4') = -4.9 \to 36 - 25 + 17$

$\frac{1}{2}(l_5 - l_5') = -3.6 \to 13 - 24 + 67$

$\frac{1}{2}(l_6 - l_6') = -0.5 \to -12 + 34 + 57$

$\frac{1}{2}(l_7 - l_7') = 3.4 \to -23 + 14 + 56$

Conclude: The two most important main effects are 1 and 5 and their interaction is probably the cause of the large effect 15 − 27

+ 46 ← − 16.2. The factory owner would therefore do well to keep the foreman on the job at all times in addition to piping in music.

The experiment also indicates that women pack faster than men, that high temperature slows packing, and that packers 25 or over pack faster than those under 25.

5.28. The missing run from the incomplete design can be "estimated" by making the tentative assumption that a (relatively) high-order interaction estimate is zero. We first complete the design by adding a row of signs $+ - + - - +$ which is chosen to give four pluses and four minuses in each column.

We now inspect the design matrix to find three columns that form a full factorial, for example, columns 1, 2, and 3. It is then seen that $4 = 123$, $5 = 23$, and $6 = 13$. Only the 12 column remains for estimation purposes and it will lead to an estimate of the combination of effects $(12 + 34 + 56)$ assuming all interactions involving three or more factors are zero. All the other six columns provide estimates of (main effect plus two-factor interactions) combinations.

The estimate of $(12 + 34 + 56)$ would be zero if the missing observation m were such that

$$\tfrac{1}{4}(29 - 27 - 42 - 0 + 30 + 0 + 39 - m) = 0,$$

that is, if $m = 29$.

We now evaluate effects other than $(12 + 34 + 56)$ which we have "sacrificed." The defining relation of the design is

$$\mathbf{I = 1234 = 235 = 136 = 145 = 246 = 1256 = 3456.}$$

We obtain

$$1 + 36 + 45 \leftarrow \quad 1.0,$$

$$2 + 35 + 46 \leftarrow \; - 21.0,$$

$$3 + 25 + 16 \leftarrow \; - \; 1.5,$$

$$4 + 15 + 26 \leftarrow \; - \; 8.5,$$

$$5 + 23 + 14 \leftarrow \quad 0.5,$$

$$6 + 24 + 13 \leftarrow \; - 19.5.$$

In view of the comparatively large sizes of the second, fourth, and sixth of these, a sensible course of action would be to "switch all signs," to perform the 2_{III}^{6-3} fraction $\mathbf{I = 1234 = - 235 = - 136}$, and to estimate main effects clear by combining both fractions.

CHAPTER 6

Note: For steepest ascent answers, $\mathbf{b} = (b_1, b_2, \ldots, b_k)'$ and does not contain b_0.

6.1. A suitable coding is $x_1 = (X_1 - 90)/10$, $x_2 = (X_2 - 20)/10$.

$$\hat{y} = 11.60 - 8.5x_1 + 6x_2.$$

Points on the path of steepest ascent have coordinates $(x_1, x_2) = (-8.5\theta, 6\theta)$ for various θ values or $(X_1, X_2) = (90 - 85\theta, 20 + 60\theta)$. Thus, all the runs given are on this path; take $\theta = 0.3, 0.5, 0.6, 0.7, 0.55, 0.65$ in turn.

$$\hat{y} = 70 - 1.5x_1, \text{ where now } x_1 = (X_1 - 39)/4.25, \ x_2 = (X_2 - 56)/3.$$

Source	SS	df	MS	F
b_1	9	1		
b_2	0	1		
Lack of fit	41	2	20.50	30.75
Pure error	2	3	0.67	
Total, corrected	52	7		

The lack of fit test F value exceeds the upper 5% F point of 9.55. It would be best to fit a second-order surface at this stage, as a diagram (not provided here) well illustrates.

6.2. $\hat{y} = 12 - 8x_1 + 5.5x_2$. Points $(x_1, x_2) = (-8\theta, 5.5\theta)$, that is, points $(X_1, X_2) = (99 - 80\theta, 17 + 110\theta)$ are on the path of steepest ascent. No value of θ gives $(59, 67)$ so the answer is no.

6.3. $\mathbf{a} = \begin{bmatrix} 4 \\ 5 \\ 6 \end{bmatrix}$ $\mathbf{b} = \begin{bmatrix} 0.25 \\ 1.00 \\ 0.50 \end{bmatrix}$ $\mathbf{e} = \mathbf{b} - \left(\dfrac{\mathbf{a'b}}{\mathbf{a'a}}\right)\mathbf{a} = \begin{bmatrix} -0.22 \\ 0.42 \\ -0.20 \end{bmatrix}$.

Restriction is $4(35 + 5x_1) + 5(15 + 5x_2) + 6(11 + 5x_3) = 371$, or $4x_1 + 5x_2 + 6x_3 = 18 = -a_0$.

$$\lambda_0 = -\frac{a_0}{\mathbf{a'b}} = 2,$$

$$\mathbf{x}_0 = \lambda_0 \begin{bmatrix} b_1 \\ b_2 \\ b_3 \end{bmatrix} = \begin{bmatrix} 0.5 \\ 2.0 \\ 1.0 \end{bmatrix}, \quad \mathbf{X}_0 = 5\mathbf{x}_0 + \begin{bmatrix} 35 \\ 15 \\ 11 \end{bmatrix} = \begin{bmatrix} 37.5 \\ 25.0 \\ 16.0 \end{bmatrix}.$$

In coded units, the new path is

$$\begin{bmatrix} 0.5 \\ 2.0 \\ 1.0 \end{bmatrix} + \mu \begin{bmatrix} -0.22 \\ 0.42 \\ -0.20 \end{bmatrix} \quad \text{or} \quad \begin{bmatrix} 35 + 5(0.5 - 0.22\mu) \\ 15 + 5(2.0 + 0.42\mu) \\ 11 + 5(1.0 - 0.20\mu) \end{bmatrix}$$

in uncoded units. This gives the point with coordinates (36.40, 27.10, 15.00) if we set $\mu = 1$. So the answer is yes.

6.4. The restriction is $(x_1 + 8) + x_2 + 14 + (\frac{1}{2}x_3 + 7) = 26.4$, or $2.6 + x_1 + x_2 + \frac{1}{2}x_3 = 0$. $\mathbf{a} = (1, 1, \frac{1}{2})'$, $\mathbf{b} = (-4.10, -9.25, 4.90)'$. $c = \mathbf{a'b}/\mathbf{a'a} = -10.9/2.25 = -4.844$. Thus $\mathbf{e} = \mathbf{b} - c\mathbf{a} = (0.744, -4.406, 7.322)$. Also, $\lambda_0 = -2.6/(-10.9) = 0.2385$, so that the point at which the path of steepest ascent is modified is $\lambda_0(-4.10, -9.25, 4.90)$, or $(-0.978, -2.206, 1.169)$. Thus the modified path is on $x_1 = 0.978 + \mu(0.744)$, $x_2 = -2.206 + \mu(-4.406)$, $x_3 = 1.169 + \mu(7.322)$, or on $(X_1, X_2, X_3) = \{7.022 + \mu(0.744), 11.794 - \mu(4.406), 7.584 + \mu(3.611)\}$. A value of $\mu = 0.114$ gives the point (7.11, 11.29, 8.00); so the answer is yes.

6.5. We have $\mathbf{a} = (1, 1, 1, 1)'$, $\mathbf{b} = (1, 2, 3, 2)'$, $\mathbf{e} = \mathbf{b} - 2\mathbf{a} = (-1, 0, 1, 0)'$. Now, $\lambda_0 = -(-4)/8 = \frac{1}{2}$. Thus in coded units the amended path has coordinates $(x_1, x_2, x_3, x_4) = \frac{1}{2}(1, 2, 3, 2) + \mu(-1, 0, 1, 0)$, or $(X_1, X_2, X_3, X_4) = \{5(0.5 - \mu) + 5, 5(1) + 5, 5(1.5 + \mu) + 5, 5(1) + 5\} = \{7.5 - 5\mu, 10, 12.5 + 5\mu, 10\}$. When $\mu = 2$, this results in the point $(-2.5, 10, 22.5, 10)$. So the answer is yes.

6.6. If $\hat{y}_1 > 21.5$, we must use (x_1, x_2) values "northeast" of the line $1.5 = x_1 + x_2$ which passes through the points (1.5, 0) and (0, 1.5). If $\hat{y}_2 < 54$, we must use (x_1, x_2) values "southwest" of the line $4 = 4x_1 + 2x_2$ which passes through the points (1, 0) and (0, 2). The lines intersect at $(\frac{1}{2}, 1)$. A pointed segment "northwest" of that point is the production region of (x_1, x_2). For example, all points $(0, x_2)$ where $1.5 \le x_2 \le 2$ lie in the production region.

CHAPTER 7

7.1. (a) $y = \beta_0 + \beta_1 x_1 + \beta_2 x_2 + \varepsilon$.

$$\mathbf{Z'Zb} \equiv \begin{bmatrix} 9 & 0 & 0 \\ 0 & 6 & 0 \\ 0 & 0 & 6 \end{bmatrix} \begin{bmatrix} b_0 \\ b_1 \\ b_2 \end{bmatrix} = \begin{bmatrix} 24.93 \\ 2.42 \\ -1.61 \end{bmatrix} \equiv \mathbf{Z'y}.$$

$$\hat{y} = 2.7700 + 0.4033x_1 - 0.2683x_2.$$

Source	SS		df	MS	F
b_0		69.0561	1		
b_1 ⎫	0.9761 ⎫	1.4081	2	0.7041	$F = 115$
b_2 ⎭	0.4320 ⎭				
Residual		0.0363	6	$s^2 = 0.0061$	
		70.5005	9		

The residuals are, in order, 10^{-2} times $-7, 4, 6, 5, 2, -14, -3, 2, 5$ (sum 0), compared with $s = 0.08$, a figure that is roughly of the same order as the standard error of the residuals.

(b) The F statistic for regression, $F = 115$ is highly significant and $R^2 = 1.4081/(1.4081 + 0.0363) = 0.9749$. All of the residuals are of the same order as $s = 0.08$ except for -0.14 which is less than $2s$, and there are no remarkable features exhibited by the residuals plot in the (x_1, x_2) plane.

Overall we conclude that a good fit to the data has been obtained.

7.2. (a)

$$\mathbf{Z} = \begin{bmatrix} 1 & -1 & -1 & 1 & 1 & 1 \\ 1 & 0 & -1 & 0 & 1 & 0 \\ 1 & 1 & -1 & 1 & 1 & -1 \\ 1 & -1 & 0 & 1 & 0 & 0 \\ 1 & 0 & 0 & 0 & 0 & 0 \\ 1 & 1 & 0 & 1 & 0 & 0 \\ 1 & -1 & 1 & 1 & 1 & -1 \\ 1 & 0 & 1 & 0 & 1 & 0 \\ 1 & 1 & 1 & 1 & 1 & 1 \end{bmatrix},$$

$$\mathbf{Z'Z} = \begin{bmatrix} 9 & 0 & 0 & 6 & 6 & 0 \\ & 6 & 0 & 0 & 0 & 0 \\ & & 6 & 0 & 0 & 0 \\ & & & 6 & 4 & 0 \\ & & & & 6 & 0 \\ \text{symmetric} & & & & & 4 \end{bmatrix}.$$

(b) Replace x_1^2 by $x_1^2 - \frac{2}{3}$ as in hint so that the new column is $\frac{1}{3}, -\frac{2}{3}, \frac{1}{3}, \frac{1}{3}, -\frac{2}{3}, \frac{1}{3}, \frac{1}{3}, -\frac{2}{3}, \frac{1}{3}$. Similarly, replacing x_2^2 by $x_2^2 - \frac{2}{3}$, we obtain a new column $\frac{1}{3}, \frac{1}{3}, \frac{1}{3}, -\frac{2}{3}, -\frac{2}{3}, -\frac{2}{3}, \frac{1}{3}, \frac{1}{3}, \frac{1}{3}$. Other columns of $\mathbf{Z}$ are as before. The "new $\mathbf{Z'Z}$" is now diagonal with diagonal entries 9, 6, 6, 2, 2, 4.

7.3. From the form described in Ex. 7.2(b) we obtain

$$\begin{bmatrix} b_0' \\ b_1 \\ b_2 \\ b_{11} \\ b_{22} \\ b_{12} \end{bmatrix} = \begin{bmatrix} \frac{1}{9} & & & & & \\ & \frac{1}{6} & & \mathbf{0} & & \\ & & \frac{1}{6} & & & \\ & & & \frac{1}{2} & & \\ & \mathbf{0} & & & \frac{1}{2} & \\ & & & & & \frac{1}{4} \end{bmatrix} \begin{bmatrix} 24.93 \\ 2.42 \\ -1.61 \\ -0.08 \\ 0.07 \\ -0.05 \end{bmatrix} = \begin{bmatrix} 2.7700 \\ 0.4033 \\ -0.2683 \\ -0.0400 \\ 0.0350 \\ -0.0125 \end{bmatrix}$$

Now $b_0' = b_0 + \frac{2}{3}b_{11} + \frac{2}{3}b_{22}$, so that $b_0 = 2.7733$. The fitted model is then $\hat{y} = 2.7733 + 0.4033x_1 - 0.2683x_2 - 0.0400x_1^2 + 0.0350x_2^2 - 0.0125x_1x_2$.

Source	SS		df		MS
b_0	69.0561		1		—
b_1	0.9761		1		0.9761
b_2	0.4320		1		0.4320
b_{11}	0.0032 ⎫		1 ⎫		
b_{22}	0.0025 ⎬ 0.0063		1 ⎬ 3		0.0021
b_{12}	0.0006 ⎭		1 ⎭		
Residual	0.0300		3		$s^2 = 0.0100$
Total	70.5005		9		

Clearly the second-order terms are not significant, so there is no point retaining them in the model. The first order test is the same as in Exercise 7.1.

7.4. In view of Exercise 7.3, we might expect to find that a transformation is unnecessary, and this is indeed the case. Numerical details follow.

x_1	x_2	y	$\hat{y}$	$y - \hat{y}$	$q = \hat{y}^2$	$q - \hat{q}$
-1	-1	2.57	2.6350	-0.0650	6.9432	
0	-1	3.08	3.0383	0.0417	9.2313	
1	-1	3.50	3.4416	0.0584	11.8446	
-1	0	2.42	2.3667	0.0533	5.6013	
0	0	2.79	2.7700	0.0200	7.6729	
1	0	3.03	3.1733	-0.1433	10.0698	
-1	1	2.07	2.0984	-0.0284	4.4033	
0	1	2.52	2.5017	0.0183	6.2585	
1	1	2.95	2.9050	0.0450	8.4390	

We obtain $\hat{q} = 7.8293 + 2.2343x_1 - 1.4864x_2$, with regression SS for this model of $\mathbf{a'Z'q} = 594.7792$. [Notation: $\mathbf{a'} = (7.8293, 2.2343, -1.4864)$, $\mathbf{Z}$ is the matrix of coefficients of $(1, x_1, x_2)$ as in Exercise 7.1, and $\mathbf{q}$ is the vector of q values.]

Thus, $\Sigma(q_u - \hat{q}_u)^2 = \Sigma q_u^2 - 594.7792 = 0.1366$. Also $\Sigma(y_u - \hat{y}_u)(q_u - \hat{q}_u) = \Sigma(y_u - \hat{y}_u)q_u = 0.0030$. Thus

$$\hat{\alpha} = \frac{(0.0030)}{0.1366} = 0.0220,$$

$$S_\alpha = \frac{(0.0030)^2}{0.1366} = 0.00007.$$

Table 7.6 can now be displayed:

Source	SS	df	MS	F
Transformation	0.00007	1	0.00007	0.01
Adjusted residual	0.03623	5	0.00725	
Residual	0.0363	6		

It is clear that no transformation is called for.

7.5. (a) $y = \beta_0 + \beta_1 x_1 + \beta_2 x_2 + \beta_{11} x_1^2 + \beta_{22} x_2^2 + \beta_{12} x_1 x_2$
$\qquad + \beta_{111} x_1^3 + \beta_{222} x_2^3 + \beta_{112} x_1^2 x_2 + \beta_{122} x_1 x_2^2 + \varepsilon.$

(b) In the appropriate Z matrix, the (iii) column $\equiv (i)$ column for $i = 1$ and 2 so that the β_{iii} are not separately estimable. However, the two (ijj) columns *cannot* be expressed as a linear combination of the other columns of Z and so both β_{122} and β_{112} *are* separately estimable.

(c) For the second-order model with third-order alternative, the alias matrix $A = (X_1'X_1)^{-1} X_1' X_2$ has the form

$$
\frac{1}{36}
\begin{bmatrix}
20 & 0 & 0 & -12 & -12 & 0 \\
0 & 6 & 0 & 0 & 0 & 0 \\
0 & 0 & 6 & 0 & 0 & 0 \\
-12 & 0 & 0 & 18 & 0 & 0 \\
-12 & 0 & 0 & 0 & 18 & 0 \\
0 & 0 & 0 & 0 & 0 & 9
\end{bmatrix}
\begin{bmatrix}
0 & 0 & 0 & 0 \\
6 & 0 & 0 & 4 \\
0 & 6 & 4 & 0 \\
0 & 0 & 0 & 0 \\
0 & 0 & 0 & 0 \\
0 & 0 & 0 & 0
\end{bmatrix}
=
\begin{bmatrix}
 & \mathbf{0} & & \\
1 & 0 & 0 & \frac{2}{3} \\
0 & 1 & \frac{2}{3} & 0 \\
 & \mathbf{0} & &
\end{bmatrix}.
$$

Thus b_1 estimates $\beta_1 + \beta_{111} + \frac{2}{3}\beta_{122}$, b_2 estimates $\beta_2 + \beta_{222} + \frac{2}{3}\beta_{112}$. The other estimates are unbiased.

(d) The question "Are your results consistent with what you found in (b)?" is a somewhat misleading one, and is asked in order to make this point: The alias structure does not in itself enable us to say whether or not a coefficient can be estimated. For example, β_1 is aliased with β_{111} (which *cannot* be estimated) and β_{122} (which *can* be estimated). This point is sometimes misunderstood.

7.6. (a) We use the formula $V(\mathbf{b}) = (Z'Z)^{-1}\sigma^2$ and then estimate σ^2 by $s^2 = 0.0061$ to get the estimated variances est $V(b_0) = 0.0061/9$, est $V(b_i) = 0.0061/6$, $i = 1, 2$. The confidence bands are given by

$$b_i \pm t(v, 1 - \tfrac{1}{2}\alpha)\{\text{est } V(b_i)\}^{1/2}$$

where v is the df of s^2 ($= 6$, here). See Eq. (3.12.9), page 76, for an alternative notation. For 95% limits $t(6, 0.975) = 2.45$ so that we obtain

$$\beta_0 : 2.77 \pm 0.06, \qquad \beta_1 : 0.40 \pm 0.08, \qquad \beta_2 : -0.27 \pm 0.08.$$

(b)   At $(x_1, x_2) = (0.7, 0.5)$, $\hat{y} = 2.9182$. est $V(\hat{y}_0) = x_0'(Z'Z)^{-1}x_0 s^2$

$$= (1, 0.7, 0.5) \begin{bmatrix} \frac{1}{9} & 0 & 0 \\ 0 & \frac{1}{6} & 0 \\ 0 & 0 & \frac{1}{6} \end{bmatrix} \begin{bmatrix} 1 \\ 0.7 \\ 0.5 \end{bmatrix} (0.0061) = 0.001430$$

$$= (0.0378)^2.$$

The 95% limits for the true mean value of y are thus: $2.9182 \pm 2.45(0.0378) = 2.92 \pm 0.09$.

(c)   The values of (β_1, β_2) in the 90% joint confidence region are bounded by the "ellipse" (which actually is a circle here)

$$[(b_1 - \beta_1), (b_2 - \beta_2)] \begin{bmatrix} 6 & 0 \\ 0 & 6 \end{bmatrix} \begin{bmatrix} b_1 - \beta_1 \\ b_2 - \beta_2 \end{bmatrix} = 2s^2 F(2, 6, 0.90);$$

see formula (3.12.5). Because the x_1 and x_2 columns are orthogonal to the column of 1's, no adjustment for the means of these columns is needed. This becomes

$$(\beta_1 - 0.4033)^2 + (\beta_2 - 0.2683)^2 = 2(0.0061)3.46/6 = 0.007035$$

$$= (0.084)^2,$$

that is, a circle of radius 0.084.

(d)   We can compare the correct 90% confidence region in (c) with the incorrect 90.25% region obtained by taking the intersection of the individual confidence intervals for β_1 and β_2. These latter are spread ± 0.078 about the estimates and so define a square of slightly smaller width than the diameter of the circle. On the whole, they would not mislead, mainly due to the lack of correlation between the estimates b_1 and b_2.

(e)

$$\text{In general } V(\hat{y}) = (1, x_1, x_2) \begin{bmatrix} \frac{1}{9} & 0 & 0 \\ 0 & \frac{1}{6} & 0 \\ 0 & 0 & \frac{1}{6} \end{bmatrix} \begin{bmatrix} 1 \\ x_1 \\ x_2 \end{bmatrix}$$

$$= \frac{1}{9} + \frac{1}{6}(x_1^2 + x_2^2).$$

This is obviously constant on circles $x_1^2 + x_2^2 = \rho^2$.

7.7. By using $(x_i^2 - \frac{2}{3})$ instead of x_i^2 in the full second-order model, we can do the least-squares fit in orthogonal form. (*Note*: This works here because of the 3^3 design used; it is not always possible to achieve orthogonalization in this way.)

The fitted model can be written

$$\hat{y} = 386.63 + 39.78x_1 + 33.89x_2 + 13.83x_3$$
$$- 205.50x_1^2 + 55.17x_2^2 - 32.67x_3^2$$
$$+ 46.42x_1x_2 - 21.17x_1x_3 + 2.50x_2x_3.$$

Residuals in order: 15, 31, -24, -49, -85, 55, 70, -64, 49,
                         51, 46, -84, -43, 35, 49, 53, -114, 7,
                         -63, 30, -3, -41, 87, -9, 6, 33, -41.

Check total of residuals is -3, which is zero within rounding error. The largest of these is -114, although it does not appear abnormally large on the overall plot of residuals. The calculations for a transformation test show SS(transformation) = 83 (1 df), SS(adjusted residual) = 76,066 (17 df) with a nonsignificant F. No transformation is needed. The 3^k algorithm leads to the following effects.

Final column of algorithm	Divisor	Mean square	Root mean square	Rank of rms	Effect name
7145	27	1,890,779	—	—	Mean
716	18	24,481	169	25	1
-3700	54	253,519	504	26	11
610	18	20,672	144	23	2
557	12	25,854	161	24	12
757	36	15,918	126	20	112
992	54	18,223	135	22	.22
-685	36	13,034	114	19	122
335	108	1,039	32	3	1122
249	18	3,445	59	15	3
-254	12	5,376	73	17	13
-804	36	17,956	134	21	113
30	12	75	9	1	23
-125	8	1,953	44	7	123
-291	24	3,528	59	16	1123
-348	36	3,364	58	14	223
217	24	1,962	44	8	1223
399	72	2,211	47	10	11223
-589	54	6,424	80	18	33
266	36	1,965	44	9	133
-292	108	789	28	2	1133
202	36	1,133	34	4	233
-265	24	2,926	54	12	1233
-479	72	3,187	56	13	11233
368	108	1,254	35	5	2233
413	72	2,369	49	11	12233
-541	216	1,355	37	6	112233

The normal plot of root mean squares indicates two things.

1. The larger effects seem to be associated with variables x_1 and x_2.
2. There appears to be bias in at least one observation.

If we take the seventeenth observation (which is associated with the largest residual) as being the possible culprit, we can replace it by a "missing value estimate" obtained in one of several possible ways. One way is to set equal to zero the highest order interaction (112233). This leads to an "estimate" for y_{17} of 497.25 instead of 362. The revised root mean squares are (3) 540 instead of 504, (4) 176, (6) 81, (7) 153, (9) 6, (19) 117 not 80, (21) 24, (22) 11, (24) 7, (25) 9, (27) set equal to zero and so dropped from the normal plot. The remainder are unchanged. A plot of the 25 revised root mean squares shows the same prominent effects as before. Further analysis is left to the reader, who should check the Davies source reference for features suppressed in our presentation of the data.

7.8. Application of the 3^k algorithm provides the figures below.

Final column of algorithm	Divisor	Mean square	Root mean square	Rank of rms	Effect name
8068	27	2,410,838	—	—	Mean
1243	18	85,836	293	26	1
−635	54	7,467	86	22	11
1048	18	61,017	247	25	2
−179	12	2,607	51	19	12
−371	36	3,823	62	21	112
−782	54	11,324	106	23	22
−95	36	251	16	12	122
253	108	593	24	16	1122
583	18	18,883	137	24	3
−31	12	80	9	7	13
13	36	5	2	3	113
56	12	261	16	13	23
−2	8	1	1	1	123
−130	24	704	27	17	1123
82	36	187	14	11	223
20	24	17	4	5	1223
52	72	38	6	6	11223
−437	54	3,536	59	20	33
−23	36	15	4	4	133
13	108	2	1	2	1133
−188	36	982	31	18	233
−44	24	81	9	8	1233
−104	72	150	12	10	11233
−206	108	393	20	15	2233
−146	72	296	17	14	12233
154	216	110	10	9	112233

The normal plot of the root mean squares shows up, as large effects (in order): 1, 2, 3, 22, 11, 112, 33, 12 and suggests that a second-order model in all three factors would do well. We can, in fact, obtain the fitted equation

$$\hat{y} = 367.48 + 69.06x_1 + 58.22x_2 + 32.39x_3$$

$$- 35.28x_1^2 - 43.44x_2^2 - 24.28x_3^2$$

$$- 14.92x_1x_2 - 2.58x_1x_3 + 4.67x_2x_3.$$

Most of these coefficients are found by dividing the first column of the table by the corresponding divisor, except for $b_0 = \bar{y} - \frac{2}{3}(b_{11} + b_{22} + b_{33})$. We can read off the SS from the MS column, because each MS has 1 df, to give this analysis of variance table:

Source	SS	df	MS	F
b_0	2,410,838	1	—	
First order	165,736	3	55,245	122.0
Second order $\|b_0$	25,275	6	4,213	9.3
Residual	7,709	17	$s^2 = 453$	
Total	2,609,558	27		

The major influences are clearly first order, although the quadratic contributions are significant also. The interaction terms are not large, comparatively.

7.9. Application of the 3^k algorithm followed by a plot on normal probability paper of root mean squares shows a well-behaved plot with largest effects being, in order, main effects 3, 1, and 2, followed after a gap, by interactions 13, 123, 23. If a second-order model is fitted we obtain

$$\hat{y} = 326.14 + 131.46x_1 + 109.43x_2 + 177.00x_3$$

$$- 28.32x_1^2 - 21.64x_2^2 + 32.74x_3^2$$

$$+ 43.58x_1x_2 + 75.47x_1x_3 + 66.03x_2x_3.$$

The first-order terms pick up $3,271,619/4,284,176 = 0.7637$ of the variation about the mean, while second-order terms with an extra SS (given b_0) of 471,778, pick up another 0.1101 for a total of 0.8738. However, a lack of fit test gives an F value of

$$\frac{\{295,990/17\}}{\{244,025/54\}} = 3.86$$

which exceeds the percentage point $F_{0.05}(17, 54) \approx 1.83$. In the circumstances, it would seem sensible to look for a transformation for y, as indications are that use of the model as fitted could lead to problems. (We do not pursue this here, but leave it to the reader.)

7.10. A full second-order model can be fitted to these data to give $\hat{y} = 219.55 + 5.42x_1 - 2.67x_2 + 75.92x_1^2 + 38.17x_2^2 + 7.13x_1x_2$ with anova table as below.

Source	SS	df	MS	F	
b_0	1,572,946.72	1			
b_1	352.08	1	352.08	1.57	not significant
b_2	85.33	1	85.33	0.38	not significant
$b_{11}|b_0$	23,053.87	1	23,053.87	102.97	significant
$b_{22}|b_0$	5,827.03	1	5,827.03	26.03	significant
b_{12}	403.13	1	403.13	1.80	not significant
Residual	2,686.84	12	$s^2 = 223.90$		
Lack of fit	1,085.34	3	361.72	2.03	not significant
Pure error	1,601.50	9	177.94		
Total	1,605,355.00	18			

$F(3,9,0.95) = 3.86$, $F(1,12,0.95) = 4.75$, and $F(1,12,0.99) = 9.33$. There is no apparent lack of fit, and the response is primarily one of pure second order in x_1 and x_2. The residuals are given below in the same order downward as the corresponding observations (and then left to right in sequence).

$$x_1 = \begin{cases} -1 \\ 0 \\ 1 \end{cases} \quad \begin{array}{ccc} -1 & x_2 = 0 & 1 \\ 0,6 & -26,0 & 14,7 \\ -2,12 & 22,-13 & 3,-22 \\ -15,-1 & 7,9 & -8,6. \end{array}$$

The residuals add to -1 (≈ 0) and show no abnormalities worth remarking. An analysis via the 3^k algorithm can also be performed. The conclusions are the same; the (122) interaction turns out to be the third largest effect.

7.11. (Partial solution).

(b) $\hat{y} = 36.25 + 3.487x_1 + 4.621x_2$. The analysis is included in (c).

(c) $\hat{y} = 7.5 + 3.487x_1 + 4.621x_2 - 5.813x_1^2 - 5.063x_2^2 + 7.5x_1x_2$.

Source	SS	df	MS	F
b_0	15,768.75	1		
First order	268.15	2	134.08	13.53
Second order$|b_0$	542.63	3	180.88	18.25
Lack of fit	42.47	3	14.16	2.50
Pure error	17.00	3	5.67	
Total	16,639.00	12		

No lack of fit is apparent, so we use $s^2 = (42.47 + 17.00)/(3 + 3) = 9.912$ in the F tests for first and second order. Both are significant at the $\alpha = 0.01$ level. When just first order is fitted, lack of fit is shown; the "second order$|b_0$" SS would be in the lack of fit for the smaller model and the lack of fit F value would be the (significant at $\alpha = 0.05$) value

$$F = \frac{\{(542.63 + 42.47)/6\}}{5.67} = 17.20.$$

(d) Adding 37 to observations 7–12 simulates the situation when there is a block effect between the two sets of runs 1–6 (block 1) and 7–12 (block 2). The blocking variable is orthogonal to the model, so exactly the same predicted equation as in (c) emerges except that b_0 is increased from 36.25 to 54.75. Only 2 df now exist for pure error, 1 df for each pair *within* a block, with pure error sum of squares $\frac{1}{2}(45 - 41)^2 + \frac{1}{2}(42 - 46)^2 = 16$. The "lost df" for pure error becomes a blocks df with SS

$$\frac{B_1^2}{6} + \frac{B_2^2}{6} - \frac{G^2}{12} = 3502.08 \ (1 \text{ df})$$

where $B_1 = 226$ is the sum of the observations in the first block and $B_2 = 431$ is the sum for the second block. $G = B_1 + B_2 = 657$ is the grand total. $SS(b_0) = 35,970.75$ now, and total SS $= 40,331$. The lack of fit SS $= 31.39$, and lack of fit is not significant. The SS for first- and second-order$|b_0$ are the same. The basic conclusions are unchanged.

7.12. $\hat{y} = 75 + 2.06x_1 + 0.85x_2 - 4x_1x_2$.

Source	SS	df	MS	F
b_0	78,750.0	1		
b_1, b_2	39.8	2	19.9	33.33
$b_{11}, b_{22}, b_{12}|b_0$	64.0	3	21.3	35.50
Blocks	350.0	1	350.0	583.33
Lack of fit	0.2	3	0.067	0.07
Pure error	4.0	4	1.0	
Total	79,208.0	14		

There is no lack of fit; $s^2 = 4.2/7 = 0.6$, with 7 df. The blocks difference is large, and worth removing. Both first and second order are significant, but second order is all interaction b_{12}. (See Section 13.7.)

7.13. *All 25 observations*:

$$\hat{y}_{(25)} = 80.59 - 2.940x_1 + 2.12x_2 - 2.157x_1^2 - 1.257x_2^2 - 3.350x_1x_2.$$

Source	SS	df	MS	F
b_1, b_2	656.90	2	328.45	30.13
$b_{11}, b_{22}, b_{12}\|b_0$	1558.61	3	519.54	47.66
Residual	207.05	19	10.90	
Total, corrected	2422.56	24		

Both first- and second-order terms are needed.

Nine selected observations:

$$\hat{y}_{(9)} = 85.78 - 2.917x_1 + 2.250x_2 - 4.021x_1^2 - 3.396x_2^2 - 7.250x_1x_2.$$

Source	SS	df	MS	F
b_1, b_2	162.83	2	81.42	69.19
$b_{11}, b_{22}, b_{12}\|b_0$	436.53	3	145.51	123.66
Residual	3.53	3	1.18	
Total, corrected	602.89	8		

Both first- and second-order terms are needed.

Table A7.13 shows that the nine point fit does very well at predicting at the nine data locations but tends to extrapolate badly at the corners due to its greater curvature. Because the 9-point fit is performed with six parameters and is over a smaller region than the 25-point fit, these features are not surprising ones.

7.14. $\hat{y} = 150.84 + 2.47x_1 - 2.06x_2 - 16.07x_1^2 - 16.30x_2^2 + 6.00x_1x_2.$

Source	SS	df	MS	F
b_1, b_2	87.90	2	43.95	2.24
$b_{11}, b_{22}, b_{12}\|b_0$	3328.62	3	1109.54	56.57
Lack of fit	100.67	3	33.557	5.92
Pure error	17.00	3	5.667	
Total, corrected	3534.19	11		

TABLE A7.13. **Fitted values from the 25- and 9-point fits. The two $\hat{y}_{(9)}$ columns are, respectively, data point predictions and predictions at the unused 16 other locations.**

y	$\hat{y}_{(25)}$	$\hat{y}_{(9)}$	$\hat{y}_{(9)}$
60	55		28
67	68		55
76	78	76	
82	85		89
91	90		95
64	65		52
71	75	72	
80	81		85
90	86	91	
90	87		90
68	71	68	
77	77		80
87	81	86	
83	81		85
77	80	77	
71	73		75
80	75	80	
76	75		79
70	73	70	
65	68		55
72	70		75
69	69		73
64	66	64	
59	60		48
55	52		26

Lack of fit is not significant; $s^2 = 117.67/6 = 19.612$. Second-order terms are significant, so the nonsignificant first-order terms must be retained.

The contours of $V = NV\{\hat{y}(\mathbf{x})\}/\sigma^2$ are given by $V = 2.979 + 0.630x_1^2x_2^2 - 1.383r^2 + 1.579r^4$, where $r^2 = x_1^2 + x_2^2$; see the diagram. Note that, if the axial points had been at distance $2^{1/2} = 1.414$, rather than 1.5, the contours would have been perfect circles.

[Additional analysis:

$$\frac{(\hat{y}_{\max} - \hat{y}_{\min})}{(ps^2/n)^{1/2}} = \frac{(150.84 - 107.94)}{\{6(19.612)/12\}^{1/2}} = 13.70,$$

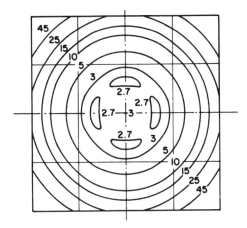

(see Section 8.2 for explanation), so that canonical reduction is justified. The reduced form is $\hat{y} = 150.98 - 13.183\tilde{X}_1^2 - 19.187\tilde{X}_2^2$, a simple maximum. The center of the system is close to the origin at $(0.067, -0.051)$. See Chapter 11.]

7.15. A cautious approach is needed here. A first reaction is to fit a second-order surface to the 20 readings as though they were 20 independent data points. (If this is done, lack of fit is exhibited when the lack of fit mean square is compared with a "pure error" mean square based on 11 df and computed from the eight pairs and the four center point readings.) However, the presentation of the data makes it clear that we simply have 10 pairs of duplicate ash readings which probably do not reflect the full experimental error within a pair; moreover, members of each pair are not independent. For this reason, we shall analyze instead the 10 averages (of the pairs), and treat only the averages for runs 1 and 8 as true pure error repeat runs.

The averages are 212, 90, 214, 91, 219.5, 55, 171.5, 165, 139.5, 160.5. The fitted second-order model is $\hat{y} = 149.915 - 57.853x_1 - 0.794x_2 - 5.780x_1^2 + 7.998x_2^2 - 0.250x_1x_2$.

Source	SS	df	MS	F
b_1, b_2	28,454.5	2	14,227.3	174.14
$b_{11}, b_{22}, b_{12}\lvert b_0$	970.3	3	323.4	3.96 not significant
Lack of fit	106.3	3	35.4	0.16 not significant
Pure error	220.5	1	220.5	
Total, corrected	29,751.6	9		

Now, no lack of fit is indicated; $s^2 = (106.3 + 220.5)/(3 + 1) = 81.7$. Evidently, second-order terms are not needed. If we refit a first-order model, we find that the x_2 term is not needed either, and the model $\hat{y} = 151.8 - 57.853x_1$ appears to adequately describe the data. The residuals indicate nothing to remark upon. Substituting for x_1 in terms of temperature gives the fitted equation $\hat{y} = 541.7 - 0.50307T$. If temperature were difficult to control, so that the settings were unreliable, there would clearly be a problem in predicting dry ash value from this equation. The data thus support the authors' conclusion (p. 44) that "It is recommended that dry ash measurements be eliminated as a control procedure except under special circumstances."

7.16. The two surfaces are very similar, indicating stability from run to run. Two possible directions for increased yields are revealed.

7.17. $\hat{y} = 58.35 - 5.482x_1 - 23.81x_2 + 0.1332x_1^2 + 3.882x_2^2 + 0.3849x_1x_2$.

Source	SS	df	MS	F
First order$\|b_0$	564.66	2	282.33	423.92
Second order$\|b_0$, first	1.13	3	0.377	0.57 not significant
Lack of fit	2.48	3	0.827	1.28 not significant
Pure error	17.50	27	0.648	
Total, corrected	585.77	35		

Lack of fit is not significant, and the residuals are unremarkable; $s^2 = (2.48 + 17.50)/(3 + 27) = 0.666$. A full second-order model is not needed. Refitting terms of up to first order provides $\hat{y} = 36.466 - 3.2337x_1 - 6.9835x_2$. Now, $s^2 = 21.11/33 = 0.64$, and standard error$(b_0) = 1.406$, standard error$(b_1) = 0.1101$, standard error$(b_2) = 0.6116$. All coefficients are necessary and this is the model of choice.

7.18. If $G > 1.055$ is substituted in equation (2), we obtain $K < 113.2$. Moreover, we must have $N \geq 0$ and $P \geq 0$.

Figure A7.18 is reproduced from the first source reference with slight modifications. Figure A7.18a shows the surface cut by the $K = 114$ plane, (114 being the next integer up from 113.2 and close enough for our purposes). Figure A7.18b shows the $N = 0$ plane intersecting the surface. Figure A7.18c shows both planes with the $N = 0$ plane regarded as transparent. It is clear from the diagram that the response decreases as N (> 0) increases and as K (< 114) decreases; this is easily substantiated by seeing that the derivatives of $\hat{Y}$ with respect to N and to $-K$ (note the minus direction, which will provide some sign changes) are both < 0 for the ranges of (N, P, K) involved. Thus the maximum response, subject to our limiting conditions, must occur at the intersection of the planes $N = 0$ and $K = 114$. Substituting these values into (1) we obtain

$$\hat{Y}(0, P, 114) = 191.4 + 4.55P - 0.0936P^2 \qquad (4)$$

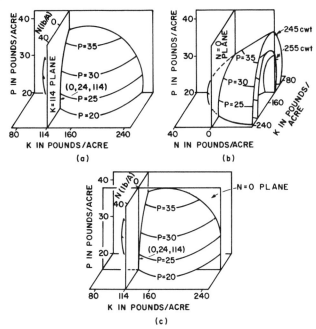

FIGURE A7.18. Constant yield surfaces showing combinations of N, P, and K fertilization which produce a particular yield. (a) and (c) show a "front" view of the 245 cwt. potatoes per acre surface, while (b) shows a "side" view of the 245 and 255 cwt. potatoes per acre surfaces. The planes K = 114 and N = 0 shown cut off the region within which the conditional maximum yield must be sought.

which is maximized at $P = -4.55/\{2(-0.0936)\} = 24.3$ or 24 if we round off to the nearest integer. So the best (N, P, K) combination is at the point $(0, 24, 114)$.

We now check if the point $(0, 24, 114)$ lies on or within (3). At the point in question, the left hand side of (3) has the value 2.669 compared with the right-hand side value of 2.618, so the ratio $(2.669/2.618)^{1/2} = 1.01$ indicates the best point is just outside (1% of the radius outside) the region of experimentation.

Note that the best point on (3) cannot be obtained by substituting $N = 0$, $K = 144$ into (3), because the line so defined does not intersect the sphere (3); the solution is imaginary. We would need to minimize (1) subject to (i) left-hand side of (3) $\leq c^2$, (ii) $N \geq 0$, (iii) $K \leq 114$. This is not an easy problem to solve but could be tackled via the Kuhn–Tucker theorem; see, for example, *Optimization by Vector Space Methods*, by D. G. Luenberger, published by J. Wiley & Sons, Inc., in 1969, pp. 247–253.

7.19. First set of data: $\hat{y} = 14.93 + 0.44046x_1 - 0.49573x_2 - 0.97305x_3 + 0.00994x_3^2 - 0.11748x_1x_3 + 0.07219x_2x_3 + 0.01164x_1x_3^2 - 0.00541x_2x_3^2$.

The pure error is very tiny; $s_e^2 = 0.003018$ (14 df). The residual MS = 0.0076 (40 df). Lack of fit MS = $0.262/26 = 0.0101$. Lack of fit $F = 0.0101/0.003018 = 3.347$ which is significant. However the model explains $R^2 = 0.9989$ of the variation about the mean. This is one of those puzzling situations where we might suspect the pure error is "too small." Derringer concludes that while the model "could stand further refinement," it remains a useful predictive tool.

Second set of data: $\hat{y} = 14.19 + 0.5766x_1 - 0.4868x_2 - 0.7944x_3 - 0.0040x_3^2 - 0.2036x_1x_3 + 0.0187x_2x_3 + 0.0206x_1x_3^2 + 0.0030x_2x_3^2$. The pure error is $s_e^2 = 0.00528/5 = 0.00106$, again tiny. The residual MS = $0.0399/11 = 0.0036$. Lack of fit MS = $(0.0399 - 0.00528)/(11 - 5) = 0.00577$ so that the lack of fit $F = 0.00577/0.00106 = 5.44$, which is significant. However, once again, the model explains nearly all the variation in the data about the mean, because $R^2 = 0.9991$. Thus similar conclusions to those for the first set of data may be drawn. It will be found that the model $\hat{y} = 14.26 + 0.5766x_1 - 0.3397x_2 - 0.8304x_3 - 0.2036x_1x_3 + 0.0206x_1x_3^3$, picks up an amount $R^2 = 0.9977$ of the variation about the mean. The terms dropped are the ones that would have been nonsignificant in the full model had not the lack of fit present rendered all the t tests invalid.

(*Note*: Because Derringer has used uncoded ξ_1 and ξ_2 values, his equations are transformed versions of ours, obtained by setting $\xi_1 = 15x_1 + 60$, $\xi_2 = 15x_2 + 21$. For applications of the models, see the original paper.)

7.20. The region $\hat{y}_1 \geq 80$ is inside an ellipse. The region $\hat{y}_2 \leq 76$ is outside a pair of saddle contours. There is no common region. Other predictor variables would

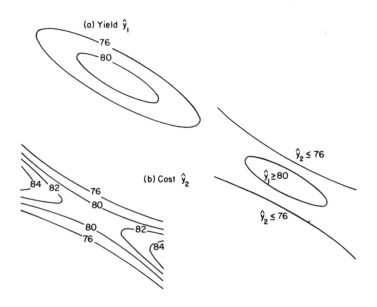

have to be considered to try to find a common region elsewhere. The diagrams make the above solution clear, but it can also be shown algebraically.

7.21. $\hat{y} = 15.8 + 0.525x_1 + 0.925x_2 + 3.808x_3$
$+ 0.019x_1^2 - 0.194x_2^2 + 0.075x_3^2$
$+ 0.7x_1x_2 + 0.025x_1x_3 - 0.025x_2x_3.$

No lack of fit is indicated and second-order terms are not significant as shown in the analysis of variance table. The reduced first-order equation is $\hat{y} = 15.8 + 0.525x_1 + 0.925x_2 + 3.808x_3$. Recomputing $s^2 = 12.359/20 = 0.61795$ we obtain the standard errors of b_1, b_2, b_3 as $(0.61795/16)^{1/2} = 0.20$, so that all terms are needed.

Source	SS	df	MS	F
b_0	5915.760	1		
First order	375.201	3	125.067	302.09
Second order$|b_0$	6.558	6	1.093	2.64 not significant
Lack of fit	3.308	5	0.661	2.39 not significant
Pure error	2.493	9	0.277	
Total	6303.320	24		

$s^2 = (3.308 + 2.493)/(5 + 9) = 0.414$; this value is used to test for first- and second-order terms in the table.

7.22. The blocking variable is not needed as it removes very little of the variation, so we drop it immediately. The second-order fit is

$$\hat{y} = 6.5961 + 0.2641x_1 + 0.1735x_2 + 0.0781x_3$$

$$-0.00403x_1^2 - 0.01402x_2^2 - 0.02507x_3^2$$

$$-0.02128x_1x_2 - 0.01258x_1x_3 - 0.01821x_2x_3.$$

The corresponding analysis of variance table is

Source	SS	df	MS	F
First order	3.3905	3	1.1302	
Second order$|b_0$	0.0320	6	0.0053	
Lack of fit	0.0081	20	0.0004	3.84
Pure error	0.0019	18	0.0001	
Total, corrected	3.4325	47		

There is lack of fit. Nevertheless, $R^2 = 0.9971$, implying that the second-order model explains nearly all the variation in the data about the mean $\bar{y}$. In such circumstances one would often use the model as a predicting tool, in spite of the lack of fit. The technically correct procedure, however, is to see if the lack of fit can be accounted for, and then to recast the model to remove it, unless it is found that the pure error is artifically small for some explainable reason.

The predicted values at the five given locations, obtained from the five corresponding $\exp(\hat{y})$ values, are 838, 707, 661, 846, and 625.

[Additional analysis:

As an exercise, we reduce the equation to the canonical form $\hat{y} = 6.088 - 0.0342\,\tilde{X}_1^2 - 0.0117\tilde{X}_2^2 + 0.0027\tilde{X}_3^2$ which is of type B (and relatively close to type C) in Table 11.1. The transformation $\tilde{\mathbf{X}} = \mathbf{M}'(\mathbf{x} - \mathbf{x}_S)$ is such that $\mathbf{x}_S = (-16.15, 19.37, -1.43)'$, so that the center is remote, and

$$\mathbf{M} = \begin{bmatrix} 0.3486 & -0.3978 & 0.8487 \\ 0.5321 & -0.6614 & -0.5286 \\ 0.7716 & 0.6359 & -0.0189 \end{bmatrix}$$

See Chapters 10 and 11 for additional explanation.]

7.23. Solution is implicit in the question.

7.24. (e) This is a personal decision. Provided there are no misgivings about the patterns exhibited by the residuals, the first-order model would be a realistic choice.

7.25. A second-order model fitted to the y_1 data reveals a significant lack of fit F value of $(101.26/5)/(0.57/3) = 106.59$. The pure error is extremely small. If it were judged artifically so, and if the results of the lack of fit test were ignored, so that $s^2 = 12.73$ were used to test coefficients, then it would be found that x_3 can be dropped and a quadratic in x_1 and x_2 provides a good fit. The appropriate technical decision, however, is that the model does not fit. For the y_2 data, the lack of fit $F = 20.97$, and $s^2 = 0.4382$, but exactly the same comments otherwise apply.

7.26. The design is not orthogonally blocked, because the ratios $\sum x_{iu}^2/(\text{number of points})$ are $16/20$ for runs 1–20, and $8/8$ for runs 21–28, and are unequal. The blocking variable suggested is orthogonal to the column of 1's arising from b_0, however. The fitted second-order model is

$$\hat{y} = 5.8142 + 0.0899z + 0.4688x_1 + 0.2655x_2 + 0.4572x_3$$

$$+ 0.1488x_4 - 0.0565x_1^2 - 0.0190x_2^2 - 0.0335x_3^2$$

$$- 0.0198x_4^2 - 0.0423x_1x_2 - 0.0838x_1x_3 - 0.0284x_1x_4$$

$$- 0.0340x_2x_3 - 0.0094x_2x_4 - 0.0284x_3x_4.$$

The model shows lack of fit, but the tiny amount of variation in the four center points is worrying, and perhaps cause for suspicion. The authors of the source paper argued that "most of the high residuals [high compared with

Source	SS × 10³	df	MS	F
b_0	910,940.2	1		
Blocks	23.5	1	23.5	
First order|blocks	12,514.7	4	3,128.68	
Second order|b_0, blocks	292.4	7	41.77	
Lack of fit	43.6	12	3.63	22.85
Pure error	0.5	3	0.16	
Total	923,814.9	28		

pure error variation]... are of no practical importance in engineering practice. Therefore, the second-order model is considered to be adequate for the present investigation." A check of the standardized residuals shows that four of the five largest in absolute value lie at star points (trials 21, 22, 27, 28). It appears possible that the x_1 and x_4 ranges are too wide for a quadratic fit to be fully appropriate. Thus with limitations in those directions, cautious interpretation of the fitted surface may not be too bad. All the x's appear to be needed.

7.27.

Estimated coefficient	$\hat{y}_0$	$\hat{y}_3$	$\hat{y}_6$	$\hat{y}_9$
b_0	74.18	35.00	25.42	17.7
b_1	46.35	−9.93	−45.69	−22.65
b_2	−13.89	−125.34[a]	−115.14[a]	−236.54[a]
b_3	196.48[a]	283.87[a]	288.19[a]	445.06[a]
b_4	−86.13[a]	−141.79[a]	−123.75[a]	−241.75[a]
b_5	47.90[a]	150.72[a]	147.29[a]	117.96
b_{11}	−13.56	29.0	43.81	81.7
b_{22}	39.19	114.0	113.3	148.0
b_{33}	103.54	37.8	38.8	93.6
b_{44}	1.69	76.5	73.8	122.3
b_{55}	22.33	52.91	56.96	65.80
b_{12}	36.17	101.07	128.44[a]	173.06[a]
b_{13}	53.90[a]	−1.01	−35.62	−14.44
b_{14}	−54.26[a]	47.26	80.00	24.31
b_{15}	21.74	−54.76	−47.81	−163.81[a]
b_{23}	3.89	−106.80[a]	−92.81	−226.62[a]
b_{24}	−28.58	39.30	29.69	153.37
b_{25}	46.68	−63.05	−97.50	−37.87
b_{34}	−106.30[a]	−169.86[a]	−161.87[a]	−279.75[a]
b_{35}	27.52	152.99[a]	142.19[a]	105.25
b_{45}	−29.14	−87.99	−110.94	−158.50

[a]Significant at $\alpha = 0.05$ level.

The table shows the estimated coefficients of second-order fitted models. The pattern of significant coefficients varies from response to response. It would now be possible to look at these four response functions directly or via canonical reduction.

The response data cover several orders of magnitude, however, suggesting that it would be worth checking for a possibly useful transformation, by the Box and Cox (1964) technique, for example. The consequent investigation shows that the $\ln y$ transformation is a reasonable choice for all four responses. Moreover, we find, from the analyses of the four second-order fits to $\ln y$, by examining the values of the t statistics stemming from the individual coefficients, that:

(a) Only terms in x_2, x_3, and x_5 appear to be needed for responses y_6 and y_9.

(b) Only terms in x_2, x_3, x_5, and $x_3 x_4$ appear to be needed for responses y_0 and y_3.

When we actually refit the model $\hat{y} = b_0 + b_2 x_2 + b_3 x_3 + b_5 x_5 + b_{34} x_3 x_4$, the term b_{34} becomes nonsignificant only for y_9 so, for uniformity, we reproduce the fitted equations with b_{34} present in all of them in a second table.

Estimated coefficient	$\ln \hat{y}_0$	$\ln \hat{y}_3$	$\ln \hat{y}_6$	$\ln \hat{y}_9$
b_0	4.4080	4.4579	4.3890	4.8040
b_2	-0.2665	-0.5076	-0.5449	-0.6175
b_3	1.4376	1.5302	1.6209	1.5608
b_5	0.3688	0.5899	0.5581	0.3257
b_{34}	-0.3605	-0.3915	-0.3457	-0.3003
R^2	0.919	0.904	0.877	0.878

Examination of the 20 estimated coefficients in this second table reveals a remarkable similarity in the coefficients line by line, and clearly calls for further conjecture and checking, preferably done with the specialist knowledge of the experimenters. An obvious question is whether or not the same model would provide a satisfactory fit to all 108 data points. Performing this fit we obtain

$$\ln \hat{y} = 4.51473 - 0.48413 x_2 + 1.53737 x_3 + 0.46064 x_5 - 0.34953 x_3 x_4$$

with $R^2 = 0.870$, a remarkably good fit in the circumstances. The analysis of variance table is as shown. All coefficients are highly significant.

Source	SS	df	MS	F
b_2	16.876	1	16.876	53.87
b_3	170.173	1	170.173	543.16
b_5	20.370	1	20.370	65.02
b_{34}	7.819	1	7.819	24.96
Residual	32.267	103	0.3133	
Total, corrected	247.505	107		

The restriction $800 \leq y_t \leq 1000$ translates into $6.68461 \leq \ln y_t \leq 6.90776$. This implies that

$$2.16988 \leq -0.48413x_2 + 1.53737x_3 + 0.46064x_5 - 0.34953x_3x_4$$

$$\leq 2.39303.$$

Points $(x_1, x_2, x_3, x_4, x_5)$ that lie between the two surfaces so defined are satisfactory ones. (For $y_t \leq 200$, we need $\ln y_t \leq 5.29832$, with similar calculations.) The model underpredicts all the y_t values at $(1, 1, 1, -1, 1)$, which is somewhat surprising. Of the 108 standardized residuals, only five exceed the range $(-2, 2)$ and these are shown in the table below, exhibiting a tendency to overprediction in four of the five cases. Further consideration needs to be given to these details. Nevertheless, all things considered, the five-parameter model does extremely well in predicting the 108 observations.

Coded x's					Value of t			
x_1	x_2	x_3	x_4	x_5	0	3	6	9
-1	-1	1	-1	1	-2.04			
0	0	1	0	0		-2.32	-2.53	
0	1	0	0	0			-2.39	
1	1	1	1	-1				2.43

It is, of course, possible that some suitable mechanistic modeling of the coefficients β_0, β_2, β_3, β_5, and β_{34} in terms of time would be even more effective. This extra pickup of additional variation would, of course, have a price in terms of extra parameters. We remark only that the assumption of quadratic functions, for example, $\beta_{2t} = \beta_{20} + \beta_{21}t + \beta_{21}t^2$, and so on, for the other coefficients, is remarkably ineffective. The reader may wish to consider other alternatives

Note: In view of the presence of the $x_3 x_4$ interaction in the final model, it might be appropriate to refit, but with an x_4 term added (x_3 is already in). One should also examine two-way tables of ln y averages at the four (± 1, ± 1) locations in the (x_3, x_4) space. A two-way table can be constructed for each response individually, and for all four together, and the nature of the $x_3 x_4$ interaction can thus be examined. We leave this further exploration to the reader.

7.28.

Estimated regression coefficients for cubic models

Estimate	$\hat{y}_1$	$\hat{y}_2$	$\hat{y}_3$	$\hat{y}_4$	$\hat{y}_5$	$\hat{y}_6$
b_0	878.34	2363.59	3803.84	666.43	10.81	15.29
b_1	220.93	708.20	582.95	-45.01	1.40	-0.60
b_2	23.02	89.74	38.47	-28.56	-2.37	-2.21
b_{11}	-8.70	1.41	-123.08	-9.17	-0.85	0.97
b_{22}	-73.40	-180.19	-203.64	5.42	0.95	5.28
b_{12}	40.17	79.03	-182.25	-22.30	-0.01	1.19
b_{111}	-42.72	-194.00	-267.07	4.01	0.05	0.29
b_{222}	87.90	184.58	114.60	-4.65	-0.90	-3.93
b_{112}	117.30	257.44	109.97	-14.29	1.16	-1.63
b_{122}	-103.34	-280.67	-237.00	19.09	-0.52	-0.10
a_1	-20.01	-87.17	15.66	7.14	1.34	0.93
a_2	-172.86	-411.44	-297.16	14.29	0.47	-0.07

The specific values of b_0, a_1, a_2 will depend on the particular blocking variable setup selected.

F ratios for lack of fit

Model order	y_1	y_2	y_3	y_4	y_5	y_6
First	40.7^b	37.9^b	13.7^a	3.5	8.5^a	110.8^b
Second	22.2^b	21.4^b	4.9	1.5	5.6	21.4^b
Third	1.9	4.4	2.2	1.3	3.4	5.3

[a] Lack of fit significant at $\alpha = 0.05$ level.
[b] Lack of fit significant at $\alpha = 0.01$ level.

The test indications are that a cubic equation is needed for responses 1, 2, and 6, a quadratic equation is adequate for responses 3 and 5 and a planar equation is adequate for response 4. (The conclusions in the paper are slightly different, reflecting practical experience of these responses.) Plots of the response surfaces may be found in the source reference.

CHAPTER 8

8.1. First plot θ as abscissa, ϕ as ordinate. Then turn the paper sideways, regard ϕ as abscissa now, and plot log ϕ as ordinate. Returning the paper to its original alignment we can now select θ, read up for ϕ, and then left for log ϕ.

8.2. Since $x^2 + y^2 = r^2$ we have $x^2 + y^2 + 4x + 6y = 12$, or $(x + 2)^2 + (y + 3)^2 = 25$, a circle with major axes parallel to the (x, y) axes, centered at $(-2, -3)$.

8.3. The fitted equation can be written as

$$\hat{y} + 3 \log \xi_3 + 5 \log \xi_4 = 0.703 - 0.517 \log \xi_3 - 0.166 \log \xi_4.$$

The anova table is

Source	SS	df	MS	F
$\hat{\gamma}$	0.2808	1		
$\hat{\delta}_1, \hat{\delta}_2 \vert \hat{\gamma}$	0.0151	2	0.007552	1.41 Not significant
Residual	0.1587	24	0.006611	
Total	0.4546	27		

So we cannot reject $H_0 : \delta_1 = \delta_2 = 0$, that is, $H_0 : \gamma_1 = -3$, $\gamma_2 = -5$.

8.4. For both y_1 and y_2, $w = \ln y$ is close to best. This gives rise to the fitted equations

$$\ln \hat{y}_1 = 4.395 + 0.513 x_1 - 0.210 x_2, \text{ and}$$

$$\ln \hat{y}_2 = 3.739 + 0.572 x_1 - 0.237 x_2.$$

Source	SS(y_1)	df$_1$	SS(y_2)	df$_2$
b_1, b_2	1.219	2	3.064	2
Lack of fit	0.018	4	0.026	6
Pure error	0.047	4	0.022	4
Total, corrected	1.284	10	3.112	12
	$s_1^2 = 0.0081$	8	$s_2^2 = 0.0049$	10

The equations explain $R_1^2 = 0.949$ and $R_2^2 = 0.985$ of the variation about respective means, there is no lack of fit indicated, and both responses need both x's. The mild irregularities in residuals plots do not seem to call for any remedial action.

8.5. $\hat{\lambda} = -0.8$. The 95% confidence interval is wide, $(-2.4, 0.7)$. Many alternative fits are thus possible. Derringer used $\lambda = 0$ to get $\ln \hat{y} = 4.468 + 0.179x_1 + 0.068x_2 + 0.032x_3$.

Source	SS	df	MS	F
b_1, b_2, b_3	0.3209	3	0.1070	54.87
Lack of fit	0.0150	7	0.0021	3.50
Pure error	0.0006	1	0.0006	
Total, corrected	0.3365	11		

There is no lack of fit and the regression is highly significant with an $R^2 = 0.954$. Predictor x_3 is not significant and, statistically, could be dropped. The ninth residual has a standardized value of -2.36 and may need additional investigation. For the F value for first order, the value $s^2 = (0.0150 + 0.0006)/(7 + 1) = 0.00195$ has been used.

CHAPTER 11

11.1. $\hat{y} = 0.410469 - 0.112913x_1 - 0.091438x_2 + 0.028766x_1^2 + 0.017328x_2^2 + 0.037593x_1 x_2$.

Source	SS	df	MS	F
b_0	6.572814	1		
First order	1.688804	2	0.8444	29.22*
Second order	0.853976	3	0.2846	9.85*
Residual	0.288915	10	0.0289	
Total	9.404509	16		

*Significant at $\alpha = 0.005$ level.

Both first- and second-order terms appear to be needed, and

$$\frac{(\hat{y}_{max} - \hat{y}_{min})}{(ps^2/n)^{1/2}} = 14.3,$$

so further interpretation appears justified. Canonical reduction produces the stationary point $x_S = (0.819, 1.750)'$, well within the experimental region and, around this point, $\hat{y} = 0.284 + 0.042694 \tilde{X}_1^2 + 0.003400 \tilde{X}_2^2$, these contours representing a simple minimum form. If low detergent consumption is desirable, the stationary point, which corresponds to temperatures of about (60°C, 62°C), is clearly best as predicted by $\hat{y}$, and elliptical contours of constant $\hat{y}$ values surround this point. A plot would show the canonical axes through $\mathbf{x}_S$ angled at 53.5° and 36.5° to the x_1 axis, elongated roughly 3.5 to 1 in the approximate NW–SE direction.

$$\tilde{X} = \begin{bmatrix} 0.803 & 0.595 \\ 0.595 & -0.803 \end{bmatrix} (\mathbf{x} - \mathbf{x}_S).$$

A plot of residuals from the quadratic fit versus z is unremarkable except that the z value (0.81) is well separated from the rest. The author of the source reference remarks that "if this lot had been scoured to approximately 0.6% grease as were the others, the detergent consumption at this point would have been even higher than indicated...". In view of the fact that at this point, $(-3, 3)$ in coded coordinates, the fitted surface is relatively high and well away from the minimum, the conclusions about desirable temperatures for low detergent consumption remain intact.

11.2.

	(a)	(b)
Center at	$(0.0266, 0.0971)$	$(1, 1)$
$\hat{y}_S$	60.015	0
λ_1	-2.128	1
λ_2	1.618	2
M	0.632 0.775	$-\frac{1}{2}\sqrt{3}$ $\frac{1}{2}$
	0.775 -0.632	$\frac{1}{2}$ $\frac{1}{2}\sqrt{3}$
Type	Saddle	Bowl

11.3. (a) Center is at $(2, 1)$. Roots are $\lambda = 12, -1$.

$$\hat{y} = 1 + 12\tilde{X}_1^2 - \tilde{X}_2^2, \qquad \tilde{X}_1 = \{3(x_1 - 2) + 2(x_2 - 1)\}/13^{1/2},$$

$$\tilde{X}_2 = \{-2(x_1 - 2) + 3(x_2 - 1)\}/13^{1/2}.$$

The surface is a saddle, broad in the $\tilde{X}_1$ direction and narrow in the $\tilde{X}_2$ direction. $\tilde{X}_1$ lies in the first quadrant of the $\{(x_1 - 2), (x_2 - 1)\}$ space and $\tilde{X}_2$ lies in the second quadrant.

(b) Center is at $(2, 0)$. Roots are $\lambda = 2, 200$

$$\hat{y} = -381 + 2\tilde{X}_1^2 + 200\tilde{X}_2^2, \qquad \tilde{X}_1 = \{(x_1 - 2) + x_2\}/2^{1/2},$$

$$\tilde{X}_2 = \{-(x_1 - 2) + x_2\}/2^{1/2}.$$

The surface is, analytically, a long, narrow, elliptical bowl but locally looks more like a stationary "river valley" rising steeply in the $\pm \tilde{X}_2$

directions and slightly in the $\pm \tilde{X}_1$ directions. $\tilde{X}_1$ and $\tilde{X}_2$ lie in the first and second quadrants of the $\{(x_1 - 2), x_2\}$ space.

(c) Center is at $(a, b) = (0.2125, 0.515)$. Roots are $\lambda = -2, 2$.

$$\hat{y} = 75.43775 - 2\tilde{X}_1^2 + 2\tilde{X}_2^2, \qquad \tilde{X}_1 = \{(x_1 - a) + (x_2 - b)\}/2^{1/2},$$

$$\tilde{X}_2 = \{-(x_1 - a) + (x_2 - b)\}/2^{1/2}.$$

The surface is a saddle, going down in the first and third quadrants of $\{(x_1 - a), (x_2 - b)\}$ and up in the second and fourth quadrants.

11.4.

	1	2	3	4
x_{1S}	0.978	0.714	0.910	119.0
x_{2S}	−1.043	1.214	0.741	99.0
$\hat{y}_S$	71.1	19.0	75.0	−206.2
λ_1	2.98	4.12	1.99	−1.01
λ_2	−0.98	−0.02	1.01	0.01
$\mathbf{m}_1$	0.937	0.277	−0.545	−0.634
	−0.350	0.961	0.838	0.773
$\mathbf{m}_2$	0.350	0.961	0.838	0.773
	0.937	−0.277	0.545	0.634

Note: $\tilde{X} = (\mathbf{m}_1, \mathbf{m}_2)'(\mathbf{x} - \mathbf{x}_S)$.
The general surface types are, respectively, (b), (d), (a), and (c) in Figure 11.1. Specific forms are determined by the sizes and signs of the λ's.

11.5. Note that the value $\alpha = 1.633 = (\frac{8}{3})^{1/2}$ would be an appropriate axial distance for orthogonal blocking if the design were to be divided into (cube plus four center points) plus (star plus two center points). We treat the design as a single block here, however.

	$\hat{Y}_1$	$\hat{Y}_2$	$\hat{Y}_3$	$\hat{Y}_4$
b_0	139.12	1261.11	400.4	68.91
b_1	16.493	268.15	−99.7	−1.41
b_2	17.880	246.50	31.4	4.32
b_3	10.906	139.48	−73.9	1.63
b_{11}	−4.009	−83.55	7.93	1.56
b_{22}	−3.447	−124.79	17.3	0.0577
b_{33}	−1.572	199.17	0.432	−0.317
b_{12}	5.125	69.38	8.75	−1.62
b_{13}	7.125	94.13	6.25	0.125
b_{23}	7.875	104.38	1.25	−0.25

		$\hat{Y}_1$	$\hat{Y}_2$	$\hat{Y}_3$	$\hat{Y}_4$
(3 df)	SS(b_i)	9,476.2	2,028,337	218,442.6	310.93
(6 df)	SS($b_{ij}|b_0$)	1,472.7	1,081,600	5,509.7	56.24
(5 df)	Lack of fit	188.0	1,030,452*	1,442.7	11.85
(5 df)	Pure error	126.8	49,941	2,800.0	4.21
(19 df)	Total, corrected	11,263.7	4,190,330	228,175.0	383.23

	$\hat{Y}_1$	$\hat{Y}_2$	$\hat{Y}_3$	$\hat{Y}_4$
$\dfrac{(\hat{Y}_{max} - \hat{Y}_{min})}{(ps^2/n)^{1/2}}$	25.2	*Lack of fit shown	22.7	18.8
x_{1S}	-1.20	—	14.85	2.79
x_{2S}	-1.37	—	-2.17	4.61
x_{3S}	-2.68	—	-18.72	1.30
$\hat{y}_S$	102.24	—	384.30	77.95
λ_1	-6.68	—	19.20	-0.40
λ_2	-6.31	—	7.22	-0.22
λ_3	3.96	—	-0.75	1.92
Shape	11.1(F)	—	11.1(D)	11.1(B)

$$\tilde{X} = M_i'(x - x_S)$$

$$M_1' = \begin{bmatrix} 0.497 & 0.485 & -0.720 \\ -0.722 & 0.692 & -0.032 \\ 0.483 & 0.536 & 0.693 \end{bmatrix}$$

$$M_3' = \begin{bmatrix} 0.383 & 0.919 & 0.094 \\ 0.849 & -0.391 & 0.355 \\ -0.363 & 0.056 & 0.930 \end{bmatrix}$$

$$M_4' = \begin{bmatrix} 0.229 & 0.613 & 0.756 \\ -0.333 & -0.681 & 0.653 \\ 0.915 & -0.401 & 0.048 \end{bmatrix}$$

Ideally, one should express the given inequalities in terms of the fitted surfaces. This would define the boundaries to four desirable regions in the x space and one should now see what, if any, region satisfies all the inequalities. More simply, we can note that the responses to compound numbers 7, 17, and 18 fall within the restrictions, indicating that feasible conditions are possible, and probably between the origin $(0,0,0)$ and the cube vertex $(-1,1,1)$. Trial values in this region may now be substituted into the response functions to see if the predicted response satisfies the restrictions.

[The original authors used a compound desirability criterion, briefly described in Section 11.7, to pick a desirable point at $(-0.05, 0.145, -0.868)$. They also found that the region around this point was insensitive to small departures from it, a good feature.] (Some details courtesy of J. C. Lu, C. M. Oliveri, and Y. C. Ki.)

11.6. The quadratic model fitted to the y_1 data shows significant lack of fit with $F = 15.23$ (df $= 5, 3$). The sixth observation is associated with a standardized residual of 2.75 and probably needs careful checking. If this observation is dropped, and the surface refitted, the lack of fit becomes nonsignificant. The fitted model is then

$$\hat{y} = 28.933 + 4.4551x_1 + 2.3337x_2 + 6.0363x_3$$

$$+ 2.7233x_1^2 - 1.1741x_2^2 + 1.6427x_3^2$$

$$+ 1.018x_1x_2 - 1.743x_1x_3 + 0.268x_2x_3.$$

Source	SS	df	MS	F
First order	812.50	3	270.83	38.52
Second order $|b_0$	185.38	6	30.90	4.39
Lack of fit	16.90	4	4.23	0.39
Pure error	32.31	3	10.77	
Total, corrected	1047.09	16		

Because there is no lack of fit, we use $s^2 = (16.90 + 32.31)/(4 + 3) = 7.03$ to test first- and second-order terms. Both sets are significant at the $\alpha = 0.05$ level.

At the additional point locations we compare the predicted $\hat{y}_1$'s and the observed y_1's as follows (in the order given in the question).

$$\hat{y}_1: \quad 21.9 \quad 16.1 \quad 29.4 \quad 10.9 \quad 20.7 \quad 4.5$$

$$y_1: \quad 33.3 \quad 16.3 \quad 31.8 \quad 28.0 \quad 17.9 \quad 22.5$$

We note that the predictions are particularly bad for the two points well outside the design region, namely the fourth and sixth. The overall message is that the fitted model is likely to be useful only over the region in which the data were taken initially.

The canonical form is $\hat{y} = 16.91 + 3.2415\tilde{X}_1^2 - 1.2613\tilde{X}_2^2 + 1.2117\tilde{X}_3^2$. The center of the system is at $x_S = (-1.682, -0.046, -2.726)'$ and the axes are rotated around this point via $\tilde{X} = M'(x - x_S)$ where

$$M = \begin{bmatrix} -0.8778 & -0.1454 & 0.4565 \\ -0.0869 & 0.9854 & 0.1467 \\ 0.4712 & -0.0891 & 0.8775 \end{bmatrix}.$$

For y_2, no lack of fit is shown for the quadratic model, but only the terms x_1 and x_3 appear effective when the estimated coefficients are compared with their standard errors. For a fit on x_1 and x_3 only, we obtain $\hat{y}_2 = 5.7339 + 0.6298x_1 + 0.7721x_3$. There is no overall lack of fit, and both coefficients are significant.

Source	SS	df	MS	F
b_1, b_3	13.5455	2	6.7727	12.28
Lack of fit	8.0021	12	0.6668	7.30
Pure error	0.2739	3	0.0913	
Total, corrected	21.8214	17		

Regression is tested using $s^2 = (8.0021 + 0.2739)/(12 + 3) = 0.5517$.

At the additional point locations, the predicted $\hat{y}_2$'s and observed y_2 values are:

$$\hat{y}_2: \quad 5.44 \quad 3.31 \quad 5.64 \quad 4.24 \quad 4.18 \quad 3.58$$

$$y_2: \quad 5.77 \quad 4.03 \quad 5.64 \quad 5.29 \quad 4.23 \quad 4.74$$

Again, prediction is worst at the two locations well outside the design region, indicating once again the dangers of extrapolation.

11.7.

$$\mathbf{X'X} = \begin{bmatrix} 15 & \mathbf{0} & 16 & 16 & 16 & \mathbf{0} \\ \mathbf{0} & 16\mathbf{I} & \mathbf{0} & \mathbf{0} & \mathbf{0} & \mathbf{0} \\ 16 & \mathbf{0} & 40 & 8 & 8 & \mathbf{0} \\ 16 & \mathbf{0} & 8 & 40 & 8 & \mathbf{0} \\ 16 & \mathbf{0} & 8 & 8 & 40 & \mathbf{0} \\ \mathbf{0} & \mathbf{0} & \mathbf{0} & \mathbf{0} & \mathbf{0} & 8\mathbf{I} \end{bmatrix},$$

$$(\mathbf{X'X})^{-1} = \begin{bmatrix} \frac{7}{9} & 0 & 0 & 0 & -\frac{2}{9} & -\frac{2}{9} & -\frac{2}{9} & 0 & 0 & 0 \\ 0 & \frac{1}{16} & & & & & & & & \\ 0 & & \frac{1}{16} & & & & & & & \\ 0 & & & \frac{1}{16} & & & & & & \\ -\frac{2}{9} & & & & \frac{26}{288} & \frac{17}{288} & \frac{17}{288} & & & \\ -\frac{2}{9} & & & & \frac{17}{288} & \frac{26}{288} & \frac{17}{288} & & & \\ -\frac{2}{9} & & & & \frac{17}{288} & \frac{17}{288} & \frac{26}{288} & & & \\ 0 & & & & & & & \frac{1}{8} & & \\ 0 & & & & & & & & \frac{1}{8} & \\ 0 & & & & & & & & & \frac{1}{8} \end{bmatrix},$$

$$\mathbf{X'y} = \begin{bmatrix} 539 \\ -21 \\ -37 \\ -17 \\ 471 \\ 395 \\ 723 \\ 73 \\ 5 \\ 7 \end{bmatrix}, \quad \mathbf{b} = \begin{bmatrix} 66.1111 \\ -1.3125 \\ -2.3125 \\ -1.0625 \\ -11.2639 \\ -13.6389 \\ -3.3889 \\ 9.125 \\ 0.625 \\ 0.875 \end{bmatrix}$$

Note: The **b** vector is not the same as the one in the original paper. We assume the quoted data have been rounded from those that determined the fitted response surface given originally. The analysis of variance table is as follows:

Source	SS	df	MS	F
b_0	19,368.07	1		
b_i	131.19	3	43.73	0.26
$b_{ii}\mid b_0$	3,123.00	3	1041.00	6.10
b_{ij}	675.38	3	225.13	1.32
$(b_{ii}, b_{ij}\mid b_0$	3,798.38	6	633.06	3.71)
Residual	853.37	5	170.67	
Total	24,151.01	15		

Comments: Only the pure second-order terms alone are significant when compared with the tabular value $F(3, 5, 0.95) = 5.41$. The ratio $6.10/5.41 = 1.13$ is very low, a bad sign (see Section 8.2). All six second-order terms are nonsignificant as a package when compared to $F(6, 5, 0.95) = 4.95$. The inevitable conclusion is that this fitted response surface is a shaky one and that canonical analysis is probably not worthwhile. Overall, this does not seem a very satisfactory fit to this set of data, in spite of an R^2 value of 0.822.

[One wonders whether application of the transformation $W = \sin^{-1}(Y/100)^{1/2}$, useful for transforming proportion or percentage type data, would improve the fit. In fact it does not, as the reader may confirm for him or herself.]

Investigating further, we see that all coefficients with a subscript 3 are small compared to their standard errors:

$b_3 = -1.0625$ has standard error of $(170.67/16)^{1/2} = 3.266$,

$b_{33} = -3.3889$ has standard error of $\{(170.67)(26/288)\}^{1/2} = 3.925$,

$b_{13} = 0.625$ has standard error of $(170.67/8)^{1/2} = 4.619$,

$b_{23} = 0.875$ has standard error of $(170.67/8)^{1/2} = 4.619$.

The standard errors are, of course, the square roots of the appropriate diagonal entries in $(\mathbf{X}'\mathbf{X})^{-1}s^2$. The above indicates that x_3 may not be needed in the model. We can check this formally by an extra sum of squares test. First we fit the reduced model

$$\hat{y} = 57.7692 - 1.3125x_1 - 2.3125x_2$$

$$-9.0481x_1^2 - 11.4231x_2^2 + 9.125x_1x_2$$

with analysis of variance:

Source	SS	df	MS	F	$F(\nu, 9, 0.95)$
b_0	19,368.07	1	—		
b_1, b_2	113.13	2	56.56	0.51	4.26
$b_{11}, b_{22}\|b_0$	2,995.79	2	1497.90	13.38	4.26
b_{12}	666.13	1	666.13	5.95	5.12
(All second order$\|b_0$	3,661.92	3	1220.64	10.90	3.86)
Residual	1,007.89	9	111.99		
	24,151.01	15			

Comparing the two tables, we find the extra sum of squares for b_3, b_{33}, b_{13}, b_{23} given the other b's is

$$1007.89 - 853.37 = 154.52,$$

$$\text{with df} = 9 - 5 = 4.$$

The F statistic for testing H_0: $\beta_3 = \beta_{33} = \beta_{13} = \beta_{23} = 0$ versus H_1: not so, is thus

$$\frac{\{154.52/4\}}{170.67} = 0.23,$$

not significant. We thus drop x_3 from the model. The resulting x_1, x_2 model is significant, but not highly so (see Section 8.2). Nevertheless, if we proceed to a canonical reduction, we obtain (see diagram)

$$\hat{y} = 58.06 - 5.520\,\tilde{X}_1^2 - 14.948\,\tilde{X}_2^2,$$

with

$$\tilde{X}_1 = -0.791(x_1 + 0.155) - 0.612(x_2 + 0.163),$$

$$\tilde{X}_2 = -0.612(x_1 + 0.155) + 0.791(x_2 + 0.163).$$

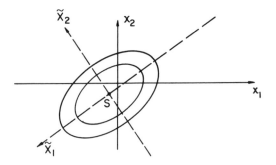

The conclusions then are that a predicted maximum response of 58.06 is obtained at a point estimated as $(x_1, x_2) = (-0.155, -0.163)$ and that the predicted response drops away from that point. In fact, a higher response of 63 was observed at $(0, 0)$. These conclusions must be hedged by a reminder of the unconvincing value of the fitted equation. One wonders if the residual mean square is "too big" but there is no way to check that in these data, because there are no repeats, a lack that is especially felt.

11.8. All three fitted equations can dispense with x_3 as indicated by an extra sum of squares F test. For the whole set of data, the reduced model is then

$$\hat{y} = 10.79 + \underset{(0.18)}{2.71x_1} + \underset{(0.18)}{1.90x_2} \underset{(0.25)}{-0.71x_1^2} \underset{(0.25)}{-0.82x_2^2} \underset{(0.25)}{+1.26x_1x_2}.$$

Standard errors are in parentheses. No lack of fit is indicated.

Source	SS	df	MS	F
b_1, b_2	493.67	2	246.84	155.62
$b_{11}, b_{22}, b_{12}\|b_0$	60.35	3	20.12	12.68
Lack of fit	46.92	27	1.74	1.89
Pure error	48.25	33	1.46	
Total, corrected	649.20	65		

$$s^2 = \frac{(46.92 + 48.25)}{(27 + 33)} = 1.5862.$$

Canonical reduction is justified in view of the fact that

$$\frac{(\hat{y}_{max} - \hat{y}_{min})}{(ps^2/n)^{1/2}} = \frac{(15.13 - 5.13)}{\{6(1.5862)/66\}^{1/2}} = 26.34.$$

The eigenvalues of

$$\mathbf{B} = \begin{bmatrix} -0.7062 & 0.62915 \\ 0.62915 & -0.82440 \end{bmatrix}$$

are -1.3972 and -0.1334. The center of the system is at $\mathbf{x}_S$ where

$$\mathbf{x}'_S = \tfrac{1}{2}(2.7117, 1.9017)\mathbf{B}^{-1} = (9.2074, 8.1804)$$

and $\hat{y}_S = b_0 + \tfrac{1}{2}\mathbf{x}'_S\mathbf{b} = 10.7853 + 20.2617 = 31.0470$. The appropriate transformation is

$$\tilde{\mathbf{X}} = \begin{bmatrix} -0.6732 & 0.7394 \\ 0.7394 & 0.6732 \end{bmatrix}(\mathbf{x} - \mathbf{x}_S)$$

and the surface form is that of an elliptical, cigar shaped in a ratio of $3.2 : 1$, mound pointing out of the positive quadrant at roughly 45 degrees. Because the center is so remote, the shape near the design center is essentially that of a rising ridge. The two other surfaces are technically saddles but with the same rising ridge characteristic near the design center, and it is this feature that dominates the interpretation of the surface. (See Figure 11.2.)

If the two designs are scaled to have the same second moment and the efficiency of the central composite design with respect to the 3^3 factorial is defined as

$$E(b) = \frac{V(b) \text{ for } 3^3 \text{ factorial}}{V(b) \text{ for central composite}} \left(\frac{27}{15}\right)$$

we find $E(b_i) = 1$, $E(b_{ii}) = 0.96$, $E(b_{ij}) = 1.32$. On balance then, the central composite design comes out slightly better. (Details courtesy of M. Lindstrom.)

11.9. All three sets of data show the same basic characteristics and we proceed here only with the full set of 50 observations, to which we fit

$$\hat{y} = 7.120 + 2.026x_1 + 1.936x_2 - 0.023x_3$$

$$-0.118x_1^2 - 1.299x_2^2 - 0.308x_3^2$$

$$+1.187x_1x_2 + 0.390x_1x_3 + 0.208x_2x_3.$$

Source	SS	df	MS	F
First order	479.11	3	159.703	61.53
Second order$\|b_0$	202.80	6	33.800	13.02
Lack of fit	71.26	25	2.850	1.31
Pure error	32.56	15	2.171	
Total, corrected	785.73	49		

$$s^2 = \frac{(71.26 + 32.56)}{(25 + 15)} = 2.5955.$$

No lack of fit is indicated, and

$$\frac{(\hat{y}_{max} - \hat{y}_{min})}{\{ps^2/n\}^{1/2}} = \frac{\{13.165 - (-1.413)\}}{\{10(2.5955)/50\}^{1/2}} = 20.24.$$

Thus, interpretation of the surface is worthwhile. The center of the system is at $\mathbf{x}_S = (-4.316, -1.490, -3.273)'$ and $\hat{y} = 1.342 - 1.546 \tilde{X}_1^2 - 0.399 \tilde{X}_2^2 + 0.220 \tilde{X}_3^2$. The transformation is

$$\tilde{\mathbf{X}} = \begin{bmatrix} 0.382 & -0.924 & 0.0176 \\ -0.362 & -0.132 & 0.923 \\ 0.851 & 0.357 & 0.385 \end{bmatrix} (\mathbf{x} - \mathbf{x}_S).$$

The cross-sectional contours in $(\tilde{X}_1, \tilde{X}_2)$ planes are ellipses, with saddles in other cross-sections. When $x_3 = 0$, we obtain the contours shown in the diagram because the center is remote from the experimental region. Similar shapes apply for other x_3 sections. The composite design is slightly more efficient than the factorial for estimating b_{ii} ($E = 1.067$) but less ($E = 0.768$) for b_{ij}. (Details courtesy of K. Kim.)

11.10. For this design we have $\Sigma_u x_{iu}^2 = 32$, $\Sigma_u x_{iu}^4 = 80$, $\Sigma_u x_{iu}^2 x_{ju}^2 = 16$.
   (a) The portion of the $\mathbf{X}'\mathbf{X}$ matrix stemming from terms in 1, x_1^2, x_2^2, and x_3^2 has inverse

$$\begin{bmatrix} 38 & 32 & 32 & 32 \\ 32 & 80 & 16 & 16 \\ 32 & 16 & 80 & 16 \\ 32 & 16 & 16 & 80 \end{bmatrix}^{-1} = \begin{bmatrix} P & Q & Q & Q \\ Q & R & S & S \\ Q & S & R & S \\ Q & S & S & R \end{bmatrix},$$

where $P = \frac{7}{74}$, $Q = -\frac{1}{27}$, $R = \frac{25}{1184}$, $S = \frac{13}{2368}$. The author's equations now follow directly from setting out $\mathbf{b} = (\mathbf{X}'\mathbf{X})^{-1}\mathbf{X}'\mathbf{y}$.
   (b) The next step is to backsolve these equations for $\mathbf{X}'\mathbf{y} = \mathbf{X}'\mathbf{Xb}$ using, successively, each set of b's given in the table. For example, for y_{14}, we

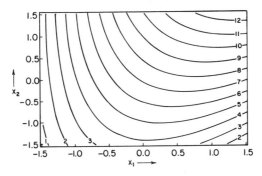

obtain:

$$
\begin{bmatrix} (0y) \\ (11y) \\ (22y) \\ (33y) \end{bmatrix} = \begin{bmatrix} 38 & 32 & 32 & 32 \\ 32 & 80 & 16 & 16 \\ 32 & 16 & 80 & 16 \\ 32 & 16 & 16 & 80 \end{bmatrix} \begin{bmatrix} 0.3021 \\ 0.1178 \\ -0.0367 \\ -0.0103 \end{bmatrix} = \begin{bmatrix} 13.7454 \\ 18.3392 \\ 8.4512 \\ 10.1408 \end{bmatrix},
$$

$$(1y) = 32(-0.3587) = -11.4784,$$

$$(2y) = 32(0.0378) = 1.2096,$$

$$(3y) = 32(0.0099) = 0.3168,$$

$$(12y) = 16(-0.0294) = -0.4704,$$

$$(13y) = 16(0.0404) = 0.6464,$$

$$(23y) = 16(0.0057) = 0.0912.$$

We actually do not need all these to fit $\hat{y} = b_0 + b_1 x + b_{11} x^2$ for which

$$
\begin{bmatrix} b_0 \\ b_1 \\ b_{11} \end{bmatrix} = \begin{bmatrix} 38 & 0 & 32 \\ 0 & 32 & 0 \\ 32 & 0 & 80 \end{bmatrix}^{-1} \begin{bmatrix} 13.7454 \\ -11.4784 \\ 18.3392 \end{bmatrix}
$$

$$
= \begin{bmatrix} 80 & 0 & -32 \\ 0 & 63 & 0 \\ -32 & 0 & 38 \end{bmatrix} (32 \times 63)^{-1} \begin{bmatrix} 13.7454 \\ -11.4784 \\ 18.3392 \end{bmatrix} = \begin{bmatrix} 0.2544 \\ -0.3587 \\ 0.1275 \end{bmatrix}.
$$

For (ii), we need

$$
\mathbf{b} = \begin{bmatrix}
38 & 0 & 0 & 0 & 32 & 0 & 0 \\
0 & 32 & 0 & 0 & & & \\
0 & 0 & 32 & 0 & & \mathbf{0} & \\
0 & 0 & 0 & 32 & & & \\
32 & & & & 80 & 0 & 0 \\
0 & & \mathbf{0} & & 0 & 16 & 0 \\
0 & & & & 0 & 0 & 16
\end{bmatrix}^{-1}
\begin{bmatrix}
(0y) \\
(1y) \\
(2y) \\
(3y) \\
(11y) \\
(12y) \\
(13y)
\end{bmatrix}
$$

From the original table we can read off $b_1 = -0.3802$, $b_2 = 0.3087$, $b_3 = 0.7634$, $b_{12} = -0.0162$, and $b_{13} = 0.1610$, all these being unchanged. For the remainder we have

$$
\begin{bmatrix} b_0' \\ b_{11}' \end{bmatrix} = \begin{bmatrix} 38 & 32 \\ 32 & 80 \end{bmatrix}^{-1} \begin{bmatrix} (0y) \\ (11y) \end{bmatrix}
$$

$$
= (32 \times 63)^{-1} \begin{bmatrix} 80 & -32 \\ -32 & 38 \end{bmatrix} \begin{bmatrix} (0y) \\ (11y) \end{bmatrix}
$$

$$
= (32 \times 63)^{-1} \begin{bmatrix} 80 & -32 & 0 & 0 \\ -32 & 38 & 0 & 0 \end{bmatrix}
$$

$$
\times \begin{bmatrix}
38 & 32 & 32 & 32 \\
32 & 80 & 16 & 16 \\
32 & 16 & 80 & 16 \\
32 & 16 & 16 & 80
\end{bmatrix}
\begin{bmatrix} b_0 \\ b_{11} \\ b_{22} \\ b_{33} \end{bmatrix}
$$

$$
= (32 \times 63)^{-1} \begin{bmatrix} 2016 & 0 & 2048 & 2048 \\ 0 & 2016 & -416 & -416 \end{bmatrix} \begin{bmatrix} b_0 \\ b_{11} \\ b_{22} \\ b_{33} \end{bmatrix}
$$

$$
\overset{.}{=} \begin{bmatrix} 1 & 0 & 1.015873 & 1.015873 \\ 0 & 1 & -0.206349 & -0.206349 \end{bmatrix} \begin{bmatrix} 2.3507 \\ -0.1622 \\ 0.0509 \\ -0.0189 \end{bmatrix}
$$

$$
= \begin{bmatrix} 2.3832 \\ -0.1688 \end{bmatrix}.
$$

For (iii), $b_0' = (0y)/38$

$$
= (38)^{-1}(38, 32, 32, 32) \begin{bmatrix} 3.0303 \\ -0.1424 \\ -0.0003 \\ 0.0948 \end{bmatrix} = 2.9900.
$$

The coefficients $b_1 = -0.0893$, $b_2 = 0.2757$, $b_3 = 0.7188$ are unchanged from before. (iv) The required calculation is exactly parallel to that of part (ii). The fitted equation that emerges is $\hat{y} = 7.4537 - 0.5334x_1 - 0.2484x_2 + 0.9903x_3 - 0.4239x_1^2 + 0.3681x_1x_2 - 0.0544x_1x_3$.

(c) The plots are straightforward and are not shown. (i) is a quadratic independent of x_3 and (iii) is a plane. Sections of (ii) and (iv) for various x_3 are quadratic curves in the (x_1, x_2) plane.

(d) In order to fit a full quadratic model in $x_1, \ldots, x_4$ we would need $(4y)$, $(i4y)$, $i = 1, 2, 3, 4$. We cannot obtain these from the information provided. Thus (e) and (f) cannot be done.

11.11. *FPP-4 responses.*

Estimate	$\hat{Y}_P$	$\hat{Y}_N$	$\hat{Y}_M$	$\hat{Y}_C$
b_0	-314	263.94	42.09	-48.35
b_1	201	-137.05	-35.03	84.61
b_2	288	-155.39	-10.65	-51.01
b_3	418	-290.75	-36.45	16.25
b_{11}	-26.1	19.45	7.60	-16.55
b_{22}	-60.9	24.03	-0.593	28.77
b_{33}	-108	77.46	10.83	1.79
b_{12}	-63.2	36.29	6.13	-11.15
b_{13}	-107	77.06	14.11	-37.70
b_{23}	-163	89.39	7.23	54.92

	$\hat{Y}_P$	$\hat{Y}_N$	$\hat{Y}_M$	$\hat{Y}_C$
(3 df) SS(first-order$|b_0$)	1476.0	361.7	15.24	2357.23
(6 df) SS(second-order$|b_0$, first)	712.0	138.9	13.54	480.53
(70 df) SS(residual)	484.6	102.9	53.18	1747.53
(79 df) SS(total, corrected)	2672.6	603.4	81.96	4585.30

	$\hat{Y}_P$	$\hat{Y}_N$	$\hat{Y}_M$	$\hat{Y}_C$
F, first order	71.07	82.0	6.7	31.5
F, second order	17.14	15.7	3.0	3.2
s^2	6.923	1.47	0.76	24.96
$\dfrac{(\hat{Y}_{max} - \hat{Y}_{min})}{(ps^2/n)^{1/2}}$	21.5	29.5	16.6	19.0
$\hat{y}_0$	81.999	1.389	3.032	9.284
λ_1	-192.3	120.6	17.66	49.29
λ_2	-8.1	3.5	1.98	-30.55
λ_3	5.3	-3.2	-1.83	-4.73

	$\hat{Y}_P$	$\hat{Y}_N$	$\hat{Y}_M$	$\hat{Y}_C$
Approximate surface type	11.1(F)	11.1(F)	11.1(F)	11.1(E)
x_{1S}	1.545-	1.081	1.616	1.913
x_{2S}	0.275-	1.001	1.053	1.001
x_{3S}	0.963	0.761	0.283	0.269

Transformation matrices $\tilde{\mathbf{X}} = \mathbf{M}'(\mathbf{x} - \mathbf{x}_S)$

$$\mathbf{M}_P = \begin{bmatrix} 0.349 & -0.743 & -0.571 \\ 0.553 & 0.655 & -0.514 \\ 0.756 & -0.137 & 0.640 \end{bmatrix},$$

$$\mathbf{M}_N = \begin{bmatrix} 0.387 & 0.782 & -0.489 \\ 0.446 & -0.623 & -0.643 \\ 0.807 & -0.031 & 0.590 \end{bmatrix},$$

$$\mathbf{M}_M = \begin{bmatrix} 0.605 & -0.774 & 0.185 \\ 0.251 & -0.035 & -0.967 \\ 0.756 & 0.632 & 0.173 \end{bmatrix},$$

$$\mathbf{M}'_C = \begin{bmatrix} 0.226 & -0.802 & -0.553 \\ 0.752 & -0.218 & 0.623 \\ -0.620 & -0.557 & 0.553 \end{bmatrix}.$$

(Details courtesy of W. Lee, R. Luneski, M. S. Chi, and S. K. Ahn.)

11.12. The center is at $(0.060, 0.215, 0.501) = \mathbf{x}'_S$. The reduced form is $\hat{y} = 68.134 - 3.190\, \tilde{X}_1^2 + 0.780\, \tilde{X}_2^2 - 0.069\, \tilde{X}_3^2$. Because λ_3 is relatively small, this is basically a form of the general nature of (E) in Figure 11.3, (E) being viewed as a limiting case of (B). The transformation needed is

$$\tilde{\mathbf{X}} = \begin{bmatrix} 0.7510 & 0.5849 & -0.3063 \\ 0.4883 & -0.8043 & -0.3386 \\ 0.4445 & -0.1047 & 0.8897 \end{bmatrix}' (\mathbf{x} - \mathbf{x}_S).$$

11.13.

Problem number	1	2	3	4
x_{1S}	0.186	0.346	-0.843	-0.196
x_{2S}	-2.431	0.320	0.501	0.430
x_{3S}	1.303	0.267	-19.482	0.244
$\hat{y}_S$	49.94	42.57	72.51	71.07
λ_1	-6.01	3.00	1.01	-3.00
λ_2	1.00	1.99	0.10	-2.00
λ_3	0.10	1.00	0.02	1.00
Approximate surface type in Table 11.1.	E	A	G	B

The transformation to canonical variables $\tilde{\mathbf{X}}$ is $\tilde{\mathbf{X}} = \mathbf{M}'_i(\mathbf{x} - \mathbf{x}_S)$ where

$\mathbf{M}_1, \ldots, \mathbf{M}_4$ are as given below.

$$\mathbf{M}_1 = \begin{bmatrix} 0.2426 & 0.8428 & -0.4804 \\ 0.4511 & 0.3404 & 0.8250 \\ 0.8588 & -0.4169 & -0.2977 \end{bmatrix},$$

$$\mathbf{M}_2 = \begin{bmatrix} 0.9459 & 0.2443 & 0.2137 \\ 0.0496 & -0.7593 & 0.6489 \\ -0.3208 & 0.6031 & 0.7303 \end{bmatrix},$$

$$\mathbf{M}_3 = \begin{bmatrix} -0.9914 & 0.1137 & -0.0654 \\ -0.1214 & -0.9842 & 0.1292 \\ 0.0497 & -0.1360 & -0.9895 \end{bmatrix},$$

$$\mathbf{M}_4 = \begin{bmatrix} 0.5552 & 0.0786 & -0.8280 \\ 0.7804 & -0.3935 & 0.4859 \\ 0.2876 & 0.9160 & 0.2798 \end{bmatrix}.$$

11.14. (a) Stationary point is $\mathbf{x}_S = (-1.37756, -2.53382, 0.18873)'$.

$$\hat{y} = 6.52 - 0.235\,\tilde{X}_1^2 + 0.0392\,\tilde{X}_2^2 + 0.1798\,\tilde{X}_3^2,$$

$$\tilde{\mathbf{X}} = \begin{bmatrix} -0.83831 & 0.44259 & 0.31834 \\ 0.49769 & 0.85956 & 0.11598 \\ 0.22188 & -0.25609 & 0.94083 \end{bmatrix}'(\mathbf{x} - \mathbf{x}_S).$$

(b) Stationary point is $\mathbf{x}_S = (2.05894, -0.04153, -0.4145)'$.

$$\hat{y} = 2.73 - 0.00645\,\tilde{X}_1^2 - 0.16118\,\tilde{X}_2^2 + 0.65004\,\tilde{X}_3^2,$$

$$\tilde{\mathbf{X}} = \begin{bmatrix} -0.95524 & -0.03588 & 0.29364 \\ 0.14772 & -0.91784 & 0.36844 \\ 0.25631 & 0.39534 & 0.88205 \end{bmatrix}(\mathbf{x} - \mathbf{x}_S).$$

11.15. $\hat{y} = 6.91082 - 0.28932\tilde{X}_1^2 + 0.07381\tilde{X}_2^2 - 0.05534\tilde{X}_3^2$.
$\mathbf{x}_S = (0.793, -0.429, -0.466)'$. $\tilde{\mathbf{X}} = \mathbf{M}'(\mathbf{x} - \mathbf{x}_S)$ where

$$\mathbf{M} = \begin{bmatrix} -0.7123 & 0.6737 & 0.1966 \\ 0.0903 & 0.3658 & -0.9263 \\ -0.6960 & -0.6421 & -0.3421 \end{bmatrix}.$$

Setting $\hat{y} > 7$ provides a quadratic restriction in x_1, x_2, x_3 and all points satisfying the restriction are suitable. In the canonical form, for example, we can set $\tilde{X}_1 = \tilde{X}_3 = 0$ and choose $\tilde{X}_2 > 1.21^{1/2} = 1.1$. This implies a requirement, for any value of δ, of

$$\mathbf{M}'(\mathbf{x} - \mathbf{x}_S) = \begin{bmatrix} 0 \\ 1.1 + \delta^2 \\ 0 \end{bmatrix}$$

or, because $\mathbf{M}^{-1} = \mathbf{M}'$, of

$$\mathbf{x} = \mathbf{x}_S + (1.1 + \delta^2) \begin{bmatrix} 0.6737 \\ 0.3658 \\ -0.6421 \end{bmatrix}$$

$$= \begin{bmatrix} 1.534 \\ -0.027 \\ -1.172 \end{bmatrix} + \delta^2 \begin{bmatrix} 0.6737 \\ 0.3658 \\ -0.6421 \end{bmatrix},$$

for any δ. Other points giving nonzero values of $\tilde{X}_1$ and $\tilde{X}_3$ are also feasible, of course.

11.16. $\hat{y} = 35.49 - 1.511x_1 + 1.284x_2 - 8.739x_3 + 4.955x_4$
$\quad\quad - 5.024x_1^2 - 2.983x_2^2 + 1.328x_3^2 - 1.198x_4^2$
$\quad\quad + 2.194x_1x_2 - 0.144x_1x_3 + 1.581x_1x_4$
$\quad\quad + 8.006x_2x_3 + 2.806x_2x_4 + 0.294x_3x_4.$

Source	SS	df	MS	F
First order	2088.6	4		
Second order$\mid b_0$	1729.3	10		
Lack of fit	2078.6	10	207.86	22.50
Pure error	46.2	5	9.24	
Total, corrected	5942.7	29		

$R^2 = 0.6475$. Lack of fit is evident. Note that in order to obtain the appropriate corrected total sum of squares we must work out the corrected sum of squares using six center point observations all equal to 32.8, and then add an $(n_0 - 1)s_e^2 = (6 - 1)9.24 = 46.2$. For reasoning, see Draper and Smith (1981, pp. 278–279). The lack of fit casts doubt on the use of the equation for predictive purposes and canonical reduction is probably not justified.

If we now fit a second-order equation to the w_i we are unable to obtain the lack of fit/pure error splitup of the residual sum of squares due to the fact that the six individual center point observations are not available. Moreover, because $R^2 = 0.6632$ is about the same size as in the previous analysis, it looks as if not much improvement has occurred.

The experimenter could be told all of the above, and be asked for the individual replicate runs at the center of the design. Were they available, lack of fit could then be tested on the $\hat{w}$ model.

11.7. $\hat{y}_1 = 8.029 + (0.7543, 0.9139, 0.2177, 0.6499)\mathbf{x} + \mathbf{x}'\mathbf{Bx}$ where $\mathbf{x} = (x_1, x_2, x_3, x_4)'$, and

$$\mathbf{B} = \begin{bmatrix} -0.203 & -0.136 & -0.0024 & -0.1256 \\ & -0.172 & 0.0066 & -0.0812 \\ & & -0.0187 & -0.0072 \\ \text{symmetric} & & & -0.118 \end{bmatrix}.$$

Source	SS	df	MS	F
First order$\|b_0$	66.24	4	16.560	67.04
Second order$\|$first, b_0	22.85	10	2.285	9.25
Lack of fit	4.27	15	0.285	2.12
Pure error	0.67	5	0.134	
Total, corrected	94.03	34		

$$s^2 = \frac{(4.27 + 0.67)}{(15 + 5)} = 0.247(20 \text{ df})$$

$$\frac{(\hat{y}_{max} - \hat{y}_{min})}{(ps^2/n)^{1/2}} = \frac{(9.61 - 2.43)}{\{14(0.247)/35\}^{1/2}} = 22.84.$$

So the surface is worthy of interpretation.

The stationary point is at $\mathbf{x}_S = (-1.43, 2.67, 0.60, 2.40)'$, and $\hat{y} = 9.55 - 0.40\tilde{X}_1^2 - 0.19\tilde{X}_2^2 - 0.07\tilde{X}_3^2 - 0.02\tilde{X}_4^2$, so that the stationary point represents a maximum. However, this point is somewhat outside the design region. Good whiteness values are obtained as we move from the origin toward the stationary point, nevertheless. The low sixteenth value has a standardized residual of -0.84 and does not appear out of line. (Details courtesy of M. Grassl.)

With observation 16:

$$\hat{y}_2 = 0.585 + (-2.8195, -1.2899, -0.2705, -1.1940)\mathbf{x} + \mathbf{x}'\mathbf{Bx},$$

$$\mathbf{B} = \begin{bmatrix} 2.043 & 0.538 & 0.141 & 0.772 \\ & 0.258 & -0.006 & 0.340 \\ & & 0.008 & 0.085 \\ \text{symmetric} & & & 0.115 \end{bmatrix}.$$

Source	SS	df	MS	F
First order$\|b_0$	141.96	4		
Second order$\|$first, b_0	318.04	10		
Lack of fit	189.34	15	12.623	51.73
Pure error	1.22	5	0.244	
Total, corrected	650.56	34		

Lack of fit is significant, so we cannot proceed with this model. The sixteenth residual is 5.33 times its standard deviation. We now reanalyze the data without this obvious outlier.

Without observation 16.

$$\hat{y}_2 = 1.399 + (-0.2254, -0.3812, 0.0165, -0.2687)\mathbf{x} + \mathbf{x}'\mathbf{B}\mathbf{x},$$

$$\mathbf{B} = \begin{bmatrix} 0.192 & -0.019 & -0.115 & 0.066 \\ & 0.034 & -0.075 & 0.115 \\ & & 0.008 & 0.011 \\ \text{symmetric} & & & 0.088 \end{bmatrix}.$$

Source	SS	df	MS	F
First order$\mid b_0$	10.12	4	2.530	2.87
Second order$\mid$first, b_0	10.09	10	1.009	1.14
Lack of fit	15.54	14	1.110	4.55
Pure error	1.22	5	0.244	
Total, corrected	36.97	33		

Lack of fit is not significant, and $s^2 = (15.54 + 1.22)/(14 + 5) = 0.882$. Neither first- nor second-order terms are significant and no further interpretation of the fitted surface is called for. The model $\hat{y}_2 = \bar{y}$ appears perfectly adequate. (Some details courtesy of D. Kim.)

$$\hat{y}_3 = 1.589 + (0.1414, 0.1125, 0.0391, 0.1406)\mathbf{x} + \mathbf{x}'\mathbf{B}\mathbf{x},$$

$$\mathbf{B} = \begin{bmatrix} -0.028 & -0.006 & 0.008 & 0.011 \\ & -0.015 & 0.006 & 0.002 \\ & & -0.004 & 0.007 \\ \text{symmetric} & & & -0.013 \end{bmatrix}.$$

Source	SS	df	MS	F
First order$\mid b_0$	3.462	4	0.8655	20.32
Second order$\mid$first, b_0	0.273	10	0.0273	0.64
Lack of fit	0.792	15	0.0528	4.40
Pure error	0.060	5	0.017	
Total, corrected	4.587	34		

Lack of fit is not significant, and $s^2 = (0.792 + 0.060)/(15 + 5) = 0.0426$. Only first-order terms are significant, so that the reduced planar model $\hat{y}_3 = 1.494 + (0.1531, 0.0875, 0.0382, 0.1384)'\mathbf{x}$ can be fitted. This shows no lack of fit, and significant first-order terms, all of which are needed. A sectional contour plot could now be constructed.

11.18. $\hat{y} = 7.277 - 0.667x_1 - 0.353x_2 - 0.092x_3 - 4.330x_4$
$\quad - 0.0342x_1^2 + 0.0225x_2^2 - 0.1075x_3^2 + 1.3993x_4^2$
$\quad - 0.0116x_1x_2 + 0.4758x_1x_3 + 0.4294x_1x_4$
$\quad - 0.0233x_2x_3 - 0.0469x_2x_4 - 0.0175x_3x_4.$

Source	SS	df	MS	F
First order	590.2	4	147.55	607.2
Second order$|b_0$	31.9	10	3.19	13.1
Residual	7.3	30	0.243	
Total, corrected	629.4	44		

All x's are needed, and so are second-order terms.

$$\frac{(\hat{y}_{max} - \hat{y}_{min})}{(ps^2/n)^{1/2}} = 39.0 \text{ so canonical reduction is worthwhile.}$$

$$\hat{y} = 8.20 + 1.432\,\tilde{X}_1^2 - 0.325\,\tilde{X}_2^2 + 0.151\,\tilde{X}_3^2 + 0.022\,\tilde{X}_4^2.$$

The coefficient of $\tilde{X}_4^2$ is close to zero. Ignoring this term leaves a surface of type 11.1(B). In theory, for reduced absorption, we follow the $\tilde{X}_2$ axis in either direction away from $\tilde{\mathbf{X}} = \mathbf{0}$. However, the stationary point is remote at $\mathbf{x}_S = (0.893, 9.899, 0.346, 1.578)'$. Thus, it is more important to plot the contours *in the design region*, for example, by setting $x_4 = -1, 0$, and 1 in succession. A plot of the response values at each level of x_4 is the next best thing. This was given in the original paper and is reproduced on page 596. The canonical transformation is

$$\tilde{\mathbf{X}} = \begin{bmatrix} 0.1477 & -0.6689 & -0.7254 & 0.0676 \\ -0.0172 & 0.0195 & 0.0714 & 0.9971 \\ 0.0173 & 0.7379 & -0.6738 & 0.0341 \\ 0.9887 & 0.0873 & 0.1214 & 0.0067 \end{bmatrix}' (\mathbf{x} - \mathbf{x}_S).$$

(Some details courtesy of H. Udomon.)

11.19. The authors provide the following fitted regression coefficients.

$b_0 = 0.845222,$

$b_1 = 0.007500,$ $\quad b_{11} = -0.025222,$

$b_2 = 0.125834,$ $\quad b_{22} = -0.091472,$

$b_3 = 0.075001,$ $\quad b_{33} = -0.016472,$

$b_4 = 0.091667,$ $\quad b_{44} = -0.063972,$

$b_5 = -0.073334,$ $\quad b_{55} = -0.046472,$

$b_{12} = -0.025000,$ $\quad b_{23} = -0.090000,$ $\quad b_{34} = -0.003750,$

$b_{13} = 0.021250,$ $\quad b_{24} = 0.010000,$ $\quad b_{35} = -0.028750,$

$b_{14} = -0.088750,$ $\quad b_{25} = -0.020000,$ $\quad b_{45} = 0.091250.$

$b_{15} = 0.053750,$

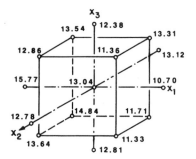

Response y when $x_4 = -1$.

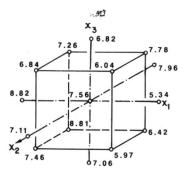

Response y when $x_4 = 0$.

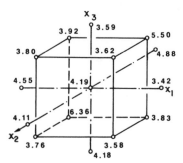

Response y when $x_4 = 1$.

596

They also provide this analysis of variance:

Source	df	MS
First order	5	0.1694
Second order$\vert b_0$	15	0.0567
Lack of fit	6	0.0156
Pure error	5	0.0001
Total, corrected	31	

Some practical difficulties now arise. Clearly there is, technically, lack of fit, which would make further analysis of the second-order model invalid. However the authors argued that the replicated center points exhibit an uncharacteristically low variation, "so the degree of bias was felt to be relatively small" (p. 320). They repooled the lack of fit and pure error sum of squares into a residual sum of squares and estimated σ^2 from that. The resulting standard errors for b_i, b_{ii}, and b_{ij} are then 0.0188, 0.0171, and 0.0231, respectively. Both first- and second-order sets of terms are significant when their mean squares are compared with $s^2 = 0.0085$, and no lack of fit is assumed.

The canonical reduction was carried out by Klein and Marshall. The stationary point is at $x'_S = (x_{1S}, x_{2S}, x_{3S}, x_{4S}, x_{5S}) = (-2.4495, 0.0429, 2.5937, 0.6375, -2.3912)$ and the canonical form is $\hat{y} = 1.0529 - 0.13168\tilde{X}_1^2 - 0.11121\tilde{X}_2^2 + 0.01996\tilde{X}_3^2 + 0.0000\tilde{X}_4^2 - 0.02069\tilde{X}_5^2$. The canonical variables are defined by $\tilde{X} = M'(x - x_S)$ where

$$M' = \begin{bmatrix} -0.37932 & 0.34843 & 0.2277 & -0.62911 & 0.53580 \\ 0.27739 & 0.82961 & 0.34969 & 0.31012 & 0.12784 \\ -0.55037 & 0.32512 & -0.63936 & 0.40256 & 0.14337 \\ 0.53409 & -0.11307 & 0.07621 & 0.20054 & 0.80990 \\ 0.25178 & -0.19807 & 0.12837 & 0.52994 & 0.77464 \end{bmatrix}.$$

Note that the predicted response is unaffected by changes in $\tilde{X}_4$, that the coefficients of $\tilde{X}_3^2$ and $\tilde{X}_5^2$ are relatively small, and that the coefficients of $\tilde{X}_1^2$, $\tilde{X}_2^2$, and $\tilde{X}_5^2$ are all negative. In the two-dimensional space defined by $\tilde{X}_1 = \tilde{X}_2 = \tilde{X}_5 = 0$, the response is $\hat{y} = 1.5029 + 0.01996\tilde{X}_3^2$ at any $\tilde{X}_4$. Thus for any chosen $\tilde{X}_4$ value, increases in $\tilde{X}_3^2$ will raise the predicted response, which has been maximized for choice of $\tilde{X}_1$, $\tilde{X}_2$, and $\tilde{X}_5$.

11.20. The fitted quadratic takes the form
$$\hat{y} = 215.7 + 7.4z_1 + 17.4z_2$$
$$+ 40x_1 - 15.08x_2 - 9.58x_3 - 9.92x_4 - 14.50x_5$$
$$- 15.53x_1^2 - 7.28x_2^2 - 7.03x_3^2 - 10.03x_4^2 - 10.03x_5^2$$
$$+ 12.13x_1x_2 + 13.13x_1x_3 + 14.38x_1x_4 + 16.75x_1x_5$$
$$+ 0.75x_2x_3 + 6.75x_2x_4 + 0.38x_2x_5$$
$$- 1.75x_3x_4 - 1.63x_3x_5 - 7.88x_4x_5.$$

Source	SS	df	MS	F
Blocks$|b_0$	1,007	2	504	
First order$|b_0$, blocks	53,471	5	10,694	
Second order$|$above	28,021	15	1,868	
Lack of fit	2,198	6	363	8.07
Pure error	227	5	45	
Total, corrected	84,924	33		

Lack of fit is significant at $\alpha = 0.05$. The source author remarked (his p. 75) that "many of the experimental errors, as determined from the replicated center point, were much smaller than previous estimates not reported here. It is suspected that nonhomogeneity of error variance is the probably cause." It is suggested earlier on the same page that "it is reasonable to expect that the difficulty in processing compounds with unusually high pigment or oil levels leads to larger error than more conventional levels of these components." Thus, a reanalysis of the data would need to take into account how the variances of the observations depended on the predictor variables. Further analysis of the model fitted above is not appropriate in the circumstances.

11.21. yield $= -85.25 - 56.88\tilde{X}_1^2 - 21.70\,\tilde{X}_2^2 + 19.70\,\tilde{X}_3^2 - 9.79\tilde{X}_4^2 - 5.23\,\tilde{X}_5^2,$

$$\tilde{X} = \begin{bmatrix} -0.37 & -0.01 & -0.10 & 0.93 & -0.02 \\ -0.52 & 0.30 & -0.64 & -0.28 & -0.39 \\ -0.77 & -0.32 & 0.43 & -0.25 & 0.25 \\ 0.05 & -0.43 & 0.28 & 0.02 & -0.86 \\ -0.09 & 0.79 & 0.53 & 0.02 & -0.22 \end{bmatrix} X + \begin{bmatrix} 0.09 \\ 1.38 \\ 0.71 \\ 0.68 \\ -1.25 \end{bmatrix},$$

$$X_S = (1.15, 1.10, 1.11, 0.49, 0.68)'.$$

In any pair of dimensions involving $\tilde{X}_3$ we have a saddle cross-section. In dimensions not involving $\tilde{X}_3$ the contours are elliptical or ellipsoidal.

purity $= 68.97 - 93.28\tilde{X}_1^2 - 25.67\tilde{X}_2^2 - 21.02\,\tilde{X}_3^2 + 7.35\tilde{X}_4^2 + 4.45\,\tilde{X}_5^2,$

$$\tilde{X} = \begin{bmatrix} -0.28 & -0.04 & 0.78 & 0.53 & -0.19 \\ -0.68 & 0.04 & -0.58 & 0.44 & -0.13 \\ -0.66 & 0.06 & 0.25 & -0.56 & 0.43 \\ -0.08 & 0.66 & 0.06 & -0.32 & -0.67 \\ -0.15 & -0.75 & 0. & -0.33 & -0.56 \end{bmatrix} X + \begin{bmatrix} -1.03 \\ 1.24 \\ 0.63 \\ 0.78 \\ 1.72 \end{bmatrix},$$

$$X_S = (1.28, 0.64, 1.31, 1.18, 1.18)'.$$

The contours are elliptical and decreasing outward in $(\tilde{X}_1, \tilde{X}_2, \tilde{X}_3)$, elliptical and increasing outward in $(\tilde{X}_4, \tilde{X}_5)$. In pairs of dimensions where one is selected from the first bracket and one from the second, the contours are saddles.

A ridge analysis can be used to show that both responses can be simultaneously improved. The sketches requested show:

(a) Two similar saddles for which improvements are obtained by moving from the origin in a south-westerly direction when X_5 is abscissa, X_4 is ordinate.

(b) and (c) show very similar characteristics to (a).

(Details courtesy of J. Tort-Martorell.)

CHAPTER 13.

13.1. To obtain the moment conditions, apply the sufficient conditions of Appendix 13B, namely moments of design equal moments of region to order $d_2 + 1$. If $d_2 = 3$, $N = 6$, we need $x_1, x_2, \ldots, x_6$ such that $\Sigma x_u = 0, \Sigma x_u^2 = 2, \Sigma x_u^3 = 0, \Sigma x_u^4 = 1.2$. The first and third of these are satisfied if we choose the x's symmetrically as $-a, -b, -c, c, b, a$, whereupon the second and fourth equations become $a^2 + b^2 + c^2 = 1$ and $a^4 + b^4 + c^4 = 0.6$. A solution is given by

$$a^2 = \tfrac{1}{2}\left\{(1 - c^2) \pm (0.2 + 2c^2 - 3c^4)^{1/2}\right\}, \qquad b^2 = 1 - c^2 - a^2,$$

for any value of c which leads to positive, nonimaginary solutions for a^2 and b^2. There are many such; for example, if $c = 0$, $a = 0.5257$, $b = 0.8507$.

13.2. The Σr_u^6 values are proportional to $142, 125, 75, 68, 62,$ and 60, respectively. Thus, the first design provides the best detectability of cubic lack of fit. However, the second design has experiments at all five levels and the detectability is quite high, so it would also be a good choice. Note that the design with two experiments at each of the five levels is not a particularly good one in the assumed circumstances, despite the fact that it might seem the natural arrangement to select.

13.3. Solution is implicit in question.

13.4. Solution is implicit in question.

13.5. This example is constructed. The surface used was

$$y = 76 + 5X_1 - X_2 + 10X_3 + X_1^2 + 0.5X_2^2 - 7X_3^2 - X_1X_2 - 1.5X_1X_3 + \varepsilon,$$

where $X_1 = \xi_1^{-1}, X_2 = \xi_2, X_3 = \xi_3^{1/2}$, and $\varepsilon \sim N(0, 4)$. Thus the estimates obtained for the second part of the question should be close to the indicated values.

CHAPTER 15.

15.1.   The variance function, obtained from the formula $V\{\hat{y}(x)\} = u'(X'X)^{-1}u\sigma^2$,
where $u = (1, x_1, x_2, x_1^2, x_2^2, x_1x_2)$, is of the form σ^2 times

$$P + \left(2Q + \frac{1}{B}\right)r^2 + Rr^4 + \frac{x_1^2x_2^2(C - 3D)}{[D(C - D)]},$$

where $P = 4\alpha^4(4 + \alpha^4)/A$, $Q = -4\alpha^4(2 + \alpha^2)/A$, $R = 4(5\alpha^4 - 4\alpha^2 + 8)/A$, $A = 2\alpha^4(16\alpha^4 - 32\alpha^2 + 64)$, $B = 4 + 2\alpha^2$, $C = 4 + 2\alpha^4$, $D = 4$, and $r^2 = x_1^2 + x_2^2$. The design is rotatable if this expression is a function of r^2 only, which happens when $C = 3D$, which implies $\alpha = 2^{1/2}$. Note that, with this value of α, the cube and star points all lie on a circle so that center points *are* essential.

15.2.   (a) $\Sigma x_{1u}^2 = 4b^2 = 9$, $\Sigma x_{1u}^4 = 4b^4 = 20.25$,
$\Sigma x_{2u}^2 = 2a^2 + 4c^2 = 9$, $\Sigma x_{2u}^4 = 2a^4 + 4c^4 = 20.25$,
$\Sigma x_{1u}^2 x_{2u}^2 = 4b^2c^2 = 6.75$.
All other sums of powers and products up to and including fourth order are zero. It is possible to estimate a full second-order model with this design provided that the $X'X$ matrix is nonsingular. In fact

$$(X'X)^{-1} = \begin{bmatrix} 9 & 0 & 0 & 9 & 9 & 0 \\ 0 & 9 & 0 & 0 & 0 & 0 \\ 0 & 0 & 9 & 0 & 0 & 0 \\ 9 & 0 & 0 & 20.25 & 6.75 & 0 \\ 9 & 0 & 0 & 6.75 & 20.25 & 0 \\ 0 & 0 & 0 & 0 & 0 & 6.75 \end{bmatrix}^{-1}$$

$$= \begin{bmatrix} \frac{1}{3} & 0 & 0 & -\frac{1}{9} & -\frac{1}{9} & 0 \\ 0 & \frac{1}{9} & 0 & 0 & 0 & 0 \\ 0 & 0 & \frac{1}{9} & 0 & 0 & 0 \\ -\frac{1}{9} & 0 & 0 & \frac{5}{54} & \frac{1}{54} & 0 \\ -\frac{1}{9} & 0 & 0 & \frac{1}{54} & \frac{5}{54} & 0 \\ 0 & 0 & 0 & 0 & 0 & \frac{4}{27} \end{bmatrix}$$

The design is rotatable, because $\Sigma x_{iu}^4 = 3\Sigma x_{1u}^2 x_{2u}^2$.

(b) $V \equiv NV\{\hat{y}(x)\}/\sigma^2 = Nx_0'(X'X)^{-1}x_0$ where

$$x_0' = \left(1, x_1, x_2, x_1^2, x_2^2, x_1x_2\right).$$

This becomes $V = 3 - r^2 + \frac{5}{6}r^4$, where $r^2 = x_1^2 + x_2^2$. The contours are circles about the origin.

(c)   If the true model contained terms $\beta_{111}x_1^3 + \beta_{122}x_1x_2^2 + \beta_{112}x_1^2x_2 + \beta_{222}x_2^3$ in addition to those of second order and below, the alias matrix would be $A = (X'X)^{-1}X'X_2$ where X_2 was a matrix with columns $x_1^3, x_1x_2^2, x_1^2x_2, x_2^3$ generated from the columns of the second-order X matrix. It can be shown that

$$A = \tfrac{1}{4}\begin{bmatrix} 0 & 0 & 0 & 0 \\ 9 & 3 & 0 & 0 \\ 0 & 0 & 3 & 1 \\ 0 & 0 & 0 & 0 \\ 0 & 0 & 0 & 0 \\ 0 & 0 & 0 & 0 \end{bmatrix}.$$

Thus

$$E(b_1) = \beta_1 + 3(3\beta_{111} + \beta_{122})/4,$$

$$E(b_2) = \beta_2 + (3\beta_{112} + \beta_{222})/4.$$

and the remaining b's are unbiased.

15.3. In general, *for a rotatable design* in k factors, $V(r) = \{ P + (2Q + 1/B)r^2 + Rr^4 \}\sigma^2$ where $P = 2(k + 2)D^2/A$, $Q = -2DB/A$, $R = \{ N(k + 1)D - (k - 1)B^2 \}/A$, $A = 2D\{ N(k + 2)D - kB^2 \}$, $B = \sum_{u=1}^{N} x_{iu}^2$, $D = \sum_{u=1}^{N} x_{iu}^2 x_{ju}^2$, and $N =$ total points in design. For the design indicated, $k = 3$, $N = 14 + n_0$, $B = 8 + 4\sqrt{2}$, $D = 8$, and so $P = 1/(n_0 + 0.011775)$, $2Q + 1/B = 0.073223 - 0.682843/(n_0 + 0.011775)$, and

$$R = \frac{(0.117157 + n_0/20)}{(n_0 + 0.011775)},$$

enabling $V(r)$ to be plotted against r for various n_0 values (suggestion). $V(0) = V(1)$ implies that $2Q + 1/B + R = 0$ which implies $n_0 = 4.58$, a result we round to the nearest value, 5. Similar work for $r^2 = B/N$ and for $r^2 = k$ provides $n_0 = 5.55$, and $n_0 = 1.48$, respectively. For additional commentary, see Draper (1982), where the 1.48 value is adjusted to 2, not 1. This is because $|V(0) - V(k^{1/2})|$ is smaller for $n_0 = 2$ than for $n_0 = 1$, even though it is zero at $n_0 = 1.48$.

15.4. Solution is implicit in the question.

15.5. Solution is implicit in the question.

15.6. For orthogonal blocking, we need, for (a) 2^{3-1} blocks and for (b) star block, respectively;

(a) $\dfrac{4}{(8 + 2\alpha^2)} = \dfrac{(4 + \tfrac{1}{2}c_0)}{(14 + c_0 + s_0)},$

(b) $\dfrac{2\alpha^2}{(8 + 2\alpha^2)} = \dfrac{(6 + s_0)}{(14 + c_0 + s_0)}.$

These are satisfied by $c_0 = 4$, $s_0 = 2$, $\alpha^2 = \frac{8}{3}$. From either equation (we really need only one), $\alpha^2 = 4(6 + s_0)/(8 + c_0)$. Some α^2 values are given below.

			s_0			
c_0	0	1	2	3	4	5
0	3	$\frac{7}{2}$	4	$\frac{9}{2}$	5	$\frac{11}{2}$
2	$\frac{12}{5}$	$\frac{14}{5}$	$\frac{16}{5}$	$\frac{18}{5}$	4	$\frac{22}{5}$
4	2	$\frac{7}{3}$	$\frac{8}{3}$	3	$\frac{10}{3}$	$\frac{11}{3}$
6	$\frac{12}{7}$	2	$\frac{16}{7}$	$\frac{18}{7}$	$\frac{20}{7}$	$\frac{22}{7}$
8	$\frac{3}{2}$	$\frac{7}{4}$	2	$\frac{9}{4}$	$\frac{5}{2}$	$\frac{11}{4}$

For rotatability as well, we need $\alpha = 2^{(k-p)/4}$, that is, $\alpha^2 = 2^{3/2}$. Substituting this in the formula gives $8 + c_0 = 2^{1/2}(6 + s_0)$ which cannot be achieved with integer values of c_0 and s_0. It follows that no design *of this form* (cube plus star plus center points) can be rotatable *and* orthogonally blocked, for $k = 3$. (It can happen for other values of k however, and there are other $k = 3$ designs for which it can be achieved. See Exercise 15.8.)

15.7. For orthogonal blocking we need

$$\frac{8}{(16 + 2\alpha^2)} = \frac{\left(8 + \frac{1}{2}c_0\right)}{(24 + c_0 + s_0)},$$

that is,

$$\alpha^2 = \frac{8(8 + s_0)}{(16 + c_0)}.$$

Clearly $c_0 = 4$, $s_0 = 2$, and $\alpha = 2$ satisfy this and the design is rotatable for this value of α. Other values of α^2 for an orthogonally blocked design are given by substituting in for s_0 and c_0. Note that when $c_0 = 2s_0$, $\alpha = 2$ so that an equal number of center points in *each* block will provided a design that is both rotatable *and* orthogonally blocked.

			s_0			
c_0	0	1	2	3	4	5
0	4	$\frac{9}{2}$	5	$\frac{11}{2}$	6	$\frac{13}{2}$
2	$\frac{32}{9}$	4	$\frac{40}{9}$	$\frac{44}{9}$	$\frac{16}{3}$	$\frac{52}{9}$
4	$\frac{16}{5}$	$\frac{18}{5}$	4	$\frac{22}{5}$	$\frac{24}{5}$	$\frac{26}{5}$
6	$\frac{32}{11}$	$\frac{36}{11}$	$\frac{40}{11}$	4	$\frac{48}{11}$	$\frac{52}{11}$
8	$\frac{8}{3}$	3	$\frac{10}{3}$	$\frac{11}{3}$	4	$\frac{13}{3}$

15.8. For rotatability, $2^{k-p} + 4\alpha^4 = 3(2^{k-p})$, that is, $\alpha^2 = 2^{(k-p-1)/2}$. For orthogonal blocking,

$$\frac{2^{k-p}/b}{2\alpha^2} = \frac{2^{k-p}/b + A}{2k + B}$$

which implies that

$$B = 2^{(k-p-1)/2}\left\{1 + \frac{Ab}{2^{k-p}}\right\} - 2k.$$

We substitute in the various values of $A = 0, 1, 2$, and of k, p, and b. The relevant solutions are those for which B is a positive integer or zero. In fact, of the 18 choices given, only the values $k = 3$, $p = 0$, $A = 2$, $b = 2$, $B = 0$ work. Thus a cube divided into two blocks, each with two center points, plus a replicated star ($\alpha = 2^{1/2}$), each star with no center points, forms a four-block orthogonally blocked rotatable design for $k = 3$ factors. (See Table 11.4.)

15.9. The general simplex design with $\sum_{u=1}^{N} x_{iu}^2 = Nc$ has $N = k + 1$ runs in k columns and the design matrix $\mathbf{D}$ is an $N \times k$ matrix of form

$$
\begin{array}{c}
\\
\mathbf{c}_1' \\
\mathbf{c}_2' \\
\mathbf{c}_3' \\
\mathbf{c}_4' \\
\\
\cdots \\
\\
\\
\\
\mathbf{c}_N'
\end{array}
\begin{array}{c}
\begin{array}{cccccccc}
x_1 & x_2 & x_3 & \cdots & x_i & \cdots & x_k \\
\end{array} \\
\left[
\begin{array}{ccccccc}
-a_1 & -a_2 & -a_3 & \cdots & -a_i & \cdots & -a_k \\
a_1 & -a_2 & -a_3 & & -a_i & & -a_k \\
0 & 2a_2 & -a_3 & & -a_i & & -a_k \\
0 & 0 & 3a_3 & & -a_i & & -a_k \\
& & & & \cdots & & \\
& & & & ia_i & & \\
& & & & 0 & & \\
& & & & \cdots & & \\
0 & 0 & 0 & & 0 & & ka_k
\end{array}
\right]
\end{array}
$$

where $a_i = \{cN/i(i+1)\}^{1/2}$. Let $\mathbf{c}_u'$ denote the u-th row. Then it can be verified that

$$\mathbf{c}_u'\mathbf{c}_v = \begin{cases} c(N-1) & \text{if} \quad u = v \\ -c & \text{if} \quad u \neq v \end{cases}$$

Each $\mathbf{c}_u$ represents a point in k-dimensional space. The distance between the *u-th* and the *v-th* is r_{uv} where

$$r_{uv}^2 = (\mathbf{c}_u - \mathbf{c}_v)'(\mathbf{c}_u - \mathbf{c}_v)$$

$$= \mathbf{c}_u'\mathbf{c}_u' - \mathbf{c}_u'\mathbf{c}_v - \mathbf{c}_v'\mathbf{c}_u + \mathbf{c}_v'\mathbf{c}_v$$

$$= c(N - 1 + 1 + 1 + N - 1)$$

$$= 2Nc.$$

The special cases follow immediately. The **X**-matrix is obtained by adjoining a column of one's to the **D** matrix.

Tables

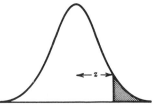

TABLE A. Tail area of unit normal distribution

z	0.00	0.01	0.02	0.03	0.04	0.05	0.06	0.07	0.08	0.09
0.0	0.5000	0.4960	0.4920	0.4880	0.4840	0.4801	0.4761	0.4721	0.4681	0.4641
0.1	0.4602	0.4562	0.4522	0.4483	0.4443	0.4404	0.4364	0.4325	0.4286	0.4247
0.2	0.4207	0.4168	0.4129	0.4090	0.4052	0.4013	0.3974	0.3936	0.3897	0.3859
0.3	0.3821	0.3783	0.3745	0.3707	0.3669	0.3632	0.3594	0.3557	0.3520	0.3483
0.4	0.3446	0.3409	0.3372	0.3336	0.3300	0.3264	0.3228	0.3192	0.3156	0.3121
0.5	0.3085	0.3050	0.3015	0.2981	0.2946	0.2912	0.2877	0.2843	0.2810	0.2776
0.6	0.2743	0.2709	0.2676	0.2643	0.2611	0.2578	0.2546	0.2514	0.2483	0.2451
0.7	0.2420	0.2389	0.2358	0.2327	0.2296	0.2266	0.2236	0.2206	0.2177	0.2148
0.8	0.2119	0.2090	0.2061	0.2033	0.2005	0.1977	0.1949	0.1922	0.1894	0.1867
0.9	0.1841	0.1814	0.1788	0.1762	0.1736	0.1711	0.1685	0.1660	0.1635	0.1611
1.0	0.1587	0.1562	0.1539	0.1515	0.1492	0.1469	0.1446	0.1423	0.1401	0.1379
1.1	0.1357	0.1335	0.1314	0.1292	0.1271	0.1251	0.1230	0.1210	0.1190	0.1170
1.2	0.1151	0.1131	0.1112	0.1093	0.1075	0.1056	0.1038	0.1020	0.1003	0.0985
1.3	0.0968	0.0951	0.0934	0.0918	0.0901	0.0885	0.0869	0.0853	0.0838	0.0823
1.4	0.0808	0.0793	0.0778	0.0764	0.0749	0.0735	0.0721	0.0708	0.0694	0.0681
1.5	0.0668	0.0655	0.0643	0.0630	0.0618	0.0606	0.0594	0.0582	0.0571	0.0559
1.6	0.0548	0.0537	0.0526	0.0516	0.0505	0.0495	0.0485	0.0475	0.0465	0.0455
1.7	0.0446	0.0436	0.0427	0.0418	0.0409	0.0401	0.0392	0.0384	0.0375	0.0367
1.8	0.0359	0.0351	0.0344	0.0336	0.0329	0.0322	0.0314	0.0307	0.0301	0.0294
1.9	0.0287	0.0281	0.0274	0.0268	0.0262	0.0256	0.0250	0.0244	0.0239	0.0233
2.0	0.0228	0.0222	0.0217	0.0212	0.0207	0.0202	0.0197	0.0192	0.0188	0.0183
2.1	0.0179	0.0174	0.0170	0.0166	0.0162	0.0158	0.0154	0.0150	0.0146	0.0143
2.2	0.0139	0.0136	0.0132	0.0129	0.0125	0.0122	0.0119	0.0116	0.0113	0.0110
2.3	0.0107	0.0104	0.0102	0.0099	0.0096	0.0094	0.0091	0.0089	0.0087	0.0084
2.4	0.0082	0.0080	0.0078	0.0075	0.0073	0.0071	0.0069	0.0068	0.0066	0.0064
2.5	0.0062	0.0060	0.0059	0.0057	0.0055	0.0054	0.0052	0.0051	0.0049	0.0048
2.6	0.0047	0.0045	0.0044	0.0043	0.0041	0.0040	0.0039	0.0038	0.0037	0.0036
2.7	0.0035	0.0034	0.0033	0.0032	0.0031	0.0030	0.0029	0.0028	0.0027	0.0026
2.8	0.0026	0.0025	0.0024	0.0023	0.0023	0.0022	0.0021	0.0021	0.0020	0.0019
2.9	0.0019	0.0018	0.0018	0.0017	0.0016	0.0016	0.0015	0.0015	0.0014	0.0014
3.0	0.0013	0.0013	0.0013	0.0012	0.0012	0.0011	0.0011	0.0011	0.0010	0.0010
3.1	0.0010	0.0009	0.0009	0.0009	0.0008	0.0008	0.0008	0.0008	0.0007	0.0007
3.2	0.0007	0.0007	0.0006	0.0006	0.0006	0.0006	0.0006	0.0005	0.0005	0.0005
3.3	0.0005	0.0005	0.0005	0.0004	0.0004	0.0004	0.0004	0.0004	0.0004	0.0003
3.4	0.0003	0.0003	0.0003	0.0003	0.0003	0.0003	0.0003	0.0003	0.0003	0.0002
3.5	0.0002	0.0002	0.0002	0.0002	0.0002	0.0002	0.0002	0.0002	0.0002	0.0002
3.6	0.0002	0.0002	0.0001	0.0001	0.0001	0.0001	0.0001	0.0001	0.0001	0.0001
3.7	0.0001	0.0001	0.0001	0.0001	0.0001	0.0001	0.0001	0.0001	0.0001	0.0001
3.8	0.0001	0.0001	0.0001	0.0001	0.0001	0.0001	0.0001	0.0001	0.0001	0.0001
3.9	0.0000	0.0000	0.0000	0.0000	0.0000	0.0000	0.0000	0.0000	0.0000	0.0000

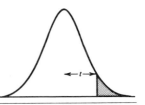

TABLE B. Probability points of the *t* distribution with *v* degrees of freedom

					tail area probability					
v	0.4	0.25	0.1	0.05	0.025	0.01	0.005	0.0025	0.001	0.0005
1	0.325	1.000	3.078	6.314	12.706	31.821	63.657	127.32	318.31	636.62
2	0.289	0.816	1.886	2.920	4.303	6.965	9.925	14.089	22.326	31.598
3	0.277	0.765	1.638	2.353	3.182	4.541	5.841	7.453	10.213	12.924
4	0.271	0.741	1.533	2.132	2.776	3.747	4.604	5.598	7.173	8.610
5	0.267	0.727	1.476	2.015	2.571	3.365	4.032	4.773	5.893	6.869
6	0.265	0.718	1.440	1.943	2.447	3.143	3.707	4.317	5.208	5.959
7	0.263	0.711	1.415	1.895	2.365	2.998	3.499	4.029	4.785	5.408
8	0.262	0.706	1.397	1.860	2.306	2.896	3.355	3.833	4.501	5.041
9	0.261	0.703	1.383	1.833	2.262	2.821	3.250	3.690	4.297	4.781
10	0.260	0.700	1.372	1.812	2.228	2.764	3.169	3.581	4.144	4.587
11	0.260	0.697	1.363	1.796	2.201	2.718	3.106	3.497	4.025	4.437
12	0.259	0.695	1.356	1.782	2.179	2.681	3.055	3.428	3.930	4.318
13	0.259	0.694	1.350	1.771	2.160	2.650	3.012	3.372	3.852	4.221
14	0.258	0.692	1.345	1.761	2.145	2.624	2.977	3.326	3.787	4.140
15	0.258	0.691	1.341	1.753	2.131	2.602	2.947	3.286	3.733	4.073
16	0.258	0.690	1.337	1.746	2.120	2.583	2.921	3.252	3.686	4.015
17	0.257	0.689	1.333	1.740	2.110	2.567	2.898	3.222	3.646	3.965
18	0.257	0.688	1.330	1.734	2.101	2.552	2.878	3.197	3.610	3.922
19	0.257	0.688	1.328	1.729	2.093	2.539	2.861	3.174	3.579	3.883
20	0.257	0.687	1.325	1.725	2.086	2.528	2.845	3.153	3.552	3.850
21	0.257	0.686	1.323	1.721	2.080	2.518	2.831	3.135	3.527	3.819
22	0.256	0.686	1.321	1.717	2.074	2.508	2.819	3.119	3.505	3.792
23	0.256	0.685	1.319	1.714	2.069	2.500	2.807	3.104	3.485	3.767
24	0.256	0.685	1.318	1.711	2.064	2.492	2.797	3.091	3.467	3.745
25	0.256	0.684	1.316	1.708	2.060	2.485	2.787	3.078	3.450	3.725
26	0.256	0.684	1.315	1.706	2.056	2.479	2.779	3.067	3.435	3.707
27	0.256	0.684	1.314	1.703	2.052	2.473	2.771	3.057	3.421	3.690
28	0.256	0.683	1.313	1.701	2.048	2.467	2.763	3.047	3.408	3.674
29	0.256	0.683	1.311	1.699	2.045	2.462	2.756	3.038	3.396	3.659
30	0.256	0.683	1.310	1.697	2.042	2.457	2.750	3.030	3.385	3.646
40	0.255	0.681	1.303	1.684	2.021	2.423	2.704	2.971	3.307	3.551
60	0.254	0.679	1.296	1.671	2.000	2.390	2.660	2.915	3.232	3.460
120	0.254	0.677	1.289	1.658	1.980	2.358	2.617	2.860	3.160	3.373
∞	0.253	0.674	1.282	1.645	1.960	2.326	2.576	2.807	3.090	3.291

Source: Taken with permission of the Biometrika Trustees from E. S. Pearson and H. O. Hartley (Eds.) (1958), *Biometrika Tables for Statisticians*. Vol. 1, Cambridge University Press. Parts of the table are also taken from Table III of Fisher and Yates: *Statistical Tables for Biological, Agricultural and Medical Research*, 5th ed., 1957, published by Longman Group Ltd., London (previously published by Oliver and Boyd, Edinburgh), by permission of the publishers.

TABLE C. Probability points of the χ^2 distribution with ν degrees of freedom

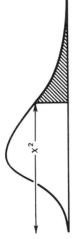

tail area probability

ν	0.995	0.99	0.975	0.95	0.9	0.75	0.5	0.25	0.1	0.05	0.025	0.01	0.005	0.001
1	—	—	—	—	0.016	0.102	0.455	1.32	2.71	3.84	5.02	6.63	7.88	10.8
2	0.010	0.020	0.051	0.103	0.211	0.575	1.39	2.77	4.61	5.99	7.38	9.21	10.6	13.8
3	0.072	0.115	0.216	0.352	0.584	1.21	2.37	4.11	6.25	7.81	9.35	11.3	12.8	16.3
4	0.207	0.297	0.484	0.711	1.06	1.92	3.36	5.39	7.78	9.49	11.1	13.3	14.9	18.5
5	0.412	0.554	0.831	1.15	1.61	2.67	4.35	6.63	9.24	11.1	12.8	15.1	16.7	20.5
6	0.676	0.872	1.24	1.64	2.20	3.45	5.35	7.84	10.6	12.6	14.4	16.8	18.5	22.5
7	0.989	1.24	1.69	2.17	2.83	4.25	6.35	9.04	12.0	14.1	16.0	18.5	20.3	24.3
8	1.34	1.65	2.18	2.73	3.49	5.07	7.34	10.2	13.4	15.5	17.5	20.1	22.0	26.1
9	1.73	2.09	2.70	3.33	4.17	5.90	8.34	11.4	14.7	16.9	19.0	21.7	23.6	27.9
10	2.16	2.56	3.25	3.94	4.87	6.74	9.34	12.5	16.0	18.3	20.5	23.2	25.2	29.6
11	2.60	3.05	3.82	4.57	5.58	7.58	10.3	13.7	17.3	19.7	21.9	24.7	26.8	31.3
12	3.07	3.57	4.40	5.23	6.30	8.44	11.3	14.8	18.5	21.0	23.3	26.2	28.3	32.9
13	3.57	4.11	5.01	5.89	7.04	9.30	12.3	16.0	19.8	22.4	24.7	27.7	29.8	34.5
14	4.07	4.66	5.63	6.57	7.79	10.2	13.3	17.1	21.1	23.7	26.1	29.1	31.3	36.1
15	4.60	5.23	6.26	7.26	8.55	11.0	14.3	18.2	22.3	25.0	27.5	30.6	32.8	37.7

df														
16	5.14	5.81	6.91	7.96	9.31	11.9	15.3	19.4	23.5	26.3	28.8	32.0	34.3	39.3
17	5.70	6.41	7.56	8.67	10.1	12.8	16.3	20.5	24.8	27.6	30.2	33.4	35.7	40.8
18	6.26	7.01	8.23	9.39	10.9	13.7	17.3	21.6	26.0	28.9	31.5	34.8	37.2	42.3
19	6.84	7.63	8.91	10.1	11.7	14.6	18.3	22.7	27.2	30.1	32.9	36.2	38.6	43.8
20	7.43	8.26	9.59	10.9	12.4	15.5	19.3	23.8	28.4	31.4	34.2	37.6	40.0	45.3
21	8.03	8.90	10.3	11.6	13.2	16.3	20.3	24.9	29.6	32.7	35.5	38.9	41.4	46.8
22	8.64	9.54	11.0	12.3	14.0	17.2	21.3	26.0	30.8	33.9	36.8	40.3	42.8	48.3
23	9.26	10.2	11.7	13.1	14.8	18.1	22.3	27.1	32.0	35.2	38.1	41.6	44.2	49.7
24	9.89	10.9	12.4	13.8	15.7	19.0	23.3	28.2	33.2	36.4	39.4	43.0	45.6	51.2
25	10.5	11.5	13.1	14.6	16.5	19.9	24.3	29.3	34.4	37.7	40.6	44.3	46.9	52.6
26	11.2	12.2	13.8	15.4	17.3	20.8	25.3	30.4	35.6	38.9	41.9	45.6	48.3	54.1
27	11.8	12.9	14.6	16.2	18.1	21.7	26.3	31.5	36.7	40.1	43.2	47.0	49.6	55.5
28	12.5	13.6	15.3	16.9	18.9	22.7	27.3	32.6	37.9	41.3	44.5	48.3	51.0	56.9
29	13.1	14.3	16.0	17.7	19.8	23.6	28.3	33.7	39.1	42.6	45.7	49.6	52.3	58.3
30	13.8	15.0	16.8	18.5	20.6	24.5	29.3	34.8	40.3	43.8	47.0	50.9	53.7	59.7

TABLE D. Percentage points of the F distribution: upper 25% points

v_2 \ v_1	1	2	3	4	5	6	7	8	9	10	12	15	20	24	30	40	60	120	∞
1	5.83	7.50	8.20	8.58	8.82	8.98	9.10	9.19	9.26	9.32	9.41	9.49	9.58	9.63	9.67	9.71	9.76	9.80	9.85
2	2.57	3.00	3.15	3.23	3.28	3.31	3.34	3.35	3.37	3.38	3.39	3.41	3.43	3.43	3.44	3.45	3.46	3.47	3.48
3	2.02	2.28	2.36	2.39	2.41	2.42	2.43	2.44	2.44	2.44	2.45	2.46	2.46	2.46	2.47	2.47	2.47	2.47	2.47
4	1.81	2.00	2.05	2.06	2.07	2.08	2.08	2.08	2.08	2.08	2.08	2.08	2.08	2.08	2.08	2.08	2.08	2.08	2.08
5	1.69	1.85	1.88	1.89	1.89	1.89	1.89	1.89	1.89	1.89	1.89	1.89	1.88	1.88	1.88	1.88	1.87	1.87	1.87
6	1.62	1.76	1.78	1.79	1.79	1.78	1.78	1.78	1.77	1.77	1.77	1.76	1.76	1.75	1.75	1.75	1.74	1.74	1.74
7	1.57	1.70	1.72	1.72	1.71	1.71	1.70	1.70	1.69	1.69	1.68	1.68	1.67	1.67	1.66	1.66	1.65	1.65	1.65
8	1.54	1.66	1.67	1.66	1.66	1.65	1.64	1.64	1.63	1.63	1.62	1.62	1.61	1.60	1.60	1.59	1.59	1.58	1.58
9	1.51	1.62	1.63	1.63	1.62	1.61	1.60	1.60	1.59	1.59	1.58	1.57	1.56	1.56	1.55	1.54	1.54	1.53	1.53
10	1.49	1.60	1.60	1.59	1.59	1.58	1.57	1.56	1.56	1.55	1.54	1.53	1.52	1.52	1.51	1.51	1.50	1.49	1.48
11	1.47	1.58	1.58	1.57	1.56	1.55	1.54	1.53	1.53	1.52	1.51	1.50	1.49	1.49	1.48	1.47	1.47	1.46	1.45
12	1.46	1.56	1.56	1.55	1.54	1.53	1.52	1.51	1.51	1.50	1.49	1.48	1.47	1.46	1.45	1.45	1.44	1.43	1.42
13	1.45	1.55	1.55	1.53	1.52	1.51	1.50	1.49	1.49	1.48	1.47	1.46	1.45	1.44	1.43	1.42	1.42	1.41	1.40
14	1.44	1.53	1.53	1.52	1.51	1.50	1.49	1.48	1.47	1.46	1.45	1.44	1.43	1.42	1.41	1.41	1.40	1.39	1.38
15	1.43	1.52	1.52	1.51	1.49	1.48	1.47	1.46	1.46	1.45	1.44	1.43	1.41	1.41	1.40	1.39	1.38	1.37	1.36
16	1.42	1.51	1.51	1.50	1.48	1.47	1.46	1.45	1.44	1.44	1.43	1.41	1.40	1.39	1.38	1.37	1.36	1.35	1.34
17	1.42	1.51	1.50	1.49	1.47	1.46	1.45	1.44	1.43	1.43	1.41	1.40	1.39	1.38	1.37	1.36	1.35	1.34	1.33
18	1.41	1.50	1.49	1.48	1.46	1.45	1.44	1.43	1.42	1.42	1.40	1.39	1.38	1.37	1.36	1.35	1.34	1.33	1.32
19	1.41	1.49	1.49	1.47	1.46	1.44	1.43	1.42	1.41	1.41	1.40	1.38	1.37	1.36	1.35	1.34	1.33	1.32	1.30
20	1.40	1.49	1.48	1.47	1.45	1.44	1.43	1.42	1.41	1.40	1.39	1.37	1.36	1.35	1.34	1.33	1.32	1.31	1.29
21	1.40	1.48	1.48	1.46	1.44	1.43	1.42	1.41	1.40	1.39	1.38	1.37	1.35	1.34	1.33	1.32	1.31	1.30	1.28
22	1.40	1.48	1.47	1.45	1.44	1.42	1.41	1.40	1.39	1.39	1.37	1.36	1.34	1.33	1.32	1.31	1.30	1.29	1.28
23	1.39	1.47	1.47	1.45	1.43	1.42	1.41	1.40	1.39	1.38	1.37	1.35	1.34	1.33	1.32	1.31	1.30	1.28	1.27
24	1.39	1.47	1.46	1.44	1.43	1.41	1.40	1.39	1.38	1.38	1.36	1.35	1.33	1.32	1.31	1.30	1.29	1.28	1.26
25	1.39	1.47	1.46	1.44	1.42	1.41	1.40	1.39	1.38	1.37	1.36	1.34	1.33	1.32	1.31	1.29	1.28	1.27	1.25
26	1.38	1.46	1.45	1.44	1.42	1.41	1.39	1.38	1.37	1.37	1.35	1.34	1.32	1.31	1.30	1.29	1.28	1.26	1.25
27	1.38	1.46	1.45	1.43	1.42	1.40	1.39	1.38	1.37	1.36	1.35	1.33	1.32	1.31	1.30	1.28	1.27	1.26	1.24
28	1.38	1.46	1.45	1.43	1.41	1.40	1.39	1.38	1.37	1.36	1.34	1.33	1.31	1.30	1.29	1.28	1.27	1.25	1.24
29	1.38	1.45	1.45	1.43	1.41	1.40	1.38	1.37	1.36	1.35	1.34	1.32	1.31	1.30	1.29	1.27	1.26	1.25	1.23
30	1.38	1.45	1.44	1.42	1.41	1.39	1.38	1.37	1.36	1.35	1.34	1.32	1.30	1.29	1.28	1.27	1.26	1.24	1.23
40	1.36	1.44	1.42	1.40	1.39	1.37	1.36	1.35	1.34	1.33	1.31	1.30	1.28	1.26	1.25	1.24	1.22	1.21	1.19
60	1.35	1.42	1.41	1.38	1.37	1.35	1.33	1.32	1.31	1.30	1.29	1.27	1.25	1.24	1.22	1.21	1.19	1.17	1.15
120	1.34	1.40	1.39	1.37	1.35	1.33	1.31	1.30	1.29	1.28	1.26	1.24	1.22	1.21	1.19	1.18	1.16	1.13	1.10
∞	1.32	1.39	1.37	1.35	1.33	1.31	1.29	1.28	1.27	1.25	1.24	1.22	1.19	1.18	1.16	1.14	1.12	1.08	1.00

Source: M. Merrington and C. M. Thompson (1943), Tables of percentage points of the inverted beta (F) distribution, *Biometrika*, 33, 73. Used by

TABLE D (continued). Percentage points of the F distribution. upper 10% points

v_2 \ v_1	1	2	3	4	5	6	7	8	9	10	12	15	20	24	30	40	60	120	∞
1	39.86	49.50	53.59	55.83	57.24	58.20	58.91	59.44	59.86	60.19	60.71	61.22	61.74	62.00	62.26	62.53	62.79	63.06	63.33
2	8.53	9.00	9.16	9.24	9.29	9.33	9.35	9.37	9.38	9.39	9.41	9.42	9.44	9.45	9.46	9.47	9.47	9.48	9.49
3	5.54	5.46	5.39	5.34	5.31	5.28	5.27	5.25	5.24	5.23	5.22	5.20	5.18	5.18	5.17	5.16	5.15	5.14	5.13
4	4.54	4.32	4.19	4.11	4.05	4.01	3.98	3.95	3.94	3.92	3.90	3.87	3.84	3.83	3.82	3.80	3.79	3.78	3.76
5	4.06	3.78	3.62	3.52	3.45	3.40	3.37	3.34	3.32	3.30	3.27	3.24	3.21	3.19	3.17	3.16	3.14	3.12	3.10
6	3.78	3.46	3.29	3.18	3.11	3.05	3.01	2.98	2.96	2.94	2.90	2.87	2.84	2.82	2.80	2.78	2.76	2.74	2.72
7	3.59	3.26	3.07	2.96	2.88	2.83	2.78	2.75	2.72	2.70	2.67	2.63	2.59	2.58	2.56	2.54	2.51	2.49	2.47
8	3.46	3.11	2.92	2.81	2.73	2.67	2.62	2.59	2.56	2.54	2.50	2.46	2.42	2.40	2.38	2.36	2.34	2.32	2.29
9	3.36	3.01	2.81	2.69	2.61	2.55	2.51	2.47	2.44	2.42	2.38	2.34	2.30	2.28	2.25	2.23	2.21	2.18	2.16
10	3.29	2.92	2.73	2.61	2.52	2.46	2.41	2.38	2.35	2.32	2.28	2.24	2.20	2.18	2.16	2.13	2.11	2.08	2.06
11	3.23	2.86	2.66	2.54	2.45	2.39	2.34	2.30	2.27	2.25	2.21	2.17	2.12	2.10	2.08	2.05	2.03	2.00	1.97
12	3.18	2.81	2.61	2.48	2.39	2.33	2.28	2.24	2.21	2.19	2.15	2.10	2.06	2.04	2.01	1.99	1.96	1.93	1.90
13	3.14	2.76	2.56	2.43	2.35	2.28	2.23	2.20	2.16	2.14	2.10	2.05	2.01	1.98	1.96	1.93	1.90	1.88	1.85
14	3.10	2.73	2.52	2.39	2.31	2.24	2.19	2.15	2.12	2.10	2.05	2.01	1.96	1.94	1.91	1.89	1.86	1.83	1.80
15	3.07	2.70	2.49	2.36	2.27	2.21	2.16	2.12	2.09	2.06	2.02	1.97	1.92	1.90	1.87	1.85	1.82	1.79	1.76
16	3.05	2.67	2.46	2.33	2.24	2.18	2.13	2.09	2.06	2.03	1.99	1.94	1.89	1.87	1.84	1.81	1.78	1.75	1.72
17	3.03	2.64	2.44	2.31	2.22	2.15	2.10	2.06	2.03	2.00	1.96	1.91	1.86	1.84	1.81	1.78	1.75	1.72	1.69
18	3.01	2.62	2.42	2.29	2.20	2.13	2.08	2.04	2.00	1.98	1.93	1.89	1.84	1.81	1.78	1.75	1.72	1.69	1.66
19	2.99	2.61	2.40	2.27	2.18	2.11	2.06	2.02	1.98	1.96	1.91	1.86	1.81	1.79	1.76	1.73	1.70	1.67	1.63
20	2.97	2.59	2.38	2.25	2.16	2.09	2.04	2.00	1.96	1.94	1.89	1.84	1.79	1.77	1.74	1.71	1.68	1.64	1.61
21	2.96	2.57	2.36	2.23	2.14	2.08	2.02	1.98	1.95	1.92	1.87	1.83	1.78	1.75	1.72	1.69	1.66	1.62	1.59
22	2.95	2.56	2.35	2.22	2.13	2.06	2.01	1.97	1.93	1.90	1.86	1.81	1.76	1.73	1.70	1.67	1.64	1.60	1.57
23	2.94	2.55	2.34	2.21	2.11	2.05	1.99	1.95	1.92	1.89	1.84	1.80	1.74	1.72	1.69	1.66	1.62	1.59	1.55
24	2.93	2.54	2.33	2.19	2.10	2.04	1.98	1.94	1.91	1.88	1.83	1.78	1.73	1.70	1.67	1.64	1.61	1.57	1.53
25	2.92	2.53	2.32	2.18	2.09	2.02	1.97	1.93	1.89	1.87	1.82	1.77	1.72	1.69	1.66	1.63	1.59	1.56	1.52
26	2.91	2.52	2.31	2.17	2.08	2.01	1.96	1.92	1.88	1.86	1.81	1.76	1.71	1.68	1.65	1.61	1.58	1.54	1.50
27	2.90	2.51	2.30	2.17	2.07	2.00	1.95	1.91	1.87	1.85	1.80	1.75	1.70	1.67	1.64	1.60	1.57	1.53	1.49
28	2.89	2.50	2.29	2.16	2.06	2.00	1.94	1.90	1.87	1.84	1.79	1.74	1.69	1.66	1.63	1.59	1.56	1.52	1.48
29	2.89	2.50	2.28	2.15	2.06	1.99	1.93	1.89	1.86	1.83	1.78	1.73	1.68	1.65	1.62	1.58	1.55	1.51	1.47
30	2.88	2.49	2.28	2.14	2.05	1.98	1.93	1.88	1.85	1.82	1.77	1.72	1.67	1.64	1.61	1.57	1.54	1.50	1.46
40	2.84	2.44	2.23	2.09	2.00	1.93	1.87	1.83	1.79	1.76	1.71	1.66	1.61	1.57	1.54	1.51	1.47	1.42	1.38
60	2.79	2.39	2.18	2.04	1.95	1.87	1.82	1.77	1.74	1.71	1.66	1.60	1.54	1.51	1.48	1.44	1.40	1.35	1.29
120	2.75	2.35	2.13	1.99	1.90	1.82	1.77	1.72	1.68	1.65	1.60	1.55	1.48	1.45	1.41	1.37	1.32	1.26	1.19
∞	2.71	2.30	2.08	1.94	1.85	1.77	1.72	1.67	1.63	1.60	1.55	1.49	1.42	1.38	1.34	1.30	1.24	1.17	1.00

Source: M. Merrington and C. M. Thompson (1943), Tables of percentage points of the inverted beta (F) distribution, *Biometrika*, 33, 73. Used by permission of the Biometrika Trustees.

TABLE D (*continued*). **Percentage points of the F distribution: upper 5% points**

v_2 \ v_1	1	2	3	4	5	6	7	8	9	10	12	15	20	24	30	40	60	120	∞
1	161.4	199.5	215.7	224.6	230.2	234.0	236.8	238.9	240.5	241.9	243.9	245.9	248.0	249.1	250.1	251.1	252.2	253.3	254.3
2	18.51	19.00	19.16	19.25	19.30	19.33	19.35	19.37	19.38	19.40	19.41	19.43	19.45	19.45	19.46	19.47	19.48	19.49	19.50
3	10.13	9.55	9.28	9.12	9.01	8.94	8.89	8.85	8.81	8.79	8.74	8.70	8.66	8.64	8.62	8.59	8.57	8.55	8.53
4	7.71	6.94	6.59	6.39	6.26	6.16	6.09	6.04	6.00	5.96	5.91	5.86	5.80	5.77	5.75	5.72	5.69	5.66	5.63
5	6.61	5.79	5.41	5.19	5.05	4.95	4.88	4.82	4.77	4.74	4.68	4.62	4.56	4.53	4.50	4.46	4.43	4.40	4.36
6	5.99	5.14	4.76	4.53	4.39	4.28	4.21	4.15	4.10	4.06	4.00	3.94	3.87	3.84	3.81	3.77	3.74	3.70	3.67
7	5.59	4.74	4.35	4.12	3.97	3.87	3.79	3.73	3.68	3.64	3.57	3.51	3.44	3.41	3.38	3.34	3.30	3.27	3.23
8	5.32	4.46	4.07	3.84	3.69	3.58	3.50	3.44	3.39	3.35	3.28	3.22	3.15	3.12	3.08	3.04	3.01	2.97	2.93
9	5.12	4.26	3.86	3.63	3.48	3.37	3.29	3.23	3.18	3.14	3.07	3.01	2.94	2.90	2.86	2.83	2.79	2.75	2.71
10	4.96	4.10	3.71	3.48	3.33	3.22	3.14	3.07	3.02	2.98	2.91	2.85	2.77	2.74	2.70	2.66	2.62	2.58	2.54
11	4.84	3.98	3.59	3.36	3.20	3.09	3.01	2.95	2.90	2.85	2.79	2.72	2.65	2.61	2.57	2.53	2.49	2.45	2.40
12	4.75	3.89	3.49	3.26	3.11	3.00	2.91	2.85	2.80	2.75	2.69	2.62	2.54	2.51	2.47	2.43	2.38	2.34	2.30
13	4.67	3.81	3.41	3.18	3.03	2.92	2.83	2.77	2.71	2.67	2.60	2.53	2.46	2.42	2.38	2.34	2.30	2.25	2.21
14	4.60	3.74	3.34	3.11	2.96	2.85	2.76	2.70	2.65	2.60	2.53	2.46	2.39	2.35	2.31	2.27	2.22	2.18	2.13
15	4.54	3.68	3.29	3.06	2.90	2.79	2.71	2.64	2.59	2.54	2.48	2.40	2.33	2.29	2.25	2.20	2.16	2.11	2.07
16	4.49	3.63	3.24	3.01	2.85	2.74	2.66	2.59	2.54	2.49	2.42	2.35	2.28	2.24	2.19	2.15	2.11	2.06	2.01
17	4.45	3.59	3.20	2.96	2.81	2.70	2.61	2.55	2.49	2.45	2.38	2.31	2.23	2.19	2.15	2.10	2.06	2.01	1.96
18	4.41	3.55	3.16	2.93	2.77	2.66	2.58	2.51	2.46	2.41	2.34	2.27	2.19	2.15	2.11	2.06	2.02	1.97	1.92
19	4.38	3.52	3.13	2.90	2.74	2.63	2.54	2.48	2.42	2.38	2.31	2.23	2.16	2.11	2.07	2.03	1.98	1.93	1.88
20	4.35	3.49	3.10	2.87	2.71	2.60	2.51	2.45	2.39	2.35	2.28	2.20	2.12	2.08	2.04	1.99	1.95	1.90	1.84
21	4.32	3.47	3.07	2.84	2.68	2.57	2.49	2.42	2.37	2.32	2.25	2.18	2.10	2.05	2.01	1.96	1.92	1.87	1.81
22	4.30	3.44	3.05	2.82	2.66	2.55	2.46	2.40	2.34	2.30	2.23	2.15	2.07	2.03	1.98	1.94	1.89	1.84	1.78
23	4.28	3.42	3.03	2.80	2.64	2.53	2.44	2.37	2.32	2.27	2.20	2.13	2.05	2.01	1.96	1.91	1.86	1.81	1.76
24	4.26	3.40	3.01	2.78	2.62	2.51	2.42	2.36	2.30	2.25	2.18	2.11	2.03	1.98	1.94	1.89	1.84	1.79	1.73
25	4.24	3.39	2.99	2.76	2.60	2.49	2.40	2.34	2.28	2.24	2.16	2.09	2.01	1.96	1.92	1.87	1.82	1.77	1.71
26	4.23	3.37	2.98	2.74	2.59	2.47	2.39	2.32	2.27	2.22	2.15	2.07	1.99	1.95	1.90	1.85	1.80	1.75	1.69
27	4.21	3.35	2.96	2.73	2.57	2.46	2.37	2.31	2.25	2.20	2.13	2.06	1.97	1.93	1.88	1.84	1.79	1.73	1.67
28	4.20	3.34	2.95	2.71	2.56	2.45	2.36	2.29	2.24	2.19	2.12	2.04	1.96	1.91	1.87	1.82	1.77	1.71	1.65
29	4.18	3.33	2.93	2.70	2.55	2.43	2.35	2.28	2.22	2.18	2.10	2.03	1.94	1.90	1.85	1.81	1.75	1.70	1.64
30	4.17	3.32	2.92	2.69	2.53	2.42	2.33	2.27	2.21	2.16	2.09	2.01	1.93	1.89	1.84	1.79	1.74	1.68	1.62
40	4.08	3.23	2.84	2.61	2.45	2.34	2.25	2.18	2.12	2.08	2.00	1.92	1.84	1.79	1.74	1.69	1.64	1.58	1.51
60	4.00	3.15	2.76	2.53	2.37	2.25	2.17	2.10	2.04	1.99	1.92	1.84	1.75	1.70	1.65	1.59	1.53	1.47	1.39
120	3.92	3.07	2.68	2.45	2.29	2.17	2.09	2.02	1.96	1.91	1.83	1.75	1.66	1.61	1.55	1.50	1.43	1.35	1.25
∞	3.84	3.00	2.60	2.37	2.21	2.10	2.01	1.94	1.88	1.83	1.75	1.67	1.57	1.52	1.46	1.39	1.32	1.22	1.00

Source: M. Merrington and C. M. Thompson (1943), Tables of percentage points of the inverted beta (F) distribution, *Biometrika*, **33**, 73. Used by

v_2 \ v_1	1	2	3	4	5	6	7	8	9	10	12	15	20	24	30	40	60	120	∞
1	4052	4999.50	5403	5625	5764	5859	5928	5982	6022	6056	6106	6157	6209	6235	6261	6287	6313	6339	6366
2	98.50	99.00	99.17	99.25	99.30	99.33	99.36	99.37	99.39	99.40	99.42	99.43	99.45	99.46	99.47	99.47	99.48	99.49	99.50
3	34.12	30.82	29.46	28.71	28.24	27.91	27.67	27.49	27.35	27.23	27.05	26.87	26.69	26.60	26.50	26.41	26.32	26.22	26.13
4	21.20	18.00	16.69	15.98	15.52	15.21	14.98	14.80	14.66	14.55	14.37	14.20	14.02	13.93	13.84	13.75	13.65	13.56	13.46
5	16.26	13.27	12.06	11.39	10.97	10.67	10.46	10.29	10.16	10.05	9.89	9.72	9.55	9.47	9.38	9.29	9.20	9.11	9.02
6	13.75	10.92	9.78	9.15	8.75	8.47	8.26	8.10	7.98	7.87	7.72	7.56	7.40	7.31	7.23	7.14	7.06	6.97	6.88
7	12.25	9.55	8.45	7.85	7.46	7.19	6.99	6.84	6.72	6.62	6.47	6.31	6.16	6.07	5.99	5.91	5.82	5.74	5.65
8	11.26	8.65	7.59	7.01	6.63	6.37	6.18	6.03	5.91	5.81	5.67	5.52	5.36	5.28	5.20	5.12	5.03	4.95	4.86
9	10.56	8.02	6.99	6.42	6.06	5.80	5.61	5.47	5.35	5.26	5.11	4.96	4.81	4.73	4.65	4.57	4.48	4.40	4.31
10	10.04	7.56	6.55	5.99	5.64	5.39	5.20	5.06	4.94	4.85	4.71	4.56	4.41	4.33	4.25	4.17	4.08	4.00	3.91
11	9.65	7.21	6.22	5.67	5.32	5.07	4.89	4.74	4.63	4.54	4.40	4.25	4.10	4.02	3.94	3.86	3.78	3.69	3.60
12	9.33	6.93	5.95	5.41	5.06	4.82	4.64	4.50	4.39	4.30	4.16	4.01	3.86	3.78	3.70	3.62	3.54	3.45	3.36
13	9.07	6.70	5.74	5.21	4.86	4.62	4.44	4.30	4.19	4.10	3.96	3.82	3.66	3.59	3.51	3.43	3.34	3.25	3.17
14	8.86	6.51	5.56	5.04	4.69	4.46	4.28	4.14	4.03	3.94	3.80	3.66	3.51	3.43	3.35	3.27	3.18	3.09	3.00
15	8.68	6.36	5.42	4.89	4.56	4.32	4.14	4.00	3.89	3.80	3.67	3.52	3.37	3.29	3.21	3.13	3.05	2.96	2.87
16	8.53	6.23	5.29	4.77	4.44	4.20	4.03	3.89	3.78	3.69	3.55	3.41	3.26	3.18	3.10	3.02	2.93	2.84	2.75
17	8.40	6.11	5.18	4.67	4.34	4.10	3.93	3.79	3.68	3.59	3.46	3.31	3.16	3.08	3.00	2.92	2.83	2.75	2.65
18	8.29	6.01	5.09	4.58	4.25	4.01	3.84	3.71	3.60	3.51	3.37	3.23	3.08	3.00	2.92	2.84	2.75	2.66	2.57
19	8.18	5.93	5.01	4.50	4.17	3.94	3.77	3.63	3.52	3.43	3.30	3.15	3.00	2.92	2.84	2.76	2.67	2.58	2.49
20	8.10	5.85	4.94	4.43	4.10	3.87	3.70	3.56	3.46	3.37	3.23	3.09	2.94	2.86	2.78	2.69	2.61	2.52	2.42
21	8.02	5.78	4.87	4.37	4.04	3.81	3.64	3.51	3.40	3.31	3.17	3.03	2.88	2.80	2.72	2.64	2.55	2.46	2.36
22	7.95	5.72	4.82	4.31	3.99	3.76	3.59	3.45	3.35	3.26	3.12	2.98	2.83	2.75	2.67	2.58	2.50	2.40	2.31
23	7.88	5.66	4.76	4.26	3.94	3.71	3.54	3.41	3.30	3.21	3.07	2.93	2.78	2.70	2.62	2.54	2.45	2.35	2.26
24	7.82	5.61	4.72	4.22	3.90	3.67	3.50	3.36	3.26	3.17	3.03	2.89	2.74	2.66	2.58	2.49	2.40	2.31	2.21
25	7.77	5.57	4.68	4.18	3.85	3.63	3.46	3.32	3.22	3.13	2.99	2.85	2.70	2.62	2.54	2.45	2.36	2.27	2.17
26	7.72	5.53	4.64	4.14	3.82	3.59	3.42	3.29	3.18	3.09	2.96	2.81	2.66	2.58	2.50	2.42	2.33	2.23	2.13
27	7.68	5.49	4.60	4.11	3.78	3.56	3.39	3.26	3.15	3.06	2.93	2.78	2.63	2.55	2.47	2.38	2.29	2.20	2.10
28	7.64	5.45	4.57	4.07	3.75	3.53	3.36	3.23	3.12	3.03	2.90	2.75	2.60	2.52	2.44	2.35	2.26	2.17	2.06
29	7.60	5.42	4.54	4.04	3.73	3.50	3.33	3.20	3.09	3.00	2.87	2.73	2.57	2.49	2.41	2.33	2.23	2.14	2.03
30	7.56	5.39	4.51	4.02	3.70	3.47	3.30	3.17	3.07	2.98	2.84	2.70	2.55	2.47	2.39	2.30	2.21	2.11	2.01
40	7.31	5.18	4.31	3.83	3.51	3.29	3.12	2.99	2.89	2.80	2.66	2.52	2.37	2.29	2.20	2.11	2.02	1.92	1.80
60	7.08	4.98	4.13	3.65	3.34	3.12	2.95	2.82	2.72	2.63	2.50	2.35	2.20	2.12	2.03	1.94	1.84	1.73	1.60
120	6.85	4.79	3.95	3.48	3.17	2.96	2.79	2.66	2.56	2.47	2.34	2.19	2.03	1.95	1.86	1.76	1.66	1.53	1.38
∞	6.63	4.61	3.78	3.32	3.02	2.80	2.64	2.51	2.41	2.32	2.18	2.04	1.88	1.79	1.70	1.59	1.47	1.32	1.00

Source: M. Merrington and C. M. Thompson (1943), Tables of percentage points of the inverted beta (F) distribution, *Biometrika,* **33**, 73. Used by permission of the Biometrika Trustees.

TABLE D (*continued*).   **Percentage points of the *F* distribution: upper 0.1% points**

v_2 \ v_1	1	2	3	4	5	6	7	8	9	10	12	15	20	24	30	40	60	120	∞
1	4053*	5000*	5404*	5625*	5764*	5859*	5929*	5981*	6023*	6056*	6107*	6158*	6209*	6235*	6261*	6287*	6313*	6340*	6366*
2	998.5	999.0	999.2	999.2	999.3	999.3	999.4	999.4	999.4	999.4	999.4	999.4	999.4	999.5	999.5	999.5	999.5	999.5	999.5
3	167.0	148.5	141.1	137.1	134.6	132.8	131.6	130.6	129.9	129.2	128.3	127.4	126.4	125.9	125.4	125.0	124.5	124.0	123.5
4	74.14	61.25	56.18	53.44	51.71	50.53	49.66	49.00	48.47	48.05	47.41	46.76	46.10	45.77	45.43	45.09	44.75	44.40	44.05
5	47.18	37.12	33.20	31.09	29.75	28.84	28.16	27.64	27.24	26.92	26.42	25.91	25.39	25.14	24.87	24.60	24.33	24.06	23.79
6	35.51	27.00	23.70	21.92	20.81	20.03	19.46	19.03	18.69	18.41	17.99	17.56	17.12	16.89	16.67	16.44	16.21	15.99	15.75
7	29.25	21.69	18.77	17.19	16.21	15.52	15.02	14.63	14.33	14.08	13.71	13.32	12.93	12.73	12.53	12.33	12.12	11.91	11.70
8	25.42	18.49	15.83	14.39	13.49	12.86	12.40	12.04	11.77	11.54	11.19	10.84	10.48	10.30	10.11	9.92	9.73	9.53	9.33
9	22.86	16.39	13.90	12.56	11.71	11.13	10.70	10.37	10.11	9.89	9.57	9.24	8.90	8.72	8.55	8.37	8.19	8.00	7.81
10	21.04	14.91	12.55	11.28	10.48	9.92	9.52	9.20	8.96	8.75	8.45	8.13	7.80	7.64	7.47	7.30	7.12	6.94	6.76
11	19.69	13.81	11.56	10.35	9.58	9.05	8.66	8.35	8.12	7.92	7.63	7.32	7.01	6.85	6.68	6.52	6.35	6.17	6.00
12	18.64	12.97	10.80	9.63	8.89	8.38	8.00	7.71	7.48	7.29	7.00	6.71	6.40	6.25	6.09	5.93	5.76	5.59	5.42
13	17.81	12.31	10.21	9.07	8.35	7.86	7.49	7.21	6.98	6.80	6.52	6.23	5.93	5.78	5.63	5.47	5.30	5.14	4.97
14	17.14	11.78	9.73	8.62	7.92	7.43	7.08	6.80	6.58	6.40	6.13	5.85	5.56	5.41	5.25	5.10	4.94	4.77	4.60
15	16.59	11.34	9.34	8.25	7.57	7.09	6.74	6.47	6.26	6.08	5.81	5.54	5.25	5.10	4.95	4.80	4.64	4.47	4.31
16	16.12	10.97	9.00	7.94	7.27	6.81	6.46	6.19	5.98	5.81	5.55	5.27	4.99	4.85	4.70	4.54	4.39	4.23	4.06
17	15.72	10.66	8.73	7.68	7.02	6.56	6.22	5.96	5.75	5.58	5.32	5.05	4.78	4.63	4.48	4.33	4.18	4.02	3.85
18	15.38	10.39	8.49	7.46	6.81	6.35	6.02	5.76	5.56	5.39	5.13	4.87	4.59	4.45	4.30	4.15	4.00	3.84	3.67
19	15.08	10.16	8.28	7.26	6.62	6.18	5.85	5.59	5.39	5.22	4.97	4.70	4.43	4.29	4.14	3.99	3.84	3.68	3.51
20	14.82	9.95	8.10	7.10	6.46	6.02	5.69	5.44	5.24	5.08	4.82	4.56	4.29	4.15	4.00	3.86	3.70	3.54	3.38
21	14.59	9.77	7.94	6.95	6.32	5.88	5.56	5.31	5.11	4.95	4.70	4.44	4.17	4.03	3.88	3.74	3.58	3.42	3.26
22	14.38	9.61	7.80	6.81	6.19	5.76	5.44	5.19	4.99	4.83	4.58	4.33	4.06	3.92	3.78	3.63	3.48	3.32	3.15
23	14.19	9.47	7.67	6.69	6.08	5.65	5.33	5.09	4.89	4.73	4.48	4.23	3.96	3.82	3.68	3.53	3.38	3.22	3.05
24	14.03	9.34	7.55	6.59	5.98	5.55	5.23	4.99	4.80	4.64	4.39	4.14	3.87	3.74	3.59	3.45	3.29	3.14	2.97
25	13.88	9.22	7.45	6.49	5.88	5.46	5.15	4.91	4.71	4.56	4.31	4.06	3.79	3.66	3.52	3.37	3.22	3.06	2.89
26	13.74	9.12	7.36	6.41	5.80	5.38	5.07	4.83	4.64	4.48	4.24	3.99	3.72	3.59	3.44	3.30	3.15	2.99	2.82
27	13.61	9.02	7.27	6.33	5.73	5.31	5.00	4.76	4.57	4.41	4.17	3.92	3.66	3.52	3.38	3.23	3.08	2.92	2.75
28	13.50	8.93	7.19	6.25	5.66	5.24	4.93	4.69	4.50	4.35	4.11	3.86	3.60	3.46	3.32	3.18	3.02	2.86	2.69
29	13.39	8.85	7.12	6.19	5.59	5.18	4.87	4.64	4.45	4.29	4.05	3.80	3.54	3.41	3.27	3.12	2.97	2.81	2.64
30	13.29	8.77	7.05	6.12	5.53	5.12	4.82	4.58	4.39	4.24	4.00	3.75	3.49	3.36	3.22	3.07	2.92	2.76	2.59
40	12.61	8.25	6.60	5.70	5.13	4.73	4.44	4.21	4.02	3.87	3.64	3.40	3.15	3.01	2.87	2.73	2.57	2.41	2.23
60	11.97	7.76	6.17	5.31	4.76	4.37	4.09	3.87	3.69	3.54	3.31	3.08	2.83	2.69	2.55	2.41	2.25	2.08	1.89
120	11.38	7.32	5.79	4.95	4.42	4.04	3.77	3.55	3.38	3.24	3.02	2.78	2.53	2.40	2.26	2.11	1.95	1.76	1.54
∞	10.83	6.91	5.42	4.62	4.10	3.74	3.47	3.27	3.10	2.96	2.74	2.51	2.27	2.13	1.99	1.84	1.66	1.45	1.00

* Multiply these entries by 100.

Source: Reproduced with permission of the Biometrika Trustees from *Biometrika Tables for Statisticians*, eds. E. S. Pearson and H. O. Hartley.

Bibliography

An Alphabetical Listing of Selected References

Abraham, T. P. and Rao, V. Y. (1966). An investigation on functional models for fertilizer-response surfaces. *J. Indian Soc. Agric. Statist.*, **18**, 45–61.

Achor, I. M., Richardson, T. and Draper, N. R. (1981). Effect of treating *Candida utilis* with acid or alkali, to remove nucleic acids, on the quality of the protein. *Agric. Food Chem.*, **29**, 27–33.

Addelman, S. (1961). Irregular fractions of the 2^n factorial experiments. *Technometrics*, **3**, 479–496.

Addelman, S. (1962a). Orthogonal main-effect plans for asymmetrical factorial experiments. *Technometrics*, **4**, 21–46.

Addelman, S. (1962b). Symmetrical and asymmetrical fractional factorial plans. *Technometrics*, **4**, 47–57.

Addelman, S. (1963). Techniques for constructing fractional replicate plans. *J. Am. Statist. Assoc.*, **58**, 45–71.

Addelman, S. (1964). Designs for the sequential application of factors. *Technometrics*, **6**, 365–370.

Addelman, S. (1969). Sequence of two-level fractional factorial plans. *Technometrics*, **11**, 477–509.

Adhikary, B. (1965). On the properties and construction of balanced block designs with variable replications. *Calcutta Statist. Assoc. Bull.*, **14**, 36–64.

Adhikary, B. (1966). Some types of m-associate P.B.I.B. association schemes. *Calcutta Statist. Assoc. Bull.*, **15**, 47–74.

Adhikary, B. (1967). Group divisible designs with variable replications. *Calcutta Statist. Assoc. Bull.*, **16**, 73–92.

Adhikary, B. (1972). A note on restricted Kronecker product method of constructing statistical designs. *Calcutta Statist. Assoc. Bull.*, **21**, 193–196.

Adhikary, B. (1973). On generalized group divisible designs. *Calcutta Statist. Assoc. Bull.*, **22**, 75–88.

Adhikary, B. and Panda, R. (1977). Restricted Kronecker product method of constructing second order rotatable designs. *Calcutta Statist. Assoc. Bull.*, **26**, 61–78.

615

Adhikary, B. and Panda, R. (1980). Construction of group divisible rotatable designs. Lecture Notes in Mathematics 885, *Combinatorics and Graph Theory Symposium Proceedings*, ISI, Calcutta, Feb. 1980, 171–184. Berlin: Springer-Verlag.

Adhikary, B. and Panda, R. (1981). Construction of GDSORD from balanced block designs with variable replication. *Calcutta Statist. Assoc. Bull.*, **30**, 129–137.

Adhikary, B. and Panda, R. (1982). On some mixed order response surface designs in sequential experiments. *Calcutta Statist. Assoc. Bull.*, **31**, 27–52.

Adhikary, B. and Panda, R. (1983). On group divisible response surface designs. *J. Statist. Planning Inference*, **7**, 387–405.

Adhikary, B. and Panda, R. (1984). Group divisible third order rotatable designs (GDTORD). *Sankhya*, **B46**, 135–146.

Adhikary, B. and Panda, R. (1985). Group divisible response surface (GDRS) designs of third order. *Calcutta Statist. Assoc. Bull.*, **34**, 75–87.

Adhikary, B. and Panda, R. (1986). Sequential rotatable designs. Unpublished ms.

Adhikary, B. and Sinha, B. K. (1976). On group divisible rotatable designs. *Calcutta Statist. Assoc. Bull.*, **25**, 79–93.

Agrawal, V. and Dey, A. (1983). Orthogonal resolution IV designs for some asymmetrical factorials. *Technometrics*, **25**, 197–199.

Aia, M. A., Goldsmith, R. L., and Mooney, R. W. (1961). Precipitating stoichiometric $CaHPO_4 \cdot 2H_2O$. *Indus. Eng. Chem.*, **53**, 55–57.

Aitchison, J. (1982). The statistical analysis of compositional data. *J. Roy. Statist. Soc.* **B44**, 139–160, *discussion* 161–177.

Aitchison, J. and Bacon-Shone, J. (1984). Log contrast models for experiments with mixtures. *Biometrika*, **71**, 323–330.

Aitken, M. and Wilson, G. T. (1980). Mixture models, outliers, and the e.m. algorithm. *Technometrics*, **22**, 325–331.

Albert, A. E. (1961). The sequential design of experiments for infinitely many states of nature. *Ann. Math. Statist.*, **32**, 774–799.

Alessi, J. and Power, J. F. (1977). Residual effects of nitrogen fertilization on dryland spring wheat in the Northern Plains I. Wheat yield and water use. *Agronomy J.*, **69**, 1007–1011.

Alessi, J. and Power, J. F. (1978). Residual effects of N fertilization on dryland spring wheat in the Northern Plains. II. Fate of fertilizer N. *Agronomy J.*, **70**, 282–286.

Allen, D. J., Reimers, H. J., Fauerstein, I. A., and Mustard, J. F. (1975). The use and analysis of multiple responses in multicompartment cellular systems. *Biometrics*, **31**, 921–929.

American Supplier Institute. (1984). *Quality Engineering using Design of Experiments: A Supplier Symposium on Taguchi Methods*, Vol. I and II. Center for Taguchi Methods, American Supplier Institute Inc., 6 Parklane Blvd., Dearborn, MI 48126.

Andersen, S. L. (1959). Statistics in the strategy of chemical experimentation. *Chem. Eng. Prog.*, **55**(4), 61–67.

Anderson, R. L. (1953). Recent advances in finding best operating conditions. *Biometrics*, **48**, 789–798.

Anderson, R. L. and Nelson, L. A. (1975). A family of models involving intersecting straight lines and concomitant experimental designs useful in evaluating response to fertilizer nutrients. *Biometrics*, **31**, 303–318.

Andrews, D. F. (1971). Significance tests based on residuals. *Biometrika*, **58**, 139–148.

Andrews, D. F., Bickel, P. J., Hampel, F. R., Huber, P. J., Rogers, W. H., and Tukey, J. W.

(1972). *Robust Estimates of Location: Survey and Advances*. Princeton: Princeton University Press.

Andrews, D. F. and Herzberg, A. M. (1979). The robustness and optimality of response surface designs. *J. Statist. Planning Inference*, 3, 249–257.

Anscombe, F. J. (1960). Rejection of outliers. *Technometrics*, 2, 123–147.

Anscombe, F. J. (1961). Examination of residuals. *Proc. Fourth Berkeley Symp. Math. Statist. Prob.*, 1, 1–36.

Anscombe, F. J. and Tukey, J. W. (1963). The examination and analysis of residuals. *Technometrics*, 5, 141–160.

Armbrust, D. V. (1968). Windblown soil abrasive injury to cotton plants. *Agronomy J.*, 60, 622–625.

Ash, A. and Hedayat, A. (1978). An introduction to design optimality with an overview of the literature. *Commun. Statist. Theory Methods*, A7, 1295–1325.

ASQC (1983). *Glossary and Tables for Statistical Quality Control*. American Society for Quality Control, Milwaukee, WI.

Atkinson, A. C. (1965). The design of several experiments to estimate a few parameters. Imperial College Diploma Dissertation.

Atkinson, A. C. (1969a). A test for discriminating between models. *Biometrika*, 56, 337–347.

Atkinson, A. C. (1969b). Constrained maximization and the design of experiments. *Technometrics*, 11, 616–618.

Atkinson, A. C. (1970a). A method of discriminating between models. *J. Roy. Statist. Soc.*, B32, 323–345, discussion 345–353.

Atkinson, A. C. (1970b). The design of experiments to estimate the slope of a response surface. *Biometrika*, 57, 319–328.

Atkinson, A. C. (1972). Planning experiments to detect inadequate regression models. *Biometrika*, 59, 275–293.

Atkinson, A. C. (1973a). Multifactor second order designs for cuboidal regions. *Biometrika*, 60, 15–19.

Atkinson, A. C. (1973b). Testing transformations to normality. *J. Roy. Statist. Soc.*, B35, 473–479.

Atkinson, A. C. (1981). A comparison of two criteria for the design of experiments for discriminating between models. *Technometrics*, 23, 301–305.

Atkinson, A. C. (1982). Developments in the design of experiments. *Int. Statist. Rev.*, 50, 161–177.

Atkinson, A. C. and Fedorov, V. V. (1975a). The design of experiments for discriminating between two rival models. *Biometrika*, 62, 57–70.

Atkinson, A. C. and Fedorov, V. V. (1975b). Optimal design: experiments for discriminating between several models. *Biometrika*, 62, 289–303.

Atkinson, A. C. and Hunter, W. G. (1968). The design of experiments for parameter estimation. *Technometrics*, 10, 271–289.

Atwood, C. L. (1969). Optimal and efficient designs of experiments. *Ann. Math. Statist.*, 40, 1570–1602.

Atwood, C. L. (1973). Sequences converging to D-optimal designs of experiments. *Ann. Statist.*, 1, 342–352.

Atwood, C. L. (1975). Estimating a response surface with an uncertain number of parameters, assuming normal errors. *J. Am. Statist. Assoc.*, **70**, 613–617.

Atwood, C. L. (1976). Convergent design sequences for sufficiently regular optimality criteria. *Ann. Statist.*, **4**, 1124–1138.

Baasel, W. D. (1965). Exploring response surfaces to establish optimum conditions. *Chem. Eng.*, **72**, October 25, 147–152.

Bacon, D. W. (1970). Making the most of a "one-shot" experiment. *Indus. Eng. Chem.*, **62**, 27–34.

Bacon, D. W. and Watts, D. G. (1971). Estimating the transition between two intersecting straight lines. *Biometrika*, **58**, 525–534.

Bacon, D. W. and Watts, D. G. (1974). Using a hyperbola as a transition model to fit two regime straight line data. *Technometrics*, **16**, 369–373.

Bailey, R. A. (1977). Patterns of confounding in factorial designs. *Biometrika*, **64**, 597–603.

Bailey, R. A., Gilchrist, F. H. L., and Patterson, H. D. (1977). Identification of effects and confounding patterns in factorial designs. *Biometrika*, **64**, 347–354.

Bain, W. A. and Batty, J. E. (1956). Inactivation of adrenaline and noradrenaline by human and other mammalian liver in vitro. *Brit. J. Pharm. Chemotherapy*, **11**, 52–57.

Baird, B. L. and Mason, D. D. (1959). Multivariable equations describing fertility-corn yield response surfaces and their agronomic and economic interpretation. *Agronomy J.*, **51**, 152–156.

Baker, F. D. and Bargmann, R. E. (1985). Orthogonal central composite designs of the third order in the evaluation of sensitivity and plant growth simulation models. *J. Amer. Statist. Assoc.*, **80**, 574–579.

Baker, R. J., Clark, M. R. B., and Nelder, J. A. (1978). The GLIM System: Release 3. London: Royal Statistical Society.

Banerjee, K. S. and Federer, W. T. (1963). On estimates for fractions of a complete factorial experiment as orthogonal linear combinations of the observations. *Ann. Math. Statist.*, **34**, 1068–1078.

Barnett, V. and Lewis, T. (1978). *Outliers in Statistical Data.* New York: Wiley.

Bates, D. M. and Watts, D. G. (1980). Relative curvature measures of nonlinearity. *J. Roy. Statist. Soc.*, **B42**, 1–16, discussion 16–25.

Bates, D. M. and Watts, D. G. (1987). *Nonlinear Regression Analysis and Its Applications.* New York: Wiley.

Baumert, L., Golomb, S. W., and Hall, M. (1962). Discovery of an Hadamard matrix of order 92. *Amer. Math. Soc. Bull.*, **68**, 237–238.

Beale, E. M. L. (1960). Confidence regions in nonlinear estimation. *J. Roy. Statist. Soc.*, **B22**, 41–76, discussion 76–88.

Beauchamp, J. J. and Cornell, R. G. (1969). Spearman simultaneous estimation for a compartmental model. *Technometrics*, **11**, 551–560.

Beaver, R. J. (1977). Weighted least squares response surface fitting in factorial paired comparisons. *Commun. Statist. Theory Methods*, **A6**, 1275–1287.

Beckman, R. J. and Cook, R. D. (1983). Outlier..........s. *Technometrics*, **25**, 119–149, discussion, 150–163. Correction, p. 390.

Behnken, D. W. (1964). Estimation of copolymer reactivity ratios: an example of nonlinear estimation. *J. Polymer Sci.*, **A2**, 645–668.

Behnken, D. W. and Draper, N. R. (1972). Residuals and their variance patterns. *Technometrics*, **14**, 101–111.

Belsley, D. A., Kuh, E., and Welsch, R. E. (1980). *Regression Diagnostics: Identifying Influential Data and Sources of Collinearity*. New York: Wiley.

Bemesderfer, J. L. (1979). Approving a process for production. *J. Qual. Technol.*, **11**, 1–12.

Berg, C. (1960). Optimization in process development. *Chem. Eng. Prog.*, **36**(8), 42–47.

Berger, P. D. (1972). On Yates' order in fractional factorial designs *Technometrics*, **14**, 971–972.

Berkson, J. (1944). Application of the logistic function to bioassay. *J. Am. Statist. Assoc.*, **39**, 357–365.

Berkson, J. (1951). Why I prefer logits to probits. *Biometrics*, **7**, 327–329.

Berry, J. W., Tucker, H., and Deutschman, A. J. (1963). Starch vinylation. *Indus. Eng. Chem. Proc. Des. Dev.* **2**(4), 318–322. (Also see **3**, 1964, 287–288.)

Bhaskar Rao, M. (1970). Balanced orthogonal designs and their application in the construction of some BIB and group divisible designs. *Sankhya A*, **32**, 439–448.

Bickel, P. J. and Herzberg, A. M. (1979). Robustness of design against autocorrelation in time I: Asymptotic theory, optimality for location and linear regression. *Ann. Statist.*, **7**, 77–95.

Bickel, P. J., Herzberg, A. M., and Schilling, M. (1981). Robustness of design against autocorrelation in time II: Numerical results for the first order autoregressive process. *J. Am. Statist. Assoc.*, **76**, 870–877.

Biles, W. E. (1975). A response surface method for experimental optimization of multi-response processes. *Ind. Eng. Chem. Proc. Des. Dev.*, **14**, 152–158.

Bisgaard, S., Hunter, W. G., and Pallesen, L. (1984). Economic selection of quality of manufactured product. *Technometrics*, **26**, 9–18.

Blight, B. J. N. and Ott, L. (1975). A Bayesian approach to model inadequacy for polynomial regression. *Biometrika*, **62**, 79–88.

Bliss, C. I. (1935a). The calculation of the dosage-mortality curve. *Ann. Appl. Biol.*, **22**, 134–167.

Bliss, C. I. (1935b). The comparison of dosage-mortality data. *Ann. Appl. Biol.*, **22**, 307–333.

Bliss, C. I. (1938). The determination of dosage-mortality curves from small numbers. *Quart. J. Pharm.*, **11**, 192–216.

Bliss, C. I. (1970). *Statistics in Biology, Volume II*. New York: McGraw-Hill.

Bloore, C. G. (1981). *A Quality Control System for the Manufacture of Spray Dried Milk Powders*. PhD. Thesis, Massey University, New Zealand.

Blum, J. R. (1954). Multidimensional stochastic approximation methods. *Ann. Math. Statist.*, **25**, 737–744.

Bohrer, R., Chow, W., Faith, R., Joshi, V. M., and Wu, C.-F. (1981). Multiple three-decision rules for factorial simple effects: Bonferroni wins again. *J. Am. Statist. Assoc.*, **76**, 119–124.

Bole, J. B. and Freyman, S. (1975). Response of irrigated field and sweet corn to nitrogen and phosphorus fertilizers in southern Alberta. *Can. J. Soil Science*, **55**, 137–143.

Bolker, H. I. (1965). Delignification by nitrogen compounds. *Ind. Eng. Chem. Prod. Res. and Devel.*, **4**, 74–79.

Bose, R. C. and Carter, R. L. (1959). Complex representation in the construction of rotatable designs. *Ann. Math. Statist.*, **30**, 771–780.

Bose, R. C., Clatworthy, W. H., and Shrikhande, S. S. (1954). Tables of partially balanced designs with two associate classes. North Carolina Agri. Expt. Station, Tech. Bull. No. 107.

Bose, R. C. and Draper, N. R. (1959). Second order rotatable designs in three dimensions. *Ann. Math. Statist.*, **30**, 1097–1112.

Box, G. E. P. (1952). Multifactor designs of first order. *Biometrika*, **39**, 49–57. (See also p. 189, note by Tocher.)

Box, G. E. P. (1953). Non-normality and tests on variances. *Biometrika*, **40**, 318–335.

Box, G. E. P. (1954). The exploration and exploitation of response surfaces: some general considerations and examples. *Biometrics*, **10**, 16–60.

Box, G. E. P. (1954). Contribution to the discussion, Symposium on Interval Estimation. *J. Roy. Statist. Soc.*, **B16**, 211–212.

Box, G. E. P. (1957). Evolutionary operation: a method for increasing industrial productivity. *Appl. Statist.*, **6**, 3–23.

Box, G. E. P. (1958). Use of statistical methods in the elucidation of basic mechanisms. *Bull. Inter. Statist. Inst.*, **36**, 215–225.

Box, G. E. P. (1959). Answer to query: replication of non-center points in the rotatable and near-rotatable central composite design. *Biometrics*, **15**, 133–135.

Box, G. E. P. (1960a). Some general considerations in process optimization. *J. Roy. Soc. Basic Eng.*, **82**, 113–119.

Box, G. E. P. (1960b). Fitting empirical data. *Annals New York Academy of Sciences*, **86**, 792–816.

Box, G. E. P. (1963). The effects of errors in the factor levels and experimental design. *Technometrics*, **5**, 247–262.

Box, G. E. P. (1964). Some notes on nonlinear estimation. University of Wisconsin–Madison Statistics Department Technical Report No. 25.

Box, G. E. P. (1966a). A simple system of evolutionary operation subject to empirical feedback. *Technometrics*, **8**, 19–26.

Box, G. E. P. (1966b). Use and abuse of regression. *Technometrics*, **8**, 625–629.

Box, G. E. P. (1967). Experimental strategy. *Proc. Sixth International Biometrics Conf.*, Sydney, Australia.

Box, G. E. P. (1968). Response surfaces. Article under "Experimental Design" in *The International Encyclopedia of the Social Sciences*, 254–259. New York: MacMillan and Free Press.

Box, G. E. P. (1976). Science and statistics. *J. Amer. Statist. Assoc.*, **71**, 791–799.

Box, G. E. P. (1980). Sampling and Bayes' inference in scientific modelling and robustness. *J. Roy. Statist. Soc.*, **A143**, 383–404, discussion 404–430.

Box, G. E. P. (1982). Choice of response surface design and alphabetic optimality. *Utilitas Mathematica*, **21B**, 11–55.

Box, G. E. P. (1985). *The Collected Works, Volumes I and II*. ed., G. C. Tiao, Belmont, CA: Wadsworth.

Box, G. E. P. and Behnken, D. W. (1960a). Some new three level designs for the study of quantitative variables. *Technometrics*, **2**, 455–475. Corrections, **3**, 1961, p. 576.

Box, G. E. P. and Behnken, D. W. (1960b). Simplex-sum designs: a class of second order rotatable designs derivable from those of first order. *Ann. Math. Statist.*, **31**, 838–864.

Box, G. E. P. and Coutie, G. A. (1956). Application of digital computers in the exploration of functional relationships. *Proc. Institution Elec. Engineers*, **103**, Part B, supplement No. 1, 100–107.

Box, G. E. P. and Cox, D. R. (1964). An analysis of transformations. *J. Roy. Statist. Soc.*, **B26**, 211–243, discussion 244–252.

Box, G. E. P. and Draper, N. R. (1959). A basis for the selection of a response surface design. *J. Am. Statist. Assoc.*, **54**, 622–654.

Box, G. E. P. and Draper, N. R. (1963). The choice of a second order rotatable design. *Biometrika*, **50**, 335–352.

Box, G. E. P. and Draper, N. R. (1965). The Bayesian estimation of common parameters from several responses. *Biometrika*, **52**, 355–365.

Box, G. E. P. and Draper, N. R. (1968). Isn't my process too variable for EVOP? *Technometrics*, **10**, 439–444.

Box, G. E. P. and Draper, N. R. (1969). *Evolutionary Operation*. New York: Wiley.

Box, G. E. P. and Draper, N. R. (1970). EVOP-makes a plant grow better. *Indus. Engr.*, April, 31–33; condensed in *Management Rev.*, July 1970, 22–25.

Box, G. E. P. and Draper, N. R. (1975). Robust designs. *Biometrika*, **62**, 347–352.

Box, G. E. P. and Draper, N. R. (1980). The variance function of the difference between two estimated responses. *J. Roy. Statist. Soc.*, **B42**, 79–82.

Box, G. E. P. and Draper, N. R. (1982a). Measures of lack of fit for response surface designs and predictor variable transformations. *Technometrics*, **24**, 1–8. See also **25**, 217.

Box, G. E. P. and Draper, N. R. (1982b). Evolutionary operation. Article in the *Encyclopedia of Statistical Sciences*, eds., N. L. Johnson and S. Kotz, Vol. 2, 564–572. New York: Wiley.

Box, G. E. P. and Fung, C. A. (1982). Some considerations in estimating data transformations. University of Wisconsin–Madison, Mathematics Research Center Technical Summary Report #2609, December.

Box, G. E. P. and Hay, W. A. (1953). A statistical design for the efficient removal of trends occurring in a comparative experiment with an application in biological assay. *Biometrics*, **9**, 304–319.

Box, G. E. P. and Hill, W. J. (1967). Discrimination among mechanistic models. *Technometrics*, **9**, 57–71.

Box, G. E. P. and Hill, W. J. (1974). Correcting inhomogeneity of variance with power transformation weighting. *Technometrics*, **16**, 385–389.

Box, G. E. P. and Hunter, J. S. (1954). A confidence region for the solution of a set of simultaneous equations with an application to experimental design. *Biometrika*, **41**, 190–199.

Box, G. E. P. and Hunter, J. S. (1957). Multifactor experimental designs for exploring response surfaces. *Ann. Math. Statist.*, **28**, 195–241.

Box, G. E. P. and Hunter, J. S. (1958). Experimental designs for the exploration and exploitation of response surfaces. In *Experimental Designs in Industry*, ed., V. Chew, 138–190. New York: Wiley.

Box, G. E. P. and Hunter, J. S. (1961a). The 2^{k-p} fractional factorial designs, I. *Technometrics*, **3**, 311–351.

Box, G. E. P. and Hunter, J. S. (1961b). The 2^{k-p} fractional factorial designs, II. *Technometrics*, **3**, 449–458.

Box, G. E. P. and Hunter, W. G. (1962). A useful method for model building. *Technometrics*, **4**, 301–318.

Box, G. E. P. and Hunter, W. G. (1965a). Sequential design of experiments for nonlinear models. *Proceedings of IBM Scientific Computing Symposium in Statistics*, 113–137.

Box, G. E. P. and Hunter, W. G. (1965b). The experimental study of physical mechanisms. *Technometrics*, **7**, 23–42.

Box, G. E. P., Hunter, W. G. and Hunter, J. S. (1978). *Statistics for Experimenters. An Introduction to Design, Data Analysis, and Model Building*. New York: Wiley.

Box, G. E. P., Hunter, W. G., MacGregor, J. F., and Erjavec, J. (1973). Some problems associated with the analysis of multiresponse data. *Technometrics*, **15**, 33–51.

Box, G. E. P. and Jenkins, G. M. (1962). Some statistical aspects of adaptive optimisation and control. *J. Roy. Statist. Soc.*, **B24**, 297–331, discussion 332–343.

Box, G. E. P. and Jenkins, G. M. (1976). *Time Series Analysis: Forecasting and Control*, 2nd ed. San Francisco: Holden-Day.

Box, G. E. P. and Kanemasu, H. (1972). Topics in model-building. Part II: On linear least squares. University of Wisconsin, Statistics Department Technical Report No. 321, November.

Box, G. E. P. and Kanemasu, H. (1984). Constrained nonlinear least squares. Article No. 17, pp. 297–318, In *Contributions to Experimental Design, Statistical Models and Genetic Statistics*. (Essays in honor of Oscar Kempthorne), ed. K. Hinkelmann. New York: Marcel Dekker.

Box, G. E. P. and Lucas, H. L. (1959). Design of experiments in nonlinear situations. *Biometrika*, **46**, 77–90.

Box, G. E. P. and Meyer, R. D. (1985). Studies in quality improvement I: dispersion effects from fractional designs. University of Wisconsin—Madison, Mathematics Research Center Technical Summary Report #2796, February. *Technometrics*, **28**, 1986, 19–27.

Box, G. E. P. and Meyer, R. D. (1985). Studies in quality improvement II: an analysis for unreplicated fractional factorials. University of Wisconsin—Madison, Mathematics Research Center Technical Summary Report #2797, March. *Technometrics*, **28**, 1986, 11–18.

Box, G. E. P. and Meyer, R. D. (1985). Analysis of unreplicated factorials allowing for faulty observations. University of Wisconsin—Madison, Mathematics Research Center Technical Summary Report #2799, March.

Box, G. E. P. and Newbold, P. (1971). Some comments on a paper by Coen, Gomme, and Kendall. *J. Roy. Statist. Soc.* **A134**, 229–240.

Box, G. E. P. and Tidwell, P. W. (1962). Transformation of the independent variables. *Technometrics*, **4**, 531–550.

Box, G. E. P. and Wetz, J. (1973). Criteria for judging adequacy of estimation by an approximating response function. University of Wisconsin Statistics Department Technical Report No. 9.

Box, G. E. P. and Wilson, K. B. (1951). On the experimental attainment of optimum conditions. *J. Roy. Statist. Soc.*, **B13**, 1–38, discussion 38–45.

Box, G. E. P. and Youle, P. V. (1955). The exploration and exploitation of response surfaces: an example of the link between the fitted surface and the basic mechanism of the system. *Biometrics*, **11**, 287–323.

Box, J. F. (1978). *R. A. Fisher: The Life of a Scientist*. New York: Wiley.

Box, M. J. (1966). A comparison of several current optimisation methods, and the use of transformations in constrained problems. *Comp. J.*, **9**, 67–77.

Box, M. J. (1968a). The use of designed experiments in nonlinear model building. In *The Future of Statistics.*, ed. D. G. Watts. New York: Academic Press.

Box, M. J. (1968b). The occurrence of replications in optimal designs of experiments to estimate parameters in nonlinear models. *J. Roy. Statist. Soc.*, **B30**, 290–302.

Box, M. J. (1969). Planning experiments to test the adequacy of nonlinear models. *Appl. Statist.*, **18**, 241–248.

Box, M. J. (1970). Some experiences with a nonlinear experimental design criterion. *Technometrics*, **12**, 569–589.

Box, M. J. (1971a). Bias in nonlinear estimation. *J. Roy. Statist. Soc.*, **B33**, 171–190, discussion 190–201.

Box, M. J. (1971b). Simplified experimental design. *Technometrics*, **13**, 19–31.

Box, M. J. (1971c). An experimental design criterion for precise estimation of a subset of the parameters of a nonlinear model. *Biometrika*, **58**, 149–153.

Box, M. J., Davies, D. and Swann, W. H. (1969). *Nonlinear Optimisation Techniques.* I.C.I. Monograph No. 5. Edinburgh: Oliver and Boyd.

Box, M. J. and Draper, N. R. (1968). Nonlinear model building under nonhomogeneous variance assumptions. I.C.I. research note 68/9.

Box, M. J. and Draper, N. R. (1971). Factorial designs, the $|X'X|$ criterion and some related matters. *Technometrics*, **13**, 731–742. Corrections, **14**, 1972, 511; **15**, 1973, 430.

Box, M. J. and Draper, N. R. (1972). Estimation and design criteria for multiresponse nonlinear models with nonhomogeneous variance. *Appl. Statist.*, **21**, 13–24.

Box, M. J. and Draper, N. R. (1974). Some minimum-point designs for second order response surfaces. *Technometrics*, **16**, 613–616.

Box, M. J., Draper, N. R., and Hunter, W. G. (1970). Missing values in multiresponse non-linear model fitting. *Technometrics*, **12**, 613–620.

Boyd, D. A. (1972). Some recent ideas on fertilizer response curves. *Proc. Ninth Int. Cong. Potash Instit.*, 461–473.

Bradley, R. A. (1958). Determination of optimum operating conditions by experimental methods, Part I. *Indus. Qual. Control*, **15**(4), 16–20.

Brannigan, M. (1981). An adaptive piecewise polynomial curve fitting procedure for data analysis. *Commun. Statist. Theory Methods*, **A10**, 1823–1848.

Brayton, R. K., Director, S. W., and Hachtel, G. D. (1980). Yield maximization and worst-case design with arbitrary statistical distributions. *IEEE Trans. Circuits and Systems*, **CAS-27**, 756–764.

Brooks, R. J. (1972). A decision theory approach to optimal regression designs. *Biometrika*, **59**, 563–571.

Brooks, S. H. (1959). A comparison of maximum seeking methods. *Operations Research*, **7**, 430–457.

Brooks, S. H. and Mickey, M. R. (1961). Optimum estimation of gradient direction in steepest ascent experiments. *Biometrics*, **17**, 48–56.

Buehler, R. J., Shah, B. V., and Kempthorne, O. (1964). Method of parallel tangents. *Chem. Eng. Prog. Symp. Ser.*, **60**(50), 1–7.

Burdick, D. S. and Naylor, T. H. (1969). Response surface methods in economics. *Rev. Internat. Statist. Inst.*, **37**, 18–35.

Cady, F. B. and Fuller, W. A. (1970). The statistics-computer interface in agronomic research. *Agronomy J.*, **62**, 599–604.

Carmer, S. G. and Jackobs, J. A. (1965). An exponential model for predicting optimum plant density and maximum corn yield. *Agronomy J.*, **57**, 241–244.

Carpenter, B. H. and Sweeney, H. C. (1965). Process improvement with "simplex" self-directing evolutionary operation. *Chem. Eng.*, **72**(14), 117–126.

Carr, J. M. and McCracken, E. A. (1960). Statistical program planning and process development. *Chem. Eng. Prog.*, **56**(11), 56–61.

Carr, N. L. (1960). Kinetics of catalytic isomerization of *n*-pentane. *Indus. Eng. Chem.*, **52**, 391–396.

Carroll, C. W. (1961). The created response surface technique for optimizing nonlinear, restrained systems. *Operations Research*, **9**, 169–185.

Carter, W. H., Jr., Chinchilli, V. M., Campbell, E. D., and Wampler, G. L. (1984). Confidence interval about the response at the stationary point of a response surface, with an application to preclinical cancer therapy. *Biometrics*, **40**, 1125–1130.

Carter, W. H., Jr. and Wampler, G. L. (1980). Survival analysis of drug combinations using a hazards model with time-dependent covariates. *Biometrics*, **36**, 537–546.

Cauchy, A. (1847). Méthode générale pour la résolution des systèmes d'equations simultanées. *Compt. rend. Acad. Sci. Paris*, **25**, 536–538.

Chaloner, K. (1984). Optimal Bayesian experimental design for linear models. *Ann. Statist.*, **12**, 283–300.

Chambers, J. M. (1973). Fitting nonlinear models: numerical techniques. *Biometrika*, **60**, 1–13.

Chang, C. D. and Kononenko, O. K. (1962). Sucrose-modified phenolic resins as plywood adhesives. *Adhesives Age*, **5**(7), July, 36–40.

Chang, C. D., Kononenko, O. K., and Franklin, R. (1960). Maximum data through a statistical design. *Indus. Eng. Chem.*, **52**, 939–942.

Chatterjee, S. K. and Mandal, N. K. (1981). Response surface designs for estimating the optimal point. *Calcutta Statist. Assoc. Bull.*, **30**, 145–169.

Cheng, C.-S. (1978a). Optimality of certain asymmetric experimental designs. *Ann. Statist.*, **6**, 1239–1261.

Cheng, C.-S. (1978b). Optimal design for the elimination of multi-way heterogeneity. *Ann. Statist.*, **6**, 1262–1272.

Cheng, C.-S. (1983). Construction of optimal balanced incomplete block designs for correlated observations. *Ann. Statist.*, **11**, 240–246.

Chernoff, H. (1953). Locally optimal designs for estimating parameters. *Ann. Math. Statist.*, **24**, 586–602.

Chernoff, H. (1959). Sequential design of experiments. *Ann. Math. Statist.*, **30**, 755–770.

Chernoff, H. (1962). Optimal accelerated life tests for estimation. *Technometrics*, **4**, 381–408.

Chew, V., ed. (1958). *Experimental Designs in Industry*. New York: Wiley.

Chow. W. M. (1962). A note on the calculation of certain constrained maxima. *Technometrics*, **4**, 135–137.

Christians, N. F., Martin, D. P. and Karnok, K. J. (1981). The interrelationship among nutrient elements applied to calcareous sand greens. *Agronomy J.*, **73**, 929–933.

Chu, C. and Hougen, O. A. (1961). Optimum design of a catalytic nitric oxide reactor. *Chem. Eng. Prog.*, **57**, June, 51–58.

Chung, K. L. (1954). On a stochastic approximation method. *Ann. Math. Statist.*, **25**, 463–483.

Claringbold, P. J. (1955). Use of the simplex design in the study of joint action of related hormones. *Biometrics*, **11**, 174–185.

Clark, V. (1965). Choice of levels in polynomial regression with one or two variables. *Technometrics*, **7**, 325–333.

Clatworthy, W. H., Connor, W. S., Deming, L. S., and Zelen, M. (1957). Fractional factorial experimental designs for factors at two levels. U. S. Dept. of Commerce, National Bureau of Standards, Applied Math. Series No. 48.

Cochran, W. G. (1973). Experiments for nonlinear functions. *J. Am. Statist. Assoc.*, **68**, 771–781.

Cochran, W. G. and Cox, G. M. (1957). *Experimental Designs*. New York: Wiley.

Cochran, W. G. and Davis, M. (1963). Sequential experiments for estimating the median lethal dose (with discussion). *Colloques Internationaux du Centre National de la Recherche Scientifique*, No. 110, Le Plan d'Experiences, Paris, 181–194.

Cochran, W. G. and Davis, M. (1964). Stochastic approximation to the median effective dose in bioassay (with discussion). In *Stochastic Models in Medicine and Biology*, ed. J. Gurland, 281–300. Madison: University of Wisconsin Press.

Cochran, W. G. and Davis, M. (1965). The Robbins-Monro method for estimating the median lethal dose. *J. Roy. Statist. Soc.*, **B27**, 28–44.

Cockerman, C. and Weir, B. S. (1977). Quadratic analyses of reciprocal crosses. *Biometrics*, **33**, 187–203.

Coen, P. J., Gomme, E. E., and Kendall, M. G. (1969). Lagged relationships in economic forecasting. *J. Roy. Statist. Soc.*, **A132**, 133–152.

Coffman, C. B. and Genter, W. A. (1977). Responses of greenhouse-grown cannabis sativa L. to nitrogen, phosphorus, and potassium. *Agronomy J.*, **69**, 832–836.

Colyer, D. and Kroth, E. H. (1968). Corn yield response and economic optima for nitrogen treatments and plant population over a seven-year period. *Agronomy J.*, **60**, 524–529.

Comer, J. R., Jr. (1964). Some stochastic approximation procedures for use in process control. *Ann. Math. Statist.*, **35**, 1136–1146.

Comer, J. R., Jr. (1965). Application of stochastic approximation procedures to process control. *J. Roy. Statist. Soc.*, **B27**, 321–331.

Connor, W. S. and Young, S. (1961). Fractional factorial designs for experiments with factors at two and three levels. U. S. Dept. of Commerce, National Bureau of Standards, Applied Math. Series No. 58.

Connor, W. S. and Zelen, M. (1959). Fractional factorial experiment designs for factors at three levels. U. S. Dept. of Commerce, National Bureau of Standards, Applied Math. Series No. 54.

Conrad, K. L. and Jones, P. R. (1965). Factorial design of experiments in ceramics, IV. Effect of composition, firing rate, and firing temperature. *Am. Ceramics Soc. Bull.*, **44**, 616–619.

Cook, R. D. (1977). Detection of influential observations in linear regression. *Technometrics*, **19**, 15–18.

Cook, R. D. and Nachtsheim, C. J. (1980). A comparison of algorithms for constructing D-optimal designs. *Technometrics*, **22**, 315–324.

Cook, R. D. and Thibodeau, L. A. (1980). Marginally restricted D-optimal designs. *J. Am. Statist. Assoc.*, **75**, 366–371.

Cook, R. D. and Weisberg, S. (1982). *Residuals and Influence in Regression*. London: Chapman and Hall.

Cooper, H. R., Hughes, I. R., and Mathews. M. E. (1977). Application of response surface methodology to the evaluation of whey protein gel systems. *N. Z. J. Dairy Sci. Technol.*, **12**, 248–252.

Cooper, R. L. (1977). Response of soybean cultivars to narrow rows and planting rates under weed-free conditions. *Agronomy J.*, **69**, 89–91.

Cornelius, P. L., Templeton, W. C., and Taylor, T. H. (1979). Curve fitting by regression on smoothed singular vectors. *Biometrics*, **35**, 849–859.

Cornell, J. A. (1975). Some comments on designs for Cox's mixture polynomials. *Technometrics*, **17**, 25–35.

Cornell, J. A. (1977). Weighted versus unweighted estimates using Scheffe's mixture model for symmetrical error variance patterns. *Technometrics*, **19**, 237–247.

Cornell, J. A. (1980). *Experiments with Mixtures: Designs, Models and the Analysis of Mixture Data*. New York: Wiley.

Cornell, J. A. and Gorman, J. W. (1978). On the detection of an additive blending component in multicomponent mixtures. *Biometrics*, **34**, 251–263.

Cornell, J. A. and Gorman, J. W. (1984). Fractional design plans for process variables in mixture experiments. *J. Qual. Technol.*, **16**, 20–38.

Cornell, J. A. and Ott, L. (1975). The use of gradients to aid in the interpretation of mixture response surfaces. *Technometrics*, **17**, 409–424.

Cornell, R. G. and Speckman, J. A. (1967). Estimation for a simple exponential model. *Biometrics*, **23**, 717–738.

Corsten, L. C. A. (1962). Balanced block designs with two different number of replicates. *Biometrics*, **18**, 499–519.

Cotter, S. C. (1979). A screening design for factorial experiments with interactions. *Biometrika*, **66**, 317–320.

Cottrell, K. M. (1971). Some uses of residuals in multiple regression. M.Sc. Thesis, University of Reading, England.

Covey-Crump, P. A. K. and Silvey, S. D. (1970). Optimal regression designs with previous observations. *Biometrika*, **57**, 551–566.

Cox, D. R. (1958). *Planning of Experiments*. New York: Wiley.

Cox, D. R. (1961). Tests of separate families of hypotheses. *Proc. Fourth Berkeley Symp. Math. Statist. Prob.*, **1**, 105–123.

Cox, D. R. (1962). Further tests on separate families of hypotheses. *J. Roy. Statist. Soc.*, **B24**, 406–424.

Cox, D. R. (1971). A note on polynomial response functions for mixtures. *Biometrika*, **58**, 155–159.

Cox, D. R. (1984). Design of experiments and regression. *J. Roy. Statist. Soc.*, **A147**, 306–315.

Cox, D. R. and Snell, E. J. (1968). A general definition of residuals. *J. Roy. Statist. Soc.*, **B30**, 248–265, discussion 265–275.

Cox, F. R. and Reid, P. H. (1965). Interaction of plant population factors and level of production on the yield and grade of peanuts. *Agronomy J.*, **57**, 455–457.

Coxeter, H. S. M. (1948). *Regular Polytopes*. London: Methuen. (A second edition was later issued by Dover Books, 61480-8.)

Cragle, R. G., Myers, R. M., Waugh, P. K., Hunter, J. S., and Anderson, R. L. (1955). The effects of various levels of sodium citrate, glycerol, and equilibration time on survival of bovine spermatozoa after storage at $-79°C$. *J. Dairy Sci.*, **38**, 508–514.

Creanga, A. and Vaduva, I. (1963). Response surfaces and the theory of regression. *Acad. R. P. Romine Stud. Cerc. Mat.*, **14**, 307–314.

Crosier, R. B. (1984). Mixture experiments: geometry and pseudocomponents. *Technometrics*, **26**, 209–216.

Crosson, L. S. and Protz, R. (1973). Prediction of soil properties from stereorthophoto measurements of landform properties. *Can. J. Soil Science*, **53**, 259–262.

Crowther, E. M. and Yates, F. (1941). Fertiliser policy in war-time. *Empire J. Exp. Agric.*, **9**, 77–97.

Curnow, R. N. (1972). The number of variables when searching for an optimum. *J. Roy. Statist. Soc.*, **B34**, 461–476, comments 477–481.

Curnow, R. N. (1973). A smooth population response curve based on an abrupt threshold and plateau model for individuals. *Biometrics*, **29**, 1–10.

Curtis, G. J. and Hornsey, K. G. (1972). Competition and yield compensation in relation to breeding sugar beet. *J. Agric. Sci., Camb.*, **79**, 115–119.

Dale, R. F. and Shaw, R. H. (1965). Effect on corn yields of moisture stress and stand at two fertility levels. *Agronomy J.*, **57**, 475–479.

Daniel, C. (1956). Fractional replication in industrial research. *Proceedings of Third Berkeley Symposium on Mathematical Statistics and Probability*, Vol. V, University of California Press, 87–98.

Daniel, C. (1958). On varying one factor at a time. *Biometrics*, **14**, 430–431.

Daniel, C. (1959). Use of half-normal plots in interpreting factorial two-level experiments. *Technometrics*, **1**, 311–341.

Daniel, C. (1962). Sequences of fractional replicates in the 2^{p-q} series. *J. Amer. Statist. Assoc.*, **57**, 403–429.

Daniel, C. (1976). *Applications of Statistics to Industrial Experimentation*. New York: Wiley.

Daniel, C. and Wilcoxon, F. (1966). Factorial 2^{p-q} plans robust against linear and quadratic trends. *Technometrics*, **8**, 259–278.

Daniel, C. and Wood, F. S. (1980). *Fitting Equations to Data*, 2nd ed. New York: Wiley.

Darroch, J. N. and Speed, T. P. (1983). Additive and multiplicative models and interactions. *Ann Statist.*, **11**, 724–738.

Darroch, J. N. and Waller, J. (1985). Additivity and interaction in three-component experiments with mixtures. *Biometrika*, **72**, 153–163.

Das, A. D. (1976). A note on the construction of asymmetrical rotatable designs with blocks. *J. Indian Soc. Agric. Statist.*, **28**, 91–96.

Das, M. N. (1960). Fractional replicates as asymmetrical factorial designs. *J. Indian Soc. Agric. Statist.*, **12**, 159–174.

Das, M. N. (1961). Construction of rotatable designs from factorial designs. *J. Indian Soc. Agric. Statist.*, **13**, 169–194.

Das, M. N. (1963). On construction of second order rotatable designs through balanced incomplete block designs with blocks of unequal sizes. *Calcutta Statist. Assoc. Bull.*, **12**, 31–46.

Das, M. N. and Dey, A. (1967). Group divisible rotatable designs. *Ann. Inst. Math. Statist. (Tokyo)*, **19**, 331–347.

Das, M. N. and Dey, A. (1970). On blocking second order rotatable designs. *Calcutta Statist. Assoc. Bull.*, **19**, 75–85.

Das, M. N. and Gill, B. S. (1974). Blocking rotatable designs for agricultural experimentation. *J. Indian Soc Agric. Statist.*, **26**, 125–138.

Das, M. N. and Giri, N. C. (1979). *Design and Analysis of Experiments*. New Delhi: Wiley Eastern.

Das, M. N. and Mehta, J. S. (1968). Asymmetric rotatable designs and orthogonal transformations. *Technometrics*, **10**, 313–322.

Das, M. N. and Narasimham, V. L. (1962). Construction of rotatable designs through balanced incomplete block designs. *Ann. Math. Statist.*, **33**, 1421–1439.

Das, M. N. and Nigam, A. K. (1966). On a method of construction of rotatable designs with smaller number of points controlling the number of levels. *Calcutta Statist. Assoc. Bull.*, **15**, 147–157.

David, H. A. (1952). Upper 5 and 1% points of the maximum F-ratio. *Biometrika*, **39**, 422–424.

David, H. A. and Arens, B. E. (1959). Optimal spacing in regression analysis. *Ann. Math. Statist.*, **30**, 1072–1081.

Davies, O. L., ed. (1954). *Design and Analysis of Industrial Experiments*. Edinburgh: Oliver and Boyd. Fourth ed., 1978, reprinted by Longman; see next reference.

Davies, O. L. and Goldsmith, P. L. (eds.) (1978). *Statistical Methods in Research and Production with Special Reference to the Chemical Industry*, 4th ed. Paperback reprint, 1984. London and New York: Longman.

Davies, O. L. and Hay, W. A. (1950). The construction and uses of fractional factorial designs in industrial research. *Biometrics*, **6**, 233–249.

DeBaun, R. M. (1956). Block effects in the determination of optimum conditions. *Biometrics*, **12**, 20–22.

DeBaun, R. M. (1959). Response surface designs for three factors at three levels. *Technometrics*, **1**, 1–8.

DeBaun, R. M. and Chew, V. (1960). Optimal allocation in regression experiments with two components of error. *Biometrics*, **16**, 451–463.

DeBaun, R. M. and Schneider, A. M. (1958). Experiences with response surface designs. In *Experimental Designs in Industry*, ed. V. Chew, 235–246. New York: Wiley.

Deckman. D. A. and Van Winkle, M. (1959). Perforated plate column studies by the Box method of experimentation. *Indus. Eng. Chem.*, **51**, 1015–1018.

De la Garza, A. (1954). Spacing of information in polynomial regression. *Ann. Math. Statist.*, **25**, 123–130.

DeLury, D. B. (1960). *Values and Integrals of the Orthogonal Polynomials up to N = 26*. University of Toronto Press, Canada.

de Mooy, C. J. and Pesek, J. (1966). Nodulation responses of soybeans to added phosphorus, potassium, and calcium salts. *Agronomy J.*, **58**, 275–280.

Dempster, A. P., Laird, N. M. and Rubin, D. B. (1977). Maximum likelihood from incomplete data via the EM algorithm. *J. Roy. Statist. Soc.*, **B34**, 1–22, discussion 22–38.

Derringer, G. C. (1969a). Sequential method for estimating response surfaces. *Indus. Eng. Chem.*, **61**, 6–13.

Derringer, G. C. (1969b). Statistically designed experiments. *Rubber Age*, **101**, November, 66–76.

Derringer, G. C. (1974a). Variable shear rate viscosity of SBR-filler-plasticizer systems. *Rubber Chemistry and Technology*, **47**, 825–836.

Derringer, G. C. (1974b). An empirical model for viscosity of filled and plasticized elastomer compounds. *J. Applied Polymer Science*, **18**, 1083–1101.

Derringer, G. C. (1983). A statistical methodology for designing elastomer formulations to meet performance specifications. *Kautschuk Gummi Kunststoffe*, **36**(5), 349–352.

Derringer, G. C. and Suich, R. (1980). Simultaneous optimization of several response variables. *J. Qual. Technol.*, **12**, 214–219.

Dey, A. (1970). On response surface designs with equi-spaced doses. *Calcutta Statist. Assoc. Bull.*, **19**, 135–144.

Dey, A. (1985). *Fractional Factorial Designs*. Calcutta: Wiley-Eastern.

Dey, A. and Das, M. N. (1970). On blocking second order rotatable designs. *Calcutta Statist. Assoc. Bull.*, **19**, 75–85.

Dey, A. and Kulshreshtha, A. C. (1973). Further second order rotatable designs. *J. Indian Soc. Agric. Statist.*, **25**, 91–96.

Dey, A. and Nigam, A. K. (1968). Group divisible rotatable designs—some further considerations. *Ann Inst. Statist. Math.* (Tokyo), **20**, 477–481.

Dickinson, A. W. (1974). Some run orders requiring a minimum number of factor level changes for the 2^4 and 2^5 main effect plans. *Technometrics*, **16**, 31–37.

Dietrich, F. H. and Marks, R. G. (1979). Analysis of a factorial quantal response assay using inverse regression. *Commun. Statist. Theory Methods*, **A8**, 85–98.

Dillon, J. L. (1977). *The Analysis of Response in Crop and Livestock Production*, 2nd ed. Oxford: Pergamon.

Dixon, L. C. W. (1972). *Nonlinear Optimisation*. London: English Universities Press.

Dixon, W. J., chief editor, with Brown, M. B., Engelman, L., Frane, J. W., Hill, M. A., Jennrich, R. I., and Toporek, J. D. (1985). *BMDP Statistical Software Manual*. Berkeley: University of California Press.

Dixon, W. J. and Mood, A. M. (1948). A method for obtaining and analysing sensitivity data. *J. Am. Statist. Assoc.*, **43**, 109–126.

Doksum, K. A. and Sievers, G. L. (1976). Plotting with confidence: graphical comparisons of two populations. *Biometrika*, **63**, 421–434.

Dowling, T. A. and Schachtman, R. H. (1975). On the relative efficiency of randomized response models. *J. Am. Statist. Assoc.*, **70**, 84–87.

Draper, N. R. (1960a). Second order rotatable designs in four or more dimensions. *Ann. Math. Statist.*, **31**, 23–33.

Draper, N. R. (1960b). Third order rotatable designs in three dimensions. *Ann. Math. Statist.*, **31**, 865–874.

Draper, N. R. (1960c). A third order rotatable design in four dimensions. *Ann. Math. Statist.*, **31**, 875–877.

Draper, N. R. (1961a). Third order rotatable designs in three dimensions: some specific designs. *Ann. Math. Statist.*, **32**, 910–913.

Draper, N. R. (1961b). Missing values in response surface designs. *Technometrics*, **3**, 389–398.

Draper, N. R. (1962). Third order designs in three factors: analysis. *Technometrics*, **4**, 219–234.

Draper, N. R. (1963). "Ridge analysis" of response surfaces. *Technometrics*, **5**, 469–479.

Draper, N. R. (1982). Center points in second-order response surface designs. *Technometrics*, **24**, 127–133.

Draper, N. R. (1983). Cubic lack of fit for three-level second order response surface designs. Letter to the Editor. *Technometrics*, **25**, 217.

Draper, N. R. (1984). Schaflian rotatability. *J. Roy. Statist. Soc.*, **B46**, 406–411.

Draper, N. R. (1985a). (1) Plackett and Burman designs. (2) Response surface designs. (3) Run. Articles in the *Encyclopedia of Statistical Sciences*, eds. N. L. Johnson and S. Kotz, Vols. 6 and 8. New York: Wiley.

Draper, N. R. (1985b). Small composite designs. *Technometrics*, **27**, 173–180.

Draper, N. R. and Beggs, W. J. (1971). Errors in the factor levels and experimental design. *Ann. Math. Statist.*, **42**, 46–58.

Draper, N. R. and Cox, D. R. (1969). On distributions and their transformation to normality. *J. Roy. Statist. Soc.*, **B31**, 472–476.

Draper, N. R. and Guttman, I. (1980). Incorporating overlap effects from neighboring units into response surface models. *Appl. Statist.*, **29**, 128–134.

Draper, N. R., Guttman, I., and Lapczak, L. (1979). Actual rejection levels in a certain stepwise test. *Comm. Statist.*, **A8**, 99–105.

Draper, N. R., Guttman, I. and Lipow, P. (1977). All-bias designs for spline functions joined at the axes. *J. Am. Statist. Assoc.*, **72**, 424–429.

Draper, N. R. and Herzberg, A. M. (1968). Further second order rotatable designs. *Ann. Math. Statist.*, **39**, 1995–2001.

Draper, N. R. and Herzberg, A. M. (1971). On lack of fit. *Technometrics*, **13**, 231–241. Correction, **14**, 1972, 245.

Draper, N. R. and Herzberg, A. M. (1973). Some designs for extrapolation outside a sphere. *J. Roy. Statist. Soc.*, **B35**, 268–276.

Draper, N. R. and Herzberg, A. M. (1979a). An investigation of first-order and second-order designs for extrapolation outside a hypersphere. *Canad. J. Statist.*, **7**, 97–101.

Draper, N. R. and Herzberg, A. M. (1979b). Designs to guard against outliers in the presence or absence of model bias. *Canad. J. Statist.*, **7**, 127–135.

Draper, N. R. and Herzberg, A. M. (1985). Fourth order rotatability. *Comm. Statist.*, **B14**(3), 515–528.

Draper, N. R. and Hunter, W. G. (1966). Design of experiments for parameter estimation in multiresponse situations. *Biometrika*, **53**, 525–533.

Draper, N. R. and Hunter, W. G. (1967a). The use of prior distributions in the design of experiments for parameter estimation in nonlinear situations. *Biometrika*, **54**, 147–153.

Draper, N. R. and Hunter, W. G. (1967b). The use of prior distributions in the design of experiments for parameter estimation in nonlinear situations: multiresponse case. *Biometrika*, **54**, 662–665.

Draper, N. R. and Hunter, W. G. (1969). Transformations: some examples revisited. *Technometrics*, **11**, 23–40.

Draper, N. R., Hunter, W. G., and Tierney, D. E. (1969a). Which product is better? *Technometrics*, **11**, 309–320.

Draper, N. R., Hunter, W. G., and Tierney, D. E. (1969b). Analyzing paired comparison tests. *J. Market. Res.*, **6**, 477–480.

Draper, N. R. and John, J. A. (1980). Testing for three or fewer outliers in two-way tables. *Technometrics*, **22**, 9–15.

Draper, N. R. and John, J. A. (1981). Influential observations and outliers in regression. *Technometrics*, **23**, 21–26.

Draper, N. R. and Joiner, B. L. (1984). Residuals with one degree of freedom. *American Statistician*, **38**, 55–57.

Draper, N. R., Kanemasu, H., and Mezaki, R. (1969). Estimating rate constants: an improvement on the time-elimination procedure. *Indus. Eng. Chem. Fundamentals*, **8**, 423–427. See also **9**, 1970, 302.

Draper, N. R. and Lawrence, W. E. (1965a). Mixture designs for three factors. *J. Roy. Statist. Soc.*, **B27**, 450–465.

Draper, N. R. and Lawrence, W. E. (1965b). Mixture designs for four factors. *J. Roy. Statist. Soc.*, **B27**, 473–478.

Draper, N. R. and Lawrence, W. E. (1965c). Designs which minimize model inadequacies; cuboidal regions of interest. *Biometrika*, **52**, 111–118.

Draper, N. R. and Lawrence, W. E. (1966). The use of second-order 'spherical' and 'cuboidal' designs in the wrong regions. *Biometrika*, **53**, 596–599.

Draper, N. R. and Lawrence, W. E. (1967). Sequential designs for spherical weight functions. *Technometrics*, **9**, 517–529.

Draper, N. R. and Mezaki, R. (1973). On the adequacy of a multi-site model suggested for the catalytic hydrogenation of olefins. Letter to the Editors, *J. Catalysis*, **28**(1), 179–181.

Draper, N. R., Mezaki, R. and Johnson, R. A. (1973). On the violation of assumptions in nonlinear least squares by interchange of response and predictor variables. *Indus. Eng. Chem. Fundamentals*, **12**, 251–254.

Draper, N. R. and Mitchell, T. J. (1967). The construction of saturated 2_R^{k-p} designs. *Ann. Math. Statist.*, **38**, 1110–1126.

Draper, N. R. and Mitchell, T. J. (1968). Construction of the set of 256-run designs of resolution ≥ 5 and the set of even 512-run designs of resolution ≥ 6 with special reference to the unique saturated designs. *Ann. Math. Statist.*, **39**, 246–255.

Draper, N. R. and Mitchell, T. J. (1970). Construction of a set of 512-run designs of resolution ≥ 5 and a set of even 1024-run designs of resolution ≥ 6. *Ann. Math. Statist.*, **41**, 876–887.

Draper, N. R. and Smith, H. (1981). *Applied Regression Analysis*, 2nd ed. New York: Wiley.

Draper, N. R. and St. John, R. C. (1977a). A mixture model with inverse terms . *Technometrics*, **19**, 37–46.

Draper, N. R. and St. John, R. C. (1977b). Designs in three and four components for mixtures models with inverse terms. *Technometrics*, **19**, 117–130.

Draper, N. R. and Stoneman, D. M. (1964). Estimating missing values in unreplicated two-level factorial and fractional factorial designs. *Biometrics*, **20**, 443–458.

Draper, N. R. and Stoneman, D. M. (1966). Alias relationships for two-level Plackett and Burman designs. *Tech. Rep. No. 96*, University of Wisconsin Statistics Department.

Draper, N. R. and Stoneman, D. M. (1968a). Response surface designs for factors at two and three levels and at two and four levels. *Technometrics*, **10**, 177–192.

Draper, N. R. and Stoneman, D. M. (1968b). Factor changes and linear trends in eight-run two-level factorial designs. *Technometrics*, **10**, 301–311.

Draper, N. R. and Van Nostrand, R. C. (1979). Ridge regression and James-Stein estimation: review and comments. *Technometrics*, **21**, 451–466.

Dupac, V. (1957). On the Kiefer-Wolfowitz approximation method (Czech) [Translated by M. D. Friedman (1960). Lincoln Lab. Rep. 22G-008]. *Casopis Pest. Mat.*, **82**, 47–75.

Dupac, V. (1965). A dynamic stochastic approximation method. *Ann. Math. Statist.*, **36**, 1695–1702.

Dupac, V. and Kral, F. (1972). Robbins-Monro procedure with both variables subject to experimental error. *Ann. Math. Statist.*, **43**, 1089–1095.

Dvoretzky, A. (1956). On stochastic approximation. *Proc. Third Berkeley Symp. Math. Statist. Prob.*, **1**, 39–55.

Dykstra, O. (1959). Partial duplication of factorial experiments. *Technometrics*, **1**, 63–75.

Dykstra, O. (1960). Partial duplication of response surface designs. *Technometrics*, **2**, 185–195.

Dykstra, O. (1971). The augmentation of experimental data to maximize $|X'X|$. *Technometrics*, **13**, 682–688.

Eccleston, J. A. (1980). Recursive techniques for the construction of experimental designs. *J. Statist. Planning Inference*, **4**, 291–297.

Ehrenfeld, S. (1955). On the efficiency of experimental designs. *Ann. Math. Statist.*, **26**, 247–255.

Eik, K. and Hanway, J. J. (1965). Some factors affecting development and longevity of leaves of corn. *Agronomy J.*, **57**, 7–12.

Elfving, G. (1952). Optimum allocation in linear regression theory. *Ann. Math. Statist.*, **23**, 255–262.

Ellerton, R. R. W. and Tsai, W-Y. (1979). Minimum bias estimation and the selection of polynomial terms for response surfaces. *Biometrics*, **35**, 631–635.

Ellis, S. R. M., Jeffreys, G. V. and Wharton, J. T. (1964). Raschig synthesis of hydrazine. Investigation of chloramine formation reaction. *Ind. Eng. Chem. Proc. Des. and Devel.*, **3**, 18–22.

Elston, R. C. (1964). On estimating time response curves. *Biometrics*, **20**, 643–647.

Elston, R. C. and Grizzle, J. E. (1962). Estimation of time response curves and their confidence bands. *Biometrics*, **18**, 148–159.

Epling, M. L. (1964). A multivariate stochastic approximation procedure. Technical Report No. 5, Department of Statistics, Stanford University.

Estes, G. O., Koch, D. W., and Breutsch, T. F. (1973). Influence of potassium nutrition on net CO_2 uptake and growth in maize. *Agronomy J*, **65**, 972–975.

Evans, J. W. (1979). Computer augmentation of experimental designs to maximize $|X'X|$. *Technometrics*, **21**, 321–330.

Evans, J. W. and Manson, A. R. (1978). Optimum experimental designs in two dimensions using minimum bias estimation. *J. Am. Statist. Assoc.*, **73**, 171–176.

Everitt, B. S. and Rushton, D. N. (1978). A method for plotting the optimum position of an array of corticol electrical phosphenes. *Biometrics*, **34**, 399–410.

Fabian, V. (1960). Stochastic approximation methods. *Czech. Math. J.*, **10**, 123–159.

Fabian, V. (1961). A stochastic approximation method for finding optimal conditions in experimental work and in self adapting systems (Czech). *Apl. Mat.*, **6**, 162–183.

Fabian, V. (1967). Stochastic approximation of minima with improved asymptotic speed. *Ann. Math. Statist.*, **38**, 191–200.

Fabian, V. (1968a). On the choice of design in stochastic approximation methods. *Ann. Math. Statist.*, **39**, 457–465.

Fabian, V. (1968b). On asymptotic normality in stochastic approximation. *Ann. Math. Statist.*, **39**, 1327–1332.

Fechner, G. T. (1860). *Elemente der Psychophysik*. Leipzig: Breitkopf und Härtel.

Federer, W. T. and Balaam, L. N. (1973). *Bibliography on Experiment and Treatment Design Pre*. 1968. Edinburgh: Oliver and Boyd.

Fedorov, V. V. (1972). *The Theory of Optimal Experiments* (Translated and edited by E. M. Klimko and W. J. Studden). New York: Academic Press.

Ferrante, G. R. (1962). Laboratory evaluation of new creaseproofing agents using statistical techniques. *Amer. Dyestuff Reporter*, **51**, January 22, 41–43.

Feuell, H. J. and Wagg, R. E. (1949). Statistical methods in detergent investigation. *Research*, **2**, 334–337.

Fieller, E. C. (1955). Some problems in interval estimation. *J. Roy. Statist. Soc.*, **B16**, 175–185.

Finney, D. J. (1945). The fractional replication of factorial arrangements. *Ann. Eugen.*, **12**, 291–301.

Finney, D. J. (1960). *Introduction to the Theory of Experimental Design*. Chicago: University of Chicago Press.

Finney, D. J. (1964). *Statistical Method in Biological Assay*, 2nd ed. London: Griffin.

Finney, D. J. (1965). The meaning of bioassay. *Biometrics*, **21**, 785–798.

Finney, D. J. (1971). *Probit Analysis*, 3rd ed. New York: Cambridge University Press.

Fisher, R. A. (1921). Studies in crop variation. I. An examination of the yield of dressed grain from Broadbalk. *J. Agric. Sci.*, **11**, 107–135.

Fisher, R. A. (1923). Studies in crop variation. II. The manurial response of different potato varieties. *J. Agric. Sci.*, **13**, 311–320.

Fisher, R. A. (1924). The influence of rainfall on the yield of wheat at Rothamsted. *Phil. Trans.*, *B*, **213**, 89–142.

Fisher, R. A. (1963). *Statistical Methods for Research Workers*, 13th ed. Edinburgh: Oliver and Boyd.

Fisher, R. A. (1966). *The Design of Experiments*, 8th ed. Edinburgh: Oliver and Boyd.

Fisk, P. R. (1967). Models of the second kind in regression analysis. *J. Roy. Statist. Soc.*, **B29**, 266–281.

Folks, J. L. (1958). Comparison of designs for exploration of response relationships. Read paper at 18th Annual Meeting, American Statistical Association at Chicago.

Fonner, D. E. Jr., Buck, J. R., and Banker, G. S. (1970). Mathematical optimisation techniques in drug product design and process analysis. *J. Pharmaceut. Sci.*, **59**, 1587–1596.

Frankel, S. A. (1961). Statistical design of experiments for process development of MBT. *Rubber Age*, **89**, 453–458.

Franklin, M. F. (1984). Constructing tables of minimum aberration p^{n-m} designs. *Technometrics*, **26**, 225–232.

Franklin, M. F. (1985). Selecting defining contrasts and confounded effects in p^{n-m} factorial experiments. *Technometrics*, **27**, 165–172.

Franklin, N. L., Pinchbeck, P. H., and Popper, F. (1956). A statistical approach to catalyst development, Part 1. The effect of process variables on the vapour phase oxidation of naphthalene. *Trans. Inst. Chem. Engrs.*, **34**, 280–293. Part II, **36**, 1958, 259–269.

Fries, A. and Hunter, W. G. (1980). Minimum aberration 2^{k-p} designs. *Technometrics*, **22**, 601–608.

Fuller, W. A. (1969). Grafted polynomials as approximating functions. *Austral. J. Agric. Econ.*, **13**, 35–46.

Gacula, M. C., Jr. and Singh, J. (1984). *Statistical Methods in Food and Consumer Research*. London: Academic Press.

Gaddum, J. H. (1933). Reports on Biological Standards III: methods of biological assay depending on a quantal response. Special report No. 183. Medical Research Council. H.M.S.O. London.

Gaffke, N. and Kraft, O. (1982). Exact *D*-optimum designs for quadratic regression. *J. Roy Statist. Soc.*, **B44**, 394–397.

Gaido, J. J. and Terhune, H. D. (1961). Evaluation of variables in the pressure-kier bleaching of cotton. *Amer. Dyestuffs Reporter*, **50**, October 16, 23–26 and 32.

Galil, Z. and Kiefer, J. (1977a). Comparison of rotatable designs for regression on balls, I (Quadratic). *J. Statist. Planning Inference*, **1**, 27–40.

Galil, Z. and Kiefer, J. (1977b). Comparison of designs for quadratic regression on cubes. *J. Statist. Planning Inference*, **1**, 121–132.

Galil, Z. and Kiefer, J. (1977c). Comparison of Box-Draper and *D*-optimum designs for experiments with mixtures. *Technometrics*, **19**, 429–440.

Galil, Z. and Kiefer, J. (1977d). Comparison of simplex designs for quadratic mixture models. *Technometrics*, **19**, 445–453.

Galil, Z. and Kiefer, J. (1979). Extrapolation designs and Φ_p-optimum designs for cubic regression on the *q*-ball. *J. Statist. Planning Inference*, **3**, 27–38.

Galil, Z. and Kiefer, J. (1980). Time and space saving computer methods, related to Mitchell's DETMAX for finding *D*-optimum designs. *Technometrics*, **22**, 302–313.

Galil, Z. and Kiefer, J. (1982). Construction methods for *D*-optimum weighing designs when $n = 3(\mod 4)$. *Ann. Statist.*, **10**, 502–510.

Gallant, A. R. and Fuller, W. A. (1973). Fitting segmented polynomial regression models whose join points have to be estimated. *J. Amer. Statist. Assoc.*, **68**, 144–147.

Gardiner, D. A. and Cowser, K. (1961). Optimization of radionuclide removal from low-level process wastes by the use of response surface methods. *Health Physics*, **5**, 70–78.

Gardiner, D. A., Grandage, A. H. E, and Hader, R. J. (1959). Third order rotatable designs for exploring response surfaces. *Ann. Math. Statist.*, **30**, 1082–1096.

Gardner, L. A., Jr. (1963). Stochastic approximation and its application of prediction and control synthesis. *International Symposium on Nonlinear Differential Equations and Nonlinear Mechanics*, 241–258. New York: Academic Press.

Gartner, N. H. (1976). On the application of a response surface methodology to traffic signal settings. *Transportation Res.*, **10**, 59–60.

Gaylor, D. W. and Merrill, J. A. (1968). Augmenting existing data in multiple regression. *Technometrics*, **10**, 73–81.

Gaylor, D. W. and Sweeney, H. C. (1965). Design for optimal prediction in simple linear regression. *J. Am. Statist. Assoc.*, **60**, 205–216.

Gaylor, V. F. and Jones, C. N. (1968). Rapid and comprehensive crude oil evaluation without distillation. *Indus. Eng. Chem. Product Res. Devel.*, **7**, 191–197.

George, K. C. and Das, M. N. (1966). A type of central composite response surface designs. *J. Indian Soc. Agric. Statist.*, **18**, 21–29.

Gill, B. S. and Das, M. N. (1974). Response surface designs for conduct of agricultural experimentation. *J. Indian Soc. Agric. Statist.*, **26**, 19–32.

Goel, B. S. and Nigam, A. K. (1979). Sequential exploration in mixture experiments. *Biometrical J.*, **21**, 277–285.

Gomez, J. et al. (1978). Agronomical applications of response surface methodology. *Agrociencia*, **32**, 125–135.

Goodman, L. A. (1975). A new model for scaling response patterns: an application of the quasi-independence concept. *J. Am. Statist. Assoc.*, **70**, 755–768.

Gopalan, R. and Dey, A. (1976). On robust experimental designs. *Sankhya*, **B38**, 297–299.

Gorman, J. W. and Cornell, J. A. (1982). A note on model reduction for experiments with both mixture components and process variables. *Technometrics*, **24**, 243–247.

Gorman, J. W. and Cornell, J. A. (1985). A note on fitting equations to freezing-point data exhibiting eutectics for binary and ternary mixture systems. *Technometrics*, **27**, 229–239.

Gorman, J. W. and Hinman, J. E. (1962). Simplex lattice designs for multicomponent systems. *Technometrics*, **4**, 463–487.

Gorman, J. W. and Toman, R. J. (1966). Selection of variables for fitting equations to data. *Technometrics*, **8**, 27–51.

Graybill, F. A. (1961). *An Introduction to Linear Statistical Models*. New York: McGraw-Hill.

Grizzle, J. E. and Allen, D. M. (1969). Analysis of growth and dose response curves. *Biometrics*, **25**, 357–381.

Grohskopf, H. (1960). Statistics in the chemical process industries, present and future. *Indus. Eng. Chem.*, **52**, 497–499.

Grove, C. C. (1965). The influence of temperature on the scouring of raw wool. *Amer. Dyestuffs Reporter*, **54**, January 4, 13–16.

Guest, P. G. (1958). The spacing of observations in polynomial regression. *Ann. Math. Statist.*, **29**, 294–299.

Guilkey, D. K. and Murphy, J. L. (1975). Directed ridge regression techniques in cases of multicollinearity. *J. Am. Statist. Assoc.*, **70**, 769–775.

Gupta, T. K. and Dey, A. (1975). On some new second order rotatable designs. *Ann. Inst. Statist. Math.*, **27**, 167–175.

Gupta, V. K. and Nigam, A. K. (1982). On a model useful for approximating fertilizer response relationships. *J. Indian Soc. Agric. Statist.*, **34**, 61–74. (December, No. 3.)

Gurney, M. and Jewett, R. S. (1975). Constructing orthogonal replications for variance estimation. *J. Am. Statist. Assoc.*, **70**, 819–821.

Guttman, I. and Smith, D. E. (1969). Investigation of rules for dealing with outliers in small samples from a normal distribution: 1. Estimation of the mean. *Technometrics*, **11**, 527–550.

Guttman, I. and Smith, D. E. (1971). Investigation of rules for dealing with outliers in small samples from a normal distribution: 2. Estimation of the variance. *Technometrics*, **13**, 101–111.

Guttman, L. and Guttman, R. (1959). An illustration of the use of stochastic approximation. *Biometrics*, **15**, 551–559.

Hackler, W. C., Kreigel, W. W. and Hader, R. J. (1956). Effect of raw-material ratios on absorption of whiteware compositions. *J. Am. Ceramic Soc.*, **39**(1), 20–25.

Hackney, H. and Jones, P. R. (1967). Response surface for dry modulus of rupture and drying shrinkage. *Amer. Ceramics. Soc. Bull.*, **46**(8), 745–749.

Hader, R. J., Harward, M. E., Mason, D. D., and Moore, D. P. (1957). An investigation of some of the relationships of copper, iron and molybdenum in the growth and nutrition of lettuce: I. Experimental design and statistical methods for characterising the response surface. *Soil Sci. Soc. Amer. Proc,,* **21**, 59–64.

Hader, R. J. and Park, S. H. (1978). Slope rotatable central composite designs. *Technometrics,* **20**, 413–417.

Hahn, G. J. (1976a). Process improvement using evolutionary operation. *Chem. Technol.,* **6**, 204–206.

Hahn, G. J. (1976b). Process improvement through simplex EVOP. *Chem. Technol.,* **6**, 343–345.

Hahn, G. J. (1984). Experimental design in the complex world. *Technometrics,* **26**, 19–31.

Hahn, G. J. (1985). More intelligent statistical software and statistical expert systems: future directions. *American Statistician,* **39**, 1–8, discussion, 8–16.

Hahn, G. J., Feder, P. I. and Meeker, W. Q. (1976). The evaluation and comparison of experimental designs for fitting regression functions. *J. Qual. Technol.,* **8**, 140–157.

Hahn, G. J., Feder, P. I. and Meeker, W. Q. (1978). Evaluating the effect of incorrect specification of a regression model. *J. Qual. Technol.,* **10**, 61–72; 93–98.

Hahn, G. J., Morgan, C. B., and Schmec, J. (1981). The analysis of a fractional factorial experiment with censored data using iterative least squares. *Technometrics,* **23**, 33–36.

Hamilton, D. C. (1986). Confidence regions for parameter subsets in nonlinear regression. *Biometrika,* **73**, 57–64.

Hamilton, D. C. and Watts, D. G. (1985). A quadratic design criterion for precise estimation in nonlinear regression models. *Technometrics,* **27**, 241–250.

Hare, L. B. (1979). Designs for mixture experiments involving process variables. *Technometrics,* **21**, 159–173.

Hare, L. B. and Brown, P. L. (1977). Plotting response surface contours for three-component mixtures. *J. Qual. Technol.,* **9**, 193–197.

Harrington, E. C. (1965). The desirability function. *Indus. Qual. Control,* **21**, 494–498.

Hartley, H. O. (1950). The maximum *F*-ratio as a short cut test for heterogeneity of variance. *Biometrika,* **37**, 308–312.

Hartley, H. O. (1959). Smallest composite designs for quadratic response surfaces. *Biometrics,* **15**, 611–624.

Hartley, H. O. (1961). The modified Gauss-Newton method for the fitting of nonlinear regression functions by least squares. *Technometrics,* **3**, 269–280.

Hartley, H. O. and Booker, A. (1965). Nonlinear least squares estimation. *Ann. Math. Statist.,* **36**, 638–650.

Hawkins, A. (1964). The use of parallel tangents in optimization. *Chem. Eng. Prog. Symp. Ser.,* **60**(50), 35–40.

Hawkins, D. F. (1964). Observations on the application of the Robbins-Monro process to sequential toxicity assays. *Brit. J. Pharmacol. and Chemotherapy,* **22**, 392–402.

Hawkins, D. M. (1980). *Identification of Outliers.* London: Chapman and Hall.

Heapt, L. A., Robertson, J. A., McBeath, D. K., von Maydell, U. M., Love, H. C., and Webster, G. R. (1976). Development of a barley yield equation for Central Alberta. 1. Effects of soil and fertilizers N and P. *Can. J. Soil Science,* **56**, 233–247.

Heapt, L. A., Robertson, J. A., McBeath, D. K., von Maydell, U. M., Love, H. C., and Webster, G. R. (1976). Development of a barley yield equation for Central Alberta. 2. Effects of soil moisture stress. *Can. J. Soil Science,* **56**, 249–256.

Hebble, T. L. and Mitchell, T. J. (1972). 'Repairing' response surface designs. *Technometrics*, **14**, 767–779.

Hedayat, A. and Afsarinejad, K. (1978). Repeated measurements designs, II. *Ann. Statist.*, **6**, 619–628.

Hedayat, A. and John, P. W. M. (1974). Resistant and susceptible BIB designs. *Ann. Statist.*, **2**, 148–158.

Hedayat, A. S. and Majumdar, D. (1985). Families of *A*-optimal block designs for comparing test treatments with a control. *Ann. Statist.*, **13**, 757–767.

Hedayat, A. and Wallis, W. D. (1978). Hadamard matrices and their applications. *Ann. Statist.*, **6**, 1184–1238.

Heller, N. B. and Staats, G. E. (1973). Response surface optimisation when experimental factors are subject to costs and constraints. *Technometrics*, **15**, 113–123.

Henderson, M. S. and Robinson, D. L. (1982). Environmental influences on fiber component concentrations of warm-season perennial grasses. *Agronomy J.*, **74**, 573–579.

Hendry, D. G., Mayo, F. R., and Schetzle, D. (1968). Stability of butadiene polyperoxide. *Indus. Eng. Chem. Product Res. Devel.*, **7**, 145–151.

Hermanson, H. P. (1961). The fertility of some Minnesota peat soils. Ph.D. Thesis, Univ. of Minnesota, 154 pp. (Univ. Microfilms, Ann Arbor, MI, No. 61-5846.)

Hermanson, H. P. (1965). Maximization of potato yield under constraint. *Agronomy J.*, **57**, 210–213.

Hermanson, H. P., Gates, C. E., Chapman, J. W., and Farnham, R. S. (1964). An agronomically useful three-factor response surface design based on dodecahedron symmetry. *Agronomy J.*, **56**, 14–17.

Herzberg, A. M. (1964). Two third order rotatable designs in four dimensions. *Ann. Math. Statist.*, **35**, 445–446.

Herzberg, A. M. (1966). Cylindrically rotatable designs. *Ann. Math. Statist.*, **37**, 242–247.

Herzberg, A. M. (1967a). Cylindrically rotatable designs of Types 1, 2, and 3. *Ann. Math. Statist.*, **38**, 167–176.

Herzberg, A. M. (1967b). The behaviour of the variance function of the difference between two estimated responses. *J. Roy. Statist. Soc.*, **B29**, 174–179.

Herzberg, A. M. (1967c). A method for the construction of second order rotatable designs in *k* dimensions. *Ann. Math. Statist.*, **38**, 177–180.

Herzberg, A. M. (1979). Are theoretical designs applicable? *Operations Research Verfahren/Methods of Operations Research*, **30**, 68–76.

Herzberg A. M. (1982a). The robust design of experiments: a review. *SERDICA Bulgaricae mathematicae publicationes*, **8**, 223–228.

Herzberg, A. M. (1982b). The design of experiments for correlated error structures: layout and robustness. *Canad. J. Statist.*, **10**, 133–138.

Herzberg, A. M. and Andrews, D. F. (1976). Some considerations in the optimal design of experiments in non-optimal situations. *J. Roy. Statist. Soc.*, **B38**, 284–289.

Herzberg, A. M. and Andrews, D. F. (1978). The robustness of chain block designs and coat-of-mail designs. *Comm. Statist*, **A7**, 479–485.

Herzberg, A. M. and Cox, D. R. (1969). Recent work on the design of experiments: a bibliography and a review. *J. Roy. Statist. Soc.*, **A132**, 29–67.

Herzberg, A. M. and Cox, D. R. (1972). Some optimal designs for interpolation and extrapolation. *Biometrika*, **59**, 551–561.

Herzberg, A. M., Garner, C. W. L., and Springer, B. G. F. (1973). Kiss-precise sequential rotatable designs. *Canad. Math. Bull.*, **16**, 207–217.

Hill, H. M. (1960). Experimental designs to adjust for time trends. *Technometrics*, **2**, 67–82.

Hill, P. D. H. (1978). A review of experimental design procedures for regression model discrimination. *Technometrics*, **20**, 15–21.

Hill, W. J. and Demler, W. R. (1970). More on planning experiments to increase research efficiency. *Indus. Eng. Chem.*, **62**, 60–65.

Hill, W. J. and Hunter, W. G. (1966). A review of response surface methodology: A literature survey. *Technometrics*, **8**, 571–590.

Hill, W. J. and Hunter, W. G. (1969). A note on designs for model discrimination: variance unknown case. *Technometrics*, **11**, 396–400.

Hillyer, M. J. and Roth, P. M. (1972). Planning of experiments when the experimental region is constrained. Application of linear transformations to factorial design. *Chem. Eng. Sci.*, **27**, 187–197.

Hinchen, J. D. (1968). Multiple regression in process development. *Technometrics*, **10**, 257–269.

Hinkley, D. V. (1969). Inference about the intersection in two phase regression. *Biometrika*, **56**, 495–504.

Hinkley, D. V. (1985). Transformation diagnostics for linear models. *Biometrika*, **72**, 487–496.

Hochberg, Y. and Quade, D. (1975). One-sided simultaneous confidence bounds on regression surfaces with intercepts. *J. Am. Statist. Assoc.*, **70**, 889–891.

Hocking, R. R. (1983). Developments in linear regression methodology: 1959–1982. *Technometrics*, **25**, 219–230; discussion 230–249.

Hodnett, G. E. (1956). The use of response curves in the analysis and planning of series of experiments with fertilisers. *Empire J. Exp. Agric.*, **24**, 205–212.

Hoel, P. G. (1958). Efficiency problems in polynomial estimation. *Ann. Math. Statist.*, **29**, 1134–1146.

Hoel, P. G. (1961). Asymptotic efficiency in polynomial estimation. *Ann. Math. Statist.*, **32**, 1042–1047.

Hoel, P. G. (1965a). Minimax designs in two-dimensional regression. *Ann. Math. Statist.*, **36**, 1097–1106.

Hoel, P. G. (1965b). Optimal designs for polynomial extrapolation. *Ann. Math. Statist.*, **36**, 1483–1493.

Hoel, P. G. (1966). A simple solution for optimal Chebyshev regression extrapolation. *Ann. Math. Statist.*, **37**, 720–725.

Hoel, P. G. (1968). On testing for the degree of a polynomial. *Technometrics*, **10**, 757–767.

Hoel, P. G. and Jennrich, R. I. (1979). Optimal designs for dose response experiments in cancer research. *Biometrika*, **66**, 307–316.

Hoel, P. G. and Levine, A. (1964). Optimal spacing and weighting in polynomial prediction. *Ann. Math. Statist.*, **35**, 1553–1560.

Hoepner, P. H. and Lutz, J. A. (1966). An application of multiple covariance analysis to the estimation of fertilizer response functions. *Agronomy J.*, **58**, 66–69.

Hoerl, A. E. (1959). Optimum solution of many variables equations. *Chem. Eng. Prog.*, **55**, 69–78.

Hoerl, A. E. (1960). Statistical analysis of an industrial production problem. *Indus. Eng. Chem.*, **52**, 513–514.

Hoerl, A. E. (1962). Application of ridge analysis to regression problems. *Chem. Eng. Prog.*, **58**, 54–59.

Hoerl, A. E. (1964). Ridge analysis. *Chem. Eng. Prog. Symp. Ser.*, **60**, 67–77.

Hoerl, A. E. and Kennard, R. W. (1970a). Ridge regression: biased estimation for nonorthogonal problems. *Technometrics*, **12**, 55–67.

Hoerl, A. E. and Kennard, R. W. (1970b). Ridge regression: applications to nonorthogonal problems. *Technometrics*, **12**, 69–82; correction **12**, 723.

Hoerl, R. W. (1985). Ridge analysis 25 years later. *Am. Statist.*, **39**, 186–192.

Holt, R. F. and Timmons, D. R. (1968). Influence of precipitation, soil water and plant population interactions on corn grain yields. *Agronomy*, **60**, 379–381.

Homme, A. C. and Othmer, D. F. (1961). Sulphuric acid...optimized condition in contact manufacturing. *Indus. Eng. Chem.*, **53**, 979–984.

Hopkins, H. S. and Jones, P. R. (1965). Factorial design of experiments in ceramics. III. Effects of firing rate, firing temperature, particle size distribution, and thickness. *Am. Ceramics Soc. Bull.*, **44**, 502–505.

Hotelling, H. (1940). The selection of variates for use in prediction with some comments on the general problem of nuisance parameters. *Ann. Math. Statist.*, **11**, 271–283.

Houtman, A. M. and Speed, T. P. (1983). Balance in designed experiments with orthogonal block structure. *Ann. Statist.*, **11**, 1069–1085.

Hozumi, K., Shinozaki, K., and Kira, T. (1958). Effect of light intensity and planting density on the growth of Hibiscus moscheutos Linn., with special reference to the interaction between two linear factors of growth. (Japanese with English summary). *Physiol. Ecol.*, **8**, 36–49.

Huda, S. (1981). A method for constructing second order rotatable designs. *Calcutta Statist. Assoc. Bull.*, **30**, 139–144.

Huda, S. (1982a). Some third-order rotatable designs in three dimensions. *Ann. Inst. Statist. Math.* (*Tokyo*), **34**, 365–371.

Huda, S. (1982b). Some third order rotatable designs. *Biom. J.*, **24**, 257–263.

Huda, S. (1982c). Cylindrically rotatable designs of type 3: Further considerations. *Biom. J.*, **24**, 469–475.

Huda, S. (1983a). Two third order rotatable designs in four dimensions. *Tech. Report No. 33/82*, Stat. Math. Division, I.S.I., Calcutta.

Huda, S. (1983b). The m-grouped cylindrically rotatable designs of types $(1, 0, m - 1)$; $(0, 1, m - 1)$; $(1, 1, m - 1)$ and $(0, 0, m)$. *Tech. Report No. 6/83*, Stat. Math. Division, I.S.I., Calcutta.

Huda, S. and Mukerjee, R. (1984). Minimizing the maximum variance of the difference between two estimated responses. *Biometrika*, **71**, 381–385.

Hunter, J. S. (1954). Searching for optimum conditions. *Trans. N.Y. Acad. Sci.*, Series 2, **17**, 124–132.

Hunter, J. S. (1956). Statistical methods for determining optimum conditions. *Trans. Tenth Annual Conv. ASQC*, 415–428.

Hunter, J. S. (1959). Determination of optimum conditions by experimental methods (in three parts). *Indus. Qual. Control*, **15**(6), 16–24; (7), 7–15; (8), 6–14.

Hunter, J. S. (1960). Some applications of statistics to experimentation. *Chem. Eng. Prog. Symp. Ser.*, *AIChE*, **56**(31), 10–26.

Hunter, J. S. (1985). Statistical design applied to product design. *J. Qual. Technol.*, **17**, 210–221.

Hunter, J. S. and Naylor, T. H. (1970). Experimental designs for computer simulation experiments. *Management Science*, **16**(7), 422–434.

Hunter, W. G., Hill, W. J. and Wichern, D. W. (1968). A joint design criterion for the dual problem of model discrimination and parameter estimation. *Technometrics*, **10**, 145–160.

Hunter, W. G. and Kittrell, J. R. (1966). EVOP—a review. *Technometrics*, **8**, 389–396.

Hunter, W. G. and Reiner, A. M. (1965). Designs for discriminating between two rival models. *Technometrics*, **7**, 307–323.

Jeffreys, G. V. and Whorton, J. T. (1965). Raschig synthesis of hydrazine. *Indus. Eng. Chem., Process Design Devel.*, **4**, 71–76.

Jeffreys, H. (1932). An alternative to the rejection of observations. *Proceedings of the Royal Society*, **A-137**, 78–87.

Jennrich, R. I. and Sampson, P. F. (1968). Application of stepwise regression to nonlinear least squares estimation. *Technometrics*, **10**, 63–72.

Jha, M. P., Kumar, A., and Bapat, B. R. (1981). Some investigations on response to fertilizer and determination of optimum dose using soil test values. *J. Indian Soc. Agric. Statist.*, **33**, 60–70.

John, J. A. and Draper, N. R. (1978). On testing for two outliers or one outlier in two-way tables. *Technometrics*, **20**, 69–78.

John, J. A. and Draper, N. R. (1980). An alternative family of transformations. *Appl. Statist.*, **29**, 190–197.

John, J. A. and Quenouille, M. H. (1977). *Experiments: Design and Analysis*. London: Griffin.

John, P. W. M. (1962). Three quarter replicates of 2^n designs. *Biometrics*, **18**, 172–184.

John, P. W. M. (1964). Blocking of $3(2^{n-k})$ designs. *Technometrics*, **6**, 371–376.

John, P. W. M. (1966). Augmenting 2^{n-1} designs. *Technometrics*, **8**, 469–480.

John, P. W. M. (1976). Robustness of balanced incomplete block designs. *Ann. Statist.*, **4**, 960–962.

John, P. W. M. (1979). Missing points in 2^n and 2^{n-k} factorial designs. *Technometrics*, **21**, 225–228.

Johnson, A. F. (1966). Properties of second order designs: Effect of transformation or truncation on prediction variance. *Appl. Statist.*, **15**, 48–50.

Johnson, M. E. and Nachtsheim, C. J. (1983). Some guidelines for constructing exact *D*-optimal designs on convex design spaces. *Technometrics*, **25**, 271–277.

Johnson, N. L. and Leone, F. C. (1977). *Statistics and Experimental Design in Engineering and the Physical Sciences*, Vols. 1 and 2, 2nd ed. New York: Wiley.

Joiner, B. L. and Campbell, C. (1976). Designing experiments when run order is important. *Technometrics*, **18**, 249–259.

Jones, B. (1976). An algorithm for deriving optimal block designs. *Technometrics*, **18**, 451–458.

Jones, B. (1979). Algorithms to search for optimum row-and-column designs. *J. Roy. Statist. Soc.*, **B41**, 210–216.

Jones, B. and Eccleston, J. A. (1980). Exchange and interchange procedures to search for optimum designs. *J. Roy. Statist. Soc.*, **B42**, 238–243.

Jones, E. R. and Mitchell, T. J. (1978). Design criteria for detecting model inadequacy. *Biometrika*, **65**, 541–551.

Jordan, L. (1977). Optimal designs for polynomial fitting. *Estadistica*, **31**, 385–394.

Jordan, L. and Kempthorne, O. (1977). Min-average bias estimable designs. *Estadistica*, **31**, 66–79.

Juusola, J. A., Bacon, D. W., and Downie, J. (1972). Sequential statistical design strategy in an experimental kinetic study. *Canad. J. Chem. Eng.*, **50**, 796–801.

Kackar, R. N. (1985). Off-line quality control, parameter design and the Taguchi method. *J. Qual. Technol.*, **17**, 176–188, discussion 189–209.

Kanemasu, H. (1979). A statistical approach to efficient use and analysis of simulation models. *Bull. Inter. Statist. Inst.*, **48**, 573–604.

Karlin, S. and Studden, W. J. (1966a). Optimal experimental designs. *Ann. Math. Statist.*, **37**, 783–815.

Karlin, S. and Studden, W. J. (1966b). *Tchebycheff Systems: With Applications in Analysis and Statistics*. New York: Wiley.

Karson, M. J. (1970). Design criterion for minimum bias estimation of response surfaces. *J. Am. Statist. Assoc.*, **65**, 1565–1572.

Karson, M. J., Manson, A. R., and Hader, R. J. (1969). Minimum bias estimation and experimental design for response surfaces. *Technometrics*, **11**, 461–475.

Katz, D. and D'Argenio, D. Z. (1983). Experimental design for estimating integrals by numerical quadrature. *Biometrics*, **39**, 621–628.

Keirstead, D. and Dekee, D. (1980). Stable bubble sensitized gel slurry explosives. *Indus. Eng. Chem. Product Res. Devel.*, **19**, 91–97.

Kempthorne, O. (1952). *The Design and Analysis of Experiments*. New York: Wiley.

Kennard, R. W. and Stone, L. (1969). Computer aided design of experiments. *Technometrics*, **11**, 137–148.

Kesten, H. (1958). Accelerated stochastic approximation. *Ann. Math. Statist.*, **29**, 41–59.

Khuri, A. I. (1984). A note on D-optimal designs for partially nonlinear regression models. *Technometrics*, **26**, 59–61.

Khuri, A. I. (1985). A test for lack of fit of a linear multiresponse model. *Technometrics*, **27**, 213–218.

Khuri, A. I. and Conlon, M. (1981). Simultaneous optimization of multiple responses represented by polynomial regression functions. *Technometrics*, **23**, 363–375.

Khuri, A. I. and Cornell, J. A. (1977). Secondary design considerations for minimum bias estimation. *Commun. Statist. Theory Methods*, **A6**, 631–647.

Khuri, A. I. and Myers, R. H. (1979). Modified ridge analysis. *Technometrics*, **21**, 467–473.

Kiefer, J. (1958). On the nonrandomised optimality and the randomised nonoptimality of symmetrical designs. *Ann. Math. Statist.*, **29**, 675–699.

Kiefer, J. (1959). Optimum experimental designs. *J. Roy. Statist. Soc.*, **B21**, 272–304, discussion 304–319.

Kiefer, J. (1960). Optimum experimental designs II, with applications to systematic and rotatable designs. *Proc. 4th Berkeley Symp. Math. Statist. Prob.*, **1**, 381–405.

Kiefer, J. (1961). Optimum designs in regression problems II. *Ann. Math. Statist.*, **32**, 298–325.

Kiefer, J. (1962a). Two more criteria equivalent to D-optimality of designs. *Ann. Math. Statist.*, **33**, 792–796.

Kiefer, J. (1962b). An extremum result. *Canad. J. Math.*, **14**, 597–601.

Kiefer, J. (1973). Optimum designs for fitting biased multiresponse surfaces. *Multivar. Anal.*, **3**,

287–297.

Kiefer, J. (1975). Optimal design: variation in structure and performance under change of criterion. *Biometrika*, **62**, 277–288.

Kiefer, J. C. (1984). The publications and writings of Jack Kiefer. *Ann. Statist.*, **12**, 424–430. ("The complete bibliography...prepared through the efforts of Roger Farrell and Ingram Olkin." The quotation is from p. 403 of the cited journal.)

Kiefer, J. C. (1985). *Jack Carl Kiefer Collected Papers*, eds. L. D. Brown, I. Olkin, J. Sacks, and H. P. Wynn, Volumes I (Statistical Inference and Probability, 1951–1963), II (Statistical Inference and Probability, 1964–1984), and III (Design of Experiments). New York: Springer-Verlag.

Kiefer, J. and Sacks, J. (1963). Asymptotically optimum sequential inference and design. *Ann. Math. Statist.*, **34**, 705–750.

Kiefer, J. and Studden, W. J. (1976). Optimal designs for large degree polynomial regression. *Ann. Statist.*, **4**, 1113–1123.

Kiefer, J. and Wolfowitz, J. (1952). Stochastic estimation of the maximum of a regression function. *Ann. Math. Statist.*, **23**, 462–466.

Kiefer, J. and Wolfowitz, J. (1959). Optimum designs in regression problems. *Ann. Math. Statist.*, **30**, 271–294.

Kiefer, J. and Wolfowitz, J. (1960). The equivalence of two extremum problems. *Canadian J. Math.*, **12**, 363–366.

Kissel, D. E. and Burnett, E. (1979). Response of Coastal bermudagrass to tillage and fertilizers on an eroded Grayland soil. *Agronomy J.*, **71**, 941–944.

Kitagawa, T. and Mitome, M. (1958). *Tables for the Design of Factorial Experiments.* New York: Dover.

Klein, I. and Marshall, D. I. (1964). Note on starch vinylation by Berry, Tucker and Deutschman. *Indus. Eng. Chem. Proc. Des. Devel.*, **3**, 287–288.

Knapp, W. R. and Knapp, J. S. (1978). Response of winter wheat to date of planting and fall fertilization. *Agronomy J.*, **70**, 1048–1052.

Koehler, T. L. (1960). How statistics apply to chemical processes. *Chem. Eng.*, **67**, December 12, 142–152.

Koshal, R. S. (1933). Application of the method of maximum likelihood to the improvement of curves fitted by the method of moments. *J. Roy. Statist. Soc.*, **A96**, 303–313.

Krewski, D. and Kovar, J. (1982). Low dose extrapolation under single parameter dose response model. *Commun. Statist.-Simul. Comp.*, **11**, 27–45.

Krzanowski, W. J. (1984). Sensitivity of principal components. *J. Roy. Statist. Soc.*, **B46**, 558–563.

Kullback, S. (1959). *Information Theory and Statistics.* New York: Wiley.

Kulshreshtha, A. C. (1969). Fitting of response surface in the presence of a concomitant variate. *Calcutta Statist. Assoc. Bull.*, **18**, 123–131.

Kunert, J. (1983). Optimal design and refinement of the linear model with applications to repeated measurements designs. *Ann. Statist.*, **11**, 247–257.

Kupper, L. L. (1972). A note on the admissibility of a response surface design. *J. Roy. Statist. Soc.*, **B34**, 28–32.

Kupper, L. L. (1973). Minimax designs for Fourier series and spherical harmonics regressions: a characterization of rotatable arrangements. *J. Roy. Statist. Soc.*, **B35**, 493–500.

Kupper, L. L. and Meydrech, E. F. (1973). A new approach to mean squared error estimation of response surfaces. *Biometrika*, **60**, 573–579.

Kupper, L. L. and Meydrech, E. F. (1974). Experimental design considerations based on a new approach to mean square error estimation of response surfaces. *J. Amer. Statist. Assoc.*, **69**, 461–463.

Lambrakis, D. P. (1968a). Experiments with mixtures: a generalisation of the simplex-lattice design. *J. Roy. Statist. Soc.*, **B30**, 123–136.

Lambrakis, D. P. (1968b). Experiments with mixtures: *p* component. *J. Roy. Statist. Soc.*, **B30**, 137–144.

Lau, T. and Studden, W. J. (1985). Optimal designs for trigonometric and polynomial regression using canonical moments. *Ann. Statist.*, **13**, 383–394.

Lawless, J. F. and Wang, P. (1976). A simulation study of ridge and other regression estimators. *Commun. Statist. Theory Methods*, **A5**, 307–323.

Laycock, P. J. and Silvey, S. D. (1968). Optimal designs in regression problems. *Biometrika*, **55**, 53–66.

Leakey, C. L. A. (1972). The effect of plant population and fertility level on yield and its components in two determinate cultivars of *Phaseolus vulgaris*. (L) Savi. *J. Agric. Sci., Camb.*, **79**, 259–267.

Leone, F. C., Nelson, L. S., and Nottingham, R. B. (1961). The folded normal distribution. *Technometrics*, **3**, 543–550.

Levenberg, K. (1944). A method for the solution of certain non-linear problems in least squares. *Q. Appl. Math.*, **2**, 164–168.

Levine, A. (1966). A problem in minimax variance polynomial extrapolation. *Ann. Math. Statist.*, **37**, 898–903.

Leyshon, A. J. and Sheard, R. W. (1974). Influence of short-term flooding on the growth and plant nutrient composition of barley. *Can. J. Soil Science*, **54**, 463–473.

Leysieffer, F. W. and Warner, S. L. (1976). Respondent jeopardy and optimal designs in randomized response models. *J. Am. Statist. Assoc.*, **71**, 649–656.

Li, C. H. (1958). Worksheet gives optimum conditions. *Chem. Eng.*, **65**, April 7, 151–156.

Li, K. (1983). Minimaxity for randomized designs: some general results. *Ann. Statist.*, **11**, 225–239.

Li, K. C. (1984). Robust regression designs when the design space consists of finitely many points. *Ann. Statist.*, **12**, 269–282.

Lin, K. M. and Kackar, R. N. (1986). Optimizing the wave soldering process. *Electronic Packaging and Production*, February, 108–115.

Lind, E. E., Goldin, J., and Hickman, J. B. (1960). Fitting yield and cost response surfaces. *Chem. Eng. Prog.*, **56**(11), 62–68.

Lind, E. E. and Young, W. R. (1965). Con-man: a 3-D device for the representation of response surfaces. *Trans. 19th. Ann. Conv., ASQC*, 545–551.

Lindley, D. V. (1968). The choice of variables in multiple regression. *J. Roy. Statist. Soc.*, **B30**, 31–53, discussion 54–66.

Lindsey, J. K. (1972). Fitting response surfaces with power transformations. *Appl. Statist.*, **21**, 234–247.

Lucas, J. M. (1974). Optimum composite designs. *Technometrics*, **16**, 561–567.

Lucas, J. M. (1976). Which response surface design is best: a performance comparison of several types of quadratic response surface designs in symmetric regions. *Technometrics*, **18**, 411–417.

Lucas, J. M. (1977). Design efficiencies for varying numbers of centre points. *Biometrika*, **64**, 145–147.

Lund, R. E. (1982). Plans for blocking and fractions of nested cube designs. *Commun. Statist. Theor. Meth.*, **A11**, 2287–2296.

Lund, R. E. and Linnell, M. G. (1982). Description and evaluation of a nested cube experimental design. *Commun. Statist. Theor. Meth.*, **A11**, 2297–2313.

Madsen, K. S. (1977). A growth curve model for studies in morphometrics. *Biometrics*, **33**, 659–669.

Maghsoodloo, S. and Hool, J. N. (1976). On response surface methodology and its computer aided teaching. *American Statistician*, **30**, 140–144.

Mallows, C. L. (1973). Some comments on C_p. *Technometrics*, **15**, 661–675.

Mandal, N. K. (1978). On estimation of the minimal point of a single factor quadratic response function. *Calcutta Statist. Assoc. Bull.*, **27**, 119–125.

Margolin, B. H. (1968). Orthogonal main effect $2^n 3^m$ designs and two-factor interaction aliasing. *Technometrics*, **10**, 559–573.

Margolin, B. H. (1969a). Results on factorial designs of resolution IV for the 2^n and $2^n 3^m$ series. *Technometrics*, **11**, 431–444.

Margolin, B. H. (1969b). Orthogonal main effect plans permitting estimation of all two factor interactions for the $2^n 3^m$ factorial series of designs. *Technometrics*, **11**, 747–762.

Margolin, B. H. (1976). Design and analysis of factorial experiments via interactive computing in APL. *Technometrics*, **18**, 135–150.

Marquardt, D. W. (1963). An algorithm for the estimation of nonlinear parameters. *Soc. Ind. Appl. Maths. J.*, **11**, 431–441.

Marquardt, D. W. and Snee, R. D. (1975). Ridge regression in practice. *American Statistician*, **29**, 3–20.

Matthews, S. (1973). The effect of time of harvest on the viability and preemergence mortality in soil of pea (*Pisum sativum L.*) seeds. *Ann. Appl. Biol.*, **73**, 211–219.

Mayer, R. P. and Stowe, R. A. (1969). Would you believe 99.9969% explained? *Indus. Eng. Chem.*, **61**, 42–46.

McKee, B. and Kshirsagar, A. M. (1982). Effect of missing plots in some response surface designs. *Commun. Statist. Theory Methods*, **11**, 1525–1549.

McLean, R. A. and Anderson, V. L. (1984). *Applied Factorial and Fractional Designs*. New York: Marcel Dekker.

Mead, R. (1970). Plant density and crop yield. *Appl. Statist.*, **19**, 64–81.

Mead, R. (1979). Competition experiments. *Biometrics*, **35**, 41–54.

Mead, R. and Freeman, K. H. (1973). An experiment game. *Appl. Statist.*, **22**, 1–6.

Mead, R. and Pike, D. J. (1975). A review of response surface methodology from a biometric viewpoint. *Biometrics.*, **31**, 803–851.

Mead, R. and Riley, J. (1981). A review of statistical ideas relevant to intercropping research. *J. Roy. Statist. Soc.*, **A144**, 462–487, discussion 487–509.

Mehta, J. S. and Das, M. N. (1968). Asymmetric rotatable designs and orthogonal transformations. *Technometrics*, **10**, 313–322.

Mendoza, J. L., Toothaker, L. E., and Crain, B. R. (1976). Necessary and sufficient conditions for F ratios in the $L \times J \times K$ factorial design with two repeated factors. *J. Am. Statist. Assoc.*, **71**, 992–993.

Meydrech, E. F. and Kupper, L. L. (1976). On quadratic approximation of a cubic response. *Commun. Statist. Theory Methods*, **A5**, 1205–1213.

Meyer, D. L. (1963). Response surface methodology in education and psychology. *J. Exper. Educ.*, **31**, 329–336.

Michaels, S. E. and Pengilly, P. J. (1963). Maximum yield for specified cost. *Appl. Statist.*, **12**, 189–193.

Miller, I. (1974). Statistical designs for experiments in combination therapy. *Cancer Chemotherapy Reports*, **4**, 151–156.

Miller, M. H. and Ashton, G. C. (1960). The influence of fertiliser placement and rate of nitrogen on fertiliser phosphorus utilisation by oats as studied using a central composite design. *Can. J. Soil Science*, **40**, 157–167.

Mirreh, H. F. and Ketcheson, J. W. (1973). Influence of soil water, matric potential, and resistance to penetration on corn root elongation. *Can. J. Soil Science*, **53**, 383–388.

Mislevy, P., Kalmbacher, R. S., and Martin, F. G. (1981). Cutting management of the tropical legume American jointvetch. *Agronomy J.*, **73**, 771–775.

Mitchell, T. J. (1974). An algorithm for the construction of *D*-optimal experimental designs. *Technometrics*, **16**, 203–210.

Mitchell, T. J. and Bayne, C. K. (1978). *D*-optimal fractions of three-level fractional designs. *Technometrics*, **20**, 369–380, discussion 381–383.

Mitra, R. K. (1981). On *G*-efficiency of some second order rotatable designs. *Biom. J.*, **23**, 749–757.

Mitscherlich, E. A. (1930). *Die Bestimmung des Dungerbedurfnisses des Bodens*. Berlin: Paul Parey.

Montgomery, D. C. (1976). *Design and Analysis of Experiments*. New York: Wiley.

Montgomery, D. C., Talavage, J. J., and Mullen, C. J. (1972). A response surface approach to improving traffic signal settings in a street network. *Transportation Res.*, **6**, 69–80.

Mooney, R. W., Comstock, A. J., Goldsmith, R. L., and Meisenhalter, G. J. (1960). Predicting chemical composition in precipitation of calcium hydrogen orthophosphate. *Indus. Eng. Chem.*, **52**(5), 427–428.

Moore, D. P., Harward, M. E., Mason, D. D., Hader, R. J., Lott, W. L. and Jackson, W. A. (1957). An investigation of some of the relationships of copper, iron and molybdenum in the growth and nutrition of lettuce. II. *Soil Sci. Soc. Amer. Proc.*, **21**, 65–74.

Moorehead, D. H. and Himmelblau, D. M. (1962). Optimization of operating conditions in a packed liquid-liquid extraction column. *Indus. Eng. Chem. Fund.*, **1**, 68–72.

Morris, M. D. and Mitchell, T. J. (1983). Two-level multifactor designs for detecting the presence of interactions. *Technometrics*, **25**, 345–355.

Morton, R. H. (1983). Response surface methodology. *Mathematical Scientist*, **8**, 31–52.

Mosby, J. F. and Albright, L. A. (1966). Alkylation of isobutane with 1-butene using sulfuric acid as catalyst at high rates of agitation. *Indus. Eng. Chem. Product Res. Devel.*, **5**, 183–190.

Moskowitz, J. R. and Chandler, J. W. (1977). Eclipse-developing products from concepts via consumer ratings. *Food Prod. Dev.*, **11**, 50–60.

Mukerjee, R. and Huda, S. (1985). Minimax second- and third-order designs to estimate the slope of a response surface. *Biometrika*, **72**, 173–178.

Mukhopadhyay, A. C. (1969). Operability region and optimum rotatable designs. *Sankhya.*, **B31**, 75–84.

Mukhopadhyay, S. (1981). Group divisible rotatable designs which minimise the mean square bias. *J. Indian Soc. Agric. Statist.*, **33**, 40–46.

Myers, R. H. (1971). *Response Surface Methodology.* Boston: Allyn and Bacon. (Reprinted by Edwards Bros., Ann Arbor, MI.)

Myers, R. H. and Carter, W. H. (1973). Response surface techniques for dual response systems. *Technometrics,* **15,** 301–317.

Myers, R. H. and Khuri, A. I. (1979). A new procedure for steepest ascent. *Commun. Statist. Theory Methods,* **A8,** 1359–1376.

Myers, R. H. and Lahoda, S. J. (1975). A generalisation of the response surface mean square error criterion with a specific application to slope. *Technometrics,* **17,** 481–486.

Nalimov, V. V., Golikova, T. I., and Mikeshina, N. G. (1970). On practical use of the concept of D-optimality. *Technometrics,* **12,** 799–812.

Narasimham, V. L., Chennarayudu, K. C., and Ramachandra Rao, P. (1983). Construction of group divisible second order rotatable designs through balanced incomplete block designs. *Proc. 37th Annual Conf. Indian Soc. Agric. Statist.,* Simla, Oct. 1983.

Narasimham, V. L. and Rao, K. N. (1980). A modified method for the construction of third order rotatable designs through complementary balanced incomplete block designs. *Proc. 2nd Annual Conf. ISTPA,* Bombay, Dec. 1980.

Narasimham, V. L., Rao, P. R., and Rao, K. N. (1983). Construction of second order rotatable designs through a pair of balanced incomplete block designs. *J. Indian Soc. Agric. Statist.,* **35,** 36–40.

National Bureau of Standards (1957). *Fractional Factorial Experimental Designs for Factors at Two Levels.* Applied Mathematics Series No. 48. Washington D.C.: U.S. Government Printing Office.

Nelder, J. A. (1961). The fitting of a generalisation of the logistic curve. *Biometrics,* **17,** 89–110.

Nelder, J. A. (1962). New kinds of systematic designs for spacing experiments. *Biometrics,* **18,** 283–307.

Nelder, J. A. (1966a). Evolutionary experimentation in agriculture. *J. Sci. Fd. Agric.,* **17,** 7–9.

Nelder, J. A. (1966b). Inverse polynomials, a useful group of multi-factor response functions. *Biometrics,* **22,** 128–141.

Nelder, J. A. (1968). Regression, model-building and invariance. *J. Roy. Statist. Soc.,* **A131,** 303–315, discussion 315–329.

Nelder, J. A. and Mead, R. (1965). A simplex method for function minimisation. *Comp. J.,* **7,** 308–313.

Nelson, L. S. (1982). Analysis of two-level factorial experiments. *J. Qual. Technol.,* **14,** 95–98.

Neuwirth, S. I. and Naphtali, L. M. (1957). New statistical method rapidly determines optimum process conditions. *Chem. Eng.,* **64,** June, 238–242.

Nigam, A. K. (1967). On third order rotatable designs with smaller number of levels. *J. Indian Soc. Agric. Statist.,* **19,** 36–41.

Nigam, A. K. (1973). Multifactor mixture experiments. *J. Roy. Statist. Soc.,* **B35,** 51–56.

Nigam, A. K. (1974). Some designs and models for mixture experiments for the sequential exploration of response surfaces. *J. Indian Soc. Agric. Statist.,* **26,** 120–124.

Nigam, A. K. (1977). A note on four and six level second order rotatable designs. *J. Indian Soc. Agric. Statist.,* **29,** 89–91.

Nigam, A. K. and Das, M. N. (1966). On a method of construction of rotatable designs with smaller number of points controlling the number of levels. *Calcutta Statist. Assoc. Bull.,* **15,** 153–174.

Nigam, A. K. and Dey, A. (1970). Four and six level second order rotatable designs. *Calcutta Statist. Assoc. Bull.*, **19**, 155–157.

Nigam, A. K., Gupta, S. C., and Gupta, S. (1983). A new algorithm for extreme vertices designs for linear mixture models. *Technometrics*, **25**, 367–371.

Nigam, A. K. and Gupta, V. K. (1985). Construction of orthogonal main-effect plans using Hadamard matrices. *Technometrics*, **27**, 37–40.

Norton, C. J. and Moss, T. E. (1963). Oxidative dealkylation of alkylaromatic hydrocarbons. *Indus. Eng. Chem. Proc. Des. and Devel.*, **2**, 140–147. Also (1964), **3**, 23–32.

Novack, J., Lynn, R. O., and Harrington, E. C. (1962). Process scale-up by sequential experimentation and mathematical optimization. *Chem. Eng. Prog.*, **58**, (2), 55–59.

Ohm, H. W. (1976). Response of 21 oat cultivars to nitrogen fertilization. *Agronomy J.*, **68**, 773–775.

Ologunde, O. O. and Sorensen, R. C. (1982). Influence of concentrations of K and Mg in nutrient solutions on sorghum. *Agronomy J.*, **74**, 41–46.

Olsson, D. M. and Nelson, L. S. (1975). The Nelder-Mead simplex procedure for function minimisation. *Technometrics*, **17**, 45–51; 393–394.

Ord, K. (1985). An interpretation of the least squares regression surface. *Amer. Statist.*, **39**, 120–123.

Ott, L. and Cornell, J. A. (1974). A comparison of methods which utilise the integrated mean square error criterion for constructing response surface designs. *Commun. Statist. Theory Methods*, **A3**, 1053–1068.

Ott, R. L. and Mendenhall, W. (1972). Designs for estimating the slope of a second order linear model. *Technometrics*, **14**, 341–354.

Ott, R. L. and Mendenhall, W. (1973). Designs for comparing the slopes of two second order response curves. *Comm. Statist.*, **1**, 243–260.

Ott, R. L. and Myers, R. H. (1968). Optimal experimental designs for estimating the independent variable in regression. *Technometrics*, **10**, 811–823.

Pablos Hach, J. L. and Castillo Morales, A. (1976). Determination of the optimal plot size using the canonical form. *Agrociencia*, **23**, 39–48.

Panda, R. (1976). Asymptotically highly efficient spring balance weighing designs under restricted setup. *Calcutta Statist. Assoc. Bull.*, **25**, 179–184.

Panton, D. (1976). Chicago Board call options as predictors of common stock price changes. *J. Econometrics*, **4**, 101–113.

Pao, T. W., Phadke, M. S., and Sherrerd, C. S. (1985). Computer response time optimization using orthogonal array experiments. *Proc. ICC, IEEE Commun. Soc.*, **2**, 890–895.

Park, S. H. (1977). Selection of polynomial terms for response surface experiments. *Biometrics*, **33**, 225–229.

Park, S. H. (1978a). Selecting contrasts among parameters in Scheffe's mixture models: Screening components and model reduction. *Technometrics*, **20**, 273–279.

Park, S. H. (1978b). Experimental designs for fitting segmented polynomial regression models. *Technometrics*, **20**, 151–154.

Park, S. H. and Kim, J. H. (1982). Axis-slope-rotatable designs for experiments with mixtures. *J. Korean Statist. Soc.*, **11**, 36–44.

Paterwardhan, V. S. and Eckert, R. E. (1981). Maximization of the conversion to m-toluenesulfonic in liquid-phase sulfonation. *Indus. Eng. Chem. Proc. Des. Devel.*, **20**, 82–85.

Patnaik, P. B. (1949). The non-central χ^2- and F-distributions and their applications. *Biometrika*, **36**, 202–232.

Pearson, E. S. and Hartley, H. O. (1958). *Biometrika Tables for Statisticians, Vol. 1*. New York: Cambridge University Press.

Perrin, E. B. (1960). Estimation of parameters in systems related to the observation by an unknown monotone transformation. Stat. Dept. Tech. Rep. No. 1., Stanford University.

Pesotan, H. and Raktoe, B. L. (1981). Further results on invariance and randomization in fractional replication. *Ann. Statist.*, **9**, 418–423.

Pesotchinsky, L. (1975). D-optimum and quasi-D-optimum second-order designs on a cube. *Biometrika*, **62**, 335–340.

Pesotchinsky, L. (1982). Optimal robust designs: linear regression in R^k. *Ann. Statist.*, **10**, 511–525.

Phadke, M. S., Kackar, R. N., Speeney, D. V., and Grieco, M. J. (1983). Off-line quality control for integrated circuit fabrication using experimental design. *Bell System Tech. J.*, **62**, 1273–1309.

Phifer, L. H. and Maginnis, J. B. (1960). Dry ashing of pulp and factors which influence it. *Tappi*, **43**, (1), 38–44.

Piepel, G. F. (1982). Measuring component effects in constrained mixture experiments. *Technometrics*, **24**, 29–39.

Piepel, G. F. (1983). Defining consistent constraint regions in mixture experiments. *Technometrics*, **25**, 97–101.

Piepel, G. F. and Cornell, J. A. (1985). Models for mixture experiments when the response depends on the total amount. *Technometrics*, **27**, 219–227.

Pike, F. R., et al. (1954). The application of statistical procedures to a study of the flooding capacity of a pulse column. *Chem. Eng. Dept. Tech. Rpt., N. C. State Coll., Raleigh*.

Pinchbeck, P. H. (1957). The kinetic implications of an empirically fitted yield surface for the vapour phase oxidation of naphthalene to phthalic anhydride. *Chem. Engr. Science.*, **6**, 105–111.

Pinthus, M. J. (1972). The effect of chlormequat seed-parent treatment on the resistance of wheat seedlings to terbutryne and simazine. *Weed Res.*, **12**, 241–247.

Plackett, R. L. (1960). *Principles of Regression Analysis*. Oxford: Oxford University Press.

Plackett, R. L. and Burman, J. P. (1946). The design of optimum multifactorial experiments. *Biometrika*, **33**, 305–325 and 328–332.

Powell, M. J. D. (1964). An efficient method for finding the minimum of a function of several variables without calculating derivatives. *Comp. J.*, **7**, 155–162.

Powell, M. J. D. (1965). A method for minimizing a sum of squares of nonlinear functions without calculating derivatives. *Comp. J.*, **8**, 303–307.

Prairie, R. R. and Zimmer, W. J. (1964). 2^p factorial experiments with the factors applied sequentially. *J. Am. Statist. Assoc.*, **59**, 1205–1216.

Prairie, R. R. and Zimmer, W. J. (1968). Fractional replications of 2^p factorial experiments with the factors applied sequentially. *J. Am. Statist. Assoc.*, **63**, 644–652.

Prasad, C. R. (1982). *Statistical Quality Control and Operational Research: 160 Case Studies in Indian Industries*. Indian Statistical Institute, Calcutta, India.

Prasad, M. (1976). Response of sugarcane to filter press mud and N, P, and K fertilizers. I. Effect on sugarcane yield and sucrose content. II. Effects on plant composition and soil chemical properties. *Agronomy J.*, **68**, 539–546.

BIBLIOGRAPHY

649

Preece, D. A. (1984). Biometry in the Third World: Science not ritual. *Biometrics*, **40**, 519–523.

Pukelsheim, F. and Titterington, D. M. (1983). General differential and Lagrangian theory for optimal experimental design. *Ann. Statist.*, **11**, 1060–1068.

Raghavarao, D. (1963). Construction of second order rotatable designs through incomplete block designs. *J. Indian Statist. Assoc.*, **1**, 221–225.

Raghavarao, D. (1971). *Constructions and Combinatorial Problems in Design of Experiments*. New York: John Wiley & Sons.

Ralston, N. and Jennrich, R. I. (1978). Derivative free nonlinear regression. *Proc. Comp. Sci. Statist. Tenth Annual Symp. on the Interface*, April 14–15, 1977; eds. D. Hogben and D. W. Fife, 312–322. Washington: National Bureau of Standards Special Publication 503.

Rao, C. R. (1947). Factorial experiments derivable from combinatorial arrangements of arrays. *J. Roy. Statist. Soc., Suppl.*, **9**, 128–139.

Rao, C. R. (1958). Some statistical methods for the comparison of growth curves. *Biometrics*, **14**, 1–17.

Rao, C. R. (1959). Some problems involving linear hypotheses in multivariate analysis. *Biometrika*, **46**, 49–58.

Rao, C. R. (1965). The theory of least squares when the parameters are stochastic, and its application to the analysis of growth curves. *Biometrika*, **52**, 447–458.

Rao, C. R. (1966). Covariate adjustment and related problems in multivariate analysis. Dayton Symposium on Multivariate Analysis. *Multivariate Analysis*, **1**, 87–103. New York: Academic Press.

Rao, C. R. (1967). Least squares theory using an estimated dispersion matrix and its application to measurement of signals. *Proc. Fifth Berkley Symp. Math. Statist. Prob.*, **1**, 355–372.

Rao, C. R. (1973). *Linear Statistical Inference and its Applications*, 2nd ed. New York: Wiley.

Read, D. R. (1954). The design of chemical experiments. *Biometrics*, **10**, 1–15.

Reed, L. J. and Berkson, J. (1929). The application of the logistic function for experimental data. *J. Phys. Chem.*, **33**, 760–799.

Reid, D. (1972). The effects of the long-term application of a wide range of nitrogen rates on the yields from perennial ryegrass swards with and without white clover. *J. Agric. Sci. Camb.*, **79**, 291–301.

Remmers, E. G. and Dunn, G. G. (1961). Process improvement of a fermentation product. *Indus. Eng. Chem.*, **53**, 743–745.

Richert, S. H., Morr, C. V., and Cooney, C. M. (1974). Effect of heat and other factors upon foaming properties of whey protein concentrates. *J. Food Sci.*, **39**, 42–48.

Robbins, H. and Monro, S. (1951). A stochastic approximation method. *Ann. Math. Statist.*, **22**, 400–407.

Robinson, P. and Nielsen, K. F. (1960). Composite designs in agricultural research. *Can. J. Soil Science*, **40**, 168–176.

Roquemore, K. G. (1976). Hybrid designs for quadratic response surfaces. *Technometrics*, **18**, 419–423; **19**, 106.

Ross, E., Damron, B. L. and Harms, R. H. (1972). The requirement for inorganic sulfate in the diet of chicks for optimum growth and feed efficiency. *Poultry Sci.*, **51**, 1606–1612.

Ross, E. G. (1961). Response surface techniques as a statistical approach to research and development in ultrasonic welding. *Trans. 15th Ann. Conv. ASQC*, 445–456.

Ross, G. J. S. (1972). Stochastic model fitting by evolutionary operation. In *Mathematical Models in Ecology*, 297–308. London: Blackwell.

Roth, P. M. and Stewart, R. A. (1969). Experimental studies with multiple responses. *Appl. Statist.*, **18**, 221–228.

Roy, S. N., Gnanadesikan, R., and Srivastava, J. N. (1970). *Analysis and Design of Certain Quantitative Multiple-Response Experiments*. Oxford: Pergamon.

Ryan, D. M. and Scott, A. J. (1980). Estimating ventilation/perfusion distributions from inert gas data: a Bayesian approach. *Biometrics*, **36**, 105–115.

Sacks, J. and Ylvisaker, D. (1966). Design for regression problems with correlated errors I. *Ann. Math. Statist.*, **37**, 66–89.

Sacks, J. and Ylvisaker, D. (1968). Design for regression problems with correlated errors II. *Ann. Math. Statist.*, **39**, 49–69.

Sacks, J. and Ylvisaker, D. (1970). Design for regression problems with correlated errors III. *Ann. Math. Statist.*, **41**, 2057–2074.

Sacks, J. and Ylvisaker, D. (1984). Some model robust designs in regression. *Ann. Statist.*, **12**, 1324–1348.

Saha, G. M. and Das, A. R. (1973). Four level second order rotatable designs from partially balanced arrays. *J. Indian Soc. Agric. Statist.*, **25**, 97–102.

Sarma, G. S. and Ravindram, M. (1975). Studies in dealkylation—a statistical analysis of process variables. *J. Indian Inst. Sci.*, **58**, 67–83.

Sasse, C. E. and Baker, D. H. (1972). The Phenylalanine and Tyrosine requirements and their interrelationship for the young chick. *Poultry Sci.*, **51**, 1531–1535.

Satyabrata, Pal (1978). Some methods of construction of second order rotatable designs. *Calcutta Statist. Assoc. Bull.*, **27**, 127–140.

Saxena, S. K. and Nigam, A. K. (1977). Restricted exploration of mixtures by symmetric-simplex design. *Technometrics*, **19**, 47–52.

Scheffé, H. (1958). Experiments with mixtures. *J. Roy. Statist. Soc.*, **B20**, 344–360.

Scheffé, H. (1959). *The Analysis of Variance*. New York: Wiley.

Scheffé, H. (1963). The simplex-centroid design for experiments with mixtures. *J. Roy. Statist. Soc.*, **B25**, 235–251, discussion 251–263.

Schmidt, R. H., Illingworth, B. L., Deng, J. C., and Cornell, J. A. (1979). Multiple regression and response surface analysis of the effects of calcium chloride and cysteine on heat-induced whey protein gelation. *J. Agric. Food Chem.*, **27**, 529–532.

Schneider, A. M. and Stockett, A. L. (1963). An experiment to select optimum conditions on the basis of arbitrary preference ratings. *Chem. Eng. Prog. Symp. Ser., AIChE*, **59**(42), 34–38.

Schoney, R. A., Bay, T. F., and Moncrief, J. F. (1981). Use of computer graphics in the development and evaluation of response surfaces. *Agronomy J.*, **73**, 437–442.

Schroeder, E. C., Nair, K. P. C., and Cardeilhac, P. T. (1972). Response of broiler chicks to a single dose of Aflatoxin. *Poultry Sci.*, **51**, 1552–1556.

Seheult, A. (1978). Minimum bias or least squares estimation. *Commun. Statist. Theory Methods*, **A7**, 277–283.

Shannon, C. E. (1948). A mathematical theory of communication. *Bell System Tech. J.*, **27**, 373–423 and 623–656.

Sharma, M. L. (1976). Interaction of water potential and temperature effects on germination of three semi-arid plant species. *Agronomy J.* **68**, 390–394.

Sheldon, F. R. (1960). Statistical techniques applied to production situations. *Indus. Eng. Chem.*, **52**, 507–509.

Shelton, J. T., Khuri, A. I., and Cornell J. A. (1983). Selecting check points for testing lack of fit in response surface models. *Technometrics*, **25**, 357–365.

Shewell, C. T. (1956). Paper studies in catalytic cracking. *Trans. Tenth Ann. Conv., ASQC*, 1–7.

Shirafuji, M. (1959). A two stage sequential design in response surface analysis. *Bull. Math. Statist.*, **8**, 115–126.

Shuster, J. J. and Dietrich, F. H. (1976). Quantal response assays by inverse regression. *Commun. Statist. Theory Methods*, **A5**, 293–305.

Sihota, S. S. and Banerjee, K. S. (1981). On the algebraic structures in the construction of confounding plans in mixed factorial designs on the lines of White and Hultquist. *J. Am. Statist. Assoc.*, **76**, 996–1101.

Silvey, S. D. (1980). *Optimal Design*. London: Chapman and Hall.

Silvey, S. D. and Titterington, D. M. (1973). A geometrical approach to optimal design theory. *Biometrika*, **60**, 21–32.

Silvey, S. D., Titterington, D. M., and Torsney, B. (1978). An algorithm for optimal designs on a finite design space. *Commun. Statist. Theory Methods*, **A7**, 1379–1389.

Singh, M. (1979). Group divisible second order rotatable designs. *Biom. J.*, **21**, 579–589.

Singh, M., Dey, A., and Mitra, R. K. (1979). An algorithm for the choice of optimal response surface designs. *J. Indian Soc. Agric. Statist.*, **31**, 50–54.

Smith, D. and Smith, R. R. (1977). Response of red clover to increasing rates of topdressed potassium fertilizer. *Agronomy J.*, **69**, 45–48.

Smith, F. B., Jr. and Shannon, D. F. (1971). An improved Marquardt procedure for nonlinear regression. *Technometrics*, **13**, 63–74.

Smith, H. and Rose, A. (1963). Subjective response in process investigations. *Indus. Eng. Chem.*, **55**, 25–28.

Snee, R. D. (1971). Design and analysis of mixture experiments. *J. Qual. Technol.*, **3**, 159–169.

Snee, R. D. (1973a). Some aspects of nonorthogonal data analysis—Part I. Developing prediction equations. *J. Qual. Technol.*, **5**, 67–69.

Snee, R. D. (1973b). Some aspects of nonorthogonal data analysis—Part II. Comparison of means. *J. Qual. Technol.*, **5**, 109–122.

Snee, R. D. (1975). Experimental designs for quadratic models in constrained mixture spaces. *Technometrics*, **17**, 149–159.

Snee, R. D. (1979). Experimental designs for mixture systems with multicomponent constraints. *Comm. Statist. Theory Methods*, **8**, 337–358.

Snee, R. D. (1981). Developing blending models for gasoline and other mixtures. *Technometrics*, **23**, 119–130.

Snee, R. D. (1983). Discussion of "Developments in linear regression methodology: 1959–1982". *Technometrics*, **25**, 230–237.

Snee, R. D. (1985). Computer-aided design of experiments—some practical experiences. *J. Qual. Technol.*, **17**, 222–236.

Snee, R. D. and Marquardt, D. W. (1974). Extreme vertices designs for linear mixture models. *Technometrics*, **16**, 533–537.

Snee, R. D. and Marquardt, D. W. (1976). Screening concepts and designs for experiments with mixtures. *Technometrics*, **18**, 19–29.

Snee, R. D. and Marquardt, D. W. (1984). Collinearity diagnostics depend on the domain of prediction, the model, and the data. *Am. Statist.*, **38**, 83–87.

Snee, R. D. and Rayner, A. A. (1982). Assessing the accuracy of mixture model regression calculations. *J. Qual. Technol*, **14**, 67–79.

Spendley, W., Hext, G. R., and Himsworth, F. R. (1962). Sequential application of simplex designs in optimisation and EVOP. *Technometrics*, **4**, 441–461.

Sprent, P. (1969). *Models in Regression and Related Topics*. London: Methuen.

Springall, A. (1973). Response surface fitting using a generalisation of the Bradley-Terry paired comparison model. *Appl. Statist.*, **22**, 59–68.

Springer, B. G. F. (1969). Numerical optimisation in the presence of variability: single factor case. *Biometrika*, **56**, 65–74.

Srivastava, I. N. and Chopra, D. V. (1971). Balanced optimal 2^m fractional factorial designs of resolution V, $m \le 6$. *Technometrics*, **13**, 257–269.

St. John, R. C. and Draper, N. R. (1975). D-optimality for regression designs: A review. *Technometrics*, **17**, 15–23.

St. John, R. C. and Draper, N. R. (1977). Designs in three and four components for mixtures models with inverse terms. *Technometrics*, **19**, 117–130.

Stablein, D. M., Carter, W. H., Jr., and Wampler, G. L. (1983). Confidence regions for constrained optima in response-surface experiments. *Biometrics*, **39**, 759–763.

Starks, T. H. (1982). A Monte Carlo evaluation of response surface analysis based on paired comparison data. *Commun. Statist. Simul. Comp.*, **11**, 603–617.

Stauber, M. S. and Burt, O. R. (1973). Implicit estimate of residual nitrogen under fertilised range conditions in the Northern Great Plains. *Agronomy J.*, **65**, 897–901.

Stein, C. (1945). A two-sample test for a linear hypothesis whose power is independent of the variance. *Ann. Math. Statist.*, **16**, 243–258.

Steinberg, D. M. (1985). Model robust response surface designs: scaling two-level factorials. *Biometrika*, **72**, 513–526.

Steinberg, D. M. and Hunter, W. G. (1984). Experimental design: review and comment. *Technometrics*, **26**, 71–97, discussion 98–130.

Stevens, W. L. (1951). Asymptotic regression. *Biometrics*, **7**, 247–267.

Stewart, T. J. (1977). A criterion for optimum design of EVOP-type experiments: Part I: Optimality criterion for quadratic response surfaces. *Computers Operat. Res.*, **4**, 181–193.

Stewart, T. J. (1978). A criterion for optimum design of EVOP-type experiments: Part II: Modified criterion for non-quadratic response surfaces. *Computers Operat. Res.*, **5**, 1–9.

Stigler, S. M. (1971). Optimal experimental design for polynomial regression. *J. Am. Statist. Assoc.*, **66**, 311–318.

Stone, R. L., Wu, S. M., and Tiemann, T. D. (1965). A statistical experimental design and analysis of the extraction of silica from quartz by digestion in sodium hydroxide solutions. *AIME Trans.*, June, 115–124.

Stoneman, D. M. (1966). Response surface designs for specified factor levels. Ph.D. Thesis. University of Wisconsin.

Stowe, R. A. and Mayer, R. P. (1969). Pitfalls of stepwise regression analysis. *Indus. Eng. Chem.*, **61**(6), 11–16.

Studden, W. J. (1968). Optimal designs on Tchebychev points. *Ann. Math. Statist.*, **39**, 1435–1447.

Studden, W. J. (1971a). Optimal designs for multivariate polynomial extrapolation. *Ann. Math. Statist.*, **42**, 828–832.

Studden, W. J. (1971b). Elfving's theorem and optimal designs for quadratic loss. *Ann. Math. Statist.*, **42**, 1613–1621.

Studden, W. J. (1982). Some robust-type D-optimal designs in polynomial regression. *J. Am. Statist. Assoc.*, **77**, 916–921.

Sullivan, T. W. and Al-Timini, A. A. (1972). Safety and toxicity of dietary organic arsenicals relative to performance of young turkeys. 3.Nitarsone. *Poultry Sci.*, **51**, 1582–1586.

Swanson, A. M., Geissler, J. J., Magnino, P. J., and Swanson, J. A. (1967). Use of statistics and computers in the selection of optimum food processing techniques. *Food Technology*, **21**(11), 99–102.

Swanson, C. L., ·Maffsiger, T., Russel, C., Hofreiter, B., and Rist, C. (1964). Xanthation of starch by a continuous process. *Indus. Eng. Chem. Devel.*, **3**, 22–27.

Taguchi, G. (1974). A new statistical analysis for clinical data, the accumulating analysis, in contrast with the chi-square test. *Saiskim-igaku (the Newest Medicine)*, **29**, 806–813.

Taguchi, G. (1976). *Experimental Designs*, 3rd ed., Vol. 1 (in Japanese). Tokyo, Japan: Maruzen Publishing Company.

Taguchi, G. (1977). *Experimental Designs*, 3rd ed., Vol. 2 (in Japanese). Tokyo, Japan: Maruzen Publishing Company.

Taguchi, G. (1978). Off-line and on-line quality control systems. *Proceedings of International Conference on Quality Control*, Tokyo, Japan.

Taguchi, G. and Phadke, M. S. (1984). Quality engineering through design optimization. Conference Record, 3, IEEE GLOBECOM 1984 Conference, 1106–1113.

Taguchi, G. and Wu, Y.-I. (1980). Introduction to off-line quality control. Central Japan Quality Control Association (available from American Supplier Institute, 6 Parklane Blvd., Dearborn, MI 48126).

Taubman, S. B. and Addelman, S. (1978). Some two-level multiple response factorial plans. *J. Am. Statist. Assoc.*, **73**, 607–612.

Terhune, H. D. (1963). A statistical analysis of the effect of four variable operating conditions on bleaching results. *Amer. Dyestuffs Reporter*, **52**(7), April 1, 33–38.

Terman, G. L., Khasawneh, F. E., Allen, S. E., and Engelstad, O. P. (1976). Yield-nutrient absorption relationships as affected by environmental growth factors. *Agronomy J.*, **68**, 107–110.

Thaker, P. J. (1962). Some infinite series of second order rotatable designs. *J. Indian Soc. Agric. Statist.*, **14**, 110–120.

Thaker, P. J. and Das, M. N. (1961). Sequential third order rotatable designs for up to eleven factors. *J. Indian Soc. Agri. Statist.*, **13**, 218–231.

Thomas, J. R. and McLean, D. M. (1967). Growth and mineral composition of squash (cucurbita pepo L.) as affected by N, P, K, and tobacco ring spot virus. *Agronomy J.*, **59**, 67–69.

Thompson, W. O. (1973). Secondary criteria in the selection of minimum bias designs in two variables. *Technometrics*, **15**, 319–328.

Tiahrt, K. J. and Weeks, P. L. (1970). A method of constrained randomisation for 2^n factorials. *Technometrics*, **12**, 471–486.

Tidwell, P. W. (1960). Chemical process improvement by response surface methods. *Indus. Eng. Chem.*, **52**, 510–512.

Titterington, D. M. (1975). Optimal design: some geometrical aspects of *D*-optimality. *Biometrika*, **62**, 313–320.

Tukey, J. W. (1949). One degree of freedom for non-additivity. *Biometrics*, **5**, 232–242.

Tukey, J. W. and Moore, P. G. (1954). Answer to query 112. *Biometrics*, **10**, 562–568.

Tyagi, B. N. (1964a). A note on the construction of a class of second order rotatable designs. *J. Indian Statist. Assoc.*, **2**, 52–54.

Tyagi, B. N. (1964b). On construction of second and third order rotatable designs through pair-wise balanced and doubly balanced designs. *Calcutta Statist. Assoc. Bull.*, **13**, 150–162.

Umland, A. W. and Smith, W. N. (1959). The use of Lagrange mutipliers with response surfaces. *Technometrics*, **1**, 289–292.

Underwood, W. M. (1962). Experimental methods for designing extrusion screws. *Chem. Eng. Prog.*, **58**(1), 59–65.

Van der Vaart, H. R. (1960). On certain types of bias in current methods of response surface estimation. *Bull. Inter. Statist. Inst.*, **37**, 191–203.

Van der Vaart, H. R. (1961). On certain characteristics of the distribution of the latent roots of a symmetric random matrix under general conditions. *Ann. Math. Statist.*, **32**, 864–873.

Vinod, H. D. (1978). A survey of ridge regression and related techniques for improvement over OLS. *Rev. Econ. Statist.*, **60**, 121–131.

Voss, R. and Pesek, J. (1965). Geometrical determination of uncontrolled-controlled factor relationships affecting crop yield. *Agronomy J.*, **57**, 460–463.

Voss, R. and Pesek, J. (1967). Yield of corn grain as affected by fertilizer rates and environmental factors. *Agronomy J.*, **59**, 567–572.

Wald, A. (1943). On the efficient design of statistical investigations. *Ann. Math. Statist.*, **14**, 134–140.

Walker, W. M. and Carmer, S. G. (1967). Determination of input levels for a selected probability of response on a curviliner regression function. *Agronomy J.*, **59**, 161–162.

Walker, W. M. and Long, O. M. (1966). Effect of selected soil fertility parameters on soybean yields. *Agronomy J.*, **58**, 403–405.

Walker, W. M. and Pesek, J. (1967), Yield of Kentucky bluegrass (poa pratensis) as a function of its percentage of nitrogen, phosphorus, and potassium. *Agronomy J.*, **59**, 44–47.

Walker, W. M., Pesek, J., and Heady, E. O. (1963). Effect of nitrogen, phosphorus and potassium fertilizer on the economics of producing bluegrass forage. *Agronomy J.*, **55**, 193–196.

Wallace, D. L. (1958). Intersection region confidence procedures with an application to the location of the maximum in quadratic regression. *Ann. Math. Statist.*, **29**, 455–475.

Warncke, D. D. and Barber, S. A. (1973). Ammonium and nitrate uptake by corn, as influenced by nitrogen concentration and NH_4^+/NO_3^- ratio. *Agronomy J.*, **65**, 950–953.

Wasan, M. T. (1969). *Stochastic Approximation*. New York: Cambridge University Press.

Webb, S. R. (1968). Saturated sequential factorial designs. *Technometrics*, **10**, 535–550.

Webb, S. R. (1971). Small incomplete factorial experiment designs for two and three level factors. *Technometrics*, **13**, 243–256.

Weissert, F. C. and Cundiff, R. R. (1963). Compounding Diene/NR blends for truck tires. *Rubber Age*, **92**, 881–887.

Welch, L. F., Adams, W. E., and Carmon, J. L. (1963). Yield response surfaces, isoquants and economic fertilizer optima for Coastal bermudagrass. *Agronomy J.*, **35**, 63–67.

Welch, L. F., Johnson, P. E., McKibben, G. E., Boone, L. V., and Pendleton, J. W. (1966). Relative efficiency of broadcast versus banded potassium for corn. *Agronomy J.*, **58**, 618–621.

Welch, W. J. (1982). Branch and bound search for experimental designs based on *D*-optimality and other criteria. *Technometrics*, **24**, 41–48.

Welch, W. J. (1984). Computer-aided design of experiments for response estimation. *Technometrics*, **26**, 217–224.

Welch, W. J. (1985). ACED: Algorithms for the construction of experimental designs. *Am. Statist.*, **39**, 146.

Westlake, W. J. (1965). Composite designs based on irregular fractions of factorials. *Biometrics*, **21**, 324–336.

Wetherill, G. B. (1963). Sequential estimation of quantal response curves. *J. Roy. Statist. Soc.*, **B25**, 1–38, discussion 39–48.

Wheeler, R. E. (1985). Regression tool kit: software for the design and analysis of regression experiments. Algorithmic tool kit: software for the algorithmic design of experiments. *Am. Statist.*, **39**, 144.

Whidden, P. (1956). Design of experiment in metals processing. *Trans. Tenth Ann. Conv.*, *ASQC*, 677–683.

Whittle, P. (1965). Some general results in sequential design. *J. Roy. Statist. Soc.*, **B27**, 371–387, discussion 387–394.

Wichern, D. W. and Churchill, G. A. (1978). A comparison of ridge estimators. *Technometrics*, **20**, 301–310.

Wilde, D. J. (1962). *Advances in Chemical Engineering Volume III. Optimization Methods*. New York: Academic Press.

Wilde, D. J. (1964). *Optimum Seeking Methods*. Englewood Cliffs, New Jersey: Prentice Hall.

Willey, R. W. and Heath, S. B. (1969). The quantitative relationships between plant population and crop yield. *Advances in Agronomy*, **21**, 281–321.

Williams, D. A. (1970). Discrimination between regression models to determine the pattern of enzyme synthesis in synchronous cell cultures. *Biometrics*, **26**, 23–32.

Williams, K. R. (1968). Designed experiments. *Rubber Age*, **100**, 65–71.

Wilson, J. H., Clowes, M. St. J., and Allison, J. C. S. (1973). Growth and yield of maize at different altitudes in Rhodesia. *Ann. Appl. Biol.*, **73**, 77–84.

Winsor, C. P. (1932). The Gompertz curve as a growth curve. *Proc. Natl. Acad. Sci.*, **18**, 1–8.

Wishart, J. (1938). Growth rate determination in nutrition studies with the bacon pig, and their analysis. *Biometrika*, **30**, 16–28.

Wishart, J. (1939). Statistical treatment of animal experiments. *J. Roy Statist. Soc.*, **B6**, 1–22.

Wood, C. L., and Cady, F. B. (1981). Intersite transfer of estimated response surfaces. *Biometrics*, **37**, 1–10.

Wood, J. T. (1974). An extension of the analysis of transformations of Box and Cox. *Appl. Statist.*, **23**, 278–283.

Wu, S. M. (1964a). Tool-life testing by response surface methodology, Part 1. *J. Eng. Indus.*, *ASME Trans.*, *Series B*, **86**, May, 105–110. Part 2, 111–116.

Wu, S. M. and Meyer, R. N. (1964b). Cutting tool temperature-predicting equation by response surface methodology. *J. Eng. Indus.*, *ASME Trans.*, *Series B*, **86**, May, 150–156.

Wu, S. M. and Meyer, R. N. (1965). A first-order five-variable cutting-tool temperature equation and chip equivalent. *J. Eng. Indus.*, *ASME Trans.*, *Series B*, **87**, November, 395–400.

Wynn, H. P. (1970). The sequential generation of *D*-optimum experimental designs. *Ann. Math. Statist.*, **41**, 1655–1664.

Wynn, H. P. (1972). Results in the theory and construction of *D*-optimum experimental designs. *J. Roy. Statist. Soc.*, **B34**, 133–147, discussion 170–185.

Wynn, H. P. (1984). Jack Kiefer's contributions to experimental design. *Ann. Statist.*, **12**, 416–423.

Yates, F. (1935). Complex experiments. *J. Roy. Statist. Soc.*, **B2**, 181–223, discussion 223–247.

Yates, F. (1937). The design and analysis of factorial experiments. London: Imperial Bureau of Soil Science.

Yates, F. (1967). A fresh look at the basic principles of the design and analysis of experiments. In *Proc. 5th Berkeley Symp. Math. Statist. Prob.*, IV, 777–790. Berkeley and Los Angeles: University of California Press.

Yates, F. (1970). *Experimental Design: Selected Papers of Frank Yates*. Darien, Connecticut: Hafner.

Index of Authors Associated with Exercises

(To facilitate agreed copyright acknowledgements, any related references are quoted on the pages indicated, as well as in the bibliography. Names may be cross-checked by page against the sequential listing by exercise number and page which follows, in order to find associated coauthors. For a listing of authors of articles mentioned in the text, see the second index. For general listings, see the third index.)

Listing by Exercise Number, Authors, Page

Index of Authors of Articles Mentioned in the Text

(Reference details are given in the Bibliography and/or in the text. For a listing of authors associated with exercises, see the foregoing index. For general listings, see the following index.)

Anderson (McLean and . . . , 1984), 147
Andrews, Bickel, Hampel, Huber, Rogers and Tukey (1972), 91
Andrews (Herzberg and . . . , 1976), 504
Anscombe (1961), 217, 219, 230
Anscombe and Tukey (1963), 217
Atkinson (1973), 216
Atwood (1969), 499

Bacon (1970), 223–226
Barella and Sust, 28, 45, 107, 205–216, 218, 231, 278–280, 286–288
Baumert, Golomb, and Hall (1961), 162
Bates and Watts (1980), 102
Bates and Watts (1987), 102
Beale (1960), 102
Behnken (Box and . . . , 1960a), 177, 223, 224, 226, 515, 520
Berger (1972), 166
Bickel (Andrews et al., 1972), 91
Box (1950), 532
Box (1952), 481, 506, 507
Box (1954), 359, 360, 368, 496
Box (1955), 191
Box (1960b), 102
Box (1964), 409
Box (1968), 499
Box (1976), 4
Box (1980), 216
Box (1982), 490
Box and Behnken (1960a), 177, 223, 224, 226, 515, 520

Box and Cox (1964), 28, 210, 218, 287, 289, 292, 293, 461, 467
Box and Draper (1959), 442
Box and Draper (1963), 442
Box and Draper (1975), 477, 504
Box and Draper (1980), 126
Box and Draper (1982a), 220
Box and Hay (1953), 481
Box and Hill, W. J. (1974), 286
Box and Hunter, J. S. (1954), 316, 341
Box and Hunter, J. S. (1957), 486, 489, 509
Box and Hunter, J. S. (1961a), 13, 457
Box and Hunter, J. S. (1961b), 13, 457
Box, Hunter, W. G., and Hunter, J. S. (1978), 12, 13, 147, 160, 161, 163–165, 167, 522
Box and Jenkins (1976), 82
Box and Kanemasu (1972), 101
Box and Kanemasu (1984), 101
Box and Lucas (1959), 496
Box and Newbold (1971), 82
Box and Tidwell (1962), 298
Box and Wetz (1973), 275, 362
Box and Wilson (1951), 200, 375, 499
Box and Youle (1955), 14, 496
Burman (Plackett and . . . , 1946), 162, 521, 522

Cochran (1973), 496
Coen, Gomme, and Kendall (1969), 82
Cowser (Gardiner and . . . , 1961), 199

659

General Index

(For authors associated with exercises, see the first index. For authors of articles mentioned in the text, see the second index.)

Absolute temperature, 408
A canonical form, 332–339, 349, 350
Adequacy of estimation, 268, 275–280
Adequacy of fit, 219, 316
Ahn, S. K., 590
Alias (or bias) matrix, 65
All-bias case, 436
All-bias design conditions, 463
All-bias designs, 434, 437
  first order, 438
  second order, 440
All-variance case, 435
All-variance designs, 435
Alphabetic optimality, 489–501
Analysis of variance, one regressor, 50
  p regressors, 63–65
  skeleton table, 442
  two regressors, 57
Answers to exercises, 529–604. *See also* Exercises
A-optimality, 491
Applications of RSM, 17–19
Approximating function, 423
Approximating polynomial, 24
Arrhenius law, 33
Arrow notation, 109, 149
Average $V(\hat{y})$, 102

Bad observations, 84, 131
Barella and Sust data, 28, 45, 107, 205–216, 218, 231, 278–280, 286–288

B canonical form, 333–339, 346
Bias and variance, 423–476
  in $E(\mathbf{b})$, 220, 232–234
  error, 208, 425
  ignoring, 435, 501
  in least squares, 66
  minimization, 437–442, 463
  (or alias) matrix, 65
  in second order model, 220–222, 317–322, 453–459
Binomial distribution, 285
Bisgaard, Søren, viii
Blocking, 313
  of 2^k, 143
  of composite designs, 509–520
  orthogonal, 503, 513–520
    conditions, 509–511
  of polymer example, 309–313
Blocks:
  pure error in, 513
  sum of squares for, 513, 515
Block-variable interaction, 461
Box–Behnken designs, 223, 226, 227, 513–520
Box and Coutie, 201
Box–Wetz check, 275–280
Burman (Plackett and . . . designs), 162, 506

Cadenza, Ozzie, 172–174
Canonical analysis, 332–373

663

~

(*continued from front*)